T0180376

Interdisciplinary Applied Mathematics

Volume 8/II

Problems in engineering, computational science, and the physical and biological sciences are using increasingly sophisticated mathematical techniques. Thus, the bridge between the mathematical sciences and other disciplines is heavily traveled. The correspondingly increased dialog between the disciplines has led to the establishment of the series: *Interdisciplinary Applied Mathematics.*

The purpose of this series is to meet the current and future needs for the interaction between various science and technology areas on the one hand and mathematics on the other. This is done, firstly, by encouraging the ways that mathematics may be applied in traditional areas, as well as point towards new and innovative areas of applications; and, secondly, by encouraging other scientific disciplines to engage in a dialog with mathematicians outlining their problems to both access new methods and suggest innovative developments within mathematics itself.

The series will consist of monographs and high-level texts from researchers working on the interplay between mathematics and other fields of science and technology.

Interdisciplinary Applied Mathematics

1. *Gutzwiller*: Chaos in Classical and Quantum Mechanics
2. *Wiggins*: Chaotic Transport in Dynamical Systems
3. *Joseph/Renardy*: Fundamentals of Two-Fluid Dynamics:
   Part I: Mathematical Theory and Applications
4. *Joseph/Renardy*: Fundamentals of Two-Fluid Dynamics:
   Part II: Lubricated Transport, Drops and Miscible Liquids
5. *Seydel*: Practical Bifurcation and Stability Analysis:
   From Equilibrium to Chaos
6. *Hornung*: Homogenization and Porous Media
7. *Simo/Hughes*: Computational Inelasticity
8. *Keener/Sneyd*: Mathematical Physiology, Second Edition:
   I: Cellular Physiology
   II: Systems Physiology
9. *Han/Reddy*: Plasticity: Mathematical Theory and Numerical Analysis
10. *Sastry*: Nonlinear Systems: Analysis, Stability, and Control
11. *McCarthy*: Geometric Design of Linkages
12. *Winfree*: The Geometry of Biological Time (Second Edition)
13. *Bleistein/Cohen/Stockwell*: Mathematics of Multidimensional Seismic Imaging, Migration, and Inversion
14. *Okubo/Levin*: Diffusion and Ecological Problems: Modern Perspectives
15. *Logan*: Transport Models in Hydrogeochemical Systems
16. *Torquato*: Random Heterogeneous Materials: Microstructure and Macroscopic Properties
17. *Murray*: Mathematical Biology: An Introduction
18. *Murray*: Mathematical Biology: Spatial Models and Biomedical Applications
19. *Kimmel/Axelrod*: Branching Processes in Biology
20. *Fall/Marland/Wagner/Tyson*: Computational Cell Biology
21. *Schlick*: Molecular Modeling and Simulation: An Interdisciplinary Guide
22. *Sahimi*: Heterogenous Materials: Linear Transport and Optical Properties (Volume I)
23. *Sahimi*: Heterogenous Materials: Non-linear and Breakdown Properties and Atomistic Modeling (Volume II)
24. *Bloch*: Nonhoionomic Mechanics and Control
25. *Beuter/Glass/Mackey/Titcombe*: Nonlinear Dynamics in Physiology and Medicine
26. *Ma/Soatto/Kosecka/Sastry*: An invitation to 3-D Vision
27. *Ewens*: Mathematical Population Genetics (Second Edition)
28. *Wyatt*: Quantum Dynamics with Trajectories
29. *Karniadakis*: Microflows and Nanoflows
30. *Macheras*: Modeling in Biopharmaceutics, Pharmacokinetics and Pharmacodynamics
31. *Samelson/Wiggins*: Lagrangian Transport in Geophysical Jets and Waves
32. *Wodarz*: Killer Cell Dynamics
33. *Pettini*: Geometry and Topology in Hamiltonian Dynamics and Statistical Mechanics
34. *Desolneux/Moisan/Morel*: From Gestalt Theory to Image Analysis

James Keener James Sneyd

Mathematical Physiology

II: Systems Physiology

Second Edition

James Keener
Department of Mathematics
University of Utah
Salt Lake City, 84112
USA
keener@math.utah.edu

James Sneyd
Department of Mathematics
University of Auckland
Private Bag 92019
Auckland, New Zealand
sneyd@math.auckland.ac.nz

ISBN 978-1-4939-3709-7 ISBN 978-0-387-79388-7 (eBook)

DOI 10.1007/978-0-387-79388-7

Softcover re-print of the Hardcover 2nd edition 2009

Printed on acid-free paper.

springer.com

To Monique,

and

To Kristine, patience personified.

Preface to the Second Edition

If, in 1998, it was presumptuous to attempt to summarize the field of mathematical physiology in a single book, it is even more so now. In the last ten years, the number of applications of mathematics to physiology has grown enormously, so that the field, large then, is now completely beyond the reach of two people, no matter how many volumes they might write.

Nevertheless, although the bulk of the field can be addressed only briefly, there are certain fundamental models on which stands a great deal of subsequent work. We believe strongly that a prerequisite for understanding modern work in mathematical physiology is an understanding of these basic models, and thus books such as this one serve a useful purpose.

With this second edition we had two major goals. The first was to expand our discussion of many of the fundamental models and principles. For example, the connection between Gibbs free energy, the equilibrium constant, and kinetic rate theory is now discussed briefly, Markov models of ion exchangers and ATPase pumps are discussed at greater length, and agonist-controlled ion channels make an appearance. We also now include some of the older models of fluid transport, respiration/perfusion, blood diseases, molecular motors, smooth muscle, neuroendocrine cells, the baroreceptor loop, tubuloglomerular oscillations, blood clotting, and the retina. In addition, we have expanded our discussion of stochastic processes to include an introduction to Markov models, the Fokker–Planck equation, the Langevin equation, and applications to such things as diffusion, and single-channel data.

Our second goal was to provide a pointer to recent work in as many areas as we can. Some chapters, such as those on calcium dynamics or the heart, close to our own fields of expertise, provide more extensive references to recent work, while in other chapters, dealing with areas in which we are less expert, the pointers are neither complete nor extensive. Nevertheless, we hope that in each chapter, enough information is given to enable the interested reader to pursue the topic further.

Of course, our survey has unavoidable omissions, some intentional, others not. We can only apologize, yet again, for these, and beg the reader's indulgence. As with the first edition, ignorance and exhaustion are the cause, although not the excuse.

Since the publication of the first edition, we have received many comments (some even polite) about mistakes and omissions, and a number of people have devoted considerable amounts of time to help us improve the book. Our particular thanks are due to Richard Bertram, Robin Callard, Erol Cerasi, Martin Falcke, Russ Hamer, Harold Layton, Ian Parker, Les Satin, Jim Selgrade and John Tyson, all of whom assisted above and beyond the call of duty. We also thank Peter Bates, Dan Beard, Andrea Ciliberto, Silvina Ponce Dawson, Charles Doering, Elan Gin, Erin Higgins, Peter Jung, Yue Xian Li, Mike Mackey, Robert Miura, Kim Montgomery, Bela Novak, Sasha Panfilov, Ed Pate, Antonio Politi, Tilak Ratnanather, Timothy Secomb, Eduardo Sontag, Mike Steel, and Wilbert van Meerwijk for their help and comments.

Finally, we thank the University of Auckland and the University of Utah for continuing to pay our salaries while we devoted large fractions of our time to writing, and we thank the Royal Society of New Zealand for the James Cook Fellowship to James Sneyd that has made it possible to complete this book in a reasonable time.

University of Utah — James Keener
University of Auckland — James Sneyd
2008

Preface to the First Edition

It can be argued that of all the biological sciences, physiology is the one in which mathematics has played the greatest role. From the work of Helmholtz and Frank in the last century through to that of Hodgkin, Huxley, and many others in this century, physiologists have repeatedly used mathematical methods and models to help their understanding of physiological processes. It might thus be expected that a close connection between applied mathematics and physiology would have developed naturally, but unfortunately, until recently, such has not been the case.

There are always barriers to communication between disciplines. Despite the quantitative nature of their subject, many physiologists seek only verbal descriptions, naming and learning the functions of an incredibly complicated array of components; often the complexity of the problem appears to preclude a mathematical description. Others want to become physicians, and so have little time for mathematics other than to learn about drug dosages, office accounting practices, and malpractice liability. Still others choose to study physiology precisely because thereby they hope not to study more mathematics, and that in itself is a significant benefit. On the other hand, many applied mathematicians are concerned with theoretical results, proving theorems and such, and prefer not to pay attention to real data or the applications of their results. Others hesitate to jump into a new discipline, with all its required background reading and its own history of modeling that must be learned.

But times are changing, and it is rapidly becoming apparent that applied mathematics and physiology have a great deal to offer one another. It is our view that teaching physiology without a mathematical description of the underlying dynamical processes is like teaching planetary motion to physicists without mentioning or using Kepler's laws; you can observe that there is a full moon every 28 days, but without Kepler's laws you cannot determine when the next total lunar or solar eclipse will be nor when

Halley's comet will return. Your head will be full of interesting and important facts, but it is difficult to organize those facts unless they are given a quantitative description. Similarly, if applied mathematicians were to ignore physiology, they would be losing the opportunity to study an extremely rich and interesting field of science.

To explain the goals of this book, it is most convenient to begin by emphasizing what this book is not; it is not a physiology book, and neither is it a mathematics book. Any reader who is seriously interested in learning physiology would be well advised to consult an introductory physiology book such as Guyton and Hall (1996) or Berne and Levy (1993), as, indeed, we ourselves have done many times. We give only a brief background for each physiological problem we discuss, certainly not enough to satisfy a real physiologist. Neither is this a book for learning mathematics. Of course, a great deal of mathematics is used throughout, but any reader who is not already familiar with the basic techniques would again be well advised to learn the material elsewhere.

Instead, this book describes work that lies on the border between mathematics and physiology; it describes ways in which mathematics may be used to give insight into physiological questions, and how physiological questions can, in turn, lead to new mathematical problems. In this sense, it is truly an interdisciplinary text, which, we hope, will be appreciated by physiologists interested in theoretical approaches to their subject as well as by mathematicians interested in learning new areas of application.

It is also an introductory survey of what a host of other people have done in employing mathematical models to describe physiological processes. It is necessarily brief, incomplete, and outdated (even before it was written), but we hope it will serve as an introduction to, and overview of, some of the most important contributions to the field. Perhaps some of the references will provide a starting point for more in-depth investigations.

Unfortunately, because of the nature of the respective disciplines, applied mathematicians who know little physiology will have an easier time with this material than will physiologists with little mathematical training. A complete understanding of all of the mathematics in this book will require a solid undergraduate training in mathematics, a fact for which we make no apology. We have made no attempt whatever to water down the models so that a lower level of mathematics could be used, but have instead used whatever mathematics the physiology demands. It would be misleading to imply that physiological modeling uses only trivial mathematics, or vice versa; the essential richness of the field results from the incorporation of complexities from both disciplines.

At the least, one needs a solid understanding of differential equations, including phase plane analysis and stability theory. To follow everything will also require an understanding of basic bifurcation theory, linear transform theory (Fourier and Laplace transforms), linear systems theory, complex variable techniques (the residue theorem), and some understanding of partial differential equations and their numerical simulation. However, for those whose mathematical background does not include all of these topics, we have included references that should help to fill the gap. We also make extensive use of asymptotic methods and perturbation theory, but include explanatory material to help the novice understand the calculations.

This book can be used in several ways. It could be used to teach a full-year course in mathematical physiology, and we have used this material in that way. The book includes enough exercises to keep even the most diligent student busy. It could also be used as a supplement to other applied mathematics, bioengineering, or physiology courses. The models and exercises given here can add considerable interest and challenge to an otherwise traditional course.

The book is divided into two parts, the first dealing with the fundamental principles of cell physiology, and the second with the physiology of systems. After an introduction to basic biochemistry and enzyme reactions, we move on to a discussion of various aspects of cell physiology, including the problem of volume control, the membrane potential, ionic flow through channels, and excitability. Chapter 5 is devoted to calcium dynamics, emphasizing the two important ways that calcium is released from stores, while cells that exhibit electrical bursting are the subject of Chapter 6. This book is not intentionally organized around mathematical techniques, but it is a happy coincidence that there is no use of partial differential equations throughout these beginning chapters.

Spatial aspects, such as synaptic transmission, gap junctions, the linear cable equation, nonlinear wave propagation in neurons, and calcium waves, are the subject of the next few chapters, and it is here that the reader first meets partial differential equations. The most mathematical sections of the book arise in the discussion of signaling in two- and three-dimensional media—readers who are less mathematically inclined may wish to skip over these sections. This section on basic physiological mechanisms ends with a discussion of the biochemistry of RNA and DNA and the biochemical regulation of cell function.

The second part of the book gives an overview of organ physiology, mostly from the human body, beginning with an introduction to electrocardiology, followed by the physiology of the circulatory system, blood, muscle, hormones, and the kidneys. Finally, we examine the digestive system, the visual system, ending with the inner ear.

While this may seem to be an enormous amount of material (and it is!), there are many physiological topics that are not discussed here. For example, there is almost no discussion of the immune system and the immune response, and so the work of Perelson, Goldstein, Wofsy, Kirschner, and others of their persuasion is absent. Another glaring omission is the wonderful work of Michael Reed and his collaborators on axonal transport; this work is discussed in detail by Edelstein-Keshet (1988). The study of the central nervous system, including fascinating topics like nervous control, learning, cognition, and memory, is touched upon only very lightly, and the field of pharmacokinetics and compartmental modeling, including the work of John Jacquez, Elliot Landaw, and others, appears not at all. Neither does the wound-healing work of Maini, Sherratt, Murray, and others, or the tumor modeling of Chaplain and his colleagues. The list could continue indefinitely. Please accept our apologies if your favorite topic (or life's work) was omitted; the reason is exhaustion, not lack of interest.

As well as noticing the omission of a number of important areas of mathematical physiology, the reader may also notice that our view of what "mathematical" means appears to be somewhat narrow as well. For example, we include very little discussion of statistical methods, stochastic models, or discrete equations, but concentrate almost wholly on continuous, deterministic approaches. We emphasize that this is not from any inherent belief in the superiority of continuous differential equations. It results rather from the unpleasant fact that choices had to be made, and when push came to shove, we chose to include work with which we were most familiar. Again, apologies are offered.

Finally, with a project of this size there is credit to be given and blame to be cast; credit to the many people, like the pioneers in the field whose work we freely borrowed, and many reviewers and coworkers (Andrew LeBeau, Matthew Wilkins, Richard Bertram, Lee Segel, Bruce Knight, John Tyson, Eric Cytrunbaum, Eric Marland, Tim Lewis, J.G.T. Sneyd, Craig Marshall) who have given invaluable advice. Particular thanks are also due to the University of Canterbury, New Zealand, where a significant portion of this book was written. Of course, as authors we accept all the blame for not getting it right, or not doing it better.

University of Utah — James Keener
University of Michigan — James Sneyd
1998

Acknowledgments

With a project of this size it is impossible to give adequate acknowledgment to everyone who contributed: My family, whose patience with me is herculean; my students, who had to tolerate my rantings, ravings, and frequent mistakes; my colleagues, from whom I learned so much and often failed to give adequate attribution. Certainly the most profound contribution to this project was from the Creator who made it all possible in the first place. I don't know how He did it, but it was a truly astounding achievement. To all involved, thanks.

University of Utah James Keener

Between the three of them, Jim Murray, Charlie Peskin and Dan Tranchina have taught me almost everything I know about mathematical physiology. This book could not have been written without them, and I thank them particularly for their, albeit unaware, contributions. Neither could this book have been written without many years of support from my parents and my wife, to whom I owe the greatest of debts.

University of Auckland James Sneyd

Table of Contents

CONTENTS, II: Systems Physiology

CONTENTS, I: Cellular Physiology

The Circulatory System

The circulatory system forms a closed loop for the flow of blood that carries oxygen from the lungs to the tissues of the body and carries carbon dioxide from the tissues back to the lungs (Figs. 11.1 and 11.2). There are two pumps to overcome the resistance and maintain a constant flow. The left heart receives oxygen-rich blood from the lungs and pumps this blood into the *systemic arteries*. The systemic arteries form a tree of progressively smaller vessels, beginning with the aorta, branching to the small arteries, then to the arterioles, and finally to the capillaries. The exchange of gases takes place in the capillaries. Leaving the systemic capillaries, the blood enters the *systemic veins*, through which it flows in vessels of progressively increasing size toward the right heart. The systemic veins consist of venules, small veins, and the venae cavae. The right heart pumps blood into the *pulmonary arteries*, which form a tree that distributes the blood to the lungs. The smallest branches of this tree are the pulmonary capillaries, where carbon dioxide leaves and oxygen enters the blood. Leaving the pulmonary capillaries, the oxygenated blood is collected by the pulmonary veins, through which it flows back to the left heart. It takes about a minute for a red blood cell to complete this circuit.

While there is an apparent structural symmetry between the pulmonary and systemic circulations, there are significant quantitative differences in pressure and blood volume. Nevertheless, the output of the right and left sides of the heart must always balance, even though the cardiac output, or total amount of blood pumped by the heart, varies widely in response to the metabolic needs of the body. One of the goals of this chapter is to understand how the cardiac output is determined and regulated in response to the metabolic needs of the body. Questions of this nature have been studied for many years, and many books have been written on the subject (see, for example, Guyton, 1963, or Reeve and Guyton, 1967. A more recent book that discusses several

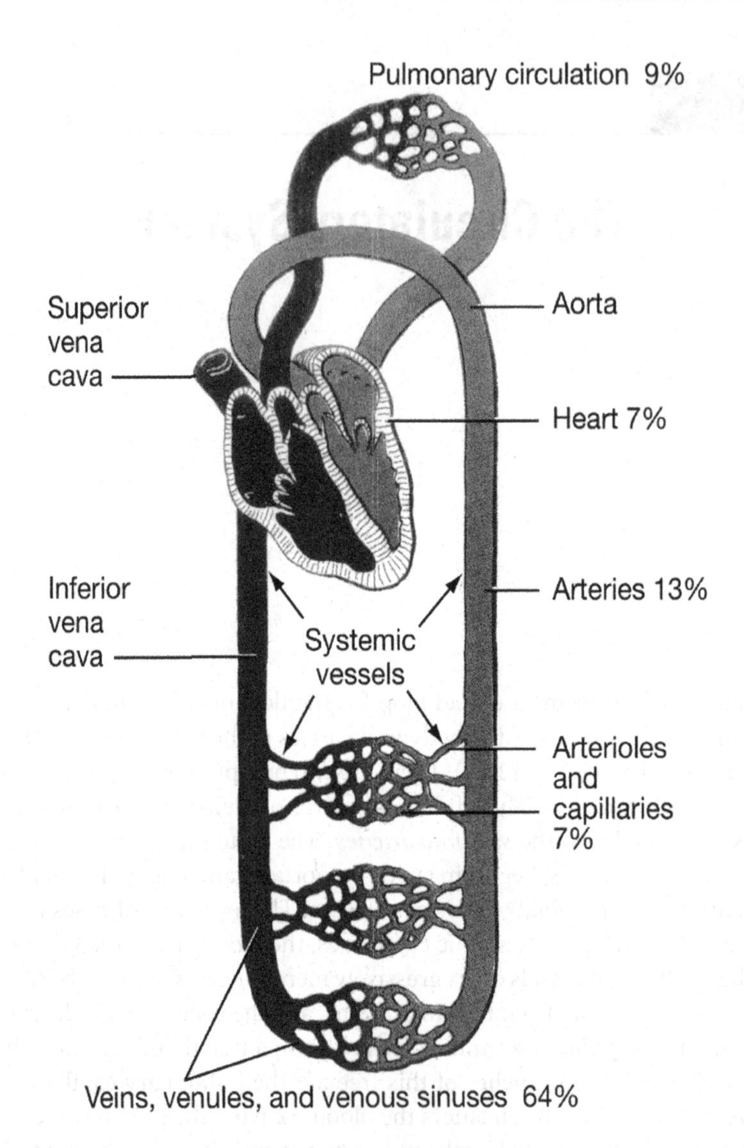

Figure 11.1 Schematic diagram of the circulatory system, showing the systemic and pulmonary circulations, the chambers of the heart, and the distribution of blood volume throughout the system. (Guyton and Hall, 1996, Fig. 14-1, p. 162.)

mathematical models of the blood and circulatory system is Ottesen et al., 2004). Here, we consider only the simplest models of the control of cardiac output.

Each beat of the heart sends a pulse of blood through the arteries, and the form of this arterial pulse changes as it moves away from the heart. An interesting problem is to understand these changes and their clinical significance in terms of the properties of the blood and the arterial walls. Again, this problem has been studied in great detail, and we present here only a brief look at the earliest and simplest models of the arterial pulse.

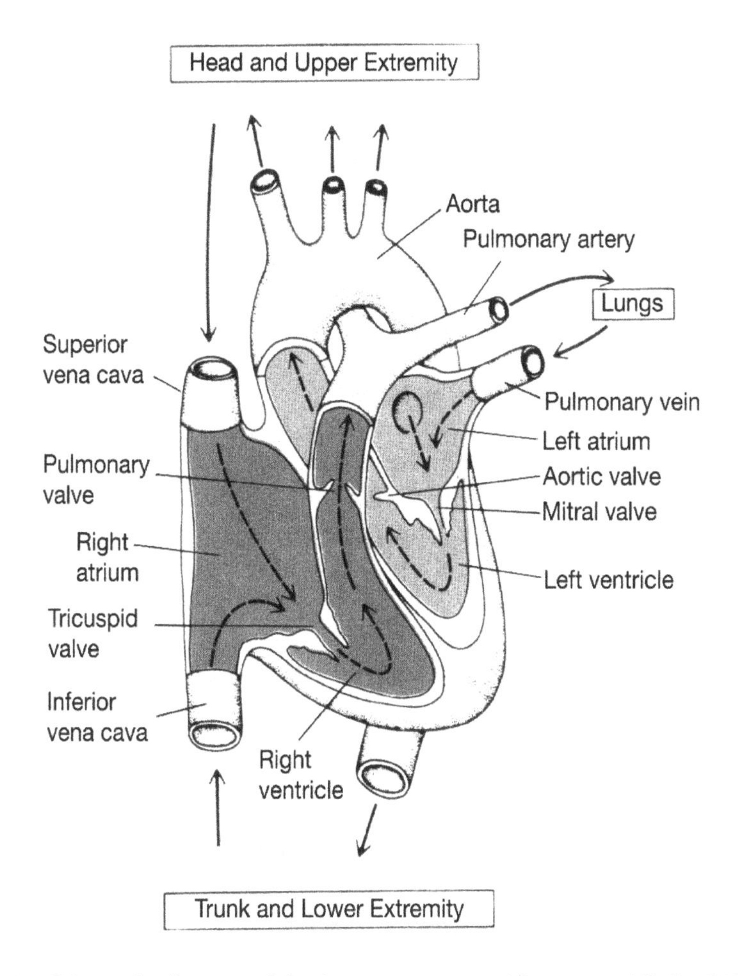

Figure 11.2 Schematic diagram of the heart as a pump. (Guyton and Hall, 1996, Fig. 9-1, p. 108.)

Similarly, we pay almost no attention to the vast body of work on the fluid mechanics of blood flow. For general discussions, the interested reader is referred to Pedley (1980), Lighthill (1975), Jaffrin and Caro (1995), and Ottesen et al. (2004).

11.1 Blood Flow

The term *blood pressure* refers to the force per unit area that the blood exerts on the walls of blood vessels. Blood pressure varies both in time and distance along the circulatory system. *Systolic pressure* is the highest surge of pressure in an artery, and results from the ejection of blood by the ventricles during ventricular contraction, or *systole*. *Diastolic pressure* is the lowest pressure reached during ventricular relaxation and filling, called *diastole*. In the aorta of a normal human, systolic pressure is about 120 mm Hg and diastolic pressure is about 80 mm Hg.

If the effects of gravity are ignored (which they are throughout this book), then one may assume that blood flows solely in response to pressure gradients. The simplest way to characterize a blood vessel is as a *resistance vessel* in which the radius is constant and the flow is linearly proportional to the pressure drop. In a linear resistance vessel, the flow, Q, is related to pressure by

$$Q = \frac{\Delta P}{R}, \tag{11.1}$$

where ΔP is the pressure drop and R is the resistance.

The relationship between resistance and radius of the vessel is dramatic. To understand this dependence suppose that a viscous fluid moves slowly and steadily through a cylindrical vessel of fixed radius. The velocity of the fluid is described by a vector $\mathbf{u}$ that has axial, radial, and angular components. Because the fluid is incompressible and fluid is conserved, it must be that

$$\nabla \cdot \mathbf{u} = 0. \tag{11.2}$$

Furthermore, because momentum is conserved, the Navier–Stokes equations (Segel, 1977) hold:

$$\rho(\mathbf{u}_t + \mathbf{u} \cdot \nabla \mathbf{u}) = -\nabla P + \mu \nabla^2 \mathbf{u}, \tag{11.3}$$

where ρ is the constant fluid density, P is the fluid pressure, and μ is the fluid viscosity. (A brief derivation of the conservation and momentum equations in the case of zero viscosity is given in Section 20.2.1.) If the flow is steady and the nonlinear terms are small compared to viscosity (in an appropriate nondimensional scaling), then

$$\mu \nabla^2 \mathbf{u} = \nabla P. \tag{11.4}$$

This simplification of the Navier–Stokes equation is called the Stokes equation.

The applicability of the Stokes equation to blood flow is suspect for several reasons. The viscosity contribution (the Laplacian) in the Stokes equation is derived from an assumed constitutive law relating stresses and strains in the fluid (Segel, 1977) that is known not to hold in fluids containing long polymers or other complicated chemical structures, including red blood cells. Furthermore, in the capillaries, the large size of the red blood cells compared to the typical diameter of a capillary suggests that a continuum description is not appropriate. However, we do not concern ourselves with these issues here and accept the Stokes equation description as adequate.

We look for a solution of the Stokes equation whose only nonzero component is the axial component. We define coordinates on the cylinder in the usual fashion, letting x denote distance along the cylinder in the axial direction and letting r denote the radial direction. The angular direction, with coordinate θ, does not enter into this analysis.

First observe that with only axial flow, the incompressibility condition (11.2) implies that $\frac{\partial u}{\partial x} = 0$, where u is the axial component of the velocity vector. Thus, u is independent of x. Then, with a steady flow, (11.4) reduces to the ordinary differential equation

$$\mu \frac{1}{r} \frac{d}{dr} \left(r \frac{d}{dr} u \right) = \frac{dP}{dx} = P_x, \tag{11.5}$$

where P_x is the axial pressure gradient along the vessel. Note also that P_x must be constant, independent of r and x. Because of viscosity, the velocity must be zero at the wall of the cylindrical vessel, $r = r_0$.

Integration of (11.5) from r to r_0 gives

$$u(r) = -\frac{P_x}{4\mu}(r_0^2 - r^2), \tag{11.6}$$

from which it follows that the total flux through the vessel is (*Poiseuille's law*)

$$Q = 2\pi \int_0^{r_0} u(r) r \, dr = -\frac{\pi P_x}{8\mu} r_0^4. \tag{11.7}$$

This illustrates that the total flow of blood through a vessel is directly proportional to the fourth power of its radius, so that the radius of the vessel is by far the most important factor in determining the rate of flow through a vessel. In terms of the cross-sectional area A_0 of the vessel, the flux through the cylinder is

$$Q = -\frac{P_x}{8\pi\mu} A_0^2, \tag{11.8}$$

while the average fluid velocity over a cross-section of the cylinder is given by

$$v = \frac{Q}{A_0} = -\frac{P_x}{8\pi\mu} A_0. \tag{11.9}$$

Note that a positive flow is defined to be in the increasing x direction, and thus a negative pressure gradient P_x drives a positive flow Q. Important controls of the circulatory system are vasodilators and vasoconstrictors, which, as their names suggest, dilate or constrict vessels and thereby adjust the vessel resistance by adjusting the radius.

If there are many parallel vessels of the same radius, then the total flux is the sum of the fluxes through the individual vessels. If there are N vessels, each of cross-sectional area A_0 and with total cross-sectional area $A = NA_0$, the total flux through the system is

$$Q = -\frac{P_x}{8\pi\mu} A_0(NA_0) = -\frac{P_x}{8\pi\mu} A_0 A, \tag{11.10}$$

and the corresponding average velocity is

$$v = \frac{Q}{A}. \tag{11.11}$$

Now, for there to be no stagnation in any portion of the systemic or pulmonary vessels, the total flux Q must be the same constant everywhere, implying that $P_x A_0 A$ must be constant. Thus, for a vessel with constant cross-sectional area, the pressure

Table 11.1 Diameter, total cross-sectional area, mean blood pressure at entrance, and mean fluid velocity of blood vessels.

Vessel	D (cm)	A (cm^2)	P(mm Hg)	v (cm/s)
Aorta	2.5	2.5	100	33
Small arteries	0.5	20	100	30
Arterioles	3×10^{-3}	40	85	15
Capillaries	6×10^{-4}	2500	30	0.03
Venules	2×10^{-3}	250	10	0.5
Small veins	0.5	80		2
Venae cavae	3.0	8	2	20

drop must be linear in distance. Furthermore, the pressure drop per unit length must be greatest in that part of the circulatory system for which A_0A is smallest.

In Table 11.1 are shown the total cross-sectional areas and pressures at entry to different components of the vascular system. The largest pressure drop occurs in the arterioles, and the pressure in the small veins and venae cavae is so low that these are often collapsed. The pressure at the capillaries must be low to keep them from bursting, since they have very thin walls. The numbers in Table 11.1 suggest that the pressure drop per unit length is greatest in the arterioles, about a factor of three times greater than in the capillaries, even though the capillaries are substantially smaller than the arterioles.

When the diameter of a vessel decreases, the velocity must increase if the flux is to remain the same. In the circulatory system, however, a decrease in vessel diameter is accompanied by an increase in total cross-sectional area (i.e., an increase in the total number of vessels), so that the velocity at the capillaries is small, even though the capillaries have very small radius. In fact, according to (11.11), the velocity in a vessel is independent of the radius of the individual vessel but depends solely on the total cross-sectional area of the collection of similar vessels. Once again, from Table 11.1 we see that the velocity drops continuously from aorta to arteries to arterioles to capillaries and then rises from capillaries to venules to veins to venae cavae. The velocity at the vena cava is about half that at the aorta.

11.2 Compliance

Because blood vessels are elastic, there is a relationship between pressure and volume. The blood vessel wall is stretched as a result of the pressure difference between the interior and exterior of the vessel. Obviously, the higher the pressure difference, the more tension in the wall. Also, the larger the radius of the vessel, the larger the tension. However, the thicker the wall the less tension there is. These three observations can be quantified via the relationship, known as Laplace's law,

$$T = \frac{Pr}{M}, \tag{11.12}$$

where T is the tension in the wall, P is the transmural pressure, r is the vessel radius and M is the wall thickness.

An application of Laplace's law arises in *dilated cardiomyopathy* in which the heart becomes greatly distended and the radius of the ventricle increases. Much larger wall tension is needed to create the same pressure during ejection of the blood, and so the dilated heart requires more energy to pump the same amount of blood compared to a heart of normal size. A new surgical procedure based on Laplace's law, called ventricular remodeling, removes portions of the ventricular wall in order to improve the function of dilated, failing hearts.

Laplace's law is the motivation for an understanding of compliance. Suppose we have an elastic vessel of volume V, with a uniform internal pressure P. The simplest assumption one can make is that V is linearly related to P, and thus

$$V = V_0 + CP, \tag{11.13}$$

for some constant C, called the *compliance* of the vessel, where V_0 is the volume of the vessel at zero pressure. Although this linear relationship is not always accurate, it is good enough for the simple models that are used here.

The compliance of the venous compartment is about 24 times as great as the compliance of the arterial system, because the veins are both larger and weaker than the arteries. It follows that large amounts of blood can be stored in the veins with only slight changes in venous pressure, so that the veins are often called storage areas. The blood vessels in the lungs are also much more compliant than the systemic arteries.

It is possible for veins and arteries to collapse and for blood flow through the vessel to cease; i.e., the radius becomes zero if the pressure is sufficiently negative. Negative pressures are possible if one takes into account that there is a fluid pressure in the body exterior to the vessels, and P actually refers to the drop in pressure across the vessel wall. The flow of whole blood is stopped at a nonzero radius, primarily because of the nonzero diameter of red blood cells. Thus, when the arterial pressure falls below about 20 mm Hg, the flow of whole blood is blocked, whereas blockage of plasma in arterioles occurs between 5 and 10 mm Hg.

Equation (11.13) is applicable when the vessel has the same internal pressure throughout. When P is not uniform, the compliance of the vessel is modeled by relating the cross-sectional area to the pressure. Again, the simplest assumption to make is that the relationship is linear (see Exercise 2), and thus

$$A = A_0 + cP, \tag{11.14}$$

for some constant c. Note that c is the compliance per unit length, since in a cylindrical vessel of length L and uniform internal pressure, $V = AL$, so that $C = cL$. However, here we refer to both C and c as compliance.

For a given flow, the pressure drop in a compliance vessel is different from the pressure drop in a resistance vessel. Further, in a compliance vessel, the flow is not a linear function of the pressure drop. From (11.8) the flux through a vessel is proportional to the product of the pressure gradient and the square of the area. Thus, for a compliance vessel,

$$8\pi\mu Q = -P_x A^2(P), \tag{11.15}$$

where $A(P)$ is the relationship between cross-sectional area and pressure for the chosen vessel. In steady state, the flux must be the same everywhere, so that

$$x = -\frac{1}{8\pi\mu Q}\int_{P_0}^{P(x)} A^2(P)\,dP \tag{11.16}$$

determines the pressure as a function of distance x. If the cross-sectional area of the vessel is given by (11.14), then the flux through a vessel of length L is related to the input pressure P_0 and output pressure P_1 by

$$RQ = \frac{1}{3\gamma}(1+\gamma P)^3\Big|_{P_1}^{P_0} \tag{11.17}$$

$$= (P_0 - P_1)\left(1 + \gamma(P_0+P_1) + \frac{\gamma^2}{3}(P_0^2 + P_0P_1 + P_1^2)\right), \tag{11.18}$$

where $R = 8\pi\mu L/A_0^2$ and $\gamma = c/A_0$. In the limit of zero compliance, this reduces to the linear ohmic law (11.1). Note that $1/\gamma$ has units of pressure, while R has units of pressure/flow. Thus, R can be interpreted as the flow resistance, as in (11.1).

Since Q is an increasing function of γ, it follows that a given flow can be driven by a smaller pressure drop in a compliance vessel than in a noncompliance vessel. This relationship is viewed graphically in Fig. 11.3, where we plot the scaled flux RQ as a function of pressure drop $\Delta P = P_0 - P_1$ for fixed γ and P_0. Clearly, the higher the compliance, the smaller the pressure drop required to drive a given fluid flux. This explains, for example, why the pressure drop in the veins can be much less than in the arteries, since the compliance of the veins is much greater than the compliance of the arteries.

We also calculate the volume of blood contained in a vessel with input pressure P_0 and output pressure P_1 to be

$$V = \int_0^L A(x)\,dx = \int_{P_0}^{P_1} A(P)x'(P)\,dP = -\frac{1}{8\pi\mu Q}\int_{P_0}^{P_1} A^3(P)\,dP. \tag{11.19}$$

For a linear compliance vessel (11.14) this is

$$\frac{V}{V_0} = \frac{3}{4}\left(\frac{(1+\gamma P)^4\big|_{P_1}^{P_0}}{(1+\gamma P)^3\big|_{P_1}^{P_0}}\right) \tag{11.20}$$

$$= 1 + \frac{\gamma}{2}(P_0+P_1) + \frac{\gamma^2}{6}(P_0-P_1)^2 + O(\gamma^3), \tag{11.21}$$

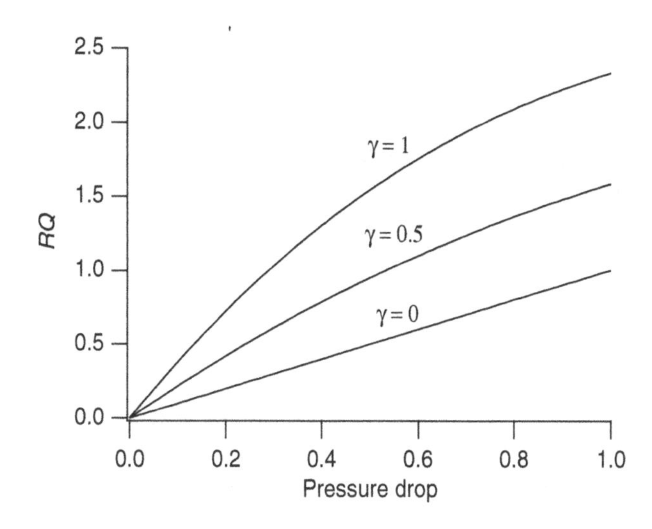

Figure 11.3 Scaled flow RQ (with units of pressure) as a function of pressure drop $\Delta P = P_0 - P_1$ for different values of compliance γ. For all curves $P_0 = 1.0$.

where V_0 is the volume of the vessel at zero pressure. It is left as an exercise (Exercise 3) to show that when there is no pressure drop across the vessel, so that $P_0 = P_1$,

$$V = V_0 + V_0 \gamma P_0 = V_0 + \frac{V_0}{A_0} cP, \tag{11.22}$$

which is the same as (11.13).

11.3 The Microcirculation and Filtration

The purpose of the circulatory system is to provide nutrients to and remove waste products from the cellular interstitium. To do so requires continuous filtration of the interstitium. This filtration is accomplished primarily at the level of capillaries, as fluid moves out of the capillaries at the arteriole end and back into the capillaries at the venous end.

The efflux or influx of fluid from or into the capillaries is determined by the local pressure differences across the capillary wall. In normal situations, the pressure drop through the capillaries is substantial, about 25 mm Hg.

A schematic diagram of the capillary network is shown in Fig. 11.4. To get some understanding of how filtration works and why a capillary pressure drop is necessary, we use a simple one-dimensional model of a capillary. We suppose that there is an influx Q_i at $x = 0$ that must be the same as the efflux at $x = L$, where L is the length of the capillary. At each point x along the capillary, there is blood flow q. The (hydrostatic) pressure P_c at each point along the capillary is determined by

$$\frac{dP_c}{dx} = -\rho q, \tag{11.23}$$

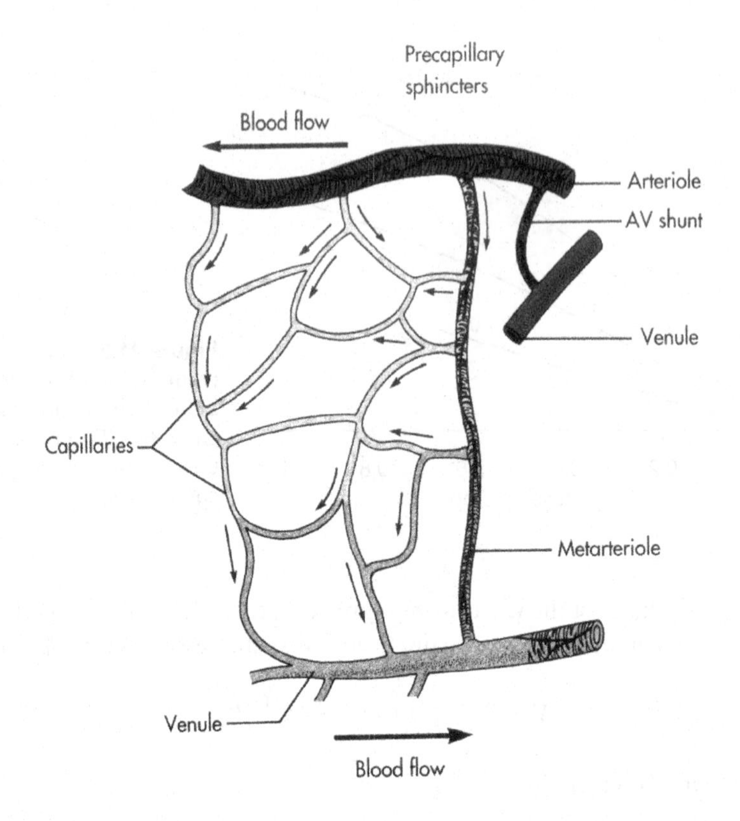

Figure 11.4 Diagram of the capillary microcirculation. (Berne and Levy, 1993, Fig. 28-1, p. 466.)

where ρ is the coefficient of capillary resistance. Flow into or out of the capillary (through the porous capillary wall) is determined by the difference between total internal pressure and total external (interstitial) pressure. The interstitial hydrostatic pressure P_i is typically about -3 mm Hg, and the interstitial fluid colloidal osmotic pressure π_i is about 8 mm Hg, while the plasma colloidal osmotic pressure π_c averages about 28 mm Hg (although it necessarily varies along the length of the capillary; see Exercise 4). Thus, the flow into the capillary is determined by

$$\frac{dq}{dx} = K_f(-P_c + P_i + \pi_c - \pi_i), \tag{11.24}$$

where K_f is the capillary filtration rate.

We assume that only P_c varies along the length of the capillary. Clearly, when P_c is high, fluid is forced out of the capillary, and when P_c drops sufficiently low, fluid is reabsorbed into the capillary.

Equations (11.23) and (11.24) form a linear system of equations, which is readily solved. We use the boundary conditions

$$q(0) = q(L) = Q_i, \tag{11.25}$$

thus giving two boundary conditions for the two linear equations. It is easiest to solve first for q. From (11.24) we see that

$$\frac{d^2q}{dx^2} = -K_f \frac{dP_c}{dx} = K_f \rho q. \tag{11.26}$$

Noting that q is the same at $x = 0$ and $x = L$, it must be symmetric about $x = \frac{L}{2}$, so that

$$q = B \cosh\left(\beta\left(x - \frac{L}{2}\right)\right), \tag{11.27}$$

where $\beta^2 = K_f \rho$. It follows easily from the boundary conditions that $B = \frac{Q_i}{\cosh(\frac{\beta L}{2})}$ and from (11.24) that

$$P_c = P_i + \pi_c - \pi_i - \frac{1}{K_f}\frac{dq}{dx} = P_i + \pi_c - \pi_i - \frac{Q_i \beta}{K_f} \frac{\sinh \beta(x - L/2)}{\cosh \frac{\beta L}{2}}. \tag{11.28}$$

In Exercise 4, this problem is studied in the phase plane, and generalized slightly to include variable osmolarity.

We can also calculate the relationship between the flux and total pressure drop along the length of the capillary. From the solution for P_c it follows that

$$\Delta P_c = P_c(0) - P_c(L) = \frac{2Q_i\beta}{K_f} \tanh \frac{\beta L}{2}, \tag{11.29}$$

and thus

$$Q_i = \frac{\Delta P_c}{R} \frac{\beta L/2}{\tanh \frac{\beta L}{2}}, \tag{11.30}$$

where $R = \rho L$ is the total resistance of the capillary. Notice the similarity of this formula to (11.1), describing the relationship between flow and pressure drop in a nonleaky vessel. Apparently, leakiness in the vessel ($\beta \neq 0$) has the effect of decreasing the overall resistance of the capillary flow.

Notice that $q(x)$ is minimal, and therefore the interstitial flow is maximal, at $x = L/2$. We define the filtration rate Q_f to be the maximal flux through the interstitium, in this case,

$$Q_f = Q_i - q(L/2). \tag{11.31}$$

It follows that

$$\frac{Q_f}{Q_i} = 1 - \operatorname{sech}\frac{\beta L}{2}. \tag{11.32}$$

The filtration rate depends on the single dimensionless parameter $\beta L = \sqrt{\rho K_f} L$. Thus filtration is enhanced in vessels that are "leaky" (large K_f) and of small radius, since vessel resistance ρ is inversely proportional to r^4.

More detailed studies of the microcirculation can be found in the work of Pries and Secomb (2000, 2005; Pries et al., 1996).

11.4 Cardiac Output

During a heartbeat cycle, the pressure and volume of the heart change in a highly specific way, most easily seen as a trajectory plotted in the pressure–volume phase plane, shown in Fig. 11.5. Notice that the pressure–volume trajectories, or "loops", are of rectangular shape, and thus the pressure and volume change at different places in the cycle. For example, on the right ascending side of the loop, the pressure increases while volume is constant, which corresponds to the ventricle contracting while the outflow valve is closed. At the top right-hand corner the outflow valve opens, and the volume decreases as the blood is pumped out at a constant pressure. The constant pressure at the top of the loop corresponds to the arterial pressure, while the constant pressure along the bottom of the loop corresponds to the venous pressure. Notice also that the top left corners of the loops lie on the same straight line. Thus, the ventricular volume at the end of a contraction (the end-systolic volume $V_{\rm ES}$) is a linear function of the arterial blood pressure. Further, in panel B of the same figure, the top left corner of the loop (the end-systolic pressure and volume) is constant, independent of the total volume of blood pumped by the ventricle, suggesting that it depends solely on the arterial pressure.

This suggests that

$$V_{\rm ES} = V_{\min} + C_s P_a, \tag{11.33}$$

where $V_{\min}$ is the intercept on the V axis, and C_s is the slope of the line connecting the three labeled points in Fig. 11.5A. The key observation is that C_s and $V_{\min}$ are independent of the arterial pressure and the end-diastolic volume, and are thus intrinsic properties of the ventricle. C_s is, in fact, the compliance of the ventricle at the end of systole. By connecting the lower right corners of the pressure–volume loops, as shown in Fig. 11.5C, we reach a similar conclusion for the end-diastolic volume.

These observations can be summarized in a simple model, in which we determine the total amount of blood pumped by the ventricle, i.e., the *cardiac output*, but ignore the time-dependent changes in volume and pressure over the beat cycle. We view the heart as a compliance vessel whose basal volume and compliance change with time. Thus,

$$V = V_0(t) + C(t)P. \tag{11.34}$$

During diastole, when the heart is filling, the heart is relaxed and compliant, so that V_0 and C are large. During this time, the aortic valve is closed, preventing backflow, so that the pressure is essentially the same as the venous pressure. During systole, the heart is contracting and much less compliant, so that C and V_0 are decreased compared to diastolic values. At this time, the mitral valve is closed, preventing backflow into the veins, so that the pressure in the heart is the same as the arterial pressure. Accordingly, the minimal volume (end systolic) is given by (11.33), and the maximal volume $V_{\rm ED}$

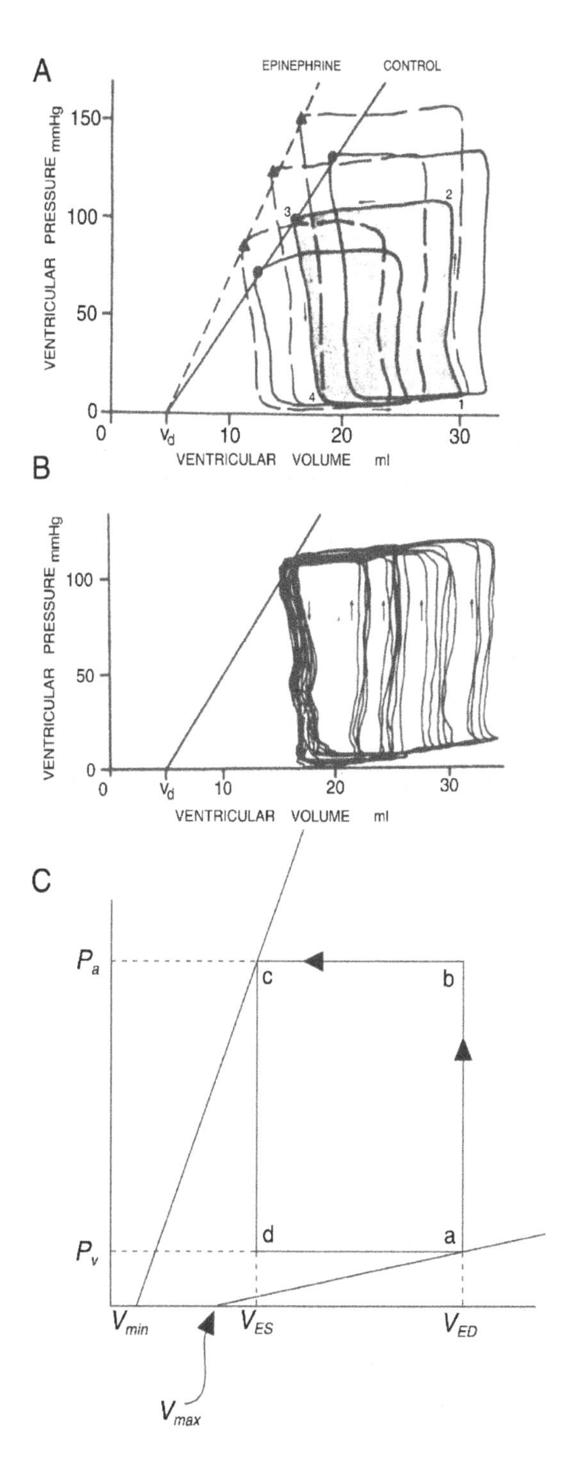

Figure 11.5 Experimental data of the pressure–volume relationship during the heartbeat cycle in the denervated left ventricle of the dog. (Sagawa et al., 1978, Fig. 11.4.) A: Three beats from different end-diastolic volumes and against different arterial pressures are shown in solid lines, with the broken lines representing the same beat cycles in the presence of adrenaline, which enhances the contraction. B: Pressure–volume loops of the same ventricle for four different end-diastolic volumes, but against the same arterial pressure. C: Schematic diagram of the pressure–volume loop (adapted from Hoppensteadt and Peskin, 2001, Fig. 1.5). a: inflow valve closes, b: outflow valve opens, c: outflow valve closes, d: inflow valve opens.

(end diastolic) is

$$V_{ED} = V_{max} + C_d P_v, \tag{11.35}$$

where P_a and P_v are the arterial and venous pressures, respectively, and C_s and C_d are the compliances of the heart during systole and diastole, respectively. This implies that the *stroke volume* is

$$V_{stroke} = V_{max} - V_{min} + C_d P_v - C_s P_a, \tag{11.36}$$

and the total cardiac output is

$$Q = FV_{stroke}, \tag{11.37}$$

where F is the heart rate in beats per unit time.

This expression of cardiac output has some features that agree with reality. For example, if C_s is small compared to C_d, then the cardiac output depends primarily on the venous pressure, or on the rate of venous return. This phenomenon, that cardiac output increases with increasing filling, is commonly referred to as *Starling's law*. While there is some decrease in output due to arterial loading, this effect is not nearly as significant as the increase in output resulting from an increase in venous pressure.

According to this formula, cardiac output is a linear function of venous pressure. This is not a terrible approximation in normal physiological ranges, although a more accurate formula would show saturation as a function of venous pressure. That is, cardiac output approaches a constant as a function of venous pressure for venous pressure above 10 mm Hg. Cardiac output also saturates at high frequencies because of inadequate fluid filling.

11.5 Circulation

11.5.1 A Simple Circulatory System

To illustrate how all the above pieces fit together to give a model of the circulatory system, consider a simple circulatory system with a single-chambered heart and a resistive closed loop (Fig. 11.6). For the resistive closed loop, we suppose that the total flux is related to the pressure drop through

$$Q = (P_a - P_v)/R, \tag{11.38}$$

so that in steady state the flux through the loop must match the cardiac output, yielding

$$Q = F(V_h + C_d P_v - C_s P_a) = (P_a - P_v)/R, \tag{11.39}$$

where $V_h = V_{max} - V_{min}$. Equation (11.39) gives a relationship between arterial and venous pressure that must be maintained in a steady-state condition. Unfortunately, these are not uniquely determined by this equation, so that this model does not have a unique solution.

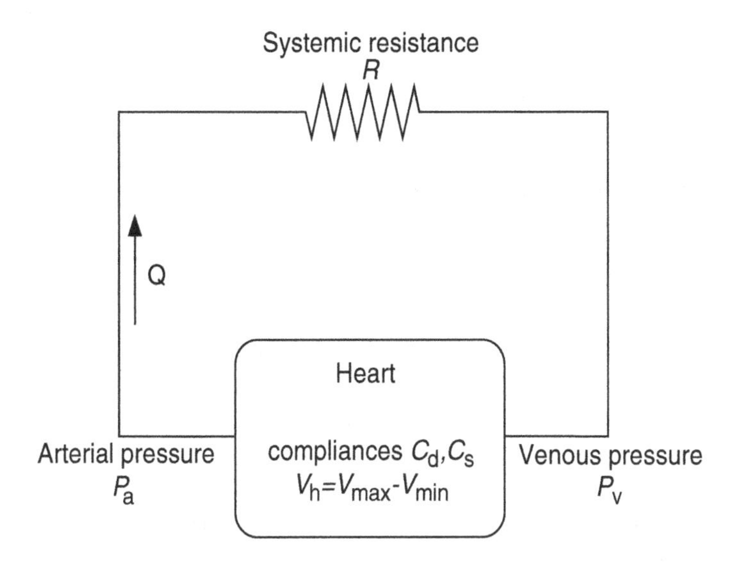

Figure 11.6 Schematic diagram of the simplest circulation model, with a single-chambered heart and a single loop.

To avoid this problem, we model the loop as a compliance vessel rather than as a simple resistance, which gives an additional relationship between pressure and total volume, thus uniquely specifying the system. Suppose that the closed circulatory loop is a compliance vessel with cross-sectional area given by (11.14). It follows from (11.17) that

$$Q = \frac{1}{3R\gamma}\left((1+\gamma P_a)^3 - (1+\gamma P_v)^3\right), \tag{11.40}$$

and the total volume of the vessel is given by

$$\frac{V}{V_0} = \frac{3}{4}\left(\frac{(1+\gamma P_a)^4 - (1+\gamma P_v)^4}{(1+\gamma P_a)^3 - (1+\gamma P_v)^3}\right). \tag{11.41}$$

Equation (11.40) is, of course, just the nonlinear version of (11.38), while (11.41) is the new equation that allows for a unique solution of the model. These two equations, together with

$$Q = F(V_h + C_d P_v - C_s P_a), \tag{11.42}$$

give a system of three equations in terms of the four unknowns Q, P_a, P_v, and V. The final equation comes from conservation of blood. Because blood is assumed to be incompressible, and because the heart chambers are assumed to have a fixed volume (as cardiac output is expressed in terms of the average output), it follows that V must be constant. The system is then completely determined. However, because it is nonlinear, a closed-form solution is not apparent, and the easiest way to obtain a solution is to solve the equations numerically.

Notice that it is also possible to use these equations to solve an inverse problem. That is, if V, Q, P_v, and P_a are known from measurements, then one could solve for the resistance R and compliances γ, C_d, and C_s.

11.5.2 A Linear Circulatory System

It is difficult to extend the analysis of the previous section to more realistic models because of the complexity of the resulting nonlinear equations. However, much can be learned using linear approximations to the governing equations. The simplest linear model is due to Guyton (1963), in which the circulatory system is represented as a closed loop with two compliance vessels and one pure resistance vessel (Fig. 11.7). The large arteries and veins are each treated as compliance vessels with linearized flow equations, and the systemic capillaries are treated as a resistance vessel with no compliance.

The equations describing this model can be conveniently divided into two groups: those describing the arterial system and those describing the venous system.

Arterial system: Cardiac output is described in terms of the compliance of the ventricle, and so

$$Q = F(C_d P_v - C_s P_a), \tag{11.43}$$

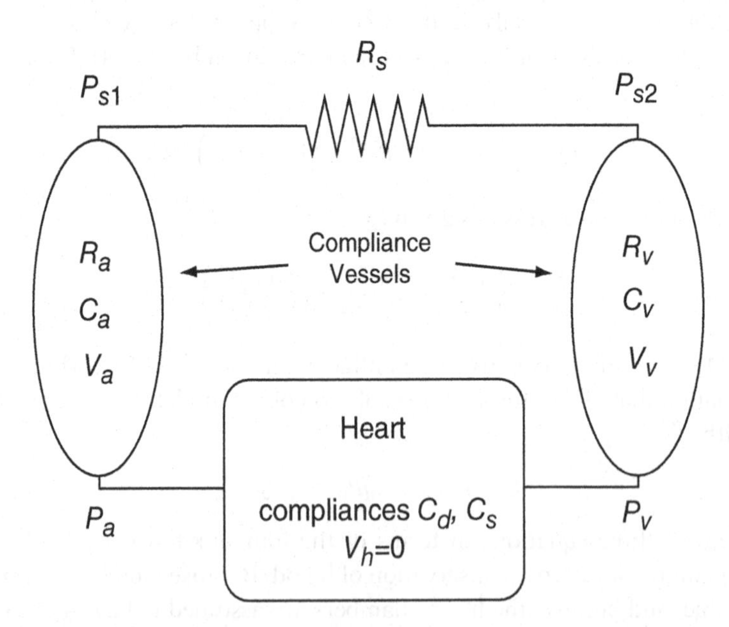

Figure 11.7 Schematic diagram of the two-compartment model of the circulation. The heart and pulmonary system are combined into a single vessel, and the systemic capillaries are modeled as a resistance vessel. The larger arteries and veins are modeled as compliance vessels.

where, for simplicity, we have assumed that $V_h = 0$, i.e., that $V_{\max} = V_{\min}$. The larger arteries are modeled as a compliance vessel, and thus, from (11.18),

$$Q = \frac{P_a - P_{s1}}{R_a}, \tag{11.44}$$

where the nonlinear terms are ignored. This approximation is reasonable if the compliance is small. The parameters are defined in Fig. 11.7. For example, P_a is the arterial pressure, while P_{s1} is the blood pressure at the (somewhat arbitrary) border between the larger arteries and the arterial capillaries, and R_a is the resistance of the larger arteries. Note that although this expression for the flux looks as though the arteries are treated as resistance vessels, this is only because the nonlinear terms are omitted. The compliance of the arteries appears in the relationship between the pressure and the volume of the arteries,

$$V_a = \frac{\gamma V_0}{2}(P_a + P_{s1}) = \frac{C_a}{2}(P_a + P_{s1}), \tag{11.45}$$

where $C_a = c_a V_0 / A_0$ is the compliance of the systemic arteries. Note that the volume of the systemic arteries is taken to be zero at zero pressure.

Venous system: We have three similar equations for the venous system, except that the equation for the cardiac output (11.43) is replaced by an equation describing the flow through the capillaries,

$$Q = \frac{P_{s1} - P_{s2}}{R_s}. \tag{11.46}$$

The remaining two equations describe the flow through the veins,

$$Q = \frac{P_{s2} - P_v}{R_v}, \tag{11.47}$$

and the volume of blood in the veins,

$$V_v = \frac{C_v}{2}(P_v + P_{s2}). \tag{11.48}$$

At this point, we have a system of six equations in seven unknowns (four pressures, two volumes, and Q). The final equation comes from conservation of volume, according to which

$$V_a + V_v = V_t, \tag{11.49}$$

where V_t is a given constant.

These seven equations, being linear, can be solved for the unknowns. However, before doing so, we make two further simplifications. First, we assume that the systolic compliance, C_s, is nearly zero, and second, that the pressure drops across the larger vessels are small, so that R_a and R_v are quite small, with the result that $P_a = P_{s1}$ and $P_v = P_{s2}$, to a good approximation. This removes two of the variables, leaving the system of five equations:

$$Q = FC_d P_v, \tag{11.50}$$

$$Q = \frac{P_a - P_v}{R_s}, \tag{11.51}$$

$$V_a = C_a P_a, \tag{11.52}$$

$$V_v = C_v P_v, \tag{11.53}$$

$$V_a + V_v = V_t. \tag{11.54}$$

The solution of this system is readily found (provided one uses symbolic manipulation software) to be

$$P_a = \frac{(1 + FC_d R_s)V_t}{C_v + (1 + FC_d R_s)C_a}, \tag{11.55}$$

$$P_v = \frac{V_t}{C_v + (1 + FC_d R_s)C_a}, \tag{11.56}$$

$$Q = \frac{FC_d V_t}{C_v + (1 + FC_d R_s)C_a}. \tag{11.57}$$

A number of qualitative features of the circulation can be seen from this solution.

1. As the heart rate increases, the arterial pressure P_a increases to a maximum of V_t/C_a, but as the heart rate falls, the arterial pressure decreases to a minimum of $V_t/(C_v + C_a)$.
2. Conversely, as the heart rate falls, the venous pressure P_v increases to a maximum of $V_t/(C_v + C_a)$. Hence, in heart failure, the arterial pressure falls and the venous pressure rises, until they are equal. With no pressure drop, there is no flow.
3. An increase in the systemic resistance, R_s, leads to a decrease in the cardiac output, an increase in the arterial pressure, and a decrease in the venous pressure.
4. Since $V_a = C_a P_a$ and $V_v = C_v P_v$, an increase in systemic resistance is accompanied by a shift in the blood volume from the venous system to the arterial system, i.e., V_v decreases, and V_a increases.

In reality, systemic resistance varies widely (decreasing, for example, during exercise), but the cardiac output compensates for this variation, keeping the arterial pressure relatively constant. Thus, the above model, which includes no control of cardiac output, needs to be modified to agree with experimental data. Later in this chapter we describe some simple models of regulation of the circulation. Before doing so, however, we consider a more complex model of the circulation, incorporating more compartments.

11.5.3 A Multicompartment Circulatory System

To construct a more detailed linear model of the circulatory system, we assume that the systemic and pulmonary loops each consist of two compliance vessels, the arterial and venous systems, connected by the capillaries, a pure resistance. Further, we assume that the heart has two chambers, the left and right hearts. A schematic diagram of the model is given in Fig. 11.8. We must write equations for the flow through each of these compartments and keep track of the total volume of blood contained in the

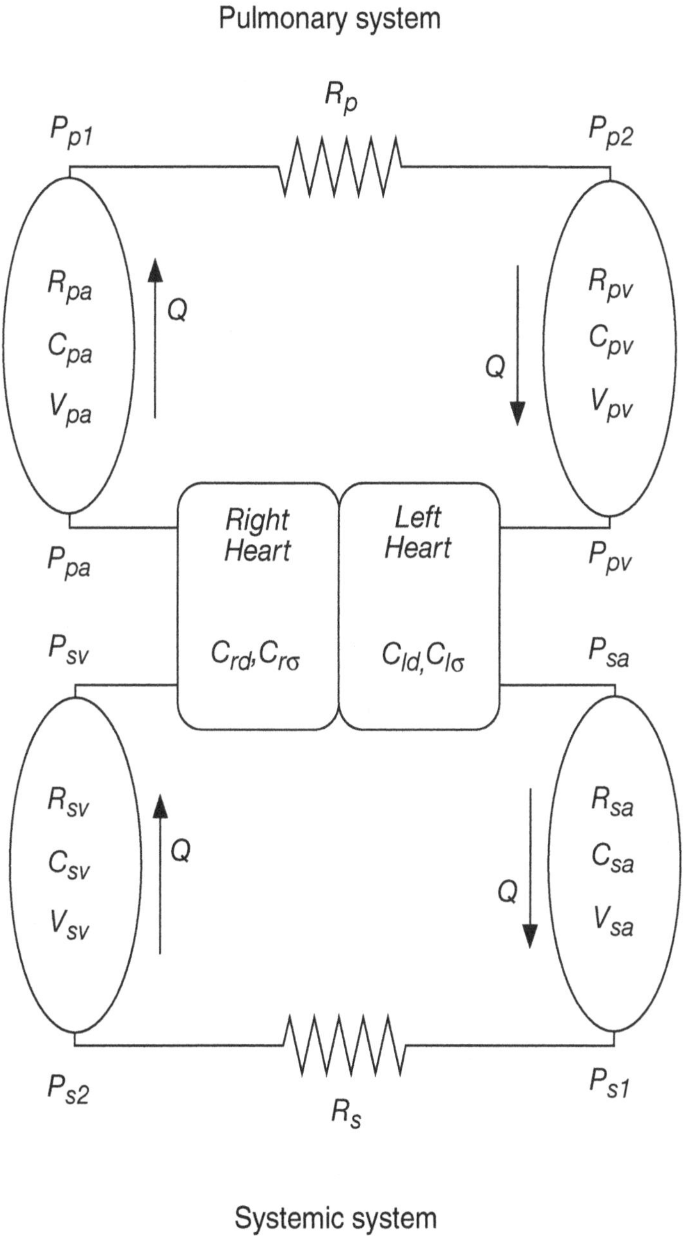

Figure 11.8 Schematic diagram of the multicompartment model of the circulation.

system. Unfortunately, the notation can be difficult to follow. We let subscripts a, v, s, and p denote, respectively, arterial, venous, systemic, and pulmonary. So, for example, P_{sa} is the pressure at the entrance to the systemic arteries, and C_{sa} is the compliance of the systemic arteries. Also, subscripts r, l, d, and σ denote, respectively, right, left, diastolic and systolic. Thus, C_{ld} denotes the diastolic compliance of the left heart. Finally, P_{s1} denotes the pressure at the ill-defined border between the systemic arteries and the systemic capillaries, with similar definitions for P_{s2}, P_{p1}, and P_{p2} as indicated in Fig. 11.8.

As before, we write the governing equations in groups.

Systemic arteries:

$$Q = \frac{P_{sa} - P_{s1}}{R_{sa}}, \tag{11.58}$$

$$Q = F(C_{ld}P_{pv} - C_{l\sigma}P_{sa}), \tag{11.59}$$

$$V_{sa} = V_0^s + \frac{C_{sa}}{2}(P_{sa} + P_{s1}). \tag{11.60}$$

Note that here it is assumed that the volume of the systemic arteries at zero pressure is V_0^s, not zero, as was assumed in the previous model.

Systemic veins:

$$Q = \frac{P_{s2} - P_{sv}}{R_{sv}}, \tag{11.61}$$

$$Q = \frac{P_{s1} - P_{s2}}{R_s}, \tag{11.62}$$

$$V_{sv} = \frac{C_{sv}}{2}(P_{sv} + P_{s2}). \tag{11.63}$$

For the venous system, it is reasonable to take the basal volume as zero, because if the blood pressure falls to zero, these vessels collapse. In the arterial system, however, such an approximation is not realistic.

Pulmonary arteries:

$$Q = \frac{P_{pa} - P_{p1}}{R_{pa}}, \tag{11.64}$$

$$Q = F(C_{rd}P_{sv} - C_{r\sigma}P_{pa}), \tag{11.65}$$

$$V_{pa} = V_0^p + \frac{C_{pa}}{2}(P_{pa} + P_{p1}). \tag{11.66}$$

Pulmonary veins:

$$Q = \frac{P_{p2} - P_{pv}}{R_{pv}}, \tag{11.67}$$

$$Q = \frac{P_{p1} - P_{p2}}{R_p}, \tag{11.68}$$

$$V_{pv} = \frac{C_{pv}}{2}(P_{pv} + P_{p2}). \tag{11.69}$$

At this stage there are 12 equations for 13 unknowns (8 pressures, 4 volumes, and Q). The final equation, as before, comes from the conservation of blood volume, whereby

$$V_{sa} + V_{sv} + V_{pa} + V_{pv} = V_t. \tag{11.70}$$

The capillary and heart volumes need not be included in this equation because they are assumed to be fixed.

This system of equations can be treated in a number of ways. First (using symbolic manipulation software), it is not difficult to find the solution directly. A second approach is to make a number of simplifying assumptions, as was done in the simpler model discussed above. For example, if it is assumed that there is no pressure drop over the arteries or veins (both pulmonary and systemic), then $P_{sa} = P_{s1}, P_{sv} = P_{s2}$, and similarly for the pulmonary equations. This removes four variables and four equations, giving a system of nine equations for the remaining nine unknowns. This variation of the model has been discussed by Hoppensteadt and Peskin (2001), and its further study is left as an exercise (Exercise 5).

A second approximation of the full system is to omit the systemic resistance and combine all the pulmonary vessels into a single compliance vessel. This results in a model consisting of only three compliance vessels (Fig. 11.9). It is left as an exercise (Exercise 6) to show that this approximation results from setting R_s and R_p approach zero and by setting $C_{pa} = C_{pv}$ and $R_{pa} = R_{pv}$. For convenience, we write $C_p = 2C_{pa}$, and $R_p = 2R_{pa}$. We also assume that the systolic compliances are negligible.

The equations governing this simplified three compartment model are as follows.

Systemic arteries:

$$Q = \frac{P_{sa} - P_s}{R_{sa}}, \tag{11.71}$$

$$Q = FC_{ld}P_{pv}, \tag{11.72}$$

$$V_{sa} = V_0^s + \frac{C_{sa}}{2}(P_{sa} + P_s). \tag{11.73}$$

Systemic veins:

$$Q = \frac{P_s - P_{sv}}{R_{sv}}, \tag{11.74}$$

$$V_{sv} = \frac{C_{sv}}{2}(P_{sv} + P_s). \tag{11.75}$$

Pulmonary system:

$$Q = \frac{P_{pa} - P_{pv}}{R_p}, \tag{11.76}$$

$$Q = FC_{rd}P_{sv}, \tag{11.77}$$

$$V_p = V_0^p + \frac{C_p}{2}(P_{pa} + P_{pv}). \tag{11.78}$$

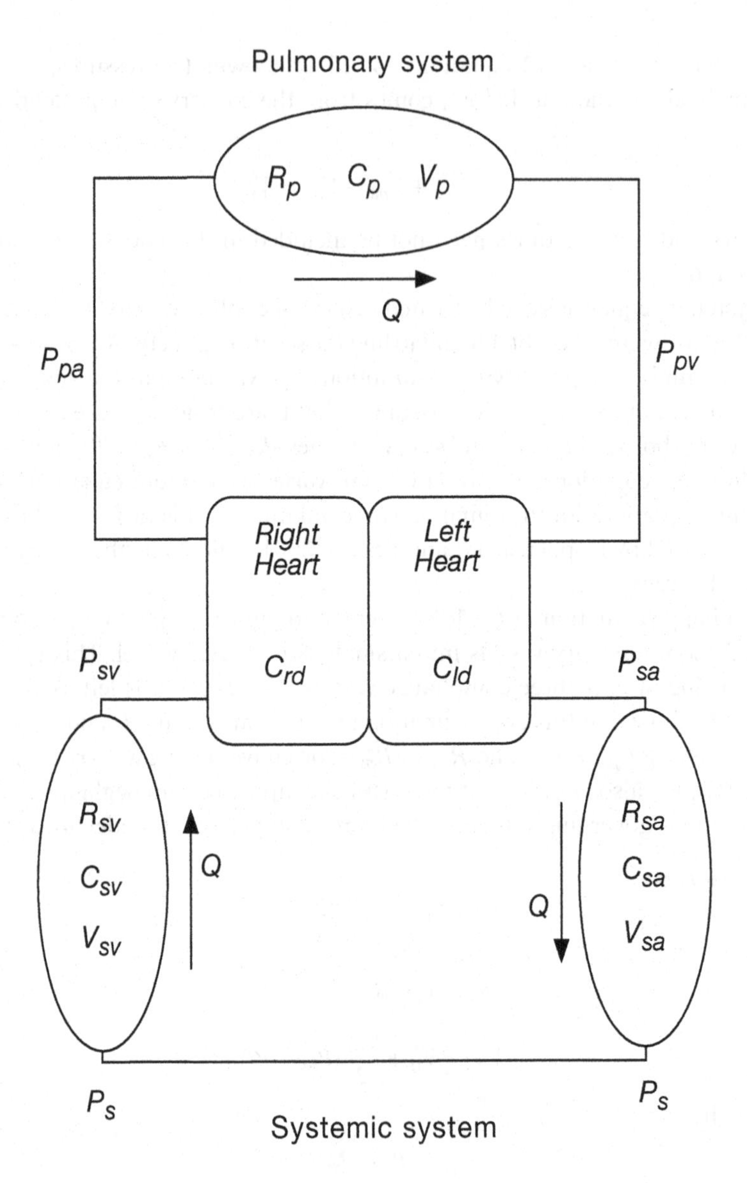

Figure 11.9 Schematic diagram of the simplified three-compartment model, obtained from the one in Fig. 11.8 by letting R_s and R_p go to zero and combining all the pulmonary vessels into one.

Conservation of volume:

$$V_{sa} + V_{sv} + V_p = V_t. \tag{11.79}$$

One can solve this linear system to obtain

$$P_{sa} = Q\left(\frac{1}{FC_{rd}} + R_{sa} + R_{sv}\right), \tag{11.80}$$

$$P_s = Q\left(\frac{1}{FC_{rd}} + R_{sv}\right), \tag{11.81}$$

$$P_{sv} = \frac{Q}{FC_{rd}}, \tag{11.82}$$

$$P_{pa} = Q\left(\frac{1}{FC_{ld}} + R_p\right), \tag{11.83}$$

$$P_{pv} = \frac{Q}{FC_{ld}}, \tag{11.84}$$

$$Q = \frac{V_e}{\alpha + \frac{C_p}{FC_{ld}} + \frac{C_{sv}+C_{sa}}{FC_{rd}}}, \tag{11.85}$$

where $\alpha = R_{sv}(C_{sa} + C_{sv}/2) + R_{sa}C_{sa}/2 + C_pR_p/2$, and $V_e = V_t - V_0^s - V_0^p$ is the excess volume beyond that which is necessary to fill the system at zero pressure. Clearly, the cardiac output saturates for large F, and

$$\lim_{F\to\infty} Q = Q_\infty = \frac{V_e}{\alpha} = \frac{2V_e}{R_{sv}(2C_{sa} + C_{sv}) + R_{sa}C_{sa} + C_pR_p}. \tag{11.86}$$

We can also see that, as in the simpler model, systemic arterial pressure P_{sa} is an increasing function of heart rate, and systemic venous pressure P_{sv} is a decreasing function of heart rate.

Cardiac output depends linearly on excess blood volume $V_t - V_0^s - V_0^p$. In trauma, if there is substantial blood loss, the cardiac output drops rapidly. If there is no compensatory control (increased heart rate or change of resistance and compliance), loss of 15–20% of blood volume over a period of half an hour is fatal. If reflexes are intact, loss of 30–40% in half an hour is fatal. Notice that in this model (taking $V_t = 5$ liters, $V_0^p + V_0^s = 1.2$ liters) a 20% loss of blood with no compensatory control leads to a 26% loss of cardiac output.

There are ten physical parameters in this system of nine equations, giving nine relationships between parameters if the solution is known from data. We take as typical volumes $V_{sa} = 1$, $V_{sv} = 3.5$, $V_p = 0.5$ (liters), and typical pressures $P_{sa} = 100$, $P_s = 30$, $P_{sv} = 2$, $P_{pa} = 15$, $P_{pv} = 5$ (mm Hg). Total cardiac output is about 5.6 liters/min, with a heart rate of 80 beats per minute and a stroke volume of 0.07 liter. Using these, we find estimates for parameters of $C_{sv} = 0.22$, $R_{sa} = 12.5$, $R_{sv} = 5.0$, $R_p = 1.78$ mm Hg/(liters/min), and $C_{ld} = 0.014$, $C_{rd} = 0.035$ (liters/mm Hg)/stroke.

As yet, C_{sa} or C_p cannot be determined because V_0^s and V_0^p are unknown. However, these quantities can be estimated using some additional information. To estimate V_0^s and C_{sa}, note that at the heart the arterial pressure varies on a beat-to-beat basis, as does the volume of blood in the arteries. During one beat, the blood ejected from the heart must be accommodated by the compliance of the arteries. Thus,

$$\Delta V = C_{sa}\Delta P, \tag{11.87}$$

where ΔV is the stroke volume, about 0.07 liter, and ΔP is the difference between systolic and diastolic pressure, about 40 mm Hg. Once C_{sa} is known ($C_{sa} = 0.0018$

liter/mm Hg), the resting volume V_0^s can be determined using the given values of volume and pressure in (11.73) as $V_0^s = 0.88$ liters. Assuming that $V_0^s + V_0^p = 1.2$ liters gives $V_0^p = 0.32$ liters. Finally, we substitute this value of V_0^p into (11.78) and solve for C_p to get $C_p = 0.018$ liters/mm Hg.

These numbers correlate with some known features of the adult circulatory system. For example, the venous system is about 24 times more compliant than the arterial system. The ratio found here is $C_{sv}/C_{sa} = 122$, which is too high, but this does not create significant errors in interpretation. Total resistance of the systemic circulation is larger than the pulmonary resistance, and the compliance of the left heart is less than that of the right heart, simply because the left ventricular wall is much thicker than the right.

To gain some understanding of the dependence of the solution on parameters, we calculate the sensitivity σ_{yx}, which is the sensitivity of dependence of the dependent variable y upon changes in the independent variable x. Thus, σ is the proportional change in y for a given proportional change in x, and so

$$\sigma_{yx} = \frac{\Delta y/y}{\Delta x/x} = \frac{x}{y}\frac{\partial y}{\partial x}, \tag{11.88}$$

in the limit as Δx goes to zero. In Table 11.2 are shown the sensitivities, expressed as percentages, for the three-compartment loop using normal parameter values, as shown in the table. For example, the first number in the first column, −2.8, is the sensitivity of the systemic arterial pressure P_{sa} to changes in the systemic arterial compliance C_{sa}.

From this table we infer some interesting features of the human circulatory system. First, the system is relatively insensitive to changes in the arterial compliance C_{sa}. In fact, compliance of the arterial system is insignificant compared to compliance of the other compartments and to a first approximation can be ignored. On the other hand, the system is strongly sensitive to changes in C_{sv}, the venous compliance. Similarly, the solution is relatively insensitive to changes in arterial resistance, R_{sa}, but is relatively sensitive to changes in venous resistance, R_{sv}. Much of the regulation of the cardiac

Table 11.2 Sensitivities for the three compartment circulatory system (expressed as percentages).

	Normal	C_{sa} 0.0018	C_{sv} 0.22	C_p 0.018	R_{sa} 12.5	R_{sv} 5	R_p 1.78	C_{ld} 0.014	C_{rd} 0.035	F 80
P_{sa}	100	−2.8	−85	−12	68	−48	−6	6	8.7	15
P_s	30	−2.8	−85	−12	−1.5	18	−6	6	4.0	10
P_{sv}	2	−2.8	−85	−12	−1.5	−76	−6	6	−89	−83
P_{pa}	15	−2.8	−85	−12	−1.5	−76	60	−27	11	−17
P_{pv}	5	−2.8	−85	−12	−1.5	−76	−6	−94	11	−83
V_{sa}	1	11	−9.2	−1.3	5.7	−3.5	−0.6	0.6	0.8	1.5
V_{sv}	3.5	−2.8	15	−12	−1.5	12	−6	6	−1.8	4.3
V_p	0.5	−1.7	−50	52	−0.9	−45	26	−26	6.3	−20
Q	5.6	−2.8	−85	−12	−1.5	−76	−6	6	11	17

systems occurs through changes in the compliance and resistance of the venous system, and this result demonstrates the efficacy of that choice.

According to the American Heart Association, cardiovascular disease, including high blood pressure, coronary heart disease, stroke and heart failure, accounted for over 36% of all deaths in the USA in 2005; coronary heart disease, caused by atherosclerosis, was the single leading cause of death. *Atherosclerosis* is a chronic inflammatory response in the walls of arteries, often called "hardening" of the arteries, and is caused by the deposition of excess cholesterol and fats in the arteries. These deposits are invaded by fibrous tissue and frequently become calcified, resulting in atherosclerotic plaques and stiffened arterial walls that can be neither constricted nor dilated. While systemic compliance is not extremely important in this model, systemic resistance is significant, and increases in systemic resistance produce increases in arterial pressure.

A person with higher than normal mean arterial pressure is said to have *hypertension*, or high blood pressure. Life expectancy is shortened substantially when mean arterial pressure is 50 percent or more above normal. The lethal effects of hypertension are

1. Increased cardiac workload, leading to congestive heart disease or coronary heart disease, often leading to a fatal heart attack;
2. Rupture of a major blood vessel in the brain (a *stroke*), resulting in paralysis, dementia, blindness, or multiple other brain disorders;
3. Multiple hemorrhages in the kidneys, leading to renal destruction and eventual kidney failure and death.

One other parameter that has an important effect is the diastolic compliance of the left heart, C_{ld}. As expected, if this compliance decreases, there is a reduction in systemic arterial pressure and a reduction in cardiac output. There is also a noticeable increase in pulmonary blood volume, V_p. Thus, left heart failure, which corresponds to a weakening of the left ventricular muscles and hence decreased cardiac efficiency and decreased compliance, results in excess fluid and fluid congestion in the lungs, known as *pulmonary edema*. Notice that in this model, failure of the left or right heart does not influence the maximal cardiac output Q_∞, although it certainly requires a higher heart rate to effect the same output. Notice, also, that with left or right heart failure, systemic volume changes little, so that one does not expect peripheral edema.

11.6 Cardiovascular Regulation

The circulatory system is equipped with a complex system for the control of arterial pressure and cardiac output. Over the short term (seconds to hours) this control occurs partly at the level of the arterioles and partly at the heart itself. Smooth muscle around the arterioles can constrict or relax, thus changing the resistance, while innervation of the heart can control the cardiac output in response to conditions in the periphery.

These short-term mechanisms dampen out the considerable variation that arises in a normal day's activity, from lying down to standing up to running one hundred meters.

There are three major types of peripheral control mechanisms:

1. Local (intrinsic) control of blood flow in the individual tissue, determined mainly by the tissue's need for blood perfusion. Local control is often called *autoregulation*.
2. Neural (extrinsic) control, by which the overall vesicular resistance and cardiac activity are controlled. The baroreceptor reflex is the most widely studied of these extrinsic mechanisms.
3. Humoral control, in which substances dissolved in the blood, such as hormones, ions, or other chemicals, cause changes in flow properties. These feedbacks are mediated by chemoreceptors.

Over the long term (days and months), blood pressure is regulated by renal mechanisms, whereby an increase in arterial pressure causes the kidneys to excrete more fluid, thus decreasing the blood volume and the arterial pressure.

Because of the intricacy of these control mechanisms, it comes as no surprise that feedback instabilities can arise, leading to periodicity in the heart beat rate and the mean arterial pressure. There are a number of different observed oscillations (although the oscillatory behavior is often highly irregular) and none is well understood. One of the most widely studied oscillations is that of heart rate, which varies periodically with respiration, such oscillations being called *respiratory sinus arrhythmia*, or RSA. The reverse is true also, with the heart rate affecting the rate of breathing. Such *cardiorespiratory coupling*, or *cardioventilatory coupling*, can result in phase locking of heart rate and breathing. Other oscillations in respiration rate, such as *Cheyne–Stokes* breathing, can also occur, and are discussed in Chapter 14.

The first observations of sinus arrhythmia were made by Ludwig. He invented the kymograph in 1846, a stylus connected to a mercury manometer connected to a rotating smoke drum, with which he could record accurately, for the first time, variables such as blood pressure, pulse rate, and respiratory frequency (Fye, 1986). Ludwig sent his original tracing of sinus arrhythmia to one of his pupils, with an inscription on the back (dated December 12, 1846) calling his observations "this first stammering of the heart and of the chest" (Lombard, 1916). Some years later, Mayer (1877) discovered arterial pressure oscillations with a frequency of about 0.1 Hz, and these were named *Mayer waves*. Oscillations in arterial pressure with a frequency around that of respiration were discovered by Traube (1865) and Hering (1869) and are called *Traube–Hering waves*. These names are not entirely consistent from author to author; some books such as Guyton and Hall (1996) describe Mayer waves and Traube–Hering waves as the same phenomenon, with the generic name *vasomotor* waves, but other authors such as Cohen and Taylor (2002) and Berne and Levy (1998) make a distinction between the frequencies of Mayer and Traube–Hering waves.

In summary, there is a confusing plethora of oscillations in heart rate, arterial pressure, and respiration, and these oscillations are almost certainly closely connected in ways we do not yet understand. Although many quantitative models of these oscillations have been constructed (Cohen and Taylor, 2002; Julien, 2005), there is no unified model that explains all of these behaviors in terms of feedback instabilities of known control mechanisms.

11.6.1 Autoregulation

Autoregulation is a local mechanism that makes flow through a tissue responsive to local oxygen demand but relatively insensitive to arterial pressure. In tissue for which the delivery of oxygen is of central importance (for example, the brain or the heart) the local blood flow is controlled to be slightly higher than required, but no higher.

In dead organs, an increase in arterial pressure produces a linear increase in blood flow, suggestive of a linear-resistance vessel. However, in normally functioning tissue, the arterial pressure can be changed over a large range with little effect on the blood flow (Fig. 11.10). For example, in muscle, with an arterial pressure between 75 mm Hg and 175 mm Hg the blood flow remains within ±10%–15% of normal.

The flow through an artery is known to be responsive to the need for oxygen. For example, an eightfold increase in metabolism produces a fourfold increase in blood flow (as shown in Fig. 11.11B). Similarly, if oxygen content falls because of anemia, high altitude, or carbon monoxide poisoning, the blood flow increases to compensate. For example, a reduction to 25% of normal oxygen saturation produces a threefold increase in blood flow, not quite enough to compensate fully for the loss (see Fig. 11.11A).

Although the mechanism for autoregulation is not completely understood, it is most likely that resistance of tissue is responsive to biochemical measures of how hard

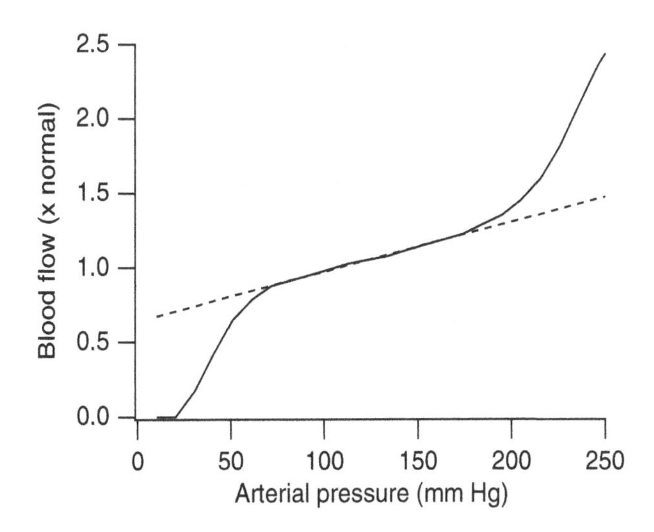

Figure 11.10 Blood flow as a function of arterial pressure if pressure is raised over a period of a few minutes. (Data (solid curve) from Guyton and Hall, 1996, Fig. 17-4, p. 203.) Dashed curve shows the result from the model (11.92).

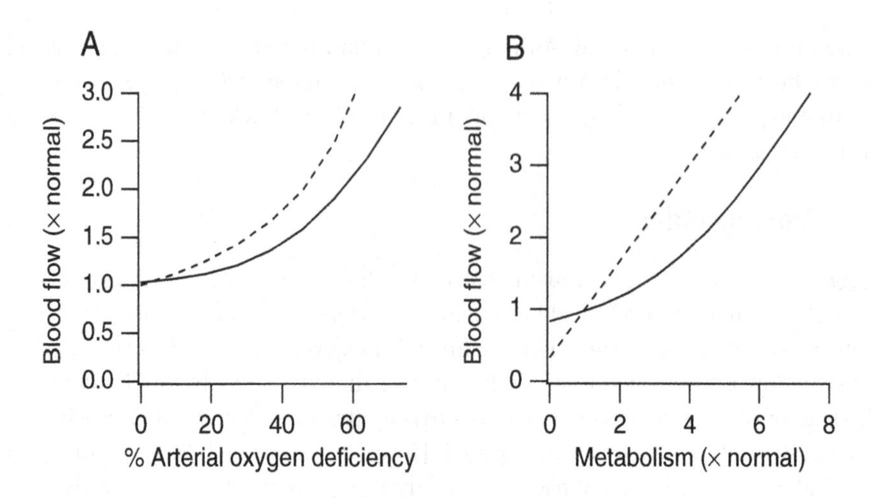

Figure 11.11 A: Blood flow as a function of the percentage of arterial oxygen deficiency, keeping arterial pressure and metabolic rate fixed. (Data (solid curve) from Guyton and Hall (1996), Fig. 17-2, p 200.) Dashed curve shows the result from the model. B: Blood flow as a function of metabolism, keeping arterial pressure and arterial oxygen content fixed at normal levels. (Data (solid curve) from Guyton and Hall, 1996, Fig. 17-1, p. 200.) Dashed curve shows the result from the model.

it is working, such as concentrations of H^+, CO_2, O_2, and lactic acid. The arterioles are highly muscular, and their diameters can change manyfold. The metarterioles (terminal arterioles) are encircled by smooth muscle fibers at intermittent points and are also used to regulate flow. However, the arterial blood has the same composition for all tissues of the body so cannot be used as a local control mechanism. This is problematic because it is the arterioles whose resistance is regulated. Here we assume that the resistance of arterioles is a function of the concentration of oxygen in the venous blood. This is possible, since arteries and veins tend to run side by side, and venous concentrations may regulate arterial resistance by release and diffusion of regulatory substances, called *vasodilators*.

An example of how this may work is as follows. If cardiac activity increases and the utilization of oxygen exceeds the supply, ATP is degraded, increasing the concentration of *adenosine*. Adenosine is a vasodilator, which leaks out of the cells into the venous flow to cause local dilation of coronary arteries. Thus, as the oxygen supply decreases, so does the arterial resistance.

A simple model of autoregulation is as follows (Huntsman et al., 1978; Hoppensteadt and Peskin, 2001). We keep track of oxygen consumption and blood flow via

$$([O_2]_a - [O_2]_v)Q = M, \tag{11.89}$$

$$P_a - P_v = RQ, \tag{11.90}$$

where $[O_2]_a$ and $[O_2]_v$ are the arterial and venous oxygen concentrations, respectively, M is the metabolic rate (oxygen consumption per unit time), P_a and P_v are the arterial and venous pressures driving the flow Q through tissue with total resistance R. In this model $[O_2]_a$ is treated as a given constant and $[O_2]_v$ as variable. Now we assume that there is some linear relationship between arterial resistance and venous oxygen content, say

$$R = R_0\left(1 + A[O_2]_v\right), \tag{11.91}$$

where $A > 0$. The assumption of linearity is reasonable in restricted ranges of oxygen content. Here, the parameter A denotes the sensitivity of resistance to oxygen; if $A = 0$, the resistance is unregulated.

These equations can be solved for the flow rate Q to get

$$Q = \frac{1}{1 + A[O_2]_a}\left(MA + \frac{P_a}{R_0}\right), \tag{11.92}$$

where it is assumed that $P_v = 0$. This is a linear relationship between flow rate and arterial pressure, which, when $A = 0$, reproduces the unregulated situation. However, with $A > 0$, the sensitivity of the flow to changes in arterial pressure varies with arterial oxygen content. Furthermore, this expression shows linear dependence of blood flow on metabolism.

We can estimate the parameters A and R_0 using the data from Fig. 11.10. In the range of pressures between 75 and 175 mm Hg, the curve is well represented by the straight line

$$\frac{Q}{Q^*} = \frac{1}{3} + \frac{2}{3}\frac{P_a}{P^*}, \tag{11.93}$$

where Q^* and P^* are the normal values of flow and pressure ($Q^* = 5.6$ liters/min, $P^* = 100$ mm Hg). Comparing this to the regulated curve (11.92) at normal values,

$$Q = \frac{1}{1 + A[O_2]_a^*}\left(M^*A + \frac{P_a}{R_0}\right), \tag{11.94}$$

where $[O_2]_a^*$ and M^* are normal values of arterial oxygen and metabolism, respectively ($M^* = Q^*([O_2]_a^* - [O_2]_v^*)$, $[O_2]_a^* = 104$ mm Hg, $[O_2]_v^* = 40$ mm Hg), we find that

$$A = \frac{Q^*}{3M^* - Q^*[O_2]_a^*}, \tag{11.95}$$

$$R_0 = \frac{P^*(3M^* - Q^*[O_2]_a^*)}{2Q^*M^*}, \tag{11.96}$$

so that

$$\frac{Q}{Q^*} = \frac{3\frac{M}{M^*}}{\left(1 + 2\frac{P_a^*}{P_a}\right)\left(3 + \frac{13}{8}\left(\frac{[O_2]_a}{[O_2]_a^*} - 1\right)\right)}, \tag{11.97}$$

using typical values for $\frac{Q^*[O_2]_a^*}{M^*} = \frac{13}{8}$.

In Fig. 11.10 are shown the data for blood flow as a function of arterial pressure, compared with the model (11.97). The good agreement over the linear range is the result of fitting.

The relationship (11.92) reproduces two other features of autoregulation that are qualitatively correct. It predicts that the flow rate increases as the arterial oxygen content decreases, and increases linearly with metabolic rate. In Fig. 11.11A is shown the blood flow plotted as a function of arterial oxygen deficiency. Here, the solid curve is taken from data, and the dashed curve is from (11.97). Similarly, in Fig. 11.11B is shown the blood flow plotted as a function of metabolic rate. As before, the solid curve is from data, and the dashed curve is from (11.97). Clearly, the model gives reasonable qualitative agreement for blood flow as a function of arterial oxygen content, and for blood flow as a function of metabolism.

11.6.2 The Baroreceptor Loop

The *baroreceptor loop* is a global feedback control mechanism using the nervous system to adjust the heart rate, the venous resistance, and thereby the venous pressure in order to maintain the arterial pressure at a given level, with the ultimate goal of regulating the cardiac output.

The need to regulate cardiac output is apparent. During exercise, when the demand for oxygen goes up, cardiac output normally rises at a linear rate, with slope about 5 (since 5 liters of blood are required to supply 1 liter of oxygen). In normal situations, the cardiac output and heart rate are roughly proportional, indicating that the stroke volume remains essentially constant. However, if heart rate is artificially driven up with a pacemaker, with no increase in oxygen consumption, then the cardiac output remains virtually the same, indicating a decrease in stroke volume. Similarly, in exercise with a fixed heart rate (set by a pacemaker), total cardiac output increases to meet the demand.

The primary nervous mechanism for the control of cardiac output is the *baroreceptor reflex* (Fig. 11.12). This reflex is initiated by stretch receptors, called *baroreceptors* or *pressoreceptors*, located in the walls of the *carotid sinus* and *aortic arch*, large arteries of the systemic circulation. A rise in arterial pressure is detected and causes a signal to be sent to the central nervous system from which feedback signals are sent through the autonomic nervous system to the circulatory system, thereby enabling the regulation of arterial pressure. For example, the baroreceptor reflex occurs when a person stands up after having been lying down. Immediately upon standing, the arterial pressure in the head and upper body falls, with dizziness or loss of consciousness a distinct possibility. The falling pressure at the baroreceptors elicits an immediate reflex, resulting in a strong sympathetic discharge throughout the entire body, thereby minimizing the decrease in blood pressure in the head. This observation suggests that the larger dinosaurs required a well-tuned baroreceptor reflex in order not to faint every time they raised their heads.

The most important part of the autonomic nervous system for regulation of the circulation is the *sympathetic nervous system*, which innervates almost all the blood

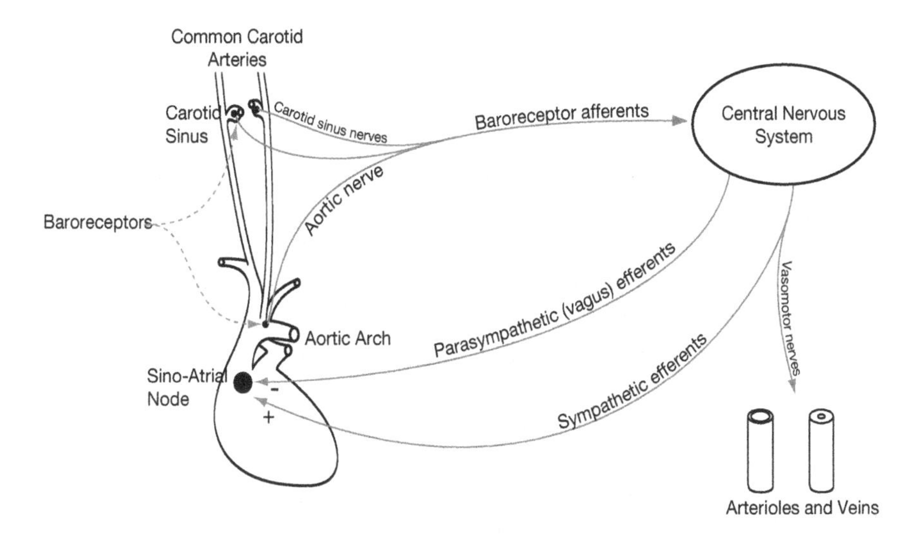

Figure 11.12 Schematic diagram of the baroreceptor reflex pathway. Adapted from Ottesen (1997), Fig. 2.

vessels, with the exception of the capillaries. The primary effects of sympathetic nervous stimulation are

1. contraction of small arteries and arterioles (by stimulation of the surrounding smooth muscle) to increase blood flow resistance and thereby decrease blood flow in the tissues;
2. constriction of the veins, thereby decreasing the amount of blood in the peripheral circulation;
3. stimulation of the heart muscle, thereby increasing both the heart rate and stroke volume.

The *parasympathetic* or *vagus* nerves have the opposite effect on the heart, decreasing the heart rate and strength of contractility. Strong sympathetic stimulation can increase the heart rate in adult humans to 180–200 beats per minute, and even as high as 220 beats per minute in young adults. Strong parasympathetic stimulation can lower the heart rate to 20–40 beats per minute and can decrease the strength of contraction by 20 to 30 percent.

The effect of the baroreceptors is to increase sympathetic stimulation and decrease parasympathetic stimulation when there is a drop in arterial pressure. This increase in sympathetic activity, in turn, increases the heart rate, the systemic resistance, and the cardiac compliances, and decreases the venous compliance. Notice from Table 11.2 that in the unregulated circulation these changes effect an increase of arterial pressure. Thus, the overall effect of the baroreceptor loop is to maintain the arterial pressure at a desired level.

The sympathetic nervous system is also stimulated by the brain vasomotor center, where an increase in carbon dioxide in the brain acts to cause a widespread vasoconstriction throughout the body. A sympathetic response is also stimulated by fright or anger, and is called the *alarm reaction*.

The sympathetic nervous system acts by three mechanisms. First, it stimulates the contraction of vessels by the release of vasoconstrictors, primarily *noradrenaline* (norepinephrine). Simultaneously, the adrenal medullae are stimulated to release adrenaline (epinephrine) and noradrenaline into the circulating blood. These two hormones are carried in the bloodstream to all parts of the body, where they act directly on blood vessels, usually to cause contraction. (Some tissues respond to adrenaline by dilation rather than constriction.) The action of secreted noradrenaline lasts about 30 minutes. Finally, sympathetic nervous activity acts to increase heart rate and heart contractility.

To include the baroreceptor loop and the sympathetic nervous system in our circulation model, we suppose that the level of sympathetic stimulation is given by S, and that S is related to the deviation of the arterial pressure P_{sa} through a simple linear relationship

$$S = S^* + \beta(P_{sa}^* - P_{sa}), \tag{11.98}$$

so that as arterial pressure decreases, sympathetic stimulation increases. Here, S^* and P_{sa}^* are "normal" values. In animal experiments, blocking all sympathetic activity leads to a drop of arterial pressure from 100 to 50 mm Hg, indicating a continuous basal level of sympathetic firing at normal pressure ($S^* \neq 0$), called *sympathetic tone*, known to be about one impulse per second.

Next, we assume that heart rate F, arterial resistance R_{sa}, and cardiac compliances C_{ld} and C_{rd} are (unspecified) increasing functions of S, while the venous compliance C_{sv} is a decreasing function of S. We account for the metabolic need of the tissue through

$$([O_2]_a - [O_2]_v)Q = M, \tag{11.99}$$

and suppose that the metabolic need is communicated to the tissue through autoregulation via

$$R_{sa} = R(S) + A[O_2]_{sv}. \tag{11.100}$$

This representation is slightly different from that in (11.91), but it has the same interpretation.

Combined with the balance equations (11.71)–(11.79), we have a closed system of equations, which can be solved to find the cardiac output Q as a function of metabolic need M. The solution is complicated, and so we leave the details to the interested reader.

However, it is useful to view the solution of these equations in a slightly different way. We suppose that the effect of baroreceptor feedback is to hold the arterial pressure fixed at some target level and to adjust other parameters such as arterial resistance, venous compliance, and heart rate so that this target pressure is maintained. Then we can view P_{sa} as a parameter of the model and let the heart rate, say, be an unknown. Thus we solve the governing equations, not for the pressures and cardiac output as functions of the heart rate, as with the unregulated flow, but for heart rate and cardiac output as functions of arterial pressure and metabolism. We then obtain

$$F = \frac{\left(\frac{1}{C_{rd}}(2C_{sv} + C_{sa}) + \frac{2}{C_{ld}}C_p\right)(AM + P_{sa}) + \frac{1}{C_{rd}}(P_{sa}C_{sa} - 2V_e)}{(2V_e - P_{sa}C_{sa})(R_{sa} + R_{sv}) - (AM + P_{sa})(C_{sv}R_{sv} + C_pR_p + C_{sa}R_{sv})}, \tag{11.101}$$

$$Q = \frac{\left(C_{ld}(2C_{sv} + C_{sa}) + 2C_{rd}C_p\right)(AM + P_{sa}) + C_{ld}(P_{sa}C_{sa} - 2V_e)}{C_{ld}\left(R_{sa}(2C_{sv} + C_{sa}) + R_{sv}C_{sv} - C_pR_p\right) + 2C_pC_{rd}(R_{sa} + R_{sv})}. \tag{11.102}$$

Although these formulas are somewhat complicated and obscure, here we see a number of features for the controlled circulation that are markedly different from those for the uncontrolled circulation. Most obvious is that heart rate and cardiac output respond to changes in metabolic need M. In fact, the cardiac output can be increased by increasing the arterial pressure or decreasing the systemic resistances.

These formulas also show some difficulties that the control system faces. Notice that there are parameter ranges for which either the numerator or denominator is negative. These are parameter values for which the solution is not valid, or, said another way, that are outside the range of physical possibility. Thus, for example, certain target pressures P_{sa} cannot be maintained if V_e is either too large or too small. Similarly (and not surprisingly), there are some large values of metabolism and pressure $(AM + P_{sa})$ that are impossible to maintain.

If the heart rate cannot be controlled by the baroreceptor loop, as for example when there is an implanted pacemaker, then F must be viewed as a tunable parameter of the model rather than an unknown. Instead, some other variables, such as cardiac compliance, are the unknowns. If we suppose that the cardiac compliances are always in the same ratio, then it is easy to see that an increase in heart rate leads to an exactly compensating decrease in compliance and stroke volume, so that the same total output is maintained.

Oscillations in the Baroreceptor Loop

One of the motivations for studying the baroreceptor loop is to try to understand the occurrence of oscillations in arterial pressure, such as the Mayer and Traube–Hering waves discussed above. This is not possible with a steady-state model like that presented in the previous section. There have been a number of mathematical models of the baroreceptor reflex that concentrate on oscillatory instabilities, the principal ones being that of de Boer et al. (1987), and that of Ottesen (1997) (later studied in a modified form by Fowler and McGuinness, 2005). Vielle (2005) has also published a model of Mayer waves, based on an earlier model of Ursino (1998, 1999). Here we present a simplified version of the Ottesen model.

We model the circulatory system as a simple lumped system, in which all the arteries are modeled by a single compliance vessel, as are all the veins, as sketched in Fig. 11.13. This is essentially a *Windkessel* model, a type of model that is discussed in more detail in Section 11.8.2.

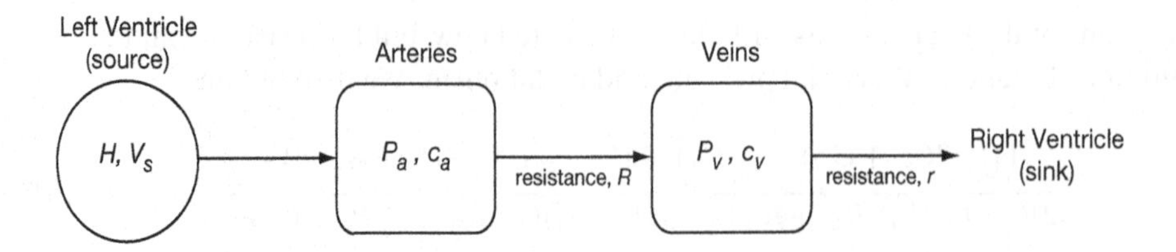

Figure 11.13 The lumped model of the circulatory system used in the model of the baroreceptor loop.

The volume of the arterial compartment is $P_a c_a$, and thus

$$\frac{d}{dt}(P_a c_a) = \frac{1}{R}(P_v - P_a) + V_s F, \tag{11.103}$$

where F is the heart rate and V_s is the stroke volume. If we assume that the arterial compliance is constant (i.e., we model only the effects of the baroreceptor loop on heart rate), then

$$c_a \frac{dP_a}{dt} = \frac{1}{R}(P_v - P_a) + V_s F. \tag{11.104}$$

Similarly,

$$c_v \frac{dP_v}{dt} = \frac{1}{R}(P_v - P_a) - \frac{1}{r} P_v. \tag{11.105}$$

We model the changes in heart rate with a sigmoidally increasing function of the arterial pressure (the action of the sympathetic efferents in Fig. 11.12) and a sigmoidally decreasing function of the arterial pressure (the parasympathetic efferents in Fig. 11.12). However, the parasympathetic system is assumed to act with a delay of τ seconds. Thus,

$$\frac{dF}{dt} = \alpha g(P_a(t - \tau)) - \beta(1 - g(P_a)), \tag{11.106}$$

where g_s is the sigmoidally decreasing function

$$g(P_a) = \frac{\mu^7}{\mu^7 + P_a^7}, \tag{11.107}$$

for some constant μ. This choice of function for the sympathetic and parasympathetic controls, although highly simplified in this version of the model, is based on experimental data and is discussed in more detail by Danielsen and Ottesen (2004).

Because P_v is so much smaller than P_a it can be neglected in the model, leaving us with the system of delay differential equations

$$c_a \frac{dP_a}{dt} = \frac{-P_a}{R} + V_s F, \tag{11.108}$$

$$\frac{dF}{dt} = \alpha g(P_a(t - \tau)) - \beta(1 - g(P_a)). \tag{11.109}$$

The (unique) steady-state solution of this system of equations is

$$\bar{P} = \mu \left(\frac{\alpha}{\beta} \right)^{\frac{1}{7}}, \tag{11.110}$$

$$\bar{F} = \frac{1}{V_s R} \bar{P}. \tag{11.111}$$

The first step to find oscillatory solutions is to determine where in parameter space this steady-state solution becomes unstable. If it becomes unstable via a change of sign of the real part of a complex eigenvalue, then we have a Hopf bifurcation and the existence of a branch of periodic solutions (although, of course, they might not be stable). To investigate stability of the steady solution we follow the usual procedure of linearizing around the steady-state solution and calculating the eigenvalues of the resultant linear system. Linearizing (11.110) and (11.111) around the steady-state solution gives

$$c_a \frac{dp}{dt} = \frac{-p}{R} + V_s f, \tag{11.112}$$

$$\frac{df}{dt} = \alpha g'(\bar{P}) p(t - \tau) + \beta g'(\bar{P}) p, \tag{11.113}$$

where $P_a = \bar{P} + p$ and $F = \bar{F} + f$, i.e., p and f are deviations from the steady state.

We now look for solutions of the form $p = p_0 e^{\lambda t}$, $f = f_0 e^{\lambda t}$. Substituting into the linearized equations gives

$$\begin{pmatrix} -1/R - c_a\lambda & V_s \\ \alpha g'(\bar{P}) e^{-\lambda t} + \beta g'(\bar{P}) & -\lambda \end{pmatrix} \begin{pmatrix} p_0 \\ f_0 \end{pmatrix} = 0. \tag{11.114}$$

This system has a nontrivial solution only if the determinant of the matrix is zero, and thus λ must be a root of the characteristic equation

$$c_a \lambda^2 + \frac{1}{R} \lambda - V_s g'(\bar{P})(\alpha e^{-\lambda t} + \beta) = 0. \tag{11.115}$$

Since $g'(\bar{P}) < 0$, all the real roots of (11.115) have negative real part. Thus, the only way the solution can become unstable is for the real part of a complex root to change sign. Thus, we look for changes of stability by setting $\lambda = iz$ and solving for z. After separating real and imaginary parts, we find

$$c_a z^2 - b^2 = a^2 \cos z\tau, \tag{11.116}$$

$$\frac{z}{R} = a^2 \sin z\tau, \tag{11.117}$$

where $a^2 = -\alpha V_s g'(\bar{P})$ and $b^2 = -\beta V_s g'(\bar{P})$. Squaring both equations and adding gives

$$c_a^2 \xi^2 + \xi \left(\frac{1}{R^2} - 2c_a b^2 \right) + b^4 - a^4, \tag{11.118}$$

where $\xi = z^2$.

Table 11.3 Parameter values of the model of oscillations in the baroreceptor loop. Taken from Ottesen (1997).

c_a = 1.55 ml/mm Hg	R = 1.05 mm Hg sec/ml
V_s = 67.9 ml	μ = 93 mm Hg
α = 0.84 sec^{-2}	β = 1.17 sec^{-2}

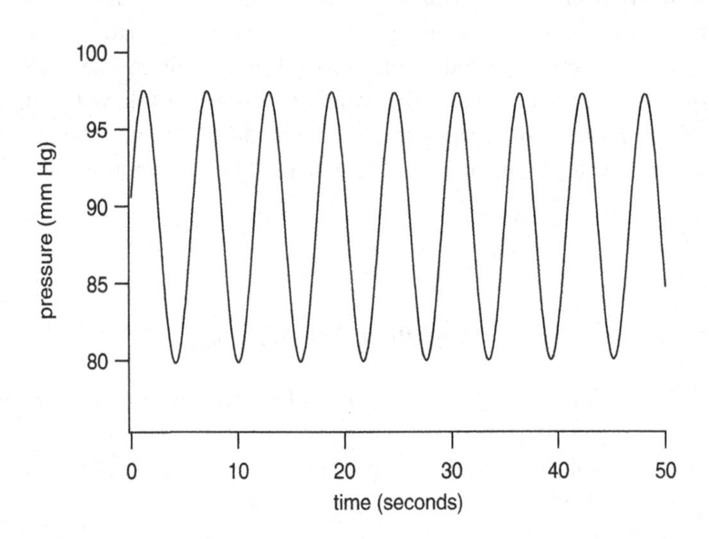

Figure 11.14 Numerical solution of the model of the baroreceptor loop for $\tau = 1.15$. The parameter values are given in Table 11.3.

As long as care is taken to select the correct root of (11.118), it is a simple matter to solve for ξ, and thus for z and τ. For the parameter values used by Ottesen (shown here in Table 11.3), $z = 1.09$ and $\tau = 1.15$. Thus, the period of the oscillation that occurs when stability is lost is $2\pi/1.09 \approx 5.8$ seconds.

Numerical solution of the differential-delay equations shows that, when $z = 1.09$ and $\tau = 1.15$, the amplitude of the periodic solution is approximately 20 mm Hg (Fig. 11.14). Although these oscillations have approximately the correct period (of around 10 seconds) and have an amplitude that is certainly consistent with Mayer waves seen experimentally (there is, in any case, enormous variability in the amplitude of Mayer waves), one should be extremely cautious in drawing any definite conclusions about the underlying mechanisms. It is clear that Mayer waves have a complex, and not well understood, genesis; feedback delays may well be one part of the story, but are certainly not the only part. Interested readers are referred to Julien (2006) and Cohen and Taylor (2005) for detailed discussions of other possibilities, and other models, including statistical approaches.

11.7 Fetal Circulation

Because the fetus receives all of its oxygen through the umbilical cord and the placenta, the lungs of the fetus are not used for gas exchange. Instead, the lungs are collapsed and have high resistance to blood flow: only 12% of the blood flow is through the lungs. This situation is reversed at birth when the newborn takes its first breath, expanding the lungs, and when the umbilical cord constricts.

Necessitated by the high resistance of the pulmonary circulation, the fetal circulatory system has a connection between the pulmonary artery and the aorta, called the *ductus arteriosus*, that shunts blood from the outflow of the right heart directly into the systemic arteries. After birth, the ductus gradually closes.

The ventricular chambers of the developing fetal heart are nearly equal in size. It is only after birth that the load on the left ventricle increases, necessitating additional growth of the left ventricular wall to accommodate an increased demand. To equalize the output of the two hearts, there is a small opening in the interatrial septum, called the *foramen ovale*. On the left side of the septum there is a small flap of tissue that allows flow from the right atrium to the left but prevents the reverse from occurring. In the fetus, this flap is open, but at birth it closes for reasons that become clear below.

To model the fetal circulatory system (Hoppensteadt and Peskin, 2001), we use the same three-compartment model as above with additional connections allowed by the ductus arteriosus and the foramen ovale (Fig. 11.15). Since there is no longer a single loop, we must keep track of the flows in each compartment. These flows are governed by the following equations:

Systemic arteries:

$$Q_s = \frac{P_{sa} - P_s}{R_{sa}}. \tag{11.119}$$

Systemic veins:

$$Q_s = \frac{P_s - P_{sv}}{R_{sv}}. \tag{11.120}$$

Pulmonary system:

$$Q_p = \frac{P_{pa} - P_{pv}}{R_p}. \tag{11.121}$$

Left heart:

$$Q_l = F(C_{ld}P_{pv} - C_{l\sigma}P_{sa}). \tag{11.122}$$

Right heart:

$$Q_r = F(C_{rd}P_{sv} - C_{r\sigma}P_{pa}). \tag{11.123}$$

The equations for the volumes are unchanged from before.

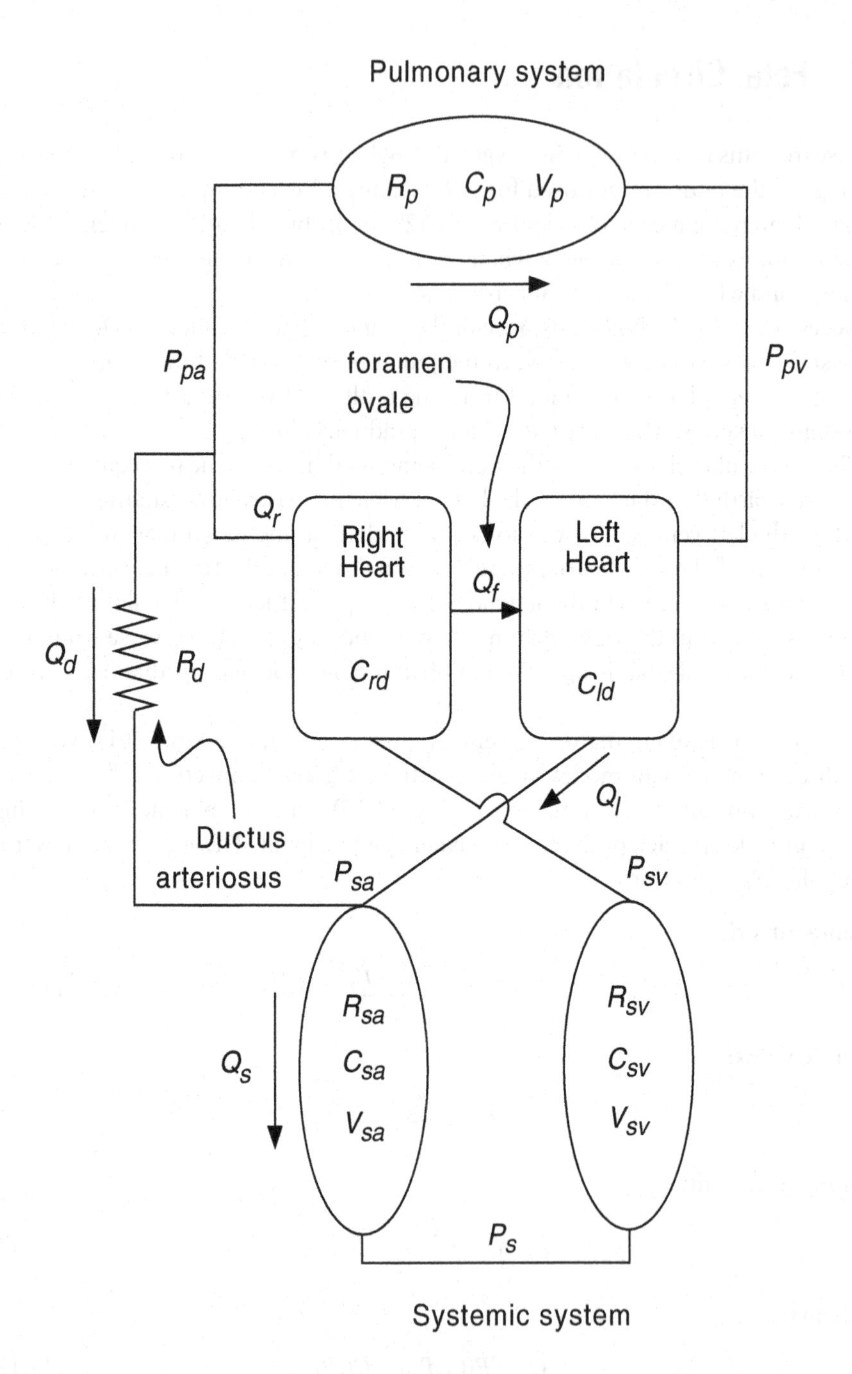

Figure 11.15 Schematic diagram of the fetal circulation. The model is based on the three-compartment model in Fig. 11.9, with additional connections to model the ductus arteriosus and the foramen ovale.

Since fluid is conserved, flow into any junction must equal the flow out. There are four junctions, and thus there are four conservation laws

$$Q_l + Q_d = Q_s, \tag{11.124}$$

$$Q_r = Q_d + Q_p, \tag{11.125}$$

$$Q_s = Q_r + Q_f, \tag{11.126}$$

$$Q_p + Q_f = Q_l, \tag{11.127}$$

where l = left heart, r = right heart, s = systemic, p = pulmonary, f = foramen ovale, d = ductus arteriosus. Notice that there are only three independent relationships here, as the first three imply the fourth, for the simple reason that total fluid is conserved. As a result, any three flows can be determined as functions of the remaining three.

A quick count shows that there are 14 variables (5 pressures, 3 volumes, and 6 flows), but only 12 equations, including the three equations for the volumes and the equation for the conservation of volume, which are not shown explicitly here. Thus, to characterize the system completely, two additional equations are needed. These are the equation for the ductus,

$$Q_d = \frac{P_{pa} - P_{sa}}{R_d}, \tag{11.128}$$

and the equation for the foramen, modeled as an ideal valve,

$$P_{sv} = P_{pv}, \quad \text{if } Q_f > 0, \tag{11.129}$$

$$Q_f = 0, \quad \text{if } P_{sv} < P_{pv}. \tag{11.130}$$

Here, there is no resistance to flow in the forward direction if the valve is open, and if the valve is closed, there is no flow in the backward direction.

There are two possible solutions. First, for the "foramen open" solution, we set $P_{sv} = P_{pv}$ and solve the governing system of equations. This yields a valid solution for all parameter values for which $Q_f > 0$. On the other hand, if we take $Q_f = 0$ (the "foramen closed" solution) and determine all the pressures, we have a valid solution for all parameter values for which $P_{sv} < P_{pv}$. For any set of parameter values there should be one, but only one, solution set.

We begin by looking for the "foramen open" solution. For simplicity, we again suppose that the systolic compliances are negligible, i.e., that $C_{r\sigma} = C_{l\sigma} = 0$. We find

$$Q_f = Q_r \frac{R_d(C_{ld} - C_{rd}) + R_p C_{ld} - C_{rd}(R_{sa} + R_{sv})}{C_{rd}(R_{sa} + R_{sv} + R_p + R_d)}, \tag{11.131}$$

$$Q_d = Q_r \frac{C_{ld}(R_{sv} + R_{sa}) - R_p C_{rd}}{C_{rd}(R_{sv} + R_{sa} + R_d + R_p)}, \tag{11.132}$$

$$Q_s = Q_r \frac{R_p(C_{rd} + C_{ld}) + R_d C_{ld}}{C_{rd}(R_{sa} + R_{sv} + R_p + R_d)}, \tag{11.133}$$

$$Q_p = Q_r \frac{(R_{sa} + R_{sv})(C_{rd} + C_{ld}) + R_d C_{rd}}{C_{rd}(R_{sa} + R_{sv} + R_p + R_d)}, \tag{11.134}$$

$$Q_l = Q_r \frac{C_{ld}}{C_{rd}}. \tag{11.135}$$

In the developing fetus, the left and right hearts are nearly the same. If $C_{rd} = C_{ld} = C_d$, the outputs from the left and right hearts are the same, and $Q_l = Q_r = Q$. With this simplification we find that

$$Q_f = Q_d = Q \frac{R_p - R_{sa} - R_{sv}}{R_{sa} + R_{sv} + R_p + R_d}, \tag{11.136}$$

$$Q_s = Q \frac{2R_p + R_d}{R_{sa} + R_{sv} + R_p + R_d}, \tag{11.137}$$

$$Q_p = Q \frac{2R_{sa} + 2R_{sv} + R_d}{R_{sa} + R_{sv} + R_p + R_d}. \tag{11.138}$$

As long as the pulmonary resistance is larger than the total systemic resistance, the flow Q_f is positive, as required by our initial assumption. Thus, by adjusting the pulmonary resistance R_p, the foramen and the ductus allow blood to be shunted from the lungs to the systemic circulation. Notice that in the extreme case of $R_p = \infty$, we have $Q_f = Q_d = Q, Q_p = 0, Q_s = 2Q$. In other words, if $R_p = \infty$, there is no pulmonary flow, the flow returning from the systemic circulation is equally divided between the left and right hearts for pumping, and the blood pumped by the right heart is shunted from the lungs to the systemic arteries.

At birth, the lungs fill with air and expand, dramatically decreasing the resistance of blood flow in the lungs. Simultaneously, the umbilical cord constricts, dramatically increasing the total systemic resistance $R_{sa} + R_{sv}$. When this happens, the flow through the foramen reverses, and closes the foramen. To find the flow solution in this "foramen closed" situation, we take $Q_f = 0$ and drop the restriction that $P_{sv} = P_{pv}$. It follows immediately from (11.126) and (11.127) that $Q_p = Q_l$, $Q_s = Q_r$, and $Q_d = Q_r - Q_l$. Thus, the flow through the ductus is used to balance the outputs of the left and right sides of the heart.

Furthermore,

$$P_{pv} = \frac{Q_r}{FC_{ld}}, \tag{11.139}$$

$$P_{sv} = \frac{Q_l}{FC_{rd}}, \tag{11.140}$$

$$\frac{P_{sa}}{P_{pa}} = \left(\frac{\frac{1}{FC_{rd}} + R_{sv} + R_{sa}}{\frac{1}{FC_{rd}} + R_{sa} + R_{sv} + R_d} \right) \left(\frac{\frac{1}{FC_{ld}} + R_p + R_d}{\frac{1}{FC_{ld}} + R_p} \right), \tag{11.141}$$

and

$$\frac{P_{pv}}{P_{sv}} = \frac{FC_{rd}(R_{sa} + R_{sv} + R_d) + 1}{FC_{ld}(R_p + R_d) + 1}, \tag{11.142}$$

which is greater than one (as required) as long as $R_p < R_{sa} + R_{sv}$. Thus, remarkably, as soon as the first breath is drawn and the pulmonary resistance drops, the pulmonary venous pressure exceeds the systemic venous pressure, keeping the foramen closed, allowing the skin flap to gradually grow over and seal tightly. In addition,

$$Q_d = Q_r \frac{(\frac{1}{FC_{ld}} + R_p) - (\frac{1}{FC_{ld}} + R_{sa} + R_{sv})}{\frac{1}{FC_{ld}} + R_d + R_p}. \tag{11.143}$$

Thus, immediately after birth, when the left and right heart compliances are the same, the flow through the ductus reverses direction, so that the left heart output exceeds the right heart output, with

$$\frac{Q_l}{Q_r} = \frac{\frac{1}{FC_{rd}} + R_{sa} + R_{sv} + R_d}{\frac{1}{FC_{ld}} + R_p + R_d}. \tag{11.144}$$

For reasons that are not completely understood (probably because of the increased concentration of oxygen in the blood), the ductus gradually closes, so that R_d grows, eventually to ∞. As it does so, the arterial pressure P_{sa} increases, causing the left ventricle to thicken gradually, decreasing its compliance. The end result (taking $R_d \to \infty$) is the solution of the single-loop system found in the previous section, although with parameter values that are not yet the adult values.

11.7.1 Pathophysiology of the Circulatory System

Occasionally, the heart or its associated blood vessels are malformed during fetal life, leaving the newborn infant with a defect called a *congenital anomaly*. There are three major types of congenital abnormalities:

1. A blockage, or *stenosis*, of the blood flow at some part of the heart or a major vessel.
2. An abnormality that allows blood to flow directly from the left heart or aorta to the right heart or pulmonary artery, bypassing the systemic circulation.
3. An abnormality that allows blood to flow from the right heart or pulmonary artery to the left heart or aorta, thereby bypassing the lungs.

Patent Ductus Arteriosus (PDA)

While the ductus arteriosus constricts to a small size shortly after birth, it is several months before flow is completely occluded. In about 1 out of 5500 babies, the ductus never closes, a condition known as *patent ductus arteriosus*. In a child with a patent ductus, there is a substantial backflow from the left heart into the lungs, so that the blood is well oxygenated, but there is decreased cardiac reserve and respiratory reserve, because insufficient blood is supplied to the systemic arteries. As the child grows and systemic pressure increases, the backflow through the ductus also increases, sometimes causing the diameter of the ductus to increase, thereby worsening the condition. Symptoms of patent ductus include fainting or dizziness during exercise, and there is usually hypertrophy of the left heart.

It can happen that the lungs respond to the excess pulmonary flow by increasing pulmonary resistance, thereby, according to (11.143), reversing the flow in the ductus, shunting blood from the right heart to the aorta, carrying deoxygenated blood directly into the systemic arteries.

Closed Foramen Ovale in Utero

In this situation the circulation is like the circulation after birth, except that the pulmonary resistance exceeds the systemic resistance, $R_p > R_{sa} + R_{sv}$. According to (11.144), the output of the left heart is low compared to the output of the right heart, so that development of the left heart is impaired and the right heart is overdeveloped at birth.

Atrial Septal Defect (ASD)

If the foramen does not close properly at birth, there remains a hole in the septum between the left and right atria, allowing oxygenated blood to leak from the left heart to the right heart. Assuming that the ductus closes successfully, so that $Q_d = 0$, it follows that

$$Q_p = Q_r = FC_{rd}P_{sv}, \tag{11.145}$$

$$Q_s = Q_l = FC_{ld}P_{pv}, \tag{11.146}$$

$$Q_f = Q_s - Q_r = Q_l - Q_p, \tag{11.147}$$

so that

$$\frac{Q_p}{Q_s} = \frac{Q_r}{Q_l} = \frac{C_{rd}}{C_{ld}}. \tag{11.148}$$

If the left heart has smaller compliance than the right heart, as would be true in an adult, the pulmonary flow exceeds the systemic flow.

ASD and PDA

The configuration here is the same as with the fetal circulation, except that there is no valve to prohibit flow from the left to right atrium. The solution is the "foramen open" solution, for which

$$Q_s - Q_p = Q_f + Q_d = Q_s\left(1 - \frac{R_dC_{rd} + (R_{sa} + R_{sv})(C_{ld} + C_{rd})}{R_dC_{ld} + R_p(C_{ld} + C_{rd})}\right), \tag{11.149}$$

which is negative for typical parameter values. This shows that it is possible to reduce the shunted flow and equalize the pulmonary and systemic flows by *banding* or surgically constricting the pulmonary artery, thus increasing R_p. The banding procedure works, however, only if $R_d \neq \infty$, that is, only if there is flow through the ductus. Banding has no effect in ASD when the ductus is closed, because in the limit $R_d \to \infty$, the flow through the foramen is

$$Q_f = Q_s\left(1 - \frac{C_{rd}}{C_{ld}}\right), \tag{11.150}$$

independent of R_p.

11.8 The Arterial Pulse

The above analysis treats the circulation as if the various pressures in the blood vessels are constant over time. Of course, since the heart pumps blood in a pulsatile manner, this is not the case. Each beat of the heart forms a pressure wave that travels along the arteries, changing shape as it moves away from the heart. Typical experimental data, taken from a dog artery, are shown in Fig. 11.16. It is evident that closer to the heart the pressure pulse is wider and does not have a distinct second wave, and the velocity and pressure waves have different forms. However, as the pulse moves away from the heart, the pressure wave becomes steeper, a second wave develops following the first, and the velocity profile becomes similar to the pressure profile. Since variations in the form of the arterial pulse are often used as clinical indicators (for example, the second wave is usually absent in patients with diabetes or atherosclerosis), it is important to gain an understanding of the physical mechanisms underlying the shape of the pulse in normal physiology. Models to explain the shape of the arterial pulse range from simple linear ones to complex models incorporating the tapering of the arterial walls and its branching structure (Pedley, 1980; Lighthill, 1975; Peskin, 1976), and the modern literature on models of the arterial pulse is vast. Here, we restrict our attention to only the simplest models.

11.8.1 The Conservation Laws

Consider flow in a blood vessel with cross-sectional area $A(x,t)$. For simplicity we assume that the flow is a *plug flow*, with velocity that is a scalar quantity u and is a

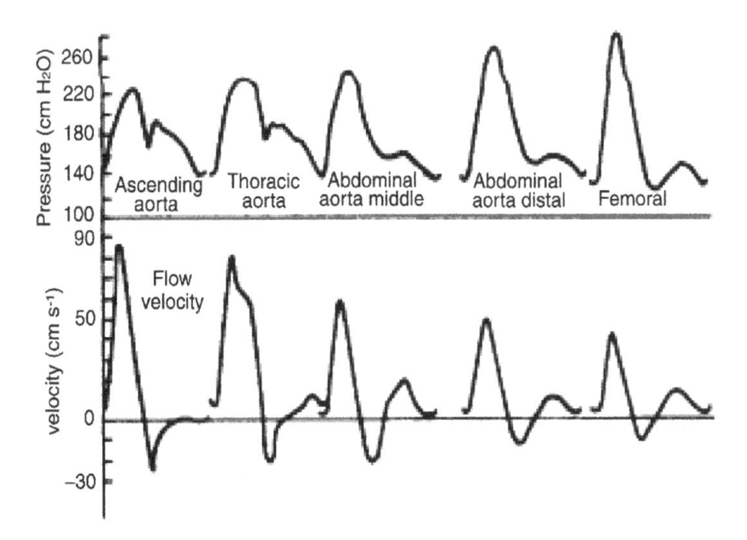

Figure 11.16 The form of the arterial pulse measured in the arteries of a dog. The top panel shows the pressure waveform, and the bottom panel shows the velocity. (Pedley, 1980, Fig. 1.14, taken from McDonald, 1974.)

function of axial distance along the vessel only. Poiseuille flow becomes plug flow in the limit of zero viscosity, and so in the following analysis we omit consideration of viscous forces. The volume of the vessel of length L is $\int_0^L A(x,t)\,dx$, and thus conservation of mass requires that

$$\frac{\partial}{\partial t}\left(\int_0^L A(x,t)\,dx\right) = u(0)A(0,t) - u(L)A(L,t). \tag{11.151}$$

Taking the partial derivative of (11.151) with respect to L and replacing L by x gives

$$A_t + (Au)_x = 0. \tag{11.152}$$

According to Newton's law, the rate of change of momentum of a fluid in some domain is equal to the net force exerted on the boundary of the domain plus the flux of momentum across the boundary. Thus, conservation of momentum requires that

$$\frac{\partial}{\partial t}\left(\rho\int_0^L A(x,t)u(x,t)\,dx\right) = \rho A(x,t)u^2(x,t)\Big|_{x=L}^{0} + P(x,t)A(x,t)\Big|_{x=L}^{0}. \tag{11.153}$$

Note that $\rho A(0,t)u(0,t)$ is the rate at which mass enters the vessel across the surface $x=0$, so that $\rho A(0,t)u^2(0,t)$ is the rate at which momentum enters the vessel across this surface. Differentiating (11.153) with respect to L and replacing L by x, we find that

$$\rho\left((Au)_t + (Au^2)_x\right) = -(PA)_x. \tag{11.154}$$

This second equation can be simplified by expanding the derivatives and using (11.152) to get

$$\rho\,(u_t + uu_x) = -P_x \tag{11.155}$$

as the equation for the conservation of momentum.

For simplicity we assume that the vessel is a linear compliance vessel with

$$A(P) = A_0 + cP. \tag{11.156}$$

One can use a more general relationship between area and pressure, but the basic conclusions of the following analysis remain unchanged. With this expression for the cross-sectional area, the equation for conservation of mass becomes

$$c(P_t + uP_x) + A(P)u_x = 0. \tag{11.157}$$

11.8.2 The Windkessel Model

One of the earliest models of the heart, dating back to the past century (Frank, 1899; translated by Sagawa et al., 1990), is the *windkessel* model, from the German word meaning an air chamber, or bellows. (The name originally arose because of the similarities between the mechanical conditions in the arterial system and the operation of the *windkessel*, or bellows, of a nineteenth-century fire engine.)

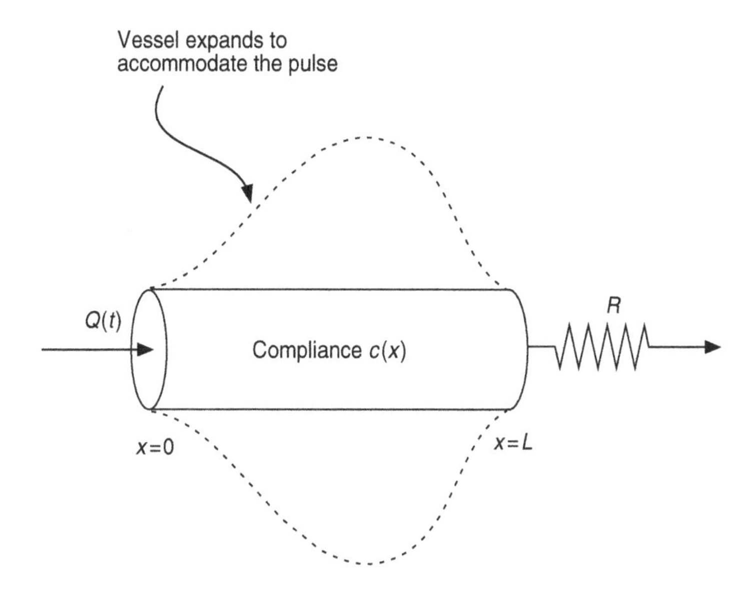

Figure 11.17 Schematic diagram of the *windkessel* model.

The *windkessel* model is obtained from (11.155) and (11.157) by letting $\rho \to 0$, in which case $P_x = 0$, so that the pressure is a function only of time. Thus, we model the greater arteries as a compliance vessel, extending from $x = 0$ to $x = L$, with time-varying pressure and volume; inflow at $x = 0$ is from the heart, and outflow at $x = L$ is into the peripheral arterial system (Fig. 11.17). However, although the pressure is uniform inside the vessel, the compliance is not, so that the cross-sectional area of the vessel varies with x, the distance along the vessel. In fact, we assume that $c(0) = c(L) = 0$, but that the compliance $c(x)$ is nonzero inside the vessel. Finally, we assume that the outflow from the vessel is into the peripheral system which is modeled as a simple resistance vessel with resistance R.

From (11.157) it follows that

$$u_x(x,t) = \frac{-c(x)P_t}{A_0 + c(x)P}, \tag{11.158}$$

and, integrating from $x = 0$ to $x = L$, we get

$$A_0 u(0,t) = \theta(P)P_t + A_0 u(L,t), \tag{11.159}$$

where

$$\theta(P) = \int_0^L \frac{A_0 c(x)}{A_0 + c(x)P}\, dx. \tag{11.160}$$

We denote $A_0 u(0,t)$ as $Q(t)$, since it is the flow into the vessel from the heart. Further, the outflow from the vessel, $A_0 u(L,t)$, must match the flow through the peripheral system, which we write as P/R (assuming the pressure drop across the peripheral system is also P). Hence, we have the differential equation for P,

$$Q(t) = \theta(P)P_t + \frac{P}{R}. \tag{11.161}$$

From this differential equation we learn that when the heart ejects blood, in which case $Q(t)$ increases quickly, there is a corresponding increase in P (i.e., the vessel fills up and expands). When the flow stops, Q is zero, and the pressure decreases to zero according to $P_t = -P/(\theta(P)R)$. This demonstrates that the major arteries act as a bellows, initially inflating to accommodate the blood from the heart and then contracting to pump the blood through the periphery.

11.8.3 A Small-Amplitude Pressure Wave

If p aan u are small, so that all nonlinear terms in (11.155) and (11.157) can be ignored, we then obtain the linear system

$$\rho u_t + P_x = 0, \tag{11.162}$$

$$cP_t + A_0 u_x = 0. \tag{11.163}$$

By cross-differentiation, we can eliminate u and find a single equation for P, namely

$$P_{tt} = \frac{A_0}{c\rho} P_{xx}, \tag{11.164}$$

which is well known as the *wave equation*.

Solutions of the wave equation include traveling wave solutions, which are functions whose shape is invariant but that move at the velocity $s = \sqrt{\frac{A_0}{c\rho}}$. For arteries, this velocity is on the order of 4 m/s, as can be verified by comparing the arrival times of the pressure pulse at the carotid artery in the neck and at the posterior tibial artery at the ankle.

The general solution of the wave equation (11.164) can be written in the form

$$P(x,t) = f(t - x/s) + g(t + x/s), \tag{11.165}$$

where f and g are arbitrary functions. Note that $f(t - x/s)$ denotes a wave with profile $f(x)$ traveling from left to right, while $g(t+x/s)$ denotes a wave traveling in the opposite direction. It follows that the general solution for u is

$$u = \frac{1}{\rho s}[f(t - x/s) - g(t + x/s)]. \tag{11.166}$$

11.8.4 Shock Waves in the Aorta

Although the linear wave equation can be used to gain an understanding of many features of the arterial pulse, such as reflected waves and waves in an arterial network (Lighthill, 1975), there are experimental indications that nonlinear effects are also important (Anliker et al., 1971a,b). One particular nonlinear effect that we investigate here is the steepening of the wave front as it moves away from the heart. If the wave front becomes too steep, the top of the front overtakes the bottom, and a shock, or discontinuity, forms, a solution typical of hyperbolic equations. Of course, a true shock is not possible, as blood viscosity and the elastic properties of the arterial wall preclude the formation of a discontinuous solution. Nevertheless, it might be possible to generate very steep pressure gradients within the aorta.

Under normal conditions, no such shocks develop. However, in conditions where the aorta does not function properly, allowing considerable backflow into the heart, the heart compensates by an increase in the ejection volume, thus generating pressure waves that are steeper and stronger than those observed normally. Furthermore, the *pistol-shot* phenomenon, a loud cracking sound heard through a stethoscope placed at the radial or femoral artery, often occurs in patients with aortic insufficiency. It has been postulated that the pistol-shot is the result of the formation of a shock wave within the artery, a shock wave that is possible because of the increased amplitude of the pressure pulse.

To model this phenomenon, recall that the governing equations are

$$c(P_t + uP_x) + A(P)u_x = 0, \tag{11.167}$$

$$\rho(u_t + uu_x) + P_x = 0, \tag{11.168}$$

which can be written in the form

$$w_t + Bw_x = 0, \tag{11.169}$$

where

$$w = \begin{pmatrix} u \\ P \end{pmatrix} \tag{11.170}$$

and

$$B = \begin{pmatrix} u & \frac{1}{\rho} \\ \frac{A(P)}{c} & u \end{pmatrix}. \tag{11.171}$$

Using the method of characteristics (Whitham 1974; Pedley, 1980; Peskin, 1976), we can determine some qualitative features of the solution. Roughly speaking, a characteristic is a curve C in the (x,t) plane along which information about the solution propagates. For example, the equation $u_t + cu_x = 0$ has solutions of the form $u(x,t) = U(x - ct)$, so that information about the solution propagates along curves $x - ct = \text{constant}$ in the (x,t) plane. Similarly, characteristics for the wave equation (11.164) are curves of the form $t \pm x/s = \text{constant}$, because it is along these curves that information about the solution travels.

To find characteristic curves, we look for curves in x,t along which the original partial differential equation behaves like an ordinary differential equation. Suppose a characteristic curve C is defined by

$$x = x(\lambda), \quad t = \lambda. \tag{11.172}$$

Derivatives of functions $w(x,t)$ along this curve are given by

$$\frac{dw}{d\lambda} = w_t + w_x \frac{dx}{d\lambda}. \tag{11.173}$$

Notice that with $dx/d\lambda = c$, the partial differential equation $u_t + cu_x = 0$ reduces to the simple ordinary differential equation $u_\lambda = 0$. Thus, curves with $dx/dt = c$ are characteristic curves for this simple equation.

To reduce the system (11.169) to characteristic form, we try to find appropriate linear combinations of the equations that transform the system to an ordinary differential equation. Thus, suppose the matrix B has a left eigenvector ξ^T with corresponding eigenvalue s, so that $\xi^T B = s\xi^T$. We multiply (11.169) by ξ^T, and find that with the identification $dx/d\lambda = s$,

$$0 = \xi^T(w_t + Bw_x) = \xi^T(w_t + sw_x) = \xi^T w_\lambda. \tag{11.174}$$

In other words, along the curve $dx/dt = s$, the original system of equations reduces to the simple ordinary differential equation $\xi^T w_\lambda = 0$.

It is an easy matter to determine that the eigenvalues of B are

$$s = u \pm K(P), \tag{11.175}$$

where

$$K(P) = \sqrt{\frac{A(P)}{\rho c}}, \tag{11.176}$$

with corresponding left eigenvector

$$\xi^T = (\xi_1, \xi_2) = (\rho K(P), \pm 1). \tag{11.177}$$

It follows from $\xi^T w_\lambda = 0$ that

$$u_\lambda \pm \frac{1}{\rho K(P)} P_\lambda = 0 \tag{11.178}$$

along the characteristic curve $dx/d\lambda = u \pm K(P)$, which we denote by $C_\pm$. Now, notice that, since $A(P)$ is linear in P,

$$\frac{1}{\rho K(P)} = 2\frac{d}{dP}K(P), \tag{11.179}$$

so that

$$\frac{d}{d\lambda}(u \pm 2K(P)) = 0. \tag{11.180}$$

In other words, $u + 2K(P)$ is conserved (remains constant) along C_+, the characteristic curve with slope $dx/dt = u + K(P)$, and $u - 2K(P)$ is conserved along C_-, the characteristic curve with slope $dx/dt = u - K(P)$.

Now, to see how this reduction allows us to solve a specific problem, consider the region $x \geq 0$, $t \geq 0$ with $u(0,t) > 0$ specified. For example, $u(0,t)$ could be the velocity

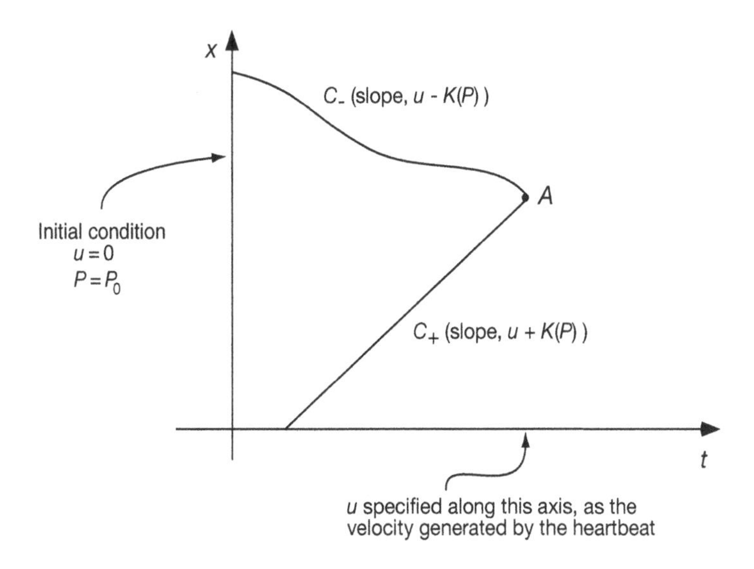

Figure 11.18 Diagram of the characteristics of the arterial pulse equations in the (t, x) plane.

pulse generated by a single heartbeat. We suppose that initially, $u(x, 0) = 0, P(x, 0) = P_0$ for all $x \geq 0$, where P_0 is the diastolic pressure.

Pick any point A in the region $x \geq 0, t \geq 0$ (Fig. 11.18). There are two characteristics passing through A, one, C_+, with positive slope $u + K(P)$ and one, C_-, with negative slope $u - K(P)$. (Here and in the following we assume that u is small enough so that C_- always has negative slope.) Following C_- up and to the left, we see that it intersects the vertical axis, where $u = 0$ and $P = P_0$ (because of the specified initial data). Since the quantity $u - 2K(P)$ is conserved on C_-, it must be that $u - 2K(P) = -2K(P_0)$ at the point A. Since A is arbitrary, it follows that $u = 2K(P) - 2K(P_0)$ everywhere in the first quadrant. Thus, $u + 2K(P) = 4K(P) - 2K(P_0)$ is constant along C_+. Hence, $K(P)$ is constant along C_+, as are both P and u, so that C_+ is a straight line. The slope of C_+ is the value of $u + K(P)$ at the intersection of C_+ with the horizontal axis.

To be specific, suppose $u(0, t)$ first increases and then decreases as a function of t, as shown in Fig. 11.19A. (In this figure $u(0, t)$ is shown as piecewise linear, but this is simply for ease of illustration.) To be consistent, since $u = 2K(P) - 2K(P_0)$ everywhere in the first quadrant, $K(P(0, t)) = K(P_0) + \frac{1}{2}u(0, t)$, so that the slope of the C_+ characteristics is $s(t) = \frac{3}{2}u(0, t) + K(P_0)$, which also increases and then decreases as a function of t. With increasing slopes, the characteristics converge, resulting in a steepening of the wave front. If characteristics meet, the solution is not uniquely defined by this method, and shocks develop.

The place a shock first develops can be found by determining the points of intersection of the characteristics. Suppose we have two C_+ characteristics, one emanating from the t-axis at $t = t_1$, described by $x = s(t_1)(t - t_1)$, and the other emanating at $t = t_2$,

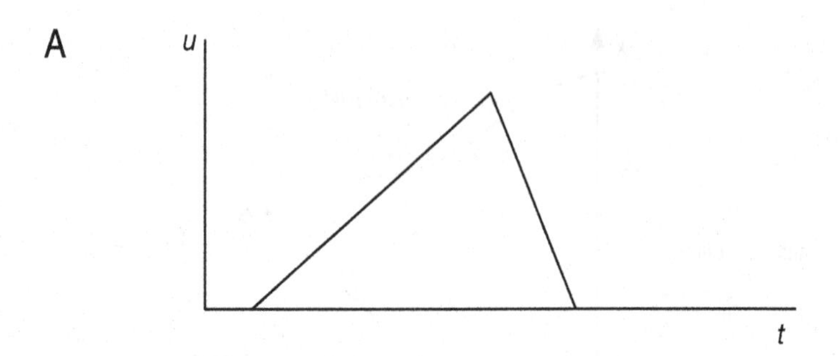

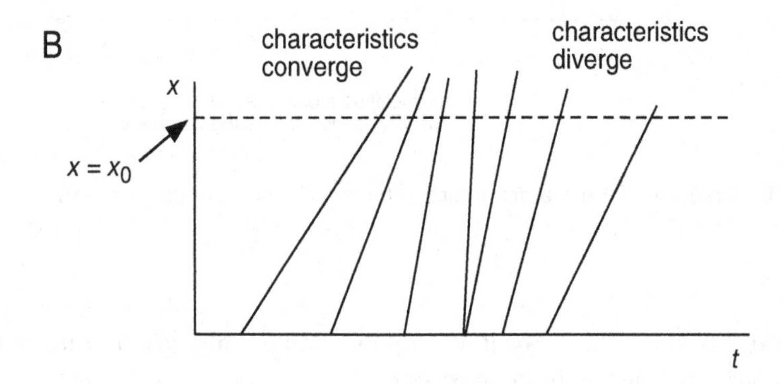

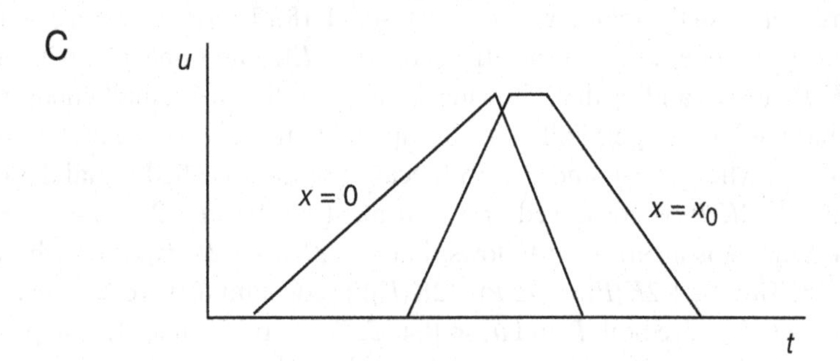

Figure 11.19 A: Sketch of $u(0, t)$. B: Characteristics generated by $u(0, t)$ in the previous figure. C: Plots of $u(x, t)$ for $x = 0$ and $x = x_0$, obtained by taking cross-sections for a fixed x (as indicated by the dotted line in B).

described by $x = s(t_2)(t - t_2)$. They intersect at any point (t_i, x_i), where

$$t_i = \frac{s(t_2)t_2 - s(t_1)t_1}{s(t_2) - s(t_1)}, \tag{11.181}$$

$$x_i = \frac{s(t_2)s(t_1)(t_2 - t_1)}{s(t_2) - s(t_1)}. \tag{11.182}$$

In the limit $t_2 \to t_1$,

$$t_i = \frac{s(t)}{s'(t)} + t, \tag{11.183}$$

$$x_i = \frac{s^2(t)}{s'(t)}, \tag{11.184}$$

which defines parametrically the envelope of intersection points as a function of t, the time of origin of one of the characteristics.

The important point to note is that the first point of shock formation occurs where x_i is smallest. In other words, for data with $s'(t)$ large, the shock develops quickly, and close to $x = 0$. Thus, generally speaking, the steeper the pulse generated by the heart, the sooner and closer a shock forms. This may explain why the pistol-shot occurs in patients with aortic insufficiency but not in other individuals. Using numerical simulations of the model equations, Anliker et al. (1971a,b) have shown that under conditions of aortic insufficiency, a steep pressure gradient can develop within 40 cm of the heart, well within the physiological range.

It is also noteworthy that the slope s depends on diastolic pressure P_0 through $s = \frac{3}{2}u+K(P_0)$. Thus, a decrease in $K(P_0)$, caused either by a decrease of P_0 or a decrease of the function $A(P)$, leads to a decrease of the first location of shock formation x_i.

Notice also that if $s'(t) < 0$, so that $u(0,t)$ is decreasing, no shock can form for positive x. This can also be seen from Fig. 11.19, since, if $u(0,t)$ is a decreasing function of t, the characteristics fan out and do not intersect for positive x.

11.9 EXERCISES

1. Equation (11.7) was derived assuming that the radius of the vessel and pressure drop along the vessel were constant, but then it was used in (11.15) as if the radius was variable. Under what conditions is this a reasonable approximation?

2. Suppose the circumference of a circular vessel at zero pressure is L_0 and is linearly related to the wall tension T via $L = L_0 + \kappa T$. Use Laplace's law to show that the compliance of the vessel is $c = A_0 \frac{\kappa}{\pi M}$ where M is the vessel wall thickness.

3. By taking the limit of (11.20) as $P_1 \to P_0$, derive (11.22).

4. (a) Choose some parameter values and plot the solution to the filtration equations, (11.27) and (11.28), in the q,P_c phase plane. What is the relationship between the nullcline $q' = 0$ and the solution?

   (b) Modify the model of capillary filtration by allowing the plasma osmotic pressure to vary along the capillary distance. Show that $\pi_c = RTc_c \frac{Q_i}{q}$, where c_c is the concentration of osmolites at $x = 0$, i.e., in the influx.

   (c) For the same parameter values as in the first part of this question, calculate and plot the solution numerically.

   (d) If incoming pressure is unchanged from the first model, what is the effect of osmotic pressure on the filtration rate?

(e) What changes must be made to the incoming pressure and the length of the capillary to maintain the same filtration rate? Hint: Study the phase portrait for this system of equations.

5. Simplify the six-compartment model of the circulation by assuming that there are no pressure drops over the arterial and venous systems (either systemic or pulmonary), and thus, for example, $P_{sa} = P_{s1}$. Assume also that the systolic compliances are negligible (why?). Solve the resultant equations and compare with the behavior of the three-compartment model presented in the text. How do the parameter values change? Are the sensitivities altered? (Calculate the sensitivities using a symbolic manipulation program.)

6. Show that (11.71)–(11.79) can be derived from (11.58)–(11.70) by letting R_s and R_p approach zero and by letting $C_{pa} = C_{pv}$ and $R_{pa} = R_{pv}$.

7. In the three-compartment circulatory model the base volume of the pulmonary circulation V_0^p was calculated using the constraint $V_0^s + V_0^p = 1.2$ liters. Using the fact that the systolic pressure in the pulmonary artery is about 22 mm Hg and the diastolic pressure is about 7 mm Hg, calculate a new value for the volume of the pulmonary circulation. How does this change to the model affect the results?

8. Find the pressures as a function of cardiac output assuming $V_l \neq 0, V_r \neq 0$, where V_l is the basal volume of the left heart, and similarly for V_r (cf. equation (11.39)). Show that

$$Q_\infty = \frac{2C_p V_l/C_{ld} + 2(C_{sv} + C_{sa})V_r/C_{rd} + 2(V_t - V_0^p - V_0^s)}{C_p R_p + C_{sv}R_{sv} + C_{sa}(R_{sa} + 2R_{sv})}. \tag{11.185}$$

9. Explore the behavior of the autoregulation model with $R = R_0(1 + a[O_2]_v)/(1 + b[O_2]_v)$.

10. What symptoms in the circulation would you predict from anemia?
Hint: Anemia refers simply to an insufficient quantity of red blood cells, which results in decreased resistance and oxygen-carrying capacity of the blood.

11. In the fetal circulation, a portion of the systemic flow is diverted to the placenta at the placental arteries for oxygen exchange. Suppose that blood entering the systemic veins from the systemic organs and tissues is completely deoxygenated, as is blood leaving the lungs to enter the left heart. Suppose also that blood leaving the placenta entering the systemic veins is fully oxygenated. What percentage of the total cardiac output should be sent through the placenta in order to maximize the total oxygen exchange?

12. In the model of autoregulation P_a and P_v are given. More realistically, they would be determined, in part at least, by R_s. Construct a more detailed model of autoregulation, including the effects of R_s on the pressures, and show how the arterial and venous pressures, and the cardiac output, depend on M and A.

13. Derive a simplified *windkessel* model by starting with a single vessel with volume $V(t) = V_0 + CP(t)$. Assume that the flow leaves through a resistance R and that there is an inflow (from the heart) of $Q(t)$. Derive the differential equation for P and compare it to (11.161).

14. Frank (1899) described a method whereby the flux of blood out of the heart could be estimated from a knowledge of the pressure pulse, even when the arterial resistance is unknown. Starting with (11.161), assume that during the second part of the arterial pulse, $Q(t) \equiv 0$. Write down equations for the first and second parts of the pulse, eliminate R, and find an expression for Q. Give a graphical interpretation of the expression for Q.

15. In the model of the arterial pulse, set $u(0,t) = at$ for some constant a, and determine the curve in the (t,x) plane along which characteristics form an envelope. Determine the first value of x at which a shock forms.

CHAPTER 12

The Heart

Of all of the human organs, the heart is in some sense the simplest. All it has to do is pump blood by contracting and expanding about 2.5 billion times during the lifetime of its owner. The heart is also one of the most studied organs of the body, probably because heart failure, either mechanical or electrical, remains the number one cause of death in the Western world.

The heart is a four-chambered pump, consisting of two pumps arranged in series. As is described in Chapter 11, one pump (the right heart) drives blood through the lungs (the pulmonary circulation) and then back to the heart, while the other pump (the left heart) drives the oxygenated blood around the body (the systemic circulation). Coordination of the mechanical activity of the heart is provided by an electrical signal, which is the topic of study in this chapter.

Cardiac tissue is a syncytium of cardiac muscle cells, each of which is contractile in much the same way as is skeletal muscle (Chapter 15), although there are important differences between the two types of striated muscle. Cardiac cells perform two functions in that they are both excitable and contractile. They are excitable, enabling action potentials to propagate, and the action potential causes the cells to contract, thereby enabling the pumping of blood. The electrical activity of the heart is initiated in a collection of cells known as the *sinoatrial node* (SA node) located just below the superior vena cava on the right atrium. The cells in the SA node are autonomous oscillators. The action potential that is generated by the SA node is then propagated through the atria by the atrial cells.

The atria and ventricles are separated by a septum composed of nonexcitable cells, which normally acts as a barrier to conduction of action potentials. There is one pathway for the action potential to continue propagation and that is through another collection of cells, known as the *atrioventricular node* (AV node), located at the base of the atria.

Conduction through the AV node is quite slow, but when the action potential exits the AV node, it propagates through a specialized collection of fibers called the *bundle of HIS*, which is composed of Purkinje fibers. The Purkinje fiber network spreads via tree-like branching into the left and right *bundle branches* throughout the interior of the ventricles, ending on the *endocardial surface* of the ventricles. As action potentials emerge from the Purkinje fiber–muscle junctions, they activate the ventricular muscle and propagate through the ventricular wall outward to the epicardial surface. A schematic diagram of the cardiac conduction system is shown in Fig. 12.1.

It should be apparent from this introduction that there are a multitude of features of the heart to study. First of all, there is the excitable and contractile function of individual cells, there is the collective oscillatory and pacemaker activity of SA nodal cells, there is one-dimensional action potential propagation along Purkinje fibers, there is higher-dimensional propagation in the atrial and ventricular muscle, and, finally, there is the collective behavior of all of these as a unit. In what follows, we describe models of all of these different features of the cardiac system. However, we start with a description of cardiac electrical activity at the organ level that uses the simplest mathematical models and ideas.

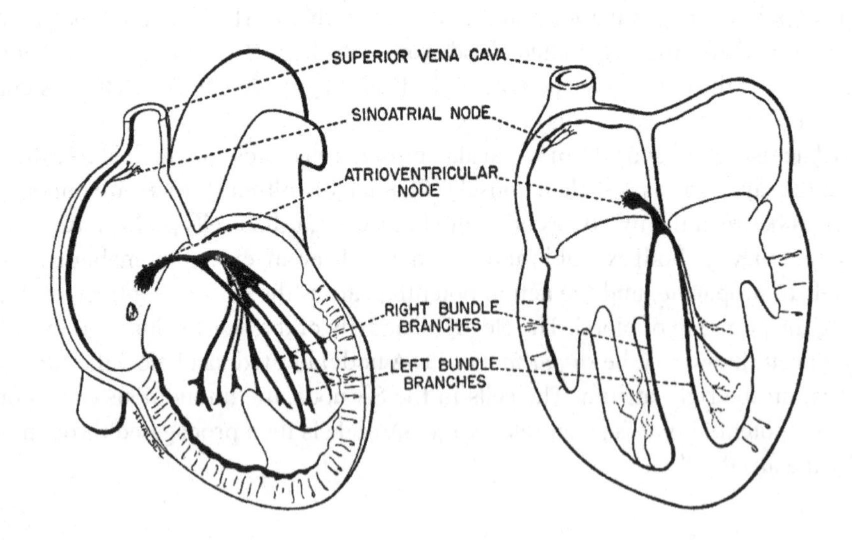

Figure 12.1 Schematic diagram of the cardiac conduction system. (Rushmer, 1976, Fig. 3-9, p. 87).

12.1 The Electrocardiogram

12.1.1 The Scalar ECG

One of the oldest and most important tools for evaluating the status of the heart and the cardiac conduction system is the *electrocardiogram* (ECG). It has been known since 1877, when the first ECG recording was made, that the action potential of the heart generates an electrical potential field that can be measured on the body surface. When an action potential is spreading through cardiac tissue, there is a wave front surface across which the membrane potential experiences a sharp increase. Along the same wave front, the extracellular potential experiences a sharp decrease. From a distance, this sharp decrease in potential looks like a Heaviside jump in potential. This rapid change in extracellular potential results from a current source (or sink) because ions are moving into or out of the extracellular space as transmembrane currents.

The body is a *volume conductor*, so when there is a current source somewhere in the body, such as during action potential spread, currents spread throughout the body. Although the corresponding voltage potential is quite weak, no larger than 4 mV, potential differences can be measured between any two points on the body using a sufficiently sensitive voltmeter. Potential differences are observed whenever the current sources are sufficiently strong. There are three such events. When the action potential is spreading across the atria, there is a measurable signal, called the *P wave*. When the action potential is propagating through the wall of the ventricles, there is the largest of all deflections, called the *QRS complex*. Finally, the recovery of ventricular tissue is seen on the ECG as the *T wave*. (The action potential typical of myocardial cells is shown in the upper panel of Fig. 12.2. In these cells the action potential has a rapid upstroke and a relatively rapid recovery separated by a plateau phase of about 300 ms. The physiological basis of the plateau phase, which does not exist in neural cells, is described later in this chapter.) The recovery of the atria is too weak to be detected on the ECG. Similarly, SA nodal firing, AV nodal conduction, and Purkinje network propagation are not detected on the normal body surface ECG because they do not involve sufficient tissue mass or generate enough extracellular current. In Fig. 12.2 is shown a sketch of a typical single electrical ECG event, and a continuous recording is shown in Fig. 12.3a. In hospitals, ECG recordings are made routinely using oscilloscopes, or, if a permanent record is required, on a continuous roll of paper. The paper speed is standardized at 25 mm per second, with a vertical scale of 1 mV per cm, and the paper is marked with a lined grid of 1 mm and darkened lines with 0.5 cm spacing.

The most important use of the single-lead ECG is to detect abnormalities of rhythm. For example, a continuous oscillatory P wave pattern suggests *atrial flutter* (Fig. 12.3b) or *atrial fibrillation* (Fig. 12.3c). A rapid repetition of QRS complexes is *ventricular tachycardia* (Fig. 12.3d), and a highly irregular pattern of ventricular activation is called *ventricular fibrillation* (Fig. 12.3e). The normal appearance of P waves with a few

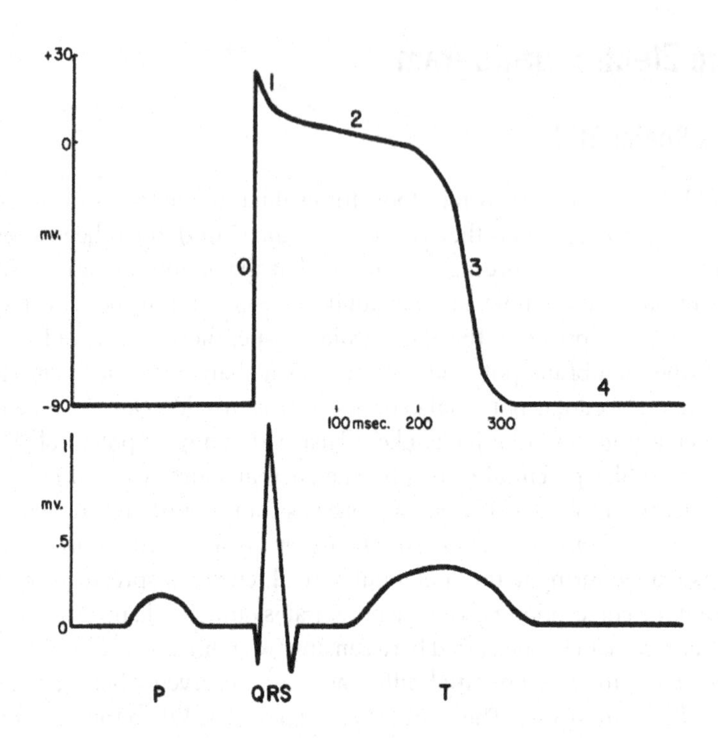

Figure 12.2 Cellular transmembrane potential and electrocardiogram. The upper tracing represents the transmembrane potential of a single ventricular myocyte and the lower tracing shows the body surface potential during the same electrical event. The numbers on the upper tracing designate phases in the action potential cycle: 0: the upstroke, 1: the brief spike, 2: the plateau, 3: the rapid recovery, 4: resting potential. (Rushmer, 1976, Fig. 8-4, p. 286.)

skipped QRS complexes implies a conduction failure in the vicinity of the AV node. Broadening of the QRS complex suggests that propagation is slower than normal, possibly because of conduction failure in the Purkinje network (Fig. 12.4). Spontaneously appearing extra deflections correspond to *extrasystoles*, arising from sources other than the SA or AV nodes.

12.1.2 The Vector ECG

There is much more information contained in the ECG than is available from a single lead. Some of this information can be extracted from the *vector electrocardiogram*. The mathematical basis for the vector ECG comes from an understanding of the nature of a volume conductor. The human body is an inhomogeneous volume conductor, meaning that it is composed of electrically conductive material. If we assume that biological tissue is ohmic, there is a linear relationship between current and potential

$$I = -\sigma \nabla \phi. \tag{12.1}$$

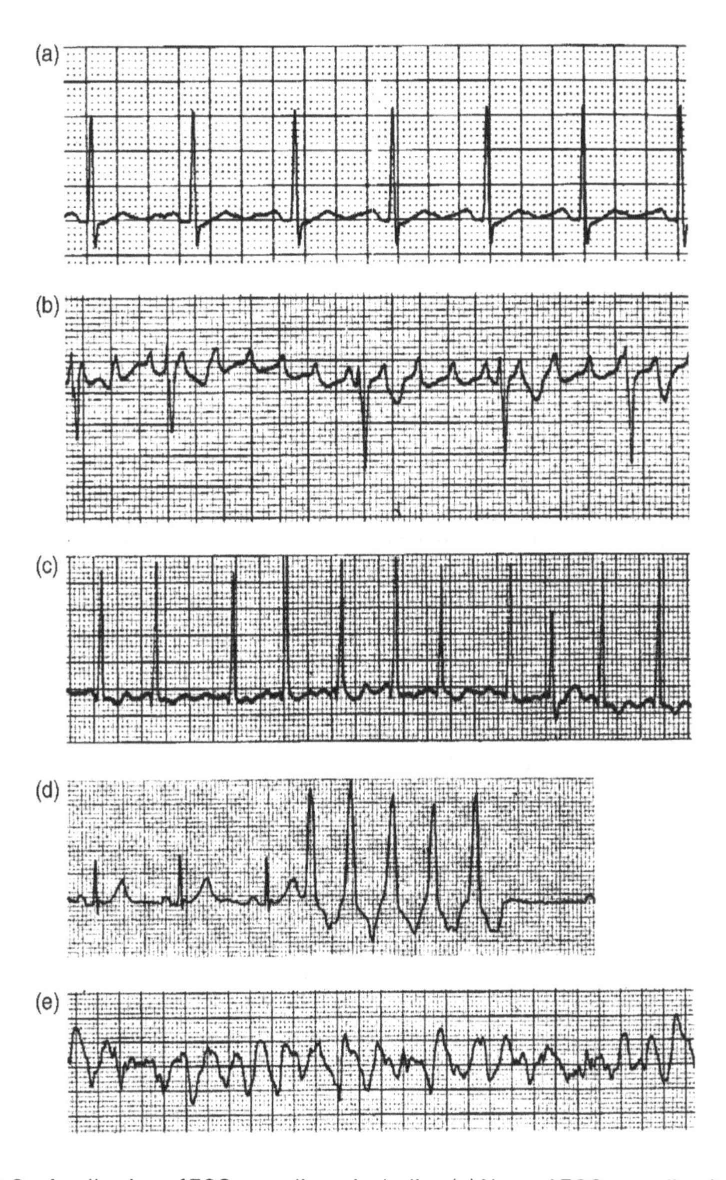

Figure 12.3 A collection of ECG recordings, including (a) Normal ECG recording (lead II) from a sedated 18-year-old male (JPK's son). (b) Atrial flutter showing rapid, periodic P waves, only some of which lead to QRS complexes. (Rushmer, 1976, Fig. 8-29, p. 316.) (c) Atrial fibrillation showing rapid, nonperiodic atrial activity and irregular QRS complexes. (Rushmer, 1976, Fig. 8-28, p. 315.) (d) (Monomorphic) ventricular tachycardia in which ventricular activity is rapid and regular (nearly periodic). (Davis et al., 1985, Fig. 17-24, p. 346.) (e) Ventricular fibrillation in which ventricular activity is rapid and irregular. (Rushmer, 1976, Fig. 8-30, p. 317.)

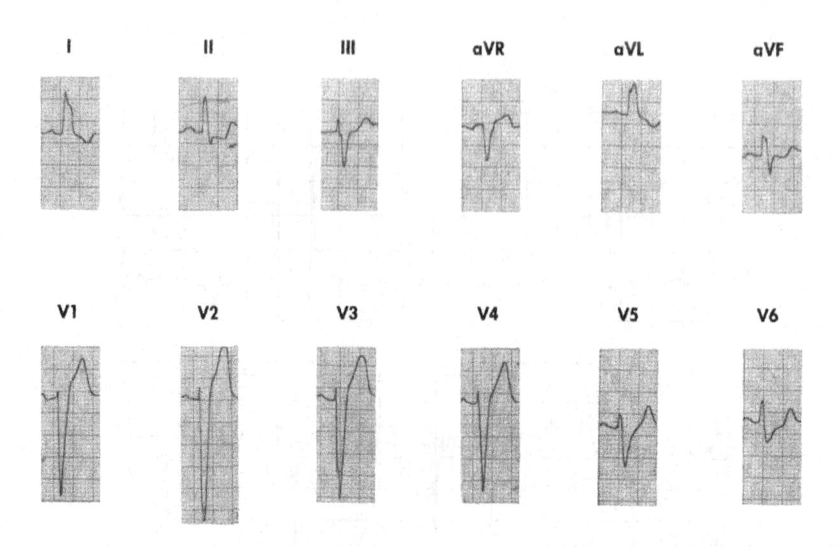

Figure 12.4 ECG from the twelve standard leads, showing left bundle branch block (LBBB), diagnosed as such because of the lengthened QRS complex (0.12 ms), a splitting of the QRS complex in leads V_1 through V_4 into two signals, and a leftward deflection of the heart vector, indicated, for example, by the amplitude shift in lead V_6. (Rushmer, 1976, Fig 8-46, p. 338.)

The conductivity tensor σ is inhomogeneous, because it is different for bone, lung, blood, etc., and it is anisotropic, because of muscle fiber striation, for example. Obviously, current is conserved, so that

$$\nabla \cdot I = -\nabla \cdot (\sigma \nabla \phi) = S, \tag{12.2}$$

where S represents all current sources.

The most significant current source in the human body is the spreading action potential wave front in the heart. The spreading cardiac action potential is well approximated as a surface of current dipoles. The rapid increase in membrane potential (of about 100 mV) translates into an extracellular decrease of about 40 mV that extends spatially over a distance (the wave front thickness) of about 0.5 mm. If the exact location and strength of this dipole surface and the conductivity tensor for the entire body were known, then we could (in principle) solve the Poisson equation (12.2) to find the body surface potential at all times during the cardiac cycle. This problem is unsolved, and is known as the *forward problem of electrocardiography*.

What we would really like to know is the operator, say T, called a *transfer function*, that solves (12.2) and yields the *body surface potential* ϕ_B, denoted by

$$\phi_B(t) = T \cdot S(t). \tag{12.3}$$

Even more useful, if the transfer function T were known, one could determine the sources by inverting the forward problem

$$S(t) = T^{-1} \cdot \phi_B(t). \tag{12.4}$$

This problem, known as the *inverse problem of electrocardiography*, is even harder to solve than the forward problem, because it is a numerically unstable mathematical problem.

Since these problems are yet unsolved, we do well to make some simplifications. Our first simplification is to view the action potential upstroke surface as a single current dipole, known as the *heart dipole vector*. We define the heart dipole vector as

$$\mathbf{H}(t) = \int_V \mathbf{J}\, dV, \tag{12.5}$$

where $\mathbf{J}$ represents the dipole density at each point of the heart, and V is the heart volume. The heart dipole vector is assumed to be located at a fixed point in space, changing only in orientation and strength as a function of time.

Next we assume that the volume conductor is homogeneous and infinite with unit conductance. Then, from standard potential theory (see Exercise 1), at any point x in space,

$$\phi(x,t) = \frac{\mathbf{H}(t)\cdot x}{4\pi |x|^3}, \tag{12.6}$$

where the dipole is assumed to be located at the origin. Thus, at each point on the body surface,

$$\phi_B(x,t) = l_x \cdot \mathbf{H}(t), \tag{12.7}$$

where l_x is a vector, called the *lead vector*, associated with the electrode lead at position x. Of course, for a real person, the lead vector is not exactly $\frac{x}{4\pi|x|^3}$, and some other method must be used to determine l_x. However, (12.7) suggests that we can think of the body surface potential as the dot product of some vector l_x with the heart vector $\mathbf{H}(t)$, and that l_x has more or less the same orientation as a vector from the heart to the point on the body where the recording is made.

Since $\mathbf{H}$ is a three-dimensional vector, if we have three leads with linearly independent lead vectors, then three copies of (12.7) yields a matrix equation that can be inverted to find $\mathbf{H}(t)$ uniquely. In other words, if our goal is to determine $\mathbf{H}(t)$, then knowledge of the full transfer function is not necessary. In fact, additional measurements from other leads should give redundant information.

Of course, the information from additional leads is not redundant, but it is nearly so. Estimates are that a good three-lead system can account for 85% of the information concerning the nature of the dipole sources. Discrepancies occur because the sources are not exactly consolidated into a single dipole, or because the lead vectors are not known with great accuracy, and so on. However, for clinical purposes, the information gleaned from this simple approximation is remarkably useful and accurate.

The next simplification is to standardize the position of the body-surface recordings and to determine the associated lead vectors. Then, with experience, a clinician can recognize features of the heart vector by looking at recordings of the potential at the leads. Or sophisticated (and expensive) equipment can be built that inverts the lead vector matrix and displays the heart vector on a CRT display device.

Cardiologists have settled on 12 standard leads. The first three were established by Einthoven, the "father of electrocardiography" (1860–1927, inventor of the string galvanometer in 1905, 1924 Nobel Prize in Physiology or Medicine) and are still used today. These are the left arm (LA), the right arm (RA), and the left leg (LL). One cannot measure absolute potentials, but only potential differences. There are three ways to measure potential differences with these three leads, namely,

$$V_{\mathrm{I}} = \phi_{\mathrm{LA}} - \phi_{\mathrm{RA}}, \tag{12.8}$$

$$V_{\mathrm{II}} = \phi_{\mathrm{LL}} - \phi_{\mathrm{RA}}, \tag{12.9}$$

$$V_{\mathrm{III}} = \phi_{\mathrm{LL}} - \phi_{\mathrm{LA}}, \tag{12.10}$$

and of course, since the potential drop around any closed loop is zero,

$$V_{\mathrm{I}} + V_{\mathrm{III}} = V_{\mathrm{II}}. \tag{12.11}$$

With these three differences, there are three lead vectors associated with the orientation of the leads, and the potential difference is the amplitude of the projection of the heart vector **H** onto the corresponding lead vector. Thus, $L_j = l_j \cdot \mathbf{H}$, and $V_j = |L_j|$ for $j =$ I, II, III.

Einthoven hypothesized that the lead vectors associated with readings $V_{\mathrm{I}}, V_{\mathrm{II}}, V_{\mathrm{III}}$ form an equilateral triangle in the vertical, frontal plane of the body, given by the unit vectors (ignoring an amplitude scale factor) $l_{\mathrm{I}} = (1, 0, 0)$, and $l_{\mathrm{II}} = (\frac{1}{2}, \frac{1}{2}\sqrt{3}, 0)$. The Einthoven triangle is shown in Fig. 12.5. Here the unit coordinate vector (1, 0, 0) is horizontal from right arm to left arm, (0, 1, 0) is vertical pointing downward, and the vector (0, 0, 1) is the third coordinate in a right-handed system, pointing in the posterior direction, from the front to back of the chest. Associated with the frontal plane is a polar coordinate system, centered at the heart, with angle $\theta = 0$ along the x axis, and $\theta = 90°$ vertically downward along the positive y axis.

Of course, the lead vectors of Einthoven are not very accurate. Experiments to measure the lead vectors in a model of the human torso filled with electrolytes produced

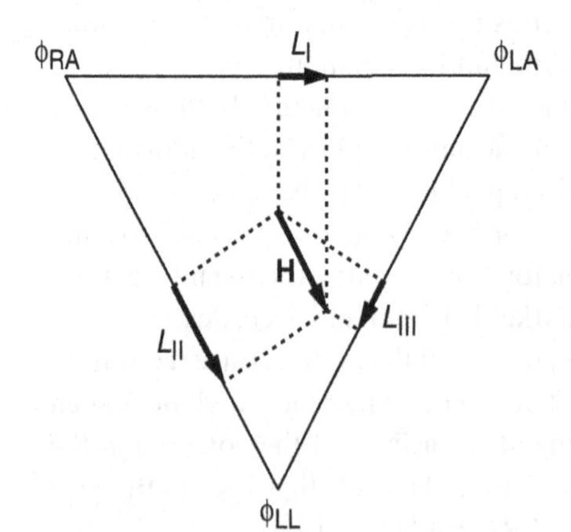

Figure 12.5 The Einthoven triangle showing a typical heart vector **H** and associated lead vectors $L_{\mathrm{I}}, L_{\mathrm{II}}$ and L_{III}. Because the body is approximately planar, the lead vectors are assumed to be in the frontal plane.

measured lead vectors $l_{\mathrm{I}} = (0.923, -0.298, 0.241)$, and $l_{\mathrm{II}} = (0.202, 0.972, -0.121)$ (Burger and van Milaan, 1946, 1947, 1948), which are not in the frontal plane. These lead vectors are known as the *Burger triangle*.

It is fairly easy to glean information about the direction of the heart vector by recognizing the information that is contained in (12.7). The vector ECG is actually a time-varying vector loop (shown in front, top, and side views in Fig. 12.7), and deducing time-dependent information is best done with an oscilloscope. However, one can estimate the mean direction of the vector by estimating the mean amplitude of a wave and then using (12.7) to estimate the mean heart vector. The mean (or time average) of the QRS complex is approximately proportional to the sum of the (positive) maximum and the (negative) minimum.

Since the lead voltage is a dot product of two vectors, a change in mean amplitude of a particular wave suggests either a change in amplitude of the heart vector or a change in direction of the heart vector. For example, the normal QRS and T wave mean dipoles are oriented about 45° below horizontal to the left (see Exercise 5). This is close to orthogonal to the lead vector l_{III}, and more or less aligned with lead vector l_{II}. Thus, on a normal ECG, we expect the mean amplitude of a QRS to be small in lead III, large in lead II, and intermediate to these two in lead I. Shifts in these relative amplitudes suggest a shift in the orientation of the heart dipole. For example, an increase in the relative amplitude of the potential difference at lead III and a decrease in amplitude at lead II suggests a shift of the heart vector to the right, away from the left, suggesting a malfunction of the conduction in the left heart.

Although two orthogonal lead vectors suffice to determine the orientation of the heart vector in the vertical plane, for ease of interpretation it is helpful to have more leads. For this reason, there are three additional leads on the frontal plane that are used clinically. To create these leads one connects two of the three Einthoven leads to a central point with 5000 Ω resistors to create a single terminal that is relatively indifferent to changes in potential and then takes the difference between this central potential and the remaining electrode of the Einthoven triangle. These measurements are denoted by aVR, aVL, or aVF, when the third unipolar lead is the right arm, the left arm, or the left foot, respectively. The initial "a" is used to denote an *augmented* unipolar limb lead.

For standard cardiographic interpretation the lead vectors for leads I, aVR^-, II, aVF, III, and aVL^- are assumed to divide the frontal plane into equal 30° sectors. For example, l_{I} is horizontal, l_{aVR^-} is declined at 30°, while l_{aVF} is vertical, etc. The superscript for aVR^- denotes the negative direction of the lead vector l_{aVR} (Fig. 12.6).

With these six leads, vector interpretation of the frontal plane orientation of the heart dipole is fast. One looks for the leads with the largest and smallest deflections, and surmises that the lead vector with largest mean amplitude is most parallel to the heart dipole, and the lead vector with the smallest mean deflection is nearly orthogonal to the heart dipole. Thus, in the normal heart situations, readings at leads II and aVR should be the largest in mean amplitude, with positive deflection at lead II, and negative deflection at lead aVR, while the mean deflections from leads III and aVL should be

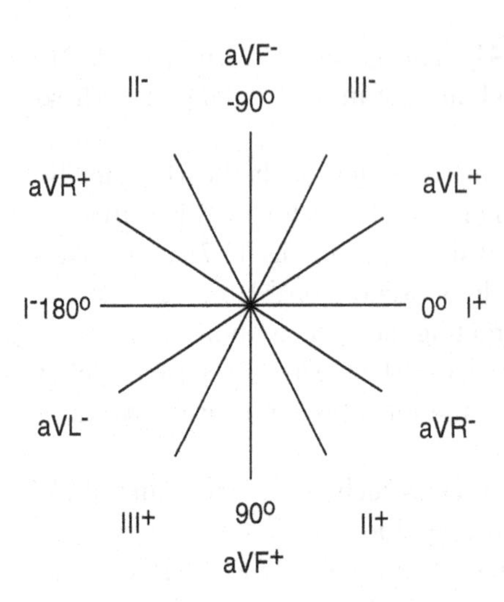

Figure 12.6 The standard six leads for the electrocardiogram (and their negatives).

the smallest, being the closest to orthogonal to the normal heart dipole (Fig. 12.7). Deviations from this suggest conduction abnormalities.

Six additional leads have been established to obtain the orientation of the heart dipole vector in a horizontal plane. For these leads, the three leads of Einthoven are connected with three 5000 Ω resistors to form a "zero reference," called the *central terminal of Wilson*. This is compared to a unipolar electrode reading taken from six different locations on the chest. These are denoted by $V_1, V_2, \ldots, V_6$ and are located on the right side of the sternum (V_1), the left side of the sternum (V_2) between the third and fourth ribs, and proceeding around the left chest following just below the fourth rib, ending on the side of the chest directly under the armpit (V_6) (Fig. 12.8).

While a detailed discussion of interpretation of a vector ECG is beyond the scope of this text, there are several features of cardiac conduction that are easy to recognize. Notice from Fig. 12.7 that the normal T wave and the normal QRS complex deflect in the same direction on leads I, II, and aVR (up on I and II, down on aVR). However, the QRS complex corresponds to the upstroke and the T wave to the downstroke of the action potential, so it must be that the activation (upstroke) and recovery (downstroke) wave fronts propagate in opposite directions. Said another way, the most recently activated tissue is the first to recover. The reason for the retrograde propagation of the wave of recovery is not fully understood. Second, an inverted wave (i.e., inverted from what is normal) implies that either the wave is propagating in the retrograde direction, or more typically with novice medical technicians, that the leads have been inadvertently reversed (see Exercise 4).

The amplitude of the QRS complex reflects the amount of muscle mass involved in propagation. Thus, if the QRS amplitude is extraordinarily large, it suggests *ventricular hypertrophy*. If the ECG vector is leftward from normal, it suggests left ventricular hypertrophy (Fig. 12.9), while a rightward orientation suggests right ventricular hypertrophy (Fig. 12.10). On the other hand if an amplitude decrease is accompanied

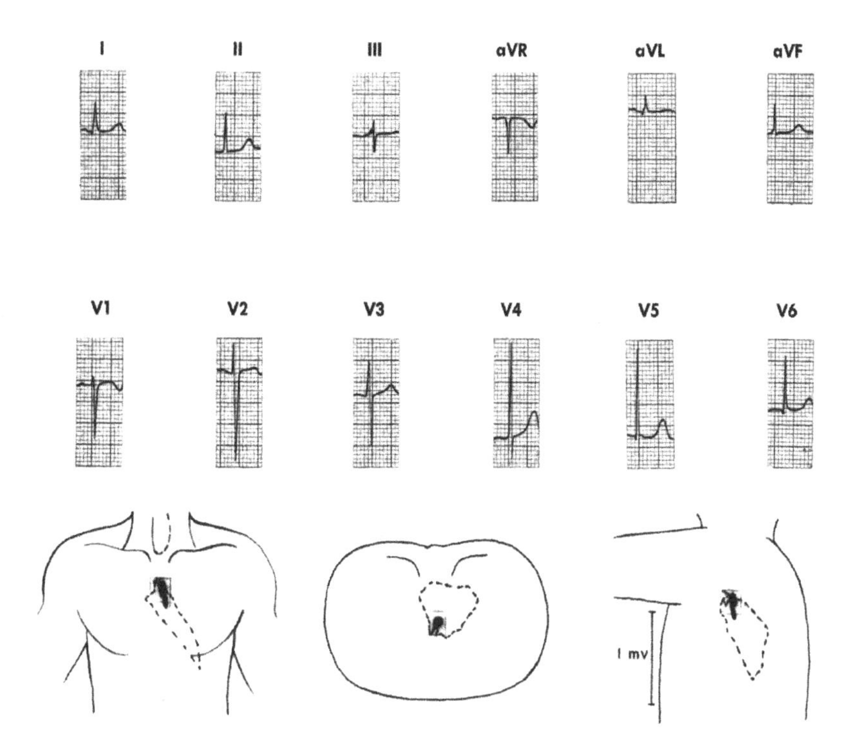

Figure 12.7 Normal ECG and VCG recording from the standard twelve leads in a nine-year old girl. (Rushmer, 1976, Fig. 8-33, p. 320, originally from Guneroth, 1965.)

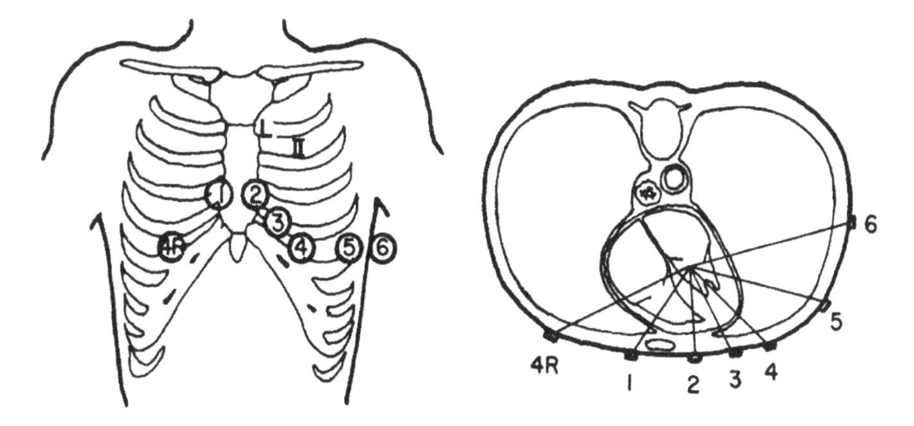

Figure 12.8 Frontal and horizontal cross-sectional views of the thorax in relation to the V-lead positions of Wilson. (Rushmer, 1976, Fig. 8-10, p. 294.)

by a rightward change in orientation, a diagnosis of *myocardial infarction* in the left ventricle is suggested, while a leftward orientation with decreased amplitude suggests a myocardial infarction of the right ventricle, as the heart vector is deflected away from the location of the infarction (see Exercises 6 and 7).

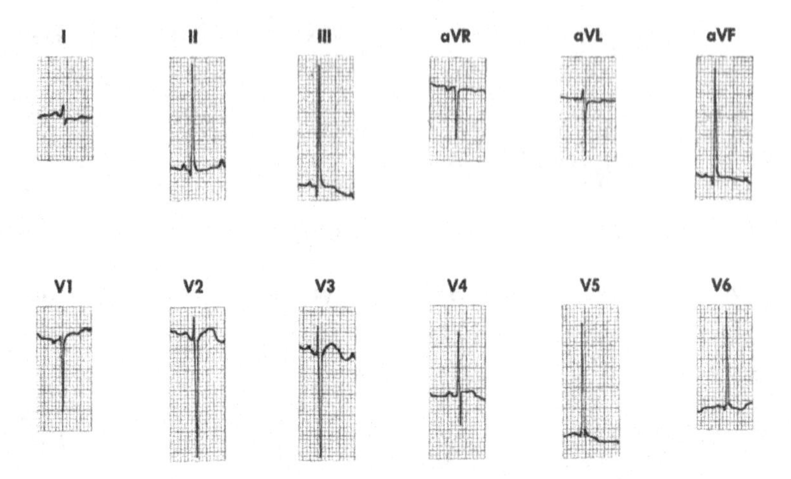

Figure 12.9 Twelve-lead ECG recording for severe left ventricular hypertrophy, particularly noticeable in leads III and aVL. (Rushmer, 1976, Fig. 8-24, p. 331, originally from Guneroth, 1965.)

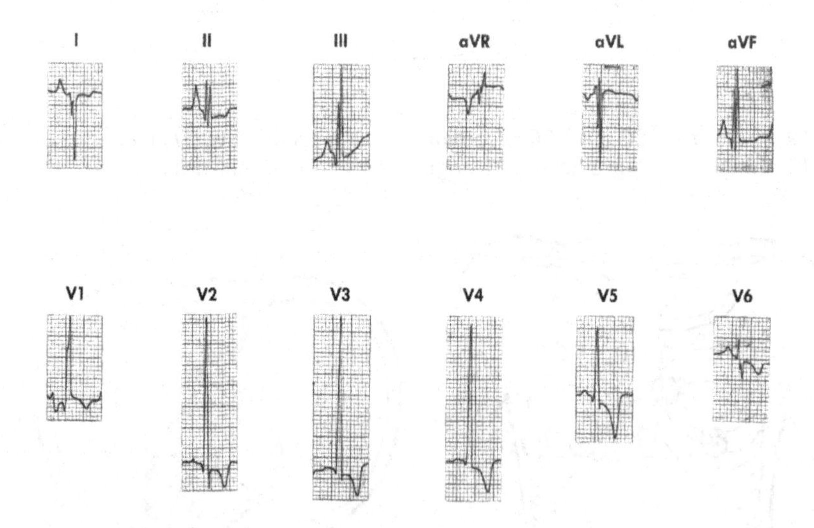

Figure 12.10 Twelve-lead ECG recording for severe right ventricular hypertrophy, particularly noticeable in leads III, V2, and V3. (Rushmer, 1976, Fig. 8-23, p. 330, originally from Guneroth, 1965.)

12.2 Cardiac Cells

We now turn our attention to developing a more detailed understanding of the physiological basis for the ECG and its abnormalities. We begin with a discussion of individual cardiac cells, and then proceed to describe spatially coupled systems of cells.

The primary cell types of cardiac cells are nodal cells (the sinoatrial (SA) and atrioventricular (AV) nodes), Purkinje fiber cells, and atrial and myocardial cells, each with a slightly different function.

The primary function of SA nodal cells is to provide a pacemaker signal for the rest of the heart. AV nodal cells transmit the electrical signal from atria to ventricles with a delay. Purkinje fiber cells are primarily for fast conduction, to activate the myocardium, and myocardial cells, both atrial and ventricular, are muscle cells and so are contractile as well as excitable.

Because of these different functions, these cell types have different action potential shapes, and all are noticeably different than the Hodgkin–Huxley action potential. The action potential for SA nodal cells is the shortest, while both Purkinje fiber cells and myocardial cells have substantially prolonged action potentials (300–400 ms compared to 3 ms for the squid axon), facilitating and controlling muscular contraction. Even within a single cell type, there can be substantial variation. For example, in the ventricles, epicardial, midmyocardial, and endocardial cells have noticeable differences in action potential duration. AV nodal cells vary substantially, to the extent that they are sometimes classified into several different subtypes. Typical action potentials for several cell types are shown in Figs. 12.11, 12.12, and 12.14, and typical ionic concentrations are given in Table 12.1.

12.2.1 Purkinje Fibers

A Phenomenological Approach

Subsequent to the work of Hodgkin and Huxley, there was substantial work done to apply their modeling approach to many different cell types, including cardiac cells. The first such model describing the action potential of a cardiac cell was proposed

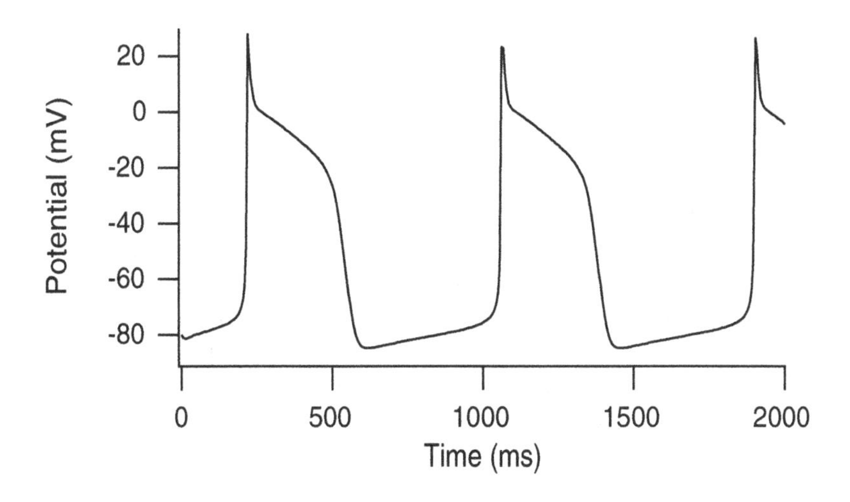

Figure 12.11 Action potential for the Noble model.

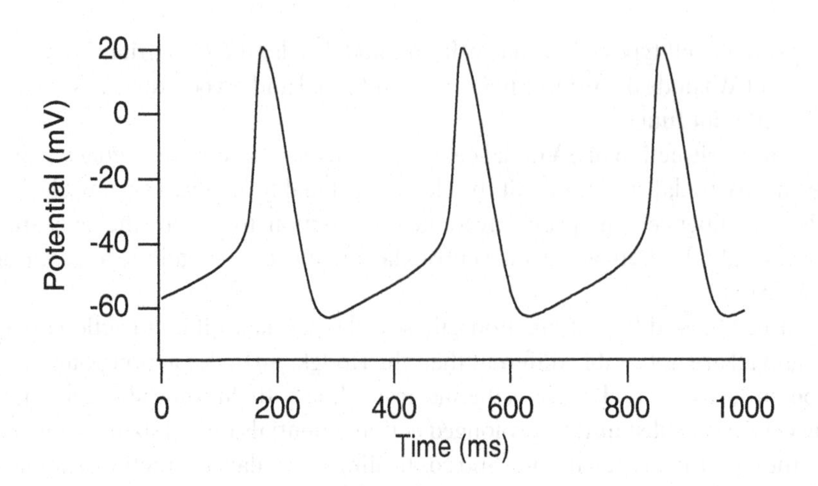

Figure 12.12 Membrane potential for the YNI model of SA nodal behavior.

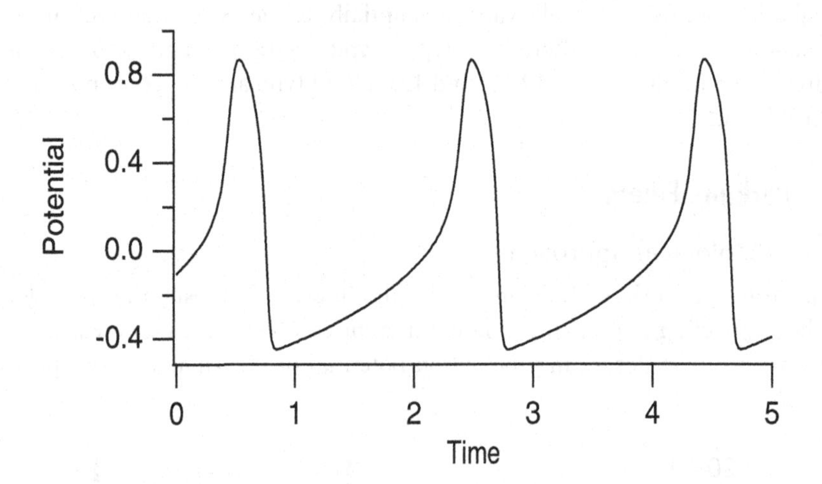

Figure 12.13 The potential $v(t)$ of the FitzHugh–Nagumo equations with $f(v,w) = v(1-v)(v-\alpha) - w$, $g(v,w) = w - \gamma v$, with $\epsilon = 0.02$, $\alpha = -0.05$, $\gamma = -0.6$.

by Noble (1962) for Purkinje fiber cells. The primary purpose of the model was to show that the action potential of a Purkinje fiber cell could be captured by a model of Hodgkin–Huxley type. The Noble model is of Hodgkin–Huxley type, expressed in terms of ionic currents and conductances. In this model there are three currents, identified as an inward Na^+ current, an outward K^+ current, and a Cl^- leak current, all of which are assumed to satisfy a linear instantaneous I–V relation,

$$I = g(V - V_{eq}). \tag{12.12}$$

For the Noble model, all changes in conductances that were measured in Na^+-deficient solutions were assumed to be for currents carried by K^+ ions and are therefore

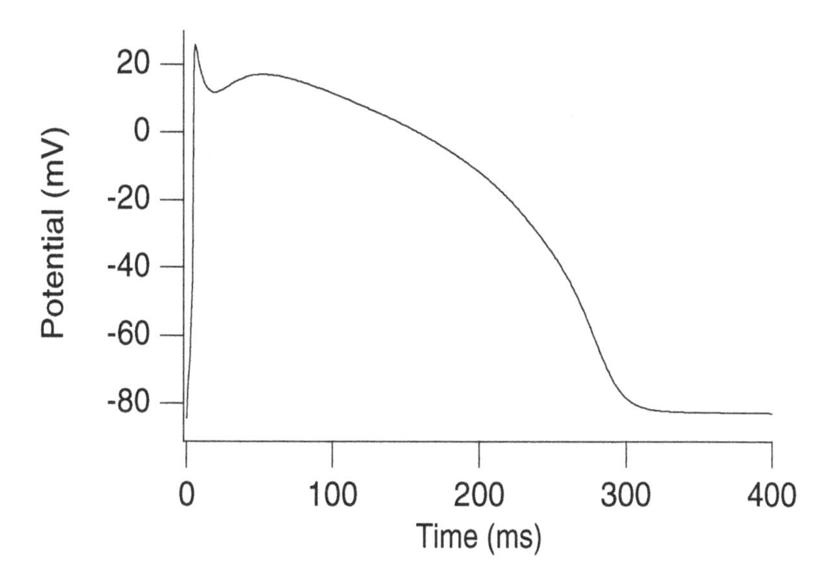

Figure 12.14 Action potential for the Beeler–Reuter equations.

Table 12.1 Ion concentrations in most cardiac cells.

Ion	Extracellular (mM)	Intracellular (mM)	Nernst Potential (mV)
Na^+	145	15	60
Cl^-	100	5	−80
K^+	4.5	160	−95
Ca^{2+}	1.8	0.0001	130
H^+	0.0001	0.0002	−18

called K^+ currents (thus, the Cl^- current is taken to be zero). While it is now known that not all these currents are carried by K^+ ions, here we follow Noble's original nomenclature.

Following the usual Hodgkin–Huxley formulation, the balance of transmembrane currents is expressed by the conservation law

$$C_m \frac{dV}{dt} + g_{\mathrm{Na}}(V - V_{\mathrm{Na}}) + (g_{\mathrm{K}_1} + g_{\mathrm{K}_2})(V - V_{\mathrm{K}}) + g_{\mathrm{an}}(V - V_{\mathrm{an}}) = I_{\mathrm{app}}, \tag{12.13}$$

where $g_{\mathrm{an}} = 0$ and $V_{\mathrm{an}} = -60$ are the conductance and equilibrium potential, respectively, for the anion Cl^- current (which therefore is not needed). In addition, $V_{\mathrm{Na}} = 40$ and $V_{\mathrm{K}} = -100$.

The Noble model assumes two different types of K^+ channels: an instantaneous, voltage-dependent, channel, and a time-dependent channel. The time-dependent K^+ channel has a similar form to the Hodgkin–Huxley K^+ channel, except that it is about

100 times slower in its response, in order to prolong the action potential plateau. This current is sometimes called the *delayed rectifier current*, because it is delayed and because it is primarily an outward (rectified) current. The conductance for this channel, g_{K_2}, depends on a time-dependent K^+ activation variable n through

$$g_{K_2} = 1.2n^4. \tag{12.14}$$

The conductance for the instantaneous channel is described empirically by

$$g_{K_1} = 1.2\exp\left(-\frac{V+90}{50}\right) + 0.015\exp\left(\frac{V+90}{60}\right). \tag{12.15}$$

The Na^+ conductance for the Noble model is of a form similar to that in the Hodgkin–Huxley equations, being

$$g_{Na} = 400m^3h + g_i, \tag{12.16}$$

where $g_i = 0.14$, with the fixed inward bias from g_i enabling a prolonged action potential without necessitating major reworking of the dynamics of h and m.

The time dependence of the variables m, n, and h is of the form

$$\frac{dw}{dt} = \alpha_w(1-w) - \beta_w w, \tag{12.17}$$

with $w = m, n$, or h, where α_w and β_w are all of the form

$$\frac{C_1\exp(\frac{V-V_0}{C_2}) + C_3(V-V_0)}{1 + C_4\exp(\frac{V-V_0}{C_5})}. \tag{12.18}$$

The constants $C_1, \ldots, C_5$ and V_0 are displayed in Table 12.2.

In the Noble model, $C_m = 12$, which is unrealistically large. This value was used because it gives a correct time scale for the length of the action potential. The choice was justified by arguing that the effective capacitance for a small bundle of cylindrical cells, for which the data were obtained, should be larger than for a single cylindrical cell, the surface area of which is only a small fraction of the total cell membrane area.

Numerical simulations show that the Noble model produces an action potential that has correct features, seen in Fig. 12.11. The sharp upstroke comes from a large, fast, inward Na^+ current, and the plateau is maintained by a continued inward Na^+ current

Table 12.2 Defining values for rate constants α and β for the Noble model.

	C_1	C_2	C_3	C_4	C_5	V_0
α_m	0	–	0.1	−1	−15	−48
β_m	0	–	−0.12	−1	5	−8
α_h	0.17	−20	0	0	–	−90
β_h	1	∞	0	1	−10	−42
α_n	0	–	0.0001	−1	−10	−50
β_n	0.002	−80	0	0	–	−90

(with conductance g_i), which nearly counterbalances the instantaneous outward K^+ current. Gradually, the slow outward K^+ current is activated, causing repolarization. A small inward Na^+ leak, called the *pacemaker current*, also allows the potential to creep upward, eventually initiating another action potential.

Because of the sharp spike at the beginning of the action potential, it is not possible to reproduce the Purkinje fiber action potential with a two-variable FitzHugh–Nagumo-type model. However, by setting $m = m_\infty(V)$, the Noble model can be reduced to a three-variable model that retains the primary qualitative features of the original model.

A Physiological Approach

While the Noble model succeeds in reproducing the Purkinje fiber action potential with a model of Hodgkin–Huxley type, the underlying physiology is incorrect, primarily because the model was constructed before data on the ionic currents were available. This lack of data was mostly because the voltage-clamp technique was not successfully applied to cardiac membrane until 1964.

The weakness of the physiology in the Noble model is exemplified by the fact that there is no current identified with Ca^{2+} ions, and the inward Na^+ current was given the dual role of generating the upstroke and maintaining the plateau.

In 1975, McAllister, Noble, and Tsien (MNT) presented an improved model for the action potential of Purkinje fibers. This model is based on a "mosaic of experimental results," because unlike the data used for the Hodgkin–Huxley equations, the required information was not obtained from a single experimental preparation. Furthermore, the model is known to have an inadequate description of the Na^+ current, so that the upstroke velocity is not accurate.

The MNT model is similar to the Noble model in that it is based on a description of transmembrane ionic currents. It is substantially more complicated than most models of its time, having nine ionic currents and nine gating variables. There are two inward currents, $I_{\rm Na}$ and $I_{\rm si}$ (called the "slow inward" current). The current $I_{\rm Na}$ resembles the Hodgkin–Huxley Na^+ current and is represented as

$$I_{\rm Na} = \bar{g}_{\rm Na} m^3 h (V - V_{\rm Na}), \tag{12.19}$$

where m and h are activation and inactivation gating variables, respectively, and $V_{\rm Na} =$ 40 mV. The inward current $I_{\rm si}$ has slower kinetics than $I_{\rm Na}$ and is carried, at least partly, by Ca^{2+} ions. This current $I_{\rm si}$ has two components and is given by

$$I_{\rm si} = (0.8df + 0.04d')(V - V_{\rm si}), \tag{12.20}$$

where $V_{\rm si}$ = 70 mV. The variables d and f are time-dependent activation and inactivation variables, respectively, while d' is only voltage-dependent, being

$$d' = \frac{1}{1 + \exp(-0.15(V + 40))}. \tag{12.21}$$

In the MNT model, there are three time-dependent outward K^+ currents, denoted by I_{K_2}, I_{x_1}, and I_{x_2}. None of these resemble the squid K^+ current from a quantitative point of view, although all are described using an activation variable and no inactivation variable. The current I_{K_2} is called the *pacemaker current* because it is responsible for periodically initiating an action potential, and it is given by

$$I_{K_2} = 2.8\bar{I}_{K_2}s, \tag{12.22}$$

where

$$\bar{I}_{K_2} = \frac{\exp(0.04(V+110)) - 1}{\exp(0.08(V+60)) + \exp(0.04(V+60))}. \tag{12.23}$$

The currents I_{x_1} and I_{x_2} are called *plateau currents* and are governed by

$$I_{x_1} = 1.2x_1\frac{\exp(0.04(V+95)) - 1}{\exp(0.04(V+45))}, \tag{12.24}$$

$$I_{x_2} = x_2(25 + 0.385V). \tag{12.25}$$

There is also a time-dependent outward current I_{Cl} carried by Cl^- ions, which is described by

$$I_{Cl} = 2.5qr(V - V_{Cl}), \tag{12.26}$$

where q and r are activation and inactivation variables, and $V_{Cl} = -70$ mV.

Finally, there are several background (leak) currents that are time-independent. There is an outward background current of K^+ ions, described by

$$I_{K_1} = \bar{I}_{K_2} + 0.2\frac{V+30}{1 - \exp(-0.04(V+30))}, \tag{12.27}$$

where $\bar{I}_{K_2}$ is given by (12.23). There is an inward background Na^+ current described by

$$I_{Na,b} = 0.105(V - 40), \tag{12.28}$$

and, finally, a background Cl^- current, given by

$$I_{Cl,b} = 0.01(V + 70). \tag{12.29}$$

All of the conductances are specified in units of mS/cm^2, and voltage is in mV. The nine gating variables $m, d, s, x_1, x_2, q, h, f$, and r all satisfy first-order differential equations of the form (12.17), where α_w and β_w are of the form (12.18). The constants $C_1, \ldots, C_5$ and V_0 are listed in Table 12.3.

The action potential for the MNT model is essentially the same as that for the Noble model, so the advantage of the MNT model is that it better isolates and depicts the activity of different channels during an action potential. Of course, because it is more complicated than the Noble model, it is also much harder to understand the model from a qualitative perspective. It therefore illustrates nicely the modeler's dilemma, the constant struggle to balance the demand for quantitative detail and qualitative understanding.

Table 12.3 Defining values for α and β for the MNT model.

	C_1	C_2	C_3	C_4	C_5	V_0
α_m	0	–	1	−1	−10	−47
β_m	40	−17.86	0	0	–	−72
α_h	0.0085	−5.43	0	0	–	−71
β_h	2.5	∞	0	1	−12.2	−10
α_d	0	–	0.002	−1	−10	−40
β_d	0.02	−11.26	0	0	–	−40
α_f	0.000987	−25	0	0	–	−60
β_f	1	∞	0	1	−11.49	−26
α_q	0	–	0.008	−1	−10	0
β_q	0.08	−11.26	0	0	–	0
α_r	0.00018	−25	0	0	–	−80
β_r	0.02	∞	0	1	−11.49	−26
α_s	0	–	0.001	−1	−5	−52
β_s	5.0×10^{-5}	−14.93	0	0	–	−52
α_{x_1}	0.0005	12.1	0	1	17.5	−50
β_{x_1}	0.0013	−16.67	0	1	−25	−20
α_{x_2}	1.27×10^{-4}	∞	0	1	−5	−19
β_{x_2}	0.0003	−16.67	0	1	−25	−20

12.2.2 Sinoatrial Node

Cells within the sinoatrial (SA) node are the primary pacemaker site within the heart. These cells are characterized as having no true resting potential, but instead generate regular, spontaneous action potentials. Unlike most other cells that elicit action potentials (e.g., nerve cells, muscle cells), the depolarizing current is carried primarily by a relatively slow, inward Ca^{2+} current instead of by fast Na^+ currents. There are, in fact, no fast Na^+ currents in SA nodal cells. Phase 0 depolarization (recall Fig. 12.2) is due primarily to increased Ca^{2+} conductance. Because the Ca^{2+} channels open more slowly than Na^+ channels (hence, the term "slow inward Ca^{2+} current"), the rate of depolarization (slope of Phase 0) is much slower than found in other cardiac cells (e.g., Purkinje cells, ventricular cells). Repolarization occurs (Phase 3) as K^+ conductance increases and Ca^{2+} conductance decreases. Spontaneous depolarization (Phase 4) is due to a fall in K^+ conductance and to a small increase in Ca^{2+} conductance. A slow inward Na^+ current also contributes to Phase 4, and is thought to be responsible for the pacemaker current. Once this spontaneous depolarization reaches threshold (about −40 mV), a new action potential is triggered.

One of the earliest models of action potential behavior for SA nodal cells is due to Yanagihara et al. (1980). As with all cardiac cell models, the YNI model is of Hodgkin–Huxley type. The YNI model includes four time-dependent currents. These are a fast inward current (incorrectly identified as a Na^+ current) I_{Na}, and the K^+ current I_K, both of which are modeled similar to the Hodgkin–Huxley currents, as well as a slow inward current I_s, and a delayed inward current activated by hyperpolarization I_h. Finally, there is a time-independent leak current I_l.

The conservation of transmembrane current takes the form

$$C_m \frac{dV}{dt} + I_{Na} + I_K + I_l + I_s + I_h = I_{app}, \tag{12.30}$$

where

$$I_{Na} = 0.5m^3h(V - 30), \tag{12.31}$$

$$I_K = 0.7p \frac{\exp(0.0277(V + 90)) - 1}{\exp(0.0277(V + 40))}, \tag{12.32}$$

$$I_l = 0.8 \left(1 - \exp\left(-\frac{V + 60}{20}\right)\right), \tag{12.33}$$

$$I_s = 12.5(0.95d + 0.05)(0.95f + 0.05) \left(\exp\left(\frac{V - 10}{15}\right) - 1\right), \tag{12.34}$$

$$I_h = 0.4q(V + 45). \tag{12.35}$$

As usual, the six gating variables m, h, p, d, f, and q satisfy first-order differential equations of the form (12.17). Some of the constants α_w and β_w can be written in the form (12.18) with constant values as shown in Table 12.4. Those that do not fit this form are

$$\alpha_p = 9 \times 10^{-3} \frac{1}{1 + \exp\left(-\frac{V+3.8}{9.71}\right)} + 6 \times 10^{-4}, \tag{12.36}$$

$$\alpha_q = 3.4 \times 10^{-4} \frac{(V + 100)}{\exp\left(\frac{V+100}{4.4}\right) - 1} + 4.95 \times 10^{-5}, \tag{12.37}$$

$$\beta_q = 5 \times 10^{-4} \frac{(V + 40)}{1 - \exp\left(-\frac{V+40}{6}\right)} + 8.45 \times 10^{-5}, \tag{12.38}$$

$$\alpha_d = 1.045 \times 10^{-2} \frac{(V + 35)}{1 - \exp\left(-\frac{V+35}{2.5}\right)} + 3.125 \times 10^{-2} \frac{V}{1 - \exp\left(-\frac{V}{4.8}\right)}, \tag{12.39}$$

$$\beta_f = 9.44 \times 10^{-4} \frac{(V + 60)}{1 + \exp\left(-\frac{V+29.5}{4.16}\right)}. \tag{12.40}$$

The behavior of the YNI equations is depicted in Fig. 12.12. The action potential is shaped similarly to the Hodgkin–Huxley action potential but is periodic in time and slower. The Na^+ current is a fast current, and there is little loss in accuracy in replacing $m(t)$ with $m_\infty(V)$. The most significant current in the YNI model is the slow inward current I_s. Not only does this current provide for most of the upstroke, it is also responsible for the oscillation, in that after repolarization by the K^+ current, the slow inward current gradually depolarizes the node until threshold is reached and an action potential is initiated.

Table 12.4 Defining values for α and β for the YNI model.

	C_1	C_2	C_3	C_4	C_5	V_0
α_m	0	—	1	−1	−10	−37
β_m	40	−17.8	0	0	—	−62
α_h	1.209×10^{-3}	−6.534	0	0	—	−20
β_h	1	∞	0	1	−10	−30
β_p	0	—	-2.25×10^{-4}	−1	13.3	−40
β_d	0	—	-4.21×10^{-3}	−1	2.5	5
α_f	0	—	-3.55×10^{-4}	−1	5.633	−20

Because the action potential of the SA node has no initial spike, it is relatively easy to replicate it using the two-variable FitzHugh–Nagumo equations. In Fig. 12.13 is shown the periodic activity of a cubic FitzHugh–Nagumo model with action potential spikes similar to those of the YNI model.

12.2.3 Ventricular Cells

The first model of the electrical behavior of ventricular myocardial cells (Beeler and Reuter, 1977) appeared shortly after the MNT equations. Like the models previously described, this model is based on data obtained from voltage-clamp experiments. The Beeler–Reuter equations are less complicated than the MNT equations, since there are only four transmembrane currents that are described, two inward currents, one fast and one slow, and two outward currents, one time-independent and one time-dependent.

As usual, there is the inward Na^+ current

$$I_{Na} = (4m^3hj + 0.003)(V - 50), \tag{12.41}$$

which is gated by the variables $m, h,$ and j. Here, Beeler and Reuter found it necessary to include the reactivation variable j, because the reactivation process is much slower than inactivation and cannot be accurately modeled with the single variable h. Thus, the Na^+ current is activated by m, inactivated by h, and reactivated by j, the slowest of the three variables. The functions h_∞ and j_∞ are identical; it is their time constants that differ. Notice also the inclusion of a Na^+ leak current; a similar Na^+ leak was included in the Noble and MNT models.

The K^+ current has two components: a time-independent current

$$I_K = 1.4\frac{\exp(0.04(V+85)) - 1}{\exp(0.08(V+53)) + \exp(0.04(V+53))} + 0.07\frac{v + 23.0}{1.0 - \exp(-0.04(V+23.0))} \tag{12.42}$$

and a time-activated outward current

$$I_x = 0.8x\frac{\exp(0.04(V+77)) - 1}{\exp(0.04(V+35))}. \tag{12.43}$$

The pacemaker K^+ current used in the MNT model is not active in myocardial tissue, which is not spontaneously oscillatory.

The primary difference between a ventricular cell and a Purkinje cell is the presence of Ca^{2+}, which is needed to activate the contractile machinery. Later in this chapter, we provide a more modern, detailed account of Ca^{2+} handling in ventricular cells. For the Beeler–Reuter equations, however, the Ca^{2+} influx is modeled by the slow inward current

$$I_s = 0.09fd(V + 82.3 + 13.0287 \ln[\mathrm{Ca}]_i), \tag{12.44}$$

activated by d and inactivated by f. Since the reversal potential for I_s is Ca^{2+}-dependent, the internal Ca^{2+} concentration must be tracked, via

$$\frac{dc}{dt} = 0.07(1 - c) - I_s, \tag{12.45}$$

where $c = 10^7[\mathrm{Ca}]_i$. Since currents are taken as positive outward, the intracellular source of Ca^{2+} is $-I_s$.

The gating variables follow (12.17), where α_w and β_w are of the form (12.18) with constants as displayed in Table 12.5. For these equations, units of V are in mV, conductances are in units of mS/cm^2, and time is measured in milliseconds (ms). A plot of the Beeler–Reuter action potential is shown in Fig. 12.14. The long plateau is maintained by the slow inward (Ca^{2+}) current, and the return to the resting potential is mediated by the slow outward K^+ current I_{x_1}.

It is a fact of modeling that most models have a relatively short lifetime. They are initially proposed because they address a particular problem or feature, but then as weaknesses become apparent, they are supplanted by a new, improved version, whose fate is ultimately similar to the model it replaces. Examples of improved ionic models

Table 12.5 Defining values for α and β for the Beeler–Reuter model.

	C_1	C_2	C_3	C_4	C_5	V_0
α_m	0	–	1	−1	−10	−47
β_m	40	−17.86	0	0	–	−72
α_h	0.126	−4	0	0	–	−77
β_h	1.7	∞	0	1	−12.2	−22.5
α_j	0.055	−4	0	1	−5	−78
β_j	0.3	∞	0	1	−10	−32
α_d	0.095	−100	0	1	−13.9	5
β_d	0.07	−58.5	0	1	20	−44
α_f	0.012	−125	0	1	6.67	−28
β_f	0.0065	−50	0	1	−5	−30
α_x	0.0005	12	0	1	17.5	−50
β_x	0.0013	−16.67	0	1	−25	−20

are those by DiFrancesco and Noble (1985) for Purkinje fiber cells and by Noble and Noble (1984) for the SA node.

Although the Beeler–Reuter model has had a fairly long and robust run of popularity, it too is gradually being replaced by updated models. One of the first modifications to the Beeler–Reuter model was of its Na^+ current. At the time the Beeler–Reuter equations were published, it was not possible to measure accurately the fast Na^+ inward current, because it activates so rapidly. As a result, all the early models (Noble, MNT, BR) used the Hodgkin–Huxley formulation of the Na^+ current. However, it is known that this does not give a sufficiently rapid upstroke for the action potential. This has little effect on the space-clamped action potential, but it has an important effect on the propagation speed for propagated action potentials.

Once appropriate data became available, it was possible to suggest an improved description of the Na^+ current. Thus, a modification of the Na^+ current was proposed by Ebihara and Johnson (1980) (EJ), which has since become the standard for most myocardial simulations.

As of this writing, the most popular detailed model of the electrical activity of ventricular myocytes is the Luo–Rudy (LR-II) model (Luo and Rudy, 1994a,b). An earlier model (LR-I, Luo and Rudy, 1991) was a direct generalization of the Beeler–Reuter model. In Fig. 12.15 is shown a diagram of all the currents included in the LR-II model.

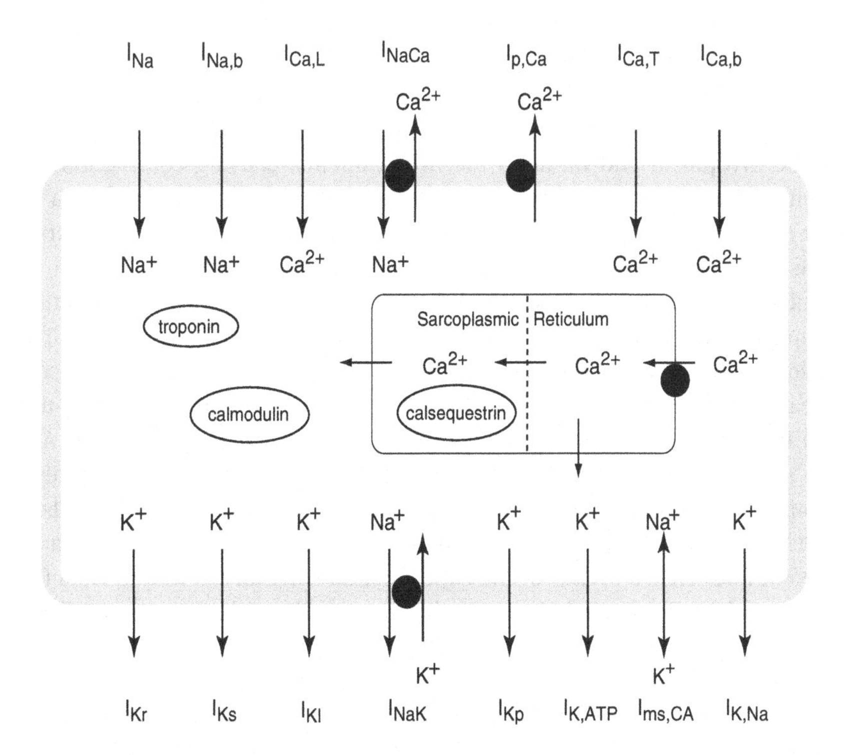

Figure 12.15 Diagram of currents included in the LR-II model.

Improvements and modifications to the LR-II model are continually being made. For example, many of the currents in the model are being given Markovian formulations, and with these it is feasible to study the effect of mutations to the channel proteins on action potential dynamics (Clancy and Rudy, 1999, 2001).

Other modifications to these models include specialization to human ventricular myocytes, as in ten Tusscher et al. (2004), or to human atrial myocytes as in Courtemanche et al. (1998). The computer codes for many models of cardiac electrical behavior can be found as part of CellML, at http://www.cellml.org.

12.2.4 Cardiac Excitation–Contraction Coupling

The final form for a myocardial ionic model has not yet been determined, since there are continual suggestions for improvements and modifications. A major difficulty with the above-mentioned models is with the Ca^{2+} current and the internal Ca^{2+} concentration. In recent years, however, there has been a virtual explosion of information regarding Ca^{2+} handling in cardiac cells, the topic to which we now turn our attention.

Excitation–contraction (EC) coupling is the process whereby an electrical stimulus is converted into muscle contraction. The basic steps of EC coupling in ventricular myocytes are shown in Fig. 12.16. Depolarization of the T-tubule by the action potential causes the opening of L-type Ca^{2+} channels (also called dihydropyridine receptors, or DHPRs) and resultant inward flow of Ca^{2+} current (I_{Ca}). The Ca^{2+} that enters the cell stimulates the release of additional Ca^{2+} from the SR via ryanodine receptors (RyR) by the process of Ca^{2+}-induced Ca^{2+} release, or CICR (Chapter 7). This Ca^{2+} diffuses through the myoplasm and binds to the myofilaments, causing contraction, before being eventually removed from the myoplasm by ATPases, which pump the Ca^{2+} into the SR or out of the cell, or by the Na^+–Ca^{2+} exchanger (NCX), which transfers Ca^{2+} to the outside of the cell.

In skeletal muscle the process is slightly different, in that the L-type channels are directly linked to the RyR; a change in conformation of the channel as a result of depolarization causes an immediate change in conformation of the RyR and consequent Ca^{2+} release. Another important difference is that, in skeletal muscle, the T-tubules penetrate into the cell in two places per sarcomere, near the junction of the A and I bands (as shown in Fig. 15.1), while in cardiac cells the T-tubules penetrate only once per sarcomere, at the Z-lines. Here we only discuss EC coupling in cardiac cells.

There is an extensive literature, both experimental and theoretical, on EC coupling. The most comprehensive review of the experimental literature is that of Bers (2001), while the modeling literature is most easily accessed through the reviews of Soeller and Cannell (2004) and Winslow et al. (2005). The shorter review by Bers (2002) is a useful introduction to the field.

Although Ca^{2+} fluxes are relatively well understood in many cell types (Chapter 7), the study of Ca^{2+} dynamics in cardiac cells is made much more difficult by the spatial aspects of the problem. Both the L-type Ca^{2+} channels and the RyR release Ca^{2+} into

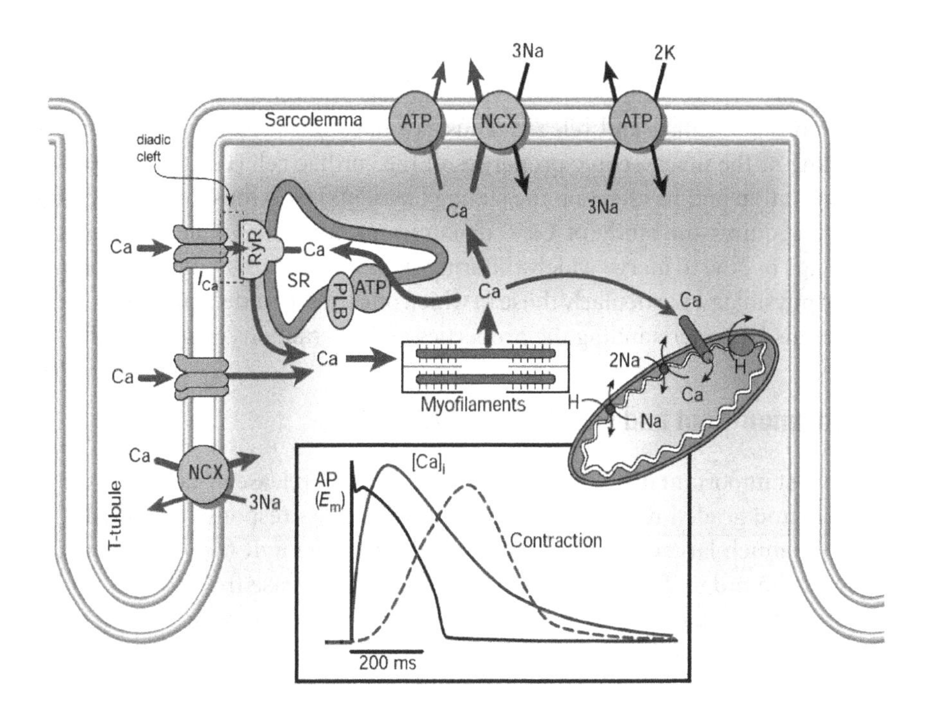

Figure 12.16 The major Ca^{2+} fluxes underlying excitation–contraction coupling in cardiac ventricular myocytes. NCX is the Na^{+}–Ca^{2+} exchanger; PLB is phospholamban; SR is the sarcoplasmic reticulum. The inset shows the time courses of the action potential (AP), the Ca^{2+} transient, and the contraction. Note that the AP happens first, followed by the Ca^{2+} transient and then by contraction. (Bers, 2002, Fig. 1: reprinted by permission from Macmillan Publishers Ltd.)

a small volume, the region between the SR and the sarcolemma. This region, called the *diadic cleft*, is only about 15 nm wide with a radius of about 200 nm (although it is not a circular region, it can be reasonably approximated as such), and so has a volume of only about 2×10^{-18} L. Calcium fluxes into such a small volume cause large spatial and temporal gradients that are impossible to measure experimentally and difficult to simulate numerically. Modeling is made even more difficult by the fact that a resting Ca^{2+} concentration of 200 nM (as is typical for the myoplasm of a ventricular myocyte) corresponds to only about 0.2 Ca^{2+} ions in the diadic cleft. In this situation, traditional deterministic and continuous models may not even be applicable. In cardiac cells the diadic clefts are separated longitudinally by approximately 2 μm, the length of a sarcomere.

During a single heartbeat, a total of about 70 μmoles of Ca^{2+} per liter cytoplasm enters the cell, with about 1% ending up as unbuffered free Ca^{2+} in the myoplasm, giving a myoplasmic concentration of around 600 nM (nmoles per liter cytoplasm). Since all this influx comes through the diadic cleft, there are clearly large and rapid changes in concentration there. Nevertheless, to understand how the Ca^{2+} transient is controlled it is vital to understand what happens inside the diadic cleft, because it is there that Ca^{2+} feeds back on the L-type channels and on the RyR to control the time course of Ca^{2+} influx and release. Thus, we have the situation in which a full understanding of the macroscopic properties of the cardiac cell (i.e., the myoplasmic Ca^{2+} concentration and its effect on the sliding filaments), and thus, ultimately, of the whole heart, requires the study of Ca^{2+} dynamics on a much smaller spatial scale. Such problems of how to merge widely differing spatial scales are at the heart of many problems in physiology, particularly those in which one tries to understand the behavior of whole organs by understanding the properties and interactions of single cells.

12.2.5 Common-Pool and Local-Control Models

Two of the most important defining characteristics of Ca^{2+} release in cardiac myocytes are high gain and graded release. High gain means that, in response to a small Ca^{2+} influx ($J_{I_{Ca}}$), a much larger amount is released through the RyR (J_{RyR}). In fact, Ca^{2+} influx is about an order of magnitude smaller than Ca^{2+} release from the SR. Graded release means that, if Ca^{2+} influx is smaller, less Ca^{2+} is released from the ER; release is a smooth and continuous function of influx. This is illustrated in Fig. 12.17. Part A of that figure shows the total L-type channel flux and the consequent RyR flux as functions of the membrane potential. The first thing to notice is that $J_{I_{Ca}}$ is a bell-shaped curve of the potential; at low V, the current through each open channel is large, but the probability of opening is small. As V increases, thus increasing the open probability of the channel, the flux increases also. However, as V gets closer to the Ca^{2+} reversal potential of the channel, the current through each open channel begins to fall, leading to a decrease in the total channel flux even though many of the channels are open.

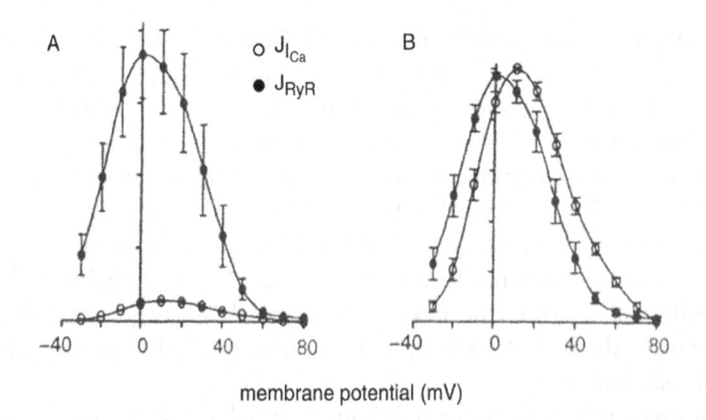

Figure 12.17 Graded release from the rat ventricular myocytes. A: the lower curve is the flux through the L-type channels ($J_{I_{Ca}}$), the upper curve is the consequent flux through the RyR (J_{RyR}). B: the same two curves plotted on a normalized scale. Adapted from Wier et al. (1994), Fig. 3, with permission from Blackwell Publishing.

As the L-type channel flux increases smoothly, so does the RyR flux, as can be seen from the upper curve in Fig. 12.17A. Interestingly, comparison of the normalized curves (Fig. 12.17B) shows that the RyR flux decreases at lower voltages than does the L-type channel flux. An explanation for this is that even while the total L-type channel current is an increasing function of voltage (due to the increased number of open channels), the flux through each individual channel decreases, leading to less effective coupling with the RyR and a resultant decrease in RyR flux.

The two requirements of high gain and graded release appear at first to be contradictory, and so require further explanation. An excitable system (Chapter 5), with a response that is approximately all-or-none, can exhibit high gain without difficulty, but in doing so cannot exhibit a graded release. Conversely, a model that exhibits graded release does not usually exhibit high gain in a stable manner. Many models of Ca^{2+} dynamics in muscle (including those used in Beeler–Reuter and LR-II models) are so-called *common-pool* models, in which it is assumed that Ca^{2+} influx and release both occur into the same well-mixed compartment. This well-mixed compartment could be the myoplasm, as in Fig. 12.18A, or a subcompartment such as the diadic cleft, as in Fig. 12.18B. Stern (1992) has shown that such common-pool models, at least in the linear regime, cannot exhibit both high gain and graded release; he introduced, instead, the concept of a *local-control* model, in which the close juxtaposition of the L-type channel

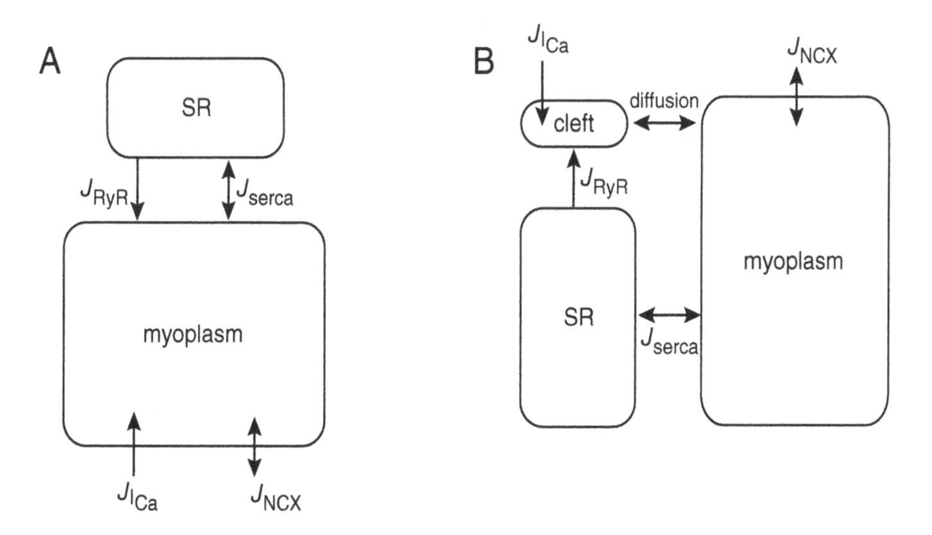

Figure 12.18 Schematic diagrams of two common-pool models. Each model consists of a number of compartments, each of which is assumed to be well mixed. A: a two-compartment model with only the SR and the myoplasm. The L-type channel current is assumed to flow directly into the myoplasm, whence it stimulates the release of further Ca^{2+} release from the SR. B: a three-compartment model in which the diadic cleft (with small volume) is included. J_{NCX} is the Na^{+}–Ca^{2+} exchange flux. Such common-pool models do not exhibit both a high gain and a graded response.

and the RyR leads to the formation of a Ca^{2+} synapse in which the opening of an RyR is controlled by the Ca^{2+} coming through a nearby L-type channel. Although each diadic cleft responds in an all-or-none manner (yielding high gain), the stochastic properties of the population of clefts can give a graded response. Since each diadic cleft is separated from its neighbors by approximately 2 μm in a longitudinal direction (the length of the sarcomere) and by about 0.8 μm in a transverse direction (Parker et al., 1996c), it can respond in a semi-independent fashion. Statistical recruitment from this large pool of locally controlled semi-independent release sites can result in graded whole-cell release, even though each individual release site is all-or-none.

Detailed models of local control, incorporating (to greater or lesser extents) the spatial geometry of a single diadic cleft and the stochastic properties of the population of clefts, have been constructed by a number of authors (Stern, 1992; Soeller and Cannell, 1997, 2002a,b; Peskoff et al., 1992; Langer and Peskoff, 1996; Stern et al., 1997; Peskoff and Langer, 1998; Smith et al., 1998; Izu et al., 2001; Greenstein and Winslow, 2002; Tameyasu, 2002; Greenstein et al., 2006). In general, these models are complex and rely on detailed numerical simulations, so we do not discuss them here. We merely comment that there are many unresolved questions in the field, which is an area of intense current research.

As we saw for models of Ca^{2+} dynamics in Chapter 7, models of EC coupling are constructed by combining individual models of each of the various Ca^{2+} fluxes, the most important of which are the L-type Ca^{2+} channel, the RyR, the SERCA pump, and the Na^+–Ca^{2+} exchanger. Here we briefly examine models of each of these fluxes except the SERCA pumps, which are discussed in detail in Chapter 7.

12.2.6 The L-type Ca^{2+} Channel

L-type Ca^{2+} channels are activated and inactivated by voltage, and inactivated by high $[Ca^{2+}]$ (Bers and Perez-Reyes, 1999). During an action potential, voltage-dependent inactivation is much less important than Ca^{2+}-dependent inactivation, and thus we ignore the former.

The most detailed model of the L-type channel is that of Jafri et al. (1998), based on an earlier model of Imredy and Yue (1994). It is a model of Monod–Wyman–Changeux type (Chapter 1), a schematic diagram of which is shown in Fig. 12.19. There are two basic conformations of the channel, called, respectively, the normal (top row) and Ca^{2+} (bottom row) conformations. State O is the only open state. Depolarization increases the rate at which the channel moves from left to right along either the normal or Ca^{2+} conformation. However, it is only from the normal conformation that the channel can change to the open state (transition N_4 to O). Binding of Ca^{2+} causes the channel to move from the normal conformation to the Ca^{2+} conformation (i.e., into a conformation from which it cannot access the open state) thus inactivating the receptor.

Good agreement with experimental data can be obtained by simpler versions of this model (Winslow et al., 2005). First, six channel states are omitted, to get the simpler

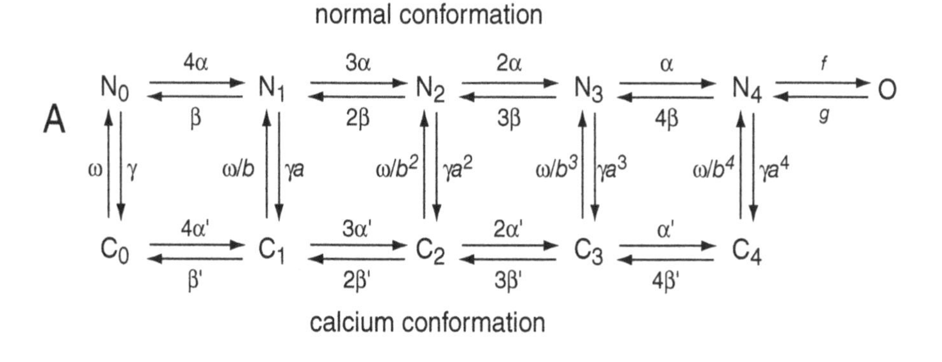

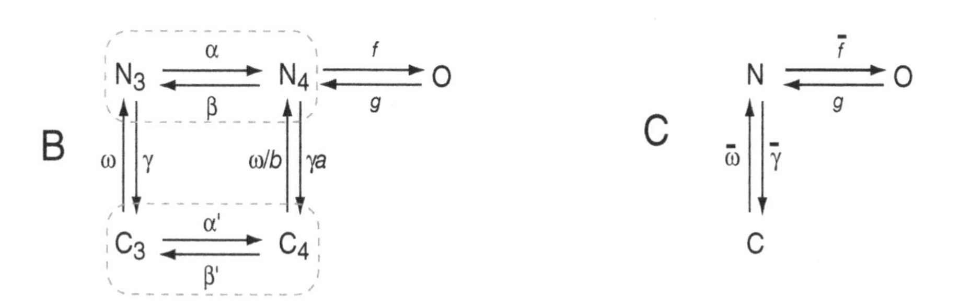

Figure 12.19 A: The full model of the L-type Ca^{2+} channel due to Jafri et al. (1998). Activation by voltage occurs because $\alpha = 2e^{0.012(V_m-35)}$ and $\beta = 0.0882e^{-0.05(V_m-35)}$. Inactivation by Ca^{2+} occurs because $\gamma = 0.44[Ca^{2+}]_{cleft}$. To satisfy detailed balance, $\alpha' = a\alpha$ and $\beta' = \beta/b$. Other parameter values are $a = 2$, $b = 1.9$, $f = 0.85$ ms^{-1}, $g = 2$ ms^{-1}, and $\omega = 0.02$ ms^{-1}. B: A simplified model obtained by truncating the leftmost three columns. C: The simplest version of the model, obtained by assuming a fast equilibrium between N_3 and N_4, and between C_3 and C_4, as shown by the dashed boxes in part B. It is left as an exercise (Exercise 8) to show that $\bar{\omega} = \omega\left(\frac{\beta'+\alpha'/b}{\alpha'+\beta'}\right)$, $\bar{\gamma} = \gamma\left(\frac{\beta+a\alpha}{\alpha+\beta}\right)$, and $\bar{f} = \frac{\alpha f}{\alpha+\beta}$.

diagram shown in Fig. 12.19B, and then the two remaining states in the normal conformation are amalgamated by assuming fast equilibrium, as are the remaining two states in the Ca^{2+} conformation (Fig. 12.19C). The details of this model simplification are left as an exercise (Exercise 8).

12.2.7 The Ryanodine Receptor

One model of the RyR due to Friel has already been described in Section 7.2.9. For this model it was assumed that activation of the RyR by Ca^{2+} is instantaneous, and that inactivation by Ca^{2+} is unimportant. However, the RyR was assumed to be activated by the cytoplasmic $[Ca^{2+}]$ and thus the spatial aspects peculiar to myocytes were not considered. Controversy surrounds the exact properties of RyR in cardiac and skeletal muscle (an excellent review is by Fill and Copello, 2002). There is widespread

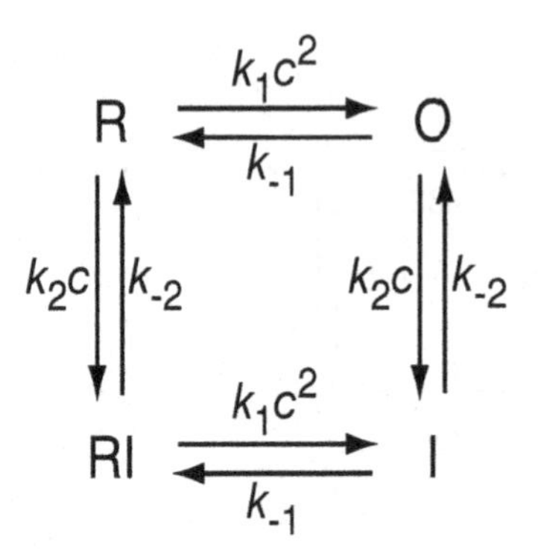

Figure 12.20 Model of the RyR due to Stern et al. (1999). R and RI are closed states, O is the open state, and I is the inactivated state. The rate constants depend on c, the $[Ca^{2+}]$ in the diadic cleft. Stern et al. used the parameters $k_1 = 35\ \mu M^{-2}\,(ms)^{-1}$, $k_{-1} = 0.06\,(ms)^{-1}$, $k_2 = 0.5\ \mu M^{-1}\,(ms)^{-1}$, $k_{-2} = 0.005\,(ms)^{-1}$.

agreement, first, that (in cardiac myocytes) Ca^{2+} is released through the RyR by a process of Ca^{2+}-induced Ca^{2+} release, giving explosive positive feedback that is crucial for EC coupling, and, second, that the RyR must respond to localized Ca^{2+} concentrations in the diadic cleft and thus have "privileged access" to the Ca^{2+} entering through the L-type channel. However, there must also be some mechanism to terminate flux through the RyR, and it is here that agreement ends. A number of mechanisms have been proposed: inactivation by diadic cleft Ca^{2+}, depletion of the SR Ca^{2+}, stochastic fluctuations resulting in RyR closure, and adaptation of the receptor. There is not enough space here to discuss in detail all the hypothetical models and the arguments for and against them. A typical model (Stern et al., 1999) is shown in Fig. 12.20. This model is based on the generic model of an adapting receptor discussed in Section 16.4. Binding of Ca^{2+} can first open the receptor (state O), but then subsequent, slower, binding of Ca^{2+} can inactivate the receptor (state I). The application of this class of models to both ryanodine and inositol trisphosphate receptors is discussed by Cheng et al. (1995) and Sachs et al. (1995).

Shannon et al. (2004) modified this basic model by assuming that the SR Ca^{2+} concentration modifies the rate constants, while a number of other groups have developed RyR models of greater or lesser complexity (Schiefer et al., 1995; Keizer and Levine, 1996; Zahradnikova and Zahradnik, 1996; Keizer and Smith, 1998; Fill et al., 2000; Sobie et al., 2002). However, there is still no consensus as to which of this multitude of models best describes RyR behavior *in vivo*.

12.2.8 The Na^+–Ca^{2+} Exchanger

All the Ca^{2+} that enters the myocyte through the L-type Ca^{2+} channel must be removed before the next heartbeat, otherwise the cell could not attain a steady state. Most of this removal is effected by the NCX, which, in rabbit, removes more than ten times the amount removed by the sarcolemmal Ca^{2+} ATPase. However, it appears that the NCX also allows for Ca^{2+} entry during the beginning of the Ca^{2+} transient. Na^+–Ca^{2+} exchange is reversible, with three Na^+ ions exchanged for one Ca^{2+} ion. Thus, the exchanger is electrogenic; its rate and direction depend on the membrane potential and the intracellular and extracellular Ca^{2+} and Na^+ concentrations. A positive membrane potential increases the exchanger's outward current, thus increasing Ca^{2+} influx, while an increase in intracellular Ca^{2+} leads to Ca^{2+} extrusion, and thus inward exchanger current.

These features can be seen in the model of the exchanger that was discussed in Section 2.4.3. In that model, the outward Ca^{2+} flux, J, through the exchanger is given by

$$J = \frac{k_1 k_2 k_3 k_4 (c_i n_e^3 - e^{\frac{FV_m}{RT}} c_e n_i^3)}{16 \text{ other terms}}, \tag{12.46}$$

where V_m is the membrane potential, n is Na^+ concentration, c is Ca^{2+} concentration, and subscripts i and e denote, respectively, internal and external concentrations. Notice that the flux of Ca^{2+} is outward whenever

$$\frac{c_i}{c_e} > e^{\frac{FV_m}{RT}} \frac{n_i^3}{n_e^3}. \tag{12.47}$$

Thus, as the membrane depolarizes, the outward flux decreases and reverses for high enough V_m, resulting in Ca^{2+} influx. This is what happens during the beginning of the action potential following depolarization. However, Ca^{2+} influx through the L-type channel and release from the RyR causes a rise in c_i that counterbalances the effects of depolarization, resulting in Ca^{2+} efflux through the exchanger.

12.3 Cellular Coupling

A single cell model, no matter how detailed or precise, can never capture the range of dynamic behaviors of cardiac tissue. This is because many of the behaviors are inherently spatio-temporal, that is, they can occur only in spatially distributed systems of cells, never in single cells.

As we will see, spatial coupling of cells has two opposite, but important features. First, coupling between cells tends to synchronize or homogenize their behavior, and second, coupling allows one excited cell to excite its neighbors, so that action potentials can propagate.

Myocardial cells are cable-like, roughly cylindrical, typically 100 μm long and 15 μm in diameter. They are packed together in a three-dimensional irregular brick-like packing, surrounded by extracellular medium (Fig. 12.21). Each cell has specialized contacts with its neighboring cells, mainly in end-to-end fashion, facilitated by a step-like surface that locks into neighboring cells. The opposing cell membranes form the *intercalated disk* structure. While the end-to-end cell membranes are typically separated by about 250 angstroms, there are places, called *junctions*, where the pre- and postjunctional membranes are fused together. The mechanical adhesion of cells is provided by adhering junctions in the intercalated disk, known as *desmosomes* or *tight junctions*. The electrical coupling of cells is provided by gap junctions (Chapter 8).

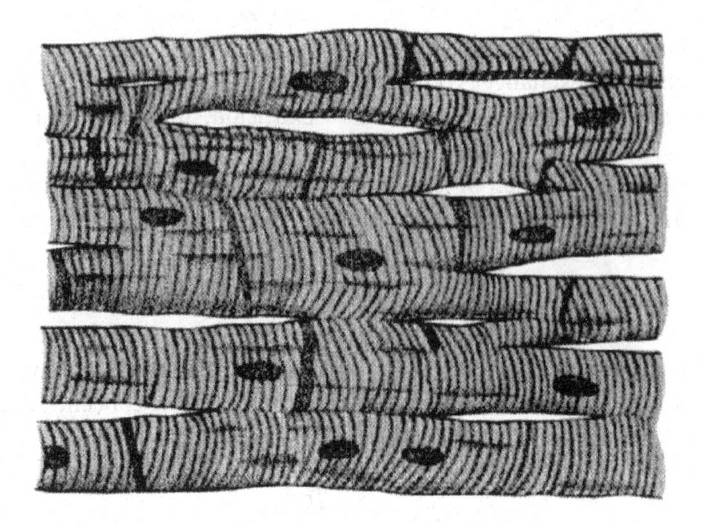

Figure 12.21 Cardiac cell structure. (Guyton and Hall, 1996, Fig. 9-2, p. 108.)

The intercellular channels provided by the gap junctions are around 20 angstroms in diameter and are characterized as "low-resistance" because the effective resistance is considerably less than what would result from two cell membranes butted together. However, compared to the intracellular cytoplasm, the gap junctions are of high resistance, simply because the cross-sectional area for electrical conduction through gap junctions is greatly reduced (about two percent of the total cross-sectional area).

12.3.1 One-Dimensional Fibers

To model a cardiac fiber, we consider a simple one-dimensional collection of cylindrical cells (with perimeter p) coupled in end-to-end fashion via gap junctions. From Chapter 8, we know that in each cell we have the cable equation,

$$p\left(C_m\frac{\partial V}{\partial t}+I_{\text{ion}}\right)=\frac{\partial}{\partial x}\left(\frac{1}{r_c}\frac{\partial V_i}{\partial x}\right)=-\frac{\partial}{\partial x}\left(\frac{1}{r_e}\frac{\partial V_e}{\partial x}\right), \tag{12.48}$$

with $r_c = \frac{R_c}{A_i}$ and $r_e = \frac{R_e}{A_e}$, where R_c and R_e are the resistivities of intracellular and extracellular space, respectively, A_i and A_e are the average cellular intracellular and extracellular cross-sectional areas, and p is the perimeter of the cell. At the ends of cells (each of length L), there is a jump in intracellular potential, but the intracellular current $-\frac{1}{r_c}\frac{\partial V_i}{\partial x}$ must be continuous. Assuming that gap junctions behave like ohmic resistors, the drop in the potential across the junction is proportional to the current through the junction,

$$\frac{[V_i]}{r_g}=\frac{1}{r_c}\frac{\partial V_i}{\partial x}, \tag{12.49}$$

where $[V_i]$ is the jump in intracellular potential across the gap junctions and r_g is the effective gap-junctional resistance. Here we are also making the assumption that there are no transmembrane currents across the ends of the cell into the small extracellular gap that separates the cells. (Recent evidence suggests that Na^+ channels may have a higher density at the ends of cells than along the cell wall (Kucera et al., 2002) so this assumption may not be valid.) The extracellular potential and current are continuous, unaffected by the gap junctions.

The time constant for this fiber is the same as (4.16). The space constant, however, is affected by the gap-junctional resistance. To find the space constant, we take $I_{\text{ion}} = V/R_m = (V_i - V_e)/R_m$ (linearized around the resting potential) and look for a geometrically decaying solution, with $V_i(x+L) = \mu V_i(x), V_e(x+L) = \mu V_e(x)$ for some constant $\mu < 1$. The constant μ relates to the space constant λ_g through $\mu = e^{-L/\lambda_g}$.

The steady-state solution of this problem can be found analytically. We suppose that for the nth cell, the solution is proportional to

$$\begin{pmatrix} V_i \\ V_e \end{pmatrix}_n = \mu^n \Phi(\mu, x) = \mu^n \begin{pmatrix} \phi_i \\ \phi_e \end{pmatrix}. \tag{12.50}$$

We also assume that the total current is zero, so that

$$\frac{1}{r_c}\frac{\partial \phi_i}{\partial x} + \frac{1}{r_e}\frac{\partial \phi_e}{\partial x} = 0, \tag{12.51}$$

and thus

$$\frac{\partial}{\partial x}\left(\frac{1}{r_c + r_e}\frac{\partial \phi}{\partial x}\right) - \frac{p\phi}{R_m} = 0, \qquad \phi = \phi_i - \phi_e. \tag{12.52}$$

It follows that

$$\phi = \alpha_1 \exp(\lambda x) + \alpha_2 \exp(-\lambda x), \tag{12.53}$$

where $\lambda^2 = \frac{p}{R_m}(r_c + r_e)$, and

$$\phi_i = \frac{r_c}{r_c + r_e}\phi(x) + \beta, \qquad \phi_e = -\frac{r_e}{r_c + r_e}\phi(x) + \beta. \tag{12.54}$$

Now there are four boundary conditions that need to be applied to determine the four unknown constants $\alpha_1, \alpha_2, \beta$ and μ. These boundary conditions are

$$\phi_e'(L) = \mu\phi_e'(0), \qquad \phi_i'(L) = \mu\phi_i'(0), \tag{12.55}$$

requiring the current to be continuous, and

$$\phi_e(L) = \mu\phi_e(0), \qquad \mu\phi_i(0) - \phi_i(L) = \frac{r_g}{r_c}\phi_i'(L), \tag{12.56}$$

requiring continuity of the extracellular potential, and the jump condition (12.49) for the intracellular potential, respectively. It is a somewhat tedious calculation (but Maple makes it easy) to determine that (up to an arbitrary scale factor)

$$\alpha_1 = \mu - \frac{1}{E}, \qquad \alpha_2 = \mu - E, \qquad \beta = 2r_e\frac{(\mu - E)(\mu - \frac{1}{E})}{\mu - 1}, \qquad E = e^{\lambda L}, \tag{12.57}$$

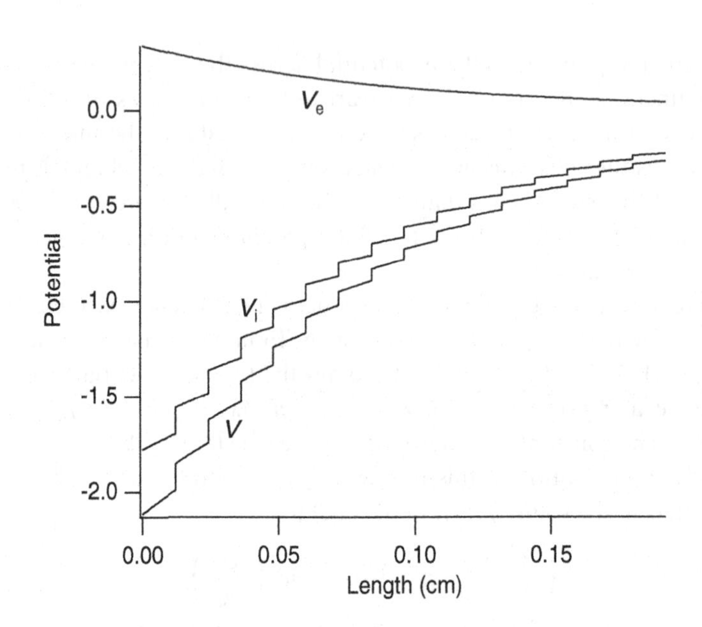

Figure 12.22 Plot of intracellular, extracellular and transmembrane potentials as functions of space, with a constant subthreshold potential maintained at a single point. For this plot cells had $L = 0.012$ cm, $A_i = 4.0 \times 10^{-6}$ cm^2, and a space constant $\lambda_g = 0.09$ cm. $R_m = 7000\,\Omega\text{cm}^2$, $R_c = 150\,\Omega\text{cm}$, $q_e = 0.5q_i$, $q_i = 5.47 \times 10^{-3}$, where $q_j = \frac{L^2 p}{R_m} r_j$, $j = i, e$. The vertical scale on this plot is arbitrary.

and, most importantly, $\mu < 1$ is a root of the (quadratic) characteristic equation

$$\frac{r_g \lambda}{r_c + r_e} = \frac{R_g}{\lambda L} = 2\frac{(\mu - \frac{1}{E})(\mu - E)}{\mu(E - \frac{1}{E})}, \tag{12.58}$$

where $R_g = \frac{Lpr_g}{R_m}$ is the effective nondimensional gap-junctional resistance.

The behavior of the extracellular, intracellular and transmembrane potentials for this solution is depicted in Fig. 12.22. As can be seen from this plot, the extracellular potential decays smoothly, but the intracellular potential decays with discrete jumps across the gap junctions. Perhaps the most important quantity here is the decay of V intracellularly relative to the decay of V due to gap junctions. If gap-junctional resistance is small then the decay of V is essentially exponential throughout the cell, whereas if gap-junctional resistance is relatively large, then the cell is nearly isopotential and most of the decay occurs across the gap junctions.

It is not possible to measure the intracellular potential with the detail shown in Fig. 12.22, because cells are usually too small to invade with multiple intracellular electrodes without irreversibly damaging the cell membrane. If the gap-junctional resistance is small compared with the cytoplasmic resistance, then (12.58) has a simplified solution. In particular, if L/λ_g is small (i.e., the length constant is much larger than the length of a single cell), then it must also be that λL is small, so that $E \approx 1 + \lambda L$ and $\mu \approx 1 - \frac{L}{\lambda_g}$. Substituting these into (12.58) gives

$$R_g \approx \left(\left(\frac{L}{\lambda_g}\right)^2 - (\lambda L)^2\right)(1 + O(\lambda L)), \tag{12.59}$$

which implies that

$$\frac{L^2}{\lambda_g^2} \approx R_g + (\lambda L)^2 = R_g + \frac{L^2 p}{R_m}(r_c + r_e). \tag{12.60}$$

The formula (12.60) is used routinely in bioengineering and in linear circuit theory; it implies that resistance along the cable is additive. This is exactly the same answer that one finds for the space constant of a uniform continuous cable if the gap-junctional resistance were uniformly distributed throughout the cytoplasm (see below). For most normal cells, the approximation (12.60) is valid. It is only when the gap-junctional resistance is excessively large, such as if there is *ischemia* or if the cells are treated with certain alcohols that block gap junctions, that this formula is substantially wrong.

However, it is also commonly asserted that gap junctions are low resistance and the bulk of the resistance is cytoplasmic in origin. To test this assertion, the effective gap-junctional resistance R_g can be calculated from (12.60). For example, using frog myocardial cells (which are longer than mammalian cells, $L = 131\,\mu$m, radius $=$ 7.5 μm), Chapman and Fry (1978) measured a space constant $\lambda_g = 0.328$ cm yielding $(\frac{L}{\lambda_g})^2 = 0.159$. From this they inferred (using $R_m = 1690\,\Omega\text{cm}^2$ and $q_e = 0$) that the effective cytoplasmic resistivity was $R_c = 588\,\Omega$cm. They were also able to measure the cytoplasmic resistivity directly, and they found $R_c = 282\,\Omega$cm so that $q_i = 0.076$. Thus, using (12.60), only 48% of the total resistivity of the cell was attributable to cytoplasmic resistance. The remaining 52% must be from gap-junctional resistance. With these numbers it is difficult to assert that gap junctions are low resistance.

The above space constant calculation represents a situation that is difficult to replicate experimentally. To do so (to enforce the no net current assumption) would require use of a bipolar electrode, with one of the poles placed intracellularly and the other placed extracellularly at a single point along the fiber, thereby holding the transmembrane potential fixed. A more realistic experiment would be to pass a current through the fiber with two extracellular electrodes at different points along the fiber. The associated mathematical problem is to solve the steady version of the cable equation (12.48) with the jump conditions (12.49) subject to the boundary conditions

$$-\frac{1}{r_e}\frac{dV_e}{dx} = I, \qquad \frac{1}{r_c}\frac{dV_i}{dx} = 0, \tag{12.61}$$

at the ends of the fiber at $x = 0, l$. These boundary conditions reflect the fact that current is injected into the extracellular space, but there is no direct current flow into the intracellular space.

The solution of this problem can also be found analytically, although it is more complicated than the previous length constant calculation. Here we state the answer, and relegate the details of the calculation to the Appendix. On the interior of cells the solution is

$$\begin{pmatrix} V_i \\ V_e \end{pmatrix}_j = \frac{1}{r_e + r_c} \begin{pmatrix} r_c \\ -r_e \end{pmatrix} V_j + \begin{pmatrix} 1 \\ 1 \end{pmatrix} (-\beta I x + \gamma_j), \tag{12.62}$$

and $V_j = (A_j + \frac{r_e}{\lambda} I) \sinh \lambda x + B_j \cosh \lambda x$, for $j = 0, 1, \ldots, N-1$, where

$$A_j = \frac{-S}{1-\mu}(\mu c_1 - c_2) f_j, \tag{12.63}$$

$$B_j = \frac{\mu c_1 - c_2}{1-\mu}\left(C f_j + \frac{1-\mu^N}{1-\mu^{2N}}\left(\mu^{j+1} + \mu^{N-j-1}\right)\right) - \frac{c_1 - \mu c_2}{1-\mu}, \tag{12.64}$$

$$f_j = \frac{1 - \mu^{N-j} - \mu^j + \mu^{N+j} + \mu^{2N-j} - \mu^{2N}}{1-\mu^{2N}}, \tag{12.65}$$

$$\gamma_{j+1} = \gamma_n - L\beta I + \frac{\beta}{r_c} K_g \left(\frac{r_e}{\lambda} I(C-1) + A_j C + B_j S\right), \tag{12.66}$$

$$C = \cosh \lambda L, \qquad S = \sinh \lambda L, \qquad c_i = \frac{S(1-\eta_i) + \eta_i K_g (1-C)}{S(2 + K_g \eta_i + K_g C)}, \tag{12.67}$$

with $\eta_1 = \frac{1}{\mu}$, $\eta_2 = \mu$, where $\mu < 1$ is the root of the characteristic equation (12.58), and $K_g = \frac{r_g \lambda}{r_c + r_e} = \frac{R_g}{\lambda L}$, $\beta = \frac{r_c r_e}{r_c + r_e}$.

In Fig. 12.23 is shown the transmembrane potential in response to a constant, but small (subthreshold), current stimulus. The current is from right to left, and at the right end the tissue is hyperpolarized, while on the left the tissue is depolarized. If this current stimulus were superthreshold, one would expect an action potential to arise on the left, and (perhaps) propagate to the right.

One can understand this profile as follows. A current that is injected into the extracellular space seeks the path of least resistance. Since there are two possible paths, extracellular space and intracellular space, total current is divided between these two spaces in such a way as to minimize the total resistance. However, to achieve this current-divided profile, current must flow into intracellular space at one end and it must exit the intracellular space at the opposite end, creating the electrode effects of hyperpolarization and depolarization, respectively. In the interior of the fiber, several space constants away from the electrodes, the current in each of the subspaces is essentially constant. However, gap-junctional resistance (or any other inhomogeneity of resistance) creates "speed bumps" that the current, while seeking the path of least resistance, circumvents by leaving or entering intracellular space, thereby creating transmembrane currents. At steady state, the membrane potential is depolarized if the transmembrane current is outward and hyperpolarized if it is inward.

The profile of transmembrane current shown in the bottom panel of Fig. 12.23 is called a sawtooth potential (Krassowska et al., 1987) and has been observed in single

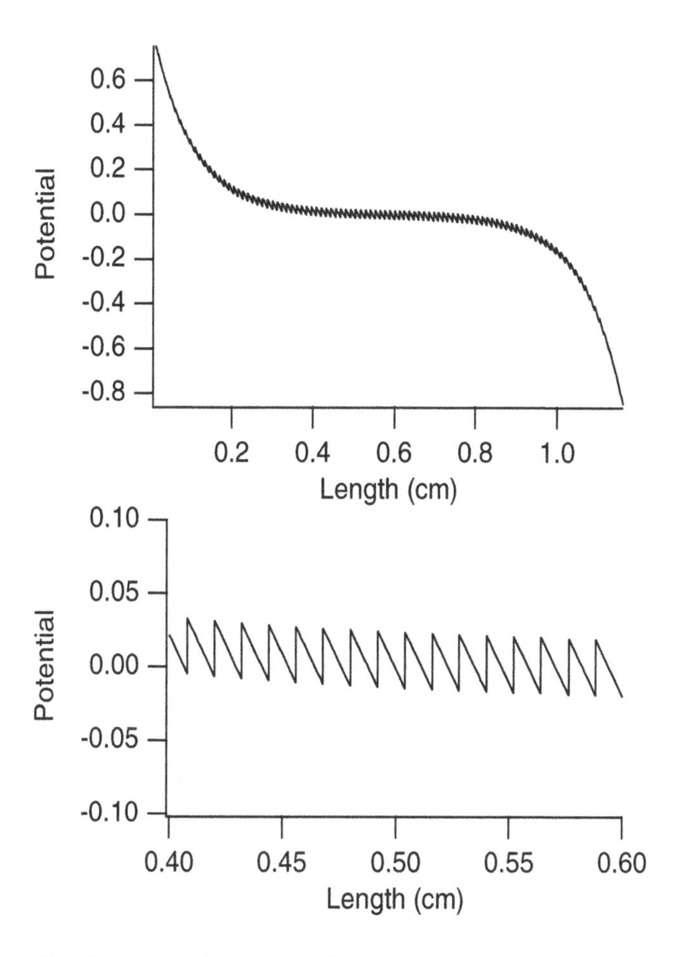

Figure 12.23 Plot of transmembrane potential as a function of space, with a constant applied current. For this plot cells had $L = 0.012$ cm, $A_i = 4.0 \times 10^{-6}$ cm^2, and a space constant $\lambda_g = 0.09$ cm. $R_m = 7000\,\Omega\text{cm}^2$, $R_c = 150\,\Omega\text{cm}$, $q_e = 0.5q_i$, $q_i = 5.47 \times 10^{-3}$, where $q_j = \frac{L^2 p}{R_m} r_j$, $j = i, e$. The vertical scale on this plot is arbitrary.

cells (Knisely et al., 1993) as well as in whole tissue (Zhou et al., 1998), although its physiological significance is debated. In other contexts, regions of localized depolarization and hyperpolarization that are spatially removed from electrodes are referred to as virtual electrodes. These are discussed further in later sections of this chapter. Regardless of the outcome of this debate, it is clear that cardiac tissue is not resistively homogeneous, and the implications of this can be profound.

Because it is impossible to make much headway studying large tissues using a spatially detailed model, some further approximations are helpful. There are two that have proven useful. The first is appropriate when the space constant is large compared to the length of a single cell.

To derive this approximation it is useful to notice that the cable equation with jump condition (12.49) is equivalent to the equation

$$p\left(C_m\frac{\partial V}{\partial t}+I_{\mathrm{ion}}\right)=\frac{\partial}{\partial x}\left(\frac{1}{r_c+\sum\delta(x-nL)r_g}\frac{\partial V_i}{\partial x}\right)=-\frac{\partial}{\partial x}\left(\frac{1}{r_e}\frac{\partial V_e}{\partial x}\right) \tag{12.68}$$

with a spatially inhomogeneous resistance. The assumption that the space constant is large compared to the length of a cell implies that the resistance $r_c+\sum\delta(x-nL)r_g$ is a rapidly varying function of space. Thus, one can use homogenization theory (Exercise 19; also see Section 7.8.2, or Keener, 1998) to replace the spatially varying resistance with a spatially averaged, effective resistance. The resulting averaged equation is

$$p\left(C_m\frac{\partial V}{\partial t}+I_{\mathrm{ion}}\right)=\frac{\partial}{\partial x}\left(\frac{1}{r_c+\frac{r_g}{L}}\frac{\partial V_i}{\partial x}\right)=-\frac{\partial}{\partial x}\left(\frac{1}{r_e}\frac{\partial V_e}{\partial x}\right). \tag{12.69}$$

Furthermore, because this is one dimensional in space, we can use that the total current is constant

$$\left(\frac{1}{r_c+\frac{r_g}{L}}\frac{\partial V_i}{\partial x}\right)+\left(\frac{1}{r_e}\frac{\partial V_e}{\partial x}\right)=I_{\mathrm{tot}}, \tag{12.70}$$

to write a single equation involving only the transmembrane potential

$$p\left(C_m\frac{\partial V}{\partial t}+I_{\mathrm{ion}}\right)=\frac{\partial}{\partial x}\left(\frac{1}{r_e+r_c+\frac{r_g}{L}}\frac{\partial V}{\partial x}\right). \tag{12.71}$$

This is the effective cable equation for cardiac fibers.

A different approximation applies when the cytoplasmic resistance r_c and extracellular resistance r_e are small compared to the gap-junctional resistance. For simplicity, we set $r_e=0$, in which case the extracellular potential is constant (so take it to be zero), and $V_i=V$. Set $x=Ly$ and rewrite the cable equation as

$$\frac{\partial^2 V}{\partial y^2}=\epsilon F(V,t), \tag{12.72}$$

subject to the jump conditions

$$\frac{\epsilon}{R_g}[V_i]=\frac{\partial V_i}{\partial y}, \tag{12.73}$$

at the integers $i=n$, where $\epsilon=\frac{pr_cL^2}{R_m}$, $R_g=\frac{r_gpL}{R_m}$ and $F(V,t)=R_mC_m\frac{\partial V}{\partial t}+R_mI_{\mathrm{ion}}$ has units of voltage.

Now we seek a power series solution of (12.72) in ϵ and find that on the interval $n<y<n+1$,

$$V(y)=V_n+a_nz_n+\frac{1}{2}\epsilon F(V_n)z_n^2, \tag{12.74}$$

where $z_n = y - n$, and V_n and a_n are constants that have yet to be determined. Of course, this implies that

$$\frac{\partial V}{\partial y} = a_n + \epsilon F(V_n) z_n. \tag{12.75}$$

Since $\frac{\partial V}{\partial y}$ must be continuous at $y = n$, it must be that

$$a_{n-1} + \epsilon F(V_{n-1}) = a_n. \tag{12.76}$$

Now substitute this solution into the jump conditions (12.73) to find that

$$\frac{1}{R_g}(V_{n+1} - 2V_n + V_{n-1}) = F(V_n, t) + \frac{\epsilon}{2R_g}(F(V_{n-1}, t) + F(V_n, t)), \tag{12.77}$$

which, to leading order in ϵ, is the discrete cable equation,

$$\frac{1}{r_g}(V_{n+1} - 2V_n + V_{n-1}) = Lp\left(C_m \frac{\partial V_n}{\partial t} + I_{\text{ion}}(V_n, t)\right), \tag{12.78}$$

where V_n is the membrane potential of the nth (isopotential) cell.

Which of these approximations is most correct?

Because it is translationally invariant, the effective cable equation implies that propagation, if it occurs, is continuous, and the propagation velocity scales with the inverse square root of the resistance

$$v = \sqrt{\frac{k}{r_c + r_e + \frac{r_g}{L}}}, \tag{12.79}$$

for some constant k. On the other hand, the discrete cable equation implies that propagation is saltatory, with individual cells becoming excited one at a time, and this excitation jumps from cell to cell in discrete fashion.

Propagation is known to be saltatory (Spach et al., 1981). Data collected on a single strand of rat myocytes (Fast and Kleber, 1993) found that an action potential required about 38 μs to travel 30 μm within a single cell, but it required 118 μs to travel 30 μm if propagation was across the ends of two cells. Hence, propagation through gap junctions led to a delay of about 80 μs. For cells of length 100 μm, this translates into a velocity of 0.8 μm/μs in the cell and 0.48 μm/μs along the fiber. If velocity within cells scales like the inverse square root of resistance, then

$$\left(\frac{0.8}{0.48}\right)^2 = \frac{r_c + r_e + \frac{r_g}{L}}{r_e + r_c}, \tag{12.80}$$

so that $\frac{r_g}{L} = 1.78(r_e + r_c)$. In other words, according to this estimate, gap-junctional resistance represents a substantial portion (here, roughly 65%) of the total resistance.

12.3.2 Propagation Failure

One-dimensional propagation in cardiac fibers is expected to occur, since the cable equation is similar to those described in Chapter 6 with excitable dynamics. It is perhaps more interesting and of greater clinical significance to understand the causes of propagation failure. The primary lesson of the discussion that follows is that inhomogeneities of resistance can cause propagation failure.

Branching

On leaving the AV node of the heart, the action potential enters the bundle of HIS. The bundle divides near the upper ventricular septum into right and left branches. The right bundle continues with little arborization toward the apex of the heart. The left bundle branch divides almost immediately into two major divisions: one anterior and superior, and the second posterior and inferior.

A bundle branch block occurs when the action potential fails to propagate through the entire branch. To understand something about the cause of bundle branch block, we consider a model of propagation in a one-dimensional fiber with a junction at which the cable properties suddenly change. Wave fronts are governed by the cable equation

$$C_m R_m \frac{\partial V}{\partial t} = \frac{R_m}{p} \frac{\partial}{\partial x}\left(\frac{A}{R_c}\frac{\partial V}{\partial x}\right) + f(V), \tag{12.81}$$

where $f(V)$ represents the fast currents, but dynamics due to slower variables are suppressed. We can also modify this equation for propagation in a bundle of fibers by letting p and A be the total membrane perimeter and cross-sectional area, respectively, for the bundle.

Suppose there is a point at which there is a sudden jump in cable properties. This could be due to a sudden change in the size of the cable, or, since this is a one-dimensional model, a splitting of the cable into two for which the total perimeter and cross-sectional area are changed. Thus, on both sides of this discontinuity, the cable equation (12.81) holds, but with different parameters p and A, say p_1, A_1 for $x < 0$, and p_2, A_2 for $x > 0$. At the junction the potential V and the axial current $\frac{A}{R_c}\frac{\partial V}{\partial x}$ must be continuous.

Using upper and lower solution techniques (Fife, 1979), one can demonstrate an important *comparison property* for the cable equation (12.81): If $V_1(x)$ and $V_2(x)$ are two functions that are ordered, with $V_1(x) \leq V_2(x)$, then the solutions of (12.81) with initial data $V_1(x)$ and $V_2(x)$, say $V_1(x,t)$ and $V_2(x,t)$ with $V_1(x,0) = V_1(x)$ and $V_2(x,0) = V_2(x)$, then $V_1(x,t) \leq V_2(x,t)$ for all time $t \geq 0$.

The importance of this theorem is that if we can establish the existence of a standing transitional profile, then traveling profiles of similar type are precluded. The standing wave is an upper bound for solutions and thereby prevents propagation (Pauwelussen, 1981).

Suppose the function $f(V)$ has three zeros, at $V = 0 < \alpha < \beta$. We look for a standing profile that connects $V = 0$ at $x = -\infty$ with $V = \beta$ at $x = \infty$. The standing profile must satisfy the ordinary differential equation

$$\frac{R_m}{p_i}\left(\frac{A_i}{R_c}V_x\right)_x + f(V) = 0, \tag{12.82}$$

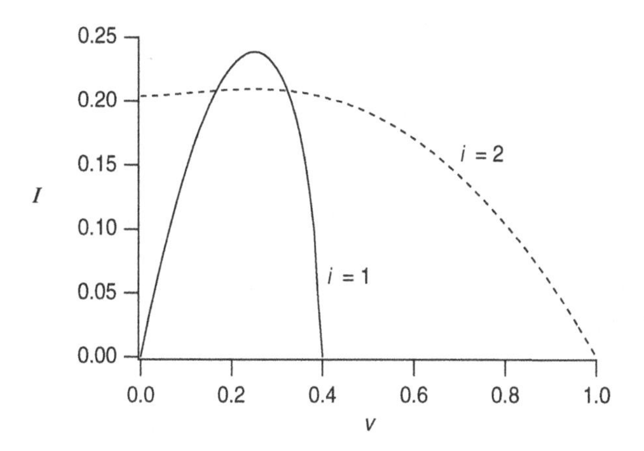

Figure 12.24 The curves (12.84) with $i = 1$ and $\frac{R_m R_c}{2A_i p_i} = 0.04$ and with $i = 2$ and $\frac{R_m R_c}{2A_i p_i} = 1.0$, for $f(v) = v(v - 1)(\alpha - v)$, $\alpha = 0.25$. An intersection of the solid curve with the dashed curve guarantees propagation failure.

with $i = 1$ for $x < 0$ and with $i = 2$ for $x > 0$. Multiplying these equations by V_x and integrating, we obtain

$$\frac{1}{2}\frac{R_m A_i}{R_c p_i} V_x^2 + F(V) = \begin{cases} 0, & \text{if } i = 1, \\ F(\beta), & \text{if } i = 2, \end{cases} \tag{12.83}$$

where $F(V) = \int_0^V f(u)\,du$. Sketches of these two curves are depicted in Fig. 12.24 in the case $F(\beta) > 0$. A connecting trajectory exists if these two curves intersect at the same level of current. We express the profiles (12.83) in terms of the axial current $I = -\frac{A}{R_c} V_x$ and obtain

$$\frac{1}{2}\frac{R_m R_c}{A_i p_i} I^2 + F(V) = \begin{cases} 0, & \text{if } i = 1, \\ F(\beta), & \text{if } i = 2. \end{cases} \tag{12.84}$$

Intersections of these two curves occur if there is a solution of

$$F(V)\left(\frac{A_1 p_1}{A_2 p_2} - 1\right) = -F(\beta), \tag{12.85}$$

with $F(V) < 0$ in the range $0 < V < \beta$. Since the minimum for $F(V)$ is at $V = \alpha$, there is a solution whenever

$$\frac{A_1 p_1}{A_2 p_2} \geq 1 - \frac{F(\beta)}{F(\alpha)}. \tag{12.86}$$

In the special case that $f(V)$ is the cubic polynomial $f(V) = AV(V - 1)(\alpha - V)$, this condition becomes

$$\frac{A_1 p_1}{A_2 p_2} \geq 1 + \frac{1 - 2\alpha}{\alpha^3(2 - \alpha)}. \tag{12.87}$$

The interpretation is clear. If at a junction along a fiber the product pA increases by a sufficient amount, as specified by (12.86), then propagation through the junction in the direction of increasing pA is not possible. Of course, this criterion for propagation block depends importantly on the excitability of the fiber as expressed through the ratio $\frac{F(\beta)}{F(\alpha)}$, and propagation failure is more likely when the fiber is less excitable. Hence propagation block is time-dependent in that if inadequate recovery time from a previous excitation is allowed, or if recovery is slowed, the likelihood of block at a junction is increased. Further analysis of problems of this type can be found in Lewis and Keener (2000a).

Gap-Junctional Coupling

We expect that gap-junctional resistance can have the similar effect of precluding propagation. To see how gap-junctional resistance affects the success or failure of propagation, we consider the idealized situation of cells of length L coupled at their ends by gap junctions, as described by (12.49) and the cable equation (12.48) with piecewise-linear ionic current

$$I_{\text{ion}}(V) = \frac{1}{R_m}\left(H(V-\alpha) - V\right). \tag{12.88}$$

This model recommends itself because it can be solved explicitly, even though it lacks quantitative reliability or a direct physiological interpretation.

As before, we look for a standing solution on the assumption that the existence of a standing solution precludes the possibility of propagation.

The method to solve this standing wave problem is similar to the method used to find the space constant in Section 12.3.1, with two important differences, namely that the solution must be bounded in both directions ($n \to \pm\infty$) rather than only one, and the transmembrane current is piecewise linear, not linear. We use solutions of the linear problem found before to construct solutions of the nonlinear problem. That is, for $n \geq 0$ we take

$$\begin{pmatrix} V_i \\ V_e \end{pmatrix}_n = A\mu^n \Phi(\mu, x), \tag{12.89}$$

and for $n < 0$ we take

$$\begin{pmatrix} V_i \\ V_e \end{pmatrix}_n = B\mu^{-n}\Phi\left(\frac{1}{\mu}, x\right) + \begin{pmatrix} 1+C \\ C \end{pmatrix}, \tag{12.90}$$

where $\mu < 1$ is a root of (12.58), and $\Phi(\mu, x)$ is specified by (12.50) and (12.54). Here A, B, and C are as yet undetermined. However, this proposed solution has the feature that $V = V_i - V_e$ approaches 0 as $n \to \infty$, it approaches 1 as $n \to -\infty$, and satisfies all junctional boundary conditions except at the junction between cell $n = -1$ and cell $n = 0$. Furthermore, this proposed solution satisfies the cable equation (12.48) with ionic current $I_{\text{ion}}(V) = -\frac{V}{R_m}$ when $n \geq 0$ and with ionic current $I_{\text{ion}}(V) = \frac{1}{R_m}(1 - V)$ when $n < 0$.

Now, to determine the coefficients A, B, and C, we require that V_e, $\frac{dV_e}{dx}$ and $\frac{dV_i}{dx}$ be continuous at the junction between cell $n = -1$ (at $x = L$) and cell $n = 0$ (at $x = 0$) and that the junctional condition (12.49) be satisfied there as well. A tedious calculation (unless it is done with Maple) yields expressions for $A = B$, and C that are too complicated to be enlightening. However, one also finds that

$$V_0(0) = \frac{1}{2}\frac{E^2 - 2\mu E + 1}{E^2 - 2\mu E + 1 + \frac{1}{2}\frac{r_g\lambda}{r_c+r_e}(E^2 - 1)}, \tag{12.91}$$

and

$$V_1(L) = 1 - V_0(0). \tag{12.92}$$

It follows that $V < \frac{1}{2}$ for cell $n = 0$ at $x = 0$ and $V > \frac{1}{2}$ for cell $n = -1$ at $x = L$ whenever r_g is positive. A plot of this solution is shown in Fig. 12.25.

Finally, to be a valid solution for the piecewise-linear ionic current (12.88), it must be that $V_0(0) < \alpha$. This leads to the condition

$$\frac{r_g\lambda}{r_c + r_e} \equiv \frac{R_g}{\lambda L} \geq \frac{1 - 2\alpha}{\alpha}\left(\frac{E - 2\mu + \frac{1}{E}}{E - \frac{1}{E}}\right). \tag{12.93}$$

where $R_g = \frac{Lpr_g}{R_m}$.

The interpretation of this formula is complicated by the fact that according to (12.58), μ is a function of R_g. However, the implication of this is that propagation failure occurs if R_g is greater than some critical gap-junctional resistance, say R_g^*, and R_g^* is that value of R_g for which (12.93) is an equality. An equivalent relationship is that

$$\frac{R_g^*}{\lambda L} = 2\frac{(\mu^* - \frac{1}{E})(\mu^* - E)}{\mu^*(E - \frac{1}{E})}, \tag{12.94}$$

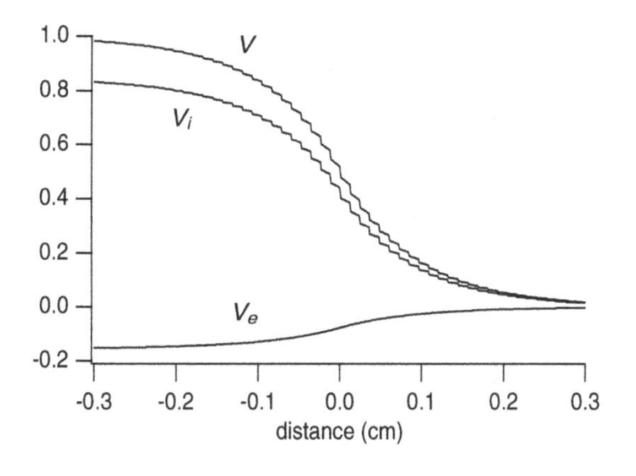

Figure 12.25 Plot of the standing wave solution for cells of length L coupled at their ends by resistive gap junctions. For this plot, cells have $L = 0.012$ cm, $A_i = 4.0 \times 10^{-6}$ cm^2, and a space constant $\lambda_g = 0.09$ cm. $R_m = 7000$ Ωcm^2, $R_i = 150$ Ωcm, $q_e = 0.5q_i$, where $q_j = \frac{L^2p}{R_m}r_j$, $j = i, e$.

where μ^* is the unique root less than $\frac{1}{E}$ of

$$\frac{1-2\alpha}{2\alpha} = \frac{(\mu^* - \frac{1}{E})(\mu^* - E)}{\mu^*(E - 2\mu^* + \frac{1}{E})}. \quad (12.95)$$

One can eliminate μ from these two equations to find the single quadratic equation for R_g^*,

$$\frac{\alpha(1-\alpha)}{(1-2\alpha)^2}\left(\frac{R_g^*}{\lambda L}\right)^2 - \left(\frac{E^2+1}{E^2-1}\right)\frac{R_g^*}{\lambda L} - 1 = 0. \quad (12.96)$$

This equation has only one positive root.

Now, with a little graphical analysis, one can show that R_g^* is a decreasing function of α for $0 < \alpha < \frac{1}{2}$. This makes good sense, since it implies that propagation is less likely to fail in a more excitable medium (with smaller α) and more likely to fail in a less excitable medium. Further, in the limit of small λL, the positive root of this quadratic equation is

$$R_g^* = \frac{(1-2\alpha)^2}{\alpha(1-\alpha)} + (\lambda L)^2\left(\frac{1-2\alpha+2\alpha^2}{2\alpha(1-\alpha)}\right) + O(\lambda L)^3. \quad (12.97)$$

On the other hand, for large λL, R_g^* is asymptotically linear, with

$$R_g^* = \lambda L\left(\frac{1-2\alpha}{1-\alpha}\right) + O(\lambda L e^{-2\lambda L}). \quad (12.98)$$

In general, one can show that for $0 < \alpha < \frac{1}{2}, R_g^*$ is an increasing function of λL. The interpretation of this is that longer cells are less likely to experience propagation failure than shorter cells with the same excitability.

Additional discussion of this problem can be found in Keener (1991b) and Keener (2000b).

12.3.3 Myocardial Tissue: The Bidomain Model

Coupling in cardiac tissue is complicated by the fact that the signal is the transmembrane potential, and this necessitates that the intracellular and extracellular spaces be continuously connected and intertwined, so that one can move continuously between any two points within one space without traversing through the opposite space. This is possible only in a three-dimensional domain.

It is impossible to write and solve equations that take into account the fine-structure details of the geometry of these two interleaving spaces. However, the microstructure can be averaged (homogenized) to yield equations that describe the potentials in an averaged, or smoothed, sense, and these are adequate for many situations (although the concerns about inhomogeneities raised by the above discussion still pertain).

In this averaged sense, we view the tissue as a two-phase medium, as if every point in space is composed of a certain fraction of intracellular space and a fraction of extracellular space. Accordingly, at each point in space there are two electrical potentials V_i and V_e, as well as two currents i_i and i_e, with subscripts i and e denoting intracellular and extracellular space, respectively.

The relationship between current and potential is ohmic,

$$i_i = -\sigma_i \nabla V_i, \quad i_e = -\sigma_e \nabla V_e, \tag{12.99}$$

where σ_i and σ_e are conductivity tensors. The principal axes of the conductivity tensors are the same, owing to the cylindrical nature of the cells, but the conductivities in these directions are possibly different. At any point in space the total current is $i_t = i_i + i_e$, and unless there are extraneous current sources, the total current is conserved, so that $\nabla \cdot i_t = 0$, or

$$\nabla \cdot (\sigma_i \nabla V_i + \sigma_e \nabla V_e) = 0. \tag{12.100}$$

At every point in space there is a membrane potential

$$V = V_i - V_e. \tag{12.101}$$

The transmembrane current i_T is the current that leaves the intracellular space to enter the extracellular space,

$$i_T = \nabla \cdot (\sigma_i \nabla V_i) = -\nabla \cdot (\sigma_e \nabla V_e). \tag{12.102}$$

For a biological membrane, the total transmembrane current is the sum of ionic and capacitive currents,

$$i_T = \chi \left(C_m \frac{\partial V}{\partial t} + I_{\text{ion}} \right) = \nabla \cdot (\sigma_i \nabla V_i). \tag{12.103}$$

Here χ is the membrane surface-to-volume ratio, needed to convert transmembrane current per unit area into transmembrane current per unit volume. In the typical scaling, $I_{\text{ion}} = -\frac{f(V)}{R_m}$.

Equation (12.103) shows how cardiac tissue is coupled, and it, together with (12.100), is called the *bidomain model*. The bidomain model was first proposed in the late 1970s by Tung and Geselowitz (Tung, 1978) and is now the generally accepted model of electrical behavior of cardiac tissue (Henriquez, 1993). However, it is not actually used that often because it is computationally expensive and there are not many situations where it matters.

Boundary conditions for the bidomain model usually assume that there is no current across the boundary that enters directly into the intracellular space, whereas if there is an injected current, it enters the tissue through the extracellular domain.

Monodomain Reduction

Equation (12.103) can be reduced to a monodomain equation for the membrane potential in one special case. Notice that (combining (12.101) and (12.99))

$$\nabla V_i = (\sigma_i + \sigma_e)^{-1} (\sigma_e \nabla V - i_t), \tag{12.104}$$

so that the balance of transmembrane currents becomes

$$\chi\left(C_m\frac{\partial V}{\partial t}+I_{\text{ion}}\right)=\nabla\cdot\left(\sigma_i(\sigma_i+\sigma_e)^{-1}\sigma_e\nabla V\right)-\nabla\cdot\sigma_i(\sigma_i+\sigma_e)^{-1}i_t. \tag{12.105}$$

Here we see that there is possibly a contribution to the transmembrane current from the divergence of the total current. We know that $\nabla\cdot i_t=0$, so this source term is zero if the matrix $\sigma_i(\sigma_i+\sigma_e)^{-1}$ is proportional to a constant multiple of the identity matrix. In other words, if the two conductivity matrices σ_i and σ_e are proportional, $\sigma_i=\alpha\sigma_e$, with α a constant, then the source term disappears, and the bidomain model reduces to the monodomain model.

$$\chi\left(C_m\frac{\partial V}{\partial t}+I_{\text{ion}}\right)=\nabla\cdot(\sigma\nabla V), \tag{12.106}$$

where $\sigma=\sigma_i(\sigma_i+\sigma_e)^{-1}\sigma_e$. When $\sigma_i=\alpha\sigma_e$, the tissue is said to have equal anisotropy ratios. A one-dimensional model with constant conductivities can always be reduced to a monodomain problem.

Plane Waves

Cardiac tissue is strongly anisotropic, with wave speeds that differ substantially depending on their direction. For example, in human myocardium, propagation is about 0.5 m/s along fibers and about 0.17 m/s transverse to fibers. To see the relationship between the wave speed and the conductivity tensor we look for plane-wave solutions of the bidomain equations. Plane waves are functions of the single variable $\xi=\mathbf{n}\cdot\mathbf{x}-ct$, where $\mathbf{n}$ is a unit vector pointing in the direction of wave-front propagation. We assume that the ionic current is such that the canonical problem

$$u''+c_0u'+f(u)=0 \tag{12.107}$$

has a wave-front solution $U(x)$ for some unique value of c_0, the value of which depends on f. The behavior of this solution was discussed in Chapter 6.

In terms of the traveling wave coordinate ξ, the bidomain equations reduce to the two ordinary differential equations

$$\frac{R_m}{\chi}\mathbf{n}\cdot\sigma_i\mathbf{n}V_i''+cC_mR_mV'+f(V)=0, \tag{12.108}$$

$$\mathbf{n}\cdot\sigma_i\mathbf{n}V_i''+\mathbf{n}\cdot\sigma_e\mathbf{n}V_e''=0. \tag{12.109}$$

Using that $V=V_i-V_e$, we find that

$$V_i'=\frac{\mathbf{n}\cdot\sigma_e\mathbf{n}}{\mathbf{n}\cdot(\sigma_i+\sigma_e)\mathbf{n}}V', \tag{12.110}$$

$$V_e'=-\frac{\mathbf{n}\cdot\sigma_i\mathbf{n}}{\mathbf{n}\cdot(\sigma_i+\sigma_e)\mathbf{n}}V', \tag{12.111}$$

and

$$\frac{R_m}{\chi}\frac{(\mathbf{n}\cdot\sigma_i\mathbf{n})(\mathbf{n}\cdot\sigma_e\mathbf{n})}{\mathbf{n}\cdot(\sigma_i+\sigma_e)\mathbf{n}}V_i''+cC_mR_mV'+f(V)=0. \tag{12.112}$$

Now we compare (12.112) with (12.107) and find that the solutions are related through

$$V(\xi) = U\left(\frac{\xi}{\Lambda}\right), \tag{12.113}$$

where $\Lambda(\mathbf{n})^2 = \frac{R_m}{\chi}\frac{(\mathbf{n}\cdot\sigma_i\mathbf{n})(\mathbf{n}\cdot\sigma_e\mathbf{n})}{\mathbf{n}\cdot(\sigma_i+\sigma_e)\mathbf{n}}$ ($\Lambda(\mathbf{n})$ is the directionally dependent space constant), and the plane-wave velocity is

$$c = \frac{\Lambda(\mathbf{n})}{C_m R_m} c_0. \tag{12.114}$$

From this we learn that the speed of propagation depends importantly on direction $\mathbf{n}$, but the membrane potential profile is independent of direction except in its spatial scale Λ. This observation allows us to estimate the coefficients of the conductivity tensors σ_i and σ_e. This we do by observing from (12.111) that the total deflection of extracellular potential is dependent on direction. If we denote the total deflection of potentials during the upstroke by ΔV and ΔV_e, then

$$r_d = \frac{\Delta V_{ed}}{\Delta V} = \left(\frac{\sigma_{id}}{\sigma_{id} + \sigma_{ed}}\right), \quad d = L, T \tag{12.115}$$

where the subscript d denotes the longitudinal (L) or the transverse (T) fiber direction and $\sigma_{id} = \mathbf{n}_d \cdot \sigma_i \mathbf{n}_d$ with $\mathbf{n}_L$ a unit vector along the fiber axis and $\mathbf{n}_T$ a unit vector transverse to the fiber axis, and similarly for σ_{ed}. It follows that

$$\frac{\sigma_{ed}}{\sigma_{id}} = \frac{1 - r_d}{r_d}. \tag{12.116}$$

Measurements on dog myocardium (Roberts and Scher, 1982) find that $\Delta V_{eL} = 74 \pm 7$ mV, $\Delta V_{eT} = 43 \pm 6$ mV. With a typical membrane potential upstroke deflection of $\Delta V = 100$ mV, it follows that

$$\frac{\sigma_{eL}}{\sigma_{iL}} = 0.35, \qquad \frac{\sigma_{eT}}{\sigma_{iT}} = 1.33, \tag{12.117}$$

implying that myocardial tissue has unequal anisotropy ratios. Combining this with the fact that the propagation speed is about three times faster in the longitudinal direction than the transverse direction leads to the conclusion that

$$\frac{\sigma_{iT}}{\sigma_{iL}} = 0.05, \qquad \frac{\sigma_{eT}}{\sigma_{iL}} = 0.07. \tag{12.118}$$

It should be pointed out that one can find a variety of estimates of the anisotropy ratios in the literature.

Virtual Electrodes

It is well accepted that cardiac tissue has unequal anisotropy ratios. Perhaps the most important consequence of unequal anisotropy is seen in the response of the membrane to a current stimulus. Recall from Fig. 12.23 that in a one-dimensional cable, the transmembrane response to a current stimulus is local depolarization or hyperpolarization with exponential decay away from the electrode site. In tissue with equal anisotropy

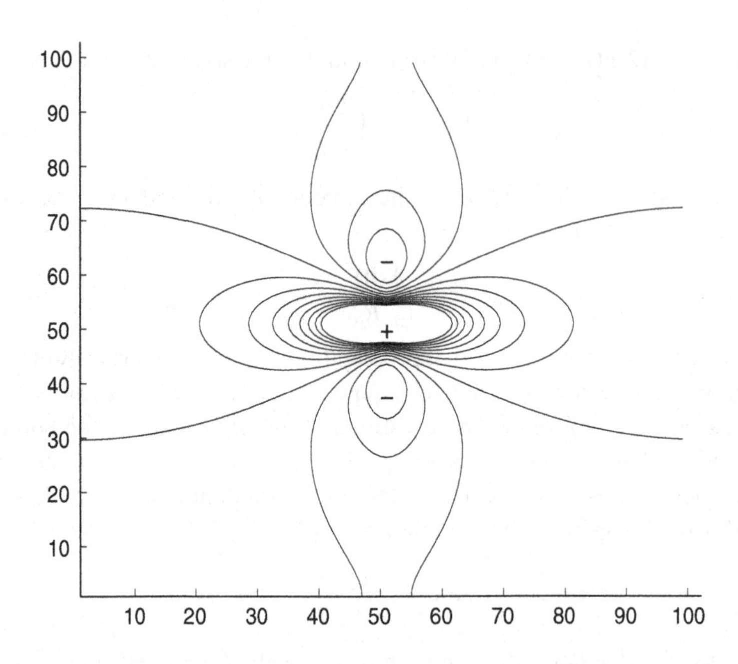

Figure 12.26 Contour plot of dogbone potential and virtual electrodes resulting from current stimulus to an anisotropic bidomain with anisotropy ratios as in (12.117) and (12.118).

ratios, the transmembrane response is roughly the same, except that the isopotential contours are elliptical, with aspect ratio given by the square root of the ratio of the principal eigenvalues of the matrix $\sigma_i(\sigma_i + \sigma_e)^{-1}\sigma_e$.

However, in tissue with unequal anisotropy ratios, the isopotential contours can be much more complicated. An example of these contours is shown in Fig. 12.26. For this plot, the passive bidomain equations (take $I_{\text{ion}} = -V/R_m$) with a constant current point source at the center of the spatial domain were solved using straightforward finite differences. The anisotropy ratios were as in (12.117) and (12.118).

These contours show two striking features. In the central region (shown with a "+"), there is a strong polarizing response, but the isopotential contours are not elliptical, but rather they are "dogbone" shaped, oriented along the axis with largest coupling. For this reason, this is referred to as the dogbone potential. Oriented orthogonal to this dogbone, there are two roughly circular regions of opposite polarity (denoted with a "−"). These are called virtual electrodes, because there is depolarization or hyperpolarization even though there is no actual electrode (Roth, 1992; Wikswo et al., 1995).

Virtual electrodes can result not only from unequal anisotropy ratios but from many other inhomogeneities of resistance, such as fiber curvature, or tissue injury (Henriquez et al. 1996, White et al. 1998). Of course, virtual electrodes were also seen above to result from inhomogeneities of resistance along cables. It is believed that virtual electrodes are important determinants of the success or failure of defibrillation, and this interesting topic is discussed below.

The Activation Sequence

Propagation in myocardial tissue is not planar and neither is it along fibers. In fact, because the initial activation of myocardial tissue occurs at the endocardial surface, myocardial propagation is primarily transverse to fibers. The fiber orientation in myocardial tissue varies through the thickness of the tissue, rotating approximately 120 degrees from epicardium to endocardium (Streeter, 1979). Additionally, the geometry of the ventricles is complicated, and the initiation of action potentials occurs at numerous places on the endocardial surface at the termini of the Purkinje fiber network.

The activation sequence is the spatial and temporal sequence in which the medium is activated by a wave initiated by the SA node. Without belaboring the details, an example of a computed action potential activation sequence for an anisotropic medium is shown in Fig. 12.27. Here is shown a sequence of wave-front surfaces at 20 ms intervals following stimulation on the top surface of a slab of tissue measuring 6 cm × 6 cm × 1 cm. The fiber orientation rotates continuously through 120 deg from top to bottom, and the velocity of propagation along fibers was taken to be three times faster than transverse to fibers. The most noticeable feature of these wave-front surfaces is the distortion from elliptical that occurs because of the rotational fiber orientation. Furthermore, there is a rotation of the elliptical axes, following the rotation of the fiber axes. However, the fastest propagation is not in the longitudinal fiber direction, as the ellipses rotate by only about 60 degrees. One can also determine (from the simulations)

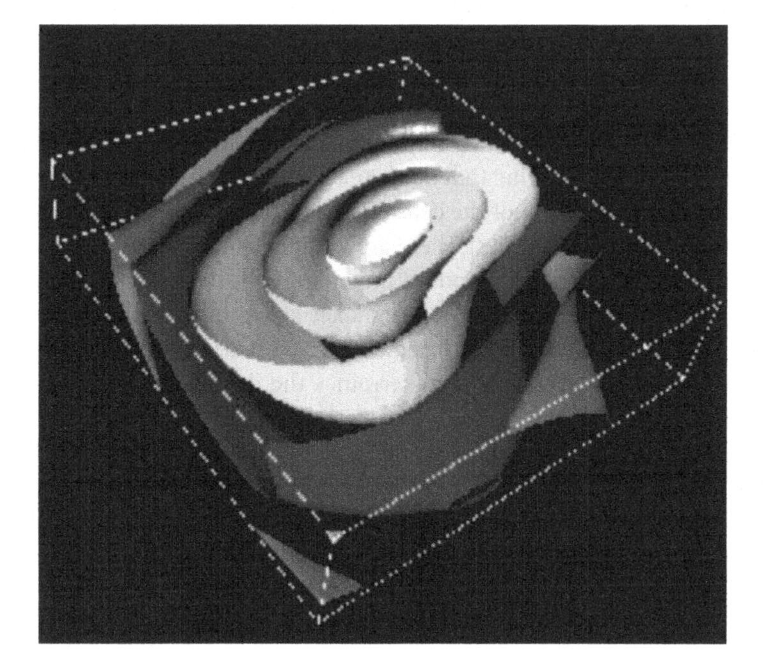

Figure 12.27 Action potential wave fronts at 20 ms intervals in a slab of tissue with rotational anisotropy.

that normal wave-front velocity is always slower than the maximal plane-wave velocity, because of curvature slowing and fiber rotation.

Simulations of this type have also been done for whole heart with realistic geometry and fiber orientation incorporated into the conductivity tensors (see for example, Keener and Panfilov (1996)). Data for the geometry and fiber orientation from Hunter's group (Nielsen et al., 1991) were incorporated into the conductivity tensors. What is noticeable in all of these simulations is that wavefronts are not elliptical, and the direction of fastest propagation is not along fibers, even though the planar velocity is fastest in the direction of fibers. In fact, significant distortions from ellipticity occur because of the variable fiber orientation.

Many of these features of propagation in three-dimensional myocardial tissue have been observed experimentally (see, for example, Taccardi et al., 1992).

12.3.4 Pacemakers

In the previous sections, the emphasis was on how gap-junctional coupling enables propagation of action potentials through cardiac tissue. In this section, we describe two other effects of gap-junctional coupling, namely synchronization and electrotonic loading.

Pacemaker Synchrony

The sinoatrial (SA) node is a clump of self-oscillatory cells located on the atrium near the superior vena cava. These cells fire regularly, initiating an action potential that propagates throughout the atrium, eventually terminating at the atrioventricular septum, or conducting into the AV node. SA nodal cells are not identical, but nevertheless fire at the same frequency. They are not synchronous in their firing (i.e., firing all at once), but they are *phase locked*, meaning that during each cycle, each cell fires once and there is a regular pattern to the firing of the cells. The variation of cellular properties in the SA node has two dominant features. There are gradual spatial gradients of the period of the oscillators and random deviations of individual cells from this average gradient.

Three questions concerning the SA pacemaker are of interest. First, since the individual cells all have different natural frequencies, what determines the frequency of the collective SA node? Second, what determines the details of the firing sequence, specifically, the location of the cell that fires earliest in the cycle and the subsequent firing order of the cells in the node? One might anticipate that the leader of the pack is the cell with the highest intrinsic frequency, but as we will see, this is not the case. Third, under what conditions does the SA node lose its ability to initiate the heartbeat (called *sinus node dysfunction*), and is it possible for other regions of (abnormal) oscillatory cells to initiate action potentials? Furthermore, since sinus node dysfunction is a potentially fatal condition, it is of clinical interest to understand how to treat this condition.

To address these questions, we suppose that the SA node is composed of self-oscillatory cells coupled together in a network, and that the action potential of the

ith cell can be described by a vector of state variables u_i, which, in the absence of coupling, has dynamics

$$\frac{du_i}{dt} = F(u_i) + \epsilon G_i(u_i). \tag{12.119}$$

Here, the term $F(u)$ represents typical dynamics applicable for every cell, and $\epsilon G_i(u)$ represents the deviation of the dynamics for the ith cell from the average. The parameter ϵ is assumed to be a small positive number, indicating that the variation among cells is small. To specify F, one might use the YNI model or the FitzHugh–Nagumo equations (as in Section 12.2), adjusted to allow for autonomous oscillations.

Next we assume that the cells are isopotential and connected discretely through resistive gap-junction coupling and that the extracellular medium is isopotential. Then, when the cells are coupled, we obtain the system of equations

$$\frac{du_i}{dt} = F(u_i) + \epsilon G_i(u_i) + \epsilon D \sum_j d_{ij}(u_j - u_i). \tag{12.120}$$

Here D is a diagonal matrix with entries of zero or one to indicate which of the state variables participate in the coupling. In neuromuscular media, only the intracellular and extracellular potentials participate in the coupling. However, because we have assumed that the extracellular potential is spatially uniform, D has only one nonzero entry, namely that one corresponding to the intracellular potential. Furthermore, since we have assumed that the extracellular potential is spatially homogeneous, the difference of intracellular potentials between two cells is the same as the difference of their transmembrane potentials.

The coefficients d_{ij} are the coupling coefficients for the network of cells, where d_{ij} is equal to the (positive) coupling strength (inversely proportional to the resistance) between cells i and j. Of course, $d_{ij} = 0$ if cells i and j are not directly coupled, and coupling is symmetric, so that $d_{ij} = d_{ji}$. Without loss of generality, $d_{ii} = 0$.

A simple example for the coupling matrix comes from considering a one-dimensional chain of cells coupled by nearest-neighbor coupling, for which $d_{i,i+1} = d_{i,i-1} = d$, the coupling strength between cells, and all other coupling coefficients are zero. The general formulation (12.120) of the problem allows us to consider a wide variety of coupling networks, including anisotropically coupled rectangular grids and hexagonal grids. The parameter ϵ scales the coupling term to indicate that the coupling is weak, so that currents through gap junctions are small compared to transmembrane currents. The evidence for weak coupling is that the wave speed in the SA node is very slow, on the order of 2–5 cm/s, compared with 50 cm/s in myocardial tissue and 100 cm/s in Purkinje fiber. (See Exercise 18 for a possible explanation of why weak coupling might be advantageous.)

Suppose the stable periodic solution of the equation $\frac{du}{dt} = F(u)$ is given by $U(t)$. Then, because ϵ is assumed to be small, one can use multiscale methods (see Section 12.7) to find that $u_i(t) = U(\omega(\epsilon)t + \delta\theta_i(t)) + O(\epsilon)$, where $\omega(\epsilon) = 1 + \epsilon\Omega_1 + O(\epsilon^2)$ and the phase shift $\delta\theta_i$ of each oscillator satisfies the equation

$$\frac{d}{dt}\delta\theta_i = \epsilon\left(\xi_i - \Omega_1 + \sum_{j\neq i} d_{ij}[h(\delta\theta_j - \delta\theta_i) - h(0)]\right). \tag{12.121}$$

The periodic coupling function h and the numbers ξ_i are specified in Section 12.7 and the scalar Ω_1 is as yet undetermined. Notice that each oscillator has frequency $\omega(\epsilon)$ and that the phase is slowly varying by comparison with the underlying oscillation and represents only the variation from the typical oscillation.

While there are many interesting questions that could be addressed at this point, of greatest interest here is to determine the *firing sequence* in a collection of phase-locked oscillators. By firing sequence, we mean the order of firing of the individual cells. If the firing sequence of cells is spatially ordered, then the firing of cells appears as a spreading wave, although it is not a propagated wave, but a *phase wave*. (It is not propagated because it would remain for some time even if coupling were set to zero. The role of coupling is merely to coordinate, not to initiate, the wavelike behavior.)

To determine the approximate firing sequence, we suppose that the cells are phase-locked and that the steady-state phase differences are not too large. This is the case for normal SA nodal cells, since all of the SA nodal cells fire within a few milliseconds of each other during an oscillatory cycle of about one second duration. If the steady-state phase differences are small enough, we can replace $h(\delta\theta_j - \delta\theta_i) - h(0)$ in (12.121) by its local linearization $h'(0)(\delta\theta_j - \delta\theta_i)$. Then the steady states of (12.121) are determined as solutions of the linear system of equations

$$\sum_j d_{ij}h'(0)(\delta\theta_j - \delta\theta_i) = \Omega_1 - \xi_i. \tag{12.122}$$

We rewrite (12.122) in matrix notation by defining a matrix A with entries $a_{ij} = d_{ij}$ if $i \neq j$ and $a_{ii} = -\sum_{j\neq i} d_{ij}$, and then (12.122) becomes

$$A\Phi = \frac{1}{h'(0)}(\vec{\Omega}_1 - \vec{\xi}), \tag{12.123}$$

where Φ is the vector with entries $\delta\theta_i$, $\vec{\Omega}_1$ is a vector with all entries Ω_1, and the entries of $\vec{\xi}$ are the numbers ξ_i.

A few observations about the matrix A are important. Notice that A is symmetric and has a nontrivial null space, since $\sum_j a_{ij} = 0$. For consistency, we must choose Ω_1 such that the sum of all rows of (12.123) is zero, so that

$$\Omega_1 = \frac{1}{N}\sum_i \xi_i. \tag{12.124}$$

Thus, the bulk frequency of the SA node is determined as the average of the frequencies of the individual oscillators. This is a purely democratic process in which one cell equals one vote, regardless of coupling strength.

Next, since all the nonzero elements of d_{ij} (and hence the off-diagonal elements of A) are positive and A has zero row sums, all the nonzero eigenvalues of A have negative real part. Furthermore, since A is real, symmetric, and nonpositive definite, it has a complete set of N mutually orthogonal, real eigenvectors, say $\{y_k\}$, with corresponding real eigenvalues λ_k. All of the eigenvalues λ_k are negative or zero. If the matrix of coupling coefficients d_{ij} is irreducible, then the constant eigenvector y_1 is the unique null vector of A, and $\lambda_k < 0$ for $k > 1$ (see Chapter 9, Exercise 9). The matrix of coupling coefficients is *irreducible* if all the cells are connected by some electrical path, so that there are no electrically isolated clumps of cells. Suppose also that the eigenvectors are ordered by increasing amplitude of the eigenvalue. The solution of (12.123) is readily expressed in terms of the eigenvectors and eigenvalues of A as

$$\Phi = -\frac{1}{h'(0)} \sum_{k \neq 1} \langle \vec{\xi}, y_k \rangle \frac{y_k}{\lambda_k}. \tag{12.125}$$

The scalar Ω_1 drops out of this expression because the eigenvector y_1 is the constant vector and $\langle y_k, y_1 \rangle = 0$ for all $k \neq 1$.

The firing sequence is now determined from Φ. That is, if $\delta\theta_k$ is the largest element of Φ, then the phase of the kth cell is the most advanced and therefore the first to fire, and so on, in decreasing order. It remains to gain some understanding of the relationship between the natural frequencies $\vec{\xi}$ and the firing sequence Φ.

The general principle of how the firing sequence is determined from the natural frequencies $\vec{\xi}$ is apparent from (12.125). The firing sequence is a superposition of the eigenvectors $\{y_k\}$ with amplitudes $\lambda_k^{-1}\langle \vec{\xi}, y_k \rangle$. Thus, eigenvector components of $\vec{\xi}$ that are most influential on the firing sequence are those components for which $\lambda_k^{-1}\langle \vec{\xi}, y_k \rangle$ is largest in amplitude. The expression (12.125) is a filter that suppresses, or filters out, certain components of $\vec{\xi}$. It follows that a single cell with high natural frequency compared to its coupled neighbors is not necessarily able to lead the firing sequence.

Equation (12.125) for the firing sequence does not give much geometrical insight. Furthermore, it is usually not a good idea to solve matrix problems such as (12.123) using eigenfunction expansions, since direct numerical methods are much faster and easier. To illustrate how (12.125) works, we consider, as an example, a two-dimensional grid of cells coupled by nearest-neighbor coupling. The natural frequencies of the cells are randomly distributed, with the fastest cells concentrated near the center of the grid. The distribution of natural frequencies is depicted in Fig. 12.28A, with darker locations representing the slowest intrinsic frequencies. The firing sequence for this collection of cells is shown in Fig. 12.28B, where cells with advanced (or largest) phase fire earliest in the firing sequence. The initiation of the firing sequence is at the site of a group of fast, but not necessarily the fastest, oscillators. Notice that the phase is smoothed, giving the appearance of wave-like motion moving from the location of largest phase to smallest phase, even though these are phase waves rather than propagated waves. In this figure, the scale of the phase variable is arbitrary.

Critical Size of a Pacemaker

The SA node is a small clump of self-oscillatory cells in a sea of excitable (but nonoscillatory) cells whose function is to initiate the cardiac action potential. SA nodal cells have

A

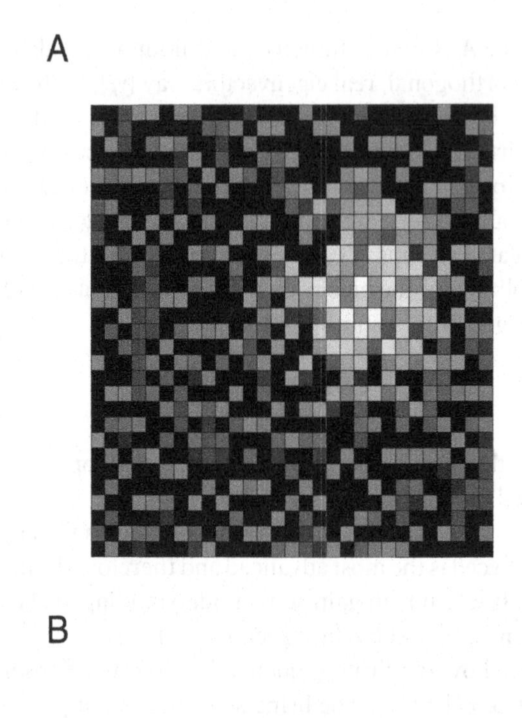

B

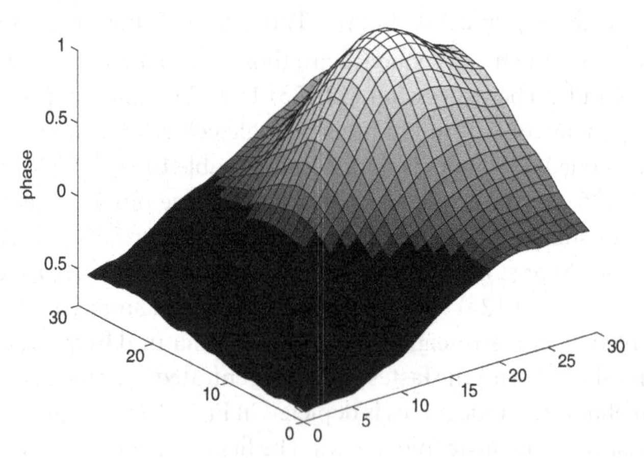

Figure 12.28 A: Natural frequencies ξ_i for a collection of oscillatory cells, lighter being faster, and darker being slower. B: Phase for oscillators in a collection of coupled cells with nearest-neighbor coupling and the natural frequencies depicted in panel A.

no contractile function and therefore no contractile machinery. Thus, when viewed in terms of contractile efficiency, SA nodal cells are a detriment to contraction and a waste of important cardiac wall space. On the other hand, the SA node cannot be too small because presumably it would not be able to generate the current necessary to entrain the rest of the heart successfully. Further it could be that surrounding resting cells drain enough current from the oscillatory cells to quench the pacemaking activity. This drain is called the electrotonic load. Clearly, it is important to have some measure of the critical size of the SA node.

An *ectopic focus* is a collection of cells other than the SA node or AV node that are normally not oscillatory but that for some reason (for example, increased extracellular K^+) become self-oscillatory and manage to entrain the surrounding tissue into a rapid beat. In some situations, particularly in people with scar tissue resulting from a previous heart attack, the appearance of an ectopic focus may be life-threatening.

To understand something about the behavior of a clump of oscillatory cells in an otherwise nonoscillatory medium, we use a simple model with FitzHugh–Nagumo dynamics,

$$\frac{\partial v}{\partial t} = \nabla^2 v + f(v) - w, \tag{12.126}$$

$$\frac{\partial w}{\partial t} = \epsilon(v - \gamma w - \alpha(r/\sigma)), \tag{12.127}$$

where v represents the membrane potential and w the recovery variable for the excitable medium. The function $f(v)$ is of typical "cubic" shape (cf. Chapter 5). The function $\alpha(r)$ is chosen to specify the intrinsic cell behavior as a function of the radial variable r. The number σ is a scale factor that measures the size of the oscillatory region. We take ϵ to be a small positive number and require $\gamma > 0, f'(v)\gamma < 1$ for all v. This requirement on γ guarantees that the steady-state solution of (12.126)–(12.127) is unique. If the domain is bounded, typical boundary conditions are Neumann (no-flux) conditions. Notice that space has been scaled to have unit space constant. We assume radial symmetry for the SA node as well as for the entire spatial domain.

When there is no spatial coupling, there are two possible types of behavior, exemplified by the phase portraits in Figs. 5.15 and 5.16. In these examples, the system has a unique steady-state solution that is globally stable (Fig. 5.15) or has an unstable steady-state solution surrounded by a stable periodic orbit (Fig. 5.16), depending on the location of the intercept of the two nullclines.

The transition from a stable to an unstable steady state is via a subcritical Hopf bifurcation. The Hopf bifurcation is readily found from standard linear analysis. Suppose v^* is the equilibrium value for v (v^* is a function of α). Then the characteristic equation for (12.126)–(12.127) (with no diffusion) is

$$f'(v^*) = \lambda + \frac{\epsilon}{\lambda + \epsilon\gamma}, \tag{12.128}$$

where λ is an eigenvalue of the linearized system. There is a Hopf bifurcation (i.e., λ is purely imaginary) when

$$f'(v^*) = \epsilon\gamma, \tag{12.129}$$

provided that $\epsilon\gamma^2 < 1$. If $f'(v^*) > \epsilon\gamma$, the steady-state solution is an unstable spiral point, whereas if $f'(v^*) < \epsilon\gamma$, the steady-state solution is linearly stable. If ϵ is small, most of the intermediate (increasing) branch of the curve $f(v)$ is unstable, with the Hopf bifurcation occurring close to the minimal and maximal points. Thus, there is a range of values of α, which we denote by $\alpha_* < \alpha < \alpha^*$, for which the steady solution is unstable.

We wish to model the physical situation in which a small collection of cells (like the SA node or an ectopic focus) is intrinsically oscillatory, while all other surrounding cells are excitable, but not oscillatory. To model this, we assume that $\alpha(r)$ is such that the steady solution is unstable for small r, but stable and excitable for large r, so that $\lim_{r\to\infty}\alpha(r) = a < \alpha_*$ and $\lim_{r\to\infty} f'(v^*(r)) < \epsilon\gamma$. As an example, we might have the bell-shaped curve

$$\alpha(r) = a + (b-a)\exp\left(-\frac{r^2}{R^2}\right), \tag{12.130}$$

$$R^2 = \log\left(\frac{b-a}{\alpha_* - a}\right), \tag{12.131}$$

with $a < \alpha_* < b < \alpha^*$. The scale factor R was chosen such that $\alpha(1) = \alpha_*$, so that cells with $r < 1$ are self-oscillatory and the cells outside unit radius are nonoscillatory.

Another way to specify $\alpha(r)$ is simply as the piecewise-constant function

$$\alpha(r) = \begin{cases} b, & \text{for} \quad 0 < r < 1, \\ a, & \text{for} \quad r > 1, \end{cases} \tag{12.132}$$

with $a < \alpha_* < b < \alpha^*$. The specification (12.132) is particularly useful when used in combination with the piecewise-linear function

$$f(v) = \begin{cases} -v, & \text{for} \quad v < \frac{1}{4}, \\ v - \frac{1}{2}, & \text{for} \quad \frac{1}{4} < v < \frac{3}{4}, \\ 1 - v, & \text{for} \quad v > \frac{3}{4}, \end{cases} \tag{12.133}$$

since then all the calculations that follow can be done explicitly (see Exercises 17 and 18).

There are two parameters whose influence we wish to understand and that we expect to be most significant, namely, a, the asymptotic value of $\alpha(r)$ as $r \to \infty$, and σ, which determines the size of the oscillatory region. Note that as a decreases, the cells become less excitable and the wavespeed of fronts decreases. We expect the behavior to be relatively insensitive to variations in b, although this should be verified as well.

With a nonuniform $\alpha(r)$, the uncoupled medium has a region of cells with unstable steady states and a region with stable steady states. With diffusive coupling, the steady-state solution is smoothed and satisfies the elliptic equation

$$\nabla^2 v + F(v,r) = 0, \tag{12.134}$$

$$F(v,r) = f(v) - w, \tag{12.135}$$

$$w = \frac{1}{\gamma}\left(v - \alpha\left(\frac{r}{\sigma}\right)\right). \tag{12.136}$$

For each r, since $f'(v)\gamma < 1$, the function $F(v,r)$ is a monotone decreasing function of v having a unique zero, say $v = v^*(r), F(v^*(r),r) = 0$. It follows that there is a unique, stable solution of (12.136), denoted by $v_0(r), w_0(r)$. In fact, this unique solution is readily found numerically as the unique steady solution of the nonlinear parabolic equation

$$\frac{\partial y}{\partial t} = \nabla^2 y + F(y,r). \tag{12.137}$$

This steady-state solution is shown in Fig. 12.29. Here are shown three different steady-state solutions of (12.126)–(12.127); the uncoupled solution (the steady states for the uncoupled medium, i.e., with no diffusive coupling), the solution for a symmetric one-dimensional medium, and the solution for a spherically symmetric three-dimensional medium. The three-dimensional solution with spherical symmetry is not much harder to find than the one-dimensional solution, because the change of variables $y = Y/r$ transforms (12.137) in three spatial dimensions into the one-dimensional problem

$$\frac{\partial Y}{\partial t} = \frac{\partial^2 Y}{\partial r^2} + rF\left(\frac{Y}{r}, r\right). \tag{12.138}$$

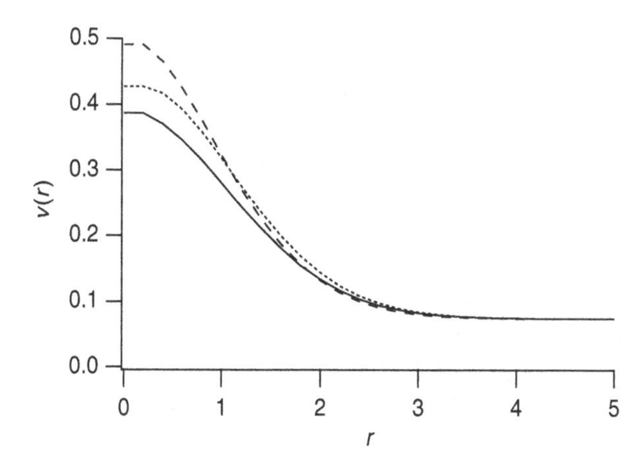

Figure 12.29 Three steady-state solutions with $f(v) = 10.0v(v-1)(0.5-v)$, $\gamma = 0.1$, and $\alpha(r)$ given by (12.130), with $a = 0.104$, $b = 0.5$, $\sigma = 2.25$. The short dashed curve shows the uncoupled solution, the long dashed curve shows the solution for a symmetric one-dimensional medium, and the solid curve shows the solution for a spherically symmetric three-dimensional medium.

Solutions of this partial differential equation are regular at the origin if we require $Y = 0$ at $r = 0$. Diffusion (i.e., electrotonic coupling) obviously smooths the steady-state solution in the oscillatory region.

The issue of collective oscillation is determined by the stability of the diffusively smoothed steady state as a solution of the partial differential equation system (12.126)–(12.127). To study the stability of the steady state, we look for a solution of (12.126)–(12.127) of the form $v(r) = v_0(r) + V(r)e^{\lambda t}, w = w_0(r) + W(r)e^{\lambda t}$ and linearize. We obtain the linear system

$$\lambda V = \nabla^2 V + f'(v_0(r))V - W, \tag{12.139}$$

$$\lambda W = \epsilon(V - \gamma W). \tag{12.140}$$

Because of the special form of this linearized system, it can be simplified to a single equation, namely,

$$\nabla^2 V + f'(v_0(r))V = \mu V, \tag{12.141}$$

where $\mu = \lambda + \frac{\epsilon}{\lambda+\epsilon\gamma}$. Equation (12.141) has a particularly nice form, being a *Schrödinger equation*. In quantum physics, the function $-f'(v_0(r))$ is the potential energy function, and the eigenvalues μ are the energy levels of bound states. In the present context, we are interested in determining the sign of the real part of λ through $\mu = \lambda + \frac{\epsilon}{\lambda+\epsilon\gamma}$. Notice that the relationship between μ and λ here is of exactly the same form as the characteristic equation for individual cells (12.128). This leads to a nice interpretation for the Schrödinger equation (12.141). Because it is a self-adjoint equation, the eigenvalues μ of (12.141) are real. Therefore, there is a Hopf bifurcation for the medium whenever $\mu = \epsilon\gamma$. The entire collection of coupled cells is stable when the largest eigenvalue satisfies $\mu < \epsilon\gamma$ and unstable if the largest eigenvalue has $\mu > \epsilon\gamma$.

In Fig. 12.30 is shown the potential function $f'(v(r))$ for the three steady profiles of Fig. 12.29. The largest eigenvalue of (12.141) represents an average over space of the influence of $f'(v_0(r))$ on the stability of the steady state. When this value is larger than $\epsilon\gamma$ (the critical slope of $f(v)$ at which Hopf bifurcations of the uncoupled system occur), then the entire medium loses stability to a Hopf bifurcation and gives rise to an oscillatory solution. The condition $\mu > \epsilon\gamma$ is therefore the condition that determines whether a region of oscillatory cells is a source of oscillation. If $\mu < \epsilon\gamma$, the oscillatory cells are held quiescent by the rest of the medium.

Some observations about the size of the eigenvalues μ are immediate. Because $\lim_{r\to\infty} v_0(r) = \lim_{r\to\infty} v^*(r)$, it follows that $f'(v_0(r)) < \epsilon\gamma$ for large r. For there to be a bounded solution of (12.141) that is exponentially decaying at $\pm\infty$, there must be a region of sinusoidal behavior in which $\mu < f'(v_0(r))$. Thus, the largest eigenvalue of (12.141) is guaranteed to be smaller than the maximum of $f'(v_0(r))$. Therefore, if $v_0(r) < \alpha_*$ (so that $f'(v_0(r)) < \epsilon\gamma$ for all r), there are no oscillatory cells, and the steady solution is stable. Furthermore, since the largest eigenvalue is strictly smaller than the maximum of $f'(v_0(r))$ and it varies continuously with changes in $v_0(r)$, there are profiles $\alpha(r)$ having a nontrivial collection of oscillatory cells that is too small to render the medium unstable. That is, there is a critical mass of oscillatory cells necessary to

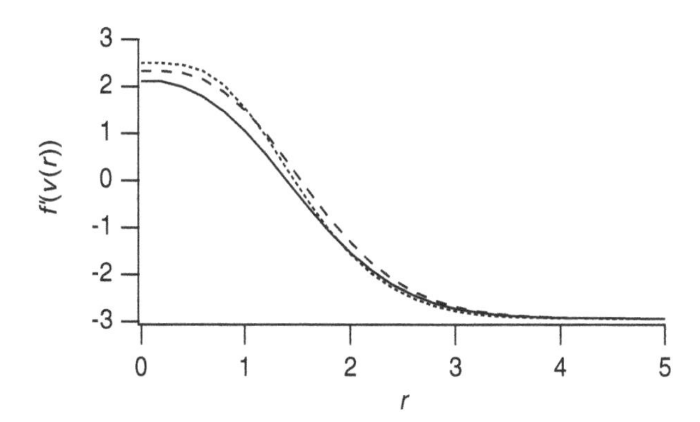

Figure 12.30 The potential function $f'(v(r))$ for three steady profiles $v(r)$. The short dashed curve corresponds to the uncoupled solution, the long dashed curve to the symmetric one-dimensional medium, and the solid curve to a spherically symmetric three-dimensional medium.

cause the medium to oscillate. Below this critical mass, the steady state is stable, and the oscillation of the oscillatory cells is quenched.

Suppose $f'(v)$ is a monotone increasing function of v in some range $v < v^+$, and suppose that $\alpha(r)$ is restricted so that $v_0(r) < v^+$ for all r. Suppose further that $\alpha(r)$ is a monotone increasing function of its asymptotic value a and a monotone decreasing function of r. Then the steady-state solution $v_0(r)$ is an increasing function (for each point r) of both a and σ. Therefore, the function $f'(v_0(r))$ is an increasing function of a and σ for all values of r, from which it follows—using standard comparison arguments for eigenfunctions (Keener, 1998, or Courant and Hilbert, 1953)—that $\mu(a, \sigma)$, the largest eigenvalue of (12.141), is an increasing function of both a and σ. As a result, if $\alpha(r)$ is restricted so that $v_0(r) < v^+$ for all r, there is a monotone decreasing function of σ, denoted by $\sigma = \Sigma(a)$, along which the largest eigenvalue $\mu(a, \sigma)$ of (12.141) is precisely $\epsilon\gamma$.

This summary statement shows that to build the SA node, one must have a sufficiently large region of oscillatory tissue, and that the critical mass requirement increases if the tissue becomes less excitable or if the coupling becomes stronger. Strong coupling inhibits oscillations, because increasing coupling increases the space constant, and σ was measured in space constant units. Therefore, an increase of the space constant increases the critical size requirement of the oscillatory region. In Fig. 12.31 is shown the critical Hopf curve $\sigma = \Sigma(a)$ for a one-dimensional domain and for a three-dimensional domain (taking $\epsilon = 0$), both found numerically.

Having established that there is a critical size for a self-oscillatory region above which oscillations occur and below which oscillations are prevented, we would like to examine the behavior of the oscillations. Two types of oscillatory behavior are possible. If the far field $r \to \infty$ is sufficiently excitable, then the oscillations of the oscillatory region excite periodic waves that propagate throughout the medium, as depicted in

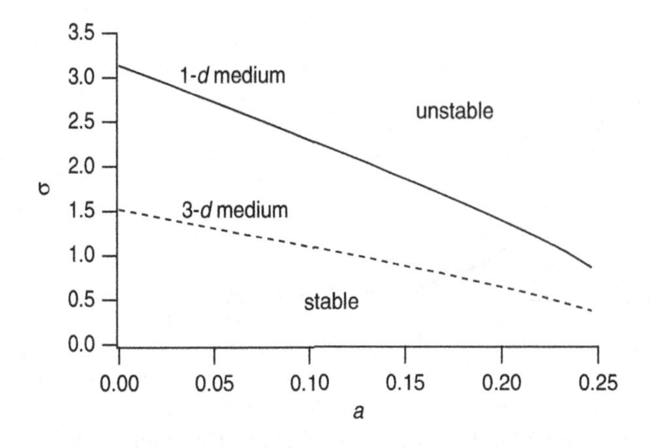

Figure 12.31 The critical curve $\sigma = \Sigma(a)$ along which there is a Hopf bifurcation for the system (12.126)–(12.127), shown solid for a one-dimensional and dashed for a three-dimensional medium.

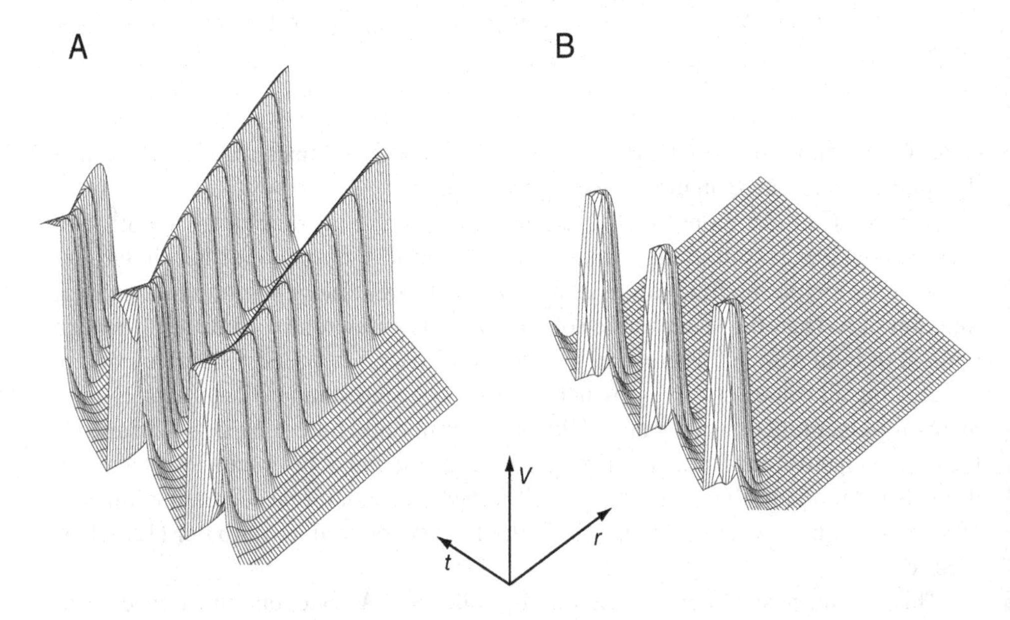

Figure 12.32 A: Waves generated by an oscillatory core that propagate into the nonoscillatory region. The nonlinearity here is the same as in Fig. 12.29 with $\epsilon = 0.1, \gamma = 0.1, a = 0.2, b = 0.5$, and $\sigma = 3.0$. B: Waves generated by an oscillatory core that fail to propagate into the nonoscillatory region. Same parameters as in A, but with $a = 0$.

Fig. 12.32A. On the other hand, it may be that there are oscillations that fail to propagate throughout the entire medium, as depicted in Fig. 12.32B. In Fig. 12.32A, the oscillatory region successfully drives oscillatory waves that propagate throughout the entire medium. Here, $a = 0.2, \sigma = 3.0$. In Fig. 12.32B, the oscillatory region is incapable of driving periodic waves into the nonoscillatory region, as the medium at infinity does not support front propagation.

The issue of whether or not the entire medium is entrained to the central oscillator is decided by the relationship between the period of the oscillator and the dispersion curve for the far medium. Roughly speaking, if the period of the central oscillator is large enough compared to the absolute refractory period of the far medium (the knee of the dispersion curve), then waves can be expected to propagate into the far field in one-to-one entrainment. On the other hand, if the frequency of the oscillation is below the knee of the dispersion curve, we expect partial or total block of propagation. Block of propagation occurs as the excitability of the far field, parameterized by a, decreases.

We can summarize how the oscillations of the medium depend on coupling strength. For a medium with fixed asymptotic excitability, if the size of the oscillatory region is large enough, there is oscillatory behavior. However, this critical mass is an increasing function of coupling strength. With sufficiently large coupling, the oscillations of any finite clump of oscillatory cells (in an infinite domain of nonoscillatory cells) are quenched. If coupling is decreased, the critical mass for oscillation decreases. Thus, any clump of oscillatory cells oscillates if coupling is weak enough. However, if coupling is too weak, then effects of discrete coupling may become important, and the oscillatory clump of cells may lose its ability to entrain the entire medium. It follows that if the medium is sufficiently excitable, there is a range of coupling strengths, bounded above and below, in which a mass of oscillatory cells entrains the medium. If the coupling is too large, the oscillations are suppressed, while if the coupling is too weak, the oscillations are localized and cannot drive oscillations in the medium far away from the oscillatory source. On the other hand, if the far region is not sufficiently excitable, then one of these two mechanisms suppresses entrainment for all coupling strengths.

12.4 Cardiac Arrhythmias

Cardiac arrhythmias are disruptions of the normal cardiac electrical cycle. They are generally of two types. There are temporal disruptions, which occur when cells act out of sequence, either by firing autonomously or by refusing to respond to a stimulus from other cells, as in AV nodal block or a bundle branch block. A collection of cells that fires autonomously is called an ectopic focus. Generally speaking, these arrhythmias cause little disruption to the ability of the heart muscle to pump blood, and so if they do not initiate some other kind of arrhythmia, are generally not life-threatening.

The second class of arrhythmias are those that are reentrant in nature and can occur only because of the spatial distribution of cardiac tissue. If they occur in the ventricles, reentrant arrhythmias are of serious concern and life-threatening, as the ability of the heart to pump blood is greatly diminished. Reentrant arrhythmias on the atria are less dangerous, since the pumping activity of the atrial muscle is not necessary to normal function with minimal physical activity, although long-lived atrial reentrant arrhythmias are known to increase the chance of strokes.

12.4.1 Cellular Arrhythmias

By a cellular arrhythmia we mean an action potential response to a stimulus protocol that is not one-to-one. Such arrhythmias are relatively easy to observe in models. Simply apply a periodic stimulus to your favorite model of cellular activity, vary the period and watch what happens. Of course, it would be nice to obtain some deeper understanding of the cause of these arrhythmias than can be provided by numerical simulation. To that end ideas of dynamical systems theory and discrete maps have been extensively used. Here we present two such examples, APD alternans and Wenckebach patterns in the AV node.

APD Alternans

Yehia et al. (1999) give a comprehensive discussion of the types of rhythms that an isolated cardiac cell can demonstrate under regular pacing. A cell isolated from rabbit ventricular muscle was stimulated periodically and the resulting sequence of action potential durations (APDs) was measured. It was found that for sufficiently large basic cycle length (BCL), after the disappearance of initial transients, each stimulus evoked the same APD. This pattern is referred to as a 1:1 rhythm - one stimulus, one response. For appropriately small BCL, the APD alternated between that of a full action potential and that of a sub-threshold response with APD essentially zero. This pattern is referred to as a 2:1 rhythm since every two stimuli evoked one super-threshold response. This 2:1 rhythm is relatively easy to understand. For large BCL, the cell has sufficient time to recover from the previous action potential before the next stimulus, but for short BCL, there is not sufficient time to recover, and the stimulus delivered during this unrecovered stage elicits a subthreshold response.

More interesting behavior occurs between these two extremes of BCL. For large stimulus amplitude, as BCL is decreased, the 1:1 rhythm is replaced by a rhythm in which there are superthreshold responses with alternating APDs. This 2:2 rhythm is referred to as APD alternans. (2:2 refers to the fact that there are 2 superthreshold responses to 2 stimuli, but the responses are not the same.)

Models of APD alternans trace back to the work of Nolasco and Dahlen (1968). The basic assumption is that APD is a function of the preceding recovery period or diastolic interval (DI). If the cell is assumed to be either excited or recovering, then $\text{BCL} = \text{APD} + \text{DI}$ where

$$\text{APD}_{n+1} = G(\text{DI}) = G(\text{BCL} - \text{APD}_n). \tag{12.142}$$

In the case of subthreshold responses, where the DI is smaller than some minimum necessary to reestablish excitability, it is assumed that the sub-threshold stimulus has no effect on the cell, but there is an extra BCL during which the cell can recover. Thus, if there are $N - 1$ unsuccessful stimuli, $\text{DI} = N \cdot \text{BCL} - \text{APD}$, so that

$$\text{APD}_{n+1} = G(\text{DI}) = G(N \cdot \text{BCL} - \text{APD}_n). \tag{12.143}$$

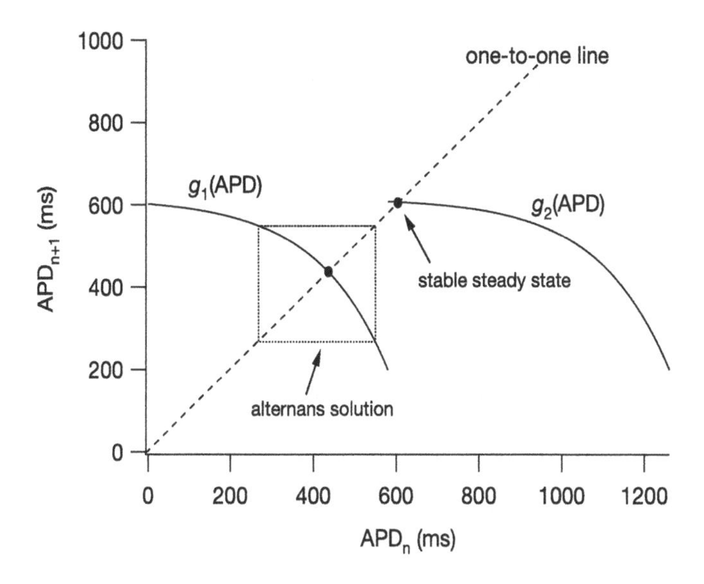

Figure 12.33 A sample APD map exhibiting alternans and bistability. Parameter values are $\mathrm{APD_{max}} = 616$, $A = 750$, $\mu = 170$, $\mathrm{DI_{min}} = 100$ and $\mathrm{BCL} = 680$, all with units of ms.

A typical choice for G is an exponential of the form

$$G(\mathrm{DI}) = \mathrm{APD_{max}} - A\exp\left(-\frac{\mathrm{DI}}{\mu}\right), \qquad \mathrm{DI} > \mathrm{DI_{min}}, \tag{12.144}$$

where $\mathrm{APD_{max}}$, A, μ, and $\mathrm{DI_{min}}$ are fit to data.

A plot of this map for typical parameter values is shown in Fig. 12.33, where $g_k(\mathrm{APD}) = G(k\cdot\mathrm{BCL} - \mathrm{APD})$. Because this is a one-dimensional map, it is quite easy to study its dynamic behavior. First note that the argument of G must always be greater than $\mathrm{DI_{min}}$. Thus, g_k is defined only if $\mathrm{APD} > k\cdot\mathrm{BCL} - \mathrm{DI_{min}}$. Specifically, the map g_1 is defined only on the interval $(0, \mathrm{BCL} - \mathrm{DI_{min}})$ which, in Fig. 12.33 is the interval $(0, 580)$. Similarly, the map g_2 is defined only on the interval $(\mathrm{BCL} - \mathrm{DI_{min}}, 2\mathrm{BCL} - \mathrm{DI_{min}})$.

Fixed points of these maps are at the intersections of the one-to-one line with the functions $g_k(\mathrm{APD})$. Stability of the fixed points is determined by the slope of $g_k(\mathrm{APD})$ at the fixed point, with stability if $|g_k'(\mathrm{APD}^*)| < 1$, and instability otherwise. Instability of the fixed point gives rise to a period two bifurcation, corresponding to alternans.

Increasing BCL shifts the curves to the right, so apparently for sufficiently large BCL there is a unique fixed point of the g_1 map, which corresponds to a 1:1 rhythm. As BCL is decreased, however, the slope of g_1 at the fixed point becomes more negative, leading, if the slope becomes less than -1, to a period two bifurcation. For the parameter values used in Fig. 12.33, there is also a region of bistability, in which stable alternans coexist with a stable 2:1 rhythm. The stable 2:1 rhythm is the steady state of the g_2 map, i.e., the point where the g_2 graph crosses the one-to-one line. This bistability has been observed experimentally (Hall et al., 1999; Yehia et al., 1999).

The observation that alternans occur in single cells when the slope of the APD curve exceeds one has lead many investigators to speculate that a steep APD restitution curve is responsible for a variety of arrhythmias in cardiac tissue as well, not only single cells. This is now known as the APD restitution hypothesis, and there is an extensive literature describing it, along with the suggestion that flattening the APD restitution curve (with drugs, for example) might prevent certain kinds of arrhythmias. However, this hypothesis remains highly conjectural. See, for example, Courtemanche et al. (1993), Cytrynbaum and Keener (2002), Qu et al. (2000), Garfinkel et al. (2000), Watanabe et al. (2001).

Discrete maps have been used with great success to study a variety of cellular arrhythmias, see for example Otani and Gilmour (1997), Hall et al. (1999), Glass et al. (1987) and Watanabe et al. (1995).

12.4.2 Atrioventricular Node—Wenckebach Rhythms

In the normal heart, the only pathway for an action potential to travel to the ventricles is through the AV node. As noted above, propagation through the AV node is quite slow compared to propagation in other cardiac cells. This slowed conduction is primarily due to a decreased density of Na^+ channels, which yields a decreased upstroke velocity, as well as a significantly decreased density of gap-junctional coupling (Pollack, 1976). With a decrease of Na^+ channel density there is also an increased likelihood of conduction failure.

Propagation failure in the AV node leads to skipped QRS complexes on the ECG, or, more prosaically, skipped heartbeats. A skipped heartbeat once in a while is not particularly dangerous, but it is certainly noticeable. During *diastole* (the period of ventricular relaxation during the heartbeat cycle), the ventricles fill with blood. Following an abnormally long diastolic period, the heart becomes enlarged, and when the next compression (*systole*) occurs, *Starling's law* (i.e., that compression is stronger when the heart is more distended initially, cf. Chapter 11) takes control, and compression is noticeably more vigorous, giving the subject a solid thump in the chest.

AV nodal conduction abnormalities are sorted into three classes. They are all readily visible from ECG recordings by looking at the time interval between the P wave and the QRS complex, i.e., the *P–R interval*. Type I AV nodal block shows itself as an increase in the P–R interval as the SA pacing rate increases. Type III AV nodal block corresponds to no AV nodal conduction whatever and total absence of a QRS complex.

Type II AV nodal block is phenomenologically the most interesting. In the simplest type, there is one QRS complex for every two P waves, a 2:1 pattern. A more complicated pattern is as follows: on the ECG (Fig. 12.34), P waves remain periodic, although the P–R interval is observed to increase gradually until one QRS complex is skipped. Following the skipped beat, the next P–R interval is quite short, but then the P–R lengthening begins again, leading to another skipped beat, and so on. A pattern with n P waves to $n-1$ QRS complexes is called an n-to-$(n-1)$ *Wenckebach pattern*, after the German cardiologist Wenckebach (1904).

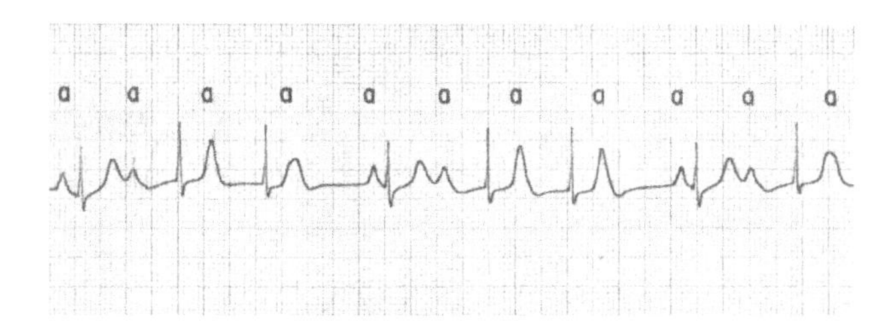

Figure 12.34 ECG recording of a Wenckebach pattern in which every fourth or fifth atrial beat is not conducted. (Rushmer, 1976, Fig. 8-24, p. 313.)

A simple mathematical description of AV nodal signal processing can be given as follows: we view the AV node as a collection of cells that fire when they are excited, which happens if their potential reaches a threshold, $\theta(t)$. Immediately after firing, the cells become refractory but then gradually recover. Effectively, at firing, the threshold increases dramatically but then decreases exponentially back to its steady-state value as recovery proceeds. This model ignores the fact that the AV node is self-oscillatory and fires without stimulus with a low frequency of 30–40 per minute. The self-oscillatory nature of the AV node becomes evident only in cases of SA nodal failure or at very low SA nodal firing rates. Thus, the model discussed here is valid at high stimulus rates (appropriate for AV nodal block) but not at low stimulus rates.

Input to the AV node comes from the action potential propagating through the atria from the SA node. The AV node experiences a periodic, time-varying potential, say $\phi(t)$. Firing occurs if the input signal reaches the threshold. Therefore, at the nth firing time, denoted by t_n,

$$\phi(t_n) = \theta(t_n). \tag{12.145}$$

Subsequent to firing, the threshold evolves according to

$$\theta(t) = \theta_0 + [\theta(t_n^+) - \theta_0]e^{-\gamma(t-t_n)}, \qquad t > t_n. \tag{12.146}$$

Note that $\theta \to \theta_0$ as $t \to \infty$, and thus θ_0 denotes the base value of the threshold. Further, $\theta = \theta(t_n^+)$ at $t = t_n$, and thus $\theta(t_n^+) - \theta(t_n^-)$ denotes the jump in the threshold caused by the firing of an action potential. To complete the model we must specify $\theta(t_n^+)$. The important feature of $\theta(t_n^+)$ is that it must have some memory, that is, depend in some way on $\theta(t_n^-)$. Therefore, we take

$$\theta(t_n^+) = \theta(t_n^-) + \Delta\theta. \tag{12.147}$$

The simple choice used here is to take $\Delta\theta$ a constant. However, consideration of the threshold in FitzHugh–Nagumo models suggests that (in a more general model) $\Delta\theta$ could also be some decreasing function of $\theta(t_n^-)$, i.e., $\Delta\theta = \Delta\theta(\theta(t_n^-)) = \Delta\theta(\phi(t_n))$, since $\phi(t_n) = \theta(t_n^-)$.

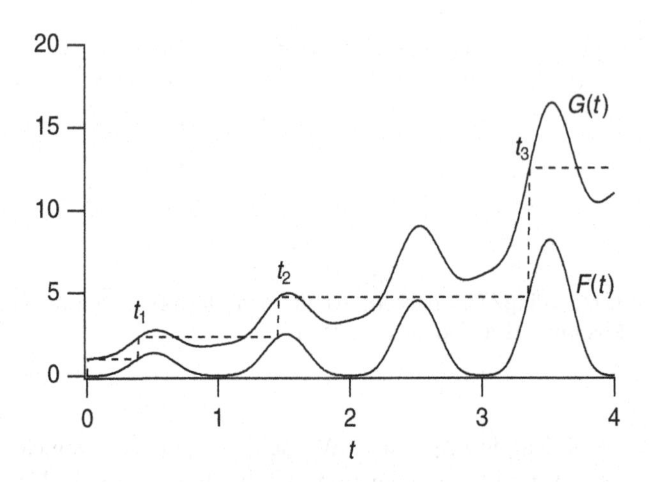

Figure 12.35 Plot of the functions $F(t)$ and $G(t)$ with $\Delta\theta = 1.0$, $\gamma = 0.6$.

Now we can find the next firing time as the smallest solution of the transcendental equation

$$\phi(t_{n+1}) = \theta_0 + [\theta(t_n^+) - \theta_0]e^{-\gamma(t_{n+1}-t_n)}. \tag{12.148}$$

Equation (12.148) can be rearranged into an equation of the form

$$F(t_{n+1}) = F(t_n) + \Delta\theta e^{\gamma t_n} = G(t_n), \tag{12.149}$$

where

$$F(t) = (\phi(t) - \theta_0)e^{\gamma t}. \tag{12.150}$$

Plots of typical functions $F(t)$ and $G(t)$ are shown in Fig. 12.35. Here we have taken $\phi(t) - \theta_0 = \sin^4(\pi t)$. The dashed lines in this figure follow a few iterates of the map.

The key observation is that the map $t_n \mapsto t_{n+1}$ as defined by (12.149) is the lift of a circle map. Before proceeding with this example, we give a brief introduction to the theory of circle maps.

The first application of circle maps to the behavior of neurons was given by Knight (1972). More detailed discussions of maps and chaos and the like with application to a wide array of biological problems can be found in Glass and Mackey (1988), Glass and Kaplan (1995), and Strogatz (1994).

Circle Maps

A circle map is a map of the circle to itself, $f : S^1 \to S^1$, but it is often easier to describe a circle map in terms of its *lift* $F : R \to R$, where F is a monotone increasing function and $F(x + 1) = F(x) + 1$. The two functions f and F are related by

$$f(x) \equiv F(x \bmod 1) \bmod 1. \tag{12.151}$$

(For convenience we normalize the circumference of the circle to be of length 1, rather than 2π.)

The primary challenge from a circle map is to determine when the behavior is periodic and to understand the possible nonperiodic behaviors. The simplest periodic behavior is a period 1 solution, say a point x_0 for which $F(x_0) = x_0 + 1$. This orbit is also said to have *rotation number* one because it rotates around the circle once on each iterate. This is also described as 1:1 phase locking between input and output. A more complicated periodic orbit would be a point x_0 and its iterates x_j with the property that $x_n = x_0 + m$. In other words, the iterates rotate around the circle m times in n iterates. The rotation number is m/n, and there is $m : n$ phase locking, with m output cycles for every n input cycles.

The key fact to understand is that the asymptotic behavior of a circle map is characterized by its *rotation number*, ρ, defined by

$$\rho = \lim_{n\to\infty} \frac{F^n(x)}{n}. \tag{12.152}$$

$F^n(x)$ is the nth iterate of the point x,

$$F^n(x) = F(F^{n-1}(x)), \tag{12.153}$$

where $F^0(x) = x$ and $F^1(x) = F(x)$.

If F is a continuous function, the rotation number has the following properties:

1. ρ exists and is independent of x.
2. ρ is rational if and only if there are periodic points.
3. If ρ is irrational, then the map F is equivalent to a rigid rotation by the amount ρ.
4. If there is a continuous family of maps F_λ, then $\rho(\lambda)$ is a continuous function of λ. Furthermore, if F_λ is a monotone increasing function of λ, then ρ is a nondecreasing function of λ.
5. Generically, if $\rho(\lambda)$ is rational at some value of λ_0, it is constant on an open interval containing λ_0.

Here is what this means in practical terms. Since ρ exists, independent of x, all orbits have the same asymptotic behavior, orbiting the circle at the same rate, independent of initial position. If ρ is rational, the asymptotic behavior is periodic, whereas if ρ is irrational, the motion is equivalent to a rigid rotation. In this case, the behavior is aperiodic, but not complicated, or "chaotic." There are no other types of behavior for a continuous circle map.

The last two features of ρ make the behavior of the orbits so unusual, being a function that is continuous, monotone nondecreasing (if F_λ is an increasing function of λ), yet locally constant at all the rational levels. Such a function is called the *Devil's staircase*. Notice that if ρ is rational on an open interval of parameter space, then phase locking is robust.

The reason for this robustness is that a periodic point with $\rho = p/q$ corresponds to a root of the equation $F^q(x) - x = p$, and roots of equations are generally, but not always, robust, or transversal (i.e., the derivative of $F^q(x) - x$ at a root is nonzero). If a root is transversal, then arbitrarily small perturbations to the equation do not destroy the root, and it persists for a range of parameter values. However, the existence of a periodic point is no guarantee that it is robust. For example, the simple shift $F(x) = x + \lambda$ has periodic points whenever λ is rational, but these periodic points are never isolated or robust.

A detailed exposition on continuous circle maps and proofs of the above statements can be found in Coddington and Levinson (1984, chapter 17).

Now we attempt to apply this theory of circle maps to (12.149). Notice that this is indeed the lift of a circle map, since if t_n and t_{n+1} satisfy (12.149), then so do $t_n + T$ and $t_{n+1} + T$. To find a circle map, we let k_n be the largest integer less than t_n/T and define $\psi_n = (t_n - k_n T)/T$. In these variables the map (12.149) can be written as

$$f(\psi_{n+1}) = (f(\psi_n) + \Delta\theta e^{\gamma T \psi_n})e^{\gamma T \Delta k_n}, \tag{12.154}$$

where

$$f(\psi) = (\Phi(\psi) - \theta_0)e^{\gamma T \psi}, \qquad \Phi(\psi) = \phi(T\psi), \tag{12.155}$$

and $\Delta k_n = k_{n+1} - k_n$.

We can make a few observations about the map $\psi_n \mapsto \psi_{n+1}$. First, and most disconcerting, the map is not continuous. In fact, it is apparent that there are values of t on the unit interval that can never be firing times. For t to be permitted as a firing time it must be the first point at which $F(t)$ reaches the level $G(t_n)$, i.e., the first time that the threshold is reached. At such a point, $F'(t) \geq 0$. Since there are regions for which $F'(t) < 0$, which can therefore never be firing times, this is a map of the unit interval *into*, but not onto, itself. However, the map $t_n \mapsto t_{n+1}$ is order preserving, since $G(t)$ is increasing whenever $F(t)$ is increasing.

Since the entire unit interval is not covered by the map, it is only necessary to examine the map on its range. Examples of the map $\psi_n \mapsto \psi_{n+1}$ are shown in Figs. 12.36–12.39. Here we have plotted the map only on the attracting range of the unit interval. These show important and typical features, namely that the map consists of either one or two continuous, monotone increasing branches. The first branch, with values above the one-to-one curve, corresponds to firing in response to the subsequent input (with $k_{n+1} = k_n + 1$), and the second, with values below the one-to-one curve, corresponds to firing after skipping one beat (with $k_{n+1} = k_n + 2$). The skipped beat occurs because when the stimulating pulse arrives, it is subthreshold and so does not evoke a response.

The sequence of figures in Figs. 12.36–12.39 is arranged according to decreasing values of γT. Note that as γ decreases, the rate of recovery from inhibition decreases. For γT sufficiently large, there is a unique fixed point, corresponding to firing in 1:1 response to the input signal. This makes intuitive sense, for when γ is large, the recovery from inhibition is fast, and thus the AV node can be driven at the frequency of the SA node. For large γT the map is relatively insensitive to changes in parameters. As γT decreases, the first branch of the map increases and the value of the fixed point

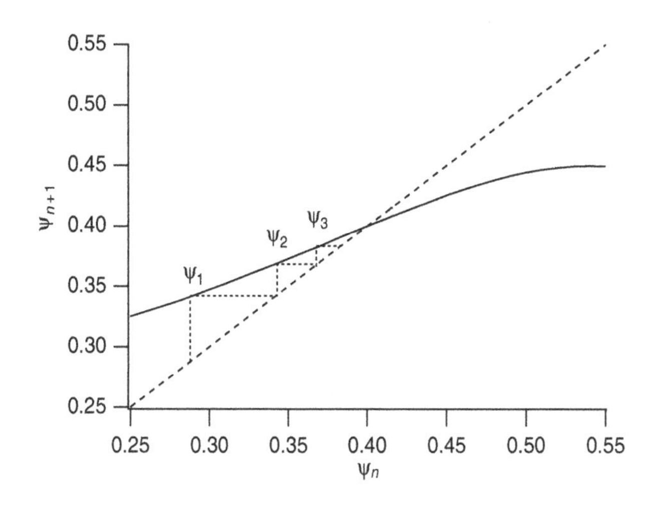

Figure 12.36 Plot of the map $\psi_n \mapsto \psi_{n+1}$ with $\Delta\theta = 1.0$, $\gamma T = 0.8$.

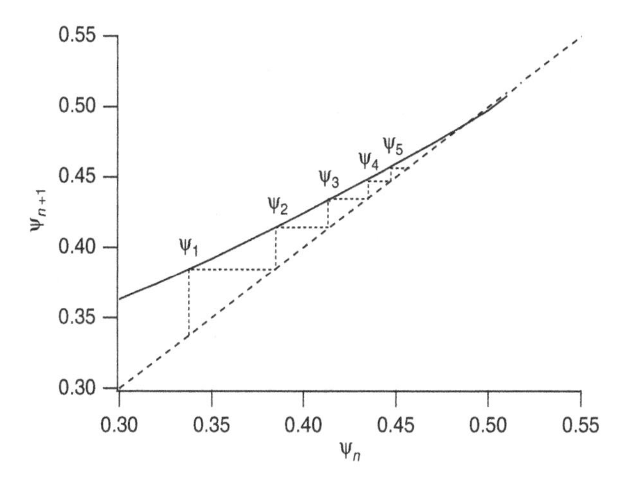

Figure 12.37 Plot of the map $\psi_n \mapsto \psi_{n+1}$ with $\Delta\theta = 1.0$, $\gamma T = 0.695$.

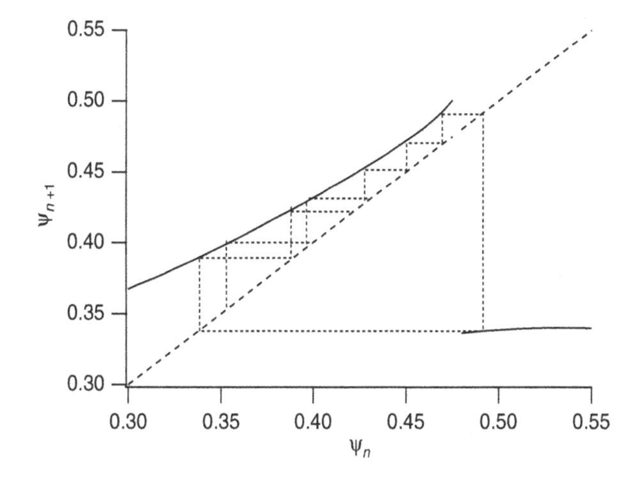

Figure 12.38 Plot of the map $\psi_n \mapsto \psi_{n+1}$ with $\Delta\theta = 1.0$, $\gamma T = 0.67$.

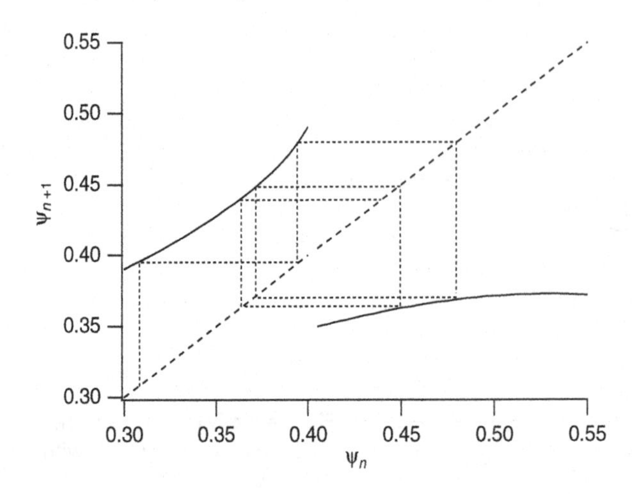

Figure 12.39 Plot of the map $\psi_n \mapsto \psi_{n+1}$ with $\Delta\theta = 1.0$, $\gamma T = 0.55$.

increases, corresponding to a somewhat delayed firing. Furthermore, because the slope of the map in the vicinity of the fixed point is close to 1, the fixed point is sensitive to changes in parameter values (depicted in Fig. 12.37), corresponding to type I AV block.

As the parameter γT is decreased further, the fixed point is lost and a second branch to the map appears (as in Fig. 12.38). Iterations show that subsequent firings become later and later in the input cycle until one beat is skipped, followed by a firing that is relatively early in the input cycle. For this region of parameter space, the map replicates the Wenckebach pattern.

Finally, as γT decreases further, the second branch "slides over" to the left and eventually intersects the one-to-one line, yielding a fixed point. This fixed point corresponds to a periodic pattern of one skipped beat for each successful firing, a two-to-one pattern, and replicates type II AV block.

The behavior of the map in the region with no fixed point can be described by the rotation number. For maps of the type (12.149) the rotation number can be defined, analogously to our earlier definition, by

$$\rho = \lim_{n\to\infty} \frac{t_n}{nT}. \tag{12.156}$$

The following features of the rotation number ρ can be verified (Keener, 1980a, 1981):

1. ρ exists and is independent of initial data.
2. ρ is a monotone decreasing function of γT.
3. ρ attains *every* rational level between 0 and 1 on an open interval of parameter space.

For continuous circle maps, it is not certain that every rational level is attained on an open interval of parameter space.

The main consequence of this result is that between 1:1 phase locking and 2:1 AV block, for every rational number there is an open interval of γT on which the rotation with that rational number is attained.

12.4.3 Reentrant Arrhythmias

A reentrant arrhythmia is a self-sustained pattern of action potential propagation that circulates around a closed path, reentering and reexiting tissue as it goes.

The simplest reentrant pattern is one for which the path of travel is one dimensional. These were first studied by Mines (1914) when he intentionally cut a ring of tissue from around the superior vena cava and managed to initiate waves that traveled in only one direction.

A classic example of a one-dimensional reentrant rhythm of clinical relevance is one in which an action potential circulates continuously between the atria and the ventricles through a loop, exiting the atria through the AV node and reentering the atria through an *accessory pathway* (or vice versa). Since conduction through the AV node is quite slow compared to other propagation, an accessory pathway that circumvents the AV node usually reveals itself on the ECG by an early, broad deflection of the QRS complex (Fig. 12.40). This deflection is broadened because it depicts propagation through myocardial tissue, which is slow compared to normal propagation through the Purkinje network. This is known clinically as Wolff–Parkinson–White (WPW) syndrome, and is life-threatening if not detected and treated. However, because the associated reentrant rhythm travels essentially along a one-dimensional pathway (at least at two points), WPW syndrome is usually curable, as cardiac surgeons can use localized radio frequency waves to burn and permanently obliterate the accessory pathway, restoring a

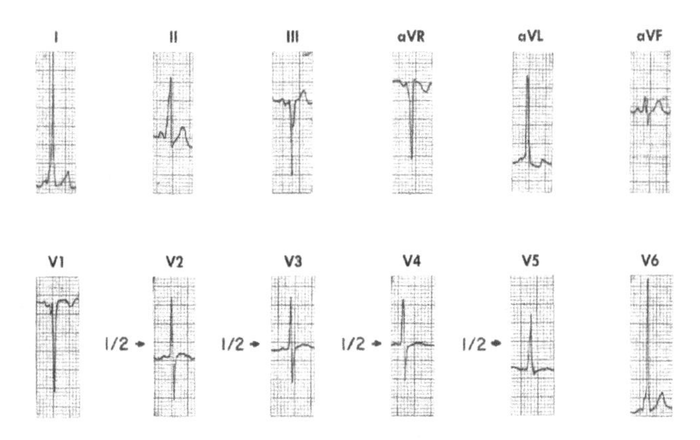

Figure 12.40 Twelve-lead ECG recording of Wolff–Parkinson–White syndrome, identifiable by the shortened P–Q interval (because the AV delay is circumvented) and the slowed QRS upstroke, particularly noticeable in leads II, aVR, and V6. (Rushmer, 1976, Fig. 8-47, p. 339; originally from Guneroth, 1965).

normal single pathway conduction and a normal ECG, and preventing the formation of a closed loop.

Reentrant patterns which are not constrained to a one-dimensional pathway are much more problematic. The two primary reentrant arrhythmias of this type are *tachycardia* and *fibrillation*. Both of these can occur on the atria (*atrial tachycardia* and *atrial fibrillation*) or in the ventricles (*ventricular tachycardia* and *ventricular fibrillation*). When they occur on the atria, they are not immediately life-threatening because the disruption of blood flow is not catastrophic (although over a long term they can cause strokes). However, when they occur on the ventricles, they are life-threatening. Ventricular fibrillation is fatal if it is not terminated quickly. Symptoms of ventricular tachycardia include dizziness or fainting, and sometimes rapid "palpitations."

Tachycardia is often classified as being either *monomorphic* or *polymorphic*, depending on the assumed morphology of the activation pattern. Monomorphic tachycardia is identified as having a simple periodic ECG, while polymorphic tachycardia is usually quasiperiodic, apparently the superposition of more than one periodic oscillation. A typical example of a polymorphic tachycardia is called *torsades de pointes*, and appears on the ECG as a rapid oscillation with slowly varying amplitude (Fig. 12.41). A vectorgram interpretation suggests a periodically rotating mean heart vector.

It is currently believed that the spatiotemporal wave pattern associated with *atrial flutter* is a spiral wave. Because the ventricles are three dimensional, a pattern that appears on the ventricular surface to be spiral-like must, in fact, correspond to a three-dimensional reentrant wave, a scroll wave. A three-dimensional view of a (numerically computed) scroll wave, corresponding to monomorphic V-tach is shown in Fig. 12.42.

Stable monomorphic ventricular tachycardia is rare, as most reentrant tachycardias degenerate into fibrillation. The likely reason for this is that there are a number of potential instabilities, although the mechanism of these instabilities has not been

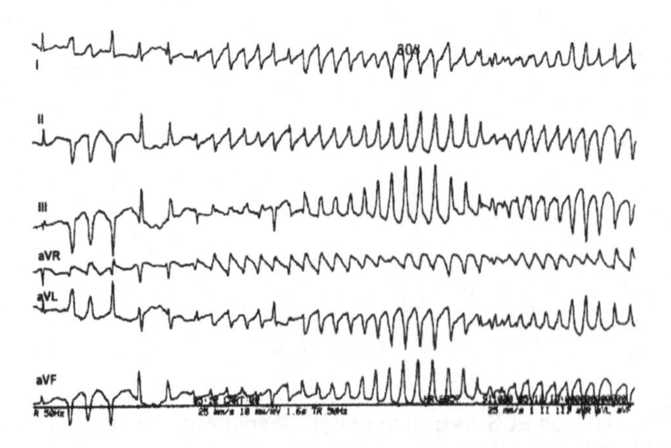

Figure 12.41 A six-lead ECG recording of *torsades de pointes*. (Zipes and Jalife, 1995, Fig. 79-1, p. 886.)

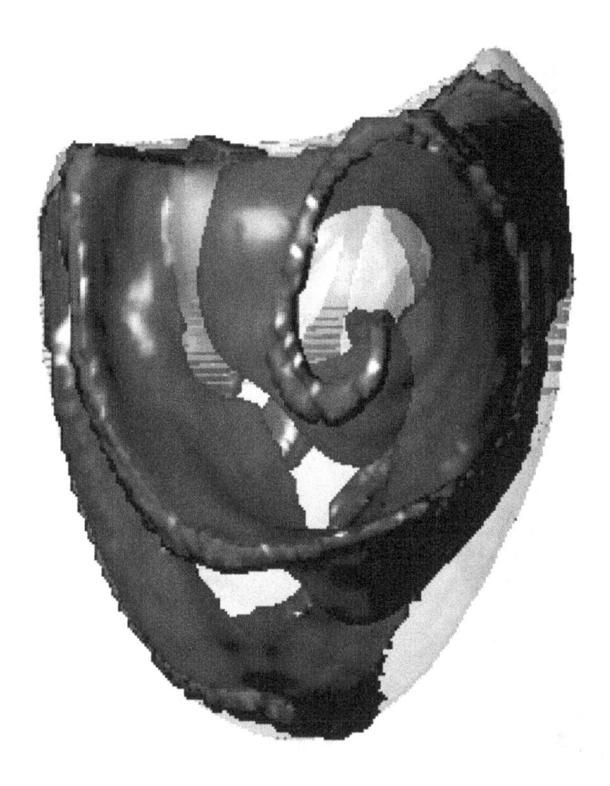

Figure 12.42 Numerically computed scroll wave in ventricular muscle. (Panfilov and Keener, 1995, p. 685, Fig. 3a.)

decisively determined. Some possibilities (including the APD alternans instability) are discussed by a number of authors (Courtemanche and Winfree, 1991; Karma, 1993, 1994; Panfilov and Holden, 1990; Panfilov and Hogeweg, 1995; Bar and Eiswirth, 1993; Courtemanche et al., 1993). It is currently believed that stable monomorphic V-tach can be maintained only by special physical structures in the cardiac tissue, such as a small conducting pathway through otherwise damaged tissue.

Fibrillation is believed to correspond to the presence of many reentrant patterns moving throughout the ventricles in continuous, perhaps erratic, fashion, leading to an uncoordinated pattern of ventricular contraction and relaxation. A surface view of a (numerically computed) fibrillatory pattern is shown in Fig. 12.43.

There is an extensive literature devoted to the mathematical study of reentrant patterns. The first such theory was due to Wiener and Rosenblueth (1946) who studied a wave circulating around an inexcitable obstacle using a simple automaton model of excitable media. More recent studies have focussed on the behavior of waves in two and three-dimensional media using systems of reaction–diffusion equations. An interesting consequence of these studies is that it has been shown that a reentrant pattern need not have a physical obstacle around which to circulate, but that there can be a spiral core that is maintained by the dynamics of the surrounding pattern, or with scroll waves, a central filament around which to circulate. This was an important observation because before this, it was generally assumed that reentrant patterns could only rotate around

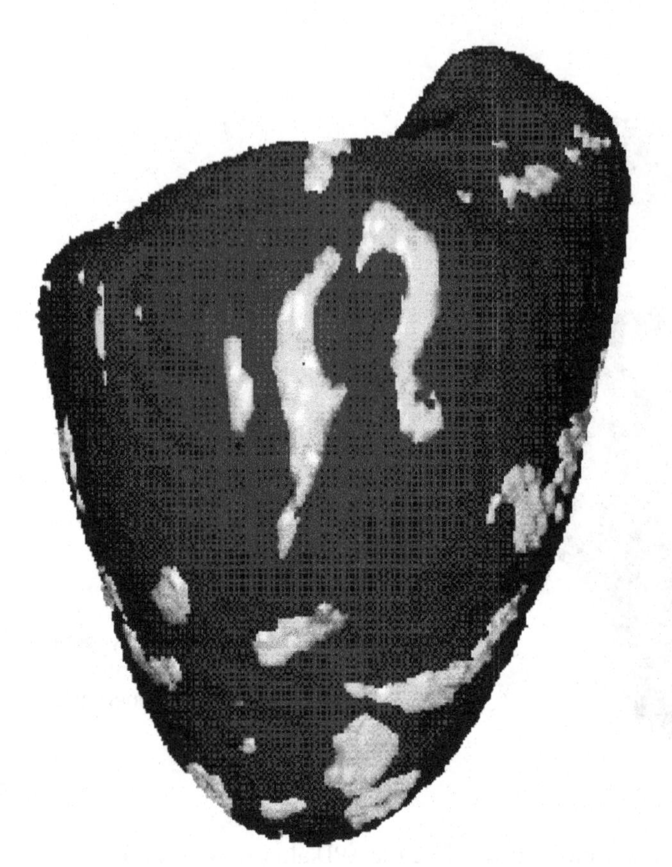

Figure 12.43 Surface view of fibrillatory reentrant activity in the ventricles (computed by A. Panfilov).

some physical or anatomical obstacle. Now there is ample experimental confirmation that anatomical obstacles are not necessary to maintain reentrant patterns.

Perhaps the most important mathematical observation about reentrant rhythms was made by Art Winfree (1987) who recognized that reentrant patterns could be characterized by their phase singularities. For any oscillatory dynamical system, one can define a phase as follows. Identify two oscillatory state variables that are not synchronous, i.e., whose extremal values occur at different times, and a phase plot of one against the other will be a closed loop, topologically equivalent to a circle. In fact, one can map the closed loop continuously to a circle, and define the phase of a point on the loop with the angle of the circle to which the point is mapped. Now this map can be continuously extended to the entire plane with the exception of a single point, the phaseless point, or phase singularity. This phaseless point is not unique because it depends on how one chooses the phase map, however, it is certain to exist. One natural way to define the phase ϕ is to take

$$\phi = \arctan\left(\frac{y - y^*}{x - x^*}\right), \tag{12.157}$$

where (x,y) is a point in the phase space, and (x^*,y^*) is a point centrally located within the closed loop.

For a spatially extended system, one can use the phase map to identify a phase for each point in space. However, if the oscillatory pattern is reentrant, that is, if it is maintained by its spatial connectivity, and is not the result of autonomous cellular oscillation, there must be phaseless points. The reason for this is as follows. A feature of an oscillatory pattern is that the variables go through one complete rotation of the phase per oscillation. For a reentrant pattern, all phases are attained somewhere in space at all times, and there is a continuous closed path in space along which the variables cycle through all the phases. However, just as there is a phaseless point at the center of all clocks, so there must be a phaseless point somewhere inside this continuous closed path that cycles through all phases. These phaseless points correspond to the center of spirals in two dimensions and the filaments of scrolls in three dimensions. Of course, there is no requirement that phaseless points be fixed in space and time, but there are precise rules regulating their creation and destruction. Phaseless points in space can be located by noticing that the closed line integral of the gradient of phase is a nonzero integer multiple of 2π if the path P encloses a phase singularity,

$$\oint_P \nabla\phi \cdot d\vec{r} = \begin{cases} 2\pi n, & n = \pm 1, \pm 2, \ldots, \text{if } P \text{ encloses a phase singularity,} \\ 0, & \text{otherwise} \end{cases} \tag{12.158}$$

Phaseless points are now well-established experimentally. There are currently two ways that experimentalists observe phaseless points in cardiac tissue. The first uses the potential $V(x,t)$ and its delay $V(x,t-\tau)$ for some fixed delay τ (Iyer and Gray, 2001). More recently, with the advent of the ability to simultaneously measure potential and intracellular Ca^{2+} concentration with fluorescent dyes, experimentalists are able to use these two variables to create a phase map.

How to Initiate a Reentrant Arrhythmia

While there is a lot of experimental and theoretical work to understand the dynamics of reentrant tachycardias and fibrillation, from a clinical point of view this is somewhat less important. It is of much greater clinical importance to understand how reentrant arrhythmias are formed, how to prevent their formation, or how to terminate them after they have formed.

The problem of how to initiate a reentrant pattern is easily understood by considering two analogies. "The Wave" (often called a "Mexican Wave" due to it first gaining a large international audience at the 1986 World Cup in Mexico City) is a peculiar behavior of football crowds around the world, in which individuals suddenly rise from their seats, wave their arms in the air and then sit down (Farkas et al., 2002). Viewed from an individual perspective there is nothing striking about this behavior (odd, yes, but striking, no). However, the collective behavior is coordinated in such a way that there appears to be a wave propagating around the stadium several times before the participants get weary or the novelty wears off.

If this behavior is viewed as analogous to an excitable system, one sees the resting phase (seated and of sound mind), or excited (standing and waving frantically). If one further assumes that no individual fan goes into the excited phase without stimulus from an excited neighbor, then one has a model of an excitable medium. Of course, in the USA the cheerleaders play the role of an autonomous oscillator that can become excited without provocation (and in some cases, with little cause).

Now, imagine what happens when the cheerleaders wish to initiate a wave. They wave their arms and cheer and in response everyone nearby does the same, initiating a wave that spreads radially outward from its source. But this does not initiate a self-maintained reentrant wave. Some additional ingredient is necessary, namely something to break the symmetry of spread, so that propagation is in one direction but not the other.

To consider a second analogy, suppose there was a large lake surrounded by very fast growing grass, and one wanted to start a grass fire that circulated around the lake continuously. Normally, if one starts a grass fire, the fire will spread, going around the lake in both directions and burning out when the flames meet on the opposite side of the lake. However, to initiate a fire that circulates continuously, two additional conditions must be met. First, there must be a breaking of the symmetry so that the fire initially spreads in only one direction, and second, the lake must be large enough so that by the time the fire has gone around the lake, the grass that was first burned has regrown sufficiently so that it is ready to burn again.

These analogies are actually quite good at describing the problem of initiation of a reentrant pattern in cardiac tissue: there must be some kind of initial stimulus, it must occur at a time and place so that propagation is not symmetric, and there must be some geometrical property that allows the wave to return to reinitiate itself. In a region with no anatomical obstacle around which to circulate, the geometrical problem (actually a topological problem) is how to create a phase singularity.

A simple mathematical model to show how this scenario might work is as follows. Suppose there is a one-dimensional closed loop conducting pathway of length L, that is stimulated by an external pacemaker (Fig. 12.44). Suppose also that somewhere in the vicinity of the stimulus site, there is a region of one way block. We know that such regions can exist, for example, at points of fiber arborization (Section 12.3.2).

Suppose that the external pacemaker fires with period T. We define the instantaneous frequency of stimulus as $\Delta T_{n+1} = t_{n+1} - t_n$, where t_n is the nth firing time of the stimulus site. Now we take a simple kinematic description of propagation along the one-way path and suppose that the speed of propagation on the path is a function of the instantaneous period, $c = c(\Delta T)$. (Typically, c is an increasing function of ΔT, since the longer the recovery time, the faster the speed of propagation.) Then the travel time around the one-way loop is $\frac{L}{c(\Delta T)}$. In cardiac tissue, the speed of an action potential is on the order of 0.5 m/s, so that travel time around the loop (perhaps a few centimeters in length) is much shorter than the period of external stimulus. Thus, the wave going around the loop typically returns to the stimulus site long before the next external stimulus arrives (i.e., we assume that $L/c < T$). If the travel time is smaller than the

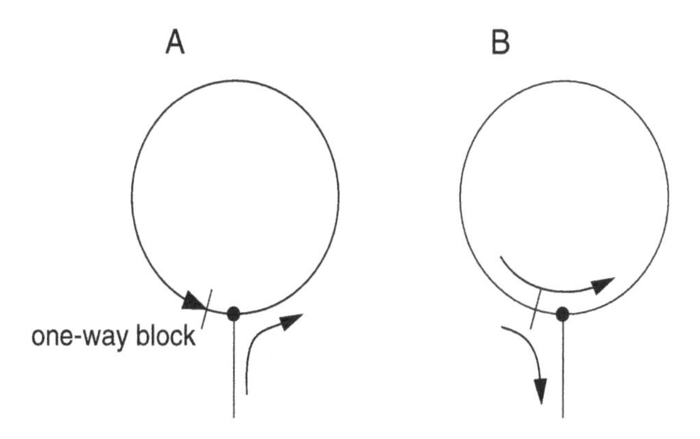

Figure 12.44 Diagram of a conducting path with one-way block, preventing conduction from right to left. A: Conduction of a stimulus around the loop until it encounters refractoriness and fails to propagate further. B: Conduction of a reentrant pattern circulating continuously around the loop and exiting via the entry pathway on every circuit.

absolute refractory period T_r of the cells, the stimulus has no effect, and the cells must await the next external stimulus before they fire, so that

$$\Delta T_{n+1} = t_{n+1} - t_n = T \tag{12.159}$$

if $\frac{L}{c(\Delta T_n)} < T_r$. On the other hand, if the travel time is larger than the T_r but smaller than T, then it stimulates the cells at the stimulus site and initiates another wave around the loop. Thus, the instantaneous period at the stimulus site is

$$\Delta T_{n+1} = t_{n+1} - t_n = \frac{L}{c(\Delta T_n)}, \tag{12.160}$$

provided that $T > \frac{L}{c(\Delta T_n)} > T_r$.

With this information, we can construct the one-dimensional map $\Delta T_n \mapsto \Delta T_{n+1}$ (shown in Fig. 12.45). There are obviously two branches for this map (shown as solid curves). Of interest are the fixed points of this map, corresponding to a periodic pattern of stimulus. The fixed point on the upper branch corresponds to the normal stimulus pattern from the external source, whereas the fixed point on the lower branch corresponds to a high-frequency reentrant pattern. The key feature of this map is that there is hysteresis between the two fixed points. In a "normal" situation (Fig. 12.45A), with L small and T large, the period is fixed at the external stimulus period T. However, as L increases or as T decreases, rendering $L > T_r c(\delta T_n)$, there is a "snap" onto the smaller-period fixed point, corresponding to initiation of a reentrant pattern (Fig. 12.45B). The pernicious nature of the reentrant pattern is demonstrated by the fact that increasing the period of the external stimulus back to previous levels does not restore the low-frequency pattern—the iterates of the map stay fixed at the lower fixed point, even though there are two possible fixed points. This is because the circulating pattern

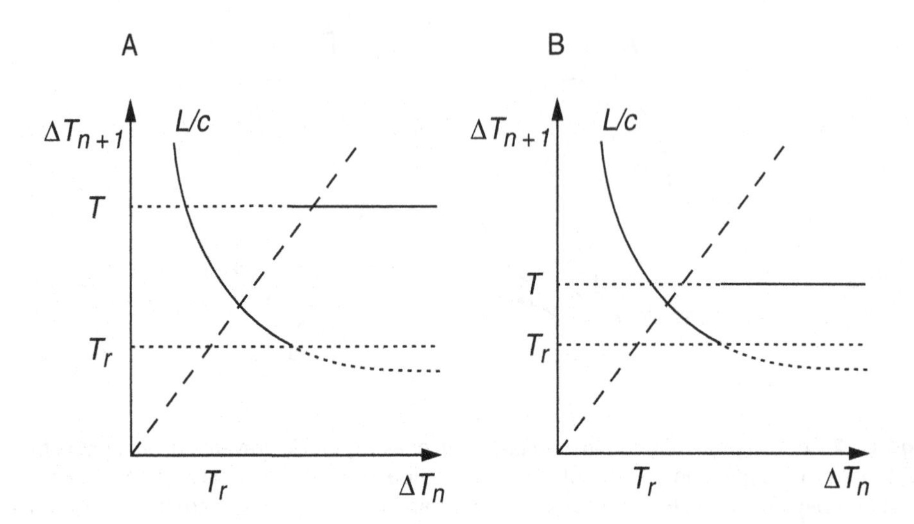

Figure 12.45 Next-interval map for a one-way conducting loop in two cases. A: With T large, so that two stable steady solutions exist. B: With T small, so that the only steady solution corresponds to reentry.

acts as a retrograde source of high-frequency stimulus on the original stimulus site, thereby masking its periodic activity.

Note that there are a number of ways that this reentrant pattern might be initiated. First, there might be a growing diseased or infarcted central region which leads to a gradual increase in L, initiating the reentrant pattern while keeping T fixed. On the other hand, a diseased central region may exist but remain static (L fixed), and the reentrant pattern is initiated following a decrease in T, for example, during strenuous exercise. Thus, a static one-way loop acts like a "period bomb" (rather than a time bomb), ready to go off whenever the period is sufficiently low.

While this simple model of reentry illustrates the basic problem, it probably has little to do with most real reentrant patterns. For example, since reentrant patterns do not require central anatomical obstacles, there must be some way that these patterns can arise in two and three-dimensional tissue without a preexisting central anatomical obstacle, or fixed region of one-way conduction block.

One of the earliest mathematical models of initiation of reentrant patterns and fibrillation to address this issue was due to Moe et al. (1964). Theirs was a finite state automaton model in which individual cells had three possible states, resting, excited and refractory, rules for how cells moved through these states in discrete time steps, and rules for how the excited state was propagated from cell to cell. The important feature of their model was that return to rest from the refractory state was spatially and randomly variable. As the spatial variability of the return from refractoriness increased, it was observed that there was an increased propensity for propagated action potentials to break up into multiple reentrant patterns reminiscent of fibrillation.

Finite state automaton models of excitable media have a rich history. They are easy to describe and understand and they are easy to simulate. However, it is important to recognize that these models are analogies to cardiac tissue, and do not have a physiological basis; it is possible (in fact, likely) that some phenomena observed in finite state automata do not occur in realistic ionic based models, and vice versa. In fact, the Moe et al. model is beset with significant difficulties and is probably not an explanation with clinical relevance. One of the difficulties of the model is that it does not employ electrotonic coupling of cells, and it is known that the crucial feature of the model, namely dispersion of refractoriness, is greatly affected by electrotonic coupling. As we saw with SA nodal cells, the effect of electrotonic coupling is to synchronize behavior, and this tends to mask cell-to-cell differences of recovery properties as well.

Nonetheless, the dispersion-of-refractoriness hypothesis, substantially modified, remains one of the best known folklore explanations for fibrillation onset in the cardiology community. Indeed, some form of dispersion of refractoriness, but not necessarily structural, is required for initiation of reentry (Weiss et al., 2006).

One of the challenges in this business is that it is difficult to design experiments in which reentrant arrhythmias are initiated in a way that is clinically relevant. It is known, however, that reentrant arrhythmias can be initiated intentionally by the correct application of point stimuli. This procedure has been described beautifully by Winfree (1987), with many gorgeous color plates, so here we content ourselves with a shorter, less colorful, verbal description of the process.

When a current is injected at some point to resting cardiac tissue, cells in the vicinity of the stimulus are depolarized (recall Fig. 12.23). If the stimulus is of sufficient amplitude and duration, the cells closest to the stimulating electrode may receive a superthreshold stimulus and become excited. Cells further away from the stimulus site receive a subthreshold stimulus, so they return to rest when the stimulus ends. At the border between subthreshold and superthreshold stimulus, a wave front is formed.

Once a transition front is formed, the local conditions of the tissue determine whether the wave moves forward or backward. That is, if the undisturbed medium is sufficiently excitable, and the initially excited domain is sufficiently large, the wave front moves outward into the unexcited region. If, however, the unaffected medium is not excitable, but partially refractory, or the excited domain is too small, the wave front recedes and collapses.

If the stimulated medium is initially uniform, these two are the only possible responses to a stimulus. However, if the state of the medium in the vicinity of the stimulating electrode is not uniform, then there is a third possible response. Suppose, for example, that there is a gradual gradient of recovery so that a portion of the stimulated region is excitable, capable of supporting wave fronts (with positive wave speed) and the remaining portion of the stimulated region cannot support wave fronts, but only wave backs (i.e., fronts with negative speed). Then, the result of the stimulus is to produce both wave fronts and wave backs.

With a mixture of wave fronts and wave backs, a portion of the wave surface expands, and a portion retracts. Allowed to continue in this way, a circular (two-dimensional) domain evolves into a linked pair of spirals, and a spherical (three-dimensional) domain evolves into a scroll. If the domain is sufficiently large, these become self-sustained reentrant patterns.

In resting tissue with no pacemaker activity, two stimuli are required to initiate a reentrant pattern. The first is required to set up a spatial gradient of recovery (dispersion of refractoriness). Then, if the timing and location of the second is within the appropriate range, a single action potential that propagates in the backward, but not forward, direction can be initiated. This window of time and space is called the *vulnerable window* or *vulnerable period*. If the tissue mass is large enough or if there is a sufficiently long closed one-dimensional path, the retrograde propagation initiates a self-sustained reentrant pattern of activation.

The mechanism of this method of initiation of reentry is well-documented experimentally (Chen et al., 1988; Frazier et al., 1988; Frazier et al., 1989).

While the Winfree scenario reliably initiates reentry, it too lacks clinical relevance because of its reliance on external stimuli. In fact, the cause of the trigger event remains a total enigma. Currently, the most popular hypothesis is that the trigger event is ectopic activity known as Early After Depolarizations (EADs) or Delayed After Depolarizations (DADs). Space does not permit us to give a full description of these events. Suffice it to say that EADs and DADs are thought to be electrical events that are triggered by spontaneous Ca^{2+} release.

While EADs and DADs are universally viewed as proarrhythmic, it has proved difficult to demonstrate a true connection between these events and initiation of reentry. For example, a common level of ventricular ectopy in patients is 2 ectopic beats per minute, or $\approx$1 million ectopic beats per year. Yet sudden cardiac death episodes in these patients occur over months to years, not minutes. In fact, medication to reduce the rate of ectopic activity has actually proven to increase mortality.

In summary, it is known what must happen in order to initiate a reentrant arrhythmia. However, why or how this happens is, for all practical purposes, unknown.

Antiarrhythmic Drugs

We close this section on arrhythmias with a brief description of antiarrhythmic drugs. Antiarrhythmic drugs are agents that have some effect on the cardiac action potential at the cellular level. The most widely used classification scheme for antiarrhythmic drugs, known as the Vaughan Williams scheme, classifies a drug based on the primary cellular mechanism of its antiarrhythmic effect. There are five main classes in the Vaughan Williams classification of antiarrhythmic agents:

The class I antiarrhythmic agents interfere with Na^+ channels. Class I agents are further grouped by the specific effect they have on Na^+ channels, and the effect they have on cardiac action potentials.

Class Ia agents block the fast Na^+ channel. Blocking this channel depresses the phase 0 depolarization (thereby reducing the maximal upstroke velocity, $V_{\max}$), which slows action potential conduction velocity. Agents in this class also cause decreased conductivity and increased refractoriness. Class Ia agents include quinidine, procainamide, and disopyramide.

Class Ib antiarrhythmic agents are Na^+ channel blockers with fast onset and offset kinetics, so that they have little or no effect at slower heart rates, and more effects at faster heart rates. These agents decrease V_{max} in partially depolarized cells with fast response action potentials, decreasing automaticity. They either do not change the action potential duration, or they may decrease the action potential duration. Class Ib agents include lidocaine, mexiletine, tocainide, and phenytoin.

Class Ic antiarrhythmic agents markedly depress the phase 0 depolarization (decreasing V_{max}). They decrease conductivity, but have a minimal effect on the action potential duration. Of the Na^+ channel blocking antiarrhythmic agents (the class I antiarrhythmic agents), the class Ic agents have the most potent Na^+ channel blocking effects. Class Ic agents include encainide, flecainide, moricizine, and propafenone.

Class II agents are beta blockers. They act by selectively blocking the effects of catecholamines at the β_1-adrenergic receptors, thereby decreasing sympathetic activity on the heart. Class II agents include esmolol, propranolol, and metoprolol.

Class III agents predominantly block K^+ channels, thereby prolonging repolarization. Since these agents do not affect the Na^+ channel, conduction velocity is not decreased. The prolongation of the action potential duration and refractory period, combined with the maintenance of normal conduction velocity, is proposed to prevent re-entrant arrhythmias. Class III agents include amiodarone, azimilide, bretylium, clofilium, dofetilide, and sotalol.

Class IV agents are slow Ca^{2+} channel blockers. They decrease conduction through the AV node. Class IV agents include verapamil and diltiazem.

Class V agents are those with other mechanisms, possibly unknown, and include digoxin and adenosine.

The history of antiarrhythmic drugs is interesting and yet tragic. One of the most interesting episodes in this saga was the CAST study (Cardiac Arrhythmia Suppression Test) which was supposed to be an 18 month study of certain type Ic antiarrhythmic drugs. Unfortunately, the test was suspended after only three months when it was realized that the patients receiving the drugs were dying from sudden cardiac death at a rate significantly higher than those receiving placebo. In other words, the antiarrhythmic drugs were actually pro-arrhythmic. A similar result was found with d-sotalol, a class III antiarrhythmic drug (Waldo et al., 1996). It is now widely recognized that all but class II antiarrhythmic drugs have pro-arrhythmic potential.

This illustrates the important difference between understanding dynamics at the cellular level and understanding the mechanisms for onset and maintenance of spatiotemporal patterns. For example, type I antiarrhythmic drugs all block Na^+ channels thereby slowing depolarization. It might seem that this would make a tachy-arrhythmia less likely because the cells are less excitable. However, it is also the case that Na^+ channel block increases the chances of propagation failure at regions of resistive inhomogeneity, perhaps increasing the possibility of formation of reentrant patterns.

Another proposal is that Na^+ channel blockers increase the size of the vulnerable window (Starmer et al., 1991, 1992).

Similarly, type III antiarrhythmic drugs block K^+ channels, lengthening the action potential. One might think that this would also increase the required size of the tissue necessary to sustain a reentrant pattern. A lengthened action potential is identified clinically as Long QT (LQT) syndrome. Seven different genetic causes of LQT syndrome have been identified, while if it is induced by nongenetic factors, it is called acquired LQT syndrome. People with LQT syndrome, whether genetic or acquired, are now known to be at risk of sudden cardiac death. And while the mechanism by which LQT increases the risk of sudden cardiac death is completely unknown, it is clear that the simple cellular-based reasoning mentioned above has no explanatory power concerning how reentrant arrhythmias are affected by these drugs.

12.5 Defibrillation

Nearly everyone who has ever watched television knows something about defibrillation. They have probably seen medical dramas where the paramedic places paddles on the chest of a man who has unexpectedly collapsed, yells "Clear!" and then a jolt of electricity shakes the body of the victim. Mysteriously, the victim revives. Since they were first made in 1947, defibrillators have saved many lives, and the recent development of implantable defibrillators will no doubt extend the lives of many people who in a previous era would have died from their first heart attack. Implanted defibrillators have become the most often used and most successful treatment for people at high risk of reentrant arrhythmias. However, implantable defibrillators also have significant drawbacks. For example, when they discharge, it is extremely painful, with the consequence that up to 40% of implantees have high levels of anxiety leading to depression.

The goal of a defibrillator is clear. Since during fibrillation different regions of tissue are in different phases of electrical activity, some excited, some refractory, some partially recovered, the purpose of defibrillation is to give an electrical impulse that stimulates the entire heart, so that the electrical activity is once again coordinated and returns to rest as a whole to await the next normal SA nodal stimulus. Said another way, the purpose is to reset the phase of each cardiac cell so that all cells are in phase, and so that there are no remaining phase singularities.

So the questions that need to be addressed are first, how does defibrillation work, and second, are there ways to improve the efficiency of defibrillators?

While it is known that defibrillation works (and is accomplished thousands of times daily around the world), the dilemma is that simple mathematical models fail to explain how this can happen, and indeed, seem to suggest that defibrillation shocks cannot achieve their goal.

To understand the dilemma, consider the numerical calculation shown in Fig. 12.46. Here is shown the result of applying a stimulus to the ends of a bidomain cable. The stimulus on the left is depolarizing and on the right is hyperpolarizing. On the left, a

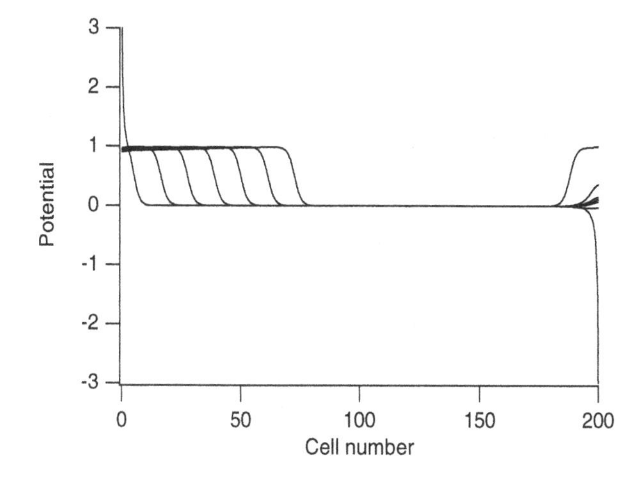

Figure 12.46 Response of a uniform cable to a stimulus of duration $t = 0.2$ applied at the ends of the cable. Traces shown start at time $t = 0.1$ and with equal time steps $\Delta t = 0.2$.

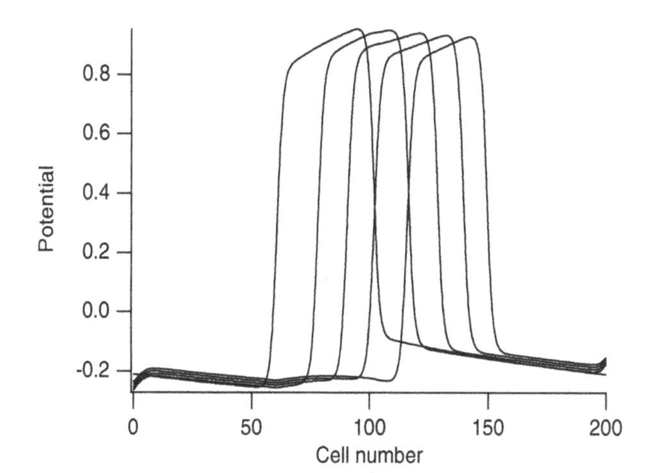

Figure 12.47 A traveling wave in a uniform cable following application of a stimulus at the ends of the cable at a time between the first and second traces.

right-moving wave is initiated almost immediately, and on the right a left-moving wave is initiated via anode break excitation (see Chapter 5, Exercise 7). The dilemma is that local stimuli can only have local effects.

A similar conclusion is drawn from Fig. 12.47. Here is shown a periodic traveling wave on a one-dimensional bidomain cable, traveling from left to right. (If the left and right ends were connected, this wave would circulate around the ring indefinitely.) What cannot be seen in this figure is that a large stimulus was applied at the ends of the cable between the first and second traces, simulating defibrillation. What can be seen from this figure is that the stimulus has essentially no effect on the traveling wave. This stimulus has no chance of defibrillating the cable, since the effects of the stimulus are localized in space.

The question, then, is how can defibrillation work, if only those regions close to the stimulating source are excited by the stimulus. The likely answer is that the medium into which the stimulus is applied has resistive inhomogeneities that are not accounted for in a uniform cable model.

To see the effect that resistive inhomogeneities might have, consider the effect of a current that is applied to a resistively inhomogeneous one-dimensional strand of cardiac tissue of length L. We take the bidomain model equations (12.48), but now assume that the conductivities $\sigma_i = A_i/R_i$ and $\sigma_e = A_e/R_e$ are continuous but nonconstant to reflect the occurrence of resistive inhomogeneities. Then, the monodomain reduction for a one-dimensional cable gives from (12.105)

$$p\left(C_m\frac{\partial V}{\partial t} - \frac{f(V)}{R_m}\right) = \frac{\partial}{\partial x}\left(\frac{\sigma_i\sigma_e}{\sigma_i+\sigma_e}\frac{\partial V}{\partial x}\right) - \frac{\partial}{\partial x}\left(\frac{\sigma_i}{\sigma_i+\sigma_e}\right)I(t), \tag{12.161}$$

where $I(t)$ is the applied current. The important observation is that if σ_i or σ_e are nonconstant, the new term in (12.161) acts as a source term everywhere throughout the medium, even though the current $I(t)$ is applied locally. The effect of resistive inhomogeneity was already evident in Fig. 12.23. Similarly, in higher dimensions, if $\sigma_i(\sigma_i+\sigma_e)^{-1}$ is inhomogeneous in space, then when the total current i_t is nonzero, there are sources and sinks of transmembrane current in the interior of the domain.

It is useful to introduce dimensionless variables $\tau = t/\tau_m, y = x/\lambda_m$, where $\tau_m = C_mR_m, \lambda_m^2 = \sigma_e DR_m/p$, and we then obtain

$$V_\tau - f(V) = \left(\frac{d}{D}V_y\right)_y - J(\tau)\left(\frac{\Sigma_e d}{\sigma_e D}\right)_y, \tag{12.162}$$

where $d = \sigma_e\sigma_i/(\sigma_i+\sigma_e)$, D^{-1} is the average value of d^{-1}, Σ_e^{-1} is the average value of σ_e^{-1}, and $J(\tau) = R_m\frac{D}{\Sigma_e}I(\tau)/(p\lambda_m)$. In addition, we have boundary conditions $V_y = -J(\tau)$ at $y=0$ and at $y = Y = L/\lambda_m$.

There are potentially many different spatial scales for resistive inhomogeneities. At the cellular level, cells are connected by gap junctions and surrounded by extracellular space containing capillaries, collagen fiber, connective tissue, etc. In addition, myocytes are assembled into layers, with extensive interlaminal clefts between these layers (Caulfield and Borg, 1979; Robinson et al., 1983; Hooks et al., 2002). At larger spatial scales, cells are organized into fibers, there is fiber branching and tapering, and the fiber orientation changes both in the longitudinal and in the transverse directions.

All of these resistive inhomogeneities produce virtual electrodes that affect the outcome of the defibrillation shock. However, here we focus on the effect of resistive inhomogeneities on the spatial scale of individual cells, and so, we suppose that σ_i and σ_e are periodic functions of y, with period $\epsilon = l/\lambda_m$, where l is the cell length. Specifically, we take

$$\sigma_j = \sigma_j\left(\frac{y}{\epsilon}\right), \qquad j = i, e, \tag{12.163}$$

with $\sigma_j(y)$ a function of period 1. Typically, ϵ is a small number, on the order of 0.1.

Now we are able to use homogenization arguments to find the effects of small scale inhomogeneities on the larger scale behavior. The result of a standard multiscale calculation gives (see Exercise 19).

$$V(z,\tau,\eta) = u_0(z,\tau,\epsilon) + \epsilon J(\tau)H(\eta) - \epsilon W(\eta)\frac{\partial u_0(z,\tau)}{\partial z}, \tag{12.164}$$

where $\eta = y/\epsilon$, $W(\eta)$ and $H(\eta)$ are periodic functions with

$$\frac{dW}{d\eta} = 1 - \frac{D}{d}, \qquad \frac{dH}{d\eta} = \frac{\Sigma_e}{\sigma_e} - \frac{D}{d}. \tag{12.165}$$

Furthermore the mean field $u_0(z,t)$ is governed by the equation

$$\frac{\partial u_0}{\partial \tau} - \overline{f(V(z,\tau,\eta))} = \frac{\partial^2 u_0}{\partial z^2}, \tag{12.166}$$

where

$$\overline{f(V(z,\tau,\eta))} = \int_0^1 f(V(z,\tau,\eta))\, d\eta. \tag{12.167}$$

The interpretation of (12.166) is significant. While a current stimulus is being applied, the response at the cellular level has an effect that is communicated to the tissue on a macroscopic scale through the nonlinearity of the ionic currents. If the current amplitude is sufficiently large (say of order $\frac{1}{\epsilon}$), the effect can be quite significant. (If the ionic current $f(V)$ were linear, the applied stimulus would have no global effect, since then $\overline{f(V(z,\tau,\eta))} = f(u_0)$.)

To get some insight into the dynamics while the stimulus is applied, we take the simple model of gap-junctional resistance

$$r_i = r_c + \frac{r_g}{l}\delta(\eta) \tag{12.168}$$

on the interval $0 \le \eta < 1$, and periodically extended from there. Here, r_c is the intracellular cytoplasmic resistance per unit length, and r_g is the gap-junctional resistance per cell. The function $\delta(\eta)$ is any positive function with small support and area one unit that represents the spatial distribution of the gap-junctional resistance, for example, the Dirac delta function.

In the specific case that $\delta(\eta)$ is the Dirac delta function, we calculate that

$$W'(\eta) = R_g(1 - \delta(\eta)), \tag{12.169}$$

where $R_g = \frac{r_g}{r_g + l(r_c + r_e)}$ is the fraction of the total resistance per unit length that is concentrated into the gap junctions. Here we used the fact that $r_i = 1/\sigma_i$. Then, $W(\eta)$ is given by

$$W(\eta) = R_g\left(\eta - \frac{1}{2}\right), 0 \le \eta < 1, \tag{12.170}$$

$W(\eta + 1) = W(\eta)$, and $H(\eta) = W(\eta)$.

The function $W(\eta)$ is a *sawtooth function*, and according to (12.164), when a stimulus is applied, the membrane potential is the sum of two components, the sawtooth function $W(\eta)$ on the spatial scale of cells and $u(y,\tau)$ on the macroscopic scale of the

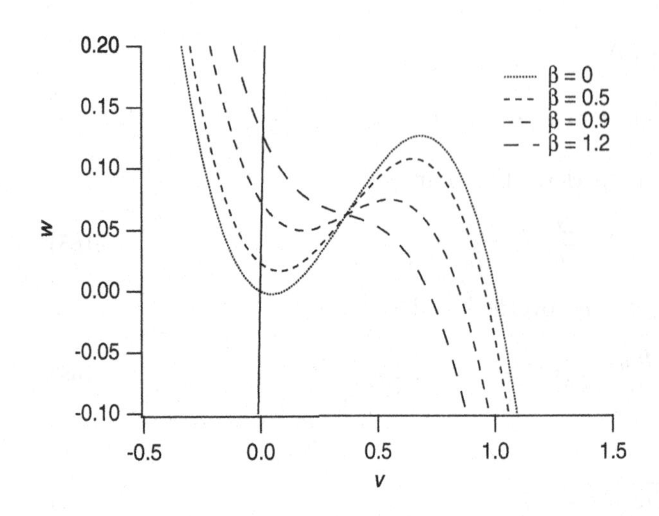

Figure 12.48 Nullclines for FitzHugh–Nagumo dynamics, $w = v(1-v)(\alpha-v)+\frac{\beta^2}{12}(1+\alpha-3v)$, $w = v/\gamma$, modified to include the effects of a current stimulus, for several values of β.

tissue. The effect of the small-scale oscillatory behavior on the larger scale problem is found by averaging, whereby

$$\overline{F}(u,\beta) = \overline{f(V(z,\tau,\eta))} = \int_{-\frac{1}{2}}^{\frac{1}{2}} f(u+\beta\eta)\,d\eta, \qquad (12.171)$$

and $\beta = R_g\epsilon(J(t) - \frac{\partial u}{\partial z})$. In other words, the effect of the current stimulus is to modify the ionic current through local averaging over the cell. The details of the structure of $W(\eta)$ are not important because they are felt only in an average sense. For the cubic model $f(V) = V(V-1)(\alpha - V)$, $\overline{F}$ can be calculated explicitly to be

$$\overline{F}(V,\beta) = f(V) + \frac{\beta^2}{12}(1+\alpha-3V). \qquad (12.172)$$

A plot of this function for different values of β is shown in Fig. 12.48.

Before examining this model for its ability to explain defibrillation, we discuss the simpler problem of direct activation of resting tissue.

12.5.1 The Direct Stimulus Threshold

Direct activation (or field stimulation) occurs if all or essentially all of the tissue is activated simultaneously without the aid of a propagated wave front. According to the model (12.166), it should be possible to stimulate cardiac tissue directly with brief stimuli of sufficiently large amplitude.

A numerical simulation demonstrating how direct stimulation can be accomplished for the bistable equation is shown in Fig. 12.49. In this simulation a one-dimensional array of 200 cells was discretized with five grid points per cell, and a brief, large current was injected at the left end and removed at the right end of the cable. In Fig. 12.46 is shown the response to the stimulus when the cable is uniform. Shown here is the membrane potential, beginning at time $t = 0.1$, and at later times with equal time steps

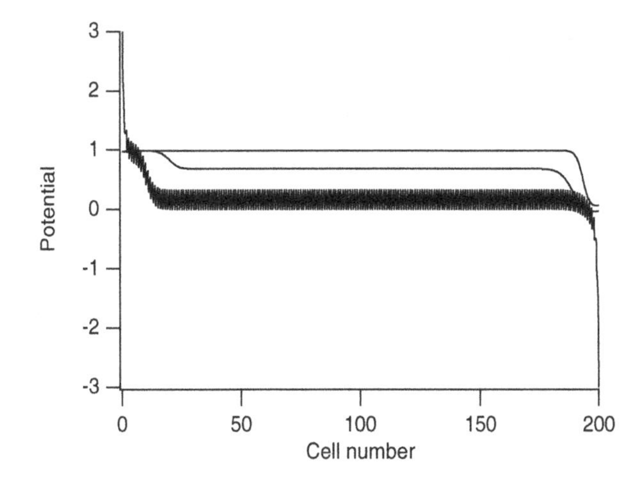

Figure 12.49 Response of a nonuniform cable with regularly spaced high-resistance nodes to a stimulus of duration $t = 0.2$ applied at the ends of the cable. The traces show the response at time $t = 0.15$, during the stimulus, and at times $t = 0.25, 0.35$ after the stimulus has terminated.

$\Delta t = 0.2$. The stimulus duration was $t = 0.2$, so its effects are seen as a depolarization on the left and hyperpolarization on the right in the first trace. As noted above, a wave is initiated from the left from superthreshold depolarization, and a wave from the right is initiated by anode break excitation.

The same stimulus protocol produces a substantially different result if the cable has nonuniform resistance. In Fig. 12.49 is shown the response of the discretized cable with high resistance at every fifth node, at times $t = 0.15, 0.25$, and 0.35, with a stimulus duration of 0.2. The first curve, at time $t = 0.15$, is blurred because the details of the membrane potential cannot be resolved on this scale. However, the overall effect of the rapid spatial oscillation is to stimulate the cable directly, as seen from the subsequent traces.

To analyze this situation, note that since direct activation occurs without the benefit of propagation, it is sufficient to ignore diffusion and the boundary conditions and simply examine the behavior of the averaged ordinary differential equation

$$\frac{dV}{d\tau} = \overline{F}(V, \beta). \tag{12.173}$$

For any resting excitable system, it is reasonable to assume that $f(V) < 0$ for $0 < V < \theta$, where θ is the threshold that must be exceeded to stimulate an action potential. To directly stimulate a medium that is initially at rest with a constant stimulus, one must apply the stimulus until $V > \theta$. The minimal time to accomplish this is given by the strength–duration relationship,

$$T = \int_0^\theta \frac{dV}{\overline{F}(V, \beta)}. \tag{12.174}$$

Clearly, this expression is meaningful only if β is sufficiently large that $\overline{F}(V, \beta) > 0$ on the interval $0 < V < \theta$. In other words, there is a minimal stimulus level (a threshold) below which the medium cannot be directly stimulated.

12.5.2 The Defibrillation Threshold

While its threshold cannot be calculated in the same way as for direct stimulus, the mechanism of defibrillation can be understood from simple phase-plane arguments. To study defibrillation, we must include the dynamics of recovery in our model equations. Thus, for purposes of illustration, we take FitzHugh–Nagumo dynamics

$$I_{\text{ion}}(v,w) = -f(v) + w, \tag{12.175}$$

$$w_\tau = g(v,w), \tag{12.176}$$

with $f(v) = v(v-1)(\alpha - v)$, with parameters chosen so that reentrant waves are persistent. This could mean that there is a stable spiral solution, or it could mean that the spiral solution is unstable but some nonperiodic reentrant motion is persistent. Either way, our goal is to demonstrate that there is a threshold for the stimulating current above which reentrant waves are terminated.

The mechanism of defibrillation is easiest to understand for a periodic wave on a one-dimensional ring, but the idea is similar for higher-dimensional reentrant patterns. For a one-dimensional ring, the phase-portrait projection of a rotating periodic wave is a closed loop. From singular perturbation theory, we know that this loop clings to the leftmost and rightmost branches of the nullcline $w = f(v)$ and has two rapid transitions connecting these branches, and these correspond to wave fronts and wave backs.

According to this model, the effect of a stimulus is to temporarily change the v nullclines and thereby to change the shape of the closed loop. After the stimulus has ended, the distorted closed loop will either go back to a closed loop, or it will collapse to a single point on the phase portrait and return to the rest point. If the latter occurs, the medium has been "defibrillated."

Clearly, if β is small and the periodic oscillation is robust, then the slight perturbation is insufficient to destroy it. On the other hand, if β is large enough, then the change is substantial and collapse may result.

There are two ways that this collapse can occur. First, and easiest to understand, if the nullcline for nonzero β is a monotone curve (as in Fig. 12.48 with $\beta = 1.2$), then the open loop collapses rapidly to a double cover of the single curve $w = \overline{F}(v,\beta)$, from where it further collapses to a single point in phase space. For the specific cubic model, this occurs if $\beta^2 > \frac{4}{3}(1 - \alpha + \alpha^2)$.

The v nullcline need not be monotone to effect a collapse of the periodic loop. In fact, as β increases, the negative-resistance region of the nullcline becomes smaller, and the periodic loop changes shape into a loop with small "thickness" (i.e., with little separation between the front and the back) and with fronts and backs that move with nearly zero speed. Indeed, if the distorted front is at a large enough w "level" (in the sense of singular perturbation theory), it cannot propagate at all and stalls, leading to a collapse of the wave. Another way to explain this is to say that with large enough β, the "excitable gap" between the refractory tail and the excitation front is excited, pushing the wave front forward (in space) as far as possible into the refractory region ahead of it, thereby causing it to stall, and eventually to collapse.

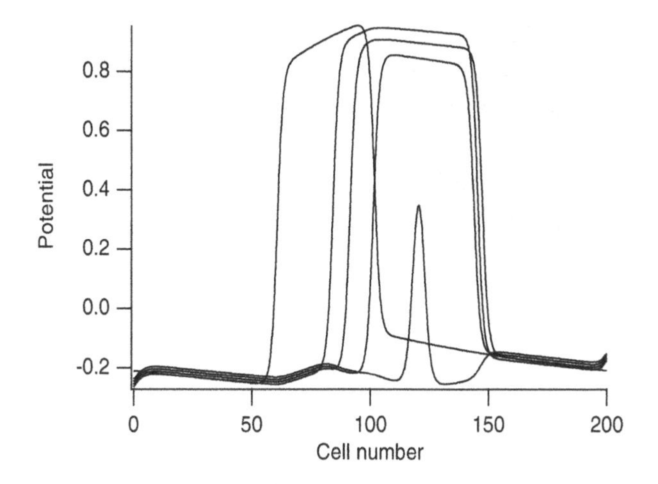

Figure 12.50 A traveling wave in a nonuniform cable following application of a stimulus at the boundary.

This scenario can be seen in Fig. 12.50, where the results of numerical simulations for a one-dimensional nonuniform cable are shown. Earlier, in Fig. 12.47, we showed a wave, propagating to the right, at equal time steps of $\Delta t = 0.55$. To simulate a reentrant arrhythmia, this wave was chosen so that were this cable a closed loop, the wave would circulate around the loop indefinitely without change of shape. It is not apparent from this previous figure that a stimulus of duration $t = 0.3$ was applied at the ends of the cable between the first and second traces, because in a uniform cable a stimulus at the boundary has little effect on the interior of the medium.

In Fig. 12.50 is shown exactly the same sequence of events for a nonuniform cable. This time, however, the applied stimulus (between the first and second traces) induces a rapidly oscillating membrane potential on the spatial scale of cells (not shown), which has the average effect (because of nonlinearity) of "pushing" the action potential forward as far as possible. This new front cannot propagate forward because it has been pushed into its refractory tail and has stalled. In fact, the direction of propagation reverses, and the action potential collapses as the front and back move toward each other.

To illustrate further this mechanism of defibrillation in a two-dimensional domain, numerical simulations were performed using a standard two-variable model of an excitable medium and using the full bidomain model derived in Section 12.8 (Keener and Panfilov, 1996). Parameters for the excitable dynamics were chosen such that spirals are not stable, but exhibit breakup and develop into "chaotic" reentrant patterns (Panfilov and Hogeweg, 1995), thereby giving a reasonable model of cardiac fibrillation (see Chapter 6, Exercise 26).

Some time after initiating a reentrant wave pattern, a constant stimulus (of duration about 2.5 times the duration of the action potential upstroke) was applied uniformly to the sides of the rectangular domain. Because the stimulus was applied uniformly along the sides of the domain, the stimulus parameter β was constant throughout the medium.

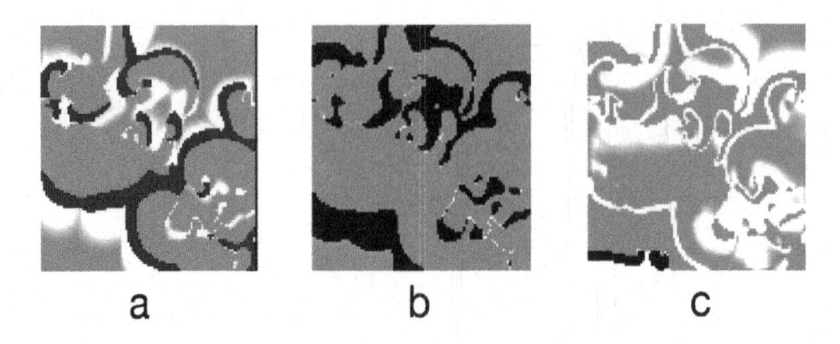

Figure 12.51 Successful defibrillation of a two-dimensional region.

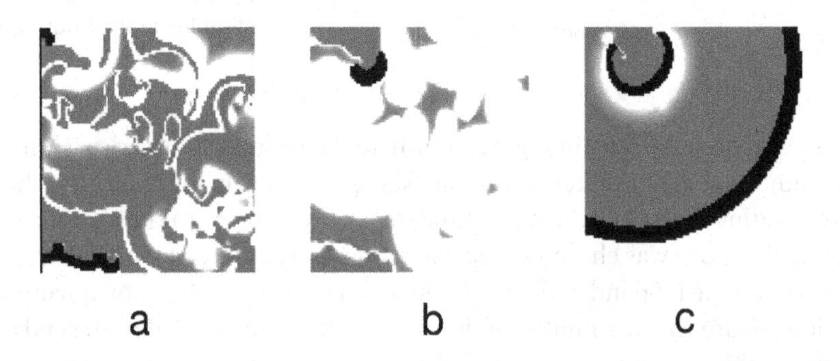

Figure 12.52 Unsuccessful defibrillation of a two-dimensional region.

In Fig. 12.51a is shown an irregular reentrant pattern just before the stimulus is applied. In this picture, the darkest regions are excited tissue, white denotes recovered and excitable tissue, and gray is refractory tissue. Following a stimulus (with $\beta = 0.86$), the excited region expanded to include essentially all of the recovered tissue, as the excitable gap was eliminated by a polarizing stimulus. Shortly thereafter (Fig. 12.51c, $t = 12$) the activation collapsed, leaving behind only recovered or refractory tissue, which shortly thereafter returned to uniform rest. The extensive patterning seen in this last figure shows a mixture of refractory and recovered tissue, but since it contains no excited tissue (except a small patch at the lower left corner that is propagating out of the domain), it cannot become re-excited, but must return to rest. (The similar fates of recovered and refractory regions can be seen in Fig. 12.52b, where the patterning has nearly disappeared.)

Defibrillation is unsuccessful with a smaller stimulus $\beta = 0.84$. The pattern at time $t = 12$ is shown in Fig. 12.52a and is similar to Fig. 12.51c, which was successful. Here, however, after the stimulus and subsequent collapse of much of the excitation, one small excited spot remains at the upper left-hand corner of the medium, which eventually evolves into a double spiral pattern (Fig. 12.52b,c), reestablishing a reentrant arrhythmia.

The primary mechanism of defibrillation still remains uncertain and debated. In this discussion we have focussed on the role of small scale resistive inhomogeneities in defibrillation, and we have ignored the effect of larger scale inhomogeneities. However, there is also a significant body of numerical work exploring the role of large scale virtual electrodes in defibrillation (Efimov et al., 1998, 2000a,b; Anderson et al., 2000; Eason and Trayanova, 2002). For a theoretical discussion of defibrillation from the viewpoint of phase singularity elimination, see Keener and Cytrynbaum (2003), and Keener (2004).

12.6 Appendix: The Sawtooth Potential

In this appendix, we provide some of the details needed for the calculation of the sawtooth potential solution (12.62)–(12.67). This is a differential-difference boundary value problem that we solve in two steps. First, on the interior of cells the solution must satisfy the cable equation (12.48) with total current $-\frac{1}{r_e}\frac{dV_e}{dx} - \frac{1}{r_c}\frac{dV_i}{dx} = -I$. This solution is

$$\begin{pmatrix} V_i \\ V_e \end{pmatrix} = \frac{1}{r_e + r_c}\begin{pmatrix} r_c \\ -r_e \end{pmatrix}(A\sinh\lambda x + B\cosh\lambda x) + \begin{pmatrix} 1 \\ 1 \end{pmatrix}(-\beta I x + \gamma), \tag{12.177}$$

and $V = A\sinh\lambda x + B\cosh\lambda x$, for $0 < x < L$, with A and B yet to be determined, $\beta = \frac{r_e r_c}{r_e + r_c}$. It is convenient to set $\psi = \frac{1}{r_c}\frac{dV_i(0)}{dx}$ so that $A = a\psi + J$, where $a = \frac{r_e + r_c}{\lambda}$, and $J = \frac{r_e}{\lambda}I$. Now, we suppose that there are a total of N cells, and we denote ψ_j, B_j, and γ_j as the values of ψ, B, and γ for the jth cell, $j = 0, 1, \ldots, N-1$. The requirement that there be no intracellular current at the ends of the cable is equivalent to the requirement that $\psi_0 = \psi_N = 0$. We impose the conditions that the current and extracellular potentials be continuous and that the jump condition is satisfied at the ends of the cells, and find that

$$\begin{pmatrix} a\psi_{n+1} \\ B_{n+1} \end{pmatrix} = \begin{pmatrix} C & S \\ S + K_g C & C + K_g S \end{pmatrix}\begin{pmatrix} a\psi_n \\ B_n \end{pmatrix} - J\begin{pmatrix} 1 - C \\ K_g(1-C) - S \end{pmatrix}, \tag{12.178}$$

and

$$\gamma_{n+1} = \gamma_n - \lambda L\frac{\beta}{r_e}J + \frac{\beta}{r_c}K_g(J(C-1) + a\psi_n C + B_n S), \tag{12.179}$$

where $K_g = \frac{r_g\lambda}{r_e + r_c}$, $C = \cosh\lambda L$, and $S = \sinh\lambda L$.

This is a system of difference equations that we wish to solve subject to the boundary conditions $\psi_0 = \psi_N = 0$. To solve these equations, observe that the matrix has eigenvalues, $\eta_1 = \frac{1}{\mu} > 1$, $\eta_2 = \mu$, where μ is the root of (12.58) with $\mu < 1$. Further, the corresponding right and left eigenvectors are

$$x_i = \begin{pmatrix} S \\ \eta_i - C \end{pmatrix}, \qquad y_i = \begin{pmatrix} S + K_g C \\ \eta_i - C \end{pmatrix}. \tag{12.180}$$

We express the solution of the difference equation (12.178) as a linear combination of the eigenvectors

$$\begin{pmatrix} a\psi_n \\ B_n \end{pmatrix} = \sum_{i=1}^{2} \alpha_{i,n} x_i, \tag{12.181}$$

and write

$$\begin{pmatrix} 1-C \\ K_g(1-C)-S \end{pmatrix} = \sum_{i=1}^{2} c_i x_i. \tag{12.182}$$

This decomposes the problem (12.178) into two scalar problems

$$\alpha_{i,n+1} = \eta_i \alpha_{i,n} - Jc_i, \qquad i = 1, 2, \tag{12.183}$$

with boundary conditions $\alpha_{1,0} + \alpha_{2,0} = 0$ and $\alpha_{1,N} + \alpha_{2,N} = 0$. The solution of this scalar problem is

$$\alpha_{i,n} = (-1)^i \alpha_0 \eta_i^n - Jc_i \frac{\eta_i^n - 1}{\eta_i - 1}, \tag{12.184}$$

which automatically satisfies the boundary condition at $n = 0$. To satisfy the boundary condition at $n = N$, we must have

$$\alpha_0 = -J \frac{\mu}{1-\mu} \frac{(\mu^N - 1)}{(\mu^{2N} - 1)} \left(c_1 + c_2 \mu^{N-1} \right). \tag{12.185}$$

As a result,

$$a\psi_j = \frac{-S}{1-\mu} (c_1\mu - c_2) f(j), \tag{12.186}$$

where

$$f(j) = \frac{1 - \mu^{N-j} - \mu^j + \mu^{j+N} + \mu^{2N-j} - \mu^{2N}}{1 - \mu^{2N}} \tag{12.187}$$

and

$$B_j = \frac{c_1\mu - c_2}{1-\eta} \left(Cf(j) + \frac{1-\mu^N}{1-\mu^{2N}} \left(\mu^{j+1} + \mu^{N-j-1} \right) \right) - \frac{c_1 - c_2\mu}{1-\mu}. \tag{12.188}$$

12.7 Appendix: The Phase Equations

Because coupled oscillators arise so frequently in mathematical biology and physiology, there is an extensive literature devoted to their study. In this book, coupled oscillators play a role in the sinoatrial node and in the digestive system. An important model for the study of coupled oscillators has been the *phase equation*, an equation describing the evolution of the phases of a loosely coupled collection of similar oscillators.

In this appendix we give a derivation of the phase equation. This derivation is similar to that of Neu (1979), using perturbation techniques and a multiscale analysis. A more technical derivation is given by Ermentrout and Kopell (1984). The attempt to describe populations of oscillators in terms of phases was first made by Winfree (1967), and in the context of reaction–diffusion systems, by Ortoleva and Ross (1973, 1974). This was improved by Kuramoto (Kuramoto and Tsuzuki, 1976; Kuramoto and Yamada, 1976). Neu's derivation is also discussed by Murray (2002).

To set the stage, consider the system of equations

$$\frac{dx}{dt} = \Lambda(r)x - \omega(r)y, \tag{12.189}$$

$$\frac{dy}{dt} = \omega(r)x + \Lambda(r)y, \tag{12.190}$$

where $r^2 = x^2 + y^2$. Systems of this form are called *lambda–omega* (Λ–ω) systems, and are special because by changing to polar coordinates, $x = r\cos\theta$, $y = r\sin\theta$, (12.189) and (12.190) can be written as

$$\frac{dr}{dt} = r\Lambda(r), \tag{12.191}$$

$$\frac{d\theta}{dt} = \omega(r). \tag{12.192}$$

This system has a stable limit cycle at any radius $r > 0$ for which $\Lambda(r) = 0$, $\Lambda'(r) < 0$. The periodic solution travels around this circle with angular velocity $\omega(r)$. Starting from any given initial conditions, the solution of (12.189) and (12.190) eventually settles onto a regular oscillation with fixed amplitude and period $\frac{\omega(r)}{2\pi}$. Hence, in the limit as $t \to \infty$, the system is described completely by its angular velocity around a circle.

Now suppose that we have two similar systems, one with a limit cycle of amplitude R_1 and angular velocity ω_1, and the other with a limit cycle of amplitude R_2 and angular velocity ω_2. If there is no coupling between the systems, each oscillates at its own frequency, unaffected by the other, and in the four-dimensional phase space the solutions approaches the torus $r_1 = R_1, r_2 = R_2$, moving around the torus with angular velocities $\theta_1' = \omega_1$, $\theta_2' = \omega_2$. Since all solutions eventually end up winding around the torus, and since any solution that starts on the torus cannot leave it, the torus is called an *attracting invariant torus*. The flow on the torus can be described entirely in terms of the rates of change of θ_1 and θ_2. In this case $(\theta_1 - \theta_2)' = \omega_1 - \omega_2$, and so the phase difference increases at a constant rate. Thus, analogously to the one-dimensional system discussed above, in the limit as $t \to \infty$, the original system of four-differential equations can be reduced to a two-dimensional system describing the flow on a two-dimensional torus.

If our two similar systems are now loosely coupled, so that each oscillator has only a small effect on the other, it is reasonable to expect (and indeed it can be proved; see Rand and Holmes (1980) for a nice discussion of this, and Hirsch et al. (1977) for a proof) that the invariant torus persists, changing its shape and position by only a small amount. In this case, the longtime solutions for r_1 and r_2 remain essentially unchanged, with $r_1 = R_1 + O(\epsilon)$ and $r_2 = R_2 + O(\epsilon)$, where $\epsilon \ll 1$ is the strength of the coupling. However, the flow *on* the torus could have a drastically different nature, as the phase difference need no longer simply increase at a constant rate. Hence, although the structure of the torus is preserved, the properties of the flow on the torus are not.

In general, the flow on the torus is described by

$$\frac{d\theta_i}{dt} = \omega_i + f_i(r_1, r_2, \theta_1, \theta_2, \epsilon), \qquad i = 1, 2, \tag{12.193}$$

but since $r_1 = R_1 + O(\epsilon)$ and $r_2 = R_2 + O(\epsilon)$, to lowest order in ϵ this simplifies to

$$\frac{d\theta_i}{dt} = \omega_i + h_i(\theta_1, \theta_2, \epsilon), \qquad i = 1, 2. \tag{12.194}$$

In general, R_1 and R_2 appear in (12.194), the so-called *phase equation*, but the independent variables r_1 and r_2 do not. It follows that the full four-dimensional system that describes the two coupled oscillators can be understood in terms of a simpler system describing the flow on a two-dimensional invariant torus.

To derive the equations describing this flow on a torus, we assume that we have a coupled oscillator system that can be written in the form

$$\frac{du_i}{dt} = F(u_i) + \epsilon G_i(u_i) + \epsilon \sum_{j=1}^{N} a_{ij} H(u_j). \tag{12.195}$$

Here, u_i is the vector of state variables for the ith oscillator, the coefficients a_{ij} represent the coupling strength, and the function $H(u)$ determines the effect of coupling. For simplicity, we assume that H is independent of i and j.

To get the special case of SA nodal coupling in (12.120), we take $a_{ij} = d_{ij}$ for $i \neq j$, $a_{ii} = -\sum_{j \neq i} d_{ij}$, and $H(u) = Du$. We take this general form of H to allow for synaptic as well as diffusive coupling.

Next, we assume that when $\epsilon = 0$ we have a periodic solution, i.e., that the equation

$$\frac{du}{dt} = F(u) \tag{12.196}$$

has a stable periodic solution, $U(t)$, scaled to have period one. Note that because of the functions G_i, we are not assuming that each oscillator is identical. Thus, the natural frequency of each oscillator is close to, but not exactly, one.

The model system (12.195) is a classic problem to which the *method of averaging* or the *multiscale method* can be applied. Specifically, since ϵ is small, we expect the behavior of (12.195) to be dominated by the periodic solution $U(t)$ of the unperturbed problem and that deviations from this behavior occur on a much slower time scale. To accommodate two different time scales, we introduce two timelike variables, $\sigma = \omega(\epsilon)t$ and $\tau = \epsilon t$, as fast and slow times, respectively. Here, ω is a function, as yet unknown, of order 1. Treating σ and τ as independent variables, we find from the chain rule that

$$\frac{d}{dt} = \omega(\epsilon)\frac{\partial}{\partial\sigma} + \epsilon\frac{\partial}{\partial\tau}, \tag{12.197}$$

and accordingly, (12.195) becomes

$$\omega(\epsilon)\frac{\partial u_i}{\partial \sigma} + \epsilon\frac{\partial u_i}{\partial \tau} = F(u_i) + \epsilon G_i(u_i) + \epsilon \sum_{j=1}^{N} a_{ij}H(u_j). \tag{12.198}$$

Next we suppose that u_i and $\omega(\epsilon)$ have power series expansions in ϵ, given by

$$u_i = u_i^0 + \epsilon u_i^1 + \cdots, \qquad \omega(\epsilon) = 1 + \epsilon\Omega_1 + \cdots. \tag{12.199}$$

Note that the first term in the expansion for ω is the frequency of the unperturbed solution, U. Expanding (12.198) in powers of ϵ and gathering terms of like order, we find a hierarchy of equations, beginning with

$$\frac{\partial u_i^0}{\partial \sigma} = F(u_i^0), \tag{12.200}$$

$$\frac{\partial u_i^1}{\partial \sigma} - F_u(u_i^0)u_i^1 = G_i(u_i^0) + \sum_{j=1}^{N} a_{ij}H(u_j^0) - \Omega_1\frac{\partial u_i^0}{\partial \sigma} - \frac{\partial u_i^0}{d\tau}. \tag{12.201}$$

Equation (12.200) is easy to solve by taking

$$u_i^0 = U(\sigma + \delta\theta_i(\tau)). \tag{12.202}$$

The phase shift $\delta\theta_i(\tau)$ allows each cell to have different phase shift behavior, and it is yet to be determined.

Next, observe that $\frac{d}{d\sigma}\left(\frac{dU}{d\sigma} - F(U)\right) = \frac{\partial U'}{\partial \sigma} - F_u(u_i^0)U' = 0$, so that the operator $LU = \frac{\partial U}{\partial \sigma} - F_u(u_i^0)U$ has a null space spanned by $U'(\sigma + \delta\theta_i(\tau))$. The null space is one dimensional because the periodic solution is assumed to be stable. It follows that the adjoint operator $L^*y = -\frac{\partial y}{\partial \sigma} - F_u(u_i^0)^T y$ has a one-dimensional null space spanned by some periodic function $y = Y(\sigma + \delta\theta_i(\tau))$. (It is a consequence of Floquet theory that the Floquet multipliers of the operator L, say μ_i, and the Floquet multipliers of L^*, say μ_i^*, are multiplicative inverses, $\mu_i\mu_i^* = 1$ for all i. Since a periodic solution has Floquet multiplier 1 and there is only one periodic solution for L, there is also precisely one periodic solution for the adjoint operator L^*. See Exercise 23.) Without loss of generality we scale Y so that $\int_0^1 U'(\sigma)\cdot Y(\sigma)\,d\sigma = 1$. Therefore, for there to be a periodic solution of (12.201), the right-hand side of (12.201) must be orthogonal to the null space of the adjoint operator L^*. This requirement translates into the system of differential equations for the phase shifts

$$\frac{d}{d\tau}\delta\theta_i = \xi_i - \Omega_1 + \sum_j a_{ij}h(\delta\theta_j - \delta\theta_i), \tag{12.203}$$

where

$$\xi_i = \int_0^1 Y(\sigma)\cdot G_i(U(\sigma))\,d\sigma, \tag{12.204}$$

$$h(\phi) = \int_0^1 Y(\sigma)\cdot H(U(\sigma + \phi))\,d\sigma. \tag{12.205}$$

The numbers ξ_i are important because they determine the approximate natural (i.e., uncoupled) frequency of the ith cell. This follows from the fact that when $a_{ij} = 0$, a simple integration gives $\delta\theta_i = \epsilon t(\xi_i - \Omega_1)$, and thus

$$u_i^0 = U(\omega(\epsilon)t + \delta\theta_i) = U((1 + \epsilon\xi_i)t). \tag{12.206}$$

Hence, the uncoupled frequency of the ith cell is $2\pi(1 + \epsilon\xi_i)$. Therefore, ξ_i can (presumably) be measured or estimated without knowing the function $G_i(u)$. The function $h(\phi)$ can be determined analytically for $\Lambda - \omega$ systems (see Exercise 24) or numerically otherwise (see, for example, Exercise 22.)

The function $h(\phi)$ has an important physical interpretation, being a *phase resetting function*. That is, $h(\phi)$ shows the effect of one oscillator on another when the two have phases that differ by ϕ. For example, in the case of two identical oscillators, the two phase shifts are governed by

$$\frac{d}{d\tau}\delta\theta_1 = -\Omega_1 + a_{12}h(\delta\theta_2 - \delta\theta_1) + a_{11}h(0), \tag{12.207}$$

$$\frac{d}{d\tau}\delta\theta_2 = -\Omega_1 + a_{21}h(\delta\theta_1 - \delta\theta_2) + a_{22}h(0). \tag{12.208}$$

Thus, when $a_{12}h(\delta\theta_2 - \delta\theta_1) > 0$, the phase of oscillator 1 is advanced, while if $a_{12}h(\delta\theta_2 - \delta\theta_1) < 0$, the phase of oscillator 1 is retarded. Furthermore, it is not necessarily the case that $h(0) = 0$, so that identical oscillators with identical phases may nonetheless exert a nontrivial influence on each other.

The system of equations (12.203) can be written in terms of phase differences by defining Φ as the vector of consecutive phase differences and defining $\overline{\delta\theta}$ as the average of the phase shifts, $\overline{\delta\theta} = \frac{1}{N}\sum_{i=1}^{N}\delta\theta_i$. In terms of these variables the system (12.203) can be written in the form

$$\frac{d\Phi}{d\tau} = \Delta + C(\Phi), \tag{12.209}$$

where Δ is the vector of consecutive differences of ξ_i. This is a closed system of $N - 1$ equations. *Phase locking* is defined as the situation in which there is a stable steady solution of (12.209), a state in which the phase differences of the oscillators do not change (see Chapter 18).

12.8 Appendix: The Cardiac Bidomain Equations

The homogenization technique that was used in Section 7.8.2 to find effective diffusion coefficients for diffusing and reacting chemical species can also be used to find effective electrical properties of three-dimensional cardiac tissue.

We assume that an individual cardiac cell is some small periodic subunit Ω contained in a small rectangular box (Fig. 12.53). The rectangular box is divided into intracellular space Ω_i and extracellular space Ω_e, separated by cell membrane Γ_m. As illustrated in Fig. 12.54 the cells are connected to each other at the sides of the boxes

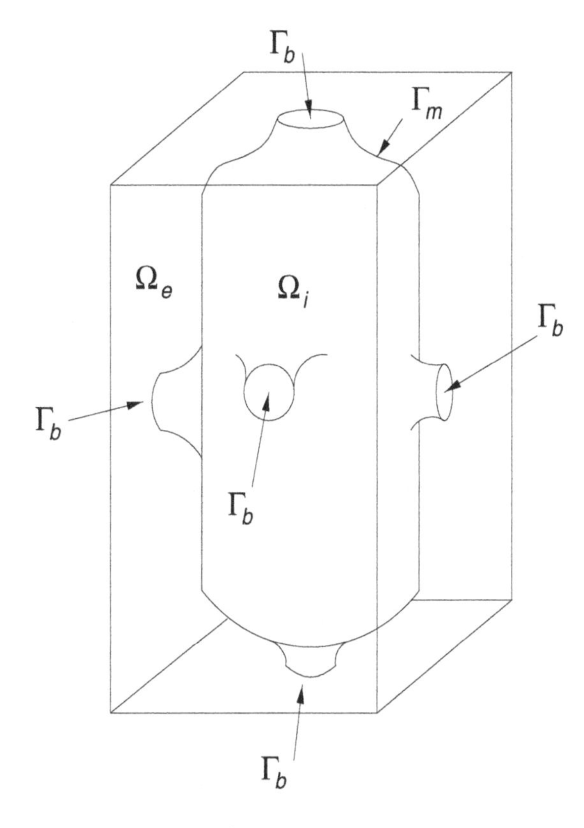

Figure 12.53 Sketch of an idealized single cell.

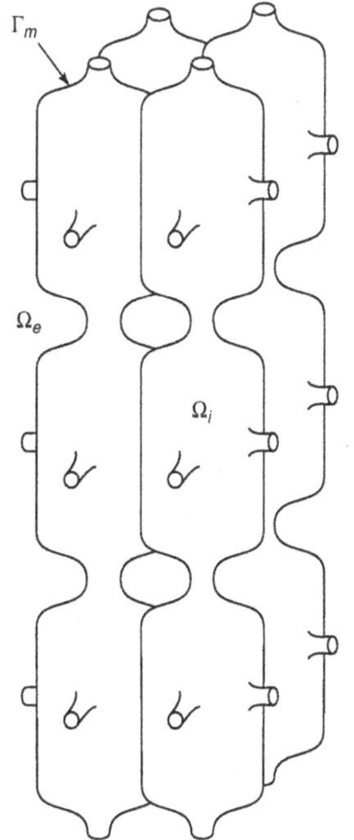

Figure 12.54 Periodic cell structure.

through gap junctions, which are simply parts of the box wall that are contiguous with intracellular space. Thus the boundary of the cellular subunit, $\partial\Gamma$, is composed of two components, cell membrane Γ_m and sides of the box Γ_b.

In either of the intracellular or extracellular spaces, currents are driven by a potential and satisfy Ohm's law $r_c i = -\nabla\phi$, where r_c is the cytoplasmic resistance (a scalar). On the interior of the region, current is conserved, so that

$$\nabla^2\phi = 0. \tag{12.210}$$

Current enters the domain only across boundaries, as a transmembrane current, according to

$$\mathbf{n}\cdot\frac{1}{r_c}\nabla\phi = I_m \tag{12.211}$$

applied in the cell membrane, denoted by Γ_m, and where $\mathbf{n}$ is the outward unit normal to the membrane boundary.

Suppose that x is the original Cartesian coordinate space. To allow for a variable fiber structure we assume that the orientation of the rectangular boxes is slowly varying (so that they are not exactly rectangular, but close enough), and that the axes of the rectangular cellular boxes form a natural "fiber" coordinate system. At each point in space the orientation of the rectangular box is determined by three orthogonal tangent vectors, forming the rows of a matrix $T(x)$. Then the fiber coordinate system is related to the original Cartesian coordinate system through

$$y = Y(x) = \int T(x)\,dx \tag{12.212}$$

and in the y-coordinate system the Laplacian operator is

$$\nabla^2\phi = \nabla_y^2\phi + \kappa\cdot\nabla_y\phi. \tag{12.213}$$

The vector κ is the curvature vector, whose components are the mean curvatures of the coordinate level surfaces. If the components of the matrix T are given by t_{ij}, then the coordinates of κ are $\kappa_j = t_{ik}\frac{\partial t_{ij}}{\partial x_k}$.

To take into account that the boundary of the cells is varying rapidly on the scale of the fiber coordinate system, we introduce the "fast" variable $\xi = \frac{y}{\epsilon}$, where ϵ is the small dimensionless parameter $\epsilon = \frac{l}{\Lambda}$, l is the length of the cell, and Λ is the natural length scale along fibers. We let $z = y$ be the slow variable and assume that κ is a function solely of z, since variations of fiber direction are not noticeable at the cellular level.

Now we apply the homogenization technique described in Chapter 7. For this particular problem, this was first done by Neu and Krassowska (1993; see also Keener and Panfilov, 1997). As the calculation here is nearly identical to that given in Section 7.8.2, we skip directly to the answer. The potential ϕ is given by

$$\phi = \Phi(z) + \epsilon W\left(\frac{z}{\epsilon}\right)\cdot T^{-1}\nabla_z\Phi(z) + O(\epsilon^2\Phi), \tag{12.214}$$

where the mean field potential Φ satisfies the averaged Poisson equation

$$\nabla \cdot (\sigma_{\text{eff}} \nabla \Phi) = -\chi \frac{1}{S_\xi} \int_{\Gamma_m} I_m(z, \xi)\, dS_\xi. \tag{12.215}$$

where $\chi = \frac{S_m}{v}$ is the cell surface area per unit volume, a quantity having units of length, and S_ξ is the surface area of Γ_m in units of ξ. $W(\xi)$ is the fundamental solution vector, periodic in ξ with zero surface average value, $\int_{\Gamma_m} W(\xi)\, d\xi = 0$, and it satisfies the vector partial differential equation

$$\nabla_\xi^2 W(\xi) = 0, \tag{12.216}$$

subject to the boundary condition

$$\mathbf{n} \cdot (\nabla_\xi W(\xi) + I) = 0, \tag{12.217}$$

on Γ_m the membrane wall. Here I is the identity matrix. Finally the effective conductivity tensor is

$$\Sigma = \frac{1}{r_c v} \int_{\Gamma_m} (\nabla_\xi W(\xi) + I)\, dV_\xi, \tag{12.218}$$

where $\sigma_{eff}(x) = T \Sigma T^{-1}$, and v is the volume of the rectangular box containing the cell.

The derivation of the bidomain model follows quickly from this. We define the potentials in the intracellular and extracellular domains as ϕ_i and ϕ_e, respectively. Then the membrane potential is the difference between the two potentials across the membrane boundary Γ_m between the two domains:

$$\phi = (\phi_i - \phi_e)|_{\Gamma_m}. \tag{12.219}$$

At each point of the cell membrane the outward transmembrane current is given by

$$I_m = C_m \frac{d\phi}{dt} + \frac{1}{R_m} f_m(\phi), \tag{12.220}$$

where C_m is the membrane capacitance and f_m/R_m represents the transmembrane ionic current. The parameter R_m is the membrane resistance.

It follows from homogenization (12.214), (12.215) that

$$\phi_i = V_i(x) + \epsilon W_i\left(\frac{x}{\epsilon}\right) \cdot T^{-1} \nabla V_i(x) + O(\epsilon^2 V_i), \tag{12.221}$$

$$\phi_e = V_e(x) + \epsilon W_e\left(\frac{x}{\epsilon}\right) \cdot T^{-1} \nabla V_e(x) + O(\epsilon^2 V_e), \tag{12.222}$$

and that $V_i(x)$ and $V_e(x)$ satisfy the averaged equations

$$\nabla \cdot (\sigma_i \nabla V_i) = -\nabla \cdot (\sigma_e \nabla V_e) = \chi \frac{1}{S_\xi} \int_{\Gamma_m} I_m(x, \xi)\, dS_\xi, \tag{12.223}$$

where I_m is the transmembrane current (positive outward). We calculate (using that $\int_{\Gamma_m} W_i dS_\xi = \int_{\Gamma_m} W_e\, dS_\xi = 0$) that

$$\frac{1}{S_\xi} \int_{\Gamma_m} I_m(x, \xi)\, dS_\xi = C_m \frac{\partial V}{\partial t} + \frac{1}{S_\xi} \int_{\Gamma_m} \frac{1}{R_m} f_m\, (V + \epsilon H(\xi, x))\, dS_\xi, \tag{12.224}$$

where

$$H(\xi,x) = W_i(\xi)\cdot T^{-1}\nabla V_i(x) - W_e(\xi)\cdot T^{-1}\nabla V_e(x), \tag{12.225}$$

$$V = V_i - V_e. \tag{12.226}$$

It follows that

$$\frac{R_m}{\chi}\nabla\cdot(\sigma_i\nabla V_i) = -\frac{R_m}{\chi}\nabla\cdot(\sigma_e\nabla V_e) = C_m R_m\frac{\partial V}{\partial t} + \frac{1}{S_m}\int_{\Gamma_m} f_m\left(V + \epsilon H(\xi,x)\right)\, dS_\xi. \tag{12.227}$$

In the limit $\epsilon = 0$, the equations (12.227) reduce to the standard bidomain model. With $\epsilon \neq 0$, this model can be used to study the effects of large defibrillating currents.

12.9 EXERCISES

1. (a) The fundamental solution of Poisson's equation in free space with a unit source at the origin ($\nabla^2\phi = -\delta(x)$) is $\phi(x) = \frac{1}{4\pi|x|}$.
   Find the solution of Poisson's equation with a source at the origin and a sink of equal strength at $x = x_1$, and let $|x_1| \to 0$. What must be assumed about the strength of the source and the sink in order to obtain a nonzero limiting potential?

   (b) Find the solution of Poisson's equation with a dipole source by solving the problem $\nabla^2\phi = \frac{1}{\epsilon}(\delta(x-\epsilon v) - \delta(x))$ with $|v| = 1$, and then taking the limit $\epsilon \to 0$.

2. Determine the heart rate for the ECG recording in Fig. 12.3a. The subject was sedated at the time this recording was made. What are the effects of the sedation?

3. Identify the different deflections in the ECG recording shown in Fig. 12.55. What can you surmise about the nature of propagation for the extra QRS complex (called an *extrasystole*). Because of the apparent periodic coupling between the normal QRS and the extrasystole, this rhythm is called a *ventricular bigeminy*.

4. (a) Suggest a diagnosis for the ECG recording in Fig. 12.56.
   Hint: What does the inverted P-wave suggest? What is the heart rate?

   (b) What possible mechanisms can account for the failure of the SA node to generate the heartbeat?

5. Estimate the mean deflection of the QRS complex in each of the six standard leads (I, II, III, aVR, aVL, aVF) in Fig. 12.7 and then estimate the mean heart vector for the normal heartbeat.

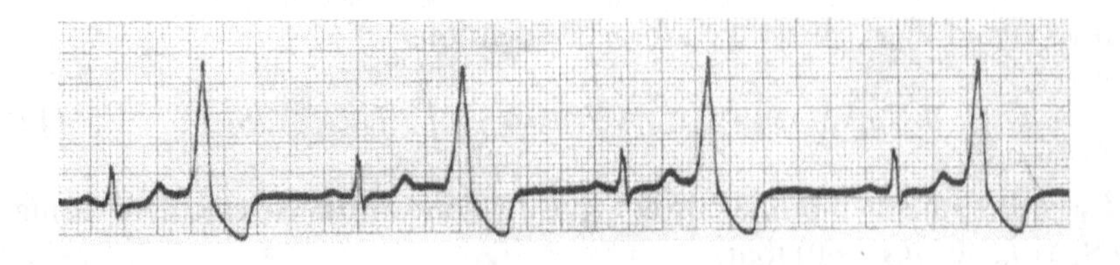

Figure 12.55 ECG recording of a ventricular extrasystole for exercise 3. (Rushmer, 1976, Fig 8-25, p. 314, originally from Guneroth, 1965.)

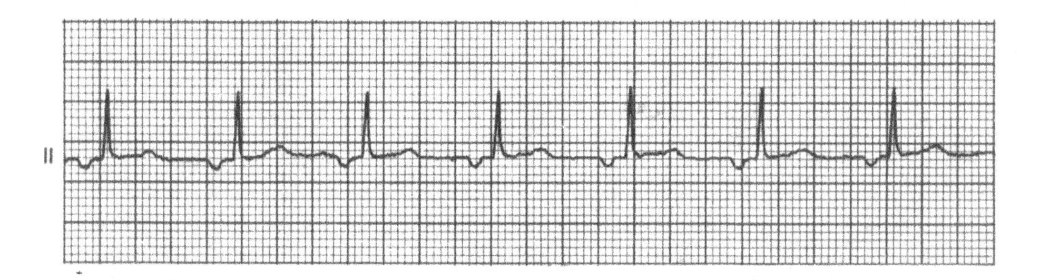

Figure 12.56 ECG for Exercise 4 (Goldberger and Goldberger, 1994, p. 45).

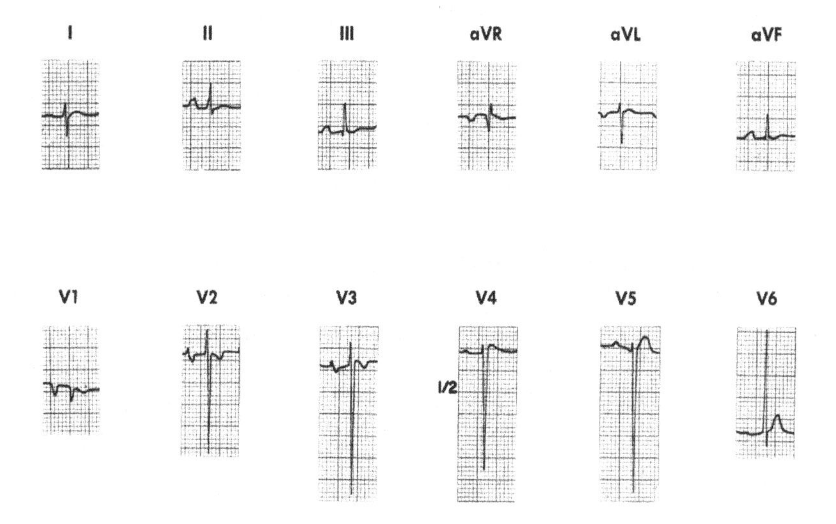

Figure 12.57 ECG for Exercise 7. Can you determine the nature of this abnormality? (Rushmer, 1976, Fig. 8-51, p. 343.)

6. Improve your skill at reading ECGs by finding the mean heart vector for the QRS complexes in Figs. 12.4, 12.9, and 12.10. Why is hypertrophy the diagnosis for Figs. 12.9 and 12.10? In what direction are the heart vectors deflected in these figures?

7. Find the mean heart vector for the ECG recording shown in Fig. 12.57. What mechanism can you suggest that accounts for this vector?

8. By assuming a fast equilibrium between states N_3 and N_4, and between states C_3 and C_4, derive the simplified reaction diagram of Fig. 12.19C for the L-type channel model.

9. Compare the solution of (12.58) with the approximation (12.60). In particular, plot $\frac{\lambda_g^2}{L^2}(Q+r)$ as a function of Q for different values of $\frac{L}{\lambda_g}$.

10. Use cable theory to estimate the effective coupling resistance for cardiac cells in the longitudinal and transverse directions. Assume that cells are 0.01 cm long, 0.00167 cm wide, $\chi = S/v = 2400$ cm^{-1}, $R_m = 7000$ Ω-cm^2, $R_c = 150$ Ω-cm, $R_e = 0$, with a longitudinal space constant of 0.09 cm and transverse space constant 0.03 cm (appropriate for canine

crista terminalis). What difference do you observe with a transverse space constant of 0.016 cm (appropriate for sheep epicardium)?

11. How should the coupling coefficient r_g be chosen so that the decay rate for the discrete linear model (12.78) with $I_{\text{ion}} = -V/R_m$ matches the space constant of the medium?

12. Using that the longitudinal and transverse cardiac action potential deflections are $\Delta V_{eL} =$ 74 mV, $\Delta V_{eT} = 43$ mV, that the membrane potential has $\Delta V = 100$ mV (independent of direction), and that the axial speed of propagation in humans is 0.5 m/s in the longitudinal fiber direction and 0.17 m/s in the transverse direction, determine the ratio of coefficients of the conductivity tensors $\frac{\sigma_{eL}}{\sigma_{iL}}, \frac{\sigma_{eT}}{\sigma_{iL}}$, and $\frac{\sigma_{iT}}{\sigma_{iL}}$. What are these ratios in dog if the ratio of longitudinal to transverse speeds is 2:1?

13. Determine the conditions under which there is propagation failure at a junction when the ionic current model is

$$I_{\text{ion}} = g_{\text{Na}} h m^2(V)(V_{\text{Na}} - V) + g_{\text{K}}(V_{\text{K}} - V), \tag{12.228}$$

where

$$m(V) = \begin{cases} 0, & V < V_{\text{K}}, \\ \frac{V - V_{\text{K}}}{V_{\text{Na}} - V_{\text{K}}}, & V_{\text{K}} < V < V_{\text{Na}}, \\ 0, & V > V_{\text{Na}} \end{cases} \tag{12.229}$$

Reasonable values for the parameters are $V_{\text{K}} = -84$ mV, $V_{\text{Na}} = 66$ mV, $g_{\text{Na}} = 3.33$ mS/cm^2, and $g_{\text{K}} = 0.66$ mS/cm^2, although these numbers are not needed to solve this problem.

14. Consider the following simple model of a forced periodic oscillator, called the *Poincaré oscillator* (also called a *radial isochron clock* or a *snap-back oscillator*; Guevara and Glass, 1982; Hoppensteadt and Keener, 1982; Keener and Glass, 1984; Glass and Kaplan, 1995). A point is moving counterclockwise around a circle of radius 1. At some point of its phase, the point is moved horizontally by an amount A and then allowed to instantly "snap back" to radius 1 moving along a radial line toward the origin (see Fig. 12.58).

    (a) Determine the *phase resetting curve* for this process. That is, given the phase θ before resetting, find the phase ϕ after resetting the clock. Plot ϕ as a function of θ for several values of A.

    (b) Show that for $A < 1$ the phase resetting function is a *type 1* map, satisfying $\phi(\theta + 2\pi) = \phi(\theta) + 2\pi$.

    (c) Show that for $A > 1$, the phase resetting function is a *type 0* map, for which $\phi(\theta + 2\pi) = \phi(\theta)$.

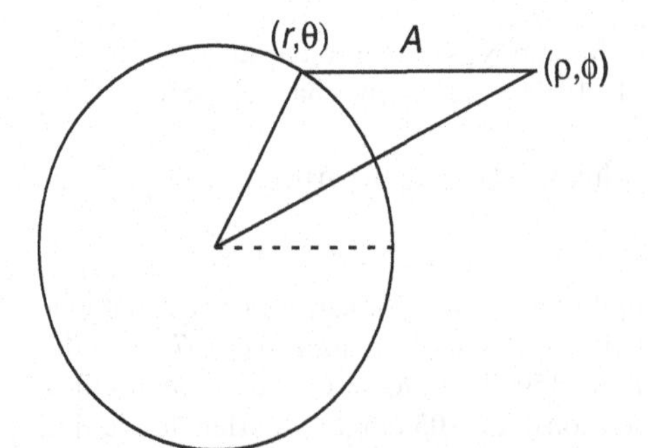

Figure 12.58 Diagram of phase resetting for the Poincaré oscillator (Exercise 14).

(d) Show that there is a *phase singularity*, that is, values of A and θ for which a new phase is not defined (Winfree, 1980).

(e) Construct a map $\theta_n \mapsto \theta_{n+1}$ by resetting the clock every T units of phase, so that

$$\theta_{n+1} = \phi(\theta_n) + T. \tag{12.230}$$

Show that for $A < 1$, this is a circle map. Determine the values of A and T for which there is one-to-one phase locking.

15. A simple model of the response of an excitable cell to a periodic stimulus is provided by the *integrate-and-fire* model (Knight, 1972; Keener, Hoppensteadt and Rinzel, 1981). In this model the membrane is assumed to be passive unless the threshold is reached, whereupon the membrane "fires" and is immediately reset to zero. The equations describing the evolution of the membrane potential are

$$\frac{dv}{dt} = -\gamma v + S(t), \tag{12.231}$$

and $v(t^+) = 0$ if $v(t) = v_T$, the threshold. We take the periodic stimulus to be a simple sinusoidal function, $S(t) = S_0 + S_m \sin(\omega t)$.

Let T_n be the time of the nth firing. Formulate the problem of determining T_{n+1} from T_n as a circle map. For what parameter values is this a continuous circle map and for what parameter values is it a discontinuous circle map? (Guevara and Glass, 1982.)

16. Suppose the membrane potential rises at a constant rate λ until it reaches a threshold $\theta(t)$, at which time the potential is reset to zero. Suppose that the threshold is the simple sinusoidal function $\theta(t) = \theta_0 + \theta_m \sin(\omega t)$. Let T_n be the time of the nth firing. Formulate the problem of determining T_{n+1} from T_n as a circle map. For what parameter values is this a continuous circle map and for what parameter values is it a discontinuous circle map? (Glass and Mackey, 1979.)

17. Suppose a constant-current stimulus I is added to a cable with FitzHugh–Nagumo dynamics and piecewise-linear function f as in (12.133).

(a) Find the steady-state solution as a function of input current I.

(b) Examine the stability of this steady-state solution. Show that the eigenvalues are eigenvalues of a Schrödinger equation with a square well potential. Find the critical Hopf bifurcation curve. Show that for ϵ sufficiently small, the solution is stable if I is small or large, but there is an intermediate range of I for which the solution is unstable (Rinzel and Keener, 1983).

Hint: Because the function $f(v)$ is piecewise linear, the potential for the Schrödinger equation is a square well potential. Solve the resulting transcendental equations numerically.

18. (a) Carry out the calculations of Section 12.3.4 for a one-dimensional piecewise-linear model (12.133) with $\alpha(r)$ specified by (12.132), $b = \frac{1}{2}$. Determine the critical stability curve.

(b) Generalize this calculation by supposing that the oscillatory cells have coupling coefficient D. What is the effect of the coupling coefficient of the oscillatory cells on the critical stability curve? Show that oscillatory behavior is more likely with weak coupling.

19. (a) Use homogenization theory to show that the averaged approximation of (12.68) is (12.69).

(b) Use homogenization to separate (12.162) into two equations on different spatial scales. Show that $V \approx u_0(z,\tau) + \epsilon J(\tau)H(\eta) - \epsilon W(\eta)\frac{\partial u_0(z,\tau)}{\partial z})$, where $W'(\eta) = 1 - \frac{D}{d(\eta)}$, $H'(\eta) - \frac{\Sigma_e}{\sigma_e} - \frac{D}{d}$, and u_0 satisfies the averaged equation (12.166). Show that this answer is valid even if $J(\tau) = O\left(\frac{1}{\epsilon}\right)$. (Hint: Use the projection method rather than a standard power series expansion.)

20. Suppose that the nonlinear function $f(V)$ for an excitable medium is well represented by

$$f(V) = V\left(\frac{V}{\theta} - 1\right), \tag{12.232}$$

at least in the vicinity of the rest point and the threshold. Find the relationship between minimal time and stimulus strength to directly stimulate the medium.
Hint: Evaluate 12.174.

21. Find the relationship between minimal time and stimulus strength to directly stimulate the medium (as in the previous problem) for the Beeler–Reuter model of myocardial tissue (Section 12.2). To do this, set $m = m_\infty(V)$ and set all other dynamic variables to their steady-state values and then evaluate (12.174).

22. Suppose a collection of FitzHugh–Nagumo oscillators described by

$$\delta\frac{dv}{dt} = f(v) - w, \tag{12.233}$$

$$\frac{dw}{dt} = v - \alpha, \tag{12.234}$$

with $f(v) = v(v-1)(\alpha - v)$ with parameter values $\delta = 0.05$, $\alpha = 0.4$, is coupled through the variable v. Calculate (numerically) the coupling function

$$h(\phi) = \int_0^P y_1(\sigma)v(\sigma + \phi)\, d\sigma, \tag{12.235}$$

where y_1 is the first component of the periodic adjoint solution.

23. (a) Consider a linear system of differential equations $\frac{dy}{dt} = A(t)y$ and the corresponding adjoint system $\frac{dv}{dt} = -A^T(t)v$, where $A(t)$ is periodic with period P. Let $Y(t)$ and $V(t)$ be matrix solutions of these equations and suppose that $V^T(0)Y(0) = I$. Show that $V^T(P)Y(P) = I$.

(b) The eigenvalues of the matrix $Y(P)Y^{-1}(0)$ are called the Floquet multipliers for the system $\frac{dy}{dt} = A(t)y$. What does the fact that $V^T(0)Y(0) = V^T(P)Y(P)$ imply about the Floquet multipliers for $\frac{dy}{dt} = A(t)y$ and its adjoint system?

24. Suppose two identical lambda–omega oscillators of the form (12.189)–(12.190) are coupled through their first variables x, with, in the notation of (12.195),

$$H = \begin{pmatrix} x \\ 0 \end{pmatrix} \tag{12.236}$$

and $a_{12} = a_{21} = \alpha$, and $a_{11} = a_{22} = 0$. Calculate the phase resetting function $h(\phi)$ and determine the stability of in-phase and out-of-phase, phase locked solutions, when $\alpha = \pm 1$.

CHAPTER 13

Blood

Blood is composed of two major ingredients: the liquid *blood plasma* and several types of cells suspended within the plasma. The cells constitute approximately 40% of the total blood volume and are grouped into three major categories: *erythrocytes* (red blood cells), *leukocytes* (white blood cells), and *thrombocytes* (platelets). The red blood cells remain within the blood vessels, and their function is to transport oxygen and carbon dioxide. The white blood cells fight infection, and thus are able to migrate out of the blood vessels and into the tissues. Platelets are not complete cells, but are small detached fragments of much larger cells called megakaryocytes. Their principal function is to aid in blood clotting.

Leukocytes themselves are subdivided into a number of categories: *granulocytes* (approximately 65%), *lymphocytes* (30%), *monocytes* (5%), and *natural killer* cells. Granulocytes are further subdivided into *neutrophils* (95%), *eosinophils*, (4%) and *basophils* (1%). Neutrophils phagocytose (i.e., ingest) and destroy small foreign bodies such as bacteria, basophils help mediate inflammatory reactions by secreting histamine, while eosinophils help to destroy parasites and modulate allergic responses.

Monocytes mature into macrophages, which help the neutrophils with the phagocytosis of foreign bodies. Macrophages are much larger than neutrophils and can phagocytose larger microorganisms. They are also responsible for removing dead or damaged cells.

Lymphocytes come in two main types, T and B, whose main functions are the production of antibodies and the removal of virus-infected cells.

The types and functions of blood cells are summarized in Table 13.1.

In this chapter we consider only a small number of mathematical questions associated with these cell types. A more general study of lymphocytes and the immune system would require an entire volume all to itself. A good basic introduction to immunology

Table 13.1 Types and functions of blood cells. The right-hand column shows typical concentrations in human blood. Adapted from Alberts et al. (1994), Table 22-1.

Type of cell	Main functions	cells/liter
Red blood cells (erythrocytes)	transport oxygen and carbon dioxide	5×10^{12}
White blood cells (leukocytes)		
Granulocytes		
Neutrophils	ingest and destroy bacteria	5×10^{9}
Eosinophils	destroy larger parasites and modulate allergic responses	2×10^{8}
Basophils	release histamine in certain immune reactions	4×10^{7}
Monocytes	become macrophages that ingest invading bacteria and foreign bodies, and remove dead or damaged cells	
Lymphocytes		
B Lymphocytes	make antibodies	2×10^{9}
T Lymphocytes	kill cells infected by virus, and regulate activities of other leukocytes	1×10^{9}
Natural killer cells (NK)	kill some tumor cells, and cells infected by virus	
Platelets	cell fragments arising from megakaryocytes in bone marrow; initiate blood clotting.	3×10^{11}

is Janeway et al. (2001), while a more detailed discussion of mathematical models and the immune system is given by Nowak and May (2000) and Perelson (2002).

13.1 Blood Plasma

The blood plasma is 89–95% water, with a variety of dissolved substances. The dissolved substances with small molecular weight include bicarbonate, Cl^-, phosphorus, Na^+, Ca^{2+}, K^+, magnesium, urea, and glucose. There are also large protein molecules including *albumin* and α-, β-, and γ-*globulins*. Of these proteins, albumin has the highest molar concentration in plasma and this makes the greatest contribution to the plasma osmotic pressure.

In addition, gases such as carbon dioxide and oxygen are dissolved in the blood plasma. For an ideal gas, the pressure, volume, and temperature are related by the ideal gas law,

$$PV = nkT, \tag{13.1}$$

where P is the pressure, V is the volume, n is the number of gas molecules, k is Boltzmann's constant, and T is temperature in Kelvin. Since concentration is $c = n/V$, for an ideal gas

$$P = ckT. \tag{13.2}$$

This representation of concentration is in units of molecules per volume, and while this seems natural, it is not the usual way that concentrations are represented. To express concentration in terms of moles per unit volume, we multiply and divide (13.2) by Avogadro's number N_A to obtain

$$P = CRT, \tag{13.3}$$

where $C = c/N_A$ and $R = kN_A$ is the universal gas constant.

Air is a mixture of different gases, with 78% nitrogen and 21% oxygen. Each of these gases contributes to the total pressure of the mixture via its *partial pressure*. The partial pressure of gas i, P_i, is defined by

$$P_i = x_i P, \tag{13.4}$$

where x_i is the mole fraction of gas i, and P is the total pressure of the gas mixture. Thus, by definition, the total pressure of a gas mixture is the sum of the partial pressures of each individual gas in the mixture. In an ideal mixture the partial pressure of gas i is equal to the pressure that gas i would exert if it alone were present.

When a gas with partial pressure P_i comes into contact with a liquid, some of the gas dissolves in the liquid. When a steady state is reached, the amount of gas dissolved in the liquid is a function of the partial pressure of the gas above the liquid. If the concentration of the dissolved gas is low enough, thus forming an ideally dilute solution, then P_i is related to the concentration of gas i by

$$c_i = \sigma_i P_i, \tag{13.5}$$

where c_i is the concentration of gas i, and σ_i is called the *solubility*. In general σ_i is a function of the temperature and the total pressure above the liquid. In Table 13.2 are shown the solubilities of important respiratory gases in blood, where it can be seen, for example, that the solubility of carbon dioxide in blood is about 20 times larger than that of oxygen.

Table 13.2 Solubility of respiratory gases in blood plasma.

Substance	σ (Molar/mm Hg)
O_2	1.4×10^{-6}
CO_2	3.3×10^{-5}
CO	1.2×10^{-6}
N_2	7×10^{-7}
He	4.8×10^{-7}

13.2 Blood Cell Production

Blood cells are produced by the bone marrow and must be continually produced over a human's lifetime. For example, one cubic millimeter of blood contains around 5 million erythrocytes. These cells have an average lifetime of 120 days and are estimated to travel through about 700 miles of blood vessels during their life span. Because of aging and rupturing, red blood cells must be constantly replaced. On average, the body must produce 3×10^9 new erythrocytes for each kilogram of body weight every day.

In a child before the age of 5, blood cells are produced in the marrow of essentially all the bones. However, with age, the marrow of the long bones becomes quite fatty and so produces no more blood cells after about age 20. In the adult, most blood cells are produced in the marrow of membranous bones, such as the vertebrae, sternum, ribs, and ilia.

In the bone marrow there are cells, called *pluripotential hematopoietic stem cells*, from which all of the cells in the circulating blood are derived. Remarkably, if endogenous hematopoiesis (blood cell production) is halted in an animal by irradiation, it can be completely restored by the addition of exogenous stem cells. As stem cells grow and reproduce, a portion of them remains exactly like the original pluripotential cells, maintaining a more or less constant supply of these cells. The larger portion of the reproduced stem cells differentiates to form other cells, called *committed stem cells*, or committed progenitors. The *in vitro* analogue of the committed stem cells are called colony-forming units (CFU). The committed stem cells produce colonies of specific types of blood cells, including erythrocytes, lymphocytes, granulocytes, monocytes, and megakaryocytes.

Although the basic outline of the process is known, the details, both of the mechanisms that control blood cell production, and the exact lineage of each cell type, remain elusive. It seems that hematopoietic stem cells pass through a series of divisions and maturational steps, at each stage of which they progressively lose their potential to change their differentiation pathway (i.e., the choice of what the cell can differentiate to becomes more limited), ending with a fully differentiated cell type such as an erythrocyte. However, it has proven difficult to separate the precursors of one cell type from those of another, leading to considerable uncertainty in the exact cell lineages. A tentative cell lineage diagram is given in Fig. 13.1.

Growth and differentiation of blood cells are controlled by an intricate array of soluble factors, or cytokines, that include hematopoietic growth and differentiation factors. However, most cytokines are known to have multiple effects and interactions, and there is no clear distinction between those that control growth and those that control differentiation. Formation of cytokines is itself controlled by factors outside the bone marrow such as, in the case of red blood cells, low oxygen concentration for an extended period of time.

The feedback system that controls red blood cell production is relatively well understood. The principal factor stimulating red blood cell production is the hormone

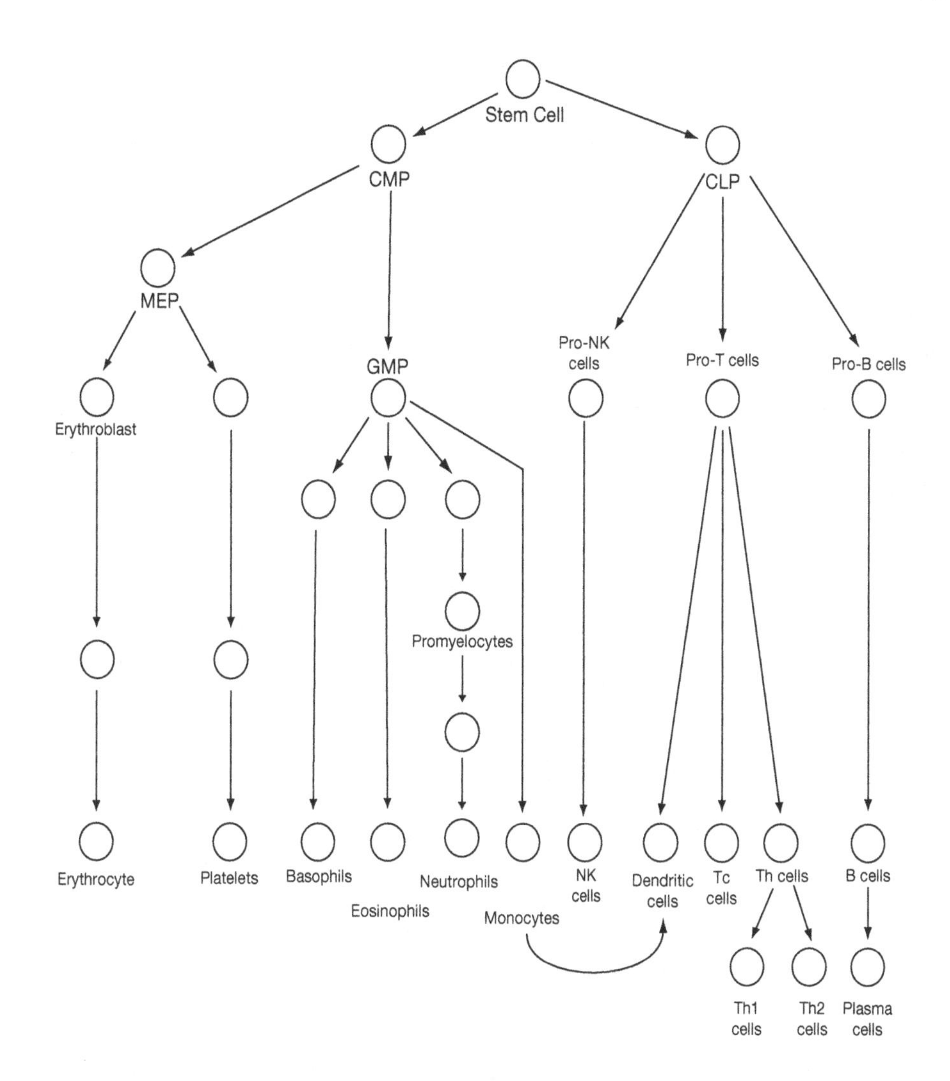

Figure 13.1 Schematic diagram of cell lineage in hematopoiesis. Stem cells differentiate into common myeloid precursors (CMP) and common lymphoid precursors (CLP). CMP then differentiate into megakaryotic/erythroid progenitors (MEP), which differentiate further into erythrocytes and platelets. CMP also give rise to granulocyte–monocyte–megakaryocyte–erythroid common progenitors (GMP), whence come granulocytes and monocytes. CLP differentiate into T lymphocytes, B lymphocytes, and natural killer cells. Th cells are helper lymphocytes, while Tc cells are cytotoxic T lymphocytes. Adapted from Kluger et al. (2004), Fig. 1.

erythropoietin. About 90% of the erythropoietin is secreted by renal tubular epithelial cells when blood is unable to deliver sufficient oxygen. The remainder is produced by other tissues (mostly the liver). When both kidneys are removed or destroyed by renal disease, the person invariably becomes anemic because of insufficient production of erythropoietin.

The role of erythropoietin in bone marrow is twofold. First, it stimulates the production of pre-erythrocytes, called *proerythroblasts,* and it also controls the speed at which the developing cells pass through the different stages. Normal production of red blood cells from stem cells takes 5–7 days, with no appearance of new cells before 5 days, even at high levels of erythropoietin. At high erythropoietin levels the rate of red blood cell production (number per unit time) can be as much as ten times normal, even though the maturation rate of an individual red blood cell varies much less.

Thus, in response to a drop of oxygen pressure in the tissues, an increased production of erythropoietin causes an increase in the rate of production of red blood cells, thus tending to restore oxygen levels. The control mechanisms that operate when the red blood cell count is too high (a condition called polycythemia or erythrocytosis) are less clear. Details of the regulatory system governing red blood cells can be found in Williams (1990), while an excellent review of much of the material discussed in this section can be found in Haurie et al. (1998).

Feedback control of the other types of blood cells is even less well understood. Production of granulocytes is controlled by granulocyte colony-stimulating factor (G-CSF), which is produced by a number of tissues, including fibroblasts, and endothelial and epithelial tissue, while megakaryocyte production is controlled, at least in part, by thrombopoietin. These feedback controls are sketched in Fig. 13.2.

13.2.1 Periodic Hematological Diseases

In most people, the production of blood cells is relatively constant. However, there are a number of pathological conditions that exhibit oscillatory behavior. The most widely studied is *cyclical neutropenia,* a condition in which the number of neutrophils periodically drops to a very low level, typically every 19 to 21 days. As it happens, the gray collie suffers from a similar disease, with a period that ranges from 11 to 15 days, and this has greatly helped the study of the disease in humans. In both humans and collies, the periodic variation in neutrophil numbers is accompanied by oscillations in the number of platelets, often the monocytes and eosinophils, and occasionally the lymphocytes. The oscillation period in these other cell types is, however, not always the same as that of the neutrophils. Some typical experimental data from nine dogs are shown in Fig. 13.3.

Another periodic blood disease is periodic chronic myelogenous leukemia (CML). Being a leukemia, this disease is characterized by the uncontrolled growth of white blood cells. However, this growth can sometimes occur in an oscillatory manner. Data showing oscillations of white blood cell count in a twelve-year-old girl with periodic CML are shown in Fig. 13.4A, and blood cell counts in cyclical neutropenia and CML are compared in Fig. 13.4B. The oscillations in CML occur with a much larger period, and generate much greater numbers of neutrophils, as expected for a leukemia.

Over the last 25 years, there has been a large number of mathematical investigations of these periodic behaviors. Early studies were those of Mackey (1978, 1979),

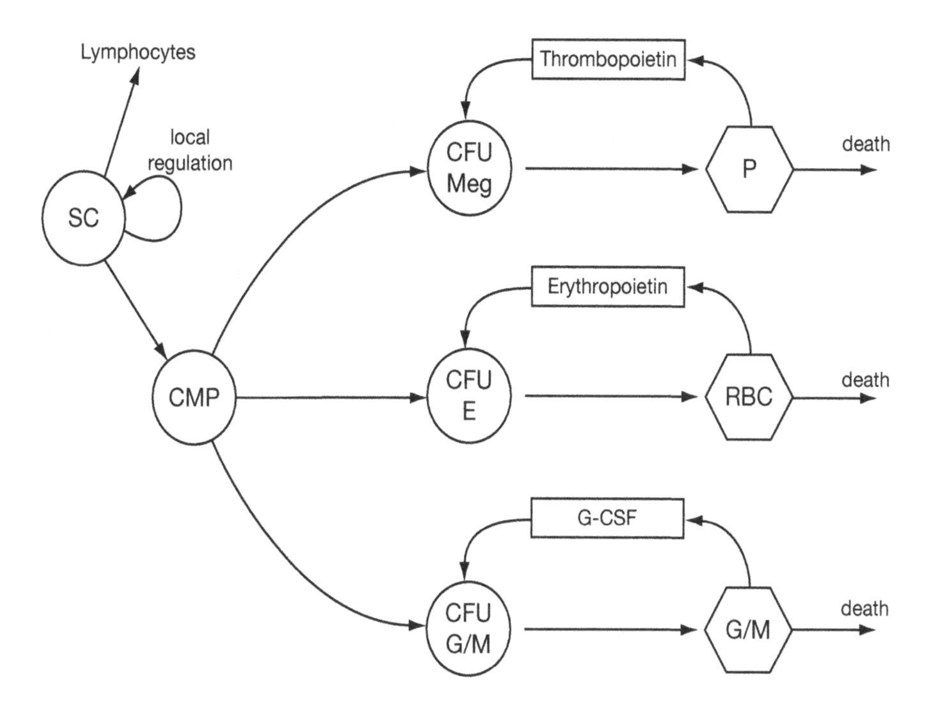

Figure 13.2 Schematic diagram of the control of blood cell production. SC: stem cells. CMP; common myeloid precursors (see Fig. 13.1). CFU: colony-forming units. E: erythrocytes. Meg: megakaryocytes. G/M: granulocytes/monocytes. P: platelets. RBC: red blood cells. The feedback loops in the diagrams are not well understood, but are certainly far more complex than shown here.

Mackey and Glass (1977), Mackey and Milton (1987), Milton and Mackey (1989), and Bélair et al. (1995), while the recent work of Mackey, Haurie, and their colleagues (Haurie et al., 1998; Hearn et al., 1998; Haurie et al., 1999; Haurie et al., 2000; Mackey et al., 2003; Colijn and Mackey, 2005a, b) have extended the earlier models into much more elaborate versions, with detailed fits to experimental data. Another of the major modeling groups studying periodic hematopoiesis is that of Wichmann, Schmitz, and their colleagues (Wichmann et al., 1988; Schmitz et al., 1990, 1993).

One of the principal goals of the recent modeling work has been to discover the site of action of the feedback that controls blood cell growth and that can lead to oscillatory behavior. Here, we first illustrate the basic modeling concepts by working through a relatively simple delay differential equation model, and then, using a more recent model, study the question of where the feedback occurs.

13.2.2 A Simple Model of Blood Cell Growth

We model the blood production cycle as follows (Belair et al., 1995). We let $n(x,t)$ be the density of blood cells at time t that are x units old, i.e., that were released into the

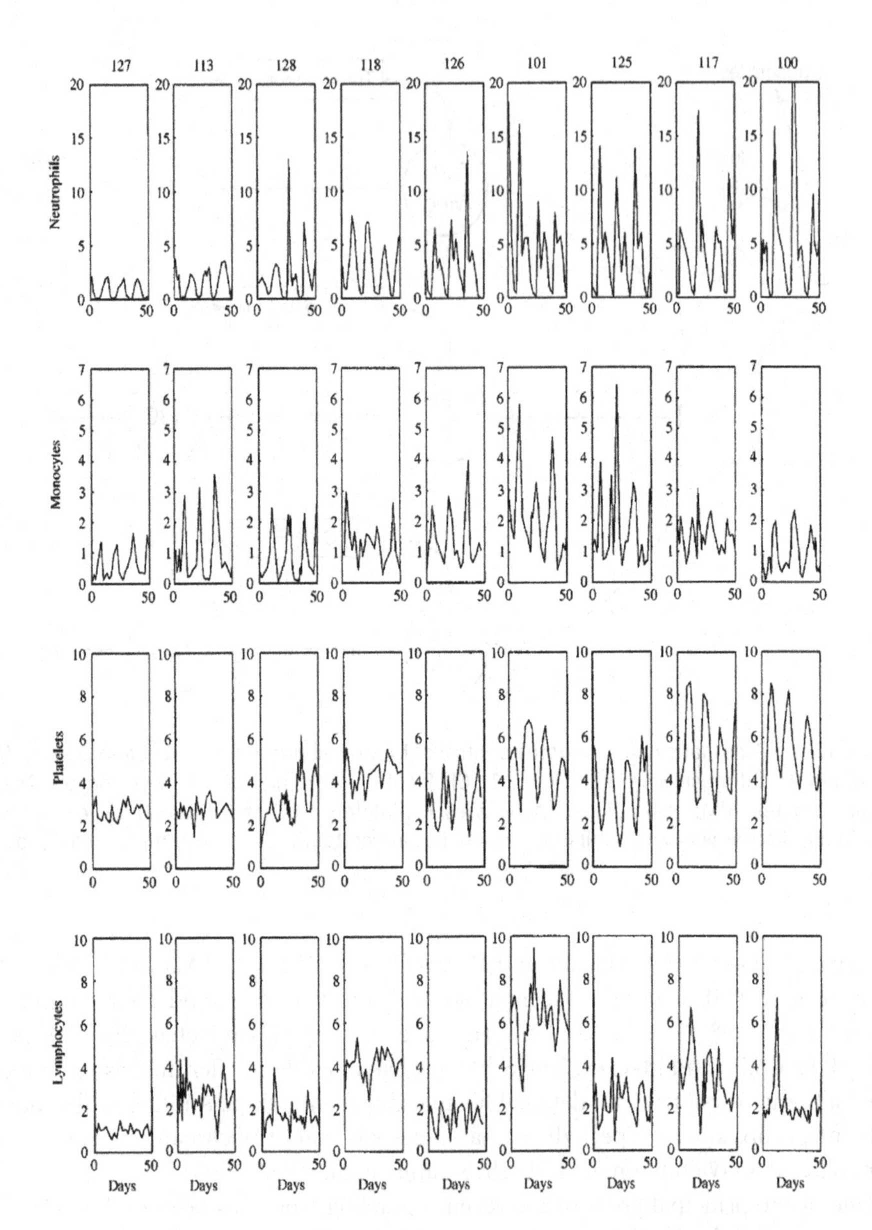

Figure 13.3 Blood counts against time (in days) for nine gray collies with cyclical neutropenia. Units: cells $\times 10^{-5}\text{mm}^{-3}$ for the platelets and cells $\times 10^{-3}\text{mm}^{-3}$ for the other cell types (Haurie et al. 2000, Fig. 1).

bloodstream at time $t-x$. We suppose that as they age, a fixed percentage of them die, but at some age X all cells die. Consider the rate of change of the total number of cells with age in the interval between $x = a$ and $x = b$. Since $n(b,t)$ is the rate at which cells leave the interval $[a,b]$ due to aging, and $n(a,t)$ is the rate at which they enter that

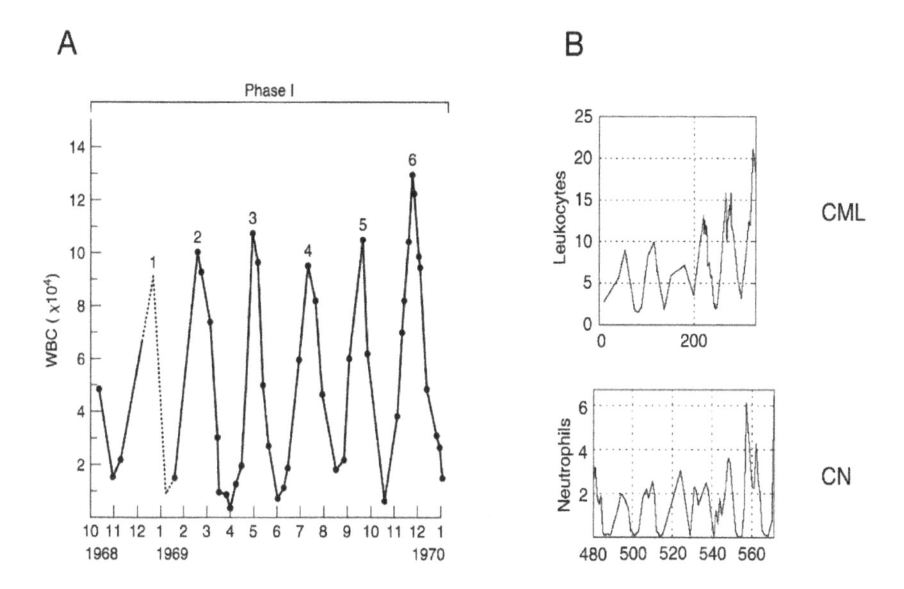

Figure 13.4 A: White blood cell count (per liter) as a function of time for a twelve-year-old girl with periodic CML. (This research was originally published in Gatti et al., 1973, Fig. 1, © the American Society of Hematology) B: Comparison of blood cell counts in CML and cyclical neutropenia (CN). The CN data are from dog 113 (Fig. 13.3, as reproduced in Colijn and Mackey, 2005b, Fig. 2), while the CML data are from Vodopick et al. (1972), as reproduced in Colijn and Mackey (2005a), Fig. 2. For both sets of data in B, the units of cell number are 10^8 per kilogram, and the time scale is days.

interval, the rate of change of total cell number is given by

$$\frac{d}{dt}\int_a^b n(x,t)\,dx = \int_a^b -\beta n(x,t)\,dx - n(b,t) + n(a,t), \tag{13.6}$$

where β is the death rate of the cells. Differentiating (13.6) with respect to b and replacing b by x gives the conservation equation

$$\frac{\partial n}{\partial t} + \frac{\partial n}{\partial x} = -\beta n. \tag{13.7}$$

In general, the death rate is expected to be a function of age, so that $\beta = \beta(x)$. However, for this model we take the death rate to be independent of age. At any given time the total number of blood cells in circulation is

$$N(t) = \int_0^X n(x,t)dx. \tag{13.8}$$

Now we suppose that the production of blood cells is controlled by N, and that once a cohort of cells is formed in the bone marrow, it emerges into the bloodstream as mature cells some fixed time d later, about 5 days. Here we ignore the fact that at high levels of feedback (for example, at high levels of erythropoietin, which occurs when oxygen levels are low), cells mature more rapidly. Thus,

$$n(0,t) = F(N(t-d)), \tag{13.9}$$

where F is some nonlinear production function that is monotone decreasing in its argument. The function F is related to the rate of secretion of growth inducer (erythropoietin, for example) in response to the blood cell population size.

The steady-state solution for this model is easy to determine. We set $\partial n/\partial t = 0$ and find that

$$n(x) = \begin{cases} n(0)e^{-\beta x}, & x < X, \\ 0, & x > X, \end{cases} \tag{13.10}$$

where $n(0)$ is yet to be determined. If $N_0 = \int_0^X n(x)\,dx$ is the total number of cells in steady state, it follows that

$$N_0 = \int_0^X n(0)e^{-\beta x}\,dx = \frac{n(0)}{\beta}(1 - e^{-\beta X}). \tag{13.11}$$

At steady state, $F(N_0) = n(0)$, and thus it follows that

$$F(N_0) = \frac{\beta N_0}{1 - e^{-\beta X}}. \tag{13.12}$$

Since $F(N_0)$ is a monotone decreasing function of N_0, (13.12) is guaranteed to have a unique solution. In fact, the solution is a monotone decreasing function of the parameter β, indicating that at higher death rates, the cell population drops while the production of cells increases. An illustration of these facts is provided by the graph in Fig. 13.5, where the two curves $F(N)$ and $\frac{\beta N}{1-e^{-\beta X}}$ are plotted as functions

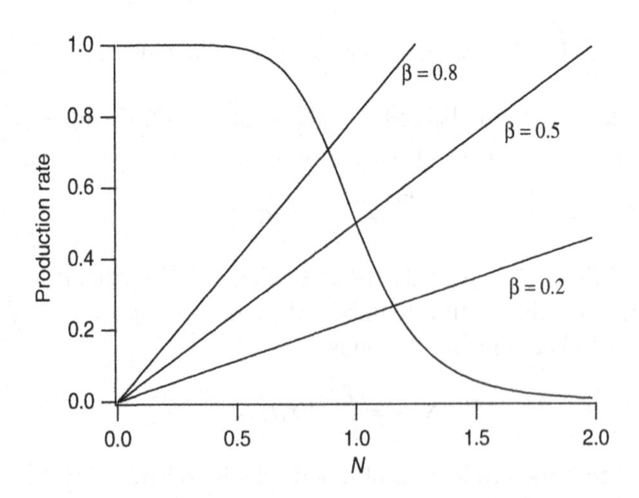

Figure 13.5 Plot of left- and right-hand sides of (13.12) for three different values of β and for $F(N) = \frac{1}{1+N^7}$ and $X = 10$.

of N. Here the function $F(N)$ is taken to be $F(N) = \frac{A}{1+N^7}$, as suggested by data from autoimmune-induced hemolytic anemia in rabbits (Belair et al., 1995).

The next interesting question to ask is whether this steady solution is stable or unstable. It is convenient to integrate the partial differential equation (13.7) to get an ordinary differential equation. Integrating (13.7) from $x = 0$ to $x = X$ gives

$$\frac{dN}{dt} + n(X,t) - n(0,t) = -\beta N. \tag{13.13}$$

Since $n(0,t) = F(N(t-d))$ and $n(X,t) = F(N(t-d-X))e^{-\beta X}$, it follows that $N(t)$ is governed by the delay differential equation

$$\frac{dN}{dt} + F(N(t-d-X))e^{-\beta X} - F(N(t-d)) = -\beta N. \tag{13.14}$$

Note that the steady state of (13.14) is (13.12), as expected.

We now linearize around the steady state, N_0, by looking for solutions of the form $N(t) = N_0(1+\epsilon e^{\lambda t})$, where $\epsilon \ll 1$. Substituting this form into (13.14) and ignoring all terms of $O(\epsilon^2)$ gives

$$\lambda + F'(N_0)e^{-\lambda(d+X)}e^{-\beta X} - F'(N_0)e^{-\lambda d} = -\beta, \tag{13.15}$$

from which it follows that

$$F'(N_0)e^{-\lambda d}\frac{1-e^{-(\lambda+\beta)X}}{\lambda+\beta} = 1. \tag{13.16}$$

The roots λ of this equation determine the stability of the linearized solution. If all the roots have negative real part, then the solution is stable, whereas if there are roots with positive real part, the steady solution is unstable. For the remainder of this discussion, we take $\beta = 0$. This implies that all cells die at exactly age X. A different simplification, taking $X \to \infty$, leads to a delay differential equation that is discussed in Chapter 14 (see also Exercise 4).

In the limit $\beta \to 0$, the characteristic equation (13.16) is

$$F'(N_0)(e^{-\lambda d} - e^{-\lambda(d+X)}) = \lambda. \tag{13.17}$$

Since $F'(N_0) < 0$, there are no positive real roots. (The root at $\lambda = 0$ is spurious.)

There is possibly one negative real root; all other roots are complex. It follows that even if the steady solution is stable, the return to steady state is oscillatory rather than monotone. Thus, following rapid disruptions of blood cell population, such as traumatic blood loss or transfusion, or a vacation at a high-altitude ski resort, the blood cell population will oscillate about its steady state.

The only possible way to have a root with positive real part is if it is complex. Furthermore, a transition from stable to unstable can occur only if a complex root changes the sign of its real part, leading to a Hopf bifurcation. If a Hopf bifurcation occurs, it does so with $\lambda = i\omega$. We substitute $\lambda = i\omega$ into (13.17) and separate this into its real and imaginary parts to obtain

$$F'(N_0)(\cos(\omega d) - \cos(\omega(d+X))) = 0, \tag{13.18}$$

$$F'(N_0)(\sin(\omega d) - \sin(\omega(d+X))) = -\omega. \tag{13.19}$$

There are two ways to solve (13.18). Because cosine is symmetric about any multiple of π, we can take $n\pi - \omega d = n\pi + \omega(d+X)$, or $\omega(2d+X) = 2n\pi$, for any positive integer n. Because cosine is 2π-periodic, we could also take $\omega X = 2n\pi$; however, since sine is also 2π-periodic, this fails to give a solution of (13.19). With $\omega(2d+X) = 2n\pi$, (13.19) becomes

$$2dF'(N_0)\sin(\omega d) = -\omega d, \tag{13.20}$$

or

$$2dF'(N_0) = -\frac{2n\pi}{2+\frac{X}{d}} \frac{1}{\sin\left(\frac{2n\pi}{2+\frac{X}{d}}\right)}. \tag{13.21}$$

Finally, we use that $F(N_0) = N_0/X$ to write

$$\frac{N_0 F'(N_0)}{F(N_0)} = -\frac{1}{2}\frac{X}{d}\frac{2n\pi}{2+\frac{X}{d}} \frac{1}{\sin\left(\frac{2n\pi}{2+\frac{X}{d}}\right)}. \tag{13.22}$$

For each integer n, this equation defines a relationship between N_0 and X/d at which there is a change of stability and thus a Hopf bifurcation. If we take F to be of the special form

$$F(x) = \frac{A}{1+x^p}, \tag{13.23}$$

we can use (13.12) (in the limit $\beta \to 0$) and (13.22) to find an analytic relationship between $dA(= dF(0))$ and X/d at which Hopf bifurcations occur (see Exercise 2). Shown in Fig. 13.6 is this curve for $n = 1$ and $p = 7$. The case $n = 1$ is the only curve of interest, since it is the first instability. That is, the steady-state solution is unstable for all the critical stability curves with $n > 1$, and so these curves do not lead to physically relevant bifurcations.

The implications of this calculation are interesting. If the nondimensional parameters X/d and $dF(0)$ are such that they lie above the curve in Fig. 13.6, then the steady solution is unstable, and a periodic or oscillatory solution is likely (but since we do not know the direction of bifurcation, this is not guaranteed). On the other hand, if these parameters lie below or to the far right of this curve, the steady solution is stable.

From this we learn that there are three mechanisms by which cell production can be destabilized, and these are by changing the maximal production rate $F(0)$, the expected lifetime X, or the production delay d. If X/d is sufficiently large (greater than approximately 14 for these parameter values), the system cannot be destabilized. However, if X/d is small enough, increasing $F(0)$ is destabilizing. Increasing d is also destabilizing. If $F(0)$ and X are held fixed, then changing d moves $y = dF(0)$ and $x = X/d$ along

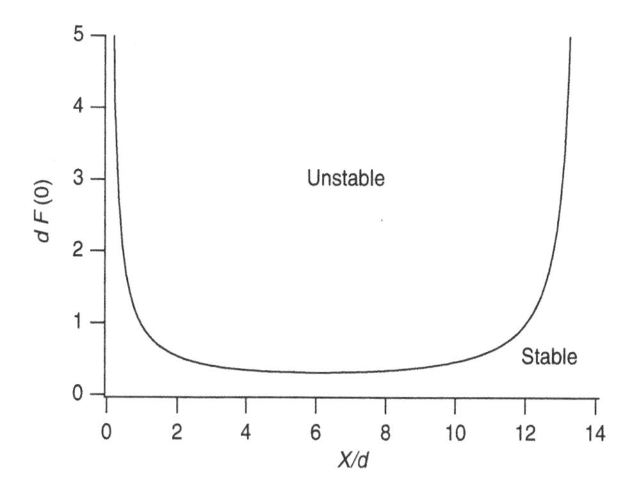

Figure 13.6 Critical stability curve (Hopf bifurcation curve) for cell growth.

the hyperbola yx = constant. Thus, decreasing d is stabilizing, since it increases X/d, moving it out of and away from the unstable region.

For normal humans, with $d = 5$ days and $X = 120$ days, there is no instability, since $X/d = 24$. However, any mechanism that substantially shortens X can have a destabilizing effect and can result in oscillatory production of blood cells (Exercise 3). Near the bifurcation, the period of oscillation is $T = \frac{2\pi}{\omega}$, where $\omega(2d+X) = 2\pi$, so that

$$T = 2d + X. \tag{13.24}$$

Thus, for example, a disorder that halves the normal lifetime of blood cells to $X = 60$ days should result in oscillatory blood cell production with a period on the order of 70 days.

13.2.3 Peripheral or Local Control?

Models of periodic hematopoietic disorders typically fall into one of two classes. Models in the first class rely on the observation that in certain disorders, many of the blood cell types show oscillatory behavior, and thus the disease is modeled as generated by local regulation of the stem cell dynamics (Fig. 13.2). In this scenario, the progenitors of each of the cell types have periodic input from the stem cells, and thus the numbers of terminally differentiated cells have the potential to oscillate (although they are not required to). Mackey (1978) and Milton and Mackey (1989) are models of this type.

The other type of model (Morley, 1979; Wichmann et al., 1988; Schmitz et al., 1990) explains the oscillations as the result of instability in the peripheral control feedback (governed by, for example, erythropoietin, or G-CSF, via the cell population; see Fig. 13.2). The model presented in the previous section is of this type.

However, a detailed analysis of peripheral control models suggests that such models, irrespective of the specific mechanisms assumed, are less likely to be an explanation for the observed oscillations (Hearn et al., 1998; Mackey et al., 2003).

To illustrate a peripheral control model, we let x be the density of neutrophils (in number of cells per microliter), let α be their death rate, and let $\mathcal{M}$ denote their rate of production. We do not specify the exact form of $\mathcal{M}$ but use only its qualitative properties. We assume that $\mathcal{M}$ depends on a weighted average of x, delayed by τ_m. Thus,

$$\frac{dx}{dt} = -\alpha x + \mathcal{M}(\tilde{x}(t - \tau_m)), \tag{13.25}$$

where τ_m is the minimal maturation delay, and

$$\tilde{x}(t) = \int_{-\infty}^{t} x(s)g(t-s)\,ds \tag{13.26}$$

is the convolution of x with weighting function g. A typical choice of g is the gamma distribution,

$$g(\tau) = \begin{cases} 0, & \tau \le 0, \\ \frac{a^{k+1}}{\Gamma(k+1)}\tau^k e^{-a\tau}, & \tau > 0. \end{cases} \tag{13.27}$$

This gives a good fit to experimental data on the distribution of maturation times, and can be further motivated by the fact that, if k is an integer, g is the solution of a kth order linear filter (see Exercise 7). For example, if $k = 0$, then $\tilde{x}$ satisfies the differential equation

$$\frac{d\tilde{x}}{dt} = x(t) - a\tilde{x}, \tag{13.28}$$

so that $\tilde{x}$ can be thought of as a substance that is produced at a rate proportional to x with a natural decay rate a.

The only additional assumption is that $\mathcal{M}$ is a monotonically nonincreasing function of $\tilde{x}$, so that there is negative feedback.

Letting x^* denote the steady-state solution, we see that

$$\alpha x^* = \mathcal{M}(x^*), \tag{13.29}$$

which, since $\mathcal{M}$ is monotonically decreasing, has a unique solution. We are interested in the stability of this solution, to determine whether it can be unstable, giving rise to oscillatory solutions.

For convenience, let $\mu = \mathcal{M}'(x^*) < 0$ and let $u = x - x^*$. Then, the linearization of (13.25) is

$$\frac{du}{dt} = -\alpha u + \mu \int_{-\infty}^{t-\tau_m} u(s)g(t-s-\tau_m)\,ds. \tag{13.30}$$

Looking for solutions of the form $u = e^{\lambda t}$ gives

$$\lambda + \alpha = \mu\left(\frac{a}{\lambda + a}\right)^{k+1} e^{-\lambda\tau_m}. \tag{13.31}$$

Solutions of this equation with the real part of λ greater than zero correspond to unstable steady states. It is left to Exercise 9 to show that, if (13.31) has a real solution, it must be negative and lie between $-a$ and $-\alpha$.

Since the only possibility for instability is when λ is complex, we set $\lambda = i\omega$ to find a possible boundary between stability and instability. This gives

$$i\omega + \alpha = \mu \left(\frac{a}{i\omega + a} \right)^{k+1} e^{-i\omega\tau_m}. \tag{13.32}$$

Our goal is to find a curve in the α, μ plane that is a possible stability boundary. We do this by separating (13.32) into real and imaginary parts and finding parametric expressions for $\alpha(\omega)$ and $\mu(\omega)$.

We begin by setting

$$e^{i\theta} = \frac{a + i\omega}{\sqrt{a^2 + \omega^2}}, \tag{13.33}$$

in which case (13.32) becomes

$$(\alpha + i\omega)(\cos[(k+1)\theta] + i \sin[(k+1)\theta]) = \mu \cos^{k+1}\theta e^{-i\omega\tau_m}. \tag{13.34}$$

Next, equate the real and imaginary parts to get the two equations

$$\alpha - \mu R \cos \omega\tau_m = \omega \tan[(k+1)\theta], \tag{13.35}$$

$$\alpha \tan[(k+1)\theta] = -\mu R \sin \omega\tau_m - \omega, \tag{13.36}$$

where

$$R = \frac{\cos^{k+1}\theta}{\cos[(k+1)\theta]}. \tag{13.37}$$

Finally, solving (13.35) and (13.36) to get α and μ as functions of ω gives

$$\alpha(\omega) = \frac{-\omega}{\tan[\omega\tau_m + (k+1)\tan^{-1}(\omega/a)]}, \tag{13.38}$$

$$\mu(\omega) = \frac{-\omega}{\cos^{k+1}[\tan^{-1}(\omega/a)] \sin[\omega\tau_m + (k+1)\tan^{-1}(\omega/a)]}. \tag{13.39}$$

Plots of μ against α are shown in Fig. 13.7, the solid curve using the parameters for normal humans, and the dashed curve using the parameters for humans with cyclical neutropenia, i.e., CN humans. Typical values for α are between the two vertical lines. The parameter values for the two different cases were estimated from experimental data on cellular maturation times (Hearn et al., 1998). In CN humans, the average maturation delay decreases from around 9.7 days to around 7.6 days, while the minimal maturation delay, τ_m, decreases also, from around 3.8 days to 1.2 days. The variance of the maturation delay also decreases, from around 16 day^2 to around 12 day^2. These changes in the maturation delay can be described (after some work) by the two sets of parameters given in the caption to Fig. 13.7.

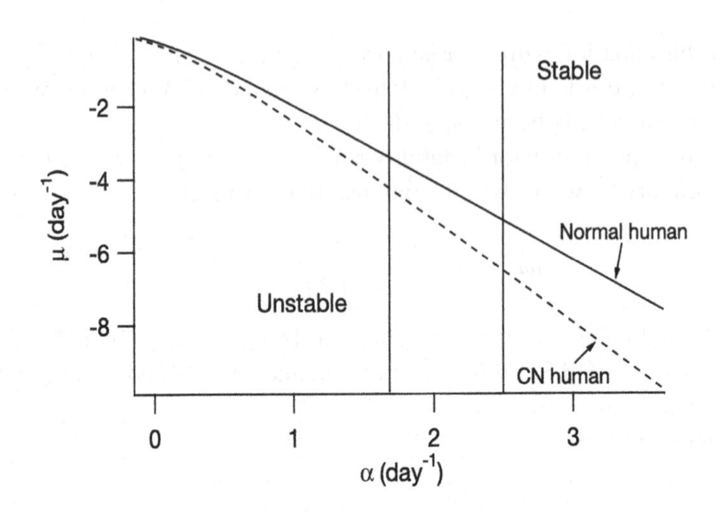

Figure 13.7 Curves that divide the μ-α plane into regions of stability and instability. Parameter values below the curves give an unstable steady state, while parameter values above the curves give a stable steady state. The two vertical lines delineate the region of reasonable values for α. Parameter values are $\tau_m = 3.8$, $a = 0.36$, $k+1 = 3.15$ (for normal humans) and $\tau_m = 1.2$, $a = 0.53$, $k+1 = 4.38$ (for CN humans).

The stability diagram in Fig. 13.7 has two interesting consequences. First, if the steady state is stable for a normal human, it is also stable for a human with cyclical neutropenia with the same values of μ and α. Thus, changes to the parameters τ, a, and m appropriate for cyclical neutropenia are not sufficient to generate instability. Furthermore, it is known experimentally that α is the same in both normal and CN humans. Hence, the only way by which the stable state can become unstable in CN humans is by a decrease in μ, i.e., a decrease in the steady-state slope of the production function.

Second, consider the period of the oscillations that arise at the Hopf bifurcations. Recall that, on the curves in Fig. 13.7, the solution has the form $e^{i\omega t}$, and thus has period $2\pi/\omega$. To get reasonable values of α (i.e., between the two vertical lines in Fig. 13.7), ω must be between approximately 0.44 and 0.46. Thus, there is only a narrow window of values for ω that give acceptable behavior, and consequently the period of the resulting oscillations is restricted to be between approximately 14.3 and 13.7 days. Since this range is considerably lower than the period of the observed oscillations, which ranges from 19 to 30 days, it suggests that this model does not capture the correct oscillatory mechanism, and thus that CN oscillations are not caused by instabilities in the peripheral control system.

This conclusion from the linear stability analysis is supported by numerical solution of the model in the nonlinear regime (Hearn et al., 1998), which shows that the period of the oscillation can be made larger only with difficulty, and at the price of having solutions which look quite different from those observed experimentally.

13.3 Erythrocytes

Erythrocytes (red blood cells) are small biconcave disks measuring about 8 μm in diameter. They are flexible, allowing them to change shape and to pass without breaking through blood vessels with diameters as small as 3 μm. Their function is the transport of oxygen from the lungs to the rest of the body, and they accomplish this with the help of a large protein molecule called *hemoglobin*, which binds oxygen in the lungs, later releasing it in tissue. Hemoglobin is the principal protein constituent of mature erythrocytes. A similar protein, *myoglobin*, is used to store and transport oxygen within muscle; mammals that dive deeply, such as whales and seals, have skeletal muscle that is especially rich in myoglobin.

13.3.1 Myoglobin and Hemoglobin

The binding of oxygen with myoglobin and hemoglobin serves as an excellent example of relatively simple chemical reactions that are of fundamental importance in blood physiology. We get some understanding of this process by examining the experimentally determined saturation function for hemoglobin and myoglobin as a function of the partial pressure of oxygen. For myoglobin the saturation curve is much like a standard Michaelis–Menten saturation function, while for hemoglobin it is sigmoidal-shaped (Fig. 13.8). From these curves we see that when the partial pressure of oxygen is at 100 mm Hg (about what it is in the lungs), hemoglobin is 97% saturated. This amount is affected only slightly by small changes in oxygen partial pressure, because at this level the saturation curve is relatively flat. In veins or tissue, however, where the partial pressure

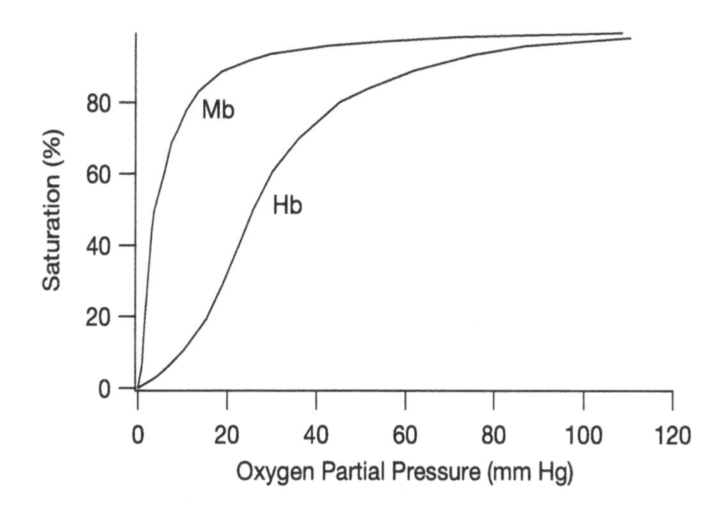

Figure 13.8 Uptake of oxygen by myoglobin and hemoglobin. (Rubinow, 1975, Fig. 2.13, p. 82, taken from Changeux, 1965.)

of oxygen is about 40 mm Hg, the saturation is about 75%. Furthermore, because this is on a steep portion of the saturation curve, if the metabolic demand for oxygen should decrease the oxygen pressure to, say, 20 mm Hg, then hemoglobin gives up its oxygen readily, reducing its saturation to about 35%. At this value of oxygen partial pressure the saturation of myoglobin is at 90%. Thus, if the tissue is muscle, oxygen is readily transferred from hemoglobin to myoglobin, since the affinity of myoglobin for oxygen is greater than that of hemoglobin.

These saturation curves are of fundamental importance to blood chemistry, so it is of interest to understand why the saturation curves of the two are as they are. We can derive models of these saturation curves from the underlying chemistry (see Section 1.4.4 on cooperativity). Myoglobin consists of a polypeptide chain and a disc-shaped molecular ring called a *heme group*, which is the active center of myoglobin. At the center of the heme group is an iron atom, which can bind with oxygen, forming oxymyoglobin. Hemoglobin consists of four such polypeptide chains (called *globin*) and four heme groups, allowing the binding of four oxygen molecules. When bound with oxygen, the iron atoms in hemoglobin and myoglobin give them their red color. Myoglobin content accounts for the difference in color between red meat such as beef, and white meat such as chicken.

A simple reaction scheme describing the binding of oxygen with myoglobin is

$$\mathrm{O_2} + \mathrm{Mb} \underset{k_-}{\overset{k_+}{\rightleftarrows}} \mathrm{MbO_2}.$$

At equilibrium $k_+[\mathrm{Mb}][\mathrm{O_2}] = k_-[\mathrm{MbO_2}]$, so that the percentage of occupied sites is

$$Y = \frac{[\mathrm{MbO_2}]}{[\mathrm{Mb}] + [\mathrm{MbO_2}]} = \frac{[\mathrm{O_2}]}{K + [\mathrm{O_2}]}, \tag{13.40}$$

where $K = k_-/k_+$.

To compare the function (13.40) with the saturation curve for myoglobin in Fig. 13.8, we must relate the oxygen concentration to the oxygen partial pressure via $[\mathrm{O_2}] = \sigma P_{\mathrm{O_2}}$. Then, (13.40) becomes

$$Y = \frac{P_{\mathrm{O_2}}}{K/\sigma + P_{\mathrm{O_2}}} = \frac{P_{\mathrm{O_2}}}{K_P + P_{\mathrm{O_2}}}, \tag{13.41}$$

and we get a good fit of the myoglobin uptake curve in Fig. 13.8 with $K_P = 2.6$ mm Hg. Notice that the equilibrium constant K_P is in units of pressure rather than concentration, as is more typical. The equilibrium constant K is related to K_P through $K = \sigma K_P$. For the myoglobin saturation curve $K = 3.7\ \mu$M; however, because it is typical to describe concentrations of dissolved gases in units of pressure, it is also typical to write the equilibrium constant K in these units as $K = 2.6\sigma$ mm Hg. A comparison of the curve (13.41) with the data is shown in Fig. 13.9.

The primary reason that the saturation curve for hemoglobin is significantly different from that for myoglobin is that it has four oxygen binding sites instead of one.

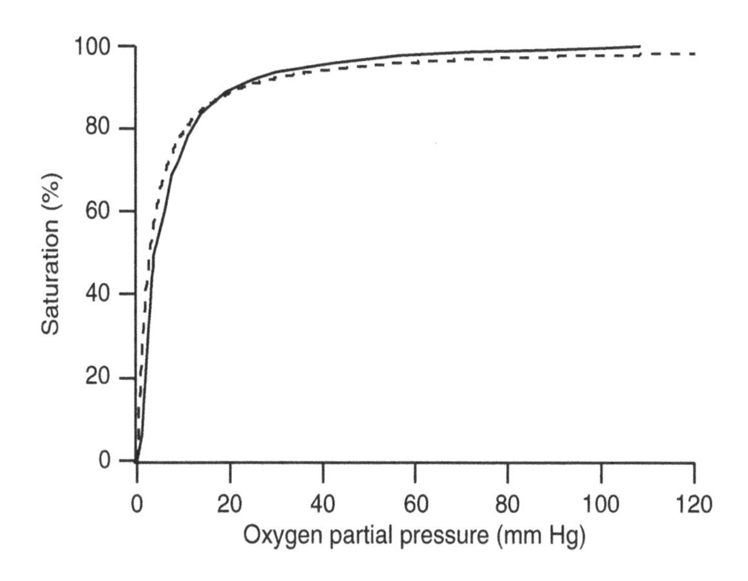

Figure 13.9 Comparison of myoglobin saturation curve (solid) with the curve (13.41) (dashed) with $K_P = 2.6$ mm Hg.

A simple kinetic scheme for the formation of oxyhemoglobin is

$$4\text{O}_2 + \text{Hb} \underset{k_-}{\overset{k_+}{\rightleftarrows}} \text{Hb(O}_2)_4,$$

with the corresponding differential equation

$$\frac{d[\text{Hb}]}{dt} = k_-[\text{Hb(O}_2)_4] - k_+[\text{Hb}][\text{O}_2]^4. \tag{13.42}$$

At steady state, the percentage of available hemoglobin sites that are bound to oxygen is

$$Y = \frac{[\text{Hb(O}_2)_4]}{[\text{Hb(O}_2)_4] + [\text{Hb}]} = \frac{[\text{O}_2]^4}{[\text{O}_2]^4 + K^4}, \tag{13.43}$$

where $K^4 = k_-/k_+$. We use the half-saturation level from the hemoglobin uptake curve in Fig. 13.8 to estimate K as $K = 26\sigma$ mm Hg.

While (13.43) (shown as a short dashed curve in Fig. 13.10) reproduces some features of the uptake curve that are qualitatively correct, it is not quantitatively accurate. In fact, one can achieve a much better fit of the data with the Hill equation

$$Y = \frac{[\text{O}_2]^n}{[\text{O}_2]^n + K^n}, \tag{13.44}$$

with $n = 2.5$ and $K = 26\sigma$ mm Hg. However, there is no adequate theoretical basis for such a model.

A better model keeps track of the elementary reactions involved in the binding process, and is given by

$$O_2 + H_{j-1} \underset{k_{-j}}{\overset{k_j}{\rightleftarrows}} H_j, \qquad j = 1, 2, 3, 4,$$

where $H_j = \mathrm{Hb}(O_2)_j$. The steady state for this reaction is attained at

$$[H_j] = \frac{k_{+j}}{k_{-j}}[H_{j-1}][O_2] = \frac{[H_{j-1}][O_2]}{K_j}, \tag{13.45}$$

and the saturation function is

$$Y = \frac{\sum_{j=0}^{4} jH_j}{4\sum_{j=0}^{4} H_j}. \tag{13.46}$$

Substituting (13.45) into (13.46) we obtain the saturation function

$$Y = \frac{\sum_{j=0}^{4} j\alpha_j [O_2]^j}{4\sum_{j=0}^{4} \alpha_j [O_2]^j}, \tag{13.47}$$

where $\alpha_j = \prod_{i=1}^{j} K_i^{-1}$, $K_j = k_{-j}/k_{+j}$, $\alpha_0 = 1$.

One can fit the saturation function (13.47) to the hemoglobin uptake curve shown in Fig. 13.8, with the result $K_1 = 45.9, K_2 = 23.9, K_3 = 243.1, K_4 = 1.52\sigma$ mm Hg (Roughton et al., 1972). The striking feature of these numbers is that K_4 is much smaller than K_1, K_2, or K_3, indicating that there is apparently a greatly enhanced affinity of oxygen for hemoglobin if three oxygen molecules are already bound to it. Hemoglobin prefers to be "filled up" with oxygen. The mechanism for this positive cooperativity is not completely understood. (If the binding sites were independent, then K_1 would be the smallest and K_4 would be the largest equilibrium constant; see Exercise 10 and Section 1.4.4.)

Notice that the affinity of oxygen for myoglobin is greater than for any of the binding sites of hemoglobin. In Fig. 13.10 is shown a comparison between the data and the approximate curves (13.43) and (13.47). The Hill equation fit (13.44) is not shown because it is nearly identical to (13.47).

The structure and function of hemoglobin has been intensively studied for over 100 years, and has motivated some of the most important biophysical models of cooperativity (Eaton et al., 1999). Some of the earliest studies were those of Christian Bohr (father of the physicist Niels Bohr) who measured the sigmoidal binding curve of hemoglobin (Bohr et al., 1904), and who also discovered that carbon dioxide lowers the oxygen binding affinity (the Bohr effect). Some years later, the physiologist Adair (Adair, 1925) discovered that hemoglobin contains four binding sites. However, the connection between the sigmoidal binding curve and the number of binding sites was not made explicit until the work of Linus Pauling (1935), who suggested that the binding of oxygen to one binding site, or subunit, could increase the binding affinity to the neighboring subunits, thus leading to a sigmoidal binding curve. The Pauling model

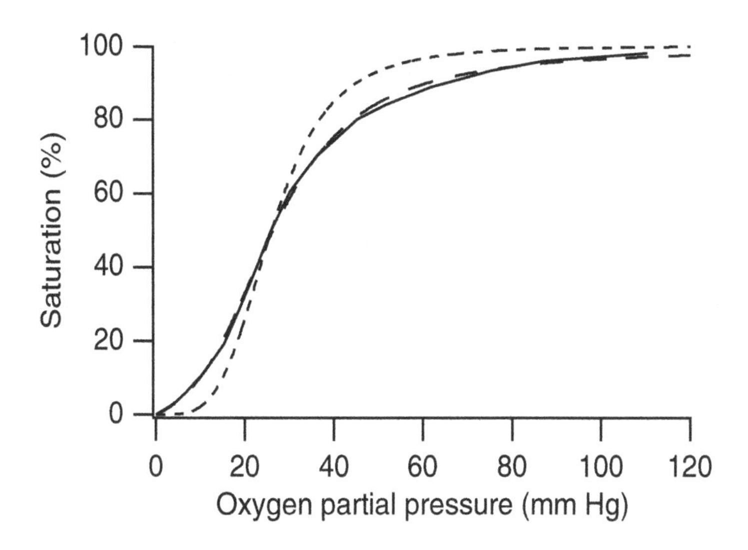

Figure 13.10 Comparison of hemoglobin saturation curve (solid) with the curves (13.43) (short dashed) and (13.47) (long dashed).

was later extended by Koshland, Nemethy and Filmer (1966; Koshland and Hamadani, 2002), giving the so-called KNF, or *sequential*, model.

However, a different model of cooperativity was proposed by Monod, Wyman, and Changeux (1965). The experimental work of Perutz et al. (1964) showed that binding of oxygen resulted in a conformational change in the hemoglobin β subunits, which moved closer together. In the Monod–Wyman–Changeux (MWC) model, hemoglobin is assumed to exist in two structural states, R (relaxed) and T (tense). The R structure has a high affinity for oxygen, while the T structure has a low affinity. As oxygen pressure increases, a greater proportion of hemoglobin exists in the R state, and thus the overall affinity for oxygen increases as oxygen is bound. The major difference from the KNF model is that, in the MWC model, all the subunits switch between the R and T states simultaneously, and are thus always in the same state. However, in the KNF model, the binding of oxygen to one subunit causes a change in the binding affinity of the neighboring subunits, rather than a change in the state of every subunit. The construction of MWC models is discussed in detail in Chapter 1 (also see Exercise 12).

Perutz (1970) proposed a specific physical mechanism that could explain the assumptions used in the MWC model. He proposed that the subunits are connected by salt bridges when hemoglobin is in the T structure. Binding of an oxygen to one subunit could break a salt bridge and destabilize the structure, thus biasing hemoglobin to the R state. The energy required to break the salt bridge upon oxygen binding serves to decrease the affinity of the T structure for oxygen.

Although many refinements of the basic MWC mechanism have been constructed, our understanding of the basic mechanism remains relatively unchanged today.

However, there is also experimental evidence for a sequential binding model of KNF type (Koshland and Hamadani, 2002), and the actual mechanism probably lies somewhere between these two extremes.

13.3.2 Hemoglobin Saturation Shifts

There are a number of factors that affect the binding of oxygen to hemoglobin, the most important of which is the hydrogen ion, which is an allosteric inhibitor of oxygen binding (Chapter 1). As is discussed in Chapter 14, the interactions between oxygen concentration and carbon dioxide concentration (which indirectly changes the hydrogen ion concentration) are important for transport of both oxygen and carbon dioxide.

Carbon monoxide combines with hemoglobin at the same binding site as oxygen (and is a competitive inhibitor), but with an affinity more than 200 times greater. Therefore the carbon monoxide saturation curve is almost identical to the oxygen saturation curve, except that the abscissa is scaled by a factor of about 200. At a carbon monoxide partial pressure of 0.5 mm Hg, and in the absence of oxygen, hemoglobin is 97% saturated with carbon monoxide. If oxygen is present at atmospheric concentrations, then it takes a carbon monoxide partial pressure of only 0.7 mm Hg (about 0.1 percent) to cause oxygen starvation in the tissues (Chapter 14).

Fetal hemoglobin, a different type of hemoglobin found in the fetus, has a considerable leftward shift for its oxygen saturation curve. This allows fetal blood to carry as much as 30% more oxygen at low oxygen partial pressures than can adult hemoglobin. This is important since the oxygen partial pressure in the fetus is always low. The left-shift of the fetal hemoglobin saturation curve is also important for the transfer of oxygen from mother to fetus.

Because it is important in the next section, here we construct a simple model of the allosteric inhibition by hydrogen ions of oxygen binding to hemoglobin. As illustrated in Fig. 13.11, we assume that the hemoglobin molecule can exist in four different states: with H^+ bound (concentration Z), with O_2 bound (concentration Y), with neither bound (concentration X), or with both bound (concentration W). This is, of course, an extreme simplification, as it ignores the cooperative nature of oxygen binding discussed in the previous section and in Chapter 1, but nevertheless the results are qualitatively correct.

Assuming that each reaction is at equilibrium, we find

$$O^4X = K_1Y, \tag{13.48}$$

$$hX = K_2Z, \tag{13.49}$$

$$O^4Z = \bar{K}_1W, \tag{13.50}$$

$$X + Y + Z + W = T_{Hb}, \tag{13.51}$$

where h denotes $[H^+]$, O denotes $[O_2]$, T_{Hb} denotes the total concentration of hemoglobin, $K_1 = k_{-1}/k_1$ and similarly for $K_2, \bar{K}_1$ and $\bar{K}_2$ (which is needed below).

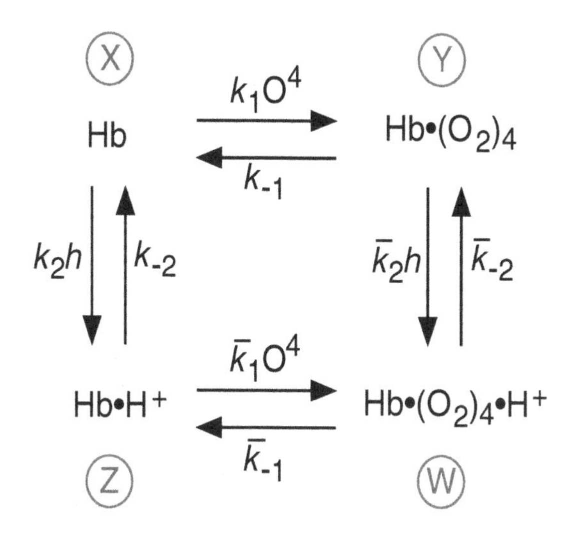

Figure 13.11 Binding diagram for the allosteric binding of hydrogen ions and oxygen to hemoglobin. We assume a single hydrogen ion binding site, and a simplified mechanism for oxygen binding. Hb denotes hemoglobin.

Solving these four equations we find

$$Y + W = \frac{O^4 T_{\mathrm{Hb}}}{\phi(h) + O^4}, \tag{13.52}$$

where

$$\phi(h) = \frac{K_1 \bar{K}_1 (K_2 + h)}{K_2 \bar{K}_1 + h K_1}. \tag{13.53}$$

We are interested in determining $Y + W$ as a function of O, since $Y + W$ is the total concentration of hemoglobin with oxygen bound, and thus plotting $Y + W$ as a function of O gives the oxygen saturation curve.

It is easily seen from (13.52) that h does not change the maximal saturation, although it shifts the mid-point of the curve. Since hydrogen ions are an allosteric inhibitor of oxygen binding, we assume that $\bar{K}_1 > K_1$. Note that, in this case, $\phi(h)$ is an increasing function of h, and thus increasing h shifts the saturation curve to the right, as expected.

Before we can discuss the importance of the allosteric effect of hydrogen ions on oxygen binding it is necessary to discuss the mechanism of carbon dioxide transport.

13.3.3 Carbon Dioxide Transport

While oxygen is taken up in the lungs and transported to the tissues, carbon dioxide must be transported in the reverse direction from the tissues to the lungs for removal from the body. In the blood, CO_2 is transported in three main forms. In venous blood a significant amount (about 6%) is present as dissolved CO_2. A slightly greater amount (about 7%) is bound to the globin part of hemoglobin as carbamino compounds, but most CO_2 (87%) is present in the form of bicarbonate ions.

In the tissues CO_2 diffuses down its concentration gradient into the plasma and into the red blood cells. In both plasma and red blood cells it combines with water to form carbonic acid (H_2CO_3), which then dissociates quickly into hydrogen ions and bicarbonate ions. Thus,

$$CO_2 + H_2O \underset{r_{-1}}{\overset{r_1}{\rightleftharpoons}} H_2CO_3 \underset{r_{-2}}{\overset{r_2}{\rightleftharpoons}} H^+ + HCO_3^-. \tag{13.54}$$

This reaction proceeds slowly in the plasma but much more rapidly in the red blood cells because of the presence there of the enzyme *carbonic anhydrase*, which increases the speed of CO_2 hydration by more than a thousand times. The H^+ formed by the dissociation of carbonic acid binds to the globin part of hemoglobin, and the bicarbonate ion diffuses into the plasma in exchange for Cl^-.

In the lungs the reaction is reversed, as CO_2 diffuses down its concentration gradient to be excreted in the alveolar air and then the expired air. It is important to emphasize that the direction of the carbonic anhydrase reaction (13.54) is determined by the local concentration of CO_2. In the tissues, $[CO_2]$ is high, which drives reaction (13.54) from left to right, thus storing CO_2 in the blood. In the lungs, $[CO_2]$ is low, driving the reaction from right to left, thus removing CO_2 from the blood. Of course, carbonic anhydrase speeds up the reaction in both directions; without this increase in speed not enough CO_2 can be stored in the blood to remove it from the body fast enough.

The importance of the allosteric effect of H^+ on oxygen binding to hemoglobin is now apparent. In the tissues, because of the high local CO_2 concentration, the hydration of CO_2 causes an increase in the local concentration of H^+ (i.e., the blood pH falls slightly, from about 7.4 to about 7.35), which in turn results in a decreased affinity of hemoglobin for oxygen, thus increasing oxygen release to the tissues. In the lungs, the reverse occurs; the low local CO_2 concentration causes a decrease in H^+ concentration which results in an increase in hemoglobin oxygen affinity, and thus increased oxygen uptake. This effect of CO_2 concentration on oxygen transport (mediated by the carbonic anhydrase reaction and hydrogen ions), is known as the *Bohr effect*.

It is interesting to note that, from the principle of detailed balance applied to the reaction scheme shown in Fig. 13.11 (i.e., from consistency of the four equilibrium equations) it must be that

$$\frac{K_1}{\bar{K}_1} = \frac{K_2}{\bar{K}_2}. \tag{13.55}$$

It follows that, if $\bar{K}_1 > K_1$ it must also be that $\bar{K}_2 > K_2$. In other words, if H^+ is an allosteric inhibitor of oxygen binding, then oxygen must also be an allosteric inhibitor of H^+ binding. Hence, as CO_2 influences oxygen transport, so too oxygen affects CO_2 transport . In the tissues, where $[O_2]$ is low, binding of H^+ to hemoglobin is enhanced. This lowers the local H^+ concentration, thus driving the carbonic anhydrase reaction from left to right, and increasing CO_2 storage. The reverse occurs at the lungs. The enhancement of CO_2 transport by low levels of oxygen is called the *Haldane effect*. Note that, according to the principle of detailed balance (at least in this simple model) the Bohr effect implies the Haldane effect, and vice versa.

To construct a mathematical model of CO_2 transport, we assume that the bicarbonate-carbon dioxide reaction is in steady state so that

$$[CO_2] = R_1 R_2 [H^+][HCO_3^-], \tag{13.56}$$

where $R_1 = r_{-1}/r_1$ and $R_2 = r_{-2}/r_2$.

Carbon dioxide enters this system from the tissues and leaves at the lungs. When it does so, bicarbonate is produced or removed. However, since the carbonic anhydrase reaction produces exactly one hydrogen ion for each bicarbonate ion it produces, and since these hydrogen ions must either be free, or bound to hemoglobin, it follows that

$$\Delta[HCO_3^-] = \Delta h + \Delta Z + \Delta W, \tag{13.57}$$

where Δ denotes the change in concentration. Hence,

$$[HCO_3^-] = h + Z + W - T_0, \tag{13.58}$$

where $T_0 = h_0 + Z_0 + W_0 - [HCO_3^-]_0$ is some reference level. In reality each hemoglobin molecule can bind many hydrogen ions, and so the conservation equation should be

$$[HCO_3^-] = h + n(Z + W) - T_0, \tag{13.59}$$

where n can be as large as 10 or 20, and $T_0 = h_0 + n(Z_0 + W_0) - [HCO_3^-]_0$. The number n is important, because without it (if $n = 0$), the pH fluctuates widely with changes in bicarbonate (Exercise 13a), whereas in normal blood, practically all the H^+ produced by the carbonic anhydrase reaction is absorbed by hemoglobin. This demonstrates the importance of hemoglobin as a hydrogen ion buffer. Note that, to be consistent, the factor n should also be included in the model of hemoglobin. However, as this would greatly increase the complexity of the binding model without adding anything fundamentally new, we include n in the bicarbonate conservation equation, but not in the binding diagram. A more accurate model gives the same qualitative result.

In addition, the oxygen and carbon dioxide concentrations in arterial and venous blood are unknowns. Their precise values are set by the rate of gas exchange in the lungs and the tissues, the rate of metabolism, and depend to some extent on the properties of the carbonic anhydrase reaction, among other things. Thus, to be strictly correct, we should not treat them as constants, but solve for them as part of a more complicated model. We omit these complications here and treat O and $[CO_2]$ as known constants, since our primary goal here is to find the other unknowns (X, Y, Z, W and h) as functions of the gas concentrations.

We now have five equations

$$O^4X = K_1Y, \tag{13.60}$$

$$hX = K_2Z, \tag{13.61}$$

$$O^4Z = \bar{K}_1W, \tag{13.62}$$

$$X + Y + Z + W = T_{Hb}, \tag{13.63}$$

$$[CO_2] = R_1R_2h[h + n(Z + W) - T_0], \tag{13.64}$$

to solve for the five unknowns. It is an easy matter to solve (13.60)–(13.63) for X, Y, Z, and W in terms of O^4, h and the other parameters, and substitute these into (13.64). This yields a single equation for $[CO_2]$ as a function of O^4 and h. This equation can be readily solved numerically for h as a function of $[CO_2]$ and O^4, from which one can determine the total amounts of carbon dioxide and oxygen in all their forms. Solution of this equation is left for the exercises (Exercise 13).

The result of this calculation shows that the Bohr effect changes the arterial concentrations of carbon dioxide and oxygen only slightly, but it increases the venous CO_2 and decreases the venous O_2 substantially, giving an increase in the total amount of these that is transported.

As a final note, the most important system for controlling the extracellular acid-base balance is the bicarbonate buffer system. Extracellular fluid contains large amounts of bicarbonate ions, mostly as $Na^+HCO_3^-$. Addition of excess H^+ ions drives the carbonic anhydrase reaction to produce carbon dioxide; the additional CO_2 produced can be removed at the lungs. Conversely, addition of a strong base and the consequent removal of H^+ results in a lowering of $[CO_2]$ and the production of additional H^+. Since the overall levels of bicarbonate and carbon dioxide are controlled by the kidneys and the lungs, respectively, this allows effective and precise control of the pH of extracellular fluid.

13.4 Leukocytes

The *leukocytes* (white blood cells) are the mobile units of the body's immune system. There are six types of white blood cells normally found in the blood. These are the *neutrophils, eosinophils, basophils, monocytes, lymphocytes*, and *natural killer cells* (see Table 13.1 and Fig. 13.1). The neutrophils, eosinophils, and basophils are called *granulocytes*, or in clinical terminology, *polymorphonuclear* (PMN) cells, because they have a granular appearance and have nuclei with a wide variety of shapes (often shaped like a bent sausage). The normal adult human has about 7000 white blood cells per microliter of blood, approximately 62% of which are neutrophils and 30% of which are lymphocytes. The granulocytes and monocytes protect the body against invading organisms mainly by ingesting them, a process called *phagocytosis*.

13.4.1 Leukocyte Chemotaxis

Leukocytes crawl about in tissue by putting out pseudopodal extensions by which they adhere to the fibrous matrix of the tissue. In uniform chemical concentrations of chemical stimulus, their motion is that of a persistent random walk. At random times they undergo random changes in direction. The *persistence time*, the average time between changes of direction, is on the order of a few minutes, and the speed of migration is on the order of 2–20 μm/min.

One important question is how leukocytes are able to find their bacterial targets. The answer is that they move preferentially in the direction of increasing chemoattractant gradients. Exactly how this is accomplished, how this should be modeled, and how well the model represents this behavior is the topic of this section.

Here we derive a simple model of directed motion in a one-dimensional medium (Tranquillo and Lauffenberger, 1987). We assume that the population of cells, c, can be subdivided into two subpopulations, $c = n^+ + n^-$, where superscripts + and − denote right-moving and left-moving cells, respectively. If v^+ is the velocity of right-moving cells, and v^- is the velocity of left-moving cells, then the flux of cells is given by

$$J_c = v^+ n^+ - v^- n^-. \tag{13.65}$$

We expect that the cell velocity should be a function only of local conditions, so that $v^+ = v^- = v$. In general, v is a function of x and t. Now we write conservation equations for the directional cell species,

$$\frac{\partial n^+}{\partial t} = -\frac{\partial (vn^+)}{\partial x} + p^- n^- - p^+ n^+, \tag{13.66}$$

$$\frac{\partial n^-}{\partial t} = \frac{\partial (vn^-)}{\partial x} + p^+ n^+ - p^- n^-, \tag{13.67}$$

where p^+ is the probability per unit time that a right-moving cell changes direction to become a left-moving cell, and p^- is the probability that a left-moving cell becomes a right-moving cell. These probabilities are also known as *turning rates*.

An equation governing the cell flux J_c is found by differentiating (13.65) and using (13.66) and (13.67), yielding

$$\frac{\partial J_c}{\partial t} - \frac{J_c}{v}\frac{\partial v}{\partial t} = -J_c(p^+ + p^-) - v\frac{\partial (vc)}{\partial x} - vc(p^+ - p^-). \tag{13.68}$$

The steady-state flux is found by setting all time derivatives equal to zero, from which we find that

$$J_c = -v^2 T_p \frac{\partial c}{\partial x} + v(p^- - p^+)T_p c - T_p v \frac{\partial v}{\partial x} c, \tag{13.69}$$

where $T_p^{-1} = p^+ + p^-$.

Now we define phenomenological population migration parameters $\mu = T_p v^2$ as the *random motility coefficient* and $V_c = T_p v(p^- - p^+)$ as the *chemotactic velocity*. Then the equilibrium flux is

$$J_c = -\mu \frac{\partial c}{\partial x} + V_c c - T_p v \frac{\partial v}{\partial x} c. \tag{13.70}$$

Finally, the total cell density is governed by the equation

$$\frac{\partial c}{\partial t} = -\frac{\partial J_c}{\partial x}. \tag{13.71}$$

The movement of cells is governed by three terms in (13.70). The first term, $-\mu \frac{\partial c}{\partial x}$, represents purely random movement of cells, since it gives a diffusive term in (13.71). The second and third terms allow for directed cell movement, since they are proportional to c. The directed motion from the second term is due to a difference in the directional change probabilities, while the directed motion in the third term is due to variation in cell speed with spatial position. The second term is called *chemotaxis*, and the third term is *chemokinesis*.

The next problem is to determine the coefficients of these movement terms and in so doing to understand more about the sensory capabilities of the cells. It is known that cell speed can vary with stimulus concentration, yielding a chemokinetic effect, and changes in the direction of movements can be biased toward attractant concentration gradients, a chemotactic response. These responses are mediated by cell surface receptors for attractant molecules that can measure the attractant concentration and its spatial gradient.

There is no a priori theory for the dependence of cell speed on attractant concentration, so it must be measured experimentally. For example, with the tripeptide attractant formyl-norleucyl-leucyl-phenylalanine (FNLLP), the data show that leukocyte velocity is a linearly increasing function of the logarithm of concentration over the range of concentrations 10^{-9} M to 10^{-6} M, with velocity about 2–5 μm/min (Zigmond et al., 1981).

Leukocytes determine the presence of an attractant when it binds to receptors on the leukocyte cell surface. When there is a spatial gradient of the attractant, there is also a spatial gradient in the concentration of bound receptors. The side of the cell that experiences a higher concentration of attractant will have a higher concentration of occupied receptors. It has been found experimentally that the fraction of leukocytes that move toward higher attractant concentrations is dependent on this gradient in receptor occupancy. The simplest reasonable expression (Zigmond, 1977) is

$$f = \frac{1}{2}\left(1 + \frac{\chi_0 \frac{\partial N_b}{\partial x}}{1 + \chi_0 \frac{\partial N_b}{\partial x}}\right), \tag{13.72}$$

where f is the fraction of cells moving toward higher concentrations, χ_0 is the chemotactic sensitivity, and N_b is the number of bound cell receptors. Notice that N_b is a function of a, the concentration of chemoattractant, and a is a function of x, so the spatial gradient of N_b is given by $\frac{\partial N_b(a)}{\partial x} = \frac{dN_b}{da}\frac{\partial a}{\partial x}$. For small gradients,

$$f \approx \frac{1}{2}\left(1 + \chi_0 \frac{dN_b}{da}\frac{\partial a}{\partial x}\right), \tag{13.73}$$

while for large gradients, $f \approx 1$. Thus in small gradients, the fraction of cells moving toward higher concentrations is linearly proportional to the gradient, and this fraction approaches 1 as the gradient increases. From data for rabbit leukocytes responding to the peptide attractant formyl-methionyl-methionyl-methionine (FMMM) it is estimated that $\chi_0 = 2 \times 10^{-5}$ cm/receptor.

In a uniform steady state (for which $\frac{\partial n}{\partial x} = 0$), $n^+p^+ = n^-p^-$, so that

$$f = \frac{n^+}{n^+ + n^-} = \left(1 + \frac{p^+}{p^-}\right)^{-1}. \tag{13.74}$$

Since $T_p = (p^- + p^+)^{-1}$, we find the chemotactic velocity to be

$$V_c = (2f - 1)v = v\frac{\chi_0 \frac{dN_b}{da}\frac{\partial a}{\partial x}}{1 + \chi_0 \frac{dN_b}{da}\frac{\partial a}{\partial x}}. \tag{13.75}$$

For a single homogeneous population of cell receptors, the number of bound receptors is related to the concentration of attractant through a Michaelis–Menten relationship

$$N_b = \frac{N_T a}{K_d + a}, \tag{13.76}$$

where K_d is the receptor dissociation constant and N_T is the total number of cell receptors.

If the function $v = v(a)$ is known, we have a complete model of the flux of cells due to an attractant concentration. In the special case that cell velocity is independent of attractant concentration, and the attractant concentration and gradient are small, this reduces to a well-known model of chemotaxis (Keller and Segel, 1971),

$$J_c = -\mu\frac{\partial c}{\partial x} + \chi c\frac{\partial a}{\partial x}, \tag{13.77}$$

where $\chi = v\chi_0 N_b'(a)$.

13.4.2 The Inflammatory Response

Leukocytes respond to a bacterial invasion by moving up a gradient of some chemical attractant produced by the bacteria and then ingesting the bacterium when it is encountered. Here we present a one-dimensional model (Alt and Lauffenberger, 1987) to determine if and when the leukocytes successfully defend against a bacterial invasion.

There are three concentrations that must be determined. These are the bacterial, attractant, and leukocyte concentrations, denoted by b, a, and c, respectively. The governing equations for these concentrations follow from the following assumptions concerning their behavior:

1. Bacteria diffuse, reproduce, and are destroyed when they come in contact with leukocytes:
$$\frac{\partial b}{\partial t} = \mu_b \frac{\partial^2 b}{\partial y^2} + (k_g - k_d c)b. \tag{13.78}$$
2. The chemoattractant is produced by bacterial metabolism and diffuses:
$$\frac{\partial a}{\partial t} = D\frac{\partial^2 a}{\partial y^2} + k_p b. \tag{13.79}$$
3. The leukocytes are chemotactically attracted to the attractant, and they die as they digest the bacteria, so that
$$\frac{\partial c}{\partial t} = -\frac{\partial J_c}{\partial y} - (g_0 + g_1 b)c. \tag{13.80}$$
For this model we assume that the leukocyte flux is given by (13.77), although more general descriptions are readily incorporated.

To specify boundary conditions we assume that $y = 0$ is the skin surface and that a blood-transporting capillary or venule lies at distance $y = L$ from the skin surface. We assume that the bacteria cannot leave the tissue domain, although the attractant may diffuse into the bloodstream. Leukocytes enter the tissue from the bloodstream at a rate proportional to the circulating leukocyte density c_b. When chemotactic attractant is present, the emigration rate increases, because leukocytes that would normally flow in the bloodstream tend to adhere to the vessel wall (*margination*) and then migrate into the interstitium. These considerations lead to the boundary conditions

$$\frac{\partial b}{\partial y} = 0 \text{ at } y = 0 \text{ and } y = L, \tag{13.81}$$

$$\frac{\partial a}{\partial y} = \begin{cases} 0, & \text{at } y = 0, \\ -h_a a, & \text{at } y = L, \end{cases} \tag{13.82}$$

$$J_c = \begin{cases} 0, & \text{at } y = 0, \\ -(h_0 + h_1 a)(c_b - c), & \text{at } y = L. \end{cases} \tag{13.83}$$

The governing equations are made dimensionless by setting $x = y/L, \tau = k_g t$, $u = c/c_b, v = b/b_0$, and $w = a/a_0$. We find that

$$\frac{\partial v}{\partial \tau} = \rho_v \frac{\partial^2 v}{\partial x^2} + (1 - \xi u)v, \tag{13.84}$$

$$\frac{\partial w}{\partial \tau} = \rho_w \left(\frac{\partial^2 w}{\partial x^2} + v\right), \tag{13.85}$$

$$\frac{\partial u}{\partial \tau} = \rho_u \left(\frac{\partial^2 u}{\partial x^2} - \alpha \frac{\partial}{\partial x}\left(u \frac{\partial w}{\partial x}\right)\right) - \gamma_0(1 + v)u, \tag{13.86}$$

where $a_0 = L^2k_pb_0/D, b_0 = g_0/g_1, \alpha = \chi a_0/\mu, \rho_v = \frac{\mu_b}{k_gL^2}, \rho_u = \frac{\mu}{k_gL^2}, \rho_w = \frac{D}{k_gL^2}, \xi = k_dc_b/k_g, \gamma_0 = g_0/k_g$.

In nondimensional form the boundary conditions become

$$\frac{\partial v}{\partial x} = 0, \text{ at } x = 0 \text{ and } x = 1, \tag{13.87}$$

$$\frac{\partial w}{\partial x} = \begin{cases} 0, & \text{at } x = 0, \\ -\sigma w, & \text{at } x = 1, \end{cases} \tag{13.88}$$

$$\rho_u\left(\frac{\partial u}{\partial x} - \alpha u\frac{\partial w}{\partial x}\right) = \begin{cases} 0, & \text{at } x = 0, \\ \gamma_0(\beta_0 + \beta_1 w)(1-u), & \text{at } x = 1, \end{cases} \tag{13.89}$$

where $\sigma = h_aL/D, \beta_0 = \frac{h_0}{g_0L}, \beta_1 = \frac{h_1a_0}{g_0L}$.

There is at least one steady-state solution for this system of equations. It is the *elimination state*, in which $v = w = 0$ and

$$u(x) = \frac{1}{A}\cosh\left(\sqrt{\frac{\gamma_0}{\rho_u}}x\right), \tag{13.90}$$

where $A = \cosh\left(\sqrt{\frac{\gamma_0}{\rho_u}}\right) + \frac{\rho_u}{\gamma_0\beta_0}\sqrt{\frac{\gamma_0}{\rho_u}}\sinh\left(\sqrt{\frac{\gamma_0}{\rho_u}}\right)$. In this state, all bacteria are eliminated, and the leukocyte density is independent of any bacterial properties. This should represent the normal state for healthy tissue. If γ_0/ρ_u is small, then this steady distribution of leukocytes is nearly constant, at level $(1 + \frac{1}{\beta_0})^{-1}$.

Bacterial diffusion is generally much smaller than the diffusion of leukocytes or of chemoattractant. Typical numbers are $D = 10^{-6}$ cm^2/s, $\mu = 10^{-7}$ cm^2/s, $\mu_b < 10^{-8}$ cm^2/s, $k_g = 0.5$ h^{-1}, and $L = 100$ μm. With these numbers, ρ_u and ρ_w are relatively large, while ρ_v is small. This leads us to consider an approximation in which bacterial diffusion is ignored, while attractant and leukocyte diffusion are viewed as fast. In this approximation, airborne bacteria can attach to the surface, but they do not move much on the time scale of leukocyte and chemoattractant motion.

Our first approximation is to ignore bacterial diffusion (take $\rho_v = 0$) and then to assume that a bacterial invasion occurs at the skin surface $x = 0$. This is a reasonable assumption for periodontal, peritoneal, and epidermal infections, which are highly localized, slowly moving infections. Then, since we neglect bacterial diffusion, we specify the bacterial distribution by

$$v(x,\tau) = V(\tau)\delta(x), \tag{13.91}$$

where $\delta(x)$ is the Dirac delta function. The governing equation for $V(\tau)$ is

$$\frac{\partial V}{\partial \tau} = (1 - \xi u(0,\tau))V. \tag{13.92}$$

Since $v = 0$ for $x > 0$, the equations for w and u simplify slightly to

$$\frac{\partial w}{\partial \tau} = \rho_w \frac{\partial^2 w}{\partial x^2}, \tag{13.93}$$

$$\frac{\partial u}{\partial \tau} = \rho_u \left(\frac{\partial^2 u}{\partial x^2} - \alpha \frac{\partial}{\partial x} \left(u \frac{\partial w}{\partial x} \right) \right) - \gamma_0 u, \tag{13.94}$$

while the effect of the bacterial concentration at the origin is reflected in the boundary conditions at $x = 0$ (found by integrating (13.85) and (13.86) "across" the origin),

$$\frac{\partial w}{\partial x} = -V, \tag{13.95}$$

$$\rho_u \left(\frac{\partial u}{\partial x} - \alpha u \frac{\partial w}{\partial x} \right) = \gamma_0 V u. \tag{13.96}$$

An identity that is important below is found by integrating (13.94) with respect to x to obtain

$$\gamma_0^{-1} \frac{dU}{dt} = -U - Vu(0,\tau) + (\beta_0 + \beta_1 w(1,\tau))(1 - u(1,\tau)), \tag{13.97}$$

where $U(\tau) = \int_0^1 u(x,\tau)dx$ is the total leukocyte population within the tissue.

Our second approximation is to assume that the chemoattractant diffusion is sufficiently large, so that the chemoattractant is in quasi-steady state,

$$\frac{\partial^2 w}{\partial x^2} = 0. \tag{13.98}$$

This implies that $w(x)$ is a linear function of x with gradient

$$\frac{\partial w}{\partial x} = -V. \tag{13.99}$$

Finally, we assume that ρ_u is large (taking $\rho_u \to \infty$), so that the leukocyte density is also in quasi-steady state with $J_c = 0$, that is,

$$\frac{\partial u}{\partial x} + \alpha V u = 0. \tag{13.100}$$

We can solve this equation and find the leukocyte spatial distribution to be

$$u(x,\tau) = U(\tau) F(\alpha V) e^{-\alpha V x}, \tag{13.101}$$

where $F(z) = \frac{z}{1-e^{-z}}$ is determined by requiring $U(\tau) = \int_0^1 u(x,\tau)dx$.

Now we are able to determine $u(0,\tau), u(1,\tau)$ from (13.101) and $w(1,\tau)$ from (13.88) and (13.99), which we substitute into the equation for total leukocyte mass (13.97) to obtain

$$\gamma_0^{-1} \frac{dU}{d\tau} = (\beta_0 + \beta V) \left(1 - UF(\alpha V) e^{-\alpha V} \right) - (VF(\alpha V) + 1)U, \tag{13.102}$$

where $\beta = \beta_1/\sigma$. Similarly, from (13.92) and (13.101), we find the equation governing V to be

$$\frac{\partial V}{\partial \tau} = V(1 - \xi UF(\alpha V)). \tag{13.103}$$

Phase-Plane Analysis

The system of equations (13.102)–(13.103) is a two-variable system of ordinary differential equations that can be studied using standard phase-plane methods. In this analysis we focus on the influence of two parameters: β, which characterizes the enhanced leukocyte emigration from the bloodstream, and α, which measures the chemotactic response of the leukocytes to the attractant.

One steady-state solution that always exists is $U = (1+\frac{1}{\beta_0})^{-1}, V = 0$. This represents the elimination state in which there are no bacteria present. Any other steady solutions that exist with $V > 0$ are compromised states in which the bacteria are allowed to persist in the tissue.

We assume that the system is at steady state at time $\tau = 0$ with $U(0) = U_0 = (1 + \frac{1}{\beta_0})^{-1}$ when a bacterial challenge with $V(0) = V_0 > 0$ is presented. We begin the analysis with simple cases for which $\alpha = 0$.

Case I: $\alpha = 0, \beta = 0$.
In this case the system reduces to

$$\gamma_0^{-1}\frac{dU}{d\tau} = \beta_0 - (\beta_0 + 1)U - VU, \tag{13.104}$$

$$\frac{\partial V}{\partial \tau} = V(1 - \xi U). \tag{13.105}$$

There are three nullclines: $\frac{dV}{d\tau} = 0$ on the vertical line $U = \frac{1}{\xi}$ and on the horizontal line $V = 0$, and $\frac{dU}{d\tau} = 0$ on the hyperbola $V = \frac{\beta_0-(\beta_0+1)U}{U}$.

Two types of behavior are possible. If $\xi U_0 < 1$, there are no steady states in the positive first quadrant. The only steady state is at $U = U_0, V = 0$. For $U \leq U_0, \frac{dV}{dt} > 0$, so that U decreases and V increases without bound. The bacterial challenge cannot be met. This situation is depicted in Fig. 13.12. In this and all the following phase portraits, the nullcline for $\frac{dV}{d\tau} = 0$ is shown as a short dashed curve, and the nullcline for $\frac{dU}{d\tau} = 0$ is shown as a long dashed curve. The solid curve shows a typical trajectory starting from initial data $U = U_0, V = V_0$.

If $\xi U_0 > 1$, there is a nontrivial steady state in the first quadrant, which is a saddle point. This means that there is a value V^* for which a trajectory starting at $U = U_0, V = V^*$ is on the stable manifold of this steady state and divides the line $U = U_0$ into two types of behavior. If $V < V^*$ initially, the trajectory evolves toward the elimination state, while if $V > V^*$ initially, the trajectory is unbounded. Thus, for large enough ξ and small enough initial bacterial population, the challenge can be withstood, but for a larger initial bacterial challenge, the bacterial population wins the competition. The number V^* is a monotone increasing function of ξ, and $\lim_{\xi\to\infty} V^* = \infty$. This follows

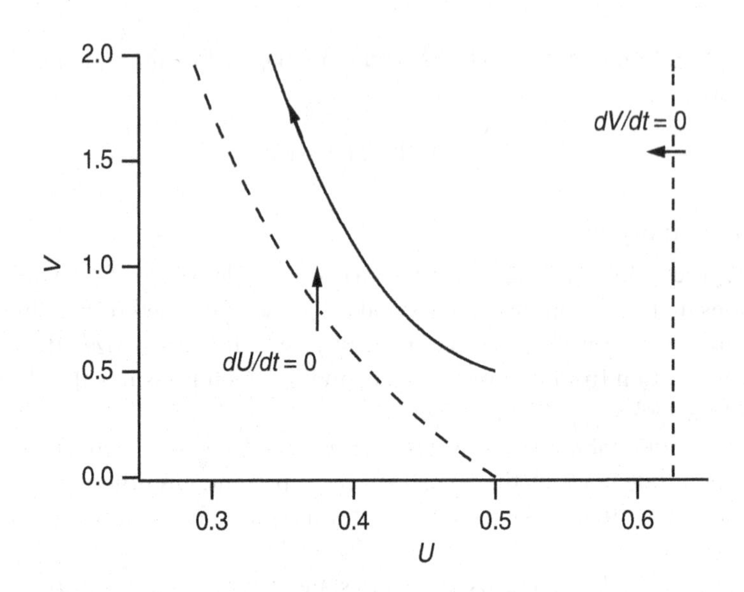

Figure 13.12 Phase portrait for the system (13.102)–(13.103) with "small" $\xi = 1.6$, "small" $\beta = 0.1$, $\alpha = 0$. Other parameters are $\beta_0 = 1.0$, $\gamma_0 = 0.2$, so that $U_0 = 0.5$.

because to the right of $U = \frac{1}{\xi}$ the stable manifold is an increasing curve as a function of U, so that V^* lies above the the value of V at the saddle point. However, as a function of ξ, the steady-state value of V is monotone increasing as ξ increases, approaching ∞ in the limit $\xi \to \infty$, so $V^* \to \infty$ as well.

The phase portrait for this situation is depicted in Fig. 13.13. In this situation the bacterial challenge is met only if ξ is large enough and V_0 is small enough, so that the leukocytes are effective killers, although with $\alpha = \beta = 0$ they are not good hunters. Note that $\xi = k_d c_b / k_g$, where k_d is the rate at which leukocytes kill bacteria, k_g is the growth rate of the bacteria, and c_b is the leukocyte density in the blood. Hence, large ξ means that leukocytes are effective killers, since they kill bacteria at a rate exceeding the growth rate of the bacteria.

Case II: $\alpha = 0, \beta > 0$.

Here, the leukocytes can respond to the bacterial challenge by enhanced emigration from the bloodstream, but they cannot localize preferentially within the tissue. The system of equations becomes

$$\gamma_0^{-1} \frac{dU}{d\tau} = (\beta_0 + \beta V)(1 - U) - (V + 1)U, \tag{13.106}$$

$$\frac{\partial V}{\partial \tau} = V(1 - \xi U). \tag{13.107}$$

The nullclines for $\frac{dV}{d\tau}$ are unchanged from above. The nullcline $\frac{dU}{d\tau} = 0$ is the hyperbola $V = \frac{\beta_0 - (\beta_0+1)U}{(\beta+1)U - \beta}$. For small β, with $\frac{\beta}{\beta+1} < U_0$, the behavior of the system changes only slightly from Case I. These phase portraits are as depicted in Figs. 13.12 and 13.13.

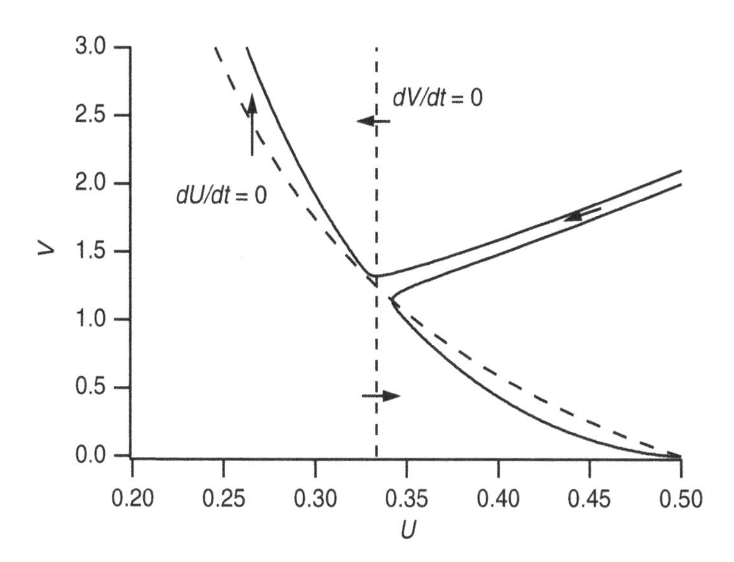

Figure 13.13 Phase portrait for the system (13.102)–(13.103) with "large" $\xi = 3.0$, "small" $\beta = 0.1$, $\alpha = 0$. Other parameters are $\beta_0 = 1.0$, $\gamma_0 = 0.2$, so that $U_0 = 0.5$.

If $\xi U_0 < 1$, the bacterial population grows without bound, whereas if $\xi U_0 > 1$, the bacterial population can be eliminated if $V < V^*$ initially. The value V^* is a monotone increasing function of β. Thus, with β small, the leukocytes have an enhanced ability to eliminate a bacterial population. In fact, if $\xi\beta > \beta + 1$ (phase portrait not shown), then $V^* = \infty$, so that a bacterial invasion of any size can be eliminated. Notice that in this case, the bacterial invasion is controlled because the leukocytes are effective killers and they effectively deploy troops to withstand the invasion. There is no mechanism making them more effective hunters.

In all of the above cases, the leukocyte population decreases initially, and if the bacterial population is controllable, the leukocyte population eventually rebounds back to normal. If β is large enough, with $\frac{\beta}{\beta+1} > U_0$, then the response to a bacterial invasion is with an initial increase in leukocyte population. If $\xi\beta < \beta + 1$, then the bacterial population is unbounded; the invasion cannot be withstood.

If $\xi\beta > \beta+1$ and $\xi U_0 < 1$, there is a nontrivial steady state in the positive first quadrant that is a stable attractor. All trajectories starting at $U = U_0$ approach this stable steady-state solution with $U > U_0$. Since $V > 0$ for this steady solution, the bacterial population is controlled but not eliminated. This situation is depicted in Fig. 13.14.

Finally, if $\xi U_0 > 1$, the leukocyte population initially increases and then decreases back to normal as the bacterial population is eliminated. This situation is depicted in Fig. 13.15.

The above information is summarized in Fig. 13.16, where four regions with differing behaviors are shown, plotted in the $(1/\beta, \xi)$ parameter space. The four regions are bounded by the curves $\xi = 1/U_0$ and $\xi = 1 + 1/\beta$ and are identified by the asymptotic

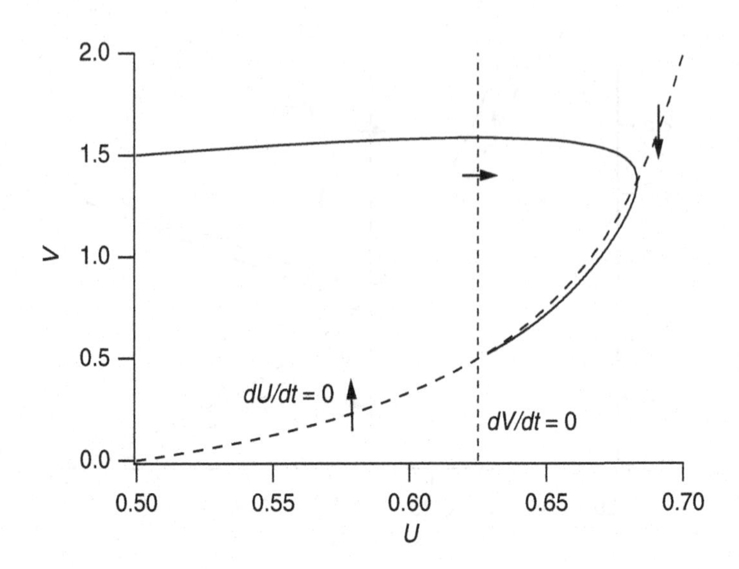

Figure 13.14 Phase portrait for the system (13.102)–(13.103) with "small" $\xi = 1.6$, "large" $\beta = 3.0$, $\alpha = 0$. Other parameters are $\beta_0 = 1.0$, $\gamma_0 = 0.2$, so that $U_0 = 0.5$.

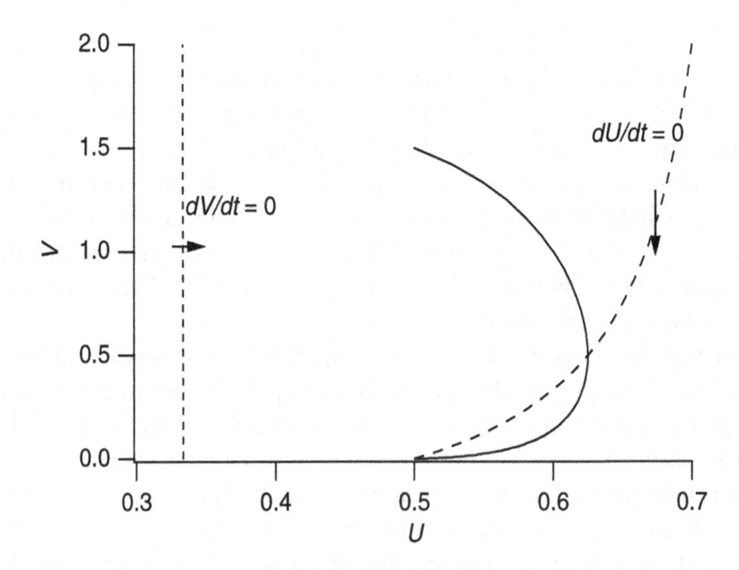

Figure 13.15 Phase portrait for the system (13.102)–(13.103) with "large" $\xi = 3.0$, "large" $\beta = 3.0$, $\alpha = 0$. Other parameters are $\beta_0 = 1.0$, $\gamma_0 = 0.2$, so that $U_0 = 0.5$.

state for $V, \lim_{\tau\to\infty} V(\tau)$. For $\xi > 1/U_0$ and $\xi > 1 + 1/\beta$, the bacteria are always eliminated. For $\xi > 1/U_0$ and $\xi < 1 + 1/\beta$, there are two possibilities, either elimination or unbounded bacterial growth, depending on the initial size of the bacterial population. For $\xi < 1/U_0$ and $\xi > 1 + 1/\beta$, the bacteria survive but are controlled at population

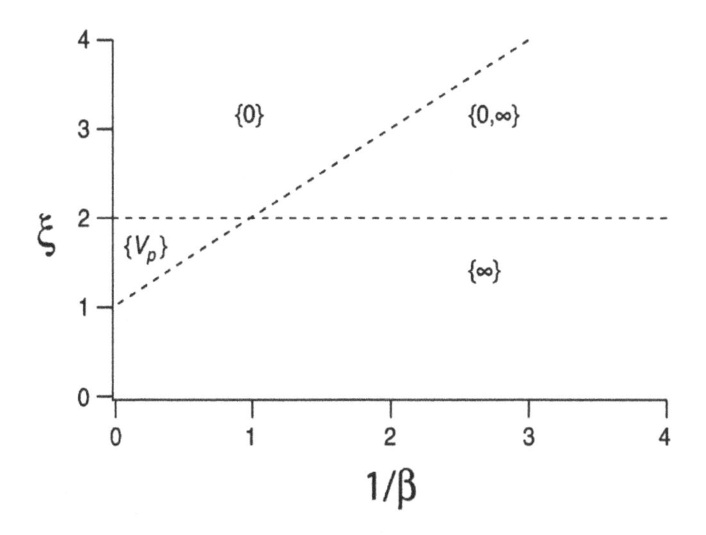

Figure 13.16 Parameter space for the system (13.102)–(13.103) with $\alpha = 0$.

size V_p, and finally, for $\xi < 1/U_0$ and $\xi < 1 + 1/\beta$, the bacterial population cannot be controlled but becomes infinite.

Case III: $\alpha > 0,\ \beta > 0$.
The primary goal of this model is to determine the effect of the chemotaxis coefficient on the performance of the leukocytes in warding off a bacterial invasion. We have seen so far that with $\alpha = 0$ there are three possible responses to an invasion. The bacteria may become unbounded, they may be controlled at a nonzero steady state, or they may be eliminated, depending on the sizes of the parameters ξ and β. With $\alpha \neq 0$, we expect control and elimination to be enhanced, if only because the bacterial growth rate is a decreasing function of α.

The effect of $\alpha \neq 0$ is seen first of all in the nullclines. The nullclines are the curves

$$\frac{dV}{d\tau} = 0: \quad U = \frac{1}{\xi F(\alpha V)}, \tag{13.108}$$

and

$$\frac{dU}{d\tau} = 0: \quad U = \frac{\beta_0 + \beta V}{(\beta_0 + \beta V)F(\alpha V)e^{-\alpha V} + VF(\alpha V) + 1}. \tag{13.109}$$

Both of these are decreasing functions of α, and both asymptote to $U = 0$ as $V \to \infty$. A steady state occurs whenever there is an intersection of these two curves. This condition we write as

$$\frac{1}{\xi} = \frac{\alpha V(\beta_0 + \beta V)}{(1 - e^{-\alpha V}) + \alpha V^2 + \alpha V e^{-\alpha V}(\beta_0 + \beta V)} = G(V). \tag{13.110}$$

One can easily see that $G(0) = U_0$ and that $\lim_{V\to\infty} G(V) = \beta$. This implies that there is an even number of roots if

$$\left(\frac{1}{\xi} - U_0\right)\left(\frac{1}{\xi} - \beta\right) > 0, \tag{13.111}$$

and an odd number of roots otherwise. An odd number of roots implies that there is at least one steady-state solution in the first quadrant; with an even number there could be no steady states. This leads to four different possible outcomes separated by the curves $\xi = \frac{1}{U_0}$ and $\xi = \frac{1}{\beta}$. These are

1. $\xi < \frac{1}{U_0}, \xi < \frac{1}{\beta}$. There can be zero or two steady states. If there are no steady states, then V becomes infinite. If there are two steady states, one of them is stable and the trajectories for sufficiently small initial bacterial populations approach the stable steady state, where they persist. We can find the boundary between these two cases by looking for a double root of (13.110). We do this by solving (13.110) and the equation $G'(V) = 0$ simultaneously. This gives a curve in the (β, ξ) parameter plane parameterized by V, as follows: For each V, β is a root of the quadratic equation

$$\alpha^2V^4\beta^2 - V(-2\alpha^2V^2\beta_0 + \alpha V - 2e^{\alpha V} + 2)\beta + \beta_0(\alpha^2V^2\beta_0 - \alpha V - 1 + e^{\alpha V}(1 - V^2\alpha)) = 0, \tag{13.112}$$

and then ξ is given by (13.110) for each V, β. It is an easy matter to determine this curve numerically. The curve is plotted in Fig. 13.17 as a solid curve, shown for the three values of $\alpha = 0.5, 0.75$, and 1.0.

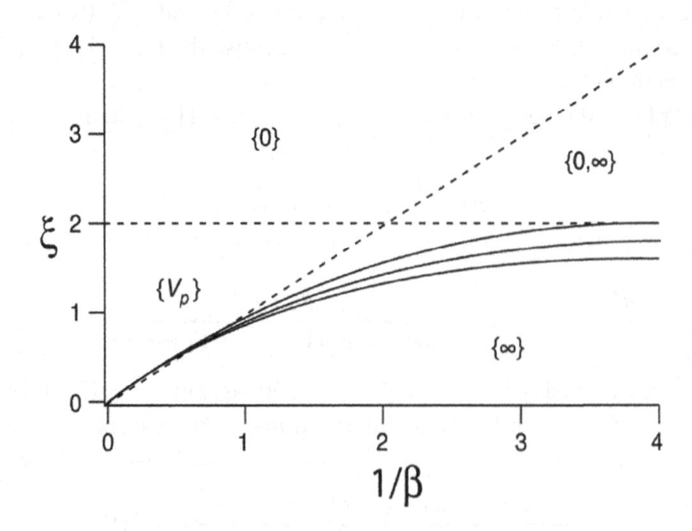

Figure 13.17 Parameter space for the system (13.102)–(13.103) with $\beta_0 = 1.0$, $\alpha > 0$.

Below this curve in the $(\frac{1}{\beta}, \xi)$ parameter space, there are no steady-state solutions. The phase portrait for this case is similar to that of Fig. 13.12 and is left as an exercise (see Exercise 15). For all trajectories starting at $U = U_0$, $V(\tau) \to \infty$.

Above the "double root" curve there are two steady solutions, one of which is stable. In this situation, some trajectories lead to persistent bacterial populations, while others (with larger initial values) become infinite. This phase portrait has similarities with Fig. 13.13 and is left as an exercise (see Exercise 15).

2. $\xi < \frac{1}{U_0}, \xi > \frac{1}{\beta}$. Here there is one stable steady state, which is a global attractor. All trajectories approach this steady state, so that the bacterial population is controlled, but it is not eliminated. The phase portrait for this case is quite similar to the previous case, except that there is only one nontrivial steady state, and no saddle point, so there is no separatrix, and all trajectories approach the persistent state.

   It should be noted that with $\xi < \frac{1}{U_0}$, the bacterial population can never be eliminated. However, with $\alpha > 0$, the population is more readily controlled than with $\alpha = 0$.
3. $\xi > \frac{1}{U_0}, \xi < \frac{1}{\beta}$. There is a single steady state in the first quadrant, which is a saddle point and which therefore divides the initial data into two types, those that are eliminated and those that become unbounded. The phase portrait for this case is similar to Fig. 13.13 and is left as an exercise (see Exercise 15).
4. $\xi > \frac{1}{U_0}, \xi > \frac{1}{\beta}$. Here there are no steady-state solutions in the positive quadrant, in which case the bacterial population is always eliminated. Here the effect of chemotaxis can be seen in the transient behavior of the leukocyte population. If the initial bacterial population is small, the leukocyte population initially increases before it decreases back to its equilibrium. If the initial bacterial population is large, then the leukocyte population initially decreases, then increases, and then finally decreases back to steady state, having eliminated the bacterial population. The phase portrait for this case has similarities with Fig. 13.15 and is left as an exercise (see Exercise 15).

In summary, to control a bacterial invasion, the leukocytes must be sufficiently lethal to the bacteria (ξ sufficiently large). They must also be able to recruit new troops, and it is advantageous that they move chemotactically, since they are more effective if $\alpha > 0$. This result is not surprising. However, the significance of this approximate analysis is that the model behaves as we had hoped, suggesting that it is a reasonable model, worthy of more detailed study and development.

13.5 Control of Lymphocyte Differentiation

The human body can develop considerable specific immunity to various kinds of invading organisms. This so-called *acquired immunity* comes in two different basic versions, both mediated by lymphocytes. First, B lymphocytes make antibodies, soluble proteins that circulate in the blood and help destroy invading organisms (humoral immunity). Second, large numbers of T lymphocytes can be activated to attack and destroy the invaders and cells infected by the invaders (cell-mediated immunity). Both types of lymphocytes are derived from common lymphoid precursors (Fig. 13.1); T lymphocytes develop from their precursors in the thymus gland (hence the nomenclature T lymphocyte), while B lymphocytes develop in the bone marrow. B lymphocytes were first discovered in birds, where they develop in an organ spectacularly named the *bursa of Fabricius*, which is not found in mammals.

One remarkable feature of acquired immunity is the combination of extreme specificity with extreme diversity. As T lymphocytes mature in the thymus they develop reactivity for specific antigens; a population of T lymphocytes contains different cells specifically targeted to each of millions of different possible antigens. B lymphocytes exhibit similar diversity and specificity, with many millions of different types of antibodies secreted by B lymphocytes that react to specific antigens. How exactly this occurs is a fascinating physiological and mathematical question, but not one that we consider further here.

T lymphocytes come in different flavors, with the major types being T helper cells, cytotoxic T cells, and regulatory T cells. As the name implies, cytotoxic T cells are designed to kill other cells, particularly those infected by invading organisms, while regulatory T cells seem to suppress the actions of both cytotoxic and helper T cells, and play a role in preventing autoimmune disease. The majority of T lymphocytes are T helper cells, which secrete a broad range of soluble mediating factors (*cytokines*) that regulate virtually all aspects of the immune response. Typical cytokines are interleukin-2 (IL2), IL3, IL4, and interferon-γ (IFNγ). Acquired immune deficiency syndrome, or AIDS, attacks the T helper cells in particular, leaving the body highly susceptible to infection.

As can be seen in Fig. 13.1, T helper (Th) cells themselves can differentiate into either Th1 or Th2 subtypes. These two subtypes mediate different responses. Th1 cells secrete IL2 and IFNγ and activate macrophages and cytotoxic T cells to destroy invading organisms; Th2 cells make IL4, IL5, and IL13, which activate mast cells, eosinophils, and B lymphocytes, leading to a humoral immune response. Clearly, making the right choice between the Th1 and Th2 pathways can be vital, and there is a complex system of controls that ensures that the appropriate response is generated.

Antigen is presented to the Th cells by an antigen presenting cell, or APC. In response to certain types of antigen, APCs secrete IL12, which promotes Th1 differentiation, via the activation of the transcription factor T-bet. T-bet itself also promotes the production of IFNγ which stimulates the APCs to make more IL12, and also stimulates the production of more T-bet. This gives an autocatalytic feedback loop, pushing more cells down the Th1 pathway. IL4, on the other hand, promotes the differentiation of Th2 cells, and, via the activation of the transcription factor GATA-3, leads to the production of further IL4 and an autocatalytic pathway pushing cells down the Th2 pathway. Crosstalk between the two pathways occurs because GATA-3

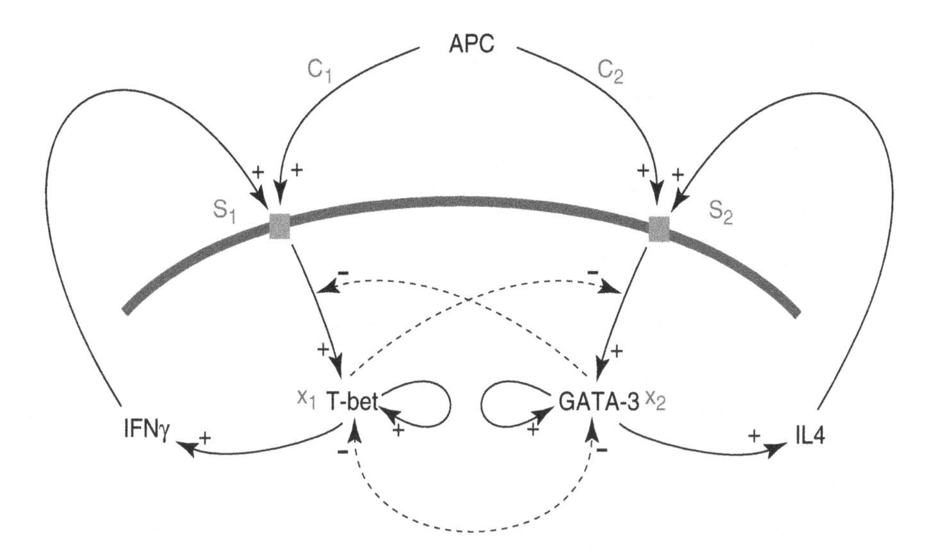

Figure 13.18 Schematic diagram of the pathways involved in the differentiation of Th1 and Th2 cells. Solid lines denote activation, while dashed lines denote inhibition. This diagram includes only a small number of the interactions in the system and is a highly simplified representation.

inhibits the expression of IL12 receptors, while T-bet inhibits GATA-3 activation by IL4. This scheme is sketched in Fig. 13.18. Of course, the actual situation is far more complicated. For example, the exact way in which T-bet and GATA-3 stimulate their own production is not clear, and neither is it known exactly how GATA-3 inhibits T-bet. Such interactions are likely to be more indirect than is shown in Fig. 13.18.

A number of groups have constructed models of Th1 and Th2 cell differentiation. One of the earliest was that of Fishman and Perelson (1993, 1994), while more recently Höfer et al. (2004) and Mariani et al. (2004) have constructed detailed models of the gene regulation networks in these cells. Here we follow the model of Yates et al. (2004).

Let x_1 and x_2 denote the concentrations (or expression levels) of T-bet and GATA-3 respectively, and let S_1 and S_2 denote the concentrations of cytokines that activate T-bet and GATA-3 respectively. We suppose that the rate of T-bet production is a saturating function of S_1, $\sigma_1 \frac{S_1}{\rho_1+S_1}$, while the rate at which it catalyzes its own production is $\alpha_1 \frac{x_1^n}{\kappa_1^n+x_1^n}$. The inhibitory effect of GATA-3 on T-bet is included by assuming that both the S_1-dependent and autocatalytic productions of T-bet are inhibited by GATA-3, via the decreasing function $\frac{\gamma_2}{\gamma_2+x_2}$. Finally, adding a background degradation term $(-\mu x_1)$ and a background production term (β_1), we get

$$\frac{dx_1}{dt} = \beta_1 - \mu x_1 + \left(\frac{\gamma_2}{\gamma_2 + x_2}\right)\left(\alpha_1 \frac{x_1^n}{\kappa_1^n + x_1^n} + \sigma_1 \frac{S_1}{\rho_1 + S_1}\right). \qquad (13.113)$$

Table 13.3 Parameters of the model of lymphocyte differentiation. These parameters, although reasonable, are not known from experimental data, but are chosen to illustrate the basic behavior of the model.

$\sigma_1, \sigma_2 = 5$ day^{-1}	$\alpha_1, \alpha_2 = 5$ day^{-1}
$\gamma_1 \quad = 1$	$\gamma_2 \quad = 0.5$
$\kappa_1, \kappa_2 = 1$	$\mu \quad = 3$ day^{-1}
$\rho_1, \rho_2 = 1$	$g \quad = 2$ day^{-1}
$\beta_1, \beta_2 = 0.05$ day^{-1}	$n \quad = 4$

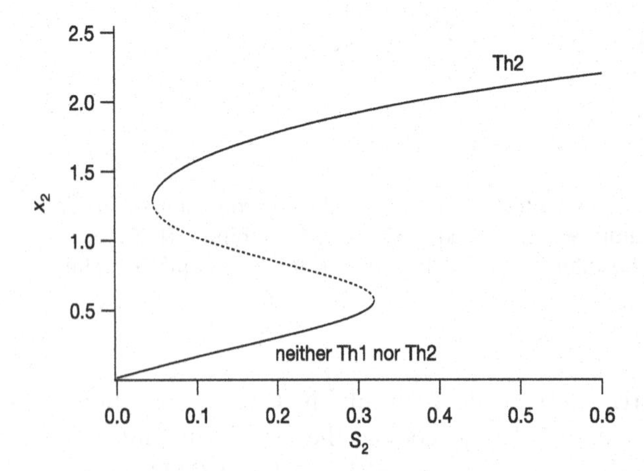

Figure 13.19 Steady-state GATA-3 concentration as a function of S_2, the level of stimulation of the GATA-3 pathway, calculated at $S_1 = 0$. Stable steady states are denoted by solid lines, unstable ones by dashed lines.

The equation for x_2 is constructed in the same way, and is essentially a mirror image of the x_1 equation:

$$\frac{dx_2}{dt} = \beta_2 - \mu x_2 + \left(\frac{\gamma_1}{\gamma_1 + x_1}\right)\left(\alpha_2 \frac{x_2^n}{\kappa_2^n + x_2^n} + \sigma_2 \frac{S_2}{\rho_2 + S_2}\right). \tag{13.114}$$

The parameters of the model are given in Table 13.3. Most of the parameters are the same for both T-bet and GATA-3 with the exception of γ_1 and γ_2. Thus, asymmetry appears only in the crosstalk terms, for which the inhibition of T-bet by GATA-3 is more sensitive (i.e., has a lower K_d) than that of GATA-3 by T-bet.

A plot of the steady-state GATA-3 concentration as a function of S_2 is shown in Fig. 13.19. As the level of stimulation of the GATA-3 pathway increases, the GATA-3 concentration undergoes a sharp increase, while a subsequent decrease in S_2 does not immediately lead to a decrease in GATA-3. This sudden jump corresponds to a change to a Th2 cell from a cell that is neither Th1 nor Th2 (i.e., one with both x_1 and x_2 low). Because of symmetry, a similar curve occurs when the steady-state is plotted against S_1, for fixed $S_2 = 0$. Such switch-like behavior and hysteresis are commonly seen in models of gene transduction pathways (see, for example, Fig. 10.5).

13.6 Clotting

13.6.1 The Clotting Cascade

The need for a clotting system is obvious. In any organism with a circulatory system, the loss of the transporters of vital metabolites and waste products has disastrous, perhaps fatal, consequences. However, the occurrence of clots in an otherwise normal circulatory system is also potentially disastrous, since it prevents a flow that is equally important to survival.

The clotting system must be fast reacting, and yet localized. Since all the ingredients for clotting are carried in the blood, there must be some control that prevents propagation. As we know from earlier chapters, a highly excitable system of diffusing species has the possibility, indeed the strong likelihood, of exhibiting traveling waves. For the clotting system, a propagating front would be as disastrous as failure of a clot to form. Thus, the challenge is to understand how a highly excitable system of reacting and diffusing chemicals is built so as not to allow uncontrolled wave propagation.

In fact, there are more than 50 substances in blood and tissue that play a role in the clotting process. Crucial to the process is the enzyme *thrombin*. Thrombin acts enzymatically on *fibrinogen*, converting it to *fibrin*, which then forms the mesh of the clot. However, this is not all, since thrombin is an extremely active enzyme, with many other regulatory roles.

Thrombin is formed when prothrombin, which is carried in the blood, is converted by an enzyme called *prothrombin activator*. Prothrombin activator is formed as the end result of two different enzymatic cascades, which are, however, closely linked. The fastest, called the *extrinsic pathway*, is initiated following tissue trauma. The second pathway, called the *intrinsic pathway*, is initiated following trauma to blood or contact of blood with collagen, or any negatively charged surface, and is not dependent on tissue trauma. However, this second pathway is much slower than the extrinsic pathway. Classic *hemophilia*, a tendency to bleed that occurs in 1 in every 10,000 males in the United States, results from a deficiency of one of the important enzymes in the intrinsic pathway.

A schematic diagram of some of the reactions involved in clotting is shown in Fig. 13.20. Although there is general agreement over the basic pathways, the exact details of the different feedbacks, and their relative importance, remains unknown, and any two papers by different groups are more than likely to contain slightly different clotting schemes. Thus, the details of Fig. 13.20 should not be taken as definitive.

Two important things may be seen immediately from Fig. 13.20. First, the intrinsic and extrinsic pathways are closely linked, and activation of one activates the other. Second, there are multiple positive feedback pathways, leading to an explosive production of fibrin once the system is triggered.

Thirteen of the important factors in the clotting cascade are denoted using Roman numerals as factors I through XIII, although for historical reasons, they also have other names. Here we retain the Roman numeral notation. Of those that have active

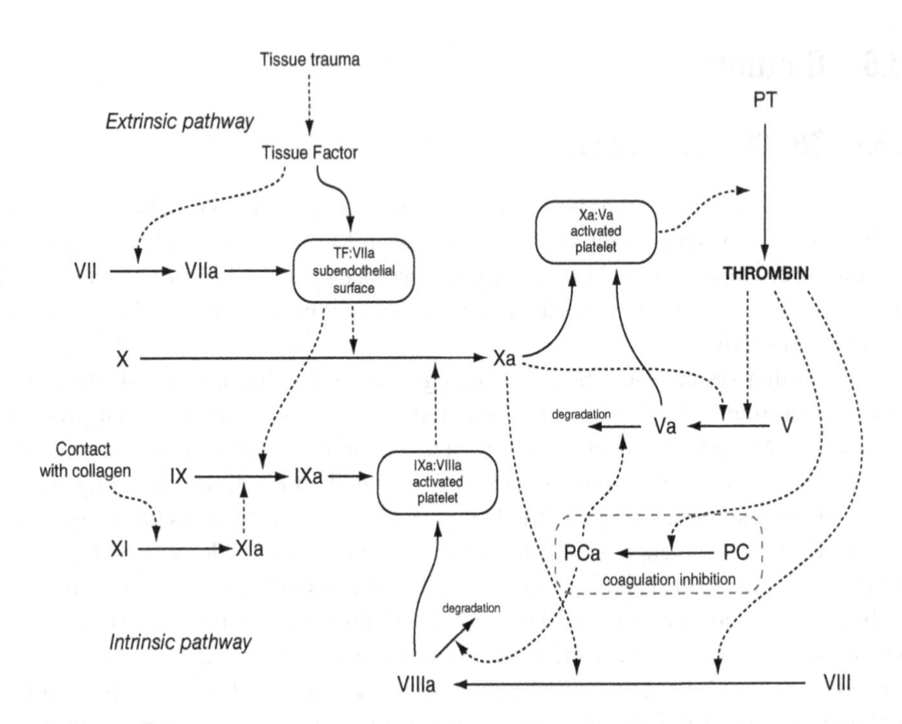

Figure 13.20 Schematic diagram of the extrinsic and intrinsic pathways for blood clotting. PT is prothrombin; PC is protein C; the factors are denoted by Roman numerals (such as VIII, IX, or X) with the activated forms denoted by a (for example, VIIIa, IXa, or Xa). Solid arrows denote conversion, while dashed arrows indicate catalysis. All the reactions, except those involving PC, activate clotting, and form a system of strong positive feedbacks. Activation of PC is the only reaction that inhibits clotting, and is thus enclosed in a dashed box. The boxes with solid lines denote complexes of activated factors that form only on the surface of the exposed subendothelium or of an activated platelet.

and inactive states, the active state is denoted by appending the letter "a" to its name. Thus, for example, Xa is the active form of factor X.

The extrinsic pathway is activated when tissue trauma causes the release of a combination of agents called, collectively, *tissue factor* (or tissue thromboplastin). Tissue factor consists primarily of certain phospholipids from the membranes of the damaged tissues, and acts enzymatically to activate factor VII, converting it to factor VIIa. Factor VIIa then acts enzymatically to activate factor X to Xa, which then combines with factor Va on the surface of an activated platelet to form prothrombin activator. As mentioned above, prothrombin activator converts prothrombin to thrombin, from which the clot eventually forms.

The speed of the extrinsic pathway is increased by various positive feedback mechanisms. First, thrombin activates factor V, thus increasing the rate of formation of prothrombin activator. Second, thrombin is one of the substances that activates platelets (described below) to make them sticky and highly reactive (although this particular feedback is not shown in Fig. 13.20). In their reactive form, platelets are a source of the phospholipids with which factor X combines to produce prothrombin activator.

The intrinsic pathway operates in a similar way. Activation of factor XI by collagen leads to the activation of factor IX, which then forms a complex with factor VIIIa on the surface of an activated platelet. This complex can then activate factor X. Again, positive feedback occurs when both Xa and thrombin activate factor VIII.

Thrombin is degraded, so that its activity is not permanent. One of the ways that thrombin is degraded is by binding to the fibrin network, so that thrombin is eventually degraded by the result of its own activity, a negative feedback loop. There are also anticoagulants, such as antithrombin III, that inactivate thrombin. Inhibition of clotting is caused by the activated form of Protein C, PCa, which increases the rate of degradation of both Va and VIIIa, thus decreasing the rate of thrombin production. Since thrombin itself catalyzes the activation of Protein C, inhibition is controlled, at least partially, by a negative feedback loop (Fig. 13.20).

Heparin is another important anticoagulant. Heparin by itself has little or no anticoagulant effect, but in complex with antithrombin III, the effectiveness of antithrombin III is increased a hundred- to a thousandfold. The concentration of heparin is normally quite slight, although it is produced in large quantities by *mast cells*, located in the connective tissue surrounding capillaries. They are especially abundant in tissue surrounding the capillaries of the lung and of the liver. This is important, because these organs receive many clots that form in the slowly moving venous blood and must be removed. Heparin is widely used in medical practice to prevent intravascular clotting.

13.6.2 Clotting Models

Early models of blood clotting focused on the initial enzyme cascades, showing how the sequence of activations could lead to enormous amplification (Hearon, 1948; Levine, 1966). In a similar style, simplified models that attempt to capture the excitability of the system include those of Jesty et al. (1993) and Beltrami and Jesty (1995).

A more complicated approach, involving large systems of differential equations, has been taken by a number of other groups (for example, Nesheim et al., 1984, 1992; Willems et al., 1991; Jones and Mann, 1994; Bungay et al., 2003). Typically, these large models are solved numerically and compared to such experimental data as the rate of rise of thrombin concentration.

Even more complex models, including the effects of blood flow on clot formation, are those of Anand et al. (2003, 2005), and those from Fogelson's group (Fogelson, 1992; Fogelson and Kuharsky, 1998; Fogelson and Tania, 2005; Fogelson and Guy, 2004). Such models are too detailed to present in full here.

13.6.3 *In Vitro* Clotting and the Spread of Inhibition

One particularly interesting model, simple enough to discuss here, is that of Ataullakhanov et al. (2002a, b), based on the earlier model of Zarnitsina et al. (1996a, b).

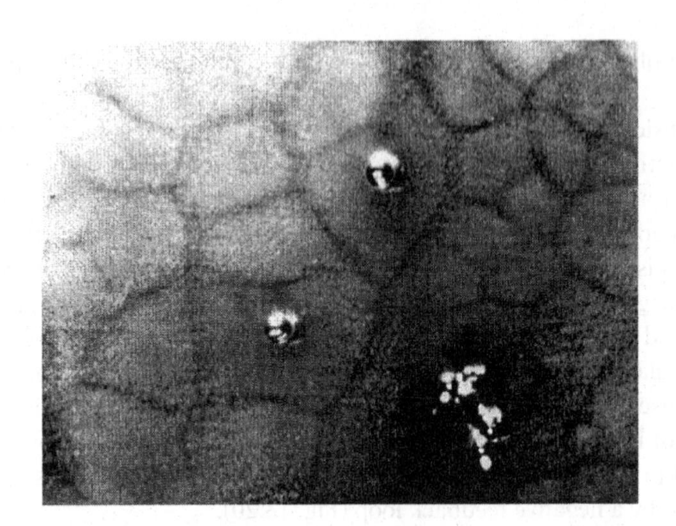

Figure 13.21 Clotting of platelet-poor plasma in a Petri dish, showing the checkerboard of clots that has formed 120 minutes after addition of the glass beads. The clots do not merge to form one large superclot, but are separated by darker channels where the plasma remains liquid for many hours. Thus, each clot appears to be surrounded by an inhibition zone. (Ataullakhanov et al., 2002a, Fig. 11b. Reproduced with permission of World Scientific.)

Using either blood, or plasma from which most of the platelets had been removed, they studied the formation of clots in Petri dishes to which small glass beads (or sometimes collagen fibers) had been added to act as clotting foci. One of their more striking observations was that the clot around each bead grows to a certain size and then stops, leaving the clot surrounded by an inhibition zone. The presence of many clotting foci can then result in a checkerboard of clots across the dish, rather than a single merged superclot. In this case, each clot is surrounded by a thin channel of blood in which clotting is inhibited, so that the blood stays liquid for many hours (Fig. 13.21). Blood clots also sometimes form in a banded structure, of similar appearance to a target pattern, but with stationary bands (Fig. 13.22).

To explain these observations, Ataullakhanov et al. proposed a nine-state model of the clotting cascade, where the nine variables are the concentrations of XIa, IXa, Xa, tissue factor, prothrombin, thrombin, VIIIa, Va, and PCa. They then simplified the model by assuming some of the reactions to be in instantaneous equilibrium, in the same manner as has appeared often throughout this book (in Chapters 1 and 2 in particular). Nondimensionalization of the equations and the omission of small terms resulted in the equations

$$\frac{\partial u_1}{\partial t} = D\frac{\partial^2 u_1}{\partial x^2} + K_1 u_1 u_2 (1 - u_1)\frac{(1 + K_2 u_1)}{(1 + K_3 u_3)} - u_1, \tag{13.115}$$

$$\frac{\partial u_2}{\partial t} = D\frac{\partial^2 u_2}{\partial x^2} + u_1 - K_4 u_2, \tag{13.116}$$

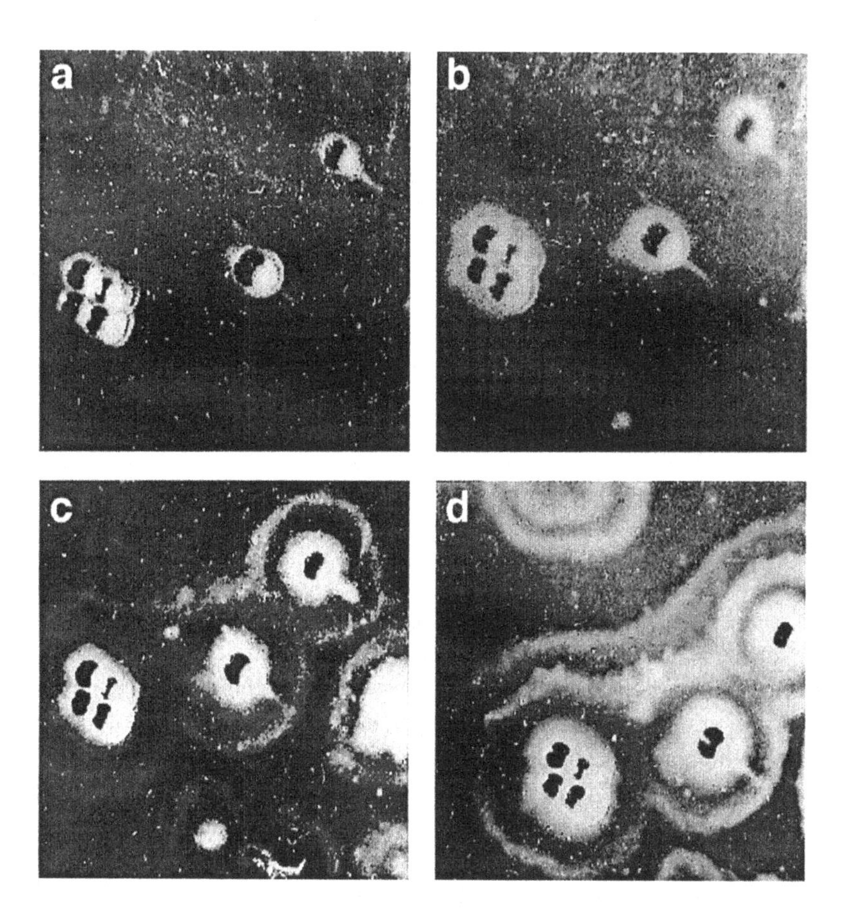

Figure 13.22 Banded clots in platelet-poor plasma *in vitro* (a) 0 minutes, (b) 12 minutes, (c) 30 minutes, and (d) 60 minutes after addition of the glass beads. (Ataullakhanov et al., 2002a, Fig. 11b. Reproduced with permission of World Scientific.)

$$\frac{\partial u_3}{\partial t} = D\frac{\partial^2 u_3}{\partial x^2} + K_5 u_1^2 - K_6 u_3. \tag{13.117}$$

Here, u_1, u_2, and u_3 are, respectively, the nondimensionalized concentrations of thrombin, factor XIa and PCa. There are a number of things to note about the structure of these equations:

- Production of u_1 is stimulated by u_1 and u_2, but inhibited by u_3.
- Production of u_2 is stimulated by u_1, and thus u_1 and u_2 together form a positive feedback loop.
- Production of u_3 is stimulated by u_1, and thus u_1 and u_3 together form a negative feedback loop.
- The concentration of u_1 is bounded by 1, but the concentrations of u_2 and u_3 are not.

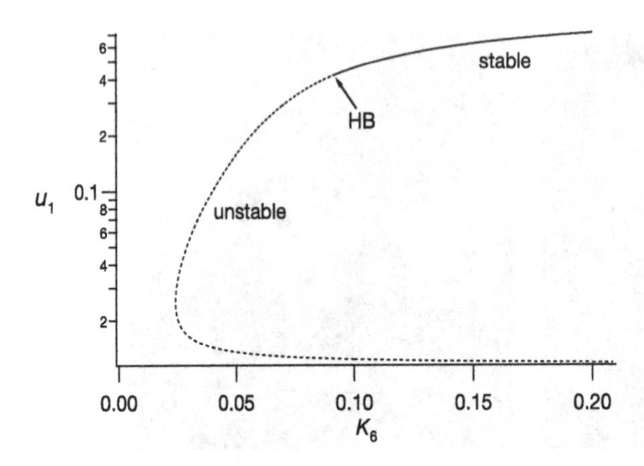

Figure 13.23 Bifurcation diagram of the clotting model (13.115)–(13.117), in the absence of diffusion. The line $u_1 = 0$ is a stable steady state for all values of K_6. Parameter values are $D = 0$, $K_1 = 6.85$, $K_2 = 11$, $K_3 = 2.36$, $K_4 = 0.087$, $K_5 = 17$.

- Each variable is degraded with first-order kinetics.
- For simplicity, each variable is assumed to have the same diffusion coefficient.

The steady states of (13.115)–(13.117) are plotted in Fig. 13.23 as functions of K_6. Since the line $u_1 = 0$ is always a stable steady state, the model has two stable steady states when K_6 is large enough. The Hopf bifurcation on the upper branch of steady states leads to a branch of periodic orbits over a narrow range of values of K_6, but since this is of little interest to the present discussion we ignore it from now on.

The model is highly excitable. Small perturbations away from the steady state $u_1 = 0$ decay back to zero, but larger perturbations cause a large and fast increase in the thrombin concentration (Exercise 16). When K_6 is less than the value at the Hopf bifurcation, this perturbation eventually dies away to zero, behaving much like an action potential (Chapter 5). When K_6 is larger than the value at the Hopf bifurcation, a superthreshold perturbation causes a transition to the higher steady state.

When diffusion is included, the model can behave much like a typical excitable system, as discussed in Chapter 6. A sufficiently large perturbation away from the steady state $u_1 = 0$ can cause a traveling wave that moves across the domain with a constant speed and shape. Because this is not at all surprising (given the excitable nature of the model) we do not show a picture of this wave.

However, for certain parameter values, something much more interesting happens, as illustrated in Fig. 13.24. Although a traveling wave appears to form in response to the superthreshold perturbation, the wave travels only a short distance before stopping. When the wave stops, it does not die away, but forms a stationary wave that is persistent and stable. This stationary solution corresponds to the experimental observation that the clot spreads only a certain distance before stopping.

Close examination of the solution shows why this happens. Immediately after the perturbation, at $t = 50$, the leading edge of the u_1 wave front (the excitation wave) is in front of the leading edge of the u_3 wave front (i.e., the inhibition wave; Fig. 13.24, upper panel). Thus, the excitation wave is able to propagate into the excitable medium. However, although the inhibition wave starts slower than the excitation wave, it eventually

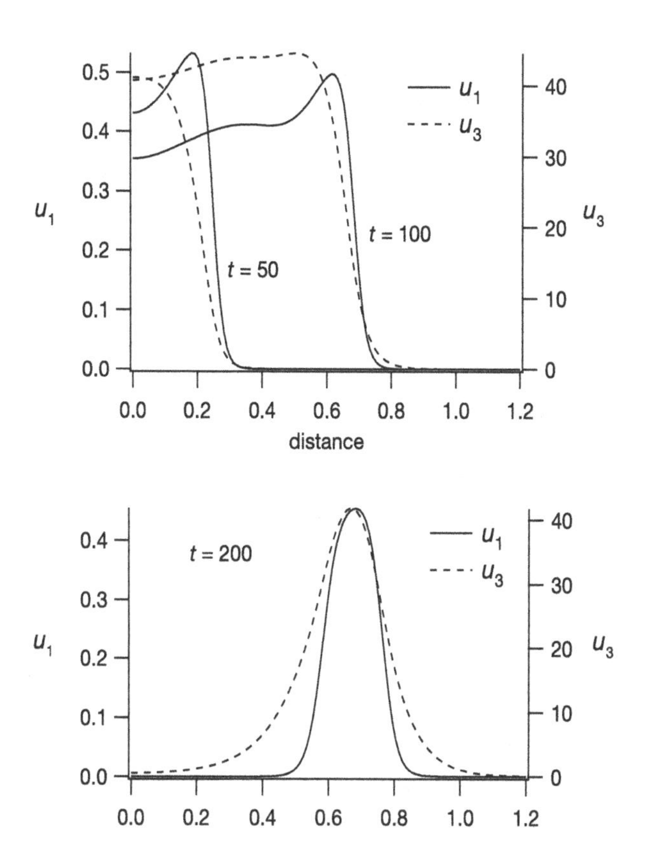

Figure 13.24 A wave of clotting in (13.115)–(13.117) (upper panel) that gradually slows down and turns into a stationary pattern (lower panel). The parameters are $K_6 = 0.066$, $D = 2.6 \times 10^{-4}$, and the times of each plot are given in the figure. The pattern in the lower panel is stable, stationary, and persistent.

catches up. By $t = 100$, the leading edge of the inhibition wave has overtaken the leading edge of the excitation wave. This slows down the excitation wave, and eventually stops it. By $t = 200$, the inhibition and excitation waves are both stationary, with the inhibition wave forming an inhibition zone around the clot (Fig. 13.24, lower panel).

Thus, this model shows how the interaction of waves of excitation and inhibition can lead to a clot that initially propagates explosively into the blood, but is eventually halted, preventing the uncontrolled spread of clotting.

13.6.4 Platelets

The experiments and model discussed in the previous section omit two important features of clotting *in vivo*: the presence of platelets and the fact that blood is often flowing past a clot. When these factors are introduced into a clotting model, the question of how clotting is inhibited and controlled becomes considerably more complicated.

As can be seen from Fig. 13.20, activation of the extrinsic pathway depends on the exposure of the subendothelial surface. However, maintenance of the clotting cascade depends on the presence of activated platelets. Thus, a better understanding of clotting *in vivo* depends on more detailed consideration of platelet dynamics, and the effects of fluid flow.

Platelets are minute round or oval disks 2 to 4 micrometers in diameter. They are formed in the bone marrow from *megakaryocytes*, which are large cells in the bone marrow that fragment into platelets. There are normally between 150,000 and 300,000 platelets per microliter of blood, constituting only a small percentage of the volume ($\approx$0.3 percent by volume).

A platelet is an active structure with a half-life of 8 to 12 days. Since platelets are cell fragments with no nucleus, they do not reproduce. A platelet normally circulates with the blood in a dormant, or inactivated, state, in which it does not adhere to other platelets or to the blood vessel wall. However, when platelets come in contact with a damaged vascular surface or sufficient chemical triggers, they become activated and change their characteristics drastically, as follows:

1. The platelet's surface membrane is altered so that the platelet becomes sticky, capable of adhering to other activated platelets or the subendothelial layer.
2. The platelet secretes chemicals, including large amounts of ADP and thrombaxane A_2, which are capable of activating other platelets.
3. The platelets change from rigid discoidal to highly deformable, extending long, thin appendages called *pseudopodia*.

An important requirement for controlled clotting is that the circulating blood must be able to build a catalytic bed in the vicinity of the injury, even though there is fluid flow. The aggregation of platelets is the means by which a catalytic reactor bed is built, and so is an important part of the process by which the flow of blood from a damaged vessel is halted.

A mathematical model of the aggregation of platelets and the formation of platelet plugs has been formulated and studied by Fogelson (1992) and Fogelson and Guy (2004). The model is a continuum model that assumes that there are concentrations of activated and nonactivated platelets. Platelets are immersed in blood and are neutrally buoyant, moving with the local fluid velocity. There is some chemical, possibly ADP, that is released by platelets when they are activated and that has the effect of stimulating nonactivated cells. Activated cells are sticky and form aggregates when they come into contact with each other.

One can write conservation equations for the density of inactivated and activated platelets. However, because this is an exercise in fluid and continuum mechanics, which is beyond the scope of this text, we do not reproduce these here. Numerical simulation of these equations then demonstrates how platelet aggregates can form in the vicinity of tissue trauma.

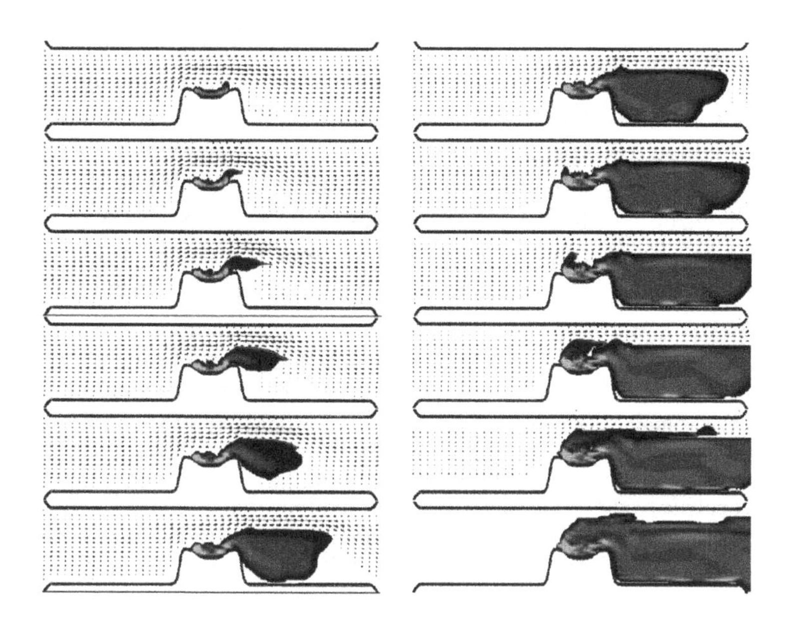

Figure 13.25 Growth of a platelet aggregate in a fluid flow. Fluid is flowing from left to right with velocity vectors shown (the velocity vectors are very small, and difficult to recognize as vectors at this resolution, appearing more like dots). Time is increasing from top left to bottom right, with the panels on the left corresponding to earlier times than the panels on the right. (Fogelson and Guy (2004), Figs. 9 and 10.)

In Fig. 13.25 is shown a series of snapshots of a (two-dimensional) fluid flow past an obstacle, the top of which is exposed subendothelial layer, and thus capable of activating platelets. The figure shows an aggregate of activated platelets growing from the obstacle, which gradually causes the occlusion of the flow (bottom right panel of Fig. 13.25).

The question remains (and is not addressed by these simulations) why the platelet system does not exhibit traveling fronts of aggregation. The putative answer is that smooth (undamaged) vascular walls are nonsticky and that they contain inhibitors of ADP, the primary factor in the activation of platelets, and inhibitors of thrombin, and these prevent the uncontrolled spread of activated platelets.

Another interesting study of clotting inhibition is that of Fogelson and Tania (2005), who propose that purely physical mechanisms can play a larger role than previously realized. In their model, there is competition between the tendency of platelets to stick to, and thus cover up, the subendothelial layer (thus inhibiting formation of the TF:VIIa complex), and the ability of thrombin to activate the platelets. It is essentially a race; if platelets cover the subendothelial layer fast enough, not enough of the IXa:VIIIa or Xa:Va complexes can be formed to maintain the clotting reaction once the TF:VIIa complex is prevented from forming. A clot is thus prevented from forming, since not enough thrombin can be produced. Formation of the IXa:VIIIa and Xa:Va complexes is also inhibited by the flow, which carries away the reactants by diffusion and convection.

However, if the blood flow is slow enough, or the formation of IXa and VIIIa fast enough, by the time the subendothelial surface is covered up by platelets, enough of the activating complexes have been formed to continue the clotting process.

In this model, purely physical factors such as occlusion of the subendothelial layer, or the transport of reactants by blood flow, are much more significant sources of inhibition than the biochemical mechanisms of inactivation by PCa or antithrombin III.

13.7 Exercises

1. What is the volume (per mole) of an ideal gas at room temperature (27° C) and 1 atm pressure? What is its volume at body temperature (98° F)?

2. Find an analytic relationship for the critical stability curve (Section 13.2.2) relating $dF(0)$ to X/d as follows: use that $F(N) = \frac{F(0)}{1+N^7}$ to solve (13.22) for N_0 as a function of X/d and then determine $dF(0)$ using that $F(N_0) = N_0/X$.

3. A deficiency of vitamin B_{12}, or folic acid, is known to cause the production of immature red blood cells with a shortened lifetime of one-half to one-third of normal. What effect does this deficiency have on the population of red blood cells?

4. Suppose $X \to \infty$ in the red blood cell production model. Show that

$$\frac{dN}{dt} = F(N(t-d)) - \beta N, \tag{13.118}$$

where $N(t) = \int_0^\infty n(x,t)dx$.

   (a) Find the stability characteristics for the steady-state solution of this equation.

   (b) Show that the period of oscillation $T = 2\pi/\omega$ at a Hopf bifurcation point is bounded between $2d$ and $4d$.

5. Suppose X is finite and $\beta = 0$ in the red blood cell production model. Show that the evolution of N is described by the delay differential equation

$$\frac{dN}{dt} = F(N(t-d)) - F(N(t-d-X)). \tag{13.119}$$

6. The maturation rate of red blood cells in bone marrow varies as a function of erythropoietin levels. Suppose that x denotes the maturity (rather than chronological age) of a red blood cell. Suppose further that cells are initially formed at maturity $x = -d$, are released into the bloodstream at maturity $x = 0$, age at the normal chronological rate, and die at age $x = X$. Suppose further that the rate of maturation G is a decreasing function of the total circulating red blood cell count N and that the rate of cell production at maturity $x = -d$ is $F(N)$.

   (a) Replace the condition (13.9) with an evolution equation of the form (13.7) to account for maturities x in the range $-d < x < 0$. Specify the boundary condition at $x = -d$ and require that the flux be continuous at $x = 0$. Show that the steady-state solution is independent of the maturation function $G(N)$.

(b) Provide a linear stability analysis for this modified model. Show that

$$\frac{dn}{dt} = \left(F'(N_0) - \frac{G'(N_0)N_0}{G(N_0)X}\right)\left(n\left(t - \frac{d}{G(N_0)}\right) - n\left(t - \frac{d}{G(N_0)} - X\right)\right), \tag{13.120}$$

where $N = N_0(1 + \delta n)$ and $\delta \ll 1$.

(c) How does the variability of G affect the stability of the steady-state solution? Does this variability make the solution more or less likely to become unstable via a Hopf bifurcation?

7. Show that the solution of the kth order linear filter

$$\frac{du_0}{dt} = -\alpha(u_0 - u_1), \tag{13.121}$$

$$\frac{du_1}{dt} = -\alpha(u_1 - u_2), \tag{13.122}$$

$$\vdots$$

$$\frac{du_k}{dt} = -\alpha(u_k - f(t)), \tag{13.123}$$

is given by the convolution

$$u_0(t) = \int_{-\infty}^{t} f(s)g(t-s)ds, \tag{13.124}$$

where $g(t)$ is the gamma distribution of order k,

$$g(t) = \frac{\alpha^{k+1}}{\Gamma(k+1)} t^k \exp(-\alpha t). \tag{13.125}$$

8. Numerically simulate (13.7) with boundary data (13.9) with parameters chosen from the stable region and from the unstable region.

9. Show that if (13.31) has a real solution, it must be negative and lie between $-a$ and $-\alpha$.

10. Suppose that a carrier (such as hemoglobin) of a molecule (such as oxygen) has n independent binding sites, with individual binding and unbinding rates k_+ and k_-. Let c_j denote the concentrations of the state with j molecules bound. Assume that concentrations are in steady state.

(a) Show that $c_j = \binom{n}{j} x^j c_0$, where $\binom{n}{j} = \frac{n!}{j!(n-j)!}$ is the *binomial coefficient*, $x = s_0/K, K = k_-/k_+$, and s_0 is the concentration of the carrier molecule.
Hint: Keep track of the total number of binding sites.

(b) Find the saturation function in the case that $n = 4$.

(c) Show that the four equilibrium constants K_1, K_2, K_3, K_4 defined in (13.45) are given by $(K_1, K_2, K_3, K_4) = K(\frac{1}{4}, \frac{2}{3}, \frac{3}{2}, 4)$.

(d) Estimate K to give a good fit of this model to the hemoglobin saturation curve. How does this curve compare with the curve (13.47)?

(e) Determine whether the hemoglobin binding sites are independent. How close are the equilibrium constants here to those found in the text?

11. Approximate numerical data for the hemoglobin saturation curve are found in Table 13.4. Fit these data to a curve of the form (13.47).

Table 13.4 Approximate numerical data for the hemoglobin saturation curve.

P_{O_2} (mm Hg)	Percent saturation
3.08	2.21
4.61	3.59
6.77	6.08
10.15	10.50
12.31	14.09
15.38	19.34
18.77	28.45
22.77	40.33
25.85	50.0
30.15	60.50
36.00	69.89
45.23	80.11
51.69	83.98
61.85	88.95
75.38	93.37
87.08	95.86
110.5	98.07

Hint: Suppose we have data points $\{x_i, y_i\}, i = 1, \ldots, n$, that we wish to fit to some function $y = f(x)$, and that the function f depends on parameters $\{\alpha_i\}, i = 1, \ldots, m$. A fit of the data is achieved when the parameters are picked such that the function

$$F = \sum_{j=1}^{n} (f(x_j) - y_j)^2 \tag{13.126}$$

is minimized. To find this fit, start with reasonable estimates for the parameters and then allow them to change dynamically (as a function of a timelike variable t) according to

$$\frac{d\alpha_k}{dt} = -\sum_{j=1}^{n} f(x_j) \frac{\partial f(x_j)}{\partial \alpha_k}. \tag{13.127}$$

With this choice,

$$\frac{dF}{dt} = \sum_{k=1}^{m} \sum_{j=1}^{n} f(x_j) \frac{\partial f(x_j)}{\partial \alpha_k} \frac{d\alpha_k}{dt} \leq 0, \tag{13.128}$$

so that F is a decreasing function of t. A fit is found when numerical integration reaches a steady-state solution of (13.127).

12. Construct a Monod–Wyman–Changeux model (Section 1.4.4) for oxygen binding to hemoglobin and determine the saturation function. Fit to the experimental data given in Table 13.4 and compare to the fit of (13.47).

13. (a) If a 25 mM solution of sodium bicarbonate is equilibrated with carbon dioxide at 40 mm Hg partial pressure, the pH is found to be 7.4. What is the equilibrium constant for the bicarbonate–carbon dioxide reaction?

    (b) If the partial pressure of carbon dioxide is increased until the pH is 6.0, what is the bicarbonate concentration? What is the carbon dioxide partial pressure at this pH?

What difference would you expect if this experiment is carried out in whole blood instead of aqueous solution?

(c) Pick reasonable values for K_1, K_2, and $\bar{K}_1$ and find the oxygen saturation curve as a function of h. (Hint: The parameters must be chosen so that hemoglobin acts as a hydrogen ion buffer at physiological concentrations. So, for example, pick K_1, K_2, and $\bar{K}_1$ so that $(\phi(h))^{1/4} = 26\ \sigma$ mm Hg at pH = 7.4. You also have to ensure that $\bar{K}_1 > K_1$ and that K_2 is not too large. Why?)

(d) Use this model to estimate the total amount of CO_2 that is transported from the tissues to the lungs and the total amount of O_2 that is transported from the lungs to the tissues. (Determine this by solving (13.60)–(13.64) numerically to find h, X, Y, Z, and W at a given $[O_2]$ and $[CO_2]$ and finding the difference between arterial and venous quantities.) Remove the Bohr and Haldane effects by setting $K_1 = \bar{K}_1$. How does this change the amount of oxygen and carbon dioxide transported? Typical parameter values are P_{CO_2} in arterial blood, 39 mm Hg; P_{CO_2} in venous blood, 46 mm Hg; P_{O_2} in arterial blood, 100 mm Hg; P_{O_2} in venous blood, 40 mm Hg; $R_1R_2 = 10^{6.1}\,M^{-1}$; $T_{Hb} = 3$ mM; $n = 10$; $[HCO_3^-] = 25$ mM, and pH = 7.4 in arterial blood.

14. Develop a detailed model of oxygen and carbon monoxide binding with hemoglobin. How can the fact that CO has 210 times the affinity for binding be used to estimate the equilibrium coefficients?

15. Sketch the phase portraits for (13.102)–(13.103) in Case III ($\alpha > 0, \beta > 0$) as follows:

(a) $\xi < \frac{1}{U_0}, \xi < \frac{1}{\beta}$. (For example,

i. $\xi = 1.0, \beta = 0.5, \alpha = 0.5, \beta_0 = 1.0, \gamma_0 = 0.2$, and

ii. $\xi = 1.5, \beta = 0.5, \alpha = 1.3, \beta_0 = 1.0, \gamma_0 = 0.2$.)

(b) $\xi < \frac{1}{U_0}, \xi > \frac{1}{\beta}$. (For example, $\xi = 1.8, \beta = 0.6, \alpha = 0.5, \beta_0 = 1.0, \gamma_0 = 0.2$.)

(c) $\xi > \frac{1}{U_0}, \xi < \frac{1}{\beta}$. (For example, $\xi = 2.2, \beta = 0.3, \alpha = 0.5, \beta_0 = 1.0, \gamma_0 = 0.2$.)

(d) $\xi > \frac{1}{U_0}, \xi > \frac{1}{\beta}$. (For example, $\xi = 2.2, \beta = 2.0, \alpha = 0.5, \beta_0 = 1.0, \gamma_0 = 0.2$.)

Locate each of these cases in Fig. 13.17.

16. Show that the clotting model of (13.115)–(13.117) is excitable. Investigate the response to perturbations away from the $u_1 = 0$ steady state. Do this for two different cases: when K_6 is less than the value at the Hopf bifurcation, and when K_6 is larger than the value at the Hopf bifurcation.

CHAPTER 14

Respiration

The respiratory system is responsible for gas transfer between the tissues and the outside air. Carbon dioxide that is produced by metabolism in the tissues must be moved by the blood to the lungs, where it is lost to the outside air, and oxygen that is supplied to the tissues must be extracted from the outside air by the lungs.

The nose, mouth, pharynx, larynx, trachea, broncheal trees, lung air sacs and respiratory muscles are the structures that make up the respiratory system (Fig. 14.1). The nasal cavities are specialized for warming and moistening inspired air and for filtering the air to remove large particles. The larynx, or "voice box," contains the vocal folds that vibrate as air passes between them to produce sounds. Below the larynx the respiratory system divides into *airways* and *alveoli*. The airways consist of a series of branching tubes that become smaller in diameter and shorter in length as they extend deeper into the lung tissue. They terminate after about 23 levels of branches in blind sacs, the alveoli. The *terminal bronchioles* represent the deepest point of the bronchial tree to which inspired air can penetrate by flowing along a pressure gradient. Beyond the terminal bronchioles, simple diffusion along concentration gradients is primarily responsible for the movement of gases.

Alveoli are thin-walled air sacs that provide the surface across which gases are exchanged (Fig. 14.2). Each lung contains about 300 million alveoli that are surrounded by respiratory membranes that serve to bring air and blood into close contact over a large surface area. In the human lung, from 70 to 140 ml of blood is spread over approximately 70 m^2.

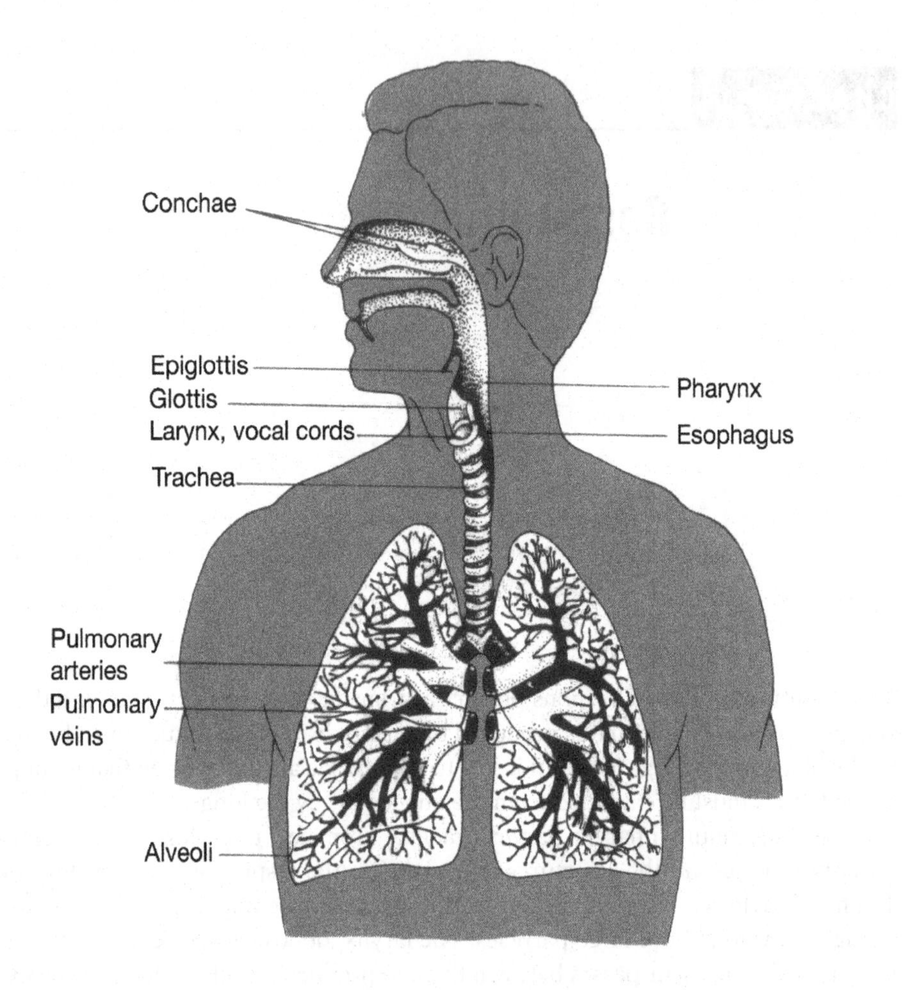

Figure 14.1 Diagram of the respiratory passages. (Guyton and Hall, 1996, Fig. 37-9, p. 486.)

14.1 Capillary–Alveoli Gas Exchange

14.1.1 Diffusion Across an Interface

Recall from Chapter 13 that the partial pressure of a gas is defined as the mole fraction of the gas multiplied by the total pressure. If a gas with partial pressure P_s is in contact with a liquid, the steady-state concentration U of gas in the liquid is given by

$$U = \sigma P_s, \tag{14.1}$$

where σ is the solubility of the gas in the liquid. Because of this, we define the partial pressure of a dissolved gas with concentration U to be U/σ.

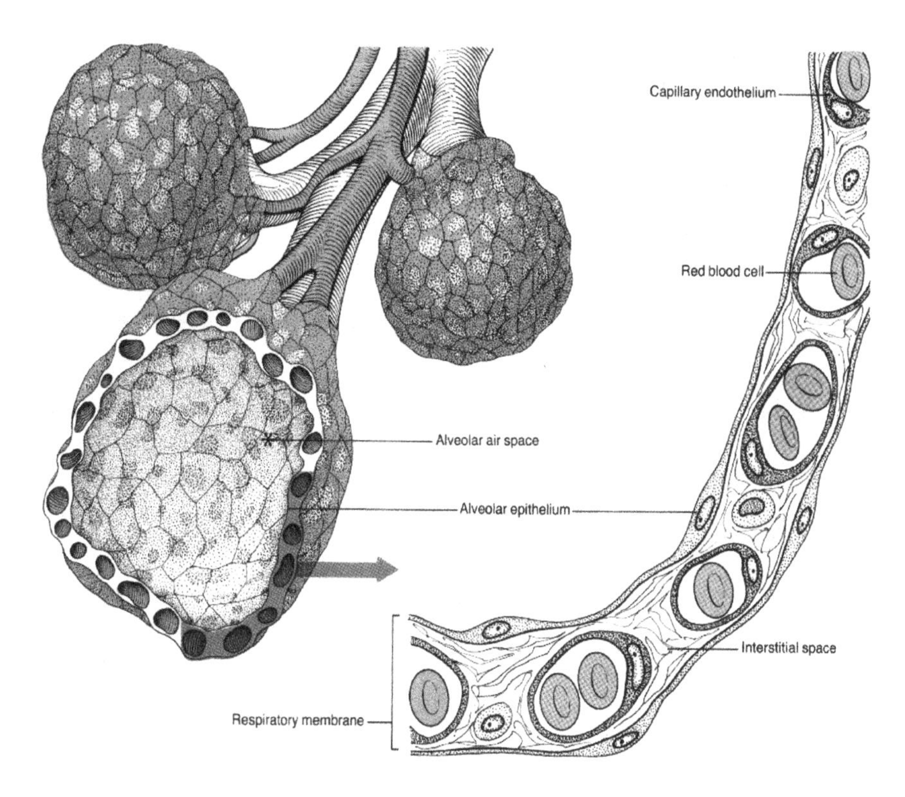

Figure 14.2 The alveoli, or air sacs, of the lung are covered by an extensive network of capillaries that form a thin layer of blood for the exchange of gases. (Davis, Holtz, and Davis, 1985, Fig. 19-4, p. 391.)

Now suppose that a gas with partial pressure P_g is brought into contact with a liquid within which that same gas is dissolved with concentration U, and thus partial pressure U/σ. If U/σ is not equal to P_g, one expects a net flow of gas across the interface. The simplest model (but not necessarily the most accurate) assumes that the flow is linearly proportional to the difference in partial pressures across the interface, and thus

$$q = D_s \left(P_g - \frac{U}{\sigma} \right), \tag{14.2}$$

where q is the net flux per unit area of the gas (positive when gas is flowing from the gaseous phase to the dissolved phase), and D_s is the *surface diffusion constant*.

14.1.2 Capillary–Alveolar Transport

To understand something about the transport of a gas across the capillary wall into the alveolar space, we begin with the simplest possible model. We suppose that a gas such as oxygen or carbon dioxide is dissolved in blood at some concentration U uniformly across the cross-section of the capillary. The blood flows along a capillary that is bounded by alveolar air space. The partial pressure of the gas in the alveolar space, P_g, is taken to be constant.

Consider a segment of the capillary, of length L, with constant cross-sectional area A and perimeter p. The total amount of the dissolved gas contained in the capillary at any time is $A\int_0^L U(x,t)\,dx$. Since mass is conserved, we have

$$\frac{d}{dt}\left(A\int_0^L U(x,t)\,dx\right) = v(0)AU(0,t) - v(L)AU(L,t) + p\int_0^L q(x,t)\,dt, \tag{14.3}$$

where $v(x)$ is the velocity of the fluid in the capillary, and q is the flux (positive inward, with units of moles per time per unit area) of gas along the boundary of the capillary. This assumes that diffusion along the length of the capillary is negligible compared to diffusion across the capillary wall. Differentiating (14.3) with respect to L and replacing L by x gives the conservation law

$$U_t + (vU)_x = \frac{pq}{A}. \tag{14.4}$$

Finally, if we assume that the flow velocity v is constant along the capillary, then using (14.2) we obtain

$$U_t + vU_x = \frac{pD_s}{A}\left(P_g - \frac{U}{\sigma}\right) = D_m(\sigma P_g - U), \tag{14.5}$$

where $D_m = \chi D_s/\sigma$, and $\chi = p/A$ is the surface-to-volume ratio. Notice that D_m has units of $(\text{time})^{-1}$, so it is the inverse of a time constant, the *membrane exchange rate*.

At steady state the conservation law (14.5) reduces to the first-order, linear ordinary differential equation

$$v\frac{dU}{dx} = D_m(\sigma P_g - U). \tag{14.6}$$

Note that, as one would expect intuitively, the rate of change of U at the steady state is inversely proportional to the fluid velocity. Now we suppose that the concentration U at the inflow $x = 0$ is fixed at U_0 (at partial pressure $P_0 = U_0/\sigma$). At steady state, the concentration at each position x is given by the exponentially decaying function

$$U(x) = \sigma P_g + (U_0 - \sigma P_g)e^{-D_m x/v}. \tag{14.7}$$

If the exposed section of the capillary has length L, the total flux of gas across the wall is $Q = p\int_0^L q\,dx = vA[U(L) - U_0]$, which is

$$Q = vA\sigma(P_g - P_0)(1 - e^{-D_m L/v}). \tag{14.8}$$

Plotted in Fig. 14.3 is the nondimensional flux

$$\bar{Q} = \frac{Q}{D_m LA\sigma(P_0 - P_g)} = \frac{v}{D_m L}(1 - e^{\frac{-D_m L}{v}}). \tag{14.9}$$

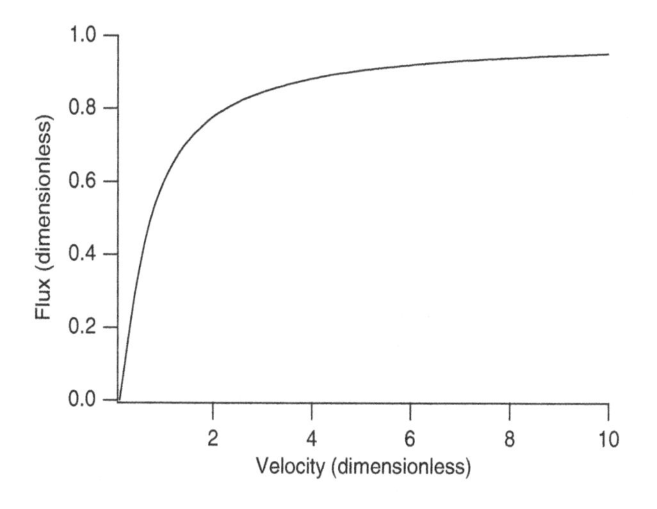

Figure 14.3 Dimensionless transmural flux $\bar{Q}$ as a function of dimensionless flow velocity $\frac{v}{D_m L}$ from (14.9).

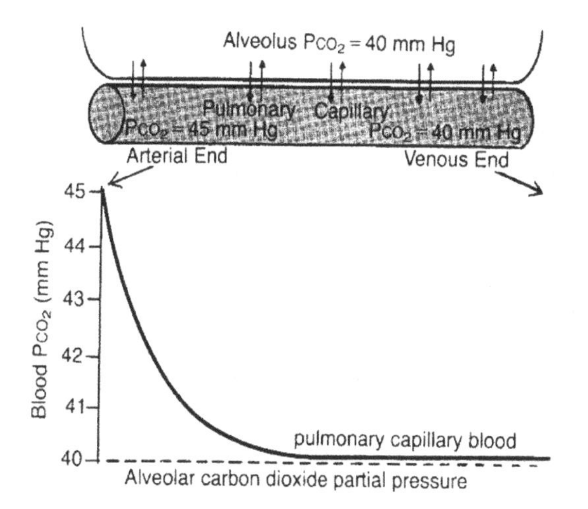

Figure 14.4 Loss of carbon dioxide from the pulmonary capillary blood into the alveolus. (The curve in this figure was constructed from data in Milhorn and Pulley, 1968). Figure from Guyton and Hall, 1996, Fig. 40-6, p. 515.)

Note that

$$Q \to vA\sigma(P_g - P_0) \tag{14.10}$$

in the limit $D_m L/v \to \infty$. Thus, an infinitely long capillary has only a finite total flux, as the dissolved gas concentration approaches the alveolar concentration along the length of the capillary.

Data on the diffusion of carbon dioxide from the pulmonary blood into the alveolus (Fig. 14.4) suggest that carbon dioxide is lost into the alveolus at an exponential rate, consistent with (14.7). Furthermore, because the solubility of carbon dioxide in water is quite high, the difference between the partial pressure for the entering blood and the alveolar air is small, about 5 mm Hg.

In contrast, the solubility of oxygen in blood is small (about 20 times smaller than carbon dioxide, see Table 13.2), and although the difference in partial pressures is

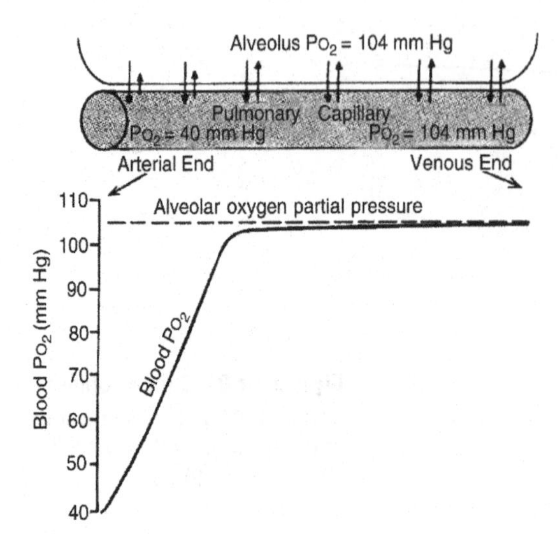

Figure 14.5 Uptake of oxygen by the pulmonary capillary blood. (The curve in this figure was constructed from data in Milhorn and Pulley, *Biophys. J.* 8:337, 1968. Figure from Guyton and Hall, 1996, Fig. 40-1, p. 514.)

larger, this is not adequate to account for the balance of oxygen inflow and carbon dioxide outflow. That is, if (14.10) is relevant, then, in order to maintain a similar transport, a decrease in σ by a factor of 20 requires a corresponding increase by a factor of 20 for the difference in partial pressures. Thus, if this is the correct mechanism for carbon dioxide and oxygen transport, the difference $P_0 - P_g$ for oxygen should be about twenty times larger than for carbon dioxide. Since $104-40 \neq 20(45-40)$ (using typical numbers from Figs. 14.4 and 14.5), there is reason to doubt this model.

Secondly, the data in Fig. 14.5 suggest that the uptake of oxygen by the capillary blood is not exponential with distance, but nearly linear for the first third of the distance, where it becomes fully saturated. We consider a model of this below. First, however, we discuss the effects of blood chemistry on gas exchange, which was ignored in the above model.

14.1.3 Carbon Dioxide Removal

Blood chemistry plays a significant role in facilitating the transport of gases between blood and alveoli. To understand something of this facilitation, we first consider a simple model of carbon dioxide transport that takes the carbon dioxide–bicarbonate chemistry into account. We assume that carbon dioxide is converted to bicarbonate via the reaction

$$CO_2 + H_2O \underset{k_{-1}}{\overset{k_1}{\rightleftharpoons}} HCO_3^- + H^+.$$

This is the carbonic anhydrase reaction discussed in Section 13.3.3. Because it has little effect, we ignore the intermediary H_2CO_3.

Now we write conservation equations for the two chemical species CO_2 and HCO_3^- (in steady state, and ignoring diffusion within the capillary) as

$$v\frac{dU}{dx} = D_{CO_2}(\sigma_{CO_2}P_{CO_2} - U) + k_{-1}[H^+]V - k_1U, \tag{14.11}$$

$$v\frac{dV}{dx} = k_1U - k_{-1}[H^+]V, \tag{14.12}$$

where $U = [CO_2]$, $V = [HCO_3^-]$. Notice that D_{CO_2} is a rate constant, similar to D_m above.

Although this is a linear problem and it can be solved exactly, it is illustrative to use an approximate, singular perturbation technique, as this technique is useful in the next section. First, notice that we can add (14.11) and (14.12) to obtain

$$v\frac{d}{dx}(U + V) = D_{CO_2}(\sigma_{CO_2}P_{CO_2} - U). \tag{14.13}$$

Now we assume that V equilibrates rapidly, so that it can be taken to be in quasi-steady state. Accordingly, we set $V = K_cU$, where $K_c = \frac{k_1}{k_{-1}[H^+]}$. It follows that, assuming that $[H^+]$ is constant,

$$v(1 + K_c)\frac{dU}{dx} = D_{CO_2}(\sigma_{CO_2}P_{CO_2} - U). \tag{14.14}$$

This equation is identical in form to (14.6). If we take the inlet conditions to be $U = U_0 = \sigma_{CO_2}P_0$ and $V = V_0 = K_cU_0$, then the total flux Q is

$$Q = vA(1 + K_c)\sigma_{CO_2}\left(P_0 - P_{CO_2}\right)\left(1 - e^{-D_{CO_2}L/(v(1+K_c))}\right), \tag{14.15}$$

and in the limit as $D_{CO_2}L/v \to \infty$,

$$Q \to vA(1 + K_c)\sigma_{CO_2}\left(P_0 - P_{CO_2}\right), \tag{14.16}$$

which is a factor of $1 + K_c$ larger than in (14.10). The only difference between this flux (14.15) and the original (14.8) is that the velocity v has been multiplied by the factor $1 + K_c$. In other words, the conversion of carbon dioxide to bicarbonate via the carbonic anhydrase reaction effectively increases the flow rate by the factor $1 + K_c$.

The equilibrium constant for the bicarbonate–carbon dioxide reaction is known to be $\log_{10}(\frac{k_1}{k_{-1}}) = -6.1$ (see also Chapter 13, Exercise 13a). Thus (since $pH = -\log_{10}[H^+]$ with $[H^+]$ in moles per liter), at $pH = 7.4$, we have $K_c = 20$, and the improvement in carbon dioxide transport because of the carbonic anhydrase reaction is substantial.

The improvement in total flux arises because the conversion of bicarbonate to carbon dioxide continually replenishes the carbon dioxide that is lost to the alveolar air. Thus, the carbon dioxide concentration in the capillary does not fall so quickly, leading to an increase in the total flux.

14.1.4 Oxygen Uptake

The chemistry for the absorption of oxygen by hemoglobin has a similar, but nonlinear, effect. We take a simple model of the chemistry of hemoglobin (discussed in Section

13.3.1), namely

$$\mathrm{Hb} + 4\mathrm{O}_2 \underset{k_{-2}}{\overset{k_2}{\rightleftharpoons}} \mathrm{Hb(O_2)_4}.$$

Of course, there are more detailed models of hemoglobin chemistry, but the qualitative behavior is affected little by these details. We write the conservation equations as

$$v\frac{dW}{dx} = D_{O_2}(\sigma_{O_2}P_{O_2} - W) + 4k_{-2}Y - 4k_2ZW^4, \tag{14.17}$$

$$v\frac{dY}{dx} = k_2ZW^4 - k_{-2}Y, \tag{14.18}$$

$$v\frac{dZ}{dx} = k_{-2}Y - k_2ZW^4, \tag{14.19}$$

where $W = [O_2]$, $Y = [Hb(O_2)_4]$, $Z = [Hb]$, and D_{O_2} is the oxygen exchange rate constant. The last of these equations is superfluous, since total hemoglobin is conserved, and so we take $Z+Y = Z_0$. Notice further that (14.17) and (14.18) can be added to obtain

$$v\frac{d}{dx}(W + 4Y) = D_{O_2}(\sigma_{O_2}P_{O_2} - W). \tag{14.20}$$

We expect oxygen uptake by hemoglobin to be fast compared to the transmural exchange, so take Y to be in quasi-steady state, setting

$$Y = Z_0\frac{W^4}{K_{O_2}^4 + W^4}, \tag{14.21}$$

where $K_{O_2}^4 = k_{-2}/k_2$. On substitution into (14.20) we find

$$v\frac{d}{dx}\left(W + 4Z_0\frac{W^4}{K_{O_2}^4 + W^4}\right) = D_{O_2}(\sigma_{O_2}P_{O_2} - W). \tag{14.22}$$

More generally, if $f(W)$ is the oxygen saturation curve for hemoglobin, then

$$v\frac{d}{dx}(W + 4Z_0f(W)) = D_{O_2}(\sigma_{O_2}P_{O_2} - W). \tag{14.23}$$

This equation is a nonlinear first-order ordinary differential equation, which, being separable, can be solved exactly. The solution is given implicitly by

$$\int_{W_1}^{W_2} \frac{1 + 4Z_0f'(W)}{\sigma_{O_2}P_{O_2} - W}\,dW = \frac{D_{O_2}L}{v}, \tag{14.24}$$

and the total flux of oxygen is given by

$$\begin{aligned} Q &= A\int_0^L D_{O_2}(\sigma_{O_2}P_{O_2} - W)\,dx \\ &= Av\int_0^L \frac{d}{dx}(W + 4Z_0f(W))\,dx \\ &= Av(W + 4Z_0f(W))\big|_{W_0}^{W_1}. \end{aligned} \tag{14.25}$$

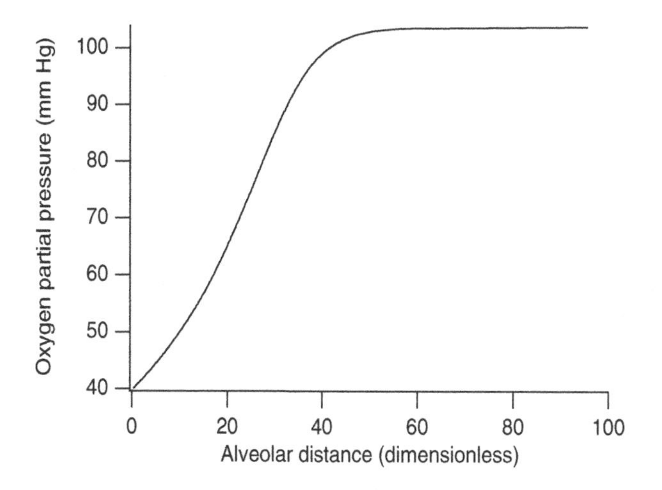

Figure 14.6 Oxygen partial pressure as a function of non-dimensional distance $\frac{D_{O_2}x}{v}$.

Finally, using (14.24) we can find the rate of oxygen uptake as a function of length along the capillary. In Fig. 14.6 is shown the partial pressure of oxygen, plotted as a function of the dimensionless distance $D_{O_2}x/v$ along the capillary. The significant observation is that oxygen partial pressure rises steeply and nearly linearly, until it saturates, comparing well with the experimental data shown in Fig. 14.5.

The exact solution (14.24) does not provide much insight. It is more useful to compare (14.23) with (14.14), in which the flux of carbon dioxide is facilitated by the factor K_c. If we rewrite (14.23) as

$$v(1 + 4Z_0 f'(W))\frac{dW}{dx} = D_{O_2}(\sigma_{O_2} P_{O_2} - W), \tag{14.26}$$

we see that there is facilitation of oxygen flux by the factor $1 + 4Z_0 f'(W)$. Clearly, the two ways to exploit this facilitation are to have a high concentration of hemoglobin and to use a saturation curve $f(W)$ with a steep slope in the range of operating values.

The maximal flux enhancement can be substantial. Letting $f(W) = W^4/(K_{O_2}^4 + W^4)$, we find that f has a maximal derivative of $1/K_{O_2}$, which occurs when $W = K_{O_2}$. Choosing $K_{O_2} = 30\sigma_{O_2}$ mm Hg (see Section 13.3.1, and Fig. 14.8), we get $K_{O_2} = 4.2 \times 10^{-2}$ mM, where the value of oxygen solubility is taken from Table 13.2.

Since the total concentration of hemoglobin in blood is around 2 mM (it varies from males to females, and from children to adults, but this is a typical concentration), we find the maximal enhancement of oxygen flux by a factor of $1+8/(4.2\times 10^{-2}) \approx 200$.

Under normal conditions the actual enhancement is, of course, less than this, as the system does not operate exclusively at the point of maximum slope of f. A more realistic estimate of the enhancement due to hemoglobin can be obtained by comparing how much oxygen is carried by the hemoglobin, and how much is carried in solution. When hemoglobin is completely saturated with oxygen, 100 mls of blood carries about 20 mls of oxygen, the so-called *20 volumes per cent*. However, assuming an alveolar oxygen partial pressure of 150 mm Hg, and an oxygen solubility of 1.4×10^{-6} moles per liter per mm Hg, it follows that, in the absence of hemoglobin, 100 mls of blood could contain only about 0.5 mls of dissolved oxygen (equivalent to a concentration of about 2.1×10^{-4} moles of oxygen per liter of blood). Thus, a more realistic estimate of the maximal enhancement due to hemoglobin is a factor of around 40.

In reality, about 97% of oxgyen is transported by hemoglobin, with the other 3% transported as dissolved oxygen, giving an enhancement factor of around 32.

There are a number of much more complicated models of gas transport and exchange in the literature. Ben-Tal (2006) reviews a range of models of gas exchange in the lungs, ranging from the simplest to the more complex, while Whiteley et al. (2001, 2002, 2003b) study a model in the same spirit as the simple model we present here, but with more accurate spatial properties, including the shape of the red blood cells. The most complex models, at least from the point of view of spatial structure, are those of Tawhai and her colleagues (Tawhai et al., 2004, 2006; Tawhai and Burrowes, 2003) who construct computational models of the entire pulmonary branching structure and study ventilation–perfusion at the level of the entire lung.

14.1.5 Carbon Monoxide Poisoning

Carbon monoxide poisoning occurs because carbon monoxide competes with oxygen for hemoglobin binding sites. The goal of this section is to see how this competition for hemoglobin affects oxygen exchange and how carbon monoxide can be eliminated from the blood.

To model this problem we assume that the chemistry for carbon monoxide binding with hemoglobin is the same as for oxygen, except that the affinity of carbon monoxide for hemoglobin is much larger (about 200 times) than the affinity of oxygen for hemoglobin. Thus,

$$\mathrm{Hb} + 4\mathrm{CO} \underset{k_{-2}}{\overset{k_2}{\rightleftharpoons}} \mathrm{Hb(CO)_4}.$$

The conservation equations for carbon monoxide gaseous exchange in the alveolus are similar in form to those for oxygen, being (ignoring diffusion of the dissolved gases)

$$v\frac{dU}{dx} = D_{\mathrm{CO}}(\sigma_{\mathrm{CO}}P_{\mathrm{CO}} - U) + 4k_{-3}S - 4k_3ZU^4, \tag{14.27}$$

$$v\frac{dS}{dx} = k_3ZU^4 - k_{-3}S, \tag{14.28}$$

where $U = [\mathrm{CO}]$, $S = [\mathrm{Hb(CO)_4}]$, $Z = [\mathrm{Hb}]$, and D_{CO} is the carbon monoxide exchange rate constant. The balance of oxygen is governed by (14.17) and (14.18). Conservation of hemoglobin implies that $Z + Y + S = Z_0$.

As before, (14.20) holds, as does

$$v\frac{d}{dx}(U + 4S) = D_{\mathrm{CO}}(\sigma_{\mathrm{CO}}P_{\mathrm{CO}} - U). \tag{14.29}$$

Now we assume that both carbon monoxide and oxygen are in quasi-steady state, so that

$$K_{CO}^4 S = ZU^4, \qquad K_{O_2}^4 Y = ZW^4, \tag{14.30}$$

where $K_{CO}^4 = k_{-3}/k_3, K_{O_2}^4 = k_{-2}/k_2$. It is convenient to introduce scaled variables w and u with $w = K_{O_2}^{-1}W, u = K_{CO}^{-1}U$. It follows from $Z + Y + S = Z_0$ that

$$S = Z_0 \frac{u^4}{1 + w^4 + u^4}, \tag{14.31}$$

$$Y = Z_0 \frac{w^4}{1 + w^4 + u^4}, \tag{14.32}$$

so that

$$v\frac{d}{dx}\left(w + 4z_0 \frac{w^4}{1 + w^4 + u^4}\right) = D_{O_2}(w^* - w), \tag{14.33}$$

$$v\frac{d}{dx}\left(u + 4\beta z_0 \frac{u^4}{1 + w^4 + u^4}\right) = D_{CO}(u^* - u), \tag{14.34}$$

where $z_0 = Z_0/K_{O_2}, \beta = K_{O_2}/K_{CO}, w^* = K_{O_2}^{-1}\sigma_{O_2}P_{O_2}, u^* = K_{CO}^{-1}\sigma_{CO}P_{CO}$.

While we cannot solve this system of differential equations explicitly, the difficulty can be readily seen. Because β is large (on the order of 200), the total carbon monoxide concentration changes as a function of x slowly, much more slowly than does the total oxygen concentration. Thus, w increases quickly to w^*, releasing some carbon monoxide as it does so, while $u + 4\beta z_0 \frac{u^4}{1+w^4+u^4}$ remains essentially fixed. As a result, in the length of the alveolus, oxygen is recharged, but very little carbon monoxide is eliminated.

The lethality of carbon monoxide can be seen from a simple steady-state analysis. Suppose that $u = u^*$ is at steady state with the environment, so that no carbon monoxide is gained or lost in the alveoli. The concentration of oxygen in the blood is proportional to $w + 4z_0 \frac{w^4}{1+w^4+u^4}$, so that the rate of oxygen transport is proportional to

$$M = w^* + 4z_0 \frac{(w^*)^4}{1 + (w^*)^4 + u^4} - w_0 - 4z_0 \frac{w_0^4}{1 + w_0^4 + u^4}, \tag{14.35}$$

where w_0 is the alveolar input level and w^* is the output level from the alveolus. When there is no carbon monoxide present (i.e., when $u = 0$), the input and output levels are 40 and 104 mm Hg, respectively, so that (with $K_{O_2} = 30\sigma$ mm Hg) $w_0 = 40/30 = 1.333, w^* = 104/30 = 3.47$. Thus, with normal metabolism, the required flow rate has $M = 53$, where we have set $z_0 = 52$. When carbon monoxide is present, this same flow rate must be maintained (as the need of the tissues for oxygen remains unchanged), but now the presence of u in the denominator changes things. Keeping M fixed, we calculate that if u is greater than 4.64, then the incoming blood has $w_0 < 0$, so that the tissue is in oxygen debt. With $\beta = 200, u = 4.64$ is equivalent to a carbon monoxide partial pressure of 0.7 mm Hg, a mere 0.1% by volume. In other words, an ambient concentration of 0.1% carbon monoxide leads to certain death because of oxygen depletion.

Since βz_0 is so large (on the order of 10^4), we can approximate the dynamics of carbon monoxide by

$$4\beta z_0 v \frac{d}{dx}\left(\frac{u^4}{1+(w^*)^4+u^4}\right) = D_{\mathrm{CO}}(u^* - u). \tag{14.36}$$

If we set

$$F = \frac{u^4}{1+(w^*)^4+u^4}, \tag{14.37}$$

we find that

$$\frac{dF}{dx} = -\frac{D_{\mathrm{CO}}}{4\beta z_0 v}(1+(w^*)^4)^{1/4}\left(\frac{F}{1-F}\right)^{1/4}, \tag{14.38}$$

where we have taken $u^* = 0$, assuming that the victim is placed in a carbon-monoxide-free environment. Clearly, the rate of carbon monoxide elimination is proportional to $(1+(w^*)^4)^{1/4}$, which for large w^* is linear in w^*. Thus, (and not surprising) the rate of carbon monoxide elimination can be increased by placing the victim in an environment of high oxygen.

In hospitals it is typical to place a carbon monoxide poisoning victim in an environment of oxygen at 2–2.5 atm. At 2 atmospheres (1 atm = 760 mm Hg), $w^* = (2 \times 760/30) = 50.7$, compared to $w^* = 3.5$ at normal oxygen levels, giving an increase in the rate of carbon monoxide elimination of about 14.

14.2 Ventilation and Perfusion

Gas exchange is mediated by the combination of ventilation of the alveoli with inspired air and the perfusion of the capillaries with blood. It is the balance of these two that determines the gas content of the lungs and of the recharged blood.

The most important early studies of ventilation and perfusion were those of Fenn et al. (1946), Rahn (1949), and Riley and Cournand (1949, 1950). A clear nonmathematical discussion is given by West (1985), who also gives a brief summary of the history of such investigations in West (2004).

To see how this balance is maintained, suppose that $\dot{V}$ is the volume flow rate of air that participates in the exchange of the alveolar content. Not all inspired air participates in this exchange, because some inspired air never reaches the terminal bronchioles. The parts of the lung that are ventilated but do not participate in gaseous exchange are called the *anatomical dead space*. In normal breathing, the total amount of inspired air is about 500 ml per breath (men 630 ml; women 390 ml). Of this, 150 ml is anatomical dead space, so only 350 ml participates in alveolar gaseous exchange. With 15 breaths per minute, $\dot{V}$ is about 5250 ml/min.

Now suppose that Q is the volume flow rate of blood into and out of the alveolar capillaries. Cardiac output is about 70 ml per beat, so at 72 beats per minute, Q is about 5000 ml/min. The ratio $\dot{V}/Q$ is called the *ventilation–perfusion ratio*, and it is the most important determinant of lung–blood gas content.

If P_i and P_a are the pressures of a gas in the inspired air and in the alveolar air, respectively (see Fig. 14.7), then the rate at which the gas is taken out of the alveolus by ventilation (in moles/sec) is

$$\dot{V}\left(\frac{P_a - P_i}{RT}\right). \tag{14.39}$$

Note that the factor RT is needed to convert pressures to concentrations. Similarly, the rate at which the gas is brought into the alveolus by the blood (again in moles/second) is given by

$$Q(c_v - c_a), \tag{14.40}$$

where c_v and c_a are the venous and alveolar gas concentrations in the blood. Since the rate at which the gas is removed by ventilation must be in balance with the rate at which it is brought in by the blood, it must be that

$$\frac{\dot{V}}{Q} = \frac{(c_v - c_a)RT}{(P_a - P_i)}. \tag{14.41}$$

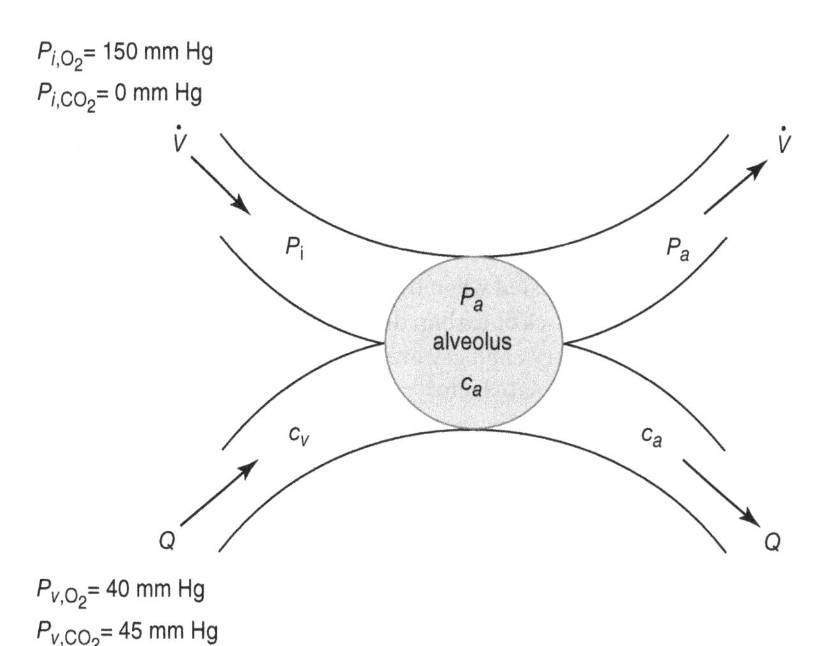

Figure 14.7 Schematic diagram of gas exchange in an alveolus.

Textbooks on respiratory physiology typically use different units from those used here, and this can be confusing for those new to the field. Since the gas in the alveolus is always saturated with water vapor and is at body temperature, the partial pressures of alveolar gas are measured at body temperature and pressure, saturated (BTPS). However, blood gas concentrations are measured in the usual way, at standard temperature and pressure, dry (STPD). To understand the difference between these units, suppose we have one mole of gas at standard temperature (273 K) and standard pressure (1 atmosphere or 760 mm Hg). Since $PV = nRT$, the STPD volume of the mole of gas, V_{STPD} is

$$V_{\mathrm{STPD}} = \frac{273R}{760}. \tag{14.42}$$

Similarly, for one mole of gas at BTPS, suppose that the total pressure, including the pressure due to water vapor, is P (in mm Hg). At body temperature (310 K), the partial pressure of water vapor is 47 mm Hg, and thus the partial pressure of the mole of gas is $P - 47$ mm Hg. Hence, the BTPS volume of the mole of gas is

$$V_{\mathrm{BTPS}} = \frac{310R}{P - 47}. \tag{14.43}$$

It follows that

$$V_{\mathrm{STPD}} = V_{\mathrm{BTPS}} \frac{273 \times (P - 47)}{310 \times 760} = \frac{P - 47}{863} V_{\mathrm{BTPS}}, \tag{14.44}$$

where the ambient pressure, P, is measured in mm Hg.

To return to gas exchange, if we were considering an inert gas that merely dissolves in blood, (14.41) would be relatively simple. The alveolar concentration c_a would be linearly related to the alveolar pressure P_a by the solubility of the gas. However, in the case of oxygen and carbon dioxide, this relationship is complicated by the blood chemistry.

It is reasonable to assume that the two most important respiratory gases, carbon dioxide and oxygen, are equilibrated when they leave the alveolus in the capillaries. In other words, the partial pressures of carbon dioxide and oxygen in the alveolus and in the blood leaving the pulmonary capillary are the same. Of course, this is not true at high perfusion rates, but it is a satisfactory assumption at normal physiological flow rates.

Because carbon dioxide is quickly converted to bicarbonate, the total blood carbon dioxide (i.e., both free and converted) is given by

$$[\mathrm{CO_2}] = \sigma_{\mathrm{CO_2}}(1 + K_c)P_{\mathrm{CO_2}}, \tag{14.45}$$

as discussed in Section 14.1.3. In other words, in the notation of Fig. 14.7, $c_a = \sigma_{\mathrm{CO_2}}(1 + K_c)P_a$ for carbon dioxide. A similar expression holds for c_v and P_v, where P_v is the pressure of carbon dioxide in the venous blood. In addition, the concentration of carbon dioxide in inspired air is very close to zero ($P_i = 0$). Thus, for carbon dioxide, the ventilation–perfusion ratio must satisfy

$$\frac{\dot{V}}{Q} = \sigma_{CO_2} RT(1 + K_c)\frac{(P_{v,CO_2} - P_{a,CO_2})}{P_{a,CO_2}}, \tag{14.46}$$

where the additional subscript CO_2 is to distinguish these pressures from those of oxygen.

Since $P_{v,CO_2} = 45$ mm Hg, we can use (14.46) to plot alveolar CO_2 pressure as a function of the ventilation–perfusion ratio (Fig. 14.8). As $\frac{\dot{V}}{Q}$ increases, the CO_2 partial pressure decreases, since removal of CO_2 from the blood becomes more efficient. When $\frac{\dot{V}}{Q} = 0$, $P_{a,CO_2} = P_{v,CO_2}$ as expected, as no CO_2 is removed from the blood when there is no ventilation.

Similarly, as described in Section 14.1.4, the total amount of oxygen in the blood, $[O_2]_t$, is given by

$$[O_2]_t = W + 4Z_0 f(W), \tag{14.47}$$

where Z_0 is the total hemoglobin concentration, and W is the concentration of unbound oxygen in the blood. In these terms, the ventilation–perfusion ratio must be

$$\frac{\dot{V}}{Q} = \left(\frac{RT}{(P_{i,O_2} - P_{a,O_2})}\right)\left(W_a - W_v + 4Z_0\left[f(W_a) - f(W_v)\right]\right), \tag{14.48}$$

where the subscripts a, i, and v have the same interpretations as above. Note that, since we assume that the partial pressure of oxygen in the alveolar air is the same as the partial pressure in the blood leaving the alveolus, we have $W_a = \sigma_{O_2} P_{a,O_2}$. Similarly, $W_v = \sigma_{O_2} P_{v,O_2}$, where $P_{v,O_2} = 40$ mm Hg is a known parameter.

A plot of the alveolar partial pressures of carbon dioxide and oxygen as a function of ventilation–perfusion ratio is shown in Fig. 14.8.

From this figure we see that the alveolar oxygen partial pressure is an increasing function of $\dot{V}/Q$, while the alveolar carbon dioxide partial pressure is a decreasing function thereof. This makes intuitive sense; the more ventilation there is compared

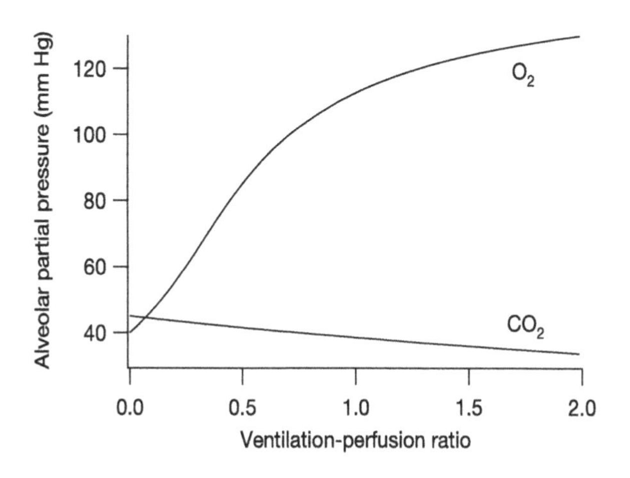

Figure 14.8 Alveolar partial pressure as a function of ventilation–perfusion ratio. Calculated using the parameters $W_v = 40\sigma_{O_2}$ mm Hg (i.e., $P_{v,O_2} = 40$ mm Hg), $K_{O_2} = 30\sigma_{O_2}$ mm Hg, $Z_0 = 2.2$ mM, $RT = 1.7 \times 10^4$ mm Hg/M, $P_{v,CO_2} = 45$ mm Hg, $K_c = 12$ and $P_{i,O_2} = 150$ mm Hg, and the function $f(W) = W^4/(K_{O_2}^4 + W^4)$.

to blood flow, the greater the opportunity for blood to pick up oxygen and lose carbon dioxide. In normal situations, the ventilation–perfusion ratio is about 1. An increase in this ratio is called *hyperventilation*, and a decrease is called *hypoventilation*. During hyperventilation, there is rapid removal of carbon dioxide, and the partial pressure of carbon dioxide in the arterial blood drops below the normal level of 40 mm Hg. This results in less carbon dioxide available for carbonic acid formation, and consequently blood pH rises above the normal level, resulting in *respiratory alkalosis*. In hyperventilation there is no substantial change in oxygen concentration because the hemoglobin is fully saturated.

The opposite situation, in which the ventilation–perfusion ratio drops, increases carbon dioxide content and decreases oxygen content of the arterial blood. The increase of carbon dioxide increases carbonic acid formation and decreases blood pH, a condition referred to as *respiratory acidosis*. To compensate for these changes, the blood gas concentration stimulates the carotid and aortic chemoreceptors to increase the rate of ventilation.

14.2.1 The Oxygen–Carbon Dioxide Diagram

Most commonly (partially for historical reasons, since this was the way it was first presented), the ventilation–perfusion ratio is discussed using the *oxygen–carbon dioxide diagram* shown in Fig. 14.9. By combining (14.46) and (14.48), we see that

$$\sigma_{\mathrm{CO}_2}(1+K_c)\frac{(P_{v,\mathrm{CO}_2}-P_{a,\mathrm{CO}_2})}{P_{a,\mathrm{CO}_2}}=\frac{W_a-W_v+4Z_0[f(W_a)-f(W_v)]}{P_{i,\mathrm{O}_2}-P_{a,\mathrm{O}_2}}, \tag{14.49}$$

from which one easily determines P_{a,CO_2} as a function of P_{a,O_2}. The resultant curve is plotted in Fig. 14.9. All possible simultaneous alveolar pressures of oxygen and carbon dioxide must lie somewhere on this curve, which is parameterized by $\dot{V}/Q$. As $\dot{V}/Q$ increases we move to the right along the curve, and in the limit as $\dot{V}/Q \to \infty$ we move to the right endpoint, where $P_{a,\mathrm{O}_2} = 150$ mm Hg, the oxygen pressure in inspired air. Similarly, as $\dot{V}/Q$ decreases we move along the curve to the left endpoint, at which point there is no ventilation and P_{a,O_2} is the venous oxygen pressure, 40 mm Hg.

14.2.2 Respiratory Exchange Ratio

Different regions of the lung lie on different points of the oxygen–carbon dioxide diagram, the actual point being determined by the local *respiratory exchange ratio*, $\mathcal{R}$, where

$$\mathcal{R}=\frac{\text{volume of carbon dioxide removed}}{\text{volume of oxygen taken up}}. \tag{14.50}$$

This ratio differs throughout the lung.

There is a close relationship between $\mathcal{R}$ and the oxygen–carbon dioxide diagram shown in Fig. 14.9. To see this, we consider curves of constant $\mathcal{R}$ for both the alveolar

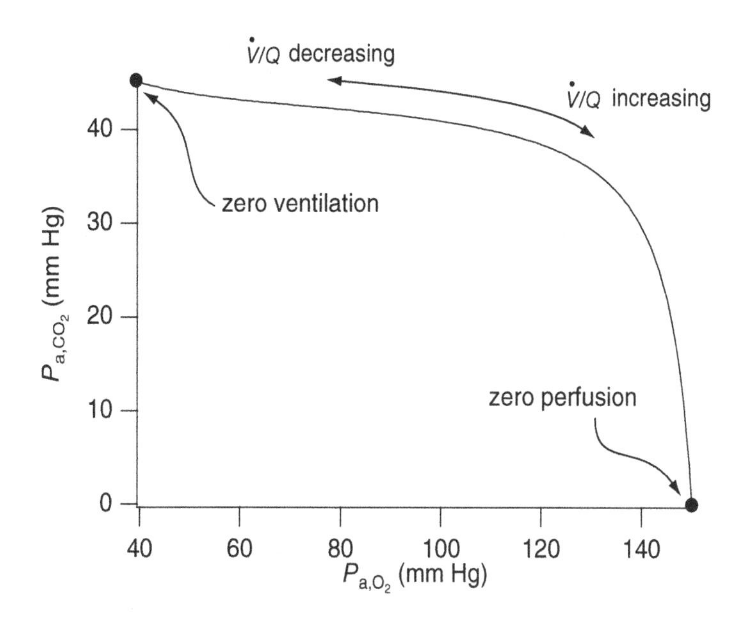

Figure 14.9 Oxygen–carbon dioxide diagram showing how the alveolar partial pressures of oxygen and carbon dioxide are related. Parameter values are the same as used for Fig. 14.8.

gas and the alveolar blood. For the alveolar gas,

$$\mathcal{R}_{\text{gas}} = \frac{P_{a,\text{CO}_2} - P_{i,\text{CO}_2}}{P_{i,\text{O}_2} - P_{a,\text{O}_2}} = \frac{P_{a,\text{CO}_2}}{P_{i,\text{O}_2} - P_{a,\text{O}_2}}. \tag{14.51}$$

Thus, in the oxygen–carbon dioxide diagram, curves of constant $\mathcal{R}$ are straight lines that emanate from the inspired air point, I (Fig. 14.10).

The total amount of oxygen taken up by the blood is $W_a + 4Z_0 f(W_a) - [W_v + 4Z_0 f(W_v)]$, while the total amount of carbon dioxide lost is $\sigma_{\text{CO}_2}(1+K_c)(P_{v,\text{CO}_2} - P_{a,\text{CO}_2})$. Thus

$$\mathcal{R}_{\text{blood}} = \frac{\sigma_{\text{CO}_2}(1 + K_c)(P_{v,\text{CO}_2} - P_{a,\text{CO}_2})}{W_a + 4Z_0 f(W_a) - [W_v + 4Z_0 f(W_v)]}, \tag{14.52}$$

where, as usual, $W_a = \sigma_{\text{O}_2} P_{a,\text{O}_2}$ and $W_v = \sigma_{\text{O}_2} P_{v,\text{O}_2}$. Curves of constant $\mathcal{R}$ are shown in Fig. 14.10. They emanate from the venous point, V, but because of the blood chemistry are nonlinear curves, not straight lines.

Since $\mathcal{R}_{\text{gas}}$ must equal $\mathcal{R}_{\text{blood}}$, it follows that the alveolar gas concentrations must lie on the intersection of the two curves $\mathcal{R}_{\text{blood}}$ and $\mathcal{R}_{\text{gas}}$. However, this intersection must lie on the oxygen–carbon dioxide curve shown in Fig. 14.9, since the equation $\mathcal{R}_{\text{gas}} = \mathcal{R}_{\text{blood}}$ is exactly the same as (14.49). Thus, the oxygen–carbon dioxide curve can be thought of as parameterized by either $\dot{V}/Q$ or $\mathcal{R}$.

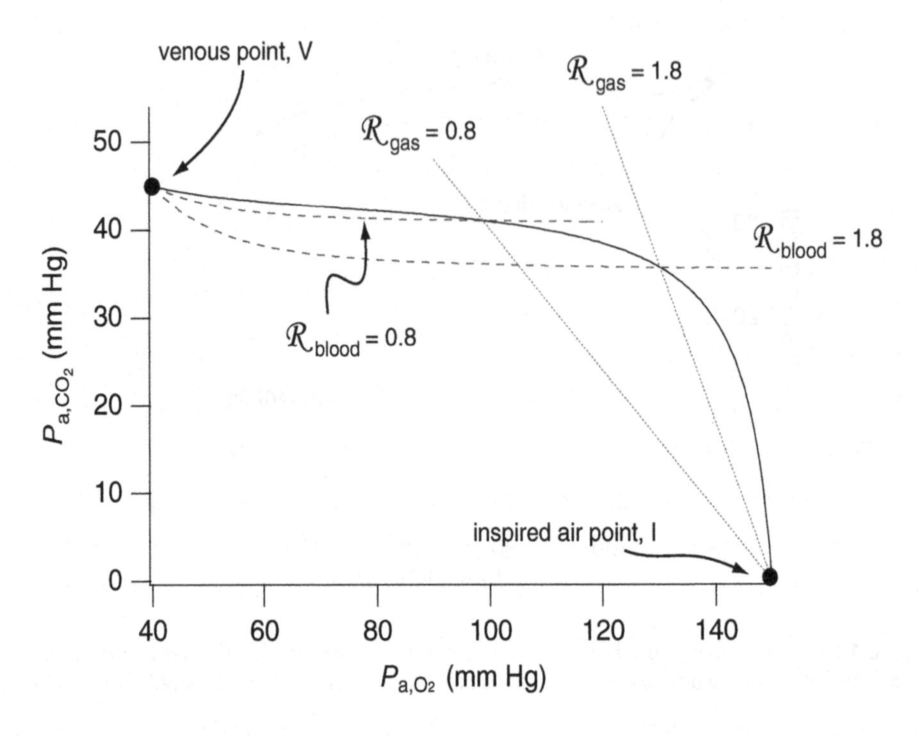

Figure 14.10 Curves of constant respiratory exchange ratio, $\mathcal{R}$, for alveolar gas and blood, superimposed on the oxygen–carbon dioxide diagram.

One way of showing the relationship between $\dot{V}/Q$ and $\mathcal{R}$ is to combine (14.48) and (14.51) to get

$$\frac{\dot{V}}{Q} = \frac{\mathcal{R}}{P_{a,\mathrm{CO}_2}} RT\Big(W_a - W_v + 4Z_0\big[f(W_a) - f(W_v)\big]\Big) \tag{14.53}$$

$$= \frac{\mathcal{R}}{P_{a,\mathrm{CO}_2}} RT(c_a - c_v), \tag{14.54}$$

where $c_a = W_a + 4Z_0 f(W_a)$ and $c_v = W_v + 4Z_0 f(W_v)$ are the total concentrations of oxygen in the alveolar blood and venous blood, respectively.

In the upper regions of the lung, $\mathcal{R}$ can be almost as large as 3, while in the lower regions, $\mathcal{R}$ can be as low as 0.6. On the level of the entire body, the ratio of total carbon dioxide produced to total oxygen used is called the *respiratory quotient*, or RQ. The oxygen that is taken in by the blood is consumed by metabolic processes to produce carbon dioxide. However, the amount of carbon dioxide produced is generally less than the amount of oxygen consumed, and thus RQ is rarely more than one. When a person uses carbohydrates for body metabolism RQ is 1.0, because one molecule of carbon dioxide is formed for every molecule of oxygen consumed. On the other hand, when oxygen reacts with fats a large share of the oxygen combines with hydrogen to form water instead of carbon dioxide. In this mode, RQ falls to as low as 0.7. For a normal person with a normal diet, RQ = 0.825 is considered normal.

14.3 Regulation of Ventilation

While the exchange of gases takes place in the lungs, the control of the rate of ventilation is accomplished in the brain. There, in the respiratory center, is located a chemosensitive area that is sensitive to the concentrations of chemicals in the blood, primarily carbon dioxide. Changes in blood P_{CO_2} are detected, and this leads to changes in the rate of breathing by activating or inhibiting the inspiratory neurons (described in Section 14.4). In Fig. 14.11 is shown the effect of carbon dioxide on ventilation rate.

To construct a model of this control, we let x denote the partial pressure of carbon dioxide in the blood. Carbon dioxide is produced at rate λ by metabolism and eliminated by ventilation at the lungs. Thus,

$$\frac{dx}{dt} = \lambda - \alpha x \dot{V}, \tag{14.55}$$

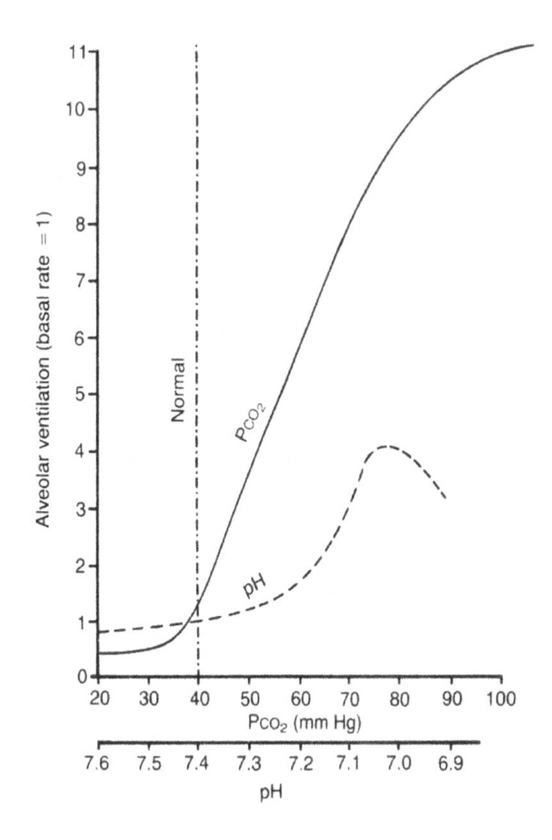

Figure 14.11 Effects of increased arterial P_{CO_2} and decreased arterial pH on the alveolar ventilation rate. (Guyton and Hall, 1996, Fig. 41-3, p. 528.)

where $\dot{V}$ is the ventilation rate, and we assume that the transport of carbon dioxide through the lungs is linearly proportional to the concentration of carbon dioxide and the ventilation rate.

Now we take the ventilation rate to be the Hill equation

$$\dot{V}(x) = V_m \frac{x^n}{\theta^n + x^n}, \tag{14.56}$$

in order to replicate the curve shown in Fig. 14.11. Furthermore, there is a substantial delay between ventilation of the blood and the measurement of P_{CO_2} at the respiratory center in the brain because the transport of blood from the lungs back to the heart and then to the brain takes time. Thus, the complete model is (Glass and Mackey, 1988)

$$\frac{dx}{dt} = \lambda - \alpha x \dot{V}(x(t-\tau)). \tag{14.57}$$

Typical parameter values for the model are given in Table 14.1.

Before proceeding further with the analysis of this equation, it is worthwhile to introduce dimensionless variables and parameters. We set $x = \theta y, t = \frac{s}{\alpha V_m}, \tau = \frac{\sigma}{\alpha V_m}$, and $\lambda = \theta \alpha V_m \beta$ and obtain

$$\frac{dy}{ds} = \beta - yF(y(s-\sigma)), \tag{14.58}$$

where F is a sigmoidal function, monotone increasing with a maximum of 1 as $y \to \infty$.

Because the function $yF(y)$ is monotone increasing in y, there is a unique steady-state solution for (14.58). Furthermore, the steady-state solution is a monotone increasing function of the parameter β, indicating that blood P_{CO_2} and ventilation increase as a function of steady metabolism. However, the dynamic situation may be quite different.

To understand more about the dynamic behavior of this equation we perform a linear stability analysis. We suppose that the steady state is $y = y^*$, and set $y = y^* + Y$, substitute into (14.58), and assume that Y is small enough so that only linear terms of the local Taylor series are necessary. The resulting linearized equation for Y is

$$\frac{dY(s)}{ds} = -F(y^*)Y(s) - y^*F'(y^*)Y(s-\sigma). \tag{14.59}$$

Solutions of exponential form $Y = Y_0 e^{\mu s}$ exist, provided that μ satisfies the characteristic equation

$$\mu + F(y^*) + y^*F'(y^*)e^{-\mu\sigma} = 0. \tag{14.60}$$

Table 14.1 Physical parameters for the Mackey–Glass model of respiratory control.

λ	= 6 mm Hg/min
V_m	= 80 liter/min
τ	= 0.25 min

Since there is a monotone relationship between y^* and β, it is convenient to view y^* as an independent parameter.

The function

$$g(y) = F(y) - yF'(y) \tag{14.61}$$

is important to the analysis that follows and has a nice geometrical interpretation. This function is constructed by drawing a straight line from the point $(y, F(y))$ to $y = 0$ with slope $F'(y)$, as illustrated in Fig. 14.12. The three functions $F(y)$, $F'(y)$, and $g(y)$ are shown in Fig. 14.13, in the case $F(y) = \frac{y^3}{1+y^3}$.

We wish to understand the behavior of the roots of the characteristic equation (14.60). First, observe that if $g(y^*)$ is positive, then all roots of (14.60) have negative real part, so that the steady solution is stable. This follows, because if the real part of μ is positive, then $|\mu + F(y^*)| > |\mu + y^*F'(y^*)| > |y^*F'(y^*)e^{-\mu\sigma}|$. Note that F, F', and y are all assumed to be positive.

The only real roots of (14.60) are negative. Thus, the only way the real part of a root can change sign is if it is complex, i.e., a Hopf bifurcation. To see whether Hopf bifurcations occur, we set $\mu = i\omega$. If this is a root of (14.60), then of necessity, $|i\omega + F(y^*)| = |y^*F'(y^*)|$, and thus $|F(y^*) + i\omega|^2 = [F(y^*)]^2 + \omega^2 = [y^*F'(y^*)]^2$. In this case, it must be that $y^*F'(y^*) > F(y^*)$, which implies that $g(y^*) < 0$. Now we split (14.60) into

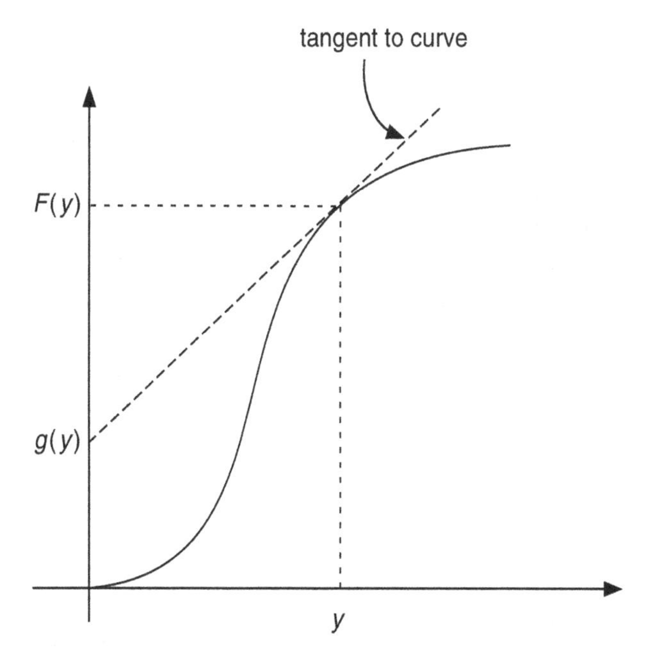

Figure 14.12 Sketch of the construction of the function $g(y) = F(y) - yF'(y)$.

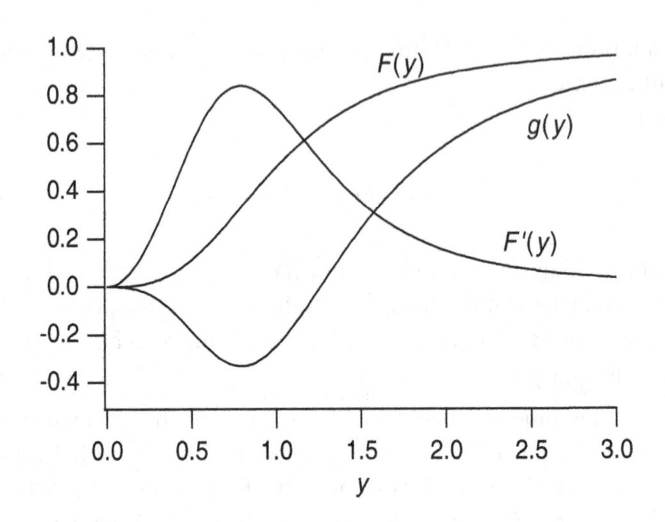

Figure 14.13 Plots of $F(y) = \frac{y^n}{1+y^n}$, $F'(y)$, and $g(y)$, with $n = 3$.

real and imaginary parts, obtaining

$$F(y^*) + y^* F'(y^*) \cos \omega\sigma = 0, \tag{14.62}$$

$$\omega - y^* F'(y^*) \sin \omega\sigma = 0. \tag{14.63}$$

It follows that $\omega = \sqrt{(y^* F'(y^*))^2 - (F(y^*))^2}$ (provided that $g(y^*) < 0$) and that

$$\tan \omega\sigma = -\frac{\omega}{F(y^*)}. \tag{14.64}$$

The smallest root of this equation lies in the interval $\frac{\pi}{2} < \omega\sigma < \pi$, and for this root,

$$\sigma = \frac{1}{\omega}\left[\pi + \tan^{-1}\left(-\frac{\omega}{F(y^*)}\right)\right]. \tag{14.65}$$

We can view this information as follows. For a given y^*, we have the frequency ω and the critical delay σ at which a Hopf bifurcation occurs. If the delay is smaller than this critical delay, then the steady solution is stable, while if the delay is larger, then the steady solution is unstable and an oscillatory solution is likely.

Plots of ω and σ are shown in Fig. 14.14. Steady solutions having σ greater than the critical value of delay (14.65) are unstable. In this case, numerical simulations show that there is a stable periodic solution of the governing equations, shown in Fig. 14.15. Here is shown the dimensionless concentration y (shown solid) and the dimensionless ventilation rate $F(y_\sigma)$ (shown dashed) as a function of time, with parameter values $\beta = 0.8, \sigma = 10.0$.

An episode of periodic fluctuation of ventilation, depicted by the periodic solution of Fig. 14.15, is called *Cheyne–Stokes breathing*. It was first described by Cheyne in 1818, but was more widely noticed after being discussed by Stokes (1854). In this condition a person breathes deeply for a short interval and then breathes slightly or not at all for an

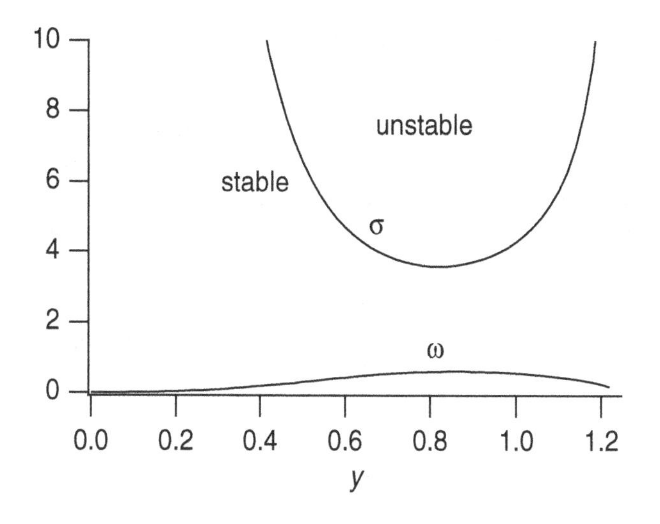

Figure 14.14 Plots of ω and σ at Hopf bifurcation points.

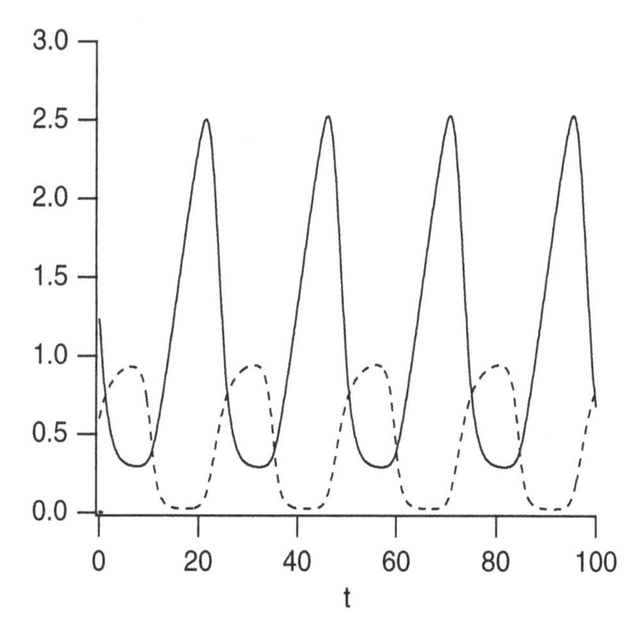

Figure 14.15 Oscillatory solution of the Mackey–Glass equation (14.58) with parameters $y^* = 0.8, \sigma = 10.0$. Carbon dioxide content is shown as a solid curve, and ventilation rate is shown as a dashed curve.

additional interval, repeating the cycle with a period of 40 to 60 seconds. Notice that, in the model above, Cheyne–Stokes breathing can be caused by an increased delay in the transport of blood to the brain or an increase in the negative feedback gain (the slope of F). The first type (delayed transport) is likely to occur in patients with chronic heart failure, and the second type (increased gain) occurs mainly in patients with brain damage, and is often a signal of impending death. Clinical observations from nine patients are discussed in Lange and Hecht (1962); all the patients in that study suffered from heart disease and seven of them had died within two years of the end of the study.

14.3.1 A More Detailed Model of Respiratory Regulation

The model discussed above, due to Mackey and Glass, is one of the simplest models of Cheyne–Stokes breathing. The most widely known model, and possibly the most physiologically realistic, is that of Grodins et al. (1967). The Grodins model describes oxygen, carbon dioxide, and nitrogen concentrations in a number of body compartments, including the lungs, the brain, and tissues, and has been widely studied (see, for example, Khoo et al., 1982; Fowler et al., 1993; Fowler and Kalamangalam, 2000, 2002; Batzel and Tran, 2000a,b,c; Whiteley et al., 2003a; Topor et al., 2004; Batzel et al., 2007). Here we present a simplified version of the full model; not only does this simplified model demonstrate reasonable oscillations, it also illustrates the principles behind the construction of the full model.

A schematic diagram of the compartments in the model is given in Fig. 14.16. For simplicity we consider only the carbon dioxide concentrations in each of these compartments. The blood flow through the lungs is Q, while the blood flow through the brain compartment is Q_B. The subscripts B and T denote, respectively, quantities in the brain and the tissues. The C's denote concentrations of carbon dioxide at the entry and exit points to the compartments, with a subscript a denoting the arterial side, and a subscript v the venous side. Let M_i denote the rate of metabolism in compartment i,

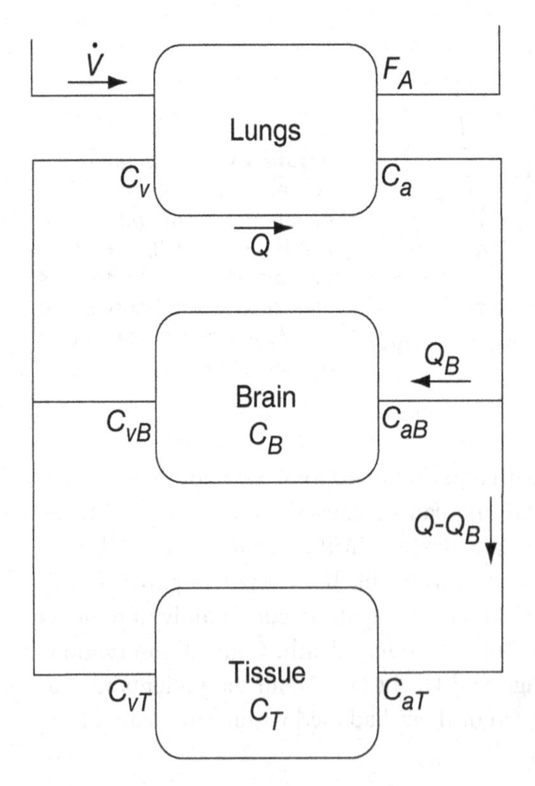

Figure 14.16 Schematic diagram of the Grodins et al. model of ventilation control. The C's denote concentrations at the entry and exit points of the compartments, while F_A is the fraction of carbon dioxide in expired alveolar air.

and let V_i denote the volume of compartment i. Then, conservation of carbon dioxide gives

$$V_B \frac{dC_B}{dt} = M_B + Q_B(C_{aB} - C_{vB}), \tag{14.66}$$

$$V_T \frac{dC_T}{dt} = M_T + (Q - Q_B)(C_{aT} - C_{vT}). \tag{14.67}$$

Similarly, if we let F denote the volume fraction of carbon dioxide, conservation of carbon dioxide at the lungs gives

$$V_L \frac{dF_A}{dt} = -\dot{V} F_A + \beta Q(C_v - C_a). \tag{14.68}$$

Since V_L and $\dot{V}$ are measured in BTPS, while the blood concentrations are measured in STPD (see (14.44)), we include the conversion factor $\beta = 863/(760 - 47)$.

First, we simplify the model by assuming that $C_{vB} = C_B$ and $C_{vT} = C_T$. In other words, the concentration that exits the brain, say, is the same as the concentration in the brain. We also assume that the concentrations C_{aB} and C_{aT} are delayed versions of C_a, and thus

$$C_{aB} = C_a(t - \tau_{aB}), \tag{14.69}$$

$$C_{aT} = C_a(t - \tau_{aT}). \tag{14.70}$$

Similarly, C_v is a delayed version of C_B and C_T, and thus

$$QC_v = Q_B C_B(t - \tau_{vB}) + (Q - Q_B) C_T(t - \tau_{vT}). \tag{14.71}$$

In a more complex version of this model, the time delays depend on the blood flow, Q, which itself is time-dependent, leading to far greater complexity. However, for simplicity, we assume both the time delays and the blood flows are constant.

We now convert the variables to partial pressures rather than concentrations. First, each carbon dioxide concentration can be converted to a partial pressure by using (14.45). To simplify the notation we write

$$C_a = K_{CO_2} P_a, \qquad C_v = K_{CO_2} P_v, \qquad \dots, \tag{14.72}$$

where $K_{CO_2} = \sigma_{CO_2}(1 + K_c)$.

Second, the fractional volume of carbon dioxide in the alveolus, F_A, is related to the alveolar partial pressure of carbon dioxide, P_A, by the relationship

$$F_A = \frac{P_A}{760 - 47}, \tag{14.73}$$

where we assume that the ambient pressure is 760 mm Hg (see (14.44)). This is because the air in the alveolus is saturated with water vapor, which (at body temperature) has a partial pressure of 47 mm Hg, as discussed in Section 14.2. Since it is reasonable to assume that the alveolar air is in equilibrium with the arterial blood, we let $P_A = P_a$.

Table 14.2 Parameters for the simplified Grodins model of respiratory control. The bracketed value for G_C is the one that was used to generate instability and oscillations (Fig. 14.17).

Q_B	= 0.75 liters/min	Q	= 6 liters/min
V_L	= 3 liters	V_B	= 1 liter
V_T	= 39 liters	K_{CO_2}	= 0.005/mm Hg
M_B	= 0.05 liters/min	M_T	= 0.182 liters/min
τ_{aB}	= 0.18 min	τ_{aT}	= 0.32 min
τ_{vT}	= 0.59 min	τ_{vB}	= 0.11 min
G_C	= 1.8 (9) liters min^{-1} (mm Hg)$^{-1}$	I_C	= 49.3 mm Hg
$\dot{V}_{\text{base}}$	= 2 liters/min		

Combining all these assumptions, (14.66)–(14.68) become

$$V_B K_{CO_2} \frac{dP_B}{dt} = M_B + Q_B K_{CO_2}(P_a(t-\tau_{aB}) - P_B), \tag{14.74}$$

$$V_T K_{CO_2} \frac{dP_T}{dt} = M_T + (Q - Q_B)K_{CO_2}(P_a(t-\tau_{aT}) - P_T), \tag{14.75}$$

$$V_L \frac{dP_a}{dt} = -\dot{V}P_a + 863 K_{CO_2} Q(P_v - P_a), \tag{14.76}$$

which we supplement with (14.71), which, in terms of partial pressures, is

$$QP_v = Q_B P_B(t - \tau_{vB}) + (Q - Q_B)P_T(t - \tau_{vT}). \tag{14.77}$$

The model parameters are given in Table 14.2.

When $\dot{V}$ is held fixed, the steady-state solution of this model is stable, and no oscillatory breathing pattern emerges. However, ventilation is dependent on the carbon dioxide concentration in the blood, and thus $\dot{V}$ is dependent on the blood pressures in the model. A number of different authors have chosen different expressions for $\dot{V}$. Here we follow Khoo et al. (1982) and assume that

$$\dot{V} = \dot{V}_{\text{base}} + G_C \max(P_B - I_C, 0), \tag{14.78}$$

where $\dot{V}_{\text{base}}$ is the base ventilation rate. If the brain carbon dioxide concentration increases above the threshold I_C, the ventilation rate increases (with slope G_C), in order to remove the excess carbon dioxide more effectively.

Typical numerical solutions are shown in Fig. 14.17. When $G_C = 1.8$, the steady state is stable and no oscillations appear. However, when $G_C = 9$ the steady state is unstable and the ventilation rate oscillates with a period close to 60 seconds.

14.4 The Respiratory Center

It is clear that control of breathing is crucial for survival. Although the rate of breathing at rest remains relatively unchanged through the lifetime of a healthy adult, the rate must respond to the differing requirements such as during exercise and sleep.

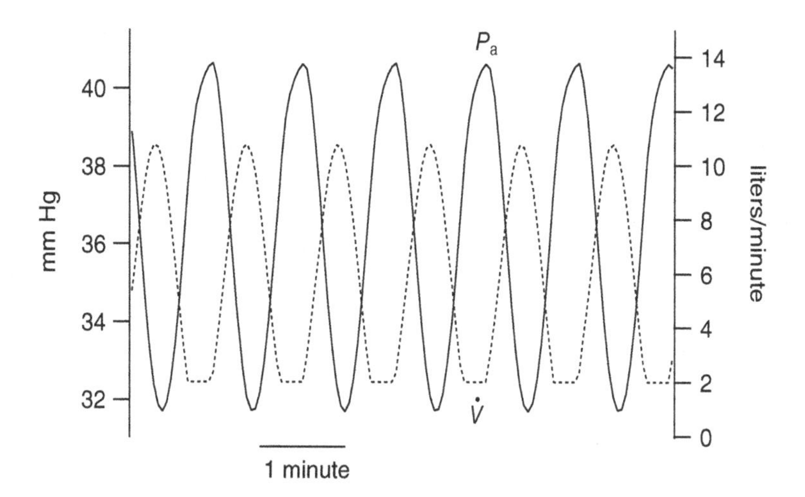

Figure 14.17 Numerical solution of the simplified Grodins model. Arterial carbon dioxide pressure, P_a, is plotted against the left axis, while ventilation ($\dot{V}$) is plotted against the right axis. This solution was calculated with $G_C = 9$. As G_C decreases, the oscillations disappear.

Furthermore, since a 70 kg adult male uses approximately 250 ml of oxygen per minute, and the body has a resevoir of only about one liter, it follows that any interruption in breathing will have catastrophic consequences, particularly for the brain, which is peculiarly susceptible to oxygen deprivation (Feldman and Del Negro, 2006).

Breathing is controlled by a neural central pattern generator that consists of multiple groups of neurons in the pons and medulla (Richter, 1996; Ramirez and Richter, 1996). The medullary network consists of three major regions; the Bötzinger complex, the pre-Bötzinger complex, and the rostral ventral respiratory group (rVRG), with each of these regions containing multiple kinds of neurons. The pontine respiratory region also contains a number of subregions, such as the Kölliker-Fuse nucleus or the parabrachial complex, but these regions play no further role in the discussion here.

The respiratory cycle is traditionally divided into three phases, inspiration, post-inspiration, and expiration, and the respiratory neurons are classified according to their behavior in different parts of the cycle (Richter, 1996; Rybak et al., 1997). So, for example, early-inspiration neurons have a burst of firing at the beginning of inspiration phase, while expiration-2 neurons have a burst of firing during the expiration phase. Ramp-inspiration neurons, located primarily in the rVRG, fire during the inspiration phase; their firing pattern is superimposed upon a rising baseline, or ramp, with a frequency that increases during the inspiration phase.

In 1991 it was discovered by Smith et al. that the primary rhythm-generating portion of the brain is located in the pre-Bötzinger complex. However, although this region is both necessary and sufficient for generating rhythmic inspiration, it is not, by itself, sufficient to explain the myriad complexities of respiration control. There seems little doubt that such control is exerted by multiple feedbacks between the various regions of the respiratory central pattern generator. Expiration, for example, results primarily from elastic recoil of the lungs and thoracic cage, but appears also to be controlled by the Bötzinger complex.

In addition to neural mechanisms operating entirely within the brain, reflex signals from the periphery also help control respiration. Located in the walls of the bronchi and bronchioles throughout the lungs are stretch receptors that transmit signals to the dorsal respiratory group. Thus, when the lungs become overly inflated, the stretch receptors activate a feedback response that switches off the inspiratory ramp and stops further inspiration. This reflex is called the *Hering–Breuer inflation reflex*.

Models of the respiratory central pattern generator tend to focus on two major questions. First, how can one best describe the behavior of individual respiratory neurons? In the most recent models (Butera et al., 1999a; Rybak et al., 1997a), individual neurons are described by models of Hodgkin–Huxley style (Chapter 5), and thus involve considerable complexity. However, to capture the different complex bursting behaviors of respiratory neurons, such complexity is unavoidable. Second, what happens when populations of such neurons are coupled in excitatory and inhibitory networks (Butera et al., 1999b; Rybak et al., 1997b; Rybak et al., 2004)? Does rhythmicity arise from the feedback interactions in such a network, or is the oscillation driven by intrinsically oscillating neurons, or pacemakers, in the pre-Bötzinger complex?

One recent proposal (Feldman and Del Negro, 2006) is that the central pattern generator consists of two coupled oscillators, one based in the pre-Bötzinger complex, the other based in the retrotrapezoid nucleus, or RTN, another region of the medulla that contains respiratory neurons. Inside each oscillator, the rhythm emerges as a result of the network connections between neurons, not from an intrinsic pacemaker, and each oscillator both inhibits and activates the other, depending on the phase of the respiratory cycle.

Another model (Smith et al., 2007) proposes that the respiratory cycle results from the interactions of neuronal populations in all the major regions, and that the pontine-medullary network has a specific spatial organization extending from the pons to the rVPN. In this model, a three-phase respiratory cycle can be obtained by a model which includes mutual inhibitory interactions between the pontine and medullary regions, a two-phase cycle can be obtained by removal of the pons, while a one-phase cycle is generated by the pre-Bötzinger complex alone. In this model there are thus three distinct oscillatory mechanisms, each of which can be uncovered by removal of portions of the network.

14.4.1 A Simple Mutual Inhibition Model

In both of the models discussed above, mutual inhibition between populations of neurons plays a major role in the generation of the respiratory cycle. It is thus important to understand, in a simpler context, how mutual inhibition can lead to oscillations. A simple qualitative model of mutual inhibition is due to von Euler (1980) and Wyman (1977), and is illustrated in Fig. 14.18. (An alternate model is suggested in Exercise 9.)

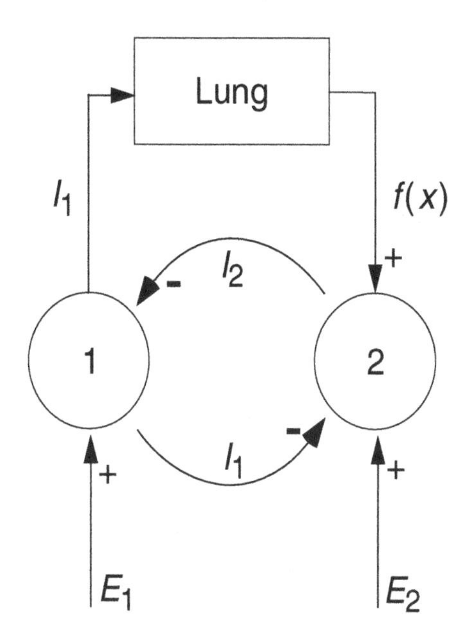

Figure 14.18 A mutual inhibition network for the control of respiration.

We suppose that there are two neurons with time-dependent outputs (their firing rates) I_1 and I_2 governed by

$$\tau_1 \frac{dI_1}{dt} + I_1 = F_1, \tag{14.79}$$

$$\tau_2 \frac{dI_2}{dt} + I_2 = F_2, \tag{14.80}$$

where F_1 and F_2 are related to the firing rates of inhibitory and excitatory inputs. For simplicity we assume that the arrangement is symmetric, so that the time constants of the neuronal output are the same, $\tau_1 = \tau_2 = \tau$. We further assume that the neurons have steady excitatory inputs, E_1 and E_2, respectively, and that they are cross-inhibited, so that the output from neuron 1 inhibits neuron 2, and vice versa. Thus we take $F_1 = F(E_1 - I_2)$ and $F_2 = F(E_2 - I_1)$. The function $F(x)$ is zero for $x < 0$ (so that the input and output are never negative), and a positive, increasing function of x for $x > 0$. Thus, we have the system of differential equations

$$\tau \frac{dI_1}{dt} + I_1 = F(E_1 - I_2), \tag{14.81}$$

$$\tau \frac{dI_2}{dt} + I_2 = F(E_2 - I_1). \tag{14.82}$$

At this point there is no feedback from the lungs.

Equations (14.81) and 14.82) are easily studied using phase-plane analysis. There are three different possible phase portraits depending on the relative sizes of E_1 and E_2, two of which are shown in Figs. 14.19. In what follows we assume that $F' > 1$ for all positive arguments, although this restriction can be weakened somewhat. If

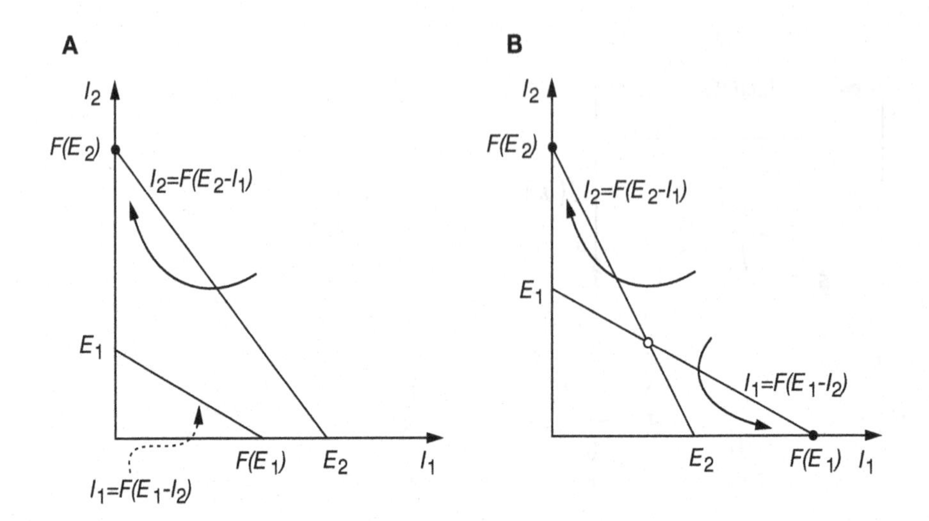

Figure 14.19 A: Phase portrait for mutual inhibition network with $E_1 < F(E_2)$ and $E_2 > F(E_1)$. B: Phase portrait for mutual inhibition network with $E_1 < F(E_2)$ and $E_2 < F(E_1)$.

E_2 is much larger than E_1, so that $E_1 < F(E_2)$ and $E_2 > F(E_1)$, then, as shown in Fig. 14.19A, there is a unique stable fixed point at $I_2 = F(E_1), I_1 = 0$, in which neuron 2 is firing and neuron 1 is quiescent. If E_1 is much larger than E_2, then the reverse is true, namely, there is a unique stable fixed point at $I_1 = F(E_2), I_2 = 0$, with neuron 1 firing and neuron 2 quiescent. There is an intermediate range of parameter values when E_1 and E_2 are similar in size, $E_1 < F(E_2)$ and $E_2 < F(E_1)$, shown in Fig. 14.19B, for which there are three steady states, the two on the axes, and one in the interior of the positive quadrant. The third (interior) steady state is a saddle point, and is therefore unstable.

This neural network exhibits hysteresis. Suppose we slowly modulate the parameter E_1. If it is initially small (compared to E_2, which is fixed at some positive level), then neuron 2 fires steadily and inhibits neuron 1. As E_1 is increased, this situation remains unchanged, even when E_1 and E_2 are of similar size, when two stable steady solutions exist. However, when E_1 becomes sufficiently large, the steady-state solution at $I_1 = F(E_2), I_2 = 0$ suddenly disappears, and the variables I_1, I_2 move to the opposite steady state at $I_2 = F(E_1), I_1 = 0$. Now if E_1 is decreased, when E_1 is small enough there is a reverse transition back to the steady state at $I_1 = F(E_2), I_2 = 0$, completing the hysteresis loop.

To use this hysteresis to control breathing, we model the diaphragm as a damped mass–spring system driven by I_1, the (firing rate) output from neuron 1, the inspiratory neuron:

$$m\frac{d^2x}{dt^2} + \mu\frac{dx}{dt} + kx = I_1. \tag{14.83}$$

We model the effect of the stretch receptors by a function $f(x)$ that is a monotone increasing function of diaphragm displacement x. The stretch receptors are assumed to excite only neuron 2, so that the output variables are governed by

$$\tau\frac{dI_1}{dt} + I_1 = F(E_1 - I_2), \tag{14.84}$$

$$\tau\frac{dI_2}{dt} + I_2 = F(E_2 - I_1 + f(x)). \tag{14.85}$$

We could allow stretch receptors to inhibit neuron 1 as well.

With this model, oscillation of the diaphragm is assured if the time constant τ is sufficiently small. The stretch receptors act to modulate the excitatory inputs, so that as the lung expands, they excite neuron 2. With $E_2 + f(x)$ sufficiently large, neuron 1, the inspiratory neuron, is switched off. With no inspiratory input, the lung relaxes, returning $f(x)$ toward zero and decreasing the excitation to neuron 2. This removes the inhibition to neuron 1 and allows it to fire once again. Thus, if parameters are adjusted properly, the hysteresis loop is exploited, and the inspiration–expiration cycle is established. The oscillations are robust and easily established.

This oscillation can be externally controlled. For example, by increasing E_2, the cycle can be stopped after expiration, whereas by increasing E_1 the inhibition of the stretch receptors can be overridden and inspiration lengthened (as in, take a deep breath, please). Decreasing E_1 shortens the inspiration time and can stop breathing altogether.

In Fig. 14.20 is shown a plot of the two inhibitory variables I_1 and I_2 (shown dashed) plotted as functions of time. Parameter values for this simulation were $\tau = 1.0, m = 0.5, \mu = 5.0, k = 1.0, E_1 = 0.5, E_2 = 0.3$. The function F was specified as $F(x) = \frac{2x^2}{0.2+x}$ for positive x and zero otherwise, and the stretch response curve was taken to be $f(x) = x^3/(1+x^3)$.

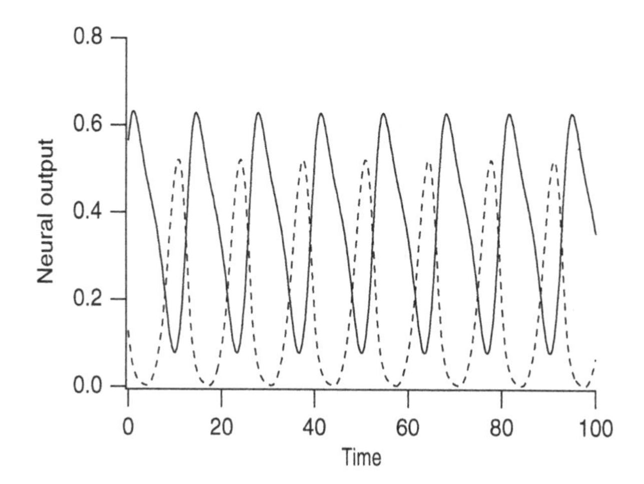

Figure 14.20 Neural output variables I_1 and I_2 (dashed) shown as functions of time.

14.5 Exercises

1. Give a "proper" mathematical derivation of (14.14) by introducing appropriate dimensionless parameters. What dimensionless parameter must be small for this approximation to be valid?
   Answer: $\epsilon = \frac{D_{CO_2}}{k_{-1}[H^+]}$.

2. (a) Develop a model of carbon dioxide and oxygen transport that includes the oxyhemoglobin buffering reaction and the effect of free hydrogen ions on the concentration of bicarbonate. Does the inclusion of proton exchange improve or hinder the rate at which oxygen and carbon dioxide are transported into or out of the blood?

   (b) Estimate the overall effect of this exchange by assuming that the pH of pulmonary venous blood is about 0.04 lower than that of arterial blood.

3. Construct a simple model of the total oxygen and total carbon monoxide in the blood. Assume that the circulatory system is a well-mixed container and that oxygen is removed by metabolism, while oxygen is added and carbon monoxide eliminated during transport through the lungs. Use the models of Section 14.1.5 to determine reasonable transfer rate functions. Estimate the parameters of the model and use numerical computations to determine the half-clearance times for elimination of carbon monoxide at different oxygen levels. How well does your model fit the experimental data shown in Table 14.3?

4. Suppose the respiratory exchange ratio is fixed. Show that there is a linear relationship between the alveolar carbon dioxide and oxygen partial pressures.

5. (a) Assume that regulatory mechanisms maintain the venous oxygen partial pressure at 40 mm Hg and the ventilation–perfusion ratio at 1. Find the alveolar P_{O_2} and the oxygen saturation leaving the alveolus as a function of atmospheric P_{O_2}.

   (b) Data are shown in Table 14.4 for breathing normal air or breathing pure oxygen. What assumption from part 5a is apparently wrong? From the data, determine the venous oxygen partial pressure.

6. (a) Devise a different model of oxygen concentration in which metabolism and the ventilation–perfusion ratio are held fixed. How do the alveolar P_{O_2} and O_2 saturation vary as a function of atmospheric pressure?

   (b) Using data from Table 14.4, estimate the altitude at which incoming alveolar blood has zero P_{O_2}, at normal metabolism.

   (c) Determine the red blood cell count (concentration of hemoglobin) that is necessary to maintain constant venous oxygen partial pressure as a function of altitude at fixed metabolism.

Table 14.3 Experimental half-clearance times for elimination of carbon monoxide from the blood (Pace et al., 1950; also see Exercise 3).

O_2 in atm	Half-clearance time (min)
0.21	249
1.0	47
2.5	22

Table 14.4 Alveolar gas concentration and oxygen saturation at different altitudes. The last column shows the alveolar P_{O_2} when breathing pure oxygen at atmospheric pressure. At this pressure, O_2 saturation is 100%. (Guyton and Hall, 1996, Table 43-1, p. 550.)

Altitude (ft)	Barometric Pressure (mm Hg)	P_{O_2} in air (mm Hg)	Alveolar P_{O_2} (in air) (mm Hg)	O_2 Saturation (in air) (%)	Alveolar P_{O_2} (in oxygen) (mm Hg)
0	760	159	104	97	673
10,000	523	110	67	90	436
20,000	349	73	40	73	262
30,000	226	47	18	24	139

7. Find the rate of carbon monoxide clearance as a function of external P_{O_2}, with fixed metabolism and ventilation.

8. (a) Determine the structure of stable steady solutions of (14.81)–(14.82) in the (E_1, E_2) parameter plane using $F(x) = \frac{2x^2}{0.2+x}$ for positive x and zero otherwise.

   (b) Numerically simulate the system of equations (14.83)–(14.85) using the parameters in the text. Plot $E_2 + f(x)$ as a function of time in the above parameter plane to see how hysteresis is exploited by this system.

9. Consider the following as a possible model of the respiratory center. Two neural FitzHugh–Nagumo oscillators have inhibitory synaptic inputs, so that

$$\frac{dv_i}{dt} = f(v_i, w_i) - s_i g_s (v_i - v_\theta), \tag{14.86}$$

$$\tau_v \frac{dw_i}{dt} = w_\infty(v) - w_i, \tag{14.87}$$

for $i = 1, 2$. The synaptic input s_i is some neurotransmitter that is released when the opposite neuron fires:

$$\frac{ds_i}{dt} = \alpha_s (1 - s_i) x_j F(v_j) - \beta_s s_i, \quad j \neq i, \tag{14.88}$$

and the amplitude of the release x_j decreases gradually when the neuron is firing, via

$$\frac{dx_i}{dt} = \alpha_x (1 - x_i) - \beta_x F(v_i) x_i. \tag{14.89}$$

   (a) Simulate this neural network with $f(v, w) = 1.35v(1 - \frac{1}{3}v^2) - w$, $w_\infty(v) = \tanh(5v)$, $F(v) = \frac{1}{2}(1 + \tanh 10v)$, and with parameters $\tau_v = 5$, $v_\theta = -2$, $\alpha_s = 0.025$, $\beta_s = 0.002$, $\alpha_x = 0.001$, $\beta_x = 0.01$, $g_s = 0.19$.

   (b) Give an approximate analysis of the fast and slow phase portraits for these equations to explain how the network works.

   (c) How does this bursting oscillator compare with those discussed in Chapter 9?

   (d) What features of this model make it a good model of the control of the respiratory system and what features are not so good?

Table 14.5 Parameter values for the bursting pacemaker neuron of Exercise 10.

$\bar{g}_{Na}$ = 28 nS	E_{Na} = 50 mV
$\bar{g}_K$ = 11.2 nS	E_K = −85 mV
$\bar{g}_{NaP}$ = 2.8 nS	C_m = 21 pF
$\bar{g}_L$ = 2.8 nS	E_L = −65 mV
θ_m = −34 mV	σ_m = −5 mV
θ_n = −29 mV	σ_n = −4 mV
θ_h = −48 mV	σ_h = 6 mV
θ_p = −40 mV	σ_p = −6 mV
$\bar{\tau}_h$ = 10,000 ms	$\bar{\tau}_n$ = 10 ms

10. A simple physiological model of bursting pacemaker neurons has been proposed by Butera et al. (1999a). The model is of standard Hodgkin–Huxley type, of the form

$$C_m \frac{dV}{dt} = -I_{NaP} - I_{Na} - I_K - I_L, \tag{14.90}$$

with ionic currents

$$I_{Na} = \bar{g}_{Na} m_\infty^3(V)(1-n)(V - E_{Na}), \tag{14.91}$$

$$I_K = \bar{g}_K n^4 (V - E_K), \tag{14.92}$$

$$I_{NaP} = \bar{g}_{NaP} p_\infty(V) h (V - E_{Na}) \tag{14.93}$$

$$I_L = \bar{g}_L (V - E_L), \tag{14.94}$$

with gating variables n and h governed by differential equations of the form

$$\tau_x(V)\frac{dx}{dt} = x_\infty(V) - x, \qquad x = n, h, \tag{14.95}$$

where

$$x_\infty(V) = \frac{1}{1 + \exp(\frac{V-\theta_x}{\sigma_x})}, \qquad \tau_x = \frac{\bar{\tau}_x}{\cosh(\frac{V-\theta_x}{2\sigma_x})}. \tag{14.96}$$

(a) Simulate this model using E_L as a control parameter (E_L in the range −65 to −54 mV). How do the burst patterns vary as a function of E_L?

(b) How does this bursting oscillator compare with those discussed in Chapter 9?

CHAPTER 15

Muscle

Muscle cells resemble nerve cells in their ability to conduct action potentials along their membrane surfaces. In addition, however, muscle cells have the ability to translate the electrical signal into a mechanical contraction, which enables the muscle cell to perform work. There are three types of muscle cells; skeletal muscle, which moves the bones of the skeleton at the joints, cardiac muscle, whose contraction enables the heart to pump blood, and smooth muscle, which is located in the walls of blood vessels and contractile visceral organs. Skeletal and cardiac muscle cells have a banded appearance under a microscope, with alternating light and dark bands, and thus they are called *striated muscle*. They have similar (though not identical) contractile mechanisms. Smooth muscle, on the other hand, is not striated, and its physiology is considerably different from the other two types of muscle.

Single skeletal muscle cells are elongated cylindrical cells with several nuclei. Each cell contains numerous cylindrical structures, called *myofibrils*, surrounded by the membranous channels of the sarcoplasmic reticulum (Fig. 15.1). Myofibrils are the functional units of skeletal muscle, containing protein filaments that make up the contractile unit. Each myofibril is segmented into numerous individual contractile units called *sarcomeres*, each about 2.5 μm long. The sarcomere, illustrated schematically in Fig. 15.2, is made up primarily of two types of parallel filaments, designated as thin and thick filaments. Viewed end on, six thin filaments are positioned around each central thick filament in a hexagonal arrangement. Viewed along its length, there are regions where thin or thick filaments are overlapping or nonoverlapping. At the end of the sarcomere is a region, called the *Z-line*, where the line filaments are anchored. Thin filaments extend from the Z-lines at each end toward the center, where they overlap with thick filaments. The regions where there is no overlap, containing only thin filaments, are called *I-bands*, and the regions containing myosin (thick) filaments

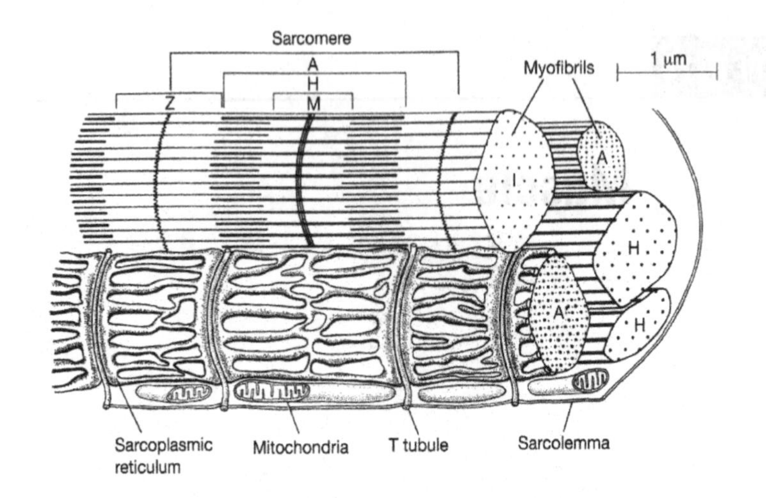

Figure 15.1 Schematic diagram of a skeletal muscle cell. (Berne and Levy, 1993, p. 283, Fig. 17-2.)

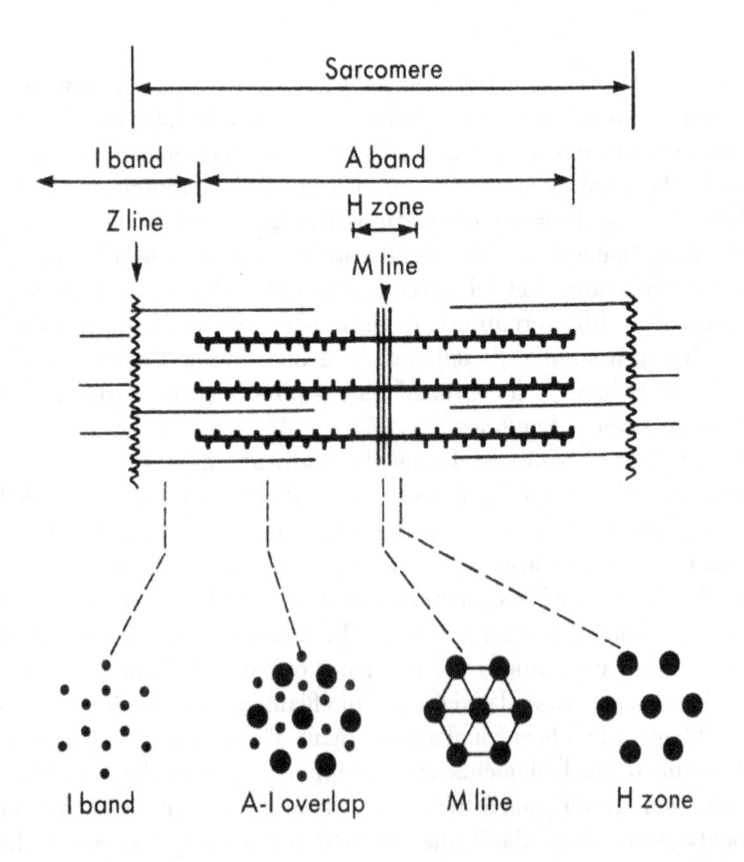

Figure 15.2 Longitudinal section (top panel) and cross-section (lower panels) of a sarcomere showing its organization into bands. (Berne and Levy, 1993, p. 283, Fig. 17-3.)

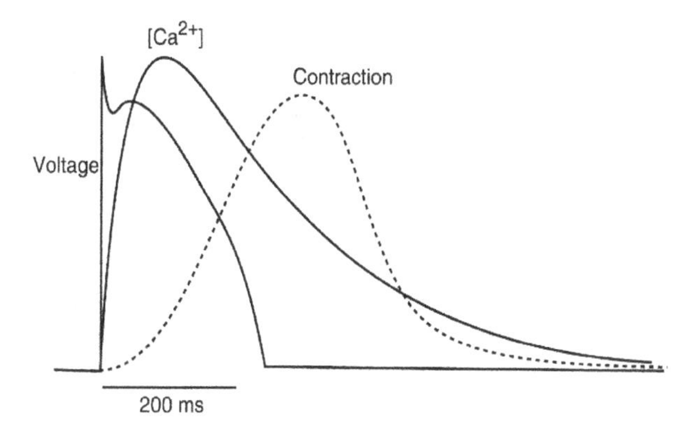

Figure 15.3 The three components of the response of a muscle. The action potential shown here is typical of those seen in cardiac ventricular myocytes (Chapter 12). Adapted from Bers (2002), Fig. 1. Adapted by permission from Macmillan Publishers Ltd.

(with some overlap with thin filaments) are called *A-bands*. The central region of the sarcomere, containing only thick filaments, is called the *H-zone*. During contraction, both the H-zone and the I-bands shorten as the overlap between thin and thick filaments increases.

In striated muscle, contraction is initiated by an action potential transmitted across a synapse from a neuron. This action potential spreads rapidly across the muscle membrane, spreading into the interior of the cell along invaginations of the cell membrane called *T-tubules*. T-tubules form a network in the cell interior, near the junction of the A- and I-bands, and increase the surface area over which the action potential can spread. They enable the action potential to reach quickly into the cell interior. (In cardiac myocytes the T-tubules penetrate into the cell at the level of the Z-lines). In cardiac muscle, voltage-gated Ca^{2+} channels are opened by the action potential, and Ca^{2+} enters the cell, initiating the release of additional Ca^{2+} from the sarcoplasmic reticulum through ryanodine receptors, or RyR (Chapters 7 and 12). In skeletal muscle there is a direct physical connection between the T-tubule and the RyR, and the action potential opens the RyR directly. In either case, the resulting high intracellular Ca^{2+} concentration causes a change in the myofilament structure that allows the thick filaments to bind and pull on the thin filaments, resulting in muscle contraction. There are thus three major components of the response of a muscle, which follow each other in sequence; the action potential, the Ca^{2+} transient, and the contraction (Fig. 15.3). The Ca^{2+}-mediated conversion of an electrical stimulus to a mechanical force is called *excitation–contraction coupling*.

15.1 Crossbridge Theory

Thick filaments contain the protein myosin, which is made up of a polypeptide chain with a globular head. These heads constitute the *crossbridges* that interact with the thin filaments to form bonds that act in ratchet-like fashion to pull on the thin filaments. In addition, the myosin heads have the ability to dephosphorylate ATP as an energy source.

Thin filaments contain the three proteins actin, tropomyosin, and troponin. Each actin monomer is approximately spherical, with a radius of about 5.5 nm, and they aggregate into a double-stranded helix, with a complete twist about every 14 monomers. Because the coil is double-stranded, this structure repeats every 7 monomers, or about every 38 nm. Tropomyosin, a rod-shaped protein, forms the backbone of the double-stranded coil. The troponin consists of a number of smaller polypeptides, which include a binding site for Ca^{2+} as well as a portion that blocks the crossbridge binding sites on the actin helix. When Ca^{2+} is bound, the confirmation of the troponin–tropomyosin complex is altered just enough to expose the crossbridge binding sites. In Fig. 15.4 we show a scale drawing of the probable way in which the actin, tropomyosin, and myosin proteins fit together.

Contraction takes place when the crossbridges bind and generate a force causing the thin filaments to slide along the thick filaments. A schematic diagram of the crossbridge reaction cycle is given in Fig. 15.5, with the accompanying physical arrangement shown in Fig. 15.6. Before binding and contraction, ATP is bound to the crossbridge heads of the myosin (M), and the concentration of Ca^{2+} is low. When the Ca^{2+} concentration increases, Ca^{2+} ions bind to the troponin–tropomyosin complex, exposing the crossbridge binding sites on the actin filament (A). Where possible, a weak bond be-

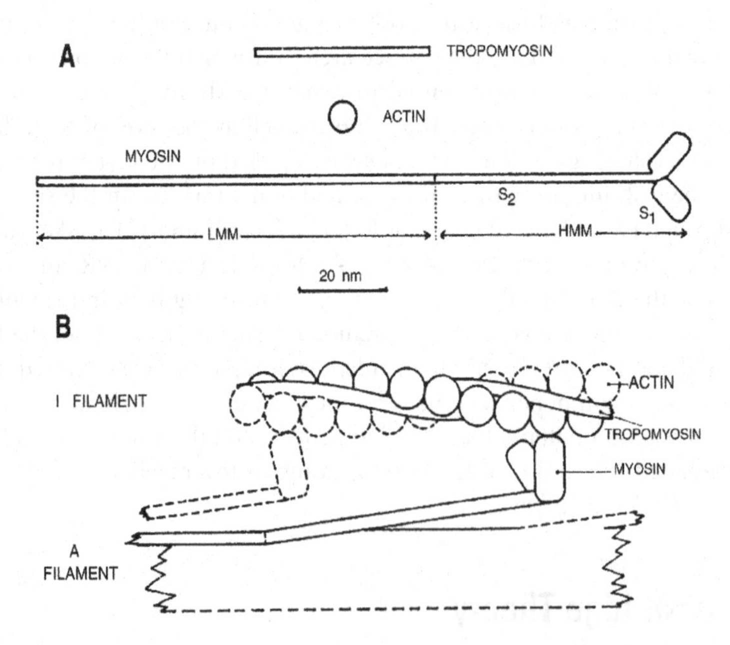

Figure 15.4 A: Scale drawing of actin, myosin, and tropomyosin proteins. B: Scale drawing of the thick and thin filaments (labeled the A and I filaments here), showing the probable way in which the actin, myosin, and tropomyosin proteins fit together. Troponin, which is bound to tropomyosin, is not included in the diagram. (White and Thorson, 1975, Fig. 9, parts A and B (i).)

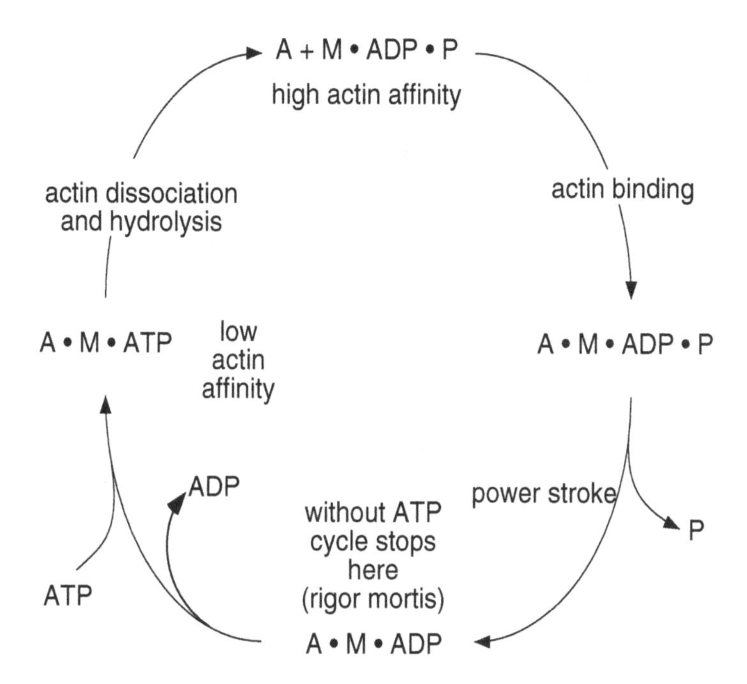

Figure 15.5 Major reaction steps in the crossbridge cycle. M denotes myosin, and A denotes actin.

tween actin and myosin is formed. Release of the phosphate changes the weak bond to a strong bond and changes the preferred configuration of the crossbridge from nearly perpendicular to a bent (foreshortened) position. While the crossbridge is in anything but this energetically preferred, bent state, there is an applied force that acts to pull the thin filament along the thick filament. The movement of the crossbridge to its newly preferred configuration is called the *power stroke*. Almost immediately upon reaching the preferred bent configuration, the crossbridge releases its ADP and binds another ATP molecule, causing dissociation from the actin binding site and return to its initial perpendicular and unbound position. ATP is then dephosphorylated, yielding ADP, phosphate, and the stored mechanical energy for the next cycle. Thus, during muscle contraction, each crossbridge cycles through sequential binding and unbinding to the actin filament.

To construct quantitative models of crossbridge binding it is necessary to know how many actin binding sites are available to a single crossbridge. One possibility is that the crossbridge must be precisely oriented to the actin binding site, and thus, in each turn of the helix, only one binding site is available to each crossbridge. In other words, from the point of view of the crossbridge, the binding sites have an effective separation of about 38 nm. Because of the physical constraints on each crossbridge, this means that at any time, there is only a single binding site available to each crossbridge. This is the assumption behind the Huxley model, described in detail below.

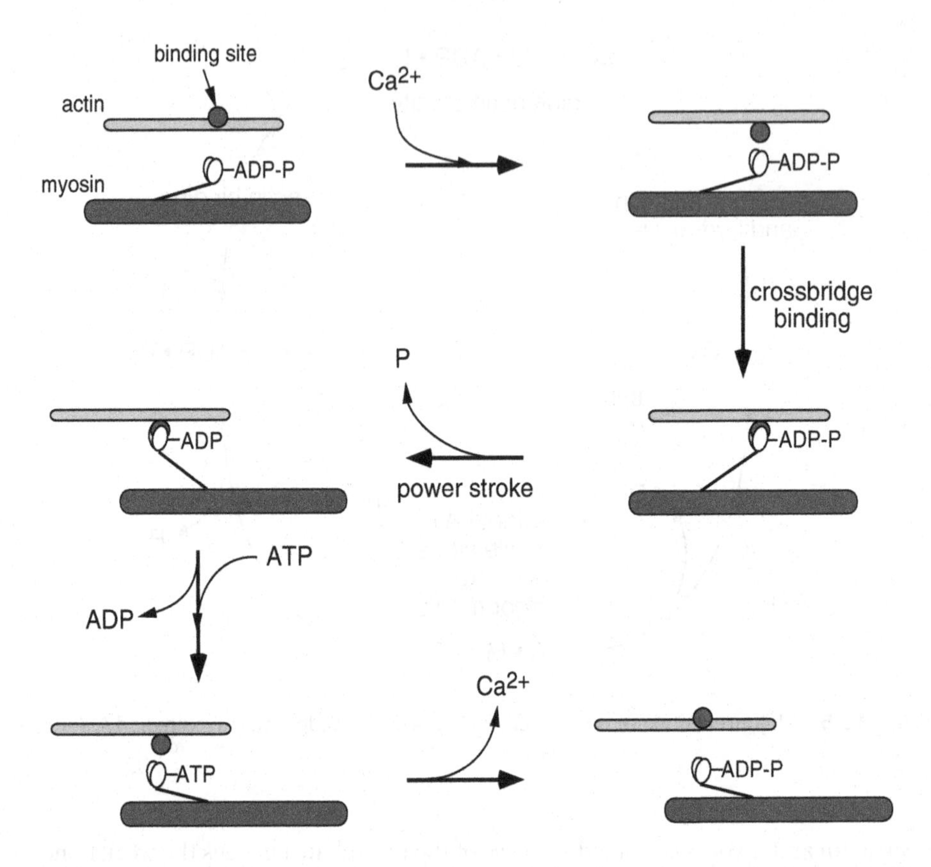

Figure 15.6 Position of crossbridge components during the major steps in the crossbridge cycle.

However, from the distribution of actin binding sites and crossbridges shown in Fig. 15.4, it is plausible that this assumption is not correct. Perhaps, depending on the flexibility of the actin filament, each crossbridge has a number of potential binding sites. In our discussion we concentrate on models of the two extreme cases: first, where each crossbridge has only a single available binding site, and second, where each crossbridge has a continuous array of available binding sites. Intermediate models, in which the crossbridge has a small number of discrete binding sites available, are considerably more complex and are mentioned only briefly.

Because of the sarcomere structure, the tension a muscle develops depends on the muscle length. In Fig. 15.7 we show a curve of isometric tension as a function of sarcomere length. By isometric tension, we mean the tension developed by a muscle when it is held at a fixed length and repeatedly stimulated (i.e., with a high-frequency periodic stimulus). Under these conditions the muscle goes into *tetanus*, a state, caused by saturating concentrations of Ca^{2+} in the sarcoplasm, in which the muscle is continually attempting to contract. Note that the muscle cannot actually contract, because it is

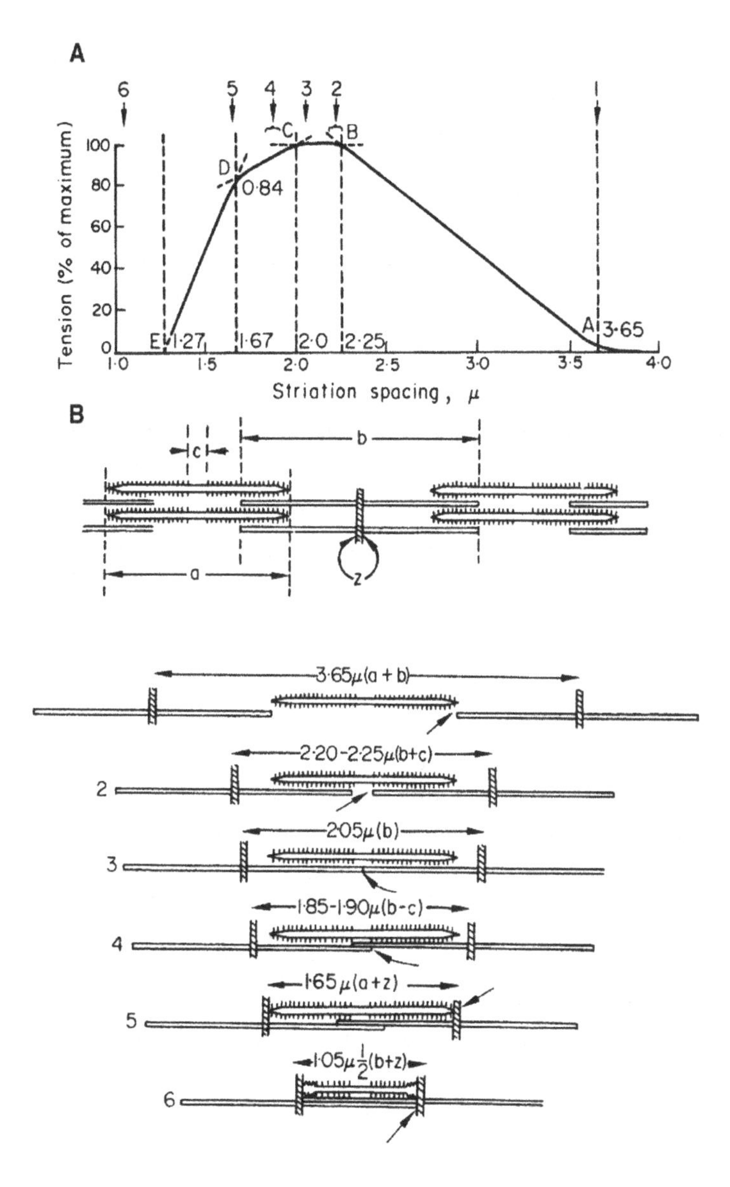

Figure 15.7 A: Isometric tension as a function of the length of the sarcomere. B: schematic diagrams of the arrangement of the thick and thin filaments for the six different places indicated in panel A. (Gordon et al., 1966, reproduced in White and Thorson, 1975, Fig. 14.)

held at constant length, although it must go through the chemistry cycle of the power stroke, since the development of tension requires that energy be consumed.

At short lengths, overlap of the thin filaments causes a drop in tension, but as this overlap decreases (as the length increases) the tension rises. However, when the length is large, there is less overlap between the thick and thin filaments, so fewer crossbridges bind, and less tension develops. When there is no overlap between the thick and thin filaments, the muscle is unable to develop any tension.

Skeletal muscle tends to operate at lengths that correspond to the peak of the isometric length–tension curve, and thus in many experimental setups the tension the muscle develops does not depend significantly on the muscle length. However, the same is not true for cardiac muscle, which considerably complicates theoretical studies of this muscle type. For these reasons we restrict our attention to models based on data from skeletal muscle. Peskin (1975) presents a detailed description of some theoretical models of cardiac muscle.

15.2 The Force–Velocity Relationship: The Hill Model

One of the earliest models of a muscle is due to A.V. Hill (1938) and was constructed before the details of the sarcomere anatomy were known. Hill observed that when a muscle contracts against a constant load (an *isotonic* contraction), the relationship between the constant rate of shortening v and the load p is well described by the *force–velocity* equation

$$(p+a)v = b(p_0 - p), \tag{15.1}$$

where a and b are constants that are determined by fitting to experimental data in a way that we discuss presently. A typical force–velocity curve is plotted in Fig. 15.8. When $v = 0$, then $p = p_0$, and thus p_0 represents the force generated by the muscle when the length is held fixed; i.e., p_0 is the *isometric* force. As discussed above, the tension generated by a skeletal muscle in isometric tetanus is approximately independent of length, and thus p_0 is also approximately independent of length. When $p = 0, v = bp_0/a$, which is the maximum speed at which a muscle is able to shorten.

In an attempt to explain these observations, we model a muscle fiber as a contractile element with the given force–velocity relationship, in series with an elastic element (Fig. 15.9). In some versions of the model a parallel elastic element is included (see Exercise 1), but as it plays no essential role in the following discussion, it is omitted here. As shown in Fig. 15.9, we let l denote the length of the contractile element, we let x denote the length of the elastic element, so that $L = l + x$ is the total length of the fiber. Then, letting v denote the velocity of contraction of the contractile element, we have

$$v = -\frac{dl}{dt}, \tag{15.2}$$

where, by assumption, v is related to the load on the muscle by the force–velocity equation (15.1). To derive a differential equation for the time dependence of p, we note that because the elastic element is in series with the contractile element, the two experience the same force. We assume that the force generated by the elastic element is a function of its length $p = P(x)$ and then use the chain rule and the force–velocity

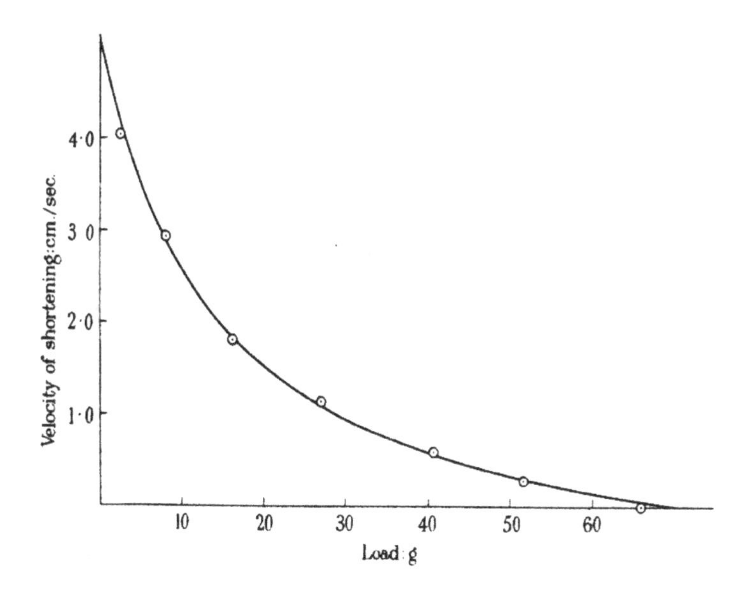

Figure 15.8 The relationship between the load on a muscle and the velocity of contraction (Hill, 1938; Fig. 12). The symbols are the data points, while the smooth curve is calculated from (15.1) using the parameter values $a = 14.35$ grams, $a/p_0 = 0.22$, $b = 1.03$ cm/s. The value of a is equivalent (in Hill's original preparation) to 357 grams (of weight) per square centimeter of muscle fiber (g-wt/cm^2). Note that if p is expressed in grams weight, so also must a be. Since Hill used a muscle of length 38 mm, the value of b is equivalent to 0.27 muscle lengths per second.

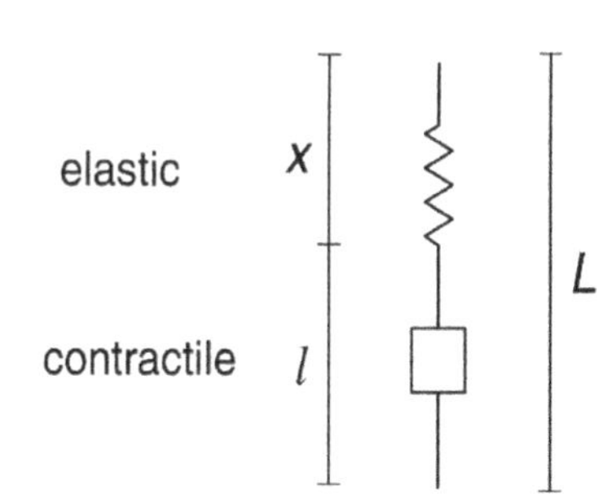

Figure 15.9 Schematic diagram of Hill's two-element model of skeletal muscle. The muscle is assumed to consist of an elastic element in series with a contractile element with a given force–velocity relationship.

equation to obtain

$$
\begin{aligned}
\frac{dp}{dt} &= \frac{dP}{dx}\frac{dx}{dt} \\
&= \frac{dP}{dx}\left[\frac{dL}{dt} - \frac{dl}{dt}\right] \\
&= \frac{dP}{dx}\left[\frac{dL}{dt} + v\right]
\end{aligned}
$$

$$= \frac{dP}{dx}\left[\frac{dL}{dt} + \frac{b(p_0 - p)}{p + a}\right]. \tag{15.3}$$

It remains to determine dP/dx.

Hill made the simplest possible assumption, that the elastic element is linear, and thus

$$P = \alpha(x - x_0), \tag{15.4}$$

where x_0 is its resting length. Thus, $dP/dx = \alpha$, and the differential equation for p is

$$\frac{dp}{dt} = \alpha\left[\frac{dL}{dt} + \frac{b(p_0 - p)}{p + a}\right]. \tag{15.5}$$

15.2.1 Fitting Data

Suppose a muscle in tetanus is held at a fixed tension until it reaches its isometric length, and then the tension is suddenly decreased and held fixed at a lower value. A typical result is shown in Fig. 15.10A, where the muscle length is plotted against time. As soon as the tension is reduced, the muscle length decreases (plotted in the vertical direction) as the elastic element contracts. After a transition period during which the length exhibits small oscillations (which are not explained by this model), the muscle decreases in length at a constant rate. Plotting the rate of decrease against the constant

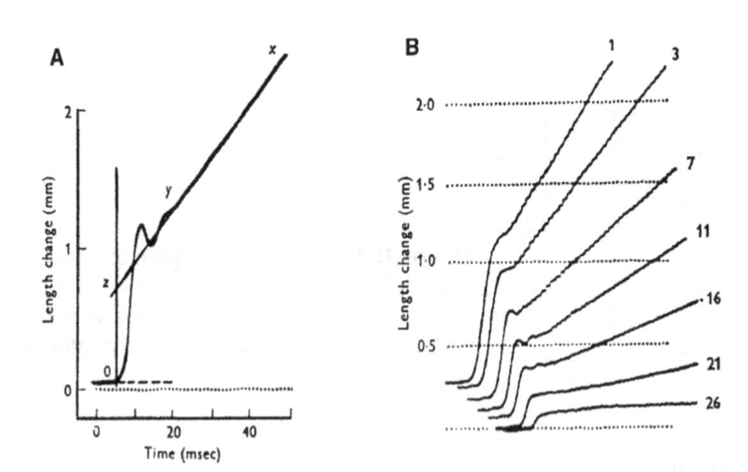

Figure 15.10 A: Plot of length change against time after a step decrease in tension. The length decreases in a sudden jump, and then, after an initial oscillatory phase, decreases at constant velocity. (Jewell and Wilkie, 1958; reproduced in White and Thorson, 1975, Fig. 5.) B: Length change as a function of time from a series of tension step experiments. The baseline of each trace has been shifted for clarity, and each dot on the horizontal axis denotes 1 ms. For each step, the value to the right of the curve denotes the final value of the tension in grams of weight (g-wt). (Jewell and Wilkie, 1958; reproduced in White and Thorson, 1975, Fig. 6.)

applied tension gives one point on the force–velocity curve. More specifically, if the tension is stepped from p_0 to p_1, the muscle contracts at the constant rate v, where

$$(p_1 + a)v = b(p_0 - p_1). \tag{15.6}$$

Repeating the experiment for tension steps of different magnitudes (shown in Fig. 15.10B) one finds a series of points on the force–velocity curve, through which one can fit the force–velocity equation to obtain values for a, b, and p_0. Note that this procedure is valid only if p_0 does not change during the course of the experiment. In other words, as the muscle shortens with constant velocity, it must be that p_0 remains unchanged. As discussed above, this is an acceptable assumption for skeletal muscle operating near the peak of the length–tension curve.

Similarly, the characteristics of the elastic element can be determined from the initial step in length. By extrapolating the line of constant speed back to the time of the tension step (the line xyz in Fig. 15.10A), one finds the distance $0z$ which is the change in length of the elastic element. This relies on the assumption that the force–velocity properties of the muscle change instantaneously with the change in tension. Knowing the change in length of the elastic element and the change in tension that produced it allows one to determine α.

15.2.2 Some Solutions of the Hill Model

Isometric Tetanus Solution

If a muscle at rest is put into tetanus by repeated stimulation, the isometric tension builds up over a period of time. Because the tension is measured isometrically, the length of the muscle does not change, and thus $dL/dt = 0$. Hence, the differential equation for the tension is

$$\frac{dp}{dt} = \alpha \left[\frac{b(p_0 - p)}{p + a} \right]. \tag{15.7}$$

Since this is a first-order differential equations, its qualitative behavior is readily understood. The right-hand side of the equation has a unique zero at $p = p_0$, and this zero is stable. Thus, all solutions of this equation must approach p_0 as $t \to \infty$.

The solution of this equation can be found explicitly, since this is a separable equation. After separation, we integrate from 0 to t and use the initial condition $p(0) = 0$ to obtain

$$-p - (p_0 + a) \log \left(\frac{p_0 - p}{p_0} \right) = \alpha b t, \tag{15.8}$$

which describes the time course of the change in tension implicitly. Notice that $p \to p_0$ as $t \to \infty$, as expected.

Release at Constant Velocity

Suppose a muscle, held originally at its isometric tension p_0, is allowed to contract with constant velocity u. It seems reasonable that the muscle tension should decrease until it reaches the value p_u determined from the force–velocity curve for a velocity u. The differential equation for p is

$$\frac{dp}{dt} = \alpha\left[-u + \frac{b(p_0 - p)}{p + a}\right], \tag{15.9}$$

with initial condition $p(0) = p_0$. As before, we assume that p_0 does not change during the course of the contraction.

Again, this is a first order differential equation whose right-hand side has a unique, stable, root at $p = p_u$, where p_u is defined by $(p_u + a)u = b(p_0 - p_u)$. The solution is found by separation to be

$$p_0 - p + (p_u + a)\log\left(\frac{p_0 - p_u}{p - p_u}\right) = \alpha t(b + u). \tag{15.10}$$

Notice that $p \to p_u$ as $t \to \infty$, as it must.

Response to a Jump in Length

Possibly the most interesting solution is the response to a step decrease in length, as this solution has been used to show that the Hill model does not provide an accurate description of all aspects of muscle behavior (Jewell and Wilkie, 1958).

First, Jewell and Wilkie determined the parameters of the Hill model by the series of experiments described above (Fig. 15.10). They then used the Hill model to predict the response of the muscle to a step decrease in muscle length. Suppose that a muscle, originally held at its isometric tension p_0, is suddenly decreased in length. One expects the muscle tension to suddenly decrease, but then slowly increase back to p_0. This is because the isometric tension is independent of length but should take some time to develop at the new length. A typical solution is sketched schematically in Fig. 15.11A.

More precisely, suppose that the length of the muscle as a function of time is given by

$$L(t) = L_1 + L_0 - L_0 H(t), \tag{15.11}$$

where $H(t)$ is the usual Heaviside function, and where L_1 and L_0 are constants, L_0 being the magnitude of the length step. Thus

$$\frac{dL}{dt} = -L_0\delta(t), \tag{15.12}$$

where $\delta(t)$ denotes the Dirac delta function. Substituting this expression into the differential equation for p gives

$$\frac{dp}{dt} = \alpha\left[-L_0\delta(t) + \frac{b(p_0 - p)}{p + a}\right], \tag{15.13}$$

$$p(0^-) = p_0. \tag{15.14}$$

Integrating (15.13) (formally) from $t = -\epsilon$ to $t = \epsilon$ and then letting $\epsilon \to 0$, we get

$$p(0^+) - p(0^-) = -\alpha L_0, \tag{15.15}$$

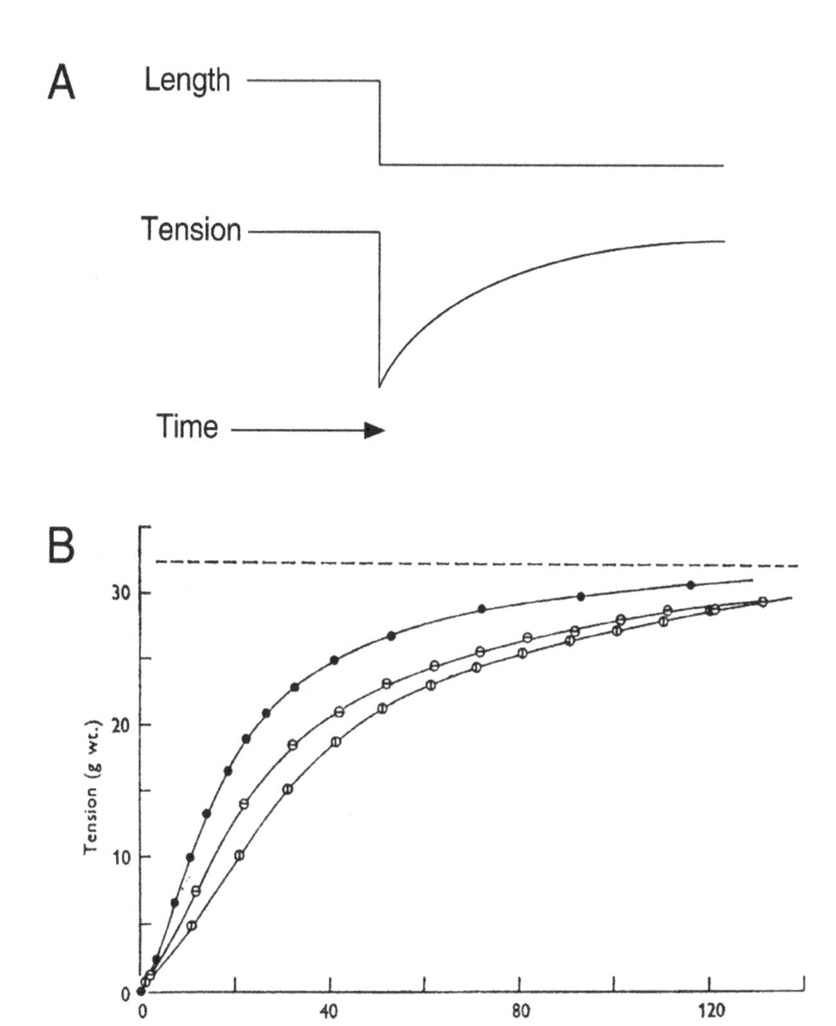

Figure 15.11 A: Schematic diagram of the response to a step decrease in length. B: Comparison of the Hill model to the data of Jewell and Wilkie (1958). The closed circles are computed from the Hill model, while the open circles are data points from two slightly different experimental procedures. (Jewel and Wilkie, 1958; reproduced in White and Thorson, 1975, Fig. 7.)

so that the delta function causes a jump of $-\alpha L_0$ in p at the origin. Thus, (15.13) and (15.14) can be rewritten as the initial value problem

$$\frac{dp}{dt} = \alpha \left[\frac{b(p_0 - p)}{p + a} \right], \qquad t > 0, \tag{15.16}$$

$$p(0) = p_0 - \alpha L_0. \tag{15.17}$$

Since we have reduced the problem to the isometric tetanus problem studied earlier (although with a different initial condition), the solution is easily calculated.

The solution calculated from the Hill model in this way does not agree with experimental observations on the tension recovery following a step decrease in length. In fact, the tension recovers less quickly than is predicted by the model, as illustrated in Fig. 15.11B. Here, the model computations (shown as closed circles), consistently lie above the data points (shown as open circles). These observations, made possible by the improvements in experimental technique in the 20 years after Hill's model was first proposed, forced the conclusion that the Hill model has serious defects. In particular, the assumption that the force–velocity relationship (15.1) is satisfied immediately after a change in tension is a probable major source of error. At the same time that Hill's model was shown to have problems, much more was being discovered about the structure of the sarcomere. This motivated the construction of a completely different type of model, based on the kinetics of the crossbridges rather than on heuristic elastic and contractile elements. The first model of this new type was due to Huxley (1957), and is the basis for the majority of subsequent models of muscle behavior.

15.3 A Simple Crossbridge Model: The Huxley Model

To formulate a mathematical model describing crossbridge interactions in a sarcomere, we suppose that a crossbridge can bind to an actin binding site at position x, where x measures the distance along the thin filament to a binding site from the crossbridge, and $x = 0$ corresponds to the position in which the bound crossbridge exerts no force during the power stroke on the thin filament (Fig. 15.12). Crossbridges can be bound to a binding site with $x > 0$, in which case they exert a contractile force, or they can be

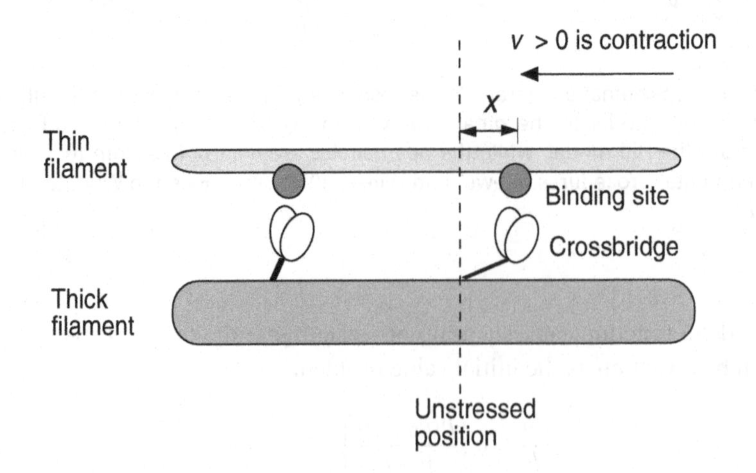

Figure 15.12 Schematic diagram of the Huxley crossbridge model.

bound to a site with $x < 0$, in which case they exert a force that opposes contraction. A crossbridge bound to a binding site at x is said to have displacement x. In his original model Huxley assumed that the actin binding sites were sufficiently far apart that each crossbridge could be bound with one and only one binding site. With this assumption, each crossbridge, whether bound or not, can be associated with a unique value of x.

Let ρ denote the number of crossbridges (either bound or unbound) with displacement x. We assume that binding is restricted to occur in some bounded interval, $-x_0 < x < x_0$, and that ρ is a constant independent of x on that interval. In other words, for each displacement x, the number of crossbridges with that displacement is conserved, and we exclude the possibility that all the crossbridges could end up, say, with the same displacement. This assumption was originally designed for the isometric case, in which the muscle has a fixed length, and is not necessarily valid under conditions where the muscle is quickly extended by an external force. For example, in response to a very fast imposed load of sufficient magnitude, each crossbridge could be quickly extended to have a large value of $x > x_0$, thus violating this conservation law, at least for a short time. However, we ignore this complication for now (but see Section 15.8.1).

Finally, we define $n(x,t)$ to be the fraction of crossbridges with displacement x that are bound.

Next, we drastically simplify the reaction mechanism, and assume that a crossbridge can be in one of two states, namely either unbound (U), or strongly bound (B) and thereby generating a force. We suppose further that the binding and unbinding of crossbridges is described by the simple reaction scheme

$$\mathrm{U} \underset{g(x)}{\overset{f(x)}{\rightleftarrows}} \mathrm{B},$$

where the rate constants are functions of the displacement x.

The conservation law for the fraction of bound crossbridges can be derived as follows. Consider all the crossbridges that are bound with displacements x between a and b. This total number is given by

$$\rho \int_a^b n(x,t)\,dx. \tag{15.18}$$

The rate of change of this total number is given by the reactions of the crossbridges as well as the fluxes across the boundaries of the interval $[a,b]$. At $x = a$, the flux of crossbridges out of the domain is $\rho v(t) n(a,t)$, while at $x = b$, the flux of crossbridges into the domain is $\rho v(t) n(b,t)$. Here $v(t)$ is the velocity of the actin filament relative to the myosin filament. For notational consistency, we assume that $v > 0$ denotes muscle contraction. Thus, conservation of crossbridges gives

$$\rho \frac{d}{dt} \int_a^b n(x,t)\,dx = \rho v(t) n(b,t) - \rho v(t) n(a,t) - \rho \int_a^b [f(x)(1 - n(x,t)) - g(x) n(x,t)]\,dx, \tag{15.19}$$

and so

$$\int_a^b \frac{\partial}{\partial t} n(x,t)\,dx = v(t)\int_a^b \frac{\partial}{\partial x} n(x,t)\,dx - \int_a^b [f(x)(1-n(x,t)) - g(x)n(x,t)]\,dx. \tag{15.20}$$

Of course, since a and b are arbitrary, the integral can be dropped, yielding the partial differential equation

$$\frac{\partial n}{\partial t} - v(t)\frac{\partial n}{\partial x} = (1-n)f(x) - ng(x). \tag{15.21}$$

This conservation law and its derivation occur many times in applications, including a number of other places in this book (see for example (2.1), (11.151), (14.3) or (17.31)).

Every time a crossbridge is bound, one ATP molecule is dephosphorylated, so the rate of energy release, ϕ, for this process is given by

$$\phi = \rho\epsilon \int_{-\infty}^{\infty} (1-n(x,t))f(x)\,dx, \tag{15.22}$$

where ϵ is the chemical energy released by one crossbridge cycle. Since n is, in general, a function of the contraction velocity, so also is ϕ. We also suppose that a bound crossbridge is like a spring, generating a restoring force $r(x)$ related to its displacement. Hence, the total force exerted by the muscle is

$$p = \rho \int_{-\infty}^{\infty} r(x)n(x,t)\,dx. \tag{15.23}$$

To find the force–velocity relationship for muscle, we assume that the fiber moves with constant velocity, and that $n(x,t)$ is equilibrated so that $\partial n/\partial t = 0$. Then, the steady distribution $n(x)$ is the solution of the first-order differential equation

$$-v\frac{dn}{dx} = (1-n)f(x) - ng(x). \tag{15.24}$$

The solution of this differential equation is easily understood. The function $n(x)$ "tracks" the quasi-steady-state solution $\frac{f(x)}{f(x)+g(x)}$ at a rate that is inversely proportional to v. Thus, if v is small, $n(x)$ is well approximated by the quasi-steady-state solution, whereas if v is large, $n(x)$ changes slowly as a function of x. From this we make two observations. First, the force is largest at small velocities. In fact, at zero velocity, the isometric force is

$$p_0 = \rho \int_{-\infty}^{\infty} r(x)\frac{f(x)}{f(x)+g(x)}\,dx. \tag{15.25}$$

Second, at large velocities, the distribution $n(x)$ has small amplitude, and so the force is small. The force decreases because the amount of time during which a crossbridge is close to a binding site is small, and so binding is less likely, with the result that a smaller fraction of crossbridges exerts a contractile force. Another factor is that at higher velocities a greater number of crossbridges are carried into the $x < 0$ region before they can dissociate, hence generating a force opposing contraction. It is intuitively reasonable that at some maximum velocity, the force generated by the crossbridges with $x < 0$

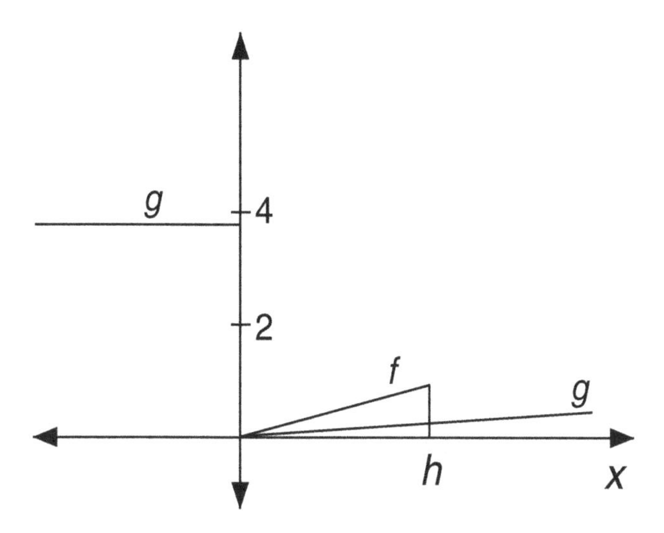

Figure 15.13 The attachment and detachment functions, *f* and *g*, in the Huxley model.

exactly balances the force generated by those with $x > 0$, at which point no tension is generated by the muscle, and the maximum velocity of shortening is attained. We have already seen that this occurs in the Hill force–velocity curve. Crossbridge theory provides an elegant explanation of this phenomenon.

To obtain quantitative formulas, one must make some reasonable guesses for the functions $f(x)$ and $g(x)$, and then calculate $n(x)$ and p numerically or analytically. Although numerical solutions can always be obtained, there are several choices of $f(x)$ and $g(x)$ for which analytical solutions are possible. The functions that Huxley chose are illustrated in Fig. 15.13 and have the form

$$f(x) = \begin{cases} 0, & x < 0, \\ f_1 x/h, & 0 \le x \le h, \\ 0, & x > h, \end{cases} \tag{15.26}$$

$$g(x) = \begin{cases} g_2, & x \le 0, \\ g_1 x/h, & x > 0. \end{cases} \tag{15.27}$$

In this model, the rate of crossbridge dissociation, g, is low when the crossbridge exerts a contractile force, but when x is negative, the crossbridge opposes contraction, and g increases. Similarly, crossbridges do not attach at a negative x ($f = 0$ when $x < 0$), and as x increases, the rate of crossbridge attachment increases as well. This ensures that crossbridge attachment contributes an overall contractile force. At some value h, the rate of crossbridge attachment falls to zero, as it is assumed that crossbridges cannot bind to a binding site that is too far away.

The steady-state solution for $n(x)$ is easily obtained by direct piecewise solution of the differential equation. Let n_I, n_{II} and n_{III} denote, respectively, the steady-state solutions in the regions $x \leq 0$, $0 < x \leq h$ and $h < x$. Then n_I is the solution of the equation

$$-v\frac{dn_I}{dx} = -g_2 n_I, \tag{15.28}$$

and thus

$$n_I = Ae^{g_2 x/v}, \tag{15.29}$$

for some constant A yet to be determined. Note that this solution is bounded as $x \to -\infty$ as it should be.

Next we solve for n_{II}, which satisfies the equation

$$-v\frac{dn_{II}}{dx} + n_{II}\left(\frac{f_1 x}{h} + \frac{g_1 x}{h}\right) = \frac{f_1 x}{h}, \tag{15.30}$$

which has solution

$$n_{II} = \frac{f_1}{f_1 + g_1} + B\exp\left(\frac{x^2(f_1 + g_1)}{2vh}\right), \tag{15.31}$$

for some constant B, also to be determined.

The only bounded solution of the equation for n_{III},

$$-v\frac{dn_{III}}{dt} = g_1\frac{x}{h}n_{III}, \tag{15.32}$$

is identically zero. This makes physical sense as well, since crossbridges can never be attached for $x > h$, if $v > 0$. Now, to find the unknown constants A and B we require that the solution be continuous at $x = 0$ and $x = h$, and thus

$$n_I(0) = n_{II}(0), \qquad n_{II}(h) = 0. \tag{15.33}$$

It follows that

$$B = -\frac{f_1}{f_1 + g_1}e^{-\phi/v}, \tag{15.34}$$

$$A = \frac{f_1}{f_1 + g_1} + B = \frac{f_1}{f_1 + g_1}(1 - e^{-\phi/v}), \tag{15.35}$$

and thus

$$n(x) = \begin{cases} F_1\left[1 - e^{-\phi/v}\right]e^{\frac{x}{2h}G_2\frac{\phi}{v}}, & x < 0, \\ F_1\left\{1 - \exp\left[\left(\frac{x^2}{h^2} - 1\right)\frac{\phi}{v}\right]\right\}, & 0 < x < h, \\ 0, & x > h, \end{cases} \tag{15.36}$$

where $\phi = (f_1 + g_1)h/2$ has units of velocity, and $F_1 = \frac{f_1}{f_1+g_1}$, and $G_2 = \frac{g_2}{f_1+g_1}$ are dimensionless. This steady solution is plotted in Fig. 15.14 for four values of v. Notice the unphysiological implication of this solution, that $n > 0$ for all $x < 0$. However, only

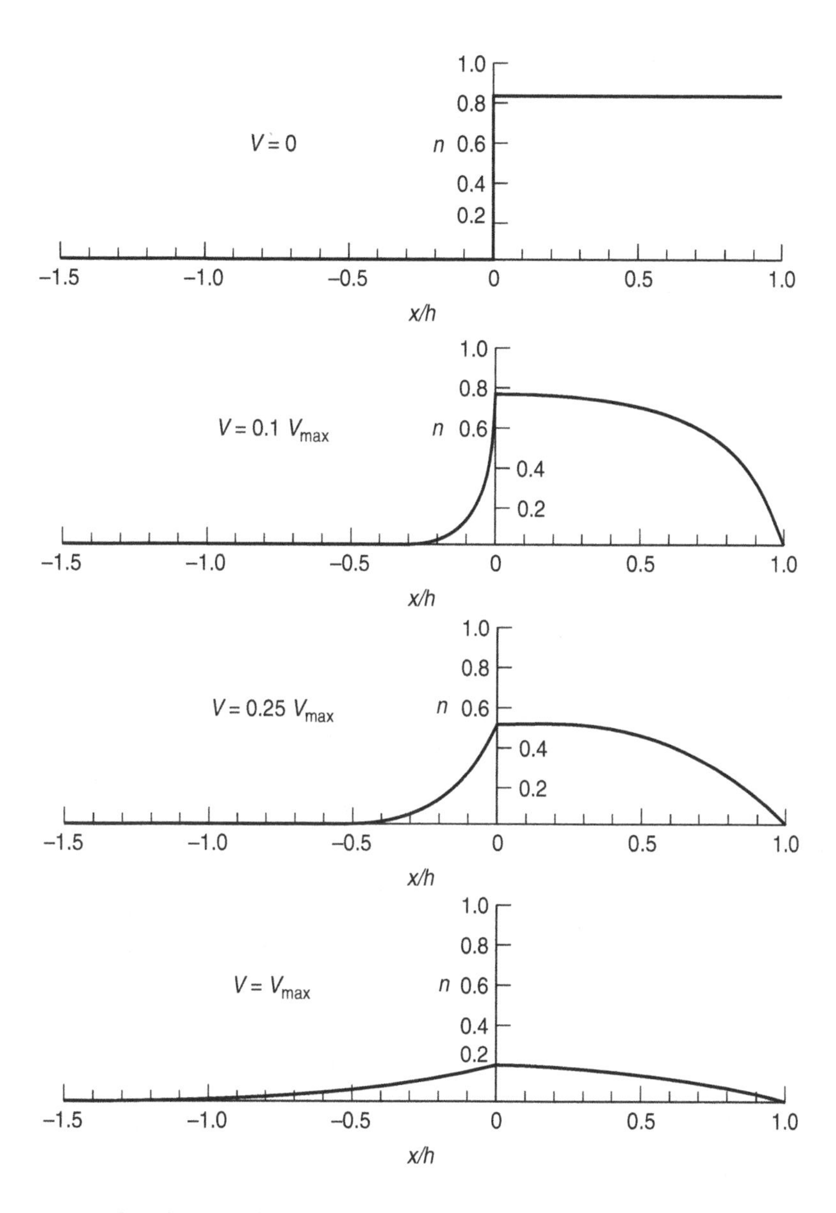

Figure 15.14 Steady-state distributions of n in the Huxley model, for different values of v, plotted as a function of dimensionless space $\frac{x}{h}$. (Huxley, 1957, Fig. 7.) The parameter values for this figure and the next were chosen by Huxley by trial and error to obtain a good fit with experimental data. The values are $F_1 = f_1/(g_1 + f_1) = 13/16$, $G_2 = g_2/(f_1 + g_1) = 3.919$.

a negligible number of crossbridges are bound at unphysiological displacements, so these have little effect on the behavior of the model.

Assuming that the crossbridge acts like a linear spring, so that $r(x) = kx$ for some constant k, the force generated by the muscle (defined by (15.23)) can be calculated

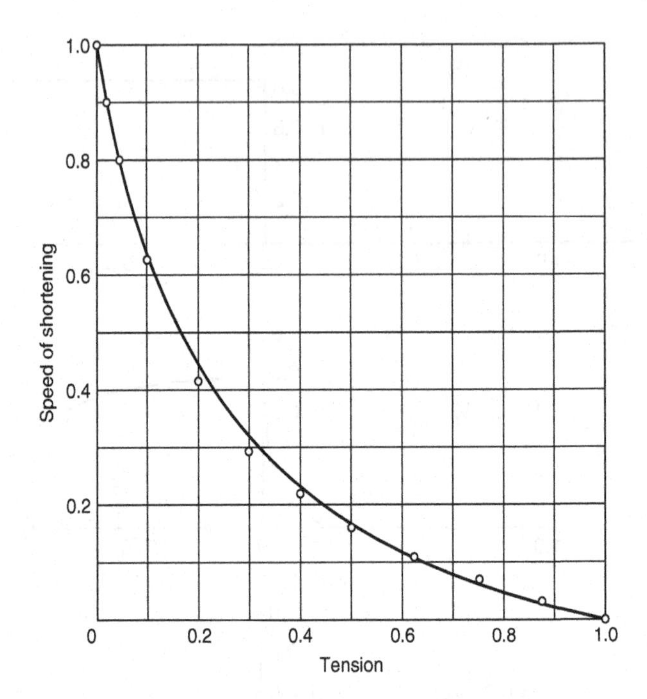

Figure 15.15 The force–velocity curve of the Huxley model (solid curve) compared to Hill's data (open circles). (Huxley, 1957, Fig. 8.) Here, p has been scaled so that $p(0) = 1$. This determines the value used for ρk. Further, the parameters have been scaled so that $v_{max} = 1$.

as a function of the velocity of contraction, and the result compared to the Hill force–velocity equation (15.1). The force–velocity equation calculated from the Huxley model is

$$p = \frac{\rho k f_1}{f_1 + g_1} \frac{h^2}{2} \left\{ 1 - \frac{v}{\phi}(1 - e^{-\phi/v}) \left(1 + \frac{1}{2G_2^2} \frac{v}{\phi} \right) \right\}, \tag{15.37}$$

which for appropriate choice of parameters gives an excellent fit to the force–velocity curve, as illustrated in Fig. 15.15.

Huxley chose the model parameters by a process of trial and error so that the rate of energy production agreed with experimental data. A good fit is obtained by choosing $F_1 = \frac{f_1}{f_1+g_1} = 13/16$ and $G_2 = \frac{g_2}{f_1+g_1} = 3.919$. One can show that for these parameter values, $4\phi \approx v_{max}$ (Exercise 5). Since in the Hill model $v_{max} = bp_0/a$, it follows that $\phi \approx \frac{bp_0}{4a}$. The parameter values that Hill obtained by fitting to data have $p_0/a \approx 4$. Hence, it follows that $\phi \approx b$. This is an elegant way of relating parameter values in the Huxley model to the parameters of the Hill model.

It is useful to see what f_1, g_1 and g_2 are in dimensional units. Brokaw (1976) used $f_1 = 65\ \text{s}^{-1}$, $g_1 = 15\ \text{s}^{-1}$, $g_2 = 313.5\ \text{s}^{-1}$ and $h = 10$ nm, to give $f_1 + g_1 = 80\ \text{s}^{-1}$ which gives $\phi = 400\ \text{nm s}^{-1}$ and thus a maximum shortening velocity of $1600\ \text{nm s}^{-1}$ per half sarcomere, and a maximum shortening velocity of 3200 nm s^{-1} per sarcomere. Assuming that a sarcomere is 2.5 μm long, each centimeter of muscle contains 4000 sarcomeres. If each sarcomere shortens at 800 nm s^{-1}, a 1 cm length of muscle shortens at 3200 × 4000 nm/s, which is 1.28 cm/s. From the caption of Fig. 15.8 we see that Hill used a piece of muscle 38 mm long, which corresponds to a maximum shortening velocity of 4.8 cm/s, in good agreement with the measured value shown in Fig. 15.8.

15.3.1 Isotonic Responses

Thus far we have shown how the Huxley model can explain the Hill force–velocity curve using crossbridge dynamics. However, for the model to give an acceptable explanation of muscle dynamics, there is a great deal of additional experimental data with which it should agree. In particular, the model should explain the response of a muscle, first, to a step change in tension (isotonic response) and second, to a step change in length (isometric response). After all, the Hill model was rejected as a satisfactory explanation because of its inability to explain all such data.

It is instructive to consider how one calculates the response of the Huxley model to a step change in tension, as the procedure is not obvious. Suppose a muscle exerts its isometric tension, p_0, at some length L. Then the steady-state crossbridge distribution is

$$n_s(x) = \frac{f(x)}{f(x) + g(x)}. \tag{15.38}$$

Now suppose the tension on the muscle is reduced to $p_1 < p_0$ so suddenly that no crossbridges are able to associate or dissociate during the reduction. In a typical experiment of Civan and Podolsky (1966), a muscle fiber of length 15,000 μm was subjected to a change in tension that changed the fiber length by less than 50 μm, a relative length change of 1/300. Hence, a typical sarcomere of length 2.5 μm changed in length by less than 10 nm, and so the length of each crossbridge was changed by less than 10 nm. A crossbridge is able to absorb such length changes without dissociating from the binding site.

Suppose the extension of each crossbridge decreases by an unknown amount ΔL, and so the crossbridge distribution suddenly changes to $n_s(x + \Delta L)$ and is no longer at steady state. The change in length is found by constraining the new tension to be p_1, and hence ΔL satisfies

$$p_1 = \int_{-\infty}^{\infty} k(x) n_s(x + \Delta L)\, dx. \tag{15.39}$$

Although (15.39) cannot in general be solved analytically, ΔL can be determined numerically, since it is easy to determine p_1 as a function of ΔL.

Following the sudden change in tension, the crossbridge population is not at steady state, so it must change according to the differential equation (15.21) with initial condition $n(x, 0) = n_s(x + \Delta L)$ and subject to the constraint that the tension is constant at $p = p_1$. However, during this evolution, v is not constant, as there is some transient behavior before the muscle reaches its steady contraction velocity (cf. Fig. 15.10).

However, we can determine an expression for $v(t)$ in terms of $n(x,t)$ that guarantees that the tension remains constant at p_1.

Since p_1 is constant, it must be that $\frac{\partial p_1}{\partial t} = 0$, or

$$0 = \int_{-\infty}^{\infty} k(x) n_t(x,t)\,dx = \int_{-\infty}^{\infty} k(x)\left(v(t)\frac{\partial n}{\partial x} + (1-n)f(x) - ng(x)\right) dx. \tag{15.40}$$

We solve this for $v(t)$ to get

$$-v(t) = \frac{\int_{-\infty}^{\infty} k(x)\left((1-n)f(x) - ng(x)\right) dx}{\int_{-\infty}^{\infty} k(x)\frac{\partial n}{\partial x}\,dx}. \tag{15.41}$$

Thus, for the tension to remain constant, the partial differential equation (15.21) must have the contraction velocity specified by (15.41).

Using a slightly different approach, Podolsky et al. (1969; Civan and Podolsky, 1966) showed that the Huxley model does not agree with experimental data in its response to a step change in tension. We saw in Fig. 15.10 that immediately after the tension reduction the muscle length changes also, and after an initial oscillatory period, the muscle contracts with a constant velocity. However, the Huxley model does not show any oscillatory behavior, the approach to constant velocity being monotonic.

Motivated by this discrepancy, Podolsky and Nolan (1972, 1973) and Podolsky et al. (1969) altered the form of the functions f and g to obtain the required oscillatory responses in the Huxley model. Of course, in Huxley's original model no physiological justification was given for the functions f and g, and modification of these functions is therefore an obvious place to start fiddling with the model to fit the data. Julian (1969) also showed that the Huxley model can be adjusted to give the correct responses to a step change in length. The details of these analyses do not concern us greatly; the main point is that Huxley's crossbridge model has enough flexibility to explain a wide array of experimental data.

15.3.2 Other Choices for Rate Functions

The simple choices for rate functions made by Huxley yield interesting analytical results. However, there are numerous other ways that the rate functions might be chosen. Suppose, for simplicity, that the rate functions $f(x)$ and $g(x)$ have nonoverlapping compact support. Where it is nonzero, we take $f(x)$ to be a constant, $f(x) = \alpha/\epsilon$, on a small interval near the maximum displacement h, say $h - \epsilon \le x \le h$, so that actin and myosin bind rapidly in a small interval near h. We expect that α depends on the local Ca^{2+} concentration. On the support of $f(x)$, $n(x) = 1 - \exp[\alpha(x-h)/(\epsilon v)]$.

The role of $g(x)$ is to break crossbridge bonds. A simple way to accomplish this is to assume that all bonds break at exactly $x = \delta < 0$, in which case

$$n(x) = \begin{cases} 1 - e^{\alpha(x-h)/\epsilon v}, & h - \epsilon \le x \le h, \\ (1 - e^{-\alpha/v}), & \delta \le x \le h - \epsilon, \\ 0, & \text{elsewhere}. \end{cases} \tag{15.42}$$

It is left as an exercise (Exercise 6) to show that in the limit as $\epsilon \to 0$ and $\delta \to 0$ the force–velocity curve for this model with a linear restoring force does not produce zero force at some positive velocity, a feature that appears in the Hill force–velocity curve.

A second option is to suppose that bonds break when $x < 0$, and to take $g(x) = \kappa/(\delta - x)$ on $\delta < x < 0$. Note that the rate of bond breakage is infinite at $x = \delta$, and thus all crossbridges are dissociated for $x < \delta$. Then

$$n(x) = \begin{cases} 1 - e^{\alpha(x-h)/\epsilon v}, & h - \epsilon \le x \le h, \\ 1 - e^{-\alpha/v}, & 0 \le x \le h - \epsilon, \\ (1 - e^{-\alpha/v})\left(1 - \frac{x}{\delta}\right)^{-\kappa/v}, & \delta \le x < 0, \\ 0, & \text{elsewhere}, \end{cases} \tag{15.43}$$

in which case $n(x)$ is a continuous function of x with compact support. Notice that this works only if $v > 0$. The model must be modified if $v < 0$ (See Exercise 9).

Another way to determine the functions f and g is to estimate the energy of the bond as a function of position, and then from Eyring rate theory to determine the rates of reaction of binding and unbinding. This is the approach followed, for example, by Pate (1997). In fact, now that the biochemistry of the crossbridge reactions is known, fairly sophisticated models of this type are possible (Marland, 1998). Other versions of the basic scheme have been constructed by Pate and Cook (1989, 1991), while the models of T.L. Hill and his colleagues (Hill 1974, 1975; Eisenberg and Hill, 1978; Eisenberg and Greene, 1980) have combined crossbridge models with detailed studies of the biochemical thermodynamics of the crossbridge cycle.

15.4 Determination of the Rate Functions

So far we have seen that an ad hoc approach to the determination of the functions f, g, and r can generate models that agree in varying degrees with experimental data. Obviously, it is desirable to find some way in which these functions can be determined more systematically, for example, to guarantee the correct form of the force–velocity curve. One way that this can be accomplished is by using a slightly different model of the crossbridge dynamics (Lacker and Peskin, 1986; Peskin, 1975, 1976).

15.4.1 A Continuous Binding Site Model

Recall that the Huxley model was based on the assumption that the actin binding sites are sufficiently separated so that each crossbridge can be associated with a unique binding site. Thus, even when a crossbridge is unbound, the distance of the crossbridge to the nearest binding site is defined. We now make the opposite assumption, that the actin binding sites are continuously distributed, so that myosin can bind anywhere along a thin filament. By analogy, one can think of the thin filament as flypaper, to which the myosin heads stick wherever they touch down. In this case, the variable x denotes the distance between the crossbridge anchor and the binding position, and the crossbridge distribution is described by a function $n(x,t)$ such that $\int_a^b n(x,t)\,dx$ is the fraction of crossbridges (at time t) that are attached with distance to the binding site x in the range $[a,b]$. Note that in this formulation, an unbound crossbridge cannot be associated with a value of x, as x is meaningful only for a bound crossbridge. The total fraction of bound crossbridges is

$$N = \int_{-\infty}^{\infty} n(x,t)\,dx < 1, \tag{15.44}$$

and the total fraction of unbound crossbridges is $1 - N$.

To derive the differential equation for n, we consider the conservation of crossbridges with x in the interval $[a,b]$. Let P denote the pool of crossbridges that are bound with $x \in [a,b]$. If the muscle is contracting at velocity $v > 0$, crossbridges move out of P at the rate $vn(a,t)$ and move into P at the rate $vn(b,t)$. Further, if f is defined such that $\int_a^b f(x)\,dx$ is the rate at which new crossbridges are formed with $x \in [a,b]$, and if $g(x)$ denotes the rate at which crossbridges with displacement x detach, then the rate of change of crossbridges is

$$\frac{d}{dt}\int_a^b n(s,t)\,ds = v[n(b,t) - n(a,t)] + (1-N)\int_a^b f(s)\,ds - \int_a^b g(s)n(s,t)\,ds. \tag{15.45}$$

Writing $v[n(b,t) - n(a,t)]$ as the integral $v\int_a^b \frac{\partial n}{\partial x}\,dx$, and noting that since a and b are arbitrary, the integrals can be dropped, we obtain

$$\frac{\partial n}{\partial t} - v(t)\frac{\partial n}{\partial x} = (1-N)f(x) - ng(x). \tag{15.46}$$

Note that in this derivation we assume that the rate of crossbridge attachment is proportional to the fraction of unattached crossbridges, $1 - N$.

The equations for the continuous binding site model are similar to those of the Huxley model, the differences being, first, that the rate of crossbridge attachment is given by $(1-N)f$ in the continuous binding site model and $(1-n)f$ in the Huxley model, and, second, that n and f have different units in the two models. In the Huxley model n is dimensionless, while in the continuous binding site model n has dimension of length^{-1}. Similarly, f has dimension of time^{-1} in the Huxley model and dimension of length^{-1} time^{-1} in the continuous binding site model.

In the following discussion we restrict our attention to a simplified version of the continuous binding site model in which all crossbridges attach at some preferred displacement, say, $x = h$. In this case $f(x) = F\delta(x - h)$, where F is the rate of crossbridge attachment. For this choice of f it is most convenient to rewrite the differential equation to incorporate crossbridge attachment as a boundary condition. We do this by integrating (15.46) from $h - \epsilon$ to $h + \epsilon$ and letting $\epsilon \to 0$. The jump in n at $x = h$ is then given by $F(1 - N)/v$, and so (15.46) can be written as

$$\frac{\partial n}{\partial t} - v(t)\frac{\partial n}{\partial x} = -ng(x), \qquad x < h, \tag{15.47}$$

$$n(h,t) = \frac{F(1-N)}{v}. \tag{15.48}$$

Although in general, N and v are functions of t, we consider only those cases in which they are constant. However, N and n are also functions of v, and we sometimes write $N(v)$ and $n(x,t;v)$ to emphasize this dependence.

15.4.2 A General Binding Site Model

Both the continuous binding site model and the Huxley model can be derived as limiting cases of a more general model (Peskin, 1975). Suppose that on the thin filament there are a discrete number of actin binding sites, with regular spacing Δx (as illustrated in Fig. 15.16). We denote the horizontal distance from the crossbridge (on the thick filament) to the kth binding site by x_k. Finally, let $n_k(t)$ denote the probability that the crossbridge is attached to site k at time t. Then, if $f(x_k)$ and $g(x_k)$ are the rates at which a crossbridge attaches and detaches respectively from the kth site, we have

$$\frac{dn_k(t)}{dt} = f(x_k)\left[1 - \sum_i n_i(t)\right] - g(x_k)n_k(t). \tag{15.49}$$

Note that the rate of crossbridge attachment is proportional to the probability that the crossbridge is not attached, $1 - \sum_i n_i(t)$.

If we now assume that the $n_k(t)$ are samples of a smooth function, so that

$$n_k(t) = n(x_k(t), t), \tag{15.50}$$

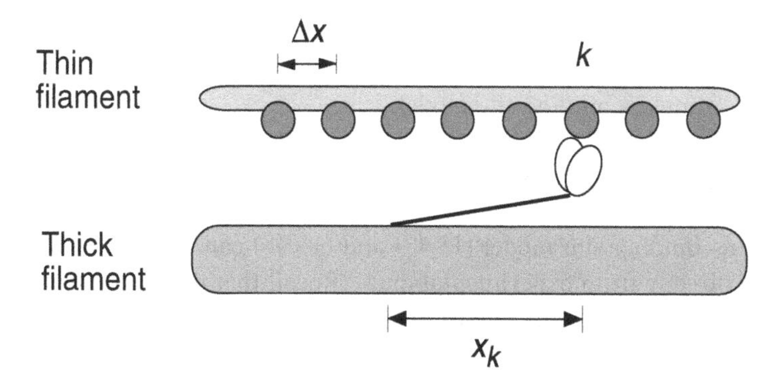

Figure 15.16 Schematic diagram of a crossbridge model with discrete binding sites. The actin binding sites are separated by a distance Δx, and x_k denotes the distance of the kth binding site from the unstressed position of the crossbridge.

it follows that

$$\frac{dn_k}{dt} = -v\frac{\partial n}{\partial x_k} + \frac{\partial n}{\partial t}, \tag{15.51}$$

where, for consistency with our assumption that v is positive for a contracting muscle, we have defined $v = -dx_k/dt$. Substituting (15.51) into (15.49) gives

$$-v\frac{\partial n}{\partial x_k} + \frac{\partial n}{\partial t} = f(x_k)\left[1 - \sum_{i=-\infty}^{\infty} n(x_k + i\Delta x, t)\right] - g(x_k)n(x_k, t). \tag{15.52}$$

Since this holds for any binding site, the subscript k may be omitted, and thus

$$-v\frac{\partial n}{\partial x} + \frac{\partial n}{\partial t} = f(x)\left[1 - \sum_{i=-\infty}^{\infty} n(x + i\Delta x, t)\right] - g(x)n. \tag{15.53}$$

By taking two different limits of (15.53) we obtain the continuous binding site and Huxley models. Suppose first that the binding sites are so widely spaced that at any given time only one is within reach of any crossbridge. This is modeled by assuming that $n(x,t) = 0$ if $|x| > \Delta x/2$. In this case, (15.53) is applicable only on the interval $|x| \leq \Delta x/2$, and with x in this interval, the only nonzero term in the sum in (15.53) is the term corresponding to $i = 0$. Thus (15.53) becomes

$$-v\frac{\partial n}{\partial x} + \frac{\partial n}{\partial t} = f(x)[1 - n(x,t)] - g(x)n(x,t), \tag{15.54}$$

which is the Huxley model.

If, however, we assume that Δx is small and if we let $n = \hat{n}\Delta x, f = \hat{f}\Delta x$, we get

$$-v\frac{\partial \hat{n}}{\partial x} + \frac{\partial \hat{n}}{\partial t} = \hat{f}(x)\left[1 - \sum_{i=-\infty}^{\infty} \hat{n}(x + i\Delta x, t)\Delta x\right] - g(x)\hat{n}. \tag{15.55}$$

In the limit as $\Delta x \to 0$ the sum becomes a Riemann integral, so that

$$-v\frac{\partial \hat{n}}{\partial x} + \frac{\partial \hat{n}}{\partial t} = \hat{f}(x)\left[1 - \int_{-\infty}^{\infty} \hat{n}(s,t)\, ds\right] - g(x)\hat{n}, \tag{15.56}$$

which is the continuous binding site model.

15.4.3 The Inverse Problem

The continuous binding site model (15.47) and (15.48) can be used to determine F, $g(x)$, and $r(x)$ directly from experimental data. (Recall that $r(x)$ is the restoring force generated by a crossbridge with extension x.) The steady-state solution of (15.47) and (15.48) can be written as

$$n(x;v) = \begin{cases} \dfrac{F[1 - N(v)]}{v} \exp\left(\displaystyle\int_x^h \frac{-g(s)}{v}\, ds\right), & x < h, \\ 0, & x > h. \end{cases} \tag{15.57}$$

Integrating (15.57) from $-\infty$ to ∞, we obtain

$$N(v) = \int_{-\infty}^{\infty} n(x; v)\, dx = \frac{F[1 - N(v)]}{v} I(v), \tag{15.58}$$

where

$$I(v) = \int_{-\infty}^{h} \exp\left(\int_{x}^{h} \frac{-g(s)}{v}\, ds \right) dx. \tag{15.59}$$

Thus, we can solve for $N(v)$ as

$$N(v) = \frac{FI(v)}{FI(v) + v}. \tag{15.60}$$

Substituting (15.60) into (15.57) we obtain an explicit solution for n,

$$n(x; v) = \frac{F}{FI(v) + v} \exp\left(\int_{x}^{h} \frac{-g(s)}{v}\, ds \right). \tag{15.61}$$

Since the average force produced by a crossbridge is

$$p(v) = \int_{-\infty}^{\infty} r(x) n(x; v)\, dx, \tag{15.62}$$

it follows that if F, g, and r are known, then (15.61) can be used to find an explicit expression for the force–velocity curve. This is the direct problem that we considered in the context of the Huxley model. Here we want to solve the inverse problem of determining $F, g,$ and r from knowledge of p. However, additional information is needed to do this.

The energy flux during constant contraction can be measured experimentally; in general it is a function of v. If we assume that the energy flux $\phi(v)$ is proportional to the rate at which crossbridges go through the cycle of binding and unbinding to the actin filament, then ϕ is proportional to the crossbridge turnover rate,

$$\phi(v) = \rho\epsilon F(1 - N(v)), \tag{15.63}$$

where ρ is the total number of crossbridges and ϵ is the energy released during each crossbridge cycle. If the fraction of attached crossbridges during isometric tetanus is known, then F can be calculated from

$$\phi_0 = \rho\epsilon F(1 - N_0), \tag{15.64}$$

where $\phi_0 = \phi(0)$ and $N_0 = N(0)$. Next, $I(v)$ can be calculated from $\phi(v)$ by substituting (15.63) into (15.60), which gives

$$I(v) = v\left(\frac{F\rho\epsilon - \phi(v)}{F\phi(v)} \right) = \frac{\rho\epsilon v[\phi_0 - (1 - N_0)\phi]}{\phi\phi_0}. \tag{15.65}$$

Hence, from experimental knowledge of N_0 and $\phi(v)$ we can calculate explicit expressions for F and $I(v)$.

To find g from $I(v)$, we define the transformation

$$y(x) = \int_x^h g(s)\,ds. \tag{15.66}$$

Since g is positive, y is a monotonic function of x and has an inverse that can be calculated explicitly. Differentiating (15.66) with respect to x, we obtain

$$\frac{dy}{dx} = -g(x), \tag{15.67}$$

from which it follows that

$$x(y) = h - \int_0^y \frac{ds}{\bar{g}(s)}, \tag{15.68}$$

where $\bar{g}$ is defined by $g(x) = g(x(y)) = \bar{g}(y)$, and where we have used the condition $y(h) = 0$, so that $x(0) = h$.

Using these definitions, and also defining $\sigma = 1/v$, we get

$$I(1/\sigma) = \int_{-\infty}^h e^{-\sigma y}\,dx \tag{15.69}$$

$$= \int_{\infty}^0 -e^{-\sigma y}\frac{dy}{g(x)} \tag{15.70}$$

$$= \int_0^\infty \frac{1}{\bar{g}(y)}e^{-\sigma y}\,dy. \tag{15.71}$$

The function $I(1/\sigma)$ is the Laplace transform of $1/\bar{g}(y)$, and so $1/\bar{g}(y)$ is obtained as the inverse Laplace transform of $I(1/\sigma)$. Furthermore, g can be obtained as a function of x, since x is defined as a function of y by (15.68). Thus, for given y we can calculate both $\bar{g}(y)$ and $x(y)$. Since $\bar{g}(y) = g(x(y))$, we thus have a parametric representation for $g(x)$.

An explicit formula for $\bar{g}(y)$ can be obtained by using the inversion formula for Laplace transforms. Thus,

$$\frac{1}{\bar{g}(y)} = \frac{1}{2\pi i}\int_{c-i\infty}^{c+i\infty} I(1/\sigma)e^{\sigma y}\,d\sigma, \tag{15.72}$$

where $c > 0$ is arbitrary.

In a similar way, $r(x)$ can be obtained from the force–velocity curve $p(v)$. It is left as an exercise to show that

$$\frac{\epsilon p(1/\sigma)}{\sigma\phi(1/\sigma)} = \int_0^\infty \frac{\bar{r}(y)}{\bar{g}(y)}e^{-\sigma y}\,dy, \tag{15.73}$$

where $\bar{r}(y) = r(x(y))$. Hence

$$\bar{r}(y) = \frac{\epsilon\bar{g}(y)}{2\pi i}\int_{c-i\infty}^{c+i\infty} \frac{p(1/\sigma)}{\sigma\phi(1/\sigma)}e^{\sigma y}\,d\sigma. \tag{15.74}$$

A Specific Example

The above analysis can be used to calculate $F, g,$ and r to fit the Hill force–velocity curve and energy flux data (also observed by Hill). First, note that the force–velocity equation (15.1) can be written in the form

$$p(v) = \frac{bp_0 - av}{v+b}. \tag{15.75}$$

Second, Hill (1938) observed that at constant rate of contraction, the heat flux $\dot{q}$ generated by a contracting muscle is linear, given by

$$\dot{q} = av + \phi_0, \tag{15.76}$$

where the constant a is the same as in the force–velocity equation, and where ϕ_0 is the energy flux at zero velocity. The energy flux is the sum of two terms: the heat flux and the power used by the muscle. The power of a muscle contracting at speed v is pv (force times velocity), and thus the energy flux, $\phi(v)$, is given by

$$\phi(v) = \dot{q} + pv = \phi_0 + \frac{bv(a+p_0)}{v+b}. \tag{15.77}$$

Substituting the expression for ϕ into (15.65) and using (15.72), we find that

$$\frac{1}{\bar{g}(y)} = \frac{\rho\epsilon}{\phi_0}\frac{1}{2\pi i}\int_{c-i\infty}^{c+i\infty}\left[\frac{\sigma_+ + N_0(\sigma-\sigma_*)}{\sigma(\sigma-\sigma_*)}\right]e^{\sigma y}\,d\sigma, \tag{15.78}$$

and

$$\bar{r}(y) = \frac{\epsilon\bar{g}(y)}{2\pi i}\int_{c-i\infty}^{c+i\infty}\left[\frac{p_0\sigma - a/b}{\sigma(\sigma-\sigma_*)}\right]e^{\sigma y}\,d\sigma, \tag{15.79}$$

where $\sigma_+ = -(a+p_0)/\phi_0$ and $\sigma_* = -1/b + \sigma_+$. These integrals can be evaluated using the contour Γ shown in Fig. 15.17.

From the residue theorem, we know that the integral around Γ is the sum of the residues inside the contour. Further, it is not difficult to see that the integral over the semicircular part of the contour goes to zero as the radius of the semicircle becomes infinite. Hence

$$2\pi i\sum\text{residues} = \int_\Gamma = \int_{c-i\infty}^{c+i\infty}. \tag{15.80}$$

Both of the integrals (15.78) and (15.79) have two simple poles inside Γ, one at $\sigma = 0$, the other at $\sigma = \sigma_*$. For the integral (15.78),

$$\text{the pole at } \sigma = 0 \text{ has residue } N_0 - \sigma_+/\sigma_*; \tag{15.81}$$

$$\text{the pole at } \sigma = \sigma_* \text{ has residue } \sigma_+e^{\sigma_* y}/\sigma_*; \tag{15.82}$$

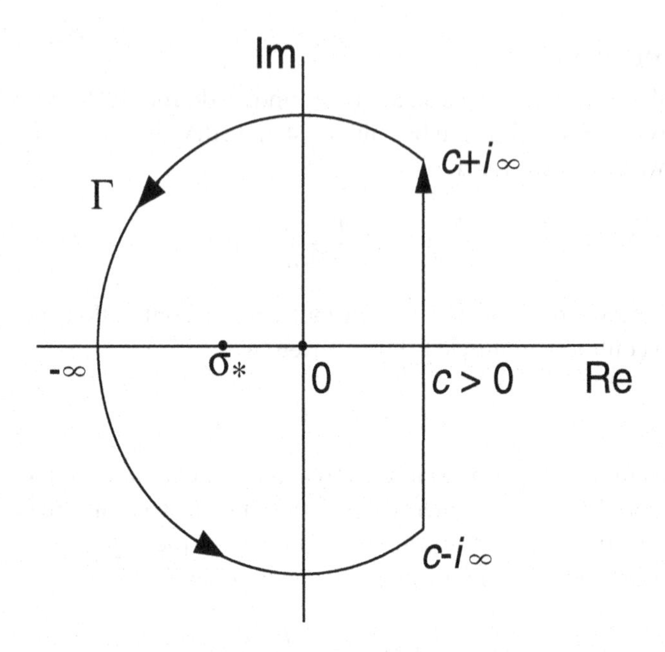

Figure 15.17 Contour for the evaluation of the path integral in the continuous binding site model. (Adapted from Lacker and Peskin, 1986, Fig. 6.)

while for the integral (15.79),

$$\text{the pole at } \sigma = 0 \text{ has residue } \frac{a}{b\sigma_*}; \tag{15.83}$$

$$\text{the pole at } \sigma = \sigma_* \text{ has residue } \left(p_0 - \frac{a}{b\sigma_*}\right) e^{\sigma_* y}. \tag{15.84}$$

Adding these residues for each integral gives, finally,

$$\frac{1}{\bar{g}(y)} = \frac{\rho\epsilon}{\phi_0}\left[N_0 + \frac{\sigma_+}{\sigma_*}(e^{\sigma_* y} - 1)\right], \tag{15.85}$$

$$\bar{r}(y) = \frac{\epsilon\bar{g}(y)}{\phi_0}\left[\frac{a}{b\sigma_*} + \left(p_0 - \frac{a}{b\sigma_*}\right)e^{\sigma_* y}\right]. \tag{15.86}$$

To calculate $x(y)$, use (15.68) from which it follows that

$$x(y) = h - \frac{\rho\epsilon}{\phi_0}\left[\left(N_0 - \frac{\sigma_+}{\sigma_*}\right)y + \frac{\sigma_+}{\sigma_*^2}(e^{\sigma_* y} - 1)\right]. \tag{15.87}$$

This gives a parametric definition of $g(x)$ and $r(x)$.

Finally, we note one important feature of this model. Each crossbridge exerts zero force at some value of $y = y_0$ such that $\bar{r}(y_0) = 0$. Solving for y_0 gives

$$y_0 = \frac{1}{\sigma_*}\ln\left(\frac{a}{a - p_0 b\sigma_*}\right). \tag{15.88}$$

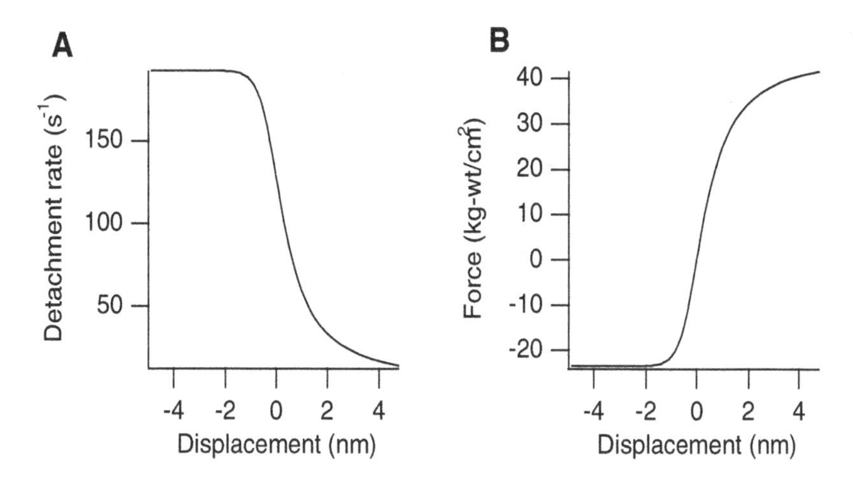

Figure 15.18 Crossbridge detachment rate (A) and force (B) in the continuous binding site model, calculated from (15.85), (15.86), and (15.87), using parameter values $p_0 = 3$ kg-wt/cm^2, $a/p_0 = 0.25$, $b = 0.325$ muscle lengths per second, $\phi_0 = ab$, $F = 125$ s^{-1}, $N_0 = 0.9$, $\phi_0/\rho\epsilon = F(1 - N_0) = 12.5$ s^{-1}. If we require that the crossbridge exerts no force at $x = 0$, then all crossbridges attach at $h = 4.78$ nm, assuming that the length of a half-sarcomere is 1.1 μm.

Hence, if we wish $x = 0$ to correspond to the equilibrium state of the crossbridge (i.e., when it exerts no force) as it was in the Huxley model, h must be chosen so that $x(y_0) = 0$, and thus

$$h = \frac{\rho\epsilon}{\phi_0}\left[\left(N_0 - \frac{\sigma_+}{\sigma_*}\right)y_0 + \frac{\sigma_+}{\sigma_*^2}(e^{\sigma_* y_0} - 1)\right]. \tag{15.89}$$

Plots of g and r are shown in Fig. 15.18. From these curves we note that as the displacement of the crossbridge becomes more negative, its probability of detachment increases, but the force it exerts decreases. This allows a high isometric force without a corresponding reduction in the maximum contraction velocity; crossbridges initially exert a large force, but tend not to be carried into the region where they oppose contraction.

15.5 The Discrete Distribution of Binding Sites

The Huxley model assumes that at any one time, each crossbridge has only a single actin binding site available for binding, while the continuous binding site model assumes the opposite, that crossbridges can bind anywhere. However, the real situation is probably something in between these two extremes. Depending on the flexibility of the actin filament, it is probable that each crossbridge has a selection of more than one binding site, but it is unlikely that the binding sites are effectively continuous (cf. Fig. 15.4). T.L. Hill (1974, 1975) has constructed a detailed series of models that treat, with varying degrees of accuracy, the intermediate case when the actin binding sites are distributed discretely but more than one is within reach of a crossbridge at any time. Detailed consideration of models of this type is left for the exercises (Exercises 7 and 8).

15.6 High Time-Resolution Data

All the models discussed so far treat crossbridge binding as a relatively simple phenomenon; either crossbridges are bound or they are not, and there is no consideration of the possibility that each crossbridge might have a number of different bound states. As we have seen, such assumptions do a good job of explaining muscle behavior on the time scale of tens of milliseconds. However, as the development of new experimental techniques allowed the measurement of muscle length and tension on much shorter time scales, the initial models were improved to take this high time-resolution data into account. One of the first models to do so was that of Huxley and Simmons (1971). The Huxley–Simmons model is quite different from models discussed above, giving a detailed description of how the force exerted by an *attached* crossbridge can vary with time over a short period, but it does not take into account the kinetics of crossbridge binding and unbinding to the thin filament.

15.6.1 High Time-Resolution Experiments

As we have already seen (Fig. 15.11B), when muscle length is decreased, the tension immediately decreases, and then, over a time period of 100 milliseconds or so, recovers to its original level. When this tension recovery is measured at a higher time resolution, it becomes apparent that there are two components of the recovery (Fig. 15.19). The initial drop in tension (which occurs simultaneously with the change in length) is followed by a rapid, partial recovery, followed in turn by a much slower complete recovery to the original tension. The slower recovery process is the one described by the other models discussed in this chapter. Typical experimental results are shown in Fig. 15.20. In this figure, T_1 denotes the value of the tension after the initial drop, and T_2 denotes the value of the tension after the initial rapid recovery. For length increases ($y > 0$) T_1 is a linear function of the change in length, while for length decreases ($y < 0$), the decrease in tension is less than might be expected from the linear relation. It is likely that because the length step is not instantaneous but takes about a millisecond to complete, when a larger length decrease is applied, the rapid recovery process has already begun to take effect by the time the length decrease has been completed. If this is true (and it appears plausible from the curves shown in Fig. 15.19), T_1 would be consistently overestimated for larger, negative, y. From the linearity of the curve for $y > 0$, it is reasonable to suppose that the relationship between T_1 and y is linear over the entire range of y, as denoted by the dashed line in the figure.

In contrast, T_2 is clearly a nonlinear function of y. For small length changes, the rapid process restores the tension to its original level, but for steps of larger length, the rapid process results in only partial recovery. The time course of the rapid recovery has

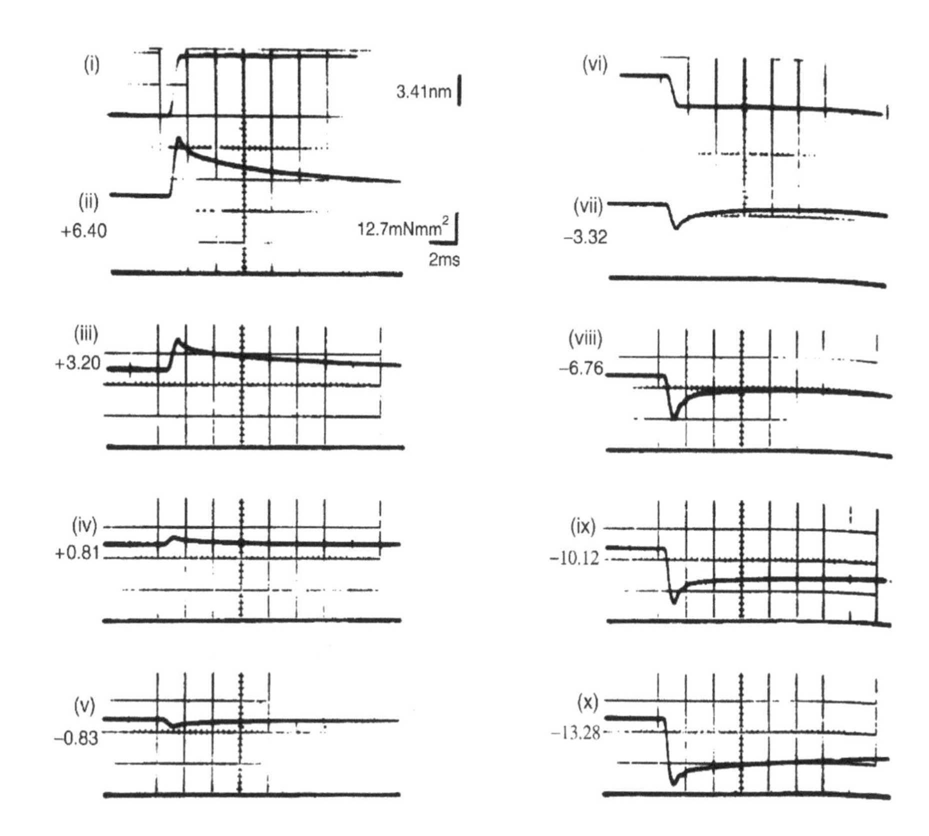

Figure 15.19 Changes in tension after a sudden stretch (ii–iv) or a sudden shortening (v, vii–x). Traces (i) and (vi) show the time course of the length change for traces (ii) and (vii), respectively. The number to the left of each record denotes the amount of the length change (in nm) per half-sarcomere. Note the high time-resolution of the measurements. (Huxley and Simmons, 1971, Fig. 2.)

a dominant rate constant, r, which is well fit by the function

$$r = \frac{r_0}{2}\left(1 + e^{-\alpha y}\right), \tag{15.90}$$

with $r_0 = 0.4$ and $\alpha = 0.5$.

15.6.2 The Model Equations

To model and give a possible explanation of the above results, we assume that a crossbridge consists of two parts: an elastic arm connected to a rotating head that can bind to the actin filament in two different configurations. Recall that in the Huxley model, tension was generated by the crossbridge length, measured as the distance from the base of the crossbridge to the binding site. Here because there are two different possible binding configurations, the tension generated by the crossbridge depends not only on the distance from the base to the binding site, but also on thé binding configuration,

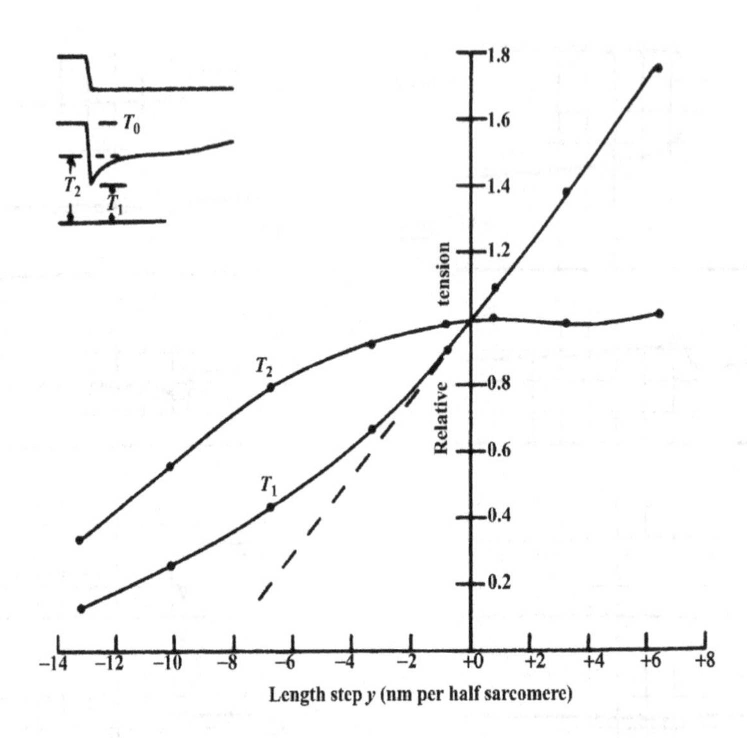

Figure 15.20 Curves of T_1 and T_2 as functions of the length step. As depicted in the inset, T_1 is the minimal tension reached during the step, while T_2 is the value of the tension reached after the quick recovery phase. The upper trace of the inset depicts the time course of the length change. (Huxley and Simmons, 1971, Fig. 3.)

with different configurations resulting in different total crossbridge lengths. The idea behind this model is that the rapid process to restore tension can be accomplished by a change of the binding configuration, without changing the overall distribution of bound and unbound crossbridges.

As illustrated in Fig. 15.21, the head of the crossbridge is assumed to contain three combining sites, M_1, M_2, M_3, each of which has the ability to bind to a corresponding site, A_1, A_2, A_3, on the actin filament. (To avoid confusion with previous terminology, the Ms and As are called combining sites, rather than binding sites.) The affinity between the combining sites is greatest for M_3A_3, and smallest for M_1A_1. As the head of the crossbridge rotates in the direction of increasing θ, it moves through the sequence of binding configurations, M_1A_1 only, M_1A_1 and M_2A_2, M_2A_2 only, M_2A_2 and M_3A_3, M_3A_3 only. During this progression the crossbridge arm is extended, and thus tension is increased. The two stable configurations of the crossbridge are those in which two consecutive combining sites are attached simultaneously. Because the binding affinity is greater for M_3A_3, the energetically most favorable position for the crossbridge head is for M_2 and M_3 to be bound to A_2 and A_3 simultaneously.

An intuitive explanation of the behavior of this model is as follows. At steady state there is a balance between the tension on the crossbridge arm and the force exerted

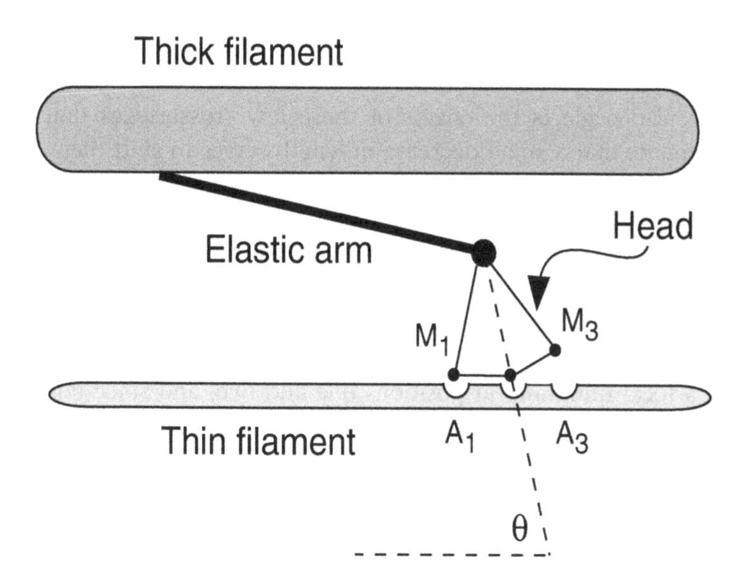

Figure 15.21 Schematic diagram of the Huxley-Simmons crossbridge model. (Adapted from Huxley and Simmons, 1971, Fig. 5).

by the head of the crossbridge. The crossbridge head, in trying to rotate to a position of lower energy, places the elastic crossbridge arm under tension, and so the crossbridge arm, in turn, exerts a contractile force on the muscle. When the muscle is held at a constant length, the sum of all the crossbridge contributions gives the isometric force. If the length of the muscle is suddenly reduced, the tension on the crossbridge arm is suddenly reduced also, and this causes the instantaneous drop in tension seen experimentally. However, over the next few milliseconds, the reduced tension on the arm allows the crossbridge head to rotate to an energetically more favorable position, thus restoring the tension on the arm, and consequently restoring the muscle tension. Hence, the instantaneous drop in tension results from the fact that the elastic crossbridge arm responds instantaneously to a change in length, while the time course of the tension recovery is governed by how fast the crossbridge head rotates, which is, in turn, governed by the kinetics of attachment and detachment of the combining sites.

It is important to note that this description relies on the assumption that during isometric tetanus, all crossbridges have a positive displacement, i.e., $x > 0$. Otherwise, if some crossbridges had $x < 0$, a shortening of the muscle fiber would *increase* the force exerted by these crossbridges, in conflict with the above interpretation. However, in the models discussed so far, this is the case. For example, in the Huxley model the isometric tetanus solution is

$$n(x) = \frac{f(x)}{f(x) + g(x)}, \tag{15.91}$$

which is zero when $f(x) = 0$. Since f is nonzero only when $x > 0$, it follows that all crossbridges have a positive displacement during isometric tetanus. Although this is the case for the Huxley model, this is not necessarily true in experimental situations or in all models.

The model also neglects the effects of those few crossbridges that have such small displacements that a small decrease in length serves to shift them to negative displacements. However, the quantitative effects of this neglect are likely to be small.

To express the model mathematically, we construct a potential-energy diagram for the crossbridge. Recall that the crossbridge head has two stable configurations, one when M_1A_1 and M_2A_2 bonds exist simultaneously, which we denote as position one, and the other when M_2A_2 and M_3A_3 bonds exist simultaneously which we denote as position two. Because these configurations are stable, the potential energy of the crossbridge head reaches a local minimum at positions one and two, and since position two is energetically favored over position one, it has a lower potential energy (see Fig. 15.22A).

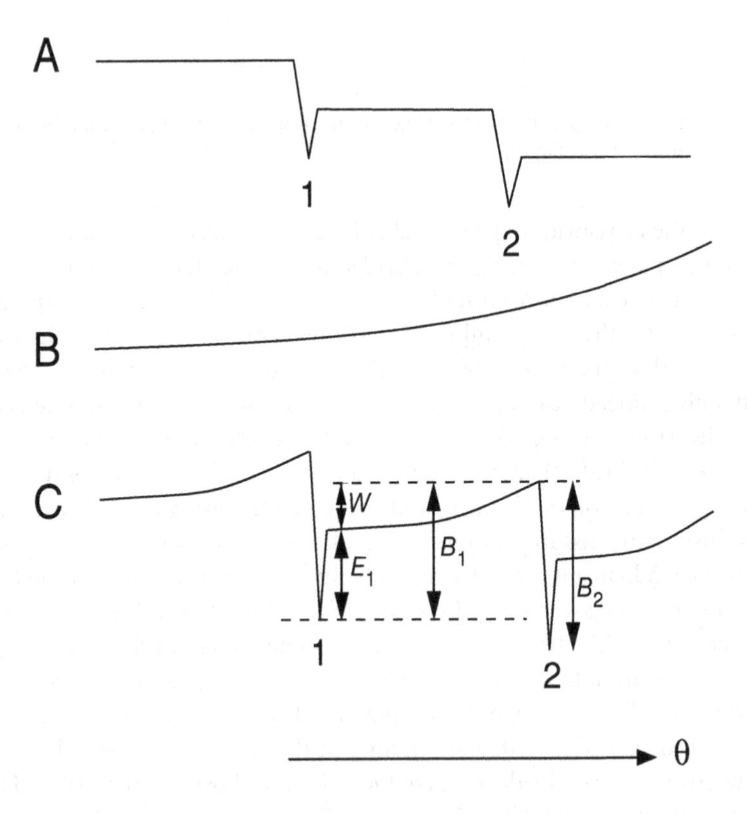

Figure 15.22 A: Potential energy of the crossbridge head. As the head rotates, with increasing θ the combining sites bind consecutively. The potential energy decreases overall from left to right, as it is assumed that M_3 and A_3 have the highest affinity. The two local minima correspond to stable configurations when two consecutive combining sites are bound simultaneously; these are called position one and position two. B: Potential energy due to the elastic energy of the crossbridge arm. C: Total potential energy of the crossbridge, showing the notation used in the model.

However, as the head rotates from position one to position two, the crossbridge arm is extended, which increases the total potential energy of the crossbridge. Thus, adding the potential energy of the crossbridge arm (Fig. 15.22B) to the potential energy of the crossbridge head, we get the total potential energy of the crossbridge (Fig. 15.22C).

Now let n_1 and $n_2 = 1 - n_1$ denote the fraction of crossbridges in positions one and two, respectively, and let y denote the displacement (i.e., length change) of the thick filament relative to the thin filament to be applied. Thus, $y = 0$ corresponds to the steady state before any length change is applied (i.e., the isometric case). Also, let y_1 and y_2 denote the lengths of the crossbridge arm when the head is at positions one and two respectively, before the length change is applied. Let $y_0 = (y_1 + y_2)/2$ (the midway position), and let $h = y_2 - y_1$. Finally, let F_1 and F_2 denote the tension in the crossbridge arm when the head is in positions one and two, respectively. Then, after the length change is applied,

$$F_1 = K(y + y_1) = K(y + y_0 - h/2), \qquad F_2 = K(y + y_2) = K(y + y_0 + h/2), \tag{15.92}$$

where K is the stiffness of the crossbridge arm, assumed to follow Hooke's law. Hence, the average tension, ϕ, on a crossbridge arm is given by

$$\phi = n_1 F_1 + n_2 F_2 = K(y + y_0 - h/2 + hn_2). \tag{15.93}$$

As the crossbridge head moves from position one to position two, the extending crossbridge arm does work, exerting an average force of approximately $(F_1+F_2)/2$ over a distance h. Thus the work, W, is given by

$$W = h\frac{F_1 + F_2}{2} = Kh(y + y_0). \tag{15.94}$$

Now, suppose that a crossbridge moves from position one to position two at kinetic rate k_+ and moves in the opposite direction at kinetic rate k_-. As with barrier models of the ionic current through a membrane channel (Chapter 3), we assume that each of these rates is an exponential function of the height of the potential-energy barrier that the crossbridge must cross in order to jump from one combining configuration to the other. With this assumption, $k_+ = \exp(\frac{-B_1}{kT})$ and $k_- = \exp(\frac{-B_2}{kT})$, where B_1 and B_2 are the barrier heights to move from configuration 1 to 2, and 2 to 1, respectively, T is the absolute temperature and k is Boltzmann's constant. Then,

$$\begin{aligned} \frac{k_+}{k_-} &= \exp\left(\frac{B_2 - B_1}{kT}\right) \\ &= \exp\left(\frac{B_2 - E_1 - W}{kT}\right) \\ &= \exp\left(\frac{B_2 - E_1 - Kh(y + y_0)}{kT}\right) \\ &= A^0 \exp\left(\frac{-Khy}{kT}\right), \end{aligned} \tag{15.95}$$

where $A^0 = \exp\left(\frac{B_2 - E_1 - Khy_0}{kT}\right)$.

When the length of the muscle is changed, the crossbridges redistribute themselves among the two configurations according to the differential equation

$$\frac{dn_2}{dt} = k_+ n_1 - k_- n_2 = k_+ - rn_2, \tag{15.96}$$

where

$$r = k_+ + k_- = k_- \left[1 + A^0 \exp\left(\frac{-yKh}{kT} \right) \right]. \tag{15.97}$$

Since r is the time constant for the redistribution of crossbridges, it follows that r is also the time constant for the development of the tension T_2 at the end of the quick recovery. Equation (15.97) is the same as (15.90) if $Kh = \alpha kT$, and $A^0 = 1$, and thus the model shows the correct time course for tension development. At steady state,

$$n_2 = \frac{k_+}{k_+ + k_-} = \frac{1}{2}\left[1 + \tanh\left(\frac{\alpha y}{2} \right) \right], \tag{15.98}$$

where we have used that $k_+/k_- = \exp(-\alpha y)$. Hence, the steady-state tension (which corresponds to T_2) is given by

$$\phi = K(y + y_0 - h/2 + hn_2) = \frac{\alpha kT}{h}\left[y_0 + y - \frac{h}{2} \tanh\left(\frac{\alpha y}{2} \right) \right]. \tag{15.99}$$

A plot of ϕ is given in Fig. 15.23, from which it is seen that the model gives an excellent qualitative description of the experimental data shown in Fig. 15.20.

In summary, although the Huxley–Simmons model does not take later events, such as crossbridge binding and unbinding, into account and is not intended to describe the full tension recovery in the manner of the Huxley model, it nevertheless provides an excellent qualitative description of the initial phase of tension recovery following a step change in length.

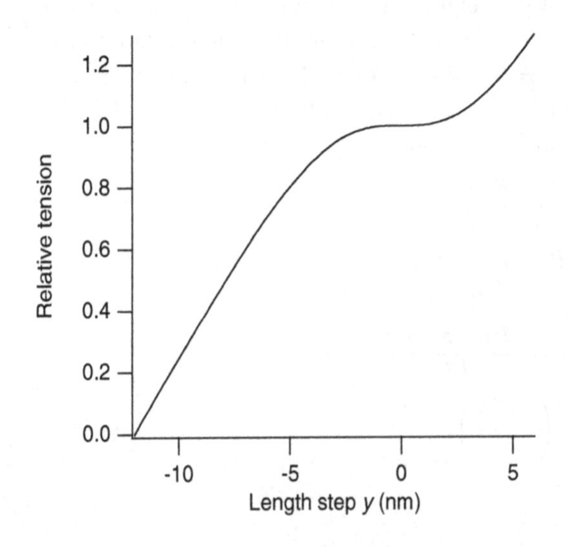

Figure 15.23 Plot of relative steady tension ϕ (15.99) in the Huxley–Simmons model. Parameter values: $\alpha = 0.5$ nm, $y_0 = 8$ nm, $h = 8$ nm. Because we plot relative tension, the numerical value of kT has no effect on the shape of the curve.

15.7 *In Vitro* Assays

Using reconstituted systems of actin and myosin (so-called *in vitro* assays), it is possible to observe single myosin molecules working their way along an actin strand (Sheetz and Spudich, 1983; Spudich et al., 1985; Uyeda et al., 1990), or single kinesin molecules moving along a microtubule (Howard et al., 1989). If a solution of myosin fragments, each containing the motor domain, is put on a nitrocellulose-coated glass surface, some of the individual myosin molecules attach to the surface. Upon addition of a solution of ATP and polymerized actin filaments, the individual actin filaments are grabbed by the myosin motor domains, which then try to move along the filament. Since the myosins are fixed but the actin filament is not, there is a resulting movement of the actin filament. Individual filaments can then be observed gliding over the field of myosin molecules. By careful regulation of the concentration of the myosin fragments, the average number of myosin molecules moving each actin strand can be controlled. Experimental results show that the speed of gliding increases with the number of myosin molecules that are moving the actin filament, and Pate and Cook (1991) have presented a simple analysis to show why this should be so.

Consider the case that the actin filament is moved by a single myosin molecule. For simplicity, assume that the spacing between actin binding sites is exactly the length of the power stroke, i.e., 10 nm, and let h denote the length of the power stroke. The speed of movement is then $h\times$ the mean cycle rate. After the myosin binds and executes a power stroke it can rebind to the actin filament only at a displacement of $x = h$ nm. Following the notation of the Huxley model (Section 15.3), the rate of rebinding is thus $f(h)$ and the rate of detachment is $g(0)$. Thus the mean time to rebind is $1/f(h)$, the mean time to detach is $1/g(0)$, and the mean cycle time is $1/f(h)+1/g(0)$. It follows that the speed of the actin filament, v, is given by

$$v = \frac{h}{\frac{1}{f(h)}+\frac{1}{g(0)}}, \tag{15.100}$$

which, using the parameter values of Brokaw (1976), $f(h) = 65\ \mathrm{s}^{-1} \ll g(0) = 313\ \mathrm{s}^{-1}$, is approximated by $v \approx f(h)h$. Thus, the rate-limiting step is the attachment of the myosin to the actin filament.

It is left as an exercise (Exercise 10) to show that, in the limit as the number of bound crossbridges goes to infinity, the gliding rate is given by

$$v \approx \frac{g_2 h}{2}, \tag{15.101}$$

where g_2 is defined by (15.27). Since, for the Huxley parameters, $f(h) < g_2/2$, it follows that the gliding rate increases as the number of bound crossbridges increases, as is observed experimentally.

15.8 Smooth Muscle

Smooth muscle is quite different, both in biochemistry and function, from skeletal or cardiac muscle. Furthermore, there is an enormous diversity of smooth muscle to match its diversity of function. Smooth muscle is found around blood vessels and airways, and is a major component of the hollow internal organs such as the gastrointestinal system (Chapter 18), the bladder, and the uterus. Unlike skeletal and cardiac muscle, smooth muscle is nonstriated, for although it contains both the actin and myosin necessary for contractile function, these proteins are not arranged in regular arrays.

Like striated muscle, smooth muscle is electrically active, with a variety of ion pumps, channels, and exchangers in the plasma membrane, that can generate action potentials. Furthermore, contraction requires a rise in the cytoplasmic Ca^{2+} concentration. However, in contrast to striated muscle, this rise in Ca^{2+} concentration often occurs via the production of IP_3, and subsequent release of Ca^{2+} from the sarcoplasmic reticulum, as described in Chapter 7.

The most important difference between smooth and striated muscle is the different way in which Ca^{2+} regulates contraction. Recall from earlier in this chapter that in striated muscle the binding of Ca^{2+} to troponin exposes the myosin binding site on the actin filament and allows the crossbridge cycle to generate force. However, smooth muscle contains no troponin. Instead, the crossbridge cycle can occur only when myosin is phosphorylated at a specific site. This phosphorylation is carried out by *myosin light chain kinase* (MLCK), which is activated by Ca^{2+}-calmodulin. Hence, a rise in Ca^{2+} leads to a rise in Ca^{2+}-calmodulin, as Ca^{2+} binds to the calmodulin, leading to increased activation of MLCK, phosphorylation of the myosin, and thus contraction.

Dephosphorylation of the myosin by *myosin light chain phosphatase* (MLCP) prevents the crossbridge cycle from occurring. However, this does not necessarily lead to relaxation of the muscle. If dephosphorylation occurs while the myosin is bound to the actin, the crossbridge enters the so-called *latch state*. While in this state, the myosin can unbind from the actin only very slowly; thus, no active force can be generated (since the crossbridge cycle cannot occur) but neither can the crossbridge relax, as the myosin remains bound. Thus, if all the crossbridges were in the latch state, the muscle would be rigid and unable to relax, but would require little energy to stay in this state. Thus, smooth muscle is able to sustain a load while using little ATP, which allows for the maintenance of organ shape and dimension without undue use of energy. The smooth muscle crossbridge cycle is illustrated in Fig. 15.24.

15.8.1 The Hai–Murphy Model

Most quantitative studies of smooth muscle are based on the work of Hai and Murphy (1989a,b; Murphy, 1994), who assumed that a crossbridge could exist in one of four forms: myosin either unattached or attached, either phosphorylated or not (Fig. 15.25).

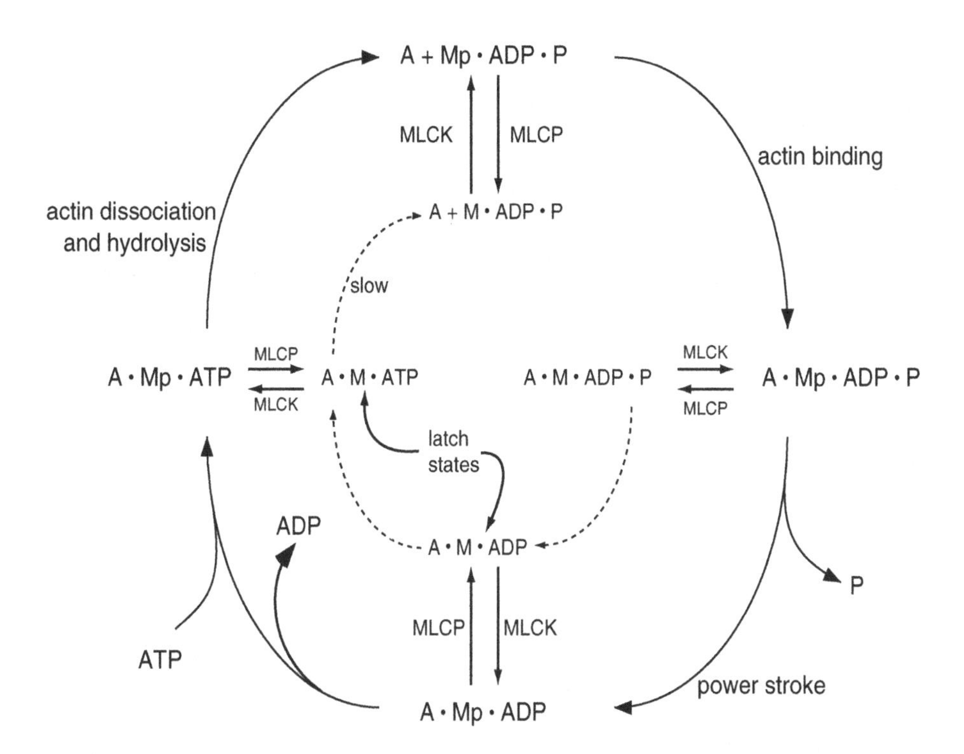

Figure 15.24 Schematic diagram of the eight-state binding model of smooth muscle. The outer ring of solid lines corresponds to the usual crossbridge cycle, but, to enter this cycle, myosin (M) must be in the phosphorylated state (Mp). The inner ring of dotted lines is incomplete, since unphosphorylated myosin cannot bind to actin. MLCK is myosin light chain kinase, MLCP is myosin light chain phosphatase. When the crossbridge is in the latch state it can neither relax quickly nor generate active force.

The rate of crossbridge phosphorylation by MLCK is denoted by k_1, and is a function of the Ca^{2+} concentration, c. The rate of dephosphorylation was assumed by Hai and Murphy to be constant, although it now seems possible that it also depends on Ca^{2+} concentration as well as the level of agonist stimulation (Wang et al., 2008). In general, unattached and attached myosin (M and AM respectively) could be phosphorylated at different rates ($k_1(c)$ and $k_5(c)$ respectively), although it is often assumed that these two rates are the same.

Mijailovich et al. (2000) extended the original Hai–Murphy model by including dependence on the crossbridge extension, in the same way as in the Huxley model (Section 15.3). Thus, the rates of attachment and detachment of phosphorylated myosin are $f(x)$ and $g(x)$ respectively, where f and g have the same shape as shown in Fig. 15.13.

However, unphosphorylated myosin cannot attach to actin, and detaches only slowly. Hence, k_7 is smaller than $g(x)$, and there is no reverse reaction.

Because k_7 is small, it is possible that some bound crossbridges experience large negative displacements. This introduces some subtleties into the model equations.

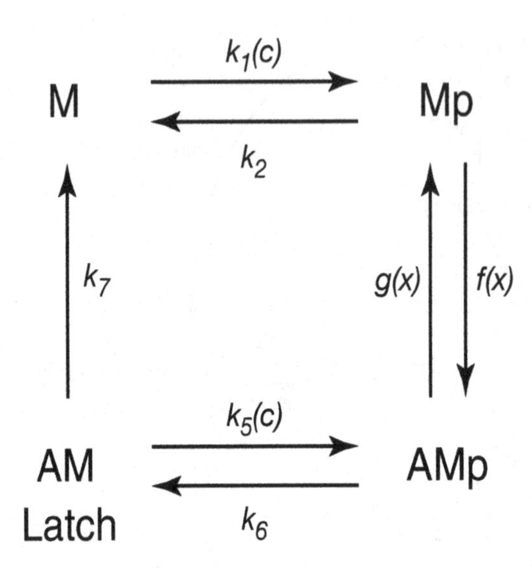

Figure 15.25 Schematic diagram of the Hai–Murphy model of smooth muscle (Hai and Murphy, 1989a,b). M – myosin; Mp – phosphorylated myosin; AMp – attached phosphorylated myosin; AM – attached myosin. The rate k_1 denotes the rate of activated myosin light chain kinase (MLCK), and is a function of the Ca^{2+} concentration (denoted by c), via Ca^{2+}-calmodulin. As in the Huxley model, x is crossbridge displacement.

Following the Huxley model, we let x be the crossbridge displacement. Then, using standard arguments, the evolution of the density of bound crossbridges is given by

$$\frac{\partial N_{\rm AMp}}{\partial t} - v(t)\frac{\partial N_{\rm AMp}}{\partial x} = k_5(c)N_{\rm AM} + f(x)N_{\rm Mp} - (k_6 + g(x))N_{\rm AMp}, \tag{15.102}$$

$$\frac{\partial N_{\rm AM}}{\partial t} - v(t)\frac{\partial N_{\rm AM}}{\partial x} = k_6 N_{\rm AMp} - (k_5(c) + k_7)N_{\rm AM}. \tag{15.103}$$

For an unbound crossbridge, the displacement is the distance to the binding site with which it could bind. Without loss of generality, we assume that $f(x)$ is nonzero only on the interval $0 < x < h$ and that Δx, the distance between binding sites, is greater than h. Thus, $N_{\rm M}$ and $N_{\rm Mp}$, the density of crossbridges in states M and Mp, respectively, which are unbound, are defined only for $0 < x < \Delta x$. However, with $v > 0$, it is possible for a bound crossbridge to have displacement anywhere in the interval $-\infty < x < h$. Furthermore, when a bound crossbridge with negative displacement becomes unbound, it becomes an unbound crossbridge with displacement $x + i\Delta x$, where i is the smallest integer for which $x + i\Delta x > 0$. A similar argument holds for $v < 0$ but with positive displacements greater than Δx. Thus, the evolution of $N_{\rm M}$ is governed by the partial differential equation

$$\frac{\partial N_{\rm M}}{\partial t} - v(t)\frac{\partial N_{\rm M}}{\partial x} = k_2 N_{\rm Mp} - k_1(c)N_{\rm M} + k_7 \sum_{i=-\infty}^{\infty} N_{\rm AM}(x - i\Delta x, t), \tag{15.104}$$

and similarly for $N_{\rm M}$,

$$\frac{\partial N_{\rm Mp}}{\partial t} - v(t)\frac{\partial N_{\rm Mp}}{\partial x} = k_1(c)N_{\rm M} - (k_2 + f(x))N_{\rm Mp} + \sum_{i=-\infty}^{\infty} g(x - i\Delta x)N_{\rm AMp}(x - i\Delta x, t). \tag{15.105}$$

In addition, the flux of unbound sites with $x = 0$ must match the flux of unbound sites with $x = \Delta x$, so that

$$\frac{\partial N_{\rm M}(0,t)}{\partial x} = \frac{\partial N_{\rm M}(\Delta x,t)}{\partial x}, \qquad \frac{\partial N_{\rm Mp}(0,t)}{\partial x} = \frac{\partial N_{\rm Mp}(\Delta x,t)}{\partial x}. \tag{15.106}$$

If we now evaluate (15.102)–(15.103) at the points $x+i\Delta x$, sum them over all i, and add the result to (15.104)–(15.105) we get the conservation law

$$N_{\rm M}(x,t) + N_{\rm Mp}(x,t) + \sum_{i=-\infty}^{\infty} \left[N_{\rm AMp}(x+i\Delta x,t) + N_{\rm AM}(x+i\Delta x,t)\right] = 1, \tag{15.107}$$

for $0 < x < \Delta x$. Under isometric conditions this reduces to $N_{\rm M}+N_{\rm Mp}+N_{\rm AMp}+N_{\rm AM} = 1$, since with $v = 0$ it is not possible to have bonds outside the range for which f is nonzero.

The Hai–Murphy crossbridge model is connected to models of intracellular Ca^{2+} dynamics via $k_1(c)$ and $k_5(c)$ (Koenigsberger et al., 2004, 2005, 2006; Fajmut et al., 2005; Payne and Stephens, 2005; Bursztyn et al., 2007, Wang et al., 2008).

15.9 Large-Scale Muscle Models

Ultimately, to model the function and properties of muscle *in situ* it is necessary to consider highly complex models that include the action potential, the Ca^{2+} transient, crossbridge and thin filament kinetics, the constitutive properties of muscle, and the consequent force and contraction that results from a train of action potentials. Furthermore, these models should use realistic geometries. Although computing facilities are not yet advanced enough to enable simulation of such detailed models in a realistic time frame, much has already been accomplished. Probably the most detailed such investigations (complementing the highly detailed electrical models of Noble's group, summarized in Noble, 2002a,b) are those from the groups of Peter Hunter and Andrew McCulloch (Guccione and McCulloch, 1993; Guccione et al., 1993; McCulloch, 1995; Hunter, 1995; Costa et al., 1996a,b; Hunter et al., 1998, 2003; Niederer et al., 2006). There is far too little space here to discuss these elaborate finite-element models in detail, so the interested reader is referred to the original works. A different approach, using the immersed boundary method rather than finite elements, is taken by Peskin and McQueen (1989, 1992; Peskin, 2002), but again there is insufficient space here to discuss this method.

15.10 Molecular Motors

In all the models described so far, the properties of the muscle result from the collective behavior of a large population of myosin molecules. However, there are many situations in which movement is achieved by many individual molecules working independently. For example, since cellular components are synthesized at sites (the Golgi apparatus, the endoplasmic reticulum, and the nucleus, for example) that are far removed from the place where those components are used (the cell membrane, the synapse, etc.) every cell relies on active transport mechanisms to distribute various cellular components.

In recent years, a lot of attention has been paid to understanding how chemical energy can be transformed into work. Of course, myosin is but one example of these molecular machines, or motors. Examples that we have seen already in this text are the ATPase ion pumps and exchangers, such as the Na^+–K^+ ATPase and the SERCA pumps. These motors reside in the cell membrane and use the chemical energy of ATP to pump ions against their gradient, thereby producing work. In the opposite direction, the best-known example is the ATP synthase molecule, which manufactures ATP from ADP and phosphate by using the energy in a transmembrane proton gradient (Wang and Oster, 1998; Elston et al., 1998; Mogilner et al., 2002).

A different mechanism (described below) is used to transport large macromolecules across a membrane, but because ATP is used, this also can be viewed as the action of a molecular motor.

Other motors are used to transport loads over larger space scales. There are three types of molecular transporters (Mallik and Gross, 2004): the myosin, kinesin and dynein motors. Each of them transports its cargo by binding the cargo to its "tail", while simultaneously stepping along a rail system (actin in the case of the myosin motor, microfilaments in the case of kinesin and dynein), in much the same way that a child swings along monkey bars. The kinesin and dynein motor proteins have two heads that cyclically bind and unbind to the rail track, but out of phase, resulting in directed motion along the track. The necessary energy comes from the hydrolysis of ATP. Movement is directed, with most kinesin-family motors transporting their cargo from the nucleus toward the cell periphery, and most dynein-family motors transporting their cargo in the opposite direction along the same microtubule system. Deterministic models of transport along axons can be found in Blum and Reed (1985, 1989), Blum et al. (1992) and Reed and Blum (1986). Here we consider only stochastic models of such transport.

In what follows we give a brief introduction to how pushing and pulling is accomplished by molecular motors. For consistency with the topics in this chapter, it would seem most reasonable to begin with a description of a myosin motor. However, from the perspective of model building and developing the appropriate mathematical framework, we begin with a description of Brownian ratchets. More detailed introductions are given by Mogilner et al. (2002) and Reimann (2002), while the mathematical theory is developed in detail by Qian (2000).

15.10.1 Brownian Ratchets

Proteins live in a world that is being continually shaken by random movements. If these random motions could be rectified, or biased, then they could be used to perform work. For example, if a particle performs an unbiased random walk on a line, then if there are no outside influences, its mean displacement is zero. However, if every time the particle moves a certain distance in one direction it is prevented by some mechanism from moving back in the opposite direction (i.e., a ratchet), then the original random motion of the particle is rectified, and the particle experiences a net drift.

This concept can be simply illustrated by the way in which a glucose transporter operates (see Chapter 2). In the absence of hexokinase, the transporter is merely an exchanger that reaches steady state when the glucose concentrations are the same on either side of the membrane. However, ATP hydrolysis and phosphorylation of glucose by hexokinase is able rectify the diffusional flux of glucose, eliminating the backflow, thus maintaining a net flux of glucose into the cytoplasm.

Another example is provided by the translocation of macromolecules through a membrane. For example, the chaperone molecule HSP-70 (heat shock protein, molecular weight 70 kilodaltons) is required for translocation of proteins into the ER and mitochondria. In the absence of HSP-70, macromolecules move through pores in the membrane by simple diffusion, moving forward or backward via a random walk. It is not certain how HSP-70 works, but one proposal (Elston, 2000) is that HSP-70, which is located inside the organelle, binds to the diffusing protein as it enters the organelle, preventing it from sliding back through the pore, thus creating a ratchet.

Another important example of a Brownian ratchet is the polymerization ratchet, a diagram of which is shown in Fig. 15.26. An actin filament that is capable of polymerizing is in contact, from time to time, with a particle, P (which may also have an applied load f). Although P is diffusing it cannot move far to the left, since the actin polymer is in the way. However, every so often, the Brownian motion of P takes it far enough to the right, providing the opportunity for another actin monomer to polymerize and extend the actin filament. The net effect is that the motion of P is biased to the right. Clearly, for this to happen a number of conditions must be satisfied: P must move randomly far enough to the right to make space for an actin monomer to join the filament, and it must stay there long enough to give actin time to bind; also, the polymerizing actin must remain strongly bound to the filament so that the process is not reversed, i.e., so

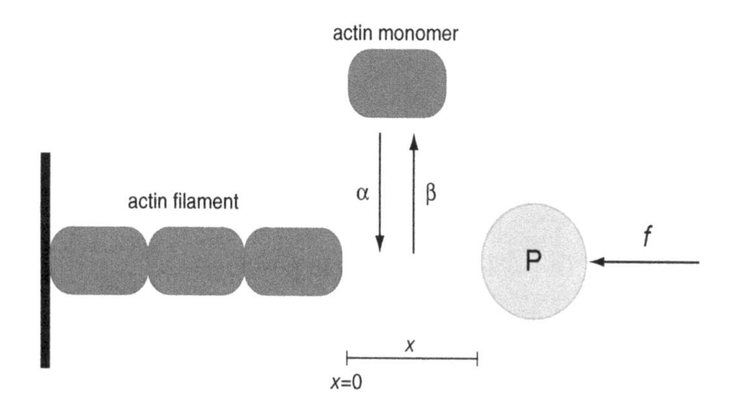

Figure 15.26 Diagram of a polymerization ratchet.

that the collision of P with the actin polymer does not cause the actin at the end to unbind.

We wish to calculate the load–velocity relationship for this system (Peskin et al., 1993). To do so, we let x denote the positive distance between the end of the actin filament and P, and let $p(x,t)$ be the probability density function for the location of the particle P. That is, $\int_a^b p(x,t)\,dx$ is the probability that the particle P lies in the interval $x \in [a,b]$.

There are only two ways that the distance between the end of the polymer and the particle can change: the particle can move to a different position or the length of the actin polymer can change. Correspondingly, $p(x,t)$ can change either because of movement of the particle or because of changes to the length of the polymer. The flux of p due to particle movement is given by

$$J = -D\left(\frac{\partial p}{\partial x} + \frac{f}{kT}p\right), \tag{15.108}$$

where k is Boltzmann's constant, T is absolute temperature, and D is the diffusion coefficient of P. This is the same as the Nernst–Planck relationship for movement of ions in an electric field, where f replaces $qz\frac{\partial\phi}{\partial x}$. Notice that the sign of f is such that it is positive to the left.

Now suppose that monomers bind at rate α and unbind at rate β. Presumably, unbinding can occur independent of x, but when it does so, the distance x is suddenly increased by the amount δ. Similarly, if binding occurs, the distance x is suddenly decreased by the amount δ. However, binding can occur only if there is enough room, i.e., only if $x > \delta$.

The rate of change of p depends on these reactions. If $x < \delta$, the rate of change of p is (in words)

$$\begin{aligned}\text{rate of change of } p = {} & \text{transport rate} \\ & + \text{binding rate at position } x+\delta \\ & - \text{unbinding rate at position } x,\end{aligned} \tag{15.109}$$

whereas, if $x > \delta$,

$$\begin{aligned}\text{rate of change of } p = {} & \text{transport rate} \\ & + \text{binding rate at position } x+\delta \\ & - \text{unbinding rate at position } x \\ & - \text{binding rate at position } x \\ & + \text{unbinding rate at position } x-\delta.\end{aligned} \tag{15.110}$$

In mathematical terms,

$$\frac{\partial p}{\partial t} = D\frac{\partial^2 p}{\partial x^2} + \frac{Df}{kT}\frac{\partial p}{\partial x} + \alpha p(x+\delta,t) - \beta p(x,t), \qquad x < \delta, \tag{15.111}$$

and

$$\frac{\partial p}{\partial t} = D\frac{\partial^2 p}{\partial x^2} + \frac{Df}{kT}\frac{\partial p}{\partial x} + \alpha[p(x+\delta,t) - p(x,t)] + \beta[p(x-\delta,t) - p(x,t)], \qquad x > \delta. \tag{15.112}$$

Since the particle cannot move across the boundary at $x = 0$, we impose the no-flux condition

$$D\frac{\partial p(0,t)}{\partial x} + \frac{Df}{kT}p(0,t) = 0. \tag{15.113}$$

Finally, we require that p be continuously differentiable at $x = \delta$.

Although this is a fairly simple example of a Brownian ratchet, its full solution is highly nontrivial. This is because the steady-state equation is a delay differential equation that must be solved on the infinite half-line, subject to boundary conditions at $x = 0$ and $x = \infty$. However, it is possible to gain an understanding of how this Brownian ratchet works in several limiting cases.

In the limit that α and β are small compared to $\frac{D}{\delta^2}$ (see Exercise 11) we can calculate the steady-state solution for p explicitly. In this case we have

$$\frac{d^2p}{dx^2} + \frac{\omega}{\delta}\frac{dp}{dx} = 0, \qquad x < \delta \text{ and } x > \delta, \tag{15.114}$$

$$\frac{dp(0)}{dx} + \frac{\omega}{\delta}p(0) = 0, \tag{15.115}$$

$$p \text{ continuous at } x = \delta, \tag{15.116}$$

where $\omega = \frac{f\delta}{kT}$. Furthermore, we require that $p \to 0$ as $x \to \infty$. This boundary value problem has solution

$$p = \frac{\omega}{\delta}e^{-\omega x/\delta}, \tag{15.117}$$

so that $\int_0^\infty p(x)\,dx = 1$.

We now can find the mean ratchet velocity, v. The proportion of positions at which a monomer can bind is $\int_\delta^\infty p(x)\,dx$, and thus the mean rate at which the length of the filament increases due to monomer binding is $\delta\alpha\int_\delta^\infty p(x)\,dx$. Similarly, all filaments can decrease in length due to unbinding of a monomer, and thus the mean rate at which the length of the filament decreases due to monomer unbinding is $\delta\beta$. Taking the difference between these two rates gives the net mean ratchet velocity

$$v = \delta\left(\alpha\int_\delta^\infty p(x)\,dx - \beta\right). \tag{15.118}$$

Substituting for p from (15.117) gives the remarkably simple expression

$$v = \delta\left[\alpha\exp\left(-\frac{f\delta}{kT}\right) - \beta\right] \tag{15.119}$$

for the force-velocity relation. The stall force, f_0, i.e., the force at which the ratchet velocity is zero, is given by

$$f_0 = -\frac{kT}{\delta}\ln\left(\frac{\beta}{\alpha}\right). \tag{15.120}$$

Note the similarity of this with the Nernst equation (2.104). Notice also that $\frac{\beta}{\alpha}$ is the equilibrium constant for binding of monomer, $\frac{\beta}{\alpha} = \exp(\frac{\Delta G^0}{RT})$, so that

$$f_0 = -\frac{\Delta G^0}{\delta N_A}. \tag{15.121}$$

Thus, the energy used to move the particle against the load is the same as the free energy of binding of the monomer. Clearly, a necessary condition for this polymerization ratchet to push a load is that the binding rate be larger than the unbinding rate.

We can also easily examine this Brownian ratchet in another limit, the fast reaction limit. We suppose that we have an ideal ratchet for which binding of monomer takes place instantly and irreversibly whenever there is sufficient space to bind, i.e., whenever x reaches δ. Thus, the mean velocity of the ratchet is δ divided by the time it takes to diffuse to the right a distance δ starting from position $x = 0$. This time is the mean first exit time $T(0)$, where $T(x)$ satisfies the differential equation

$$D\left(\frac{d^2T}{dx^2} - \frac{f}{kT}\frac{dT}{dx}\right) = -1 \tag{15.122}$$

(see Section 2.9.6), subject to boundary conditions $T'(0) = 0$ (since the boundary at $x = 0$ is a reflecting boundary) and $T(\delta) = 0$. The solution is (see Exercise 29)

$$T(0) = \frac{\delta^2}{D}\left(\frac{kT}{f\delta}\right)^2\left[\exp\left(\frac{f\delta}{kT}\right) - \frac{f\delta}{kT} - 1\right], \tag{15.123}$$

so that

$$v = \frac{D}{\delta}\frac{\left(\frac{f\delta}{kT}\right)^2}{\exp\left(\frac{f\delta}{kT}\right) - 1 - \frac{f\delta}{kT}}. \tag{15.124}$$

This load–velocity curve is a monotone decreasing function of load f. Notice, however, that for this model there is no stall velocity, because, since the bound monomer is assumed to bind irreversibly, there is the implicit assumption that the binding energy is infinite, which, of course, cannot be correct.

Peskin et al. (1993) applied this model to the study of filopod protrusion, as well as to propulsion of the bacteria *Listeria monocytogenes*, while a number of elaborations of the basic scheme have been constructed (Mogilner and Oster, 1996, 1999; Simon et al., 1992).

15.10.2 The Tilted Potential

There is another approximate model of the Brownian ratchet that has proven to be useful. Rather than thinking of binding and unbinding of a monomer, we suppose that at the fixed positions $x = x_k = k\delta$, the particle P is given a "kick" of energy ΔG, but that it diffuses freely elsewhere, against a load f. Now the probability distribution function $p(x,t)$ for the location of the particle is the solution of the Fokker–Planck equation

$$\frac{\partial p}{\partial t} = \frac{\partial}{\partial x}(V'(x)p) + D\frac{\partial^2 p}{\partial x^2}, \tag{15.125}$$

where $V(x)$ is the potential energy function with

$$V'(x) = \frac{Df}{kT} - D\frac{\Delta G}{kT}\sum_n \delta(x - n\delta). \tag{15.126}$$

The potential $V(x)$ shown plotted in Fig. 15.27 is a piecewise-linear staircase, often called a *tilted ratchet* potential.

One feature of this that is immediately obvious is that the stall load is $f = \Delta G$, because it is the load at which the potential $V(x)$ is periodic. This also makes sense intuitively because at this value the load is exactly balanced by the energy expended.

The second feature of this model is that the load–velocity relationship can be calculated explicitly. To do so we look for a steady periodic solution of the Fokker–Planck equation (15.125). For $x \neq n\delta$, $p(x)$ must satisfy the differential equation

$$\frac{Df}{kT}p + D\frac{dp}{dx} = -J, \tag{15.127}$$

while at the points $x = n\delta$, the solution experiences a jump,

$$\ln p|_{x_n^-}^{x_n^+} = \frac{\Delta G}{kT}, \tag{15.128}$$

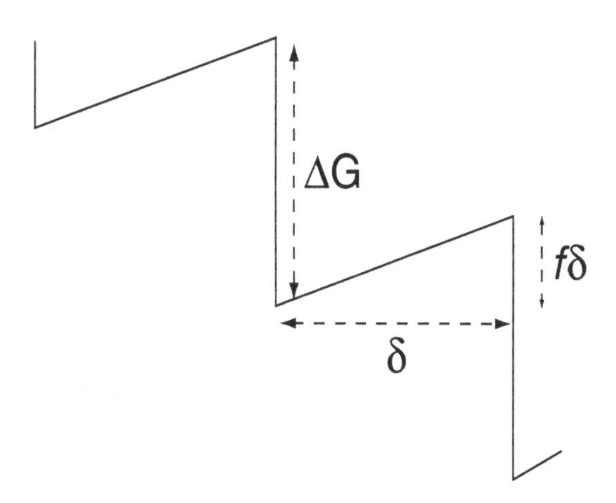

Figure 15.27 Diagram of a tilted ratchet potential (15.126).

so that

$$p(n\delta^+) = p(n\delta^-)\exp\left(\frac{\Delta G}{kT}\right). \tag{15.129}$$

For periodicity, it must be that

$$p(0) = p(\delta)\exp\left(\frac{\Delta G}{kT}\right). \tag{15.130}$$

In addition, we require $\int_0^\delta p(x)\,dx = 1$. The solution has

$$J = \frac{D}{\delta^2}\omega_l^2 \frac{1-\exp(\omega_0-\omega_l)}{(\exp(\omega_0)-1)(\exp(-\omega_l)-1)+\omega_l(\exp(\omega_0-\omega_l)-1)}, \tag{15.131}$$

where

$$\omega_0 = \frac{\Delta G}{kT}, \qquad \omega_l = \frac{f\delta}{kT}. \tag{15.132}$$

Furthermore,

$$\lim_{f\to 0} J = 2\frac{D}{\delta^2}\frac{\exp(\omega_0)-1}{\exp(\omega_0)+1}. \tag{15.133}$$

The mean velocity of the ratchet is given by

$$\left\langle \frac{dx}{dt} \right\rangle = \int_0^\delta J\,dx = \delta J. \tag{15.134}$$

A Motor Pulling Cargo

The tilted potential Brownian ratchet was motivated as a model of a polymerization ratchet that is pushing a loaded particle. However, another interpretation is that it describes the movement of a molecular motor moving along a track, pulling a load. Elston and Peskin (2000) extended this idea by assuming that the molecular motor moves in a titled potential, but that the load is attached to the motor by a spring. They then addressed the question of how the velocity of the motor with its load was affected by the flexibility of the spring.

To study this problem, suppose that the molecular motor is a molecule with diffusion coefficient D_1, the cargo molecule has diffusion coefficient D_2, and the two are connected by a spring with spring constant κ. Thus, the load on the molecular motor is not fixed but is determined by the distance between motor and cargo times the spring constant. One can readily write down the Fokker–Planck equation for the location of this motor–cargo complex. Elston and Peskin used asymptotic analysis for large and small κ to determine the velocity of the motor, but their results can also be explained using intuitive physical reasoning.

Suppose the spring is very stiff, $\kappa \to \infty$. With no flexibility, the motor–cargo complex is expected to move as a single molecule with an effective diffusion coefficient. Since (recall from Chapter 2) friction is inversely proportional to the diffusion coefficient

$f = \frac{kT}{D}$, and we expect friction to be additive, it follows that the diffusion coefficient of the complex should be

$$D_c = \frac{1}{\frac{1}{D_1} + \frac{1}{D_2}} = \frac{D_1 D_2}{D_1 + D_2}. \tag{15.135}$$

The velocity of the complex should be that for a molecule with diffusion coefficient D_c, that is,

$$v = 2\frac{D_c}{\delta}\frac{\exp(\omega_0) - 1}{\exp(\omega_0) + 1}. \tag{15.136}$$

In the opposite limit of a soft spring, $\kappa \to 0$, one expects the load to be essentially constant. This is because small fluctuations in the separation between cargo and motor change the load very little. Since the load on the cargo must match the velocity times friction coefficient, it must be that

$$F_l = \frac{kT}{D_2}v, \tag{15.137}$$

so that

$$w_l = \frac{\delta}{D_2}v. \tag{15.138}$$

This same load is experienced by the motor. Since the motor and cargo must have the same average velocity, it follows that

$$v = \frac{D_1}{\delta}\omega_l^2 \frac{1 - \exp(\omega_0 - \omega_l)}{(\exp(\omega_0) - 1)(\exp(-\omega_l) - 1) + \omega_l(\exp(\omega_0 - \omega_l) - 1)} \tag{15.139}$$

as well. Here we have two relationships between load and velocity that must both be satisfied. Since (15.138) is a monotone increasing relationship between load and velocity and (15.139) is a monotone decreasing relationship between load and velocity, there is a unique intersection of these two curves, hence a unique velocity.

Plots of the velocities in the hard-spring limit and in the soft-spring limit as a function of the parameter ω_0 are shown in Fig. 15.28. The important observation is that the velocity in the soft-spring limit is always larger than in the hard-spring limit. Thus, a flexible spring allows a motor to transport its cargo much more efficiently than does an inflexible spring.

15.10.3 Flashing Ratchets

Although the Brownian ratchet model presented above is a reasonable qualitative model of a molecular motor, it does not provide any insight into the actual physical mechanisms by which a molecular motor works.

To build a more mechanistic model of molecular motors, reconsider how an actin–myosin crossbridge works. In the Huxley model, the crossbridge can be in one of two states: bound or unbound. If it is unbound, its movement is governed by diffusion and

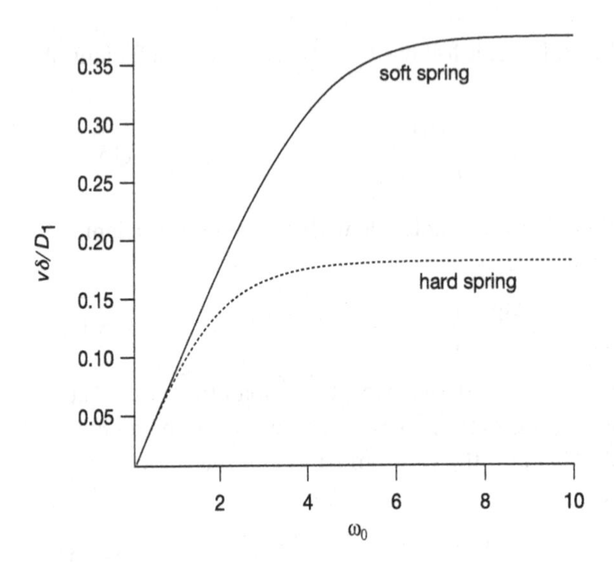

Figure 15.28 Plots of the velocities of the molecular motor–cargo complex in the hard-spring limit ($\kappa \to \infty$) and soft-spring limit ($\kappa \to 0$), plotted for $\frac{D_2}{D_1} = 0.1$.

any applied load. If it is bound, then its movement is additionally governed by a force, the strength of which is proportional to the deviation from its preferred position.

This is the idea behind a *flashing ratchet* model. We suppose there are two conformations, bound and unbound, with binding and unbinding rate constants α and β, respectively. In the unbound state the particle (or motor protein) moves by simple diffusion, while in the bound state it is also driven by the periodic potential $V(x)$. Since $V(x)$ is periodic, there can be no net movement if the protein is always bound. However, if the potential is asymmetric, then net drift is possible, even against a load.

The reason that net movement is possible can be understood as follows. When the particle is bound, it is forced to be mostly near the location of the minimum of the potential well; the deeper the potential well, the more narrowly the particle is confined, and the faster it equilibrates. However, when the particle is unbound it is free to move away from the minimum, and the longer it stays unbound, the more uniform the distribution becomes. When the particle is bound again, there is some probability that the particle will remain in the same attractive basin and there is some probability that it will be in an adjacent attractive basin. If the potential is asymmetric, then the probability that the particle falls into an adjacent attractive basin is biased, with one direction preferred over another, decided by the asymmetry of the potential. Thus, if the minimum of the potential well is biased to the left, the particle is more likely to move to the left.

The following is a simple model of an ideal flashing ratchet. We suppose that the periodic potential is deep, with each attracting basin between $n\delta$ and $(n+1)\delta$ having its minimum at $x = n\delta + a$, so that binding has the effect of localizing the particle to exactly the minimum at position $x = n\delta + a$. Suppose we start the process with the particle localized at $x = a$. Then, when unbinding occurs, the probability distribution function for the position of the particle is exactly a delta function localized at $x = a$.

Diffusion renders the distribution a Gaussian with mean $x = a$ and variance $2Dt$. When the next binding occurs, the probability of being in an attractive basin to the right, p_r, is the probability of being to the right of $x = \delta$, while the probability of moving to the left, p_l, is the probability of being to the left of $x = 0$. Thus,

$$p_r(t) = \frac{1}{2}\mathrm{erfc}\left(\frac{\delta - a}{\sqrt{2Dt}}\right), \qquad p_l(t) = \frac{1}{2}\mathrm{erfc}\left(\frac{a}{\sqrt{2Dt}}\right). \tag{15.140}$$

Since the average time spent unbound is $\frac{1}{\alpha}$ and the average time spent bound is $\frac{1}{\beta}$, the average velocity is approximately (making the approximation that the particle moves no more than one position to the left or right)

$$\begin{aligned} V &= \delta\frac{\alpha\beta}{\alpha+\beta}\left[p_r\left(\frac{1}{\alpha}\right) - p_l\left(\frac{1}{\alpha}\right)\right] \\ &= \frac{\delta\alpha}{2}\frac{\beta}{\alpha+\beta}\left[\mathrm{erfc}\left(\sqrt{\frac{\alpha\delta^2}{2D}}\left(1-\frac{a}{\delta}\right)\right) - \mathrm{erfc}\left(\sqrt{\frac{\alpha\delta^2}{2D}}\frac{a}{\delta}\right)\right] \\ &= \frac{D}{\delta}\frac{\beta}{\alpha+\beta}v\left(\frac{\alpha\delta^2}{2D}\right), \end{aligned} \tag{15.141}$$

where

$$v(\eta) = \eta\left[\mathrm{erfc}\left(\sqrt{\eta}\frac{a}{\delta}\right) - \mathrm{erfc}\left(\sqrt{\eta}\left(1-\frac{a}{\delta}\right)\right)\right]. \tag{15.142}$$

The function $v(\eta)$ is shown in Fig. 15.29 with $\frac{a}{\delta} = \frac{3}{4}$. It is a nonmonotone function of η, with a maximum at about $\eta = 11.5$.

One can easily understand that there should be an optimal rate of switching between the bound and the unbound states. If the switching occurs rapidly (compared to the rate of diffusion), then there is not much time for the distribution to spread out from the localized minimum, and hence there is little opportunity to move to an adjacent potential well. On the other hand, if switching is slow, then much of the time

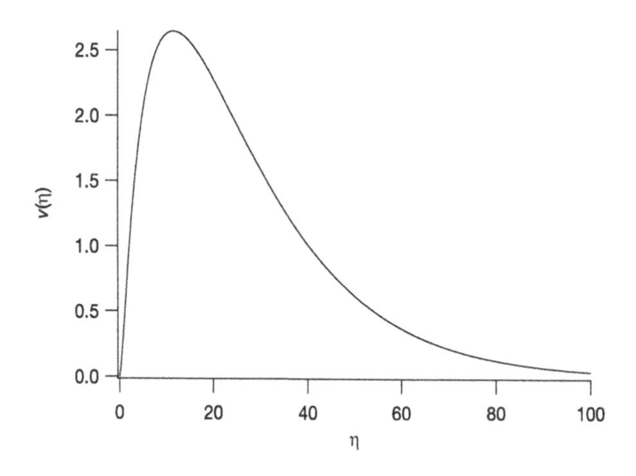

Figure 15.29 Plot of the function $v(x)$ given by (15.142), with $\frac{a}{\delta} = \frac{3}{4}$.

spent in the bound state is wasted, since once the asymmetric potential has localized the particle, there is no reason for the particle to remain bound.

It is typical in the literature on flashing ratchets that the driving potential is assumed to be asymmetric. If the binding and unbinding rates are independent of position, this is a necessary condition for net directional transport. However, if the binding and unbinding rates are spatially inhomogeneous, then it is not necessary that the potential be asymmetric. The symmetry can be broken by the arrangement of binding sites relative to the potential well locations. Thus, for example, actin–myosin crossbridge dynamics, wherein the force is assumed to be driven by a simple linear spring, have directed motion because the crossbridge binding site is shifted away from the minimum of the spring potential energy, due to the conformational change of the crossbridge.

A more detailed analysis of flashing ratchets requires examination of the appropriate Fokker–Planck equations. Suppose that there are a number of chemical states, and that the transition rates between the states are given by $k_{ij}(x)$. Further, assume that in each chemical state the molecule is subject to the potential $\psi_j(x)$. The governing Fokker–Planck equations are then given by

$$\frac{\partial p_j}{\partial t} = \frac{\partial}{\partial x}\left(D\left(\left(\frac{\psi_j'(x)}{kT} + \frac{f}{kT}\right)p_j + \frac{\partial p_j}{\partial x}\right)\right) + \sum_i k_{ij}(x)p_i - \sum_i k_{ji}p_j, \tag{15.143}$$

where $p_j(x,t)$ is the probability distribution for being in state j at position x at time t. For example, a model of the myosin motor that follows the Huxley crossbridge model would have two states, bound and unbound, with spatially periodic binding and unbinding rates $f(x)$ and $g(x)$, and with a periodic, symmetric potential $\psi(x)$.

This general framework has been used in many studies of real and generic molecular motors. See, for example, Howard (2001) or Reimann (2002) and references therein.

15.11 EXERCISES

1. Derive a differential equation for the load in a three-element Hill model with an elastic element in series with a contractile element (as shown in Fig. 15.9) in parallel with an additional elastic element of length L.

2. When a muscle cell dies, its ATP is depleted, with the result that the power stroke stalls and Ca^{2+} cannot be withdrawn using the Ca^{2+} ATPase. How does this explain rigor mortis?

3. Calculate the response of the Huxley model to a step change in length. In other words, calculate how n changes as a function of t and x, and hence calculate how the tension changes as a function of time following a step change in length.

4. Find the force-velocity relationship for the Huxley model for $v > 0$ with

$$f(x) = \begin{cases} f_{max} e^{(x-h)/\lambda}, & x < h, \\ 0, & x > h, \end{cases} \tag{15.144}$$

$$g(x) = f_{max}\left(1 - e^{(x-h)/\lambda}\right), \tag{15.145}$$

$$r(x) = r_{max} \frac{e^{(x-h)/\lambda} - \alpha}{1 - \alpha}, \tag{15.146}$$

Remark: This model reproduces the Hill force–velocity curve exactly, although it gives poor agreement with the energy flux data.

5. Show that in the Huxley model, $4\phi = v_{max}$ (at least approximately) if $g_2/(f_1 + g_1) = 3.919$. Hint: To find the maximum velocity, set $p = 0$ in (15.37). Then solve numerically for v_{max}/ϕ.

6. Assuming that the crossbridges act as linear springs, calculate the force–velocity curve for (15.42) and show that in the limit as $\epsilon \to 0$ and $\delta \to 0$ it does not produce zero force for some positive velocity. Give an intuitive explanation for this.

7. This exercise and the next are based on the discrete binding site models of T.L. Hill (1974, 1975). Suppose that each crossbridge is within reach of no more than two binding sites at one time, and that adjacent crossbridges are not within reach of the same two binding sites. Suppose also that adjacent binding sites are separated by a distance Δx, and let x denote the distance of the crossbridge from one of the binding sites, binding site 0, say. Define x such that if the crossbridge is bound to site 0 and has $x = 0$, it exerts no force. Also, let $n_i(x,t), i = 0, -1$, denote the fraction of crossbridges with displacement x that are bound to binding site i.

   (a) Show that the conservation equations are

$$-v\frac{dn_0}{dx} = f(x)[1 - n_0(x) - n_{-1}(x)] - g(x)n_0(x), \tag{15.147}$$

$$-v\frac{dn_{-1}}{dx} = f(x - \Delta x)[1 - n_0(x) - n_{-1}(x)] - g(x - \Delta x)n_{-1}(x), \tag{15.148}$$

   where, as usual, v denotes the steady contraction velocity.

   (b) Derive expressions for the isometric distributions of n_0 and n_{-1}. Compute the isometric force. Show that if the Huxley model is modified to include two binding sites, the isometric force is increased.

   (c) Compute the force–velocity curve. Hint: For each v solve the differential equations numerically, using the boundary conditions $n_0(h) = 0, n_{-1}(h+\Delta x) = 0$, then substitute the result into the expression for the force and integrate numerically.

   (d) Modify the model to include slippage of the crossbridge from one binding site to another. Show that in the limit as slippage becomes very fast, the two differential equations (15.147) and (15.148) reduce to a single equation.

8. Consider the general binding site model (15.53) that incorporates the discrete distribution of binding sites. Why is this equation much harder to integrate than the models we have discussed previously? The isometric solution is considerably easier to calculate than the

fully general solution. To do so, let $n_u(x) = 1 - \sum_{-\infty}^{\infty} n(x + i\Delta x)$. Show that

$$1 - n_u(x) = \frac{\sum_{-\infty}^{\infty} \frac{f(x+i\Delta x)}{g(x+i\Delta x)}}{1 + \sum_{-\infty}^{\infty} \frac{f(x+i\Delta x)}{g(x+i\Delta x)}}. \tag{15.149}$$

Use this to calculate the isometric solution $n(x)$.

9. A muscle fiber must be able to produce a force even at negative velocities. For example, if you slowly lower a brick onto a table, your bicep is extending and simultaneously resisting the freefall of the brick. Investigate the behavior of the Huxley model when $v < 0$.

   (a) Calculate the force-velocity curve for $v < 0$. Plot the full force-velocity curve using Huxley's parameter values.

   (b) Plot $n(x)$ for $v = -0.01, -0.1$, and -1.

   (c) Derive an expression for the maximum force as $v \to -\infty$.

10. Show that, in the limit as the number of bound crossbridges goes to infinity, the gliding rate of an actin filament (cf. Section 15.7) is given by

$$v \approx \frac{g_2 h}{2}. \tag{15.150}$$

    Hint: When many crossbridges are bound, we recover the Huxley model. Use (15.37) with $p = 0$ to estimate v when $R = \frac{f_1+g_1}{g_2}$ is small.

11. Use a formal perturbation calculation to find the load–velocity relationship for the polymerization ratchet (15.111) and (15.112) in the limit that α and β are small. That is, introduce a nondimensional scaling and define an appropriate small parameter ϵ, and then find a power series solution $p = p_0 + \epsilon p_1 + \cdots$ and the corresponding flux $J = J_0 + \epsilon J_1 + \cdots$.

12. Suppose $V(x)$ is the driving potential for a Brownian ratchet, and that $V'(x)$ is periodic with period δ. What is the stall load for the molecular motor driven by the potential $V(x)$?

13. Use a regular perturbation expansion to find the force-velocity curve for the flashing ratchet in the limit that the reaction rates are slow compared to diffusion.

CHAPTER 16

The Endocrine System

Hormones control a vast array of bodily functions, including sexual reproduction and sexual development, whole-body metabolism, blood glucose levels, plasma Ca^{2+} concentration, and growth. Hormones are produced in, and released from, diverse places, including the hypothalamus and pituitary, the adrenal gland, the thyroid gland, the testes and ovaries, and the pancreas, and they act on target cells that are often at a considerable physical distance from the site of production. Since they are carried in the bloodstream, hormones are capable of a diffuse whole-body effect, as well as a localized effect, depending on the distance between the production site and the site of action. In many ways the endocrine system is similar to the nervous system, in that it is an intercellular signaling system in which cells communicate via cellular secretions. Hormones are, in a sense, neurotransmitters that are capable of acting on target cells throughout the body, or conversely, neurotransmitters can be thought of as hormones with a localized action.

Despite the analogy with neural transmission, there are significant differences between the endocrine and nervous systems that have important ramifications for mathematical modeling. Not only is the endocrine system extremely complicated, but the data that are presently obtainable are less susceptible to quantitative analysis than, say, voltage measurements in neurons. Further, the distance between the sites of hormone production and action, and the complexities inherent in the mode of transport, make it extraordinarily difficult to construct quantitative models of hormonal control. For these reasons, models in endocrinology are less mechanistic than many of the models presented elsewhere in this book, and thus, in some ways, less realistic.

There are a number of basic types of hormones. Some, such as adrenaline and noradrenaline, originate from the amino acid tyrosine. Other, water-soluble, hormones are derived from proteins or peptides, while the *steroid* hormones are derived from cholesterol and are thus lipid-soluble. The diversity of the chemical composition of hormones results in a corresponding diversity of mechanisms of hormone action.

Steroid hormones, being lipid-soluble, diffuse across the cell membrane and bind to receptors located in the cell cytoplasm. The resultant conformational change in the receptor leads to activation of specific portions of DNA, thus initiating the transcription of RNA, eventually (possibly hours or days later) resulting in the production of specific proteins that modify cell behavior. An example of one such hormone is aldosterone, whose effect on epithelial cells is to enhance the production of ion channel proteins, rendering the cell more permeable to Na^+.

Other hormones, such as acetylcholine, act by binding to receptors located on the cell-surface membrane and causing a conformational change that results in the opening or closing of ionic channels.

Another important mechanism of hormone action is through second messengers, of which there are several examples in this book. Many hormone receptors are linked to G-proteins; binding of a hormone to the receptor results in the activation of the G-protein, and the triggering of a cascade of enzymatic reactions. For example, in the adenylate cyclase cascade, a wide variety of hormones (including adrenocorticotropin, luteinizing hormone, and vasopressin) cause activation of a G-protein, which in turn activates the membrane-bound enzyme adenylate cyclase. This activation results in an increase in the intracellular concentration of cAMP, and the consequent activation of a number of enzymes, with eventual effects on cell behavior; the specific effects depend on the cell type and the type of hormonal stimulus. In Chapter 7 we described the result of another signaling cascade, the phosphoinositide cascade, in which activation of cell-surface receptors leads to the activation of phospholipase C, the cleavage of phosphotidyl inositol 4,5-bisphosphate, and the resultant production of inositol 1,4,5-trisphosphate (IP_3) and diacylglycerol. As we saw, IP_3 releases Ca^{2+} from internal stores, and this can lead to intracellular Ca^{2+} oscillations and traveling waves.

Hormones can also act by directly converting the receptors into activated enzymes. For example, when insulin binds to a membrane receptor, the portion of the binding protein that protrudes into the cell interior becomes an activated kinase, which then promotes the phosphorylation of several substances inside the cell. The phosphorylation of proteins in the cell leads to a variety of other effects, including the enhanced uptake of glucose.

Much hormonal activity is characterized by oscillatory behavior, with the period of oscillation ranging from milliseconds (β-cell spiking) to minutes (insulin secretion) to hours (β-endorphin). In Table 16.1 are shown examples of pulsatile secretion of various hormones in humans. The pulsatility of normal hormonal activity is not well understood, but has significant implications for the treatment of hormonal abnormalities with drug therapies.

Table 16.1 Examples of pulsatile secretion of hormones in man (Brabant et al., 1992). Different values correspond to different primary sources.

Hormone	Pulses/Day
Growth hormone	9–16, 29
Prolactin	4–9, 7–22
Thyroid-stimulating hormone	6–12, 13
Adrenocorticotropic hormone	15, 54
Luteinizing hormone	7–15, 90–121
Follicle-stimulating hormone	4–16, 19
β-endorphin	13
Melatonin	18–24, 12–20
Vasopressin	12–18
Renin	6, 8–12
Parathyroid hormone	24–139, 23
Insulin	108–144, 120
Pancreatic polypeptide	96
Somatostatin	72
Glucagon	103, 144
Estradiol	8–19
Progesterone	6– 6
Testosterone	8–12, 13
Aldosterone	6, 9–12
Cortisol	15, 39

16.1 The Hypothalamus and Pituitary Gland

One of the most important components of the endocrine system is the hypothalamus and pituitary gland (Fig. 16.1A). The pituitary gland sits below the hypothalamic region of the brain and is a combination of two different types of cells. The adenohypophysis, or anterior pituitary, is made up of hormone-producing endocrine cells, while the neurohypophysis, or posterior pituitary, consists of secretory neural cells. In fact, the posterior pituitary is a collection of axons of neurons whose cell bodies lie in the hypothalamus. Hormones are synthesized at the cell body and transported to the secretory terminals, where they await release in response to an appropriate stimulus.

The pattern of blood circulation from the posterior to the anterior pituitary is a crucial part of the function of the gland (Fig. 16.1B). Hormones released from the posterior pituitary enter the inferior hypophyseal artery, whence they can travel through the portal vein to stimulate the endocrine cells of the anterior pituitary. In this way the hypothalamus can exert direct control over the pituitary gland.

Cells in the anterior pituitary can also be stimulated (or inhibited) by hormones released from other hypothalamic neurons that secrete into the superior hypophyseal artery.

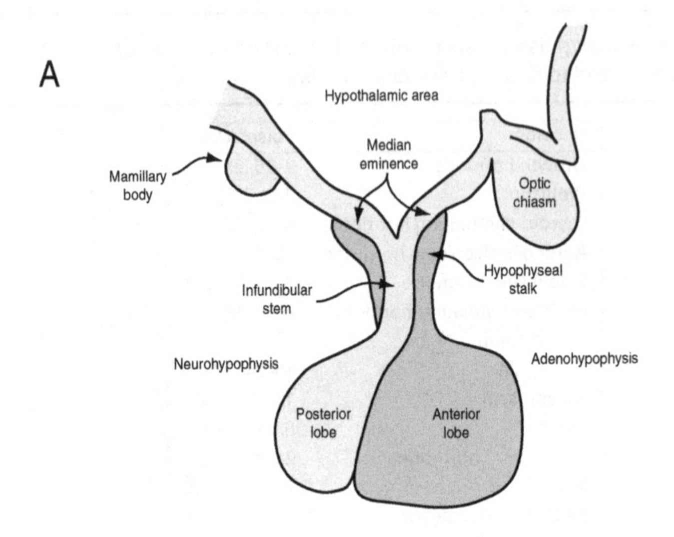

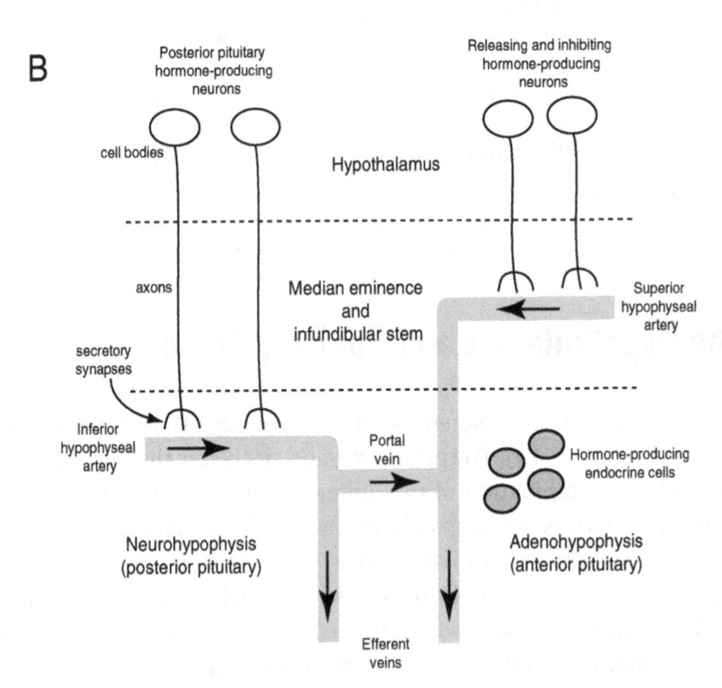

Figure 16.1 A: Schematic diagram of the hypothalamus and pituitary gland. B: Schematic diagram of the pituitary circulation. (Adapted from Berne and Levy, 1998, Figs. 49-1 and 49-2.)

Communication between the hypothalamus and pituitary is not all unidirectional. It is possible for hormones secreted from the anterior pituitary to be carried by a reverse blood flow back up to the axons in the median eminence or the cell bodies in the hypothalamus, thus forming a short feedback loop.

Almost all the hormones secreted from the anterior pituitary stimulate the secretion of hormones from target glands in the periphery, such as the thyroid and adrenal glands, or the gonads. This gives a feedforward control system: hormones secreted from hypothalamic neurons stimulate the secretion of hormones from the anterior pituitary, which in turn stimulate secretion of hormones from target glands in the periphery. For example, secretion of gonadotropin-releasing hormone (GnRH) from the hypothalamus stimulates the secretion of gonadotropins, such as luteinizing hormone (LH) or follicle-stimulating hormone (FSH), from the anterior pituitary. These in turn have multiple effects on the gonads, one example of which is stimulating the secretion of estradiol and progesterone from the ovaries.

The enormous complexity of the endocrine system arises partly because the hormones released by the peripheral glands can influence, in their turn, the secretion of the hypothalamus and the anterior pituitary. This allows for a highly complicated system of feedbacks, both positive and negative, most of which involve delays due to the circulation of the blood from one gland to another.

16.1.1 Pulsatile Secretion of Luteinizing Hormone

Luteinizing hormone and follicle-stimulating hormone, known collectively as gonadotropin, have a monthly cycle (in humans) related to ovulation, and also vary periodically on a time scale of hours. Although the precise function of these hourly variations is unclear, they occur in both males and females, and are crucial to development and maturation in both sexes. Gonadotropin is produced by the pituitary gonadotrophs in response to gonadotropin-releasing hormone (GnRH), sometimes called luteinizing-hormone-releasing hormone (LHRH), which is itself produced in the hypothalamus. Periodic variations in gonadotropin secretion are therefore the result of periodic variations in GnRH secretion. In fact, if GnRH secretion is constant rather than pulsatile, the secretion of gonadotropin is greatly reduced, and thus the pulsatility of GnRH secretion has an important regulatory function (Knobil, 1981).

This observation has been used as the basis for clinical treatments of certain reproductive disorders. In women suffering from abnormal GnRH secretion, the pulsatile administration of GnRH can, in some cases, restore normal ovulation and fertility. However, the frequency of the pulse must be controlled carefully. Wildt et al. (1981) have shown that the secretion of gonadotropin in rhesus monkeys is approximately maximized by the administration of GnRH pulses with a frequency of one per hour. If the frequency of the GnRH pulse is increased to 2 per hour, gonadotropin secretion is inhibited. Conversely, if the frequency is decreased to one pulse every three hours, the rate of secretion of follicle-stimulating hormone (FSH) increases, while the rate of secretion of luteinizing hormone (LH) decreases.

An example of pulsatile secretion of LH and testosterone in males is shown in Fig. 16.2. Although the testosterone secretion is not obviously oscillatory, the fluctuations in LH secretion clearly are.

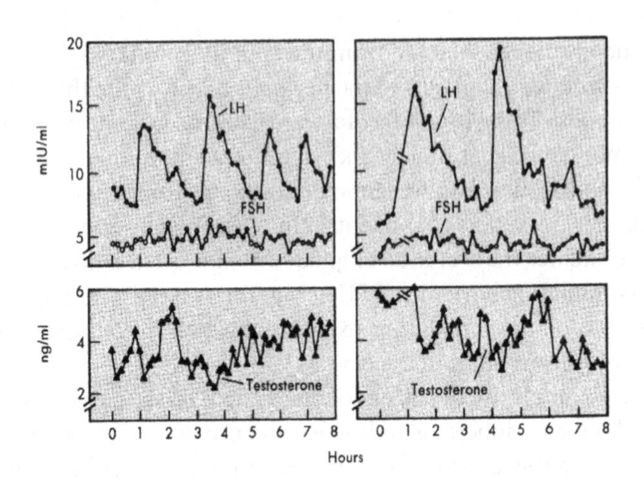

Figure 16.2 Pulsatile secretionof LH, FSH, and testosterone in men. (Berne and Levy, 1993, Fig. 48-15, p. 912.)

In males, gonadotropin stimulates the production of testosterone from the testes, while in females it stimulates the production of estradiol from the ovaries. Under some circumstances (see the model of ovulation below) estradiol can stimulate further production of gonadotropin, forming a positive feedback loop. However, estradiol can have both positive and negative feedback effects on the production of gonadotropin. In models of pulsatile testosterone and gonadotropin secretion, negative feedback from estradiol and testosterone to gonadotropin production is the important mechanism.

An early model of LH levels in the rat is that of Shotkin (1974a,b), although this model did not consider oscillatory aspects. One of the earliest models of oscillatory GnRH release is that of Smith (1980, 1983), later extended by Cartwright and Husain (1986) and by Murray (2002). Although these early models are highly simplistic and phenomenological, they illustrate the basic modeling approach often used to model pulsatile hormone release. A later model of this type is due to Liu and Deng (1991), and was the first to make a serious attempt to determine model parameters by fitting to experimental data, while similar models of the hypothalamus–pituitary–adrenal axis have been constructed by Jelić et al. (2005) and Lenbury and Pornsawad (2005).

A schematic diagram of the Smith model is shown in Fig. 16.3. LHRH, with concentration R, is produced by the hypothalamus, and stimulates the secretion of LH, with concentration L, from the pituitary. This in turn stimulates the secretion of testosterone, with concentration T, from the testes, which then decreases the rate of LHRH secretion by the hypothalamus. The simplest possible formulation of this model is obtained by assuming that all the reactions, except the feedback of T to R, are linear. In this case,

$$\dot{R} = f(T) - b_1 R, \tag{16.1}$$

$$\dot{L} = g_1 R - b_2 L, \tag{16.2}$$

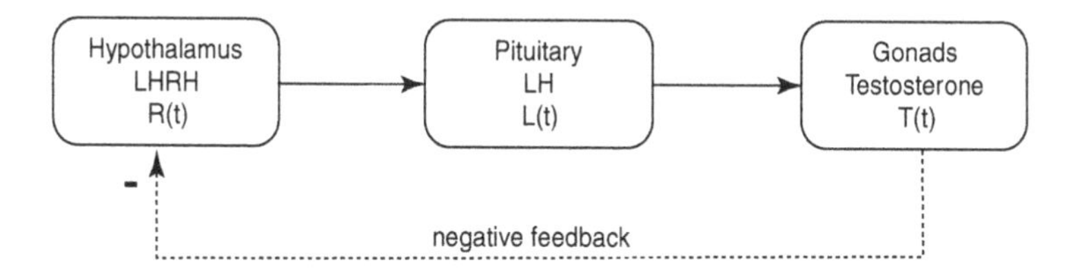

Figure 16.3 Schematic diagram of the Smith model of pulsatile LH release.

$$\dot{T} = g_2 L - b_3 T, \tag{16.3}$$

where b_i and g_i are positive constants, and $f(T)$ is a decreasing function. This is an example of a feedback repression model, a type of model that was studied in detail in the 1970s by a number of authors (Rapp, 1975,1976; Othmer, 1976; Rapp and Berridge, 1977; Hastings et al., 1977).

Because of the presence of negative feedback, it comes as no surprise that this model can exhibit stable oscillations; the exact conditions under which these occur and are stable are given by Smith (1980).

In a later modification, Murray (2002) introduced a delay into the production of testosterone, replacing the term g_2L with $g_2L(t-\tau)$, for some delay τ. They argued that the control of testosterone secretion could occur only after the LH has traveled through the blood to the testes, a process necessarily involving some delay.

Although this gives a minimal increase in model realism, the model remains too simplistic to be a useful model for comparing to data. Nevertheless, it illustrates the basic approach that has been used many times subsequently in the modeling of pulsatile hormone secretion. Typically such models (i) are compartmental, with different compartments for the various hormones and glands, (ii) involve negative or positive feedback, or both, and (iii) use discrete delays, either single or multiple, to model the time taken for the transport of hormones in the blood, and their subsequent action on sites distant from the point of action.

16.1.2 Neural Pulse Generator Models

A completely different approach to modeling pulsatile secretion of LH was taken by Brown et al. (1994), and has since been used in a number of modeling studies (Brown et al., 2004; MacGregor and Leng, 2005). In this model, pulsatile secretion of LH is assumed to be the result of pulsatile bursts of electrical activity in the hypothalamus, without any need for feedback from the peripheral gland (although such feedback is not ruled out). Experimental evidence that immortalized LHRH neurons exhibit spontaneous pulsatile electrical activity (Wetsel et al., 1992) lends considerable support to this assumption.

In this model, LHRH neurons stimulate the release of the neurotransmitter GABA from neighboring GABA neurons, which, in turn, decreases the activity of LHRH

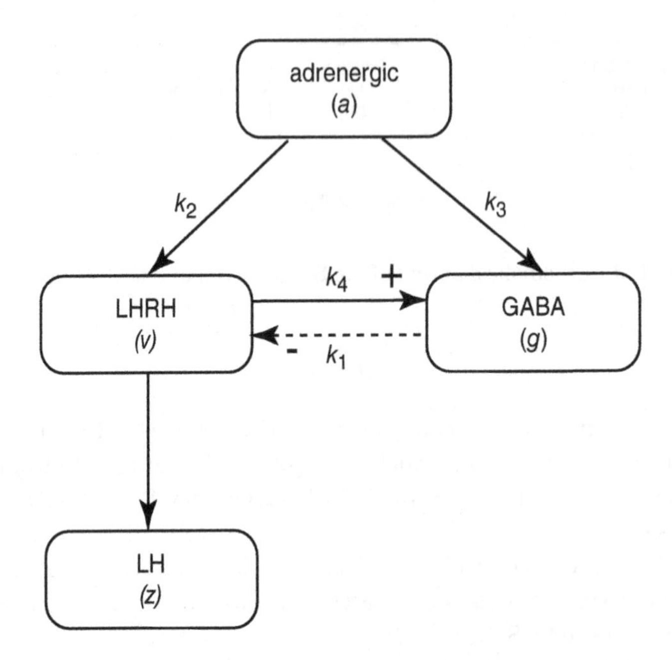

Figure 16.4 Schematic diagram of the neural model of pulsatile LH release.

neurons, thus resulting in a negative feedback loop. Both LHRH and GABA neurons are driven by adrenergic input from the neurotransmitters adrenaline and noradrenaline, which are released from neurons whose cell bodies lie at distant sites in the brainstem. Finally, the activity of LHRH neurons drives the release of LH, with no feedback from LH to LHRH or GABA.

A schematic diagram of this system is given in Fig. 16.4. Each group of neurons is modeled by a single compartment, which assumes complete synchronization within each group. The core of the model is the assumption that the activity of LHRH neurons is intrinsically bistable (see Chapter 5). Thus, if we let v denote the activity of the LHRH neurons, the neurons can be turned off, and not firing (denoted by $v = 0$), or they can be turned on and firing (denoted arbitrarily by $v = 1$). To avoid possible confusion we emphasize that the variable v does not denote the voltage of the synchronized neuron group, merely its activity state. Using the bistable equation to model the activity of the LHRH neurons, we get

$$\epsilon \frac{dv}{dt} = v(c - v)(v - 1) - k_1 g + k_2 a(t), \tag{16.4}$$

$$\frac{dg}{dt} = k_3 a(t) + k_4 v - r_g g, \tag{16.5}$$

$$\frac{dz}{dt} = p(v) - r_z z, \tag{16.6}$$

where, as indicated in Fig. 16.4, $a(t)$ denotes a time-dependent adrenergic input, g denotes the activity of GABA neurons, and z denotes the blood concentration of LH. The constants r_g and r_z denote the linear removal rate of g and z respectively. The rate of production of LH is assumed to be some increasing and bounded function of v.

This model has a structure similar to that of the FitzHugh–Nagumo equations (Chapter 5). Here, however, both the excitatory (v) and inhibitory (g) variables are driven by the same input, and the LH concentration is a filtered version of the LHRH activity. Because of this similarity with the FitzHugh–Nagumo equations, we can understand intuitively how this model behaves. If the input a is a low-frequency spike train, each peak of a gives a spike, or a series of spikes in v, since an increase in a is equivalent to an increase in injected current in the FitzHugh–Nagumo equations. However, when the frequency of a is high enough, and thus the average level of a is high enough, it is possible that this increases the level of the inhibitory variable, g, to such an extent that further firing of v is prevented. At intermediate frequencies, irregular firing patterns can develop, dependent on the exact balance between excitation and inhibition. Each burst of firing of the LHRH neurons results in a burst of LH release.

A typical solution is shown in Fig. 16.5. The input is given by

$$a(t) = \begin{cases} 3.5, & 0 < t < 0.2, \\ 0, & 0.2 \leq t \leq 0.21, \end{cases}$$

and then repeated periodically. For this input, LHRH activity is mostly inhibited, but every so often it increases for a short time, resulting in a burst of LH release. More realistic stochastic inputs were considered by Brown et al. (1994), but these give qualitatively similar results.

Although this model can reproduce patterns of pulsatile LH release that look reasonably realistic, it suffers from the fault of not providing a mechanistic explanation of the pulsatility. Essentially, the pulsatility of hormonal release is assumed to arise from an unexplained pulsatility in a neural network deeper in the brain. Nevertheless, this is as likely an explanation as any other; it is entirely plausible that the origins of pulsatile hormone release lie in neural networks in the brain rather than in an intrinsic oscillatory delayed feedback from the peripheral glands.

Pulsatile Release of Growth Hormone

In humans, and most other mammals, growth hormone (GH) is secreted in a pulsatile manner. In the male rat, which has one of the most defined secretory patterns, pulses occur every three hours, with very little secretion between the pulses. Secretion of GH from the anterior pituitary is stimulated by growth-hormone-releasing hormone (GHRH), which is secreted from the hypothalamus and is inhibited by somatostatin, which is also secreted from the hypothalamus. GH, in turn, activates the production of somatostatin, and inhibits the production of GHRH, both effects occurring with a time delay. In addition, there is evidence that somatostatin can inhibit GHRH neurons directly.

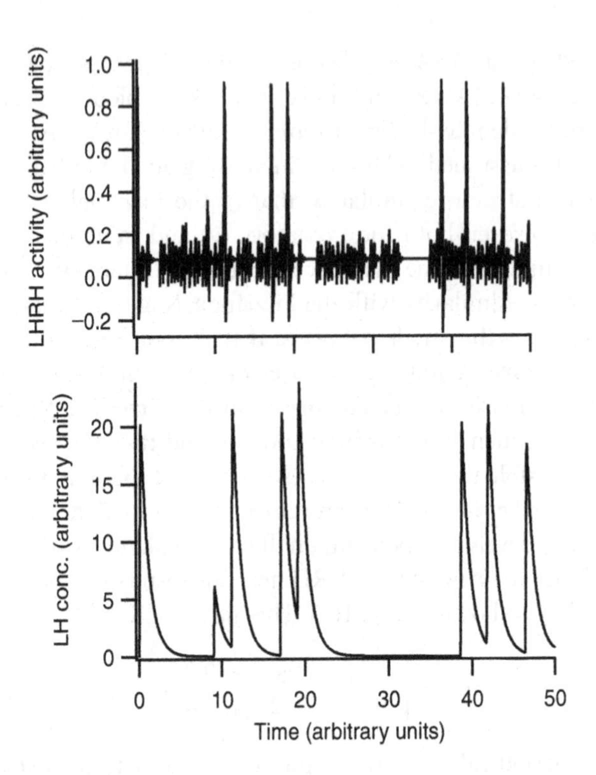

Figure 16.5 Typical results of the neural model of pulsatile LH release. Parameter values are $\epsilon = 1/200$, $c = 0.2$, $k_1 = 1$, $k_2 = 0.02$, $k_3 = 0.02$, $k_4 = 1$, $k_5 = 2.5$, $b_1 = 2$, $b_2 = 0.59$, $b_3 = 10$, $b_4 = 0.1$, $p_1 = 100$, $p_2 = 100$, $p_3 = 0.3$, $d_1 = 1$. All the units are arbitrary.

Feedback from the periphery occurs via somatomedins, secreted by the liver and other tissues, which increase the production of both GHRH and somatostatin, and decrease the rate of GH secretion from the pituitary. However, peripheral effects are not included directly in the model discussed below.

A model of this system, similar in style to the model of Section 16.1.2, was constructed and studied by Brown et al. (2004) and by MacGregor and Leng (2005). They showed how, as in the previous model, a train of input pulses to the hypothalamus could be converted into pulsatile release. They also showed how, by mimicking experimental sampling that is done only every 10 minutes or so, the output of the model could be made to look considerably more stochastic, and thus much more realistic.

A schematic diagram of the model structure, and some typical results, are shown in Fig. 16.6.

Pulsatile Prolactin Secretion

Another model that explains pulsatile hormone secretion by interactions between the hypothalamus and pituitary, without the need for feedback from the periphery, is that of Bertram et al. (2006), who modeled the mating-induced prolactin rhythm

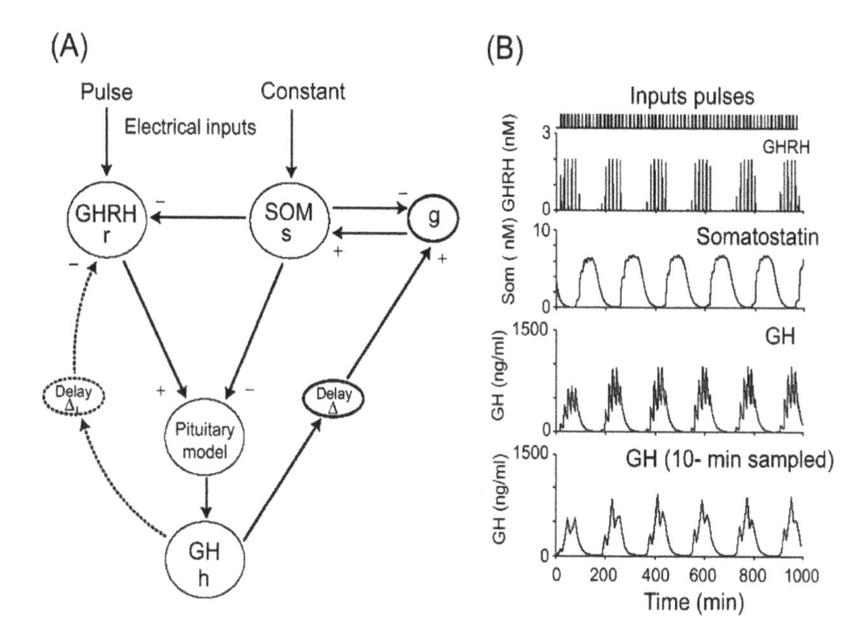

Figure 16.6 Schematic diagram (A) and some typical results (B) of the model of pulsatile GH secretion. The lowest panel of B shows how, by mimicking experimental sampling (which occurs every 10 minutes), the model results can be made to look much more realistic. Reproduced from MacGregor and Leng, 2005, Fig. 6, with permission from Blackwell Publishing.

of female rats. In response to a copulatory stimulus, or an injection of a bolus of oxytocin, the hormone prolactin is secreted by pituitary lactotrophs; in the first 10 days of pregnancy, secretion of prolactin is pulsatile, with regular morning and afternoon surges. Since these rhythms can be observed in ovariectomized rats, it is plausible that the rhythm is generated by interactions between the hypothalamus and the pituitary. Prolactin secretion is regulated principally by the hormone dopamine, which is secreted from the hypothalamus, but, because of the circadian nature of the pulsatility, is presumably also influenced by the suprachiasmatic nucleus of the hypothalamus.

The basis for pulsatility in the model is the interaction between dopamine neurons and lactotrophs. Dopamine inhibits the production of prolactin, which in turn activates, with a time delay, the production of dopamine, leading to a classic negative feedback loop, and oscillations. However, Bertram et al. proceed to build a much more complex model on this simple base; the complex model can explain much more of the experimental data, including different surges in morning and afternoon, the persistence of the oscillations after a single brief stimulus, and abrupt termination of the oscillation after 10 days. This work provides a good example of how to build more complex models from simpler ones, ensuring that, at each step, there is a compelling scientific reason for the incorporation of additional complexity.

16.2 Ovulation in Mammals

16.2.1 A Model of the Menstrual Cycle

At birth, the human ovary contains approximately 2 million ovarian *follicles*, which consist of germ cells, or oocytes, surrounded by a cluster of endocrine cells that provide an isolated and protected environment for the oocyte. In the first stage of follicle development, occurring mostly before puberty, and taking anywhere from 13 to 50 years, the cells surrounding the oocyte (the granulosa cells) divide and form several layers around the oocyte, forming the so-called secondary follicle. Subsequent to puberty, these secondary follicles form a reserve pool from which follicles are recruited to begin the second stage of development. In this second stage, follicles increase to a final size of up to 20 mm before they rupture and release the oocyte to be fertilized. The release of the oocyte is called *ovulation*.

A normal menstrual cycle for an adult woman has a period anywhere from 25 to 35 days, and is regulated by the complex interplay of a number of different hormones. In response to the release of LHRH from the hypothalamus, LH and FSH (follicle-stimulating hormone) are released from the pituitary. LH and FSH initiate follicle development and control the release of the ovarian hormones, estradiol, inhibin and progesterone. Feedback occurs because the ovarian hormones themselves affect, in multiple ways, the rate of secretion of LH and FSH.

Blood concentrations of all these hormones vary in a systematic manner throughout the menstrual cycle. During the *follicular phase* (which begins after the onset of menstrual bleeding and averages 15 days), LH and FSH levels remain fairly low, while the concentration of estradiol rises. The follicular phase is followed by the *ovulatory phase* (about 1 to 3 days long), during which there is very large increase in LH levels, the so-called LH surge, while the level of estradiol falls rapidly. The final phase, the *luteal phase* (lasting about 13 days), ends with the onset of menstrual bleeding, and is characterized by high levels of progesterone and low levels of LH and FSH.

A model of the menstrual cycle has been constructed by Selgrade, Schlosser, and their colleagues (Selgrade and Schlosser, 1999; Schlosser and Selgrade, 2000; Clark et al., 2003). It consists of two submodels, one for the pituitary and one for the ovaries. Its most prominent features are, first, the omission of a specific variable for LHRH concentration, and, second, a detailed model of the different stages of the ovary.

The Pituitary Submodel

A schematic diagram of the pituitary submodel is shown in Fig. 16.7. Both the synthesis and release of LH and FSH are modulated by estradiol and progesterone levels, but in an antagonistic manner. These interactions are not modeled in a mechanistic way, but are described by mathematical functions with the correct qualitative behavior, and thus have a limited physiological interpretation.

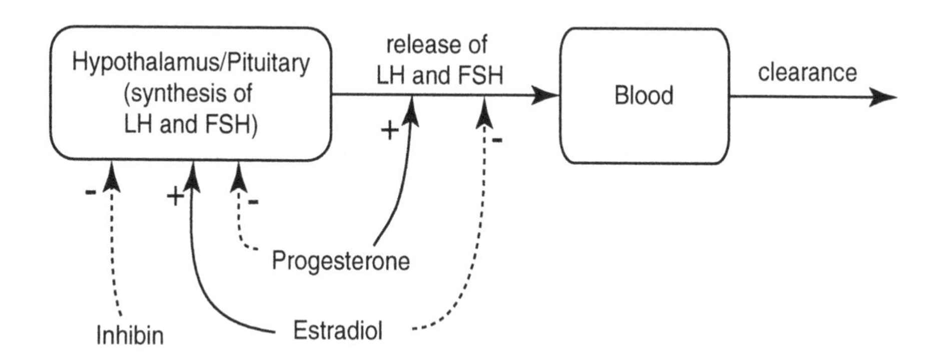

Figure 16.7 Schematic diagram of the pituitary submodel of the model of the menstrual cycle. Adapted from Clark et al., 2003, Fig. 1.

We begin by defining the variables

$$\begin{aligned} \mathrm{LH_R} &- \text{amount of LH in the releasable pool,} \\ \mathrm{LH} &- \text{blood concentration of LH,} \\ \mathrm{FSH_R} &- \text{amount of FSH in the releasable pool,} \\ \mathrm{FSH} &- \text{blood concentration of FSH,} \\ E &- \text{blood concentration of estradiol,} \\ P &- \text{blood concentration of progesterone,} \\ I &- \text{blood concentration of inhibin.} \end{aligned}$$

Then, based on the feedbacks shown in Fig. 16.7, LH and FSH production and release are modeled by the phenomenological equations

$$\frac{d}{dt}\mathrm{LH_R} = \frac{V_{0L} + \frac{V_{1L}E^8}{K_{mL}^8+E^8}}{1+P(t-\tau)/K_{iL}} - \frac{k_L(1+c_{LP}P)\mathrm{LH_R}}{1+c_{LE}E}, \tag{16.7}$$

$$\frac{d}{dt}\mathrm{LH} = \frac{1}{v}\frac{k_L(1+c_{LP}P)\mathrm{LH_R}}{1+c_{LE}E} - c_L\mathrm{LH}, \tag{16.8}$$

$$\frac{d}{dt}\mathrm{FSH_R} = \frac{V_F}{1+I(t-\tau)/K_{iF}} - \frac{k_F(1+c_{FP}P)\mathrm{FSH_R}}{1+c_{FE}E^2}, \tag{16.9}$$

$$\frac{d}{dt}\mathrm{FSH} = \frac{1}{v}\frac{k_F(1+c_{FP}P)\mathrm{FSH_R}}{1+c_{FE}E^2} - c_F\mathrm{FSH}. \tag{16.10}$$

Thus, the rate of production of $\mathrm{LH_R}$ is an increasing function of E, but a decreasing function of P, with some time delay, τ. Similarly, the rate of secretion of LH into the blood (i.e., the second term on the right of (16.7) and the first term on the right of (16.8)) is an increasing function of P, but a decreasing function of E. In (16.8) this secretion term is multiplied by the scale factor $1/v$, which takes into account the different volumes of the two compartments. The equations for $\mathrm{FSH_R}$ and FSH have the same form, with slightly different functional forms.

The Ovary Submodel

A schematic diagram of the ovary submodel is shown in Fig. 16.8. Each follicle is assumed to move through 10 different stages, with different rates of hormone production during each stage. This allows for time-dependent rates of hormone release, without the need to build in such time dependence explicitly. The passage of the follicle through the stages is described mostly by linear transitions (making many of the stages act like linear filters, as described briefly in the appendix to Chapter 19; see also Exercise 7 of this chapter, and Exercise 7 of Chapter 13), although some of the transitions are dependent on LH.

Inactive follicles are recruited (i.e., moved into the recruited follicle box, RcF) by FSH (at rate bFSH). Recruited follicles also multiply at a rate proportional to both RcF and FSH. In a similar manner, LH stimulates the movement of follicles into the secondary phase and then into the preovulatory phase, while simultaneously stimulating an increase in the number of follicles in the secondary phase by a process of positive feedback. Subsequently, the follicle moves through the different phases in a

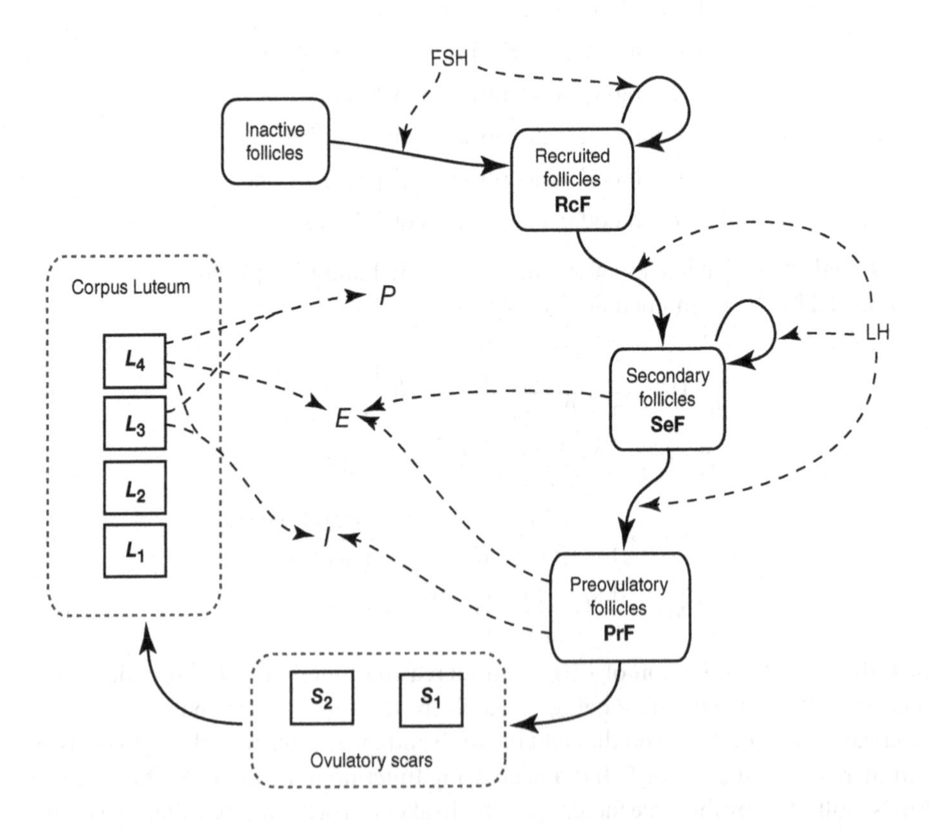

Figure 16.8 Schematic diagram of the ovary submodel of the model of the menstrual cycle. Adapted from Clark et al., 2003, Fig. 2.

linear fashion. This sequence of steps, as illustrated in Fig. 16.8, is modeled by

$$\frac{d}{dt}\text{RcF} = c\text{FSH} + (c_1\text{FSH} - c_2\text{LH}^{\alpha})\text{RcF}, \tag{16.11}$$

$$\frac{d}{dt}\text{SeF} = c_2\text{LH}^{\alpha} \cdot \text{RcF} + (c_3\text{LH}^{\beta} - c_4\text{LH})\text{SeF}, \tag{16.12}$$

$$\frac{d}{dt}\text{PrF} = c_4\text{LH} \cdot \text{SeF} - c_5\text{PrF}, \tag{16.13}$$

$$\frac{dS_1}{dt} = c_5\text{PrF} - S_1/k, \tag{16.14}$$

$$k\frac{dS_2}{dt} = S_1 - S_2, \tag{16.15}$$

$$k\frac{dL_1}{dt} = S_2 - L_1, \tag{16.16}$$

$$k\frac{dL_2}{dt} = L_1 - L_2, \tag{16.17}$$

$$k\frac{dL_3}{dt} = L_2 - L_3, \tag{16.18}$$

$$k\frac{dL_4}{dt} = L_3 - L_4. \tag{16.19}$$

To complete the model we need to specify how the levels of estradiol, progesterone, and inhibin depend on the various follicle stages. The model assumes that each hormone can be produced only at certain follicle stages, as indicated in Fig. 16.8. Thus we also have the auxiliary equations

$$E = e_0 + e_1\text{SeF} + e_2\text{PrF} + e_3L_4, \tag{16.20}$$

$$P = p(L_3 + L_4), \tag{16.21}$$

$$I = h_0 + h_1\text{PrF} + h_2L_3 + h_3L_4. \tag{16.22}$$

Comparing to Data

The model parameters were determined by fitting the model to the data of MacLachlan et al. (1990), who measured blood concentrations of the major hormones over the course of a menstrual cycle. Such a procedure does not determine the parameters to any degree of accuracy, and certainly does not demonstrate that this model is the only possible model that can reproduce the data. Nevertheless, it does show that this model is sufficient to explain many important features of the menstrual cycle.

The parameter values (slightly changed from the original values of Clark et al., 2003) are given in Tables 16.2 and 16.3, and a typical stable oscillatory solution is shown in Fig. 16.9. The LH surge can be clearly seen in the upper panel, while the typical multihumped solution for the estradiol concentration is shown in the lower panel.

Interestingly, Clark et al. found a second stable limit cycle solution of this same model, a periodic solution that has a slightly smaller period, much less variation in

Table 16.2 Parameter values for the pituitary submodel of the model of the menstrual cycle.

LH equations		
k_L	2.5	day^{-1}
c_L	14	day^{-1}
V_{0L}	1263	μg/day
V_{1L}	91000	μg/day
K_{mL}	360	ng/L
K_{iL}	31	nmol/L
c_{LE}	0.005	L/ng
c_{LP}	0.07	L/nmol
τ	2.00	day

FSH equations		
V_F	5700	μg/day
c_F	8.2	day^{-1}
k_F	7.3	day^{-1}
c_{FE}	0.16	(L/ng)^2
K_{iF}	641	U/L
c_{FP}	644	L/nmol
v	2.5	L

Table 16.3 Parameter values for the ovarian submodel and the auxiliary equations of the model of the menstrual cycle.

Ovarian equations		
b	0.004	L/day
c_1	0.006	(L/μg)/day
c_2	0.05	day^{-1}
c_3	0.004	day^{-1}
c_4	0.006	day^{-1}
c_5	1.3	day^{-1}
k	1.43	day
α	0.77	
β	0.16	

Auxiliary equations		
e_0	48	ng/L
e_1	0.1	1/kL
e_2	0.17	1/kL
e_3	0.23	1/kL
p	0.05	(nmol/L)/μg
h_0	274	U/L
h_1	0.5	(U/L)/μg
h_2	0.5	(U/L)/μg
h_3	2	(U/L)/μg

estradiol levels, lower levels of progesterone, and an elevated LH/FSH ratio. Since these features are typical of the disease polycystic ovary syndrome (PCOS), one of the leading causes of female infertility, it is possible that the model might provide a way to investigate proposed treatments. For example, Clark et al. showed that the solution can be switched from the abnormal cycle to the normal cycle by the application of a sufficient amount of exogenous progesterone, but not by the application of too small a dose. Although such conclusions are highly tentative, the possibility of using this model to test proposed treatments is intriguing.

16.2.2 The Control of Ovulation Number

One particularly remarkable feature about ovulation is that, although many follicles begin the second developmental stage, few reach full maturity and ovulate, as the rest atrophy and die. In fact, the number of oocytes reaching full maturity is carefully controlled, so that litter sizes are generally restricted to within a relatively narrow range, and different species have different typical litter sizes. For example, to quote

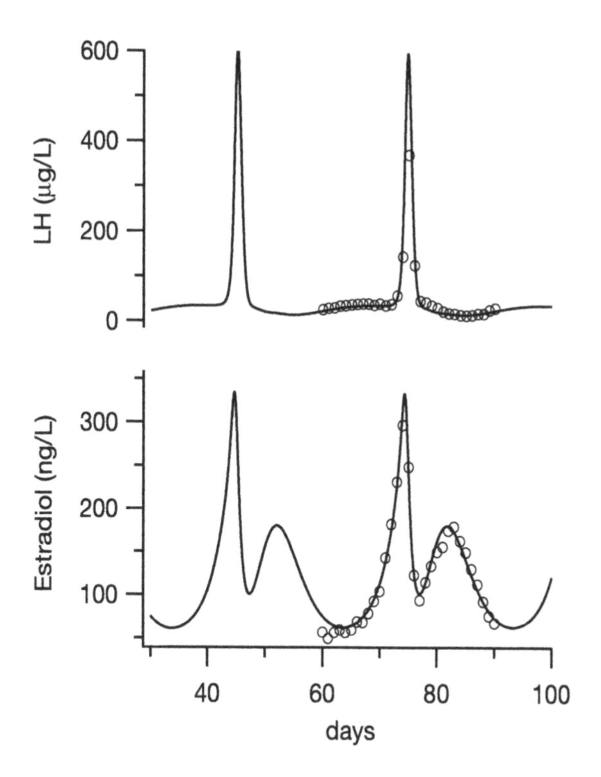

Figure 16.9 An oscillatory solution of the model of the menstrual cycle. The experimental data of MacLachlan et al. (1990) are shown as open circles.

some interesting, if not particularly useful, facts from Asdell (1946), both the dugong and llama have a typical litter size of 1, the crestless Himalayan porcupine typically gives birth to two offspring, while the dingo produces, on average, 3. Different breeds of pigs have litter sizes ranging from 6 to 11.

There must therefore be a complex process that, despite the continuous recruitment of secondary follicles into the second developmental stage, allows precise regulation of the number remaining at ovulation. Further, the temporal periodicity of ovulation is tightly controlled, with ovulation occurring at regularly spaced time intervals.

In addition to questions related to the nature of the control of ovulation, there is the question of efficiency. It appears inefficient to regulate the final number of mature follicles by initiating the growth of many and killing off most of them. One might speculate that it would be more reasonable to initiate growth in only the required number and ensure that they all progress through to ovulation.

Normal ovulation involves growth in both ovaries. However, since removal of one ovary does not change the total number of eggs released during ovulation, the control mechanism is not a local one, but a global one, known to operate through the circulatory system. Maturation of follicles is stimulated by *gonadotropin*, which is released

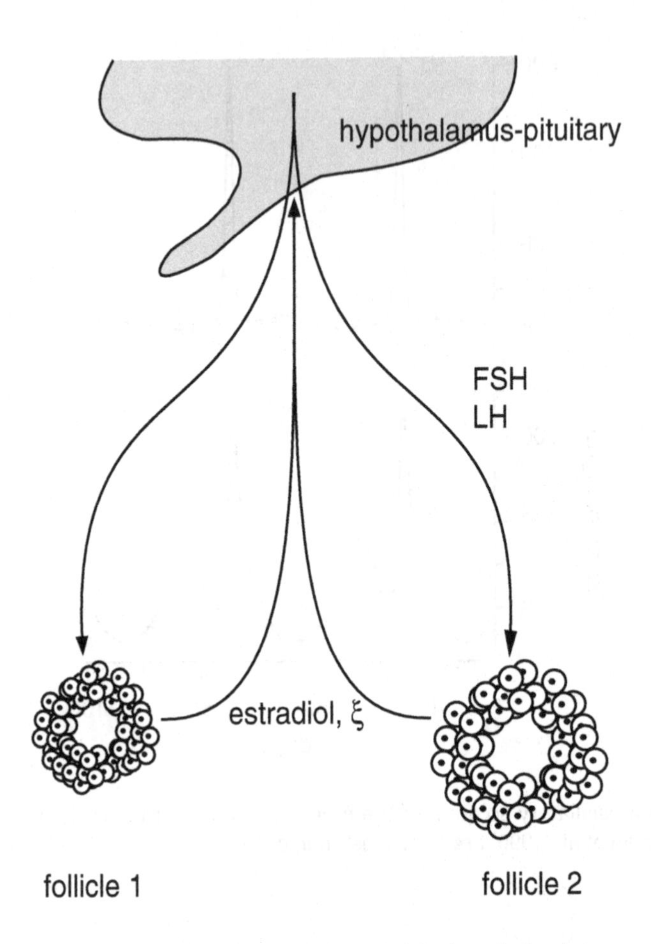

Figure 16.10 Schematic diagram of the Lacker model of ovulation in mammals. (Adapted from Lacker, 1981, Fig. 1.)

from pituitary gonadotrophs. Gonadotropin consists of two different hormones called *follicle-stimulating hormone* (FSH) and *luteinizing hormone* (LH). However, follicles themselves secrete *estradiol*, which stimulates the production of gonadotropin, forming a feedback control loop for the control of follicle maturation (Fig. 16.10).

One of the most elegant models of hormonal control is due to Lacker (1981; Lacker and Peskin, 1981; Akin and Lacker, 1984) and describes a possible mechanism by which mammals control the number of eggs released at ovulation. In the model it is assumed that each follicle interacts with other follicles only through the hormone concentrations in the bloodstream. As follicles mature they become more sensitive to gonadotropin, and their secretion of estradiol increases. The model of this feedback control loop is considerably oversimplified, since it does not incorporate a detailed mechanistic description of how estradiol production depends on gonadotropin or vice versa. However, it provides a phenomenological description of how a global interaction mechanism can be organized to give precise control over the final number of eggs reaching maturity.

The three basic assumptions of the model are that

1. The rate at which follicles secrete estradiol is a marker of follicle maturity.
2. The concentration of estradiol in the blood controls the release of FSH and LH from the pituitary.
3. The concentrations of FSH and LH control the rate of follicle maturation, and at any given instant, the response of each follicle to FSH and LH is a function of the follicle's maturity.

To express these assumptions mathematically, we define the following variables and parameters:

ξ concentration of estradiol,
γ rate of clearance of estradiol from the blood,
V plasma volume,
s_i rate of secretion of estradiol from the ith follicle,
N number of interacting follicles.

Here, all concentrations refer to serum concentrations (i.e., concentrations in the blood). Then, the rate of change of the total estradiol concentration is given by

$$V\frac{d\xi}{dt} = \sum_{i=1}^{N} s_i(t) - \gamma\xi. \tag{16.23}$$

Assuming that the rates of addition and removal of estradiol are much faster than the rate of follicle maturation, we take ξ to be at pseudo-steady state, and thus

$$\xi = \frac{1}{\gamma}\sum_{i=1}^{N} s_i(t) = \sum_{i=1}^{N} \xi_i(t), \tag{16.24}$$

where $\xi_i(t) = s_i(t)/\gamma$ is the contribution that the ith follicle makes to ξ. In general, $d\xi_i/dt$ is a function of both ξ_i and ξ, but does not depend directly on any other $\xi_j, j \neq i$. This is because we assume that local follicle–follicle interactions are not an important feature of the control mechanism, but that follicles interact only via the total estradiol concentration. Hence, the most general form of the model equations is

$$\frac{d\xi_i}{dt} = f(\xi_i, \xi), \qquad i = 1, \ldots, N. \tag{16.25}$$

The function f is called the *maturation function*. Note that the concentrations of FSH and LH do not appear explicitly, as their effect on ξ_i is modeled indirectly by assuming that $\frac{d\xi_i}{dt}$ depends on ξ.

Here we describe one particular form of the maturation function. This form is not based on experimental evidence but is chosen to give the correct behavior. Specifically, we take

$$\frac{d\xi_i}{dt} = f(\xi_i, \xi) = \xi_i\phi(\xi_i, \xi), \qquad i = 1, \ldots, N, \tag{16.26}$$

where

$$\phi(\xi_i, \xi) = 1 - (\xi - M_1\xi_i)(\xi - M_2\xi_i). \tag{16.27}$$

The constants M_1 and M_2 are parameters that are the same for every follicle, so that each follicle obeys the same developmental rules. As a function of ξ_i, for fixed ξ, ϕ is an inverted parabola with a maximum at

$$\xi_{i,\max} = \frac{\xi}{2}\left(\frac{1}{M_1} + \frac{1}{M_2}\right). \tag{16.28}$$

If ξ_i is large or small, the growth of ξ_i is negative, and so this growth rate is fastest for those follicles with maturity within a narrow range, depending on the total estradiol concentration. Thus, with a given initial distribution of follicle maturities, those with ξ_i close to $\xi_{i,\max}$ grow at the expense of the others. Further, since the growth rate f is proportional to ξ_i, the selective growth of the ith follicle leads to an autocatalytic increase in ξ_i.

Numerical Solutions

Before we study the behavior of the model analytically, it is helpful to see some typical numerical solutions. The numerical solution of (16.26)–(16.27), with $M_1 = 3.85$, and $M_2 = 15.15$, starting with a group of 10 follicles with initial maturities randomly distributed between 0 and 0.1, shows that the maturity of four or five follicles goes to infinity in finite time, while the other follicles die (Fig. 16.11). Since ovulation is triggered by high, fast-rising estradiol levels, solutions that become infinite in finite time are interpreted as ovulatory solutions. Not only do a similar number of follicles ovulate in each run, they also ovulate at the same time. Hence, ovulatory solutions for ξ_i and ξ_j, say, are ones in which $\xi_i(t)$ and $\xi_j(t) \to \infty$ as $t \to T < \infty$, with $\xi_i/\xi_j \to 1$. These numerical solutions show that the model has the correct qualitative behavior. However, analytic methods give a deeper understanding of how this control is accomplished.

Symmetric Solutions

Much of the behavior of the ovulation model can be understood by considering symmetric solutions, in which M of the follicles have the same maturity, while all others have zero maturity. Thus, $\xi_i = \xi/M, i = 1, \ldots, M$, and $\xi_i = 0, i = M+1, \ldots, N$, in which case the model simplifies to

$$\frac{d\xi}{dt} = \xi + \mu\xi^3, \tag{16.29}$$

where $\mu = -(1 - M_1/M)(1 - M_2/M)$. The solution is given implicitly by

$$\frac{\xi}{\xi_0}\sqrt{\frac{1+\mu\xi_0^2}{1+\mu\xi^2}} = e^t, \tag{16.30}$$

where $\xi_0 = \xi(0)$ is the initial value. When $\mu > 0$, $t \to \log\left(\sqrt{(1+\mu\xi_0^2)/(\mu\xi_0^2)}\right)$ as $\xi \to \infty$, while when $\mu < 0$, t blows up to infinity as $\xi \to \sqrt{-1/\mu}$ (Fig. 16.12). Thus, when $\mu > 0$,

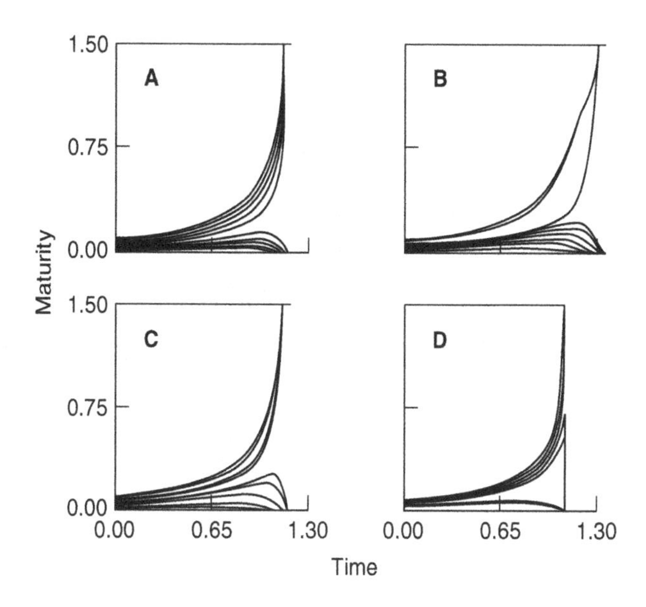

Figure 16.11 Typical numerical solutions of the ovulation model. Each numerical simulation was started with a group of 10 follicles with maturities randomly distributed between 0 and 0.1. Parameter values are $M_1 = 3.85$, $M_2 = 15.15$. In panels A and D five follicles ovulate (their maturity blows up in finite time), while in panels B and C only four ovulate. All other follicles atrophy and die. (Lacker, 1981, Fig. 7.)

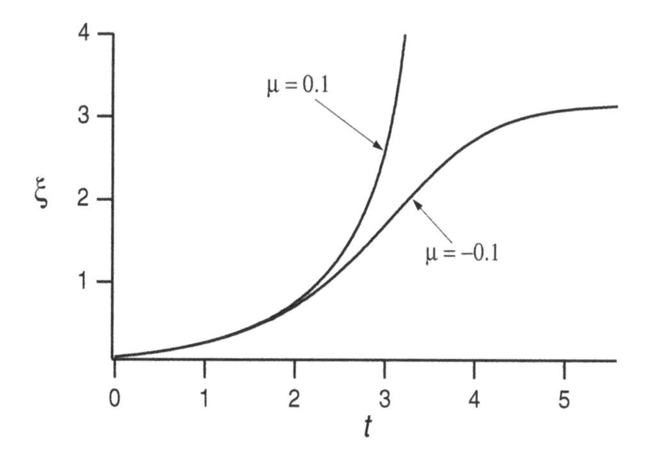

Figure 16.12 Symmetric solutions of the ovulation model for two values of μ. The initial condition was set arbitrarily at $\xi_0 = 0.1$. When $\mu > 0$, the solution blows up in finite time, while when $\mu < 0$, the solution approaches a steady state as $t \to \infty$.

ξ becomes infinite in finite time, while when $\mu < 0$, ξ goes to the steady state $\sqrt{-1/\mu}$ as $t \to \infty$. The former solution corresponds to an ovulatory solution, and the time of ovulation, T, is

$$T = \log\left(\sqrt{\frac{1+\mu\xi_0^2}{\mu\xi_0^2}}\right). \tag{16.31}$$

It follows that if M is between M_1 and M_2, all M follicles progress to ovulation at time T and the other follicles are suppressed, but if M is outside this range, all M follicles go to the steady (nonovulatory) state

$$\xi_M = \frac{1}{M}\sqrt{\frac{-1}{\mu}} = \frac{1}{\sqrt{(M-M_1)(M-M_2)}}. \tag{16.32}$$

Hence, in the symmetric case, ovulation numbers must be between M_1 and M_2.

Solutions in Phase Space

To understand these symmetric solutions more fully, and to understand how they relate to the behavior of nonsymmetric solutions, it is helpful to consider the trajectories in the N-dimensional phase space defined by $\xi_i, i = 1, \ldots, N$. Each symmetric solution lies on a line of symmetry l_M of the M-dimensional coordinate hyperplane. This is illustrated in Fig. 16.13 for the case $N = 3$: the l_1 lines are the ξ_1, ξ_2, and ξ_3 axes, the l_2 lines lie in the two-dimensional coordinate planes, and the l_3 line makes a 45 degree angle with the ξ_1, ξ_2 plane. Note that not all the l_1 and l_2 lines of symmetry are included in the diagram. When M is between M_1 and M_2, l_M contains no critical point, and any trajectory starting on l_M goes to infinity along l_M, reaching infinity in finite time T. However, when M is outside the range of M_1 and M_2, l_M contains a critical point, P_M, and solutions that start on l_M stay on l_M, approaching P_M as $t \to \infty$. In Fig. 16.13, $M_1 = 1.9$ and $M_2 = 2.9$, and so the only possible ovulation number is 2. Thus, each l_1 contains a critical point, P_1, that prevents the ovulation of single follicles, and similarly for l_3. The l_2 lines are the only lines of symmetry not containing a critical point.

The relationship between the symmetric solutions and the general solutions is most easily seen by analyzing the stability of the critical points P_M. Linearizing (16.25) around P_M gives the linear system (after rearranging the variables)

$$\frac{d\tilde{P}}{dt} = A\tilde{P}, \tag{16.33}$$

where $\tilde{P}$ is a small perturbation from P_M and

$$A = \begin{pmatrix} A_1 + B_1 & B_2 \\ 0 & A_2 \end{pmatrix}, \tag{16.34}$$

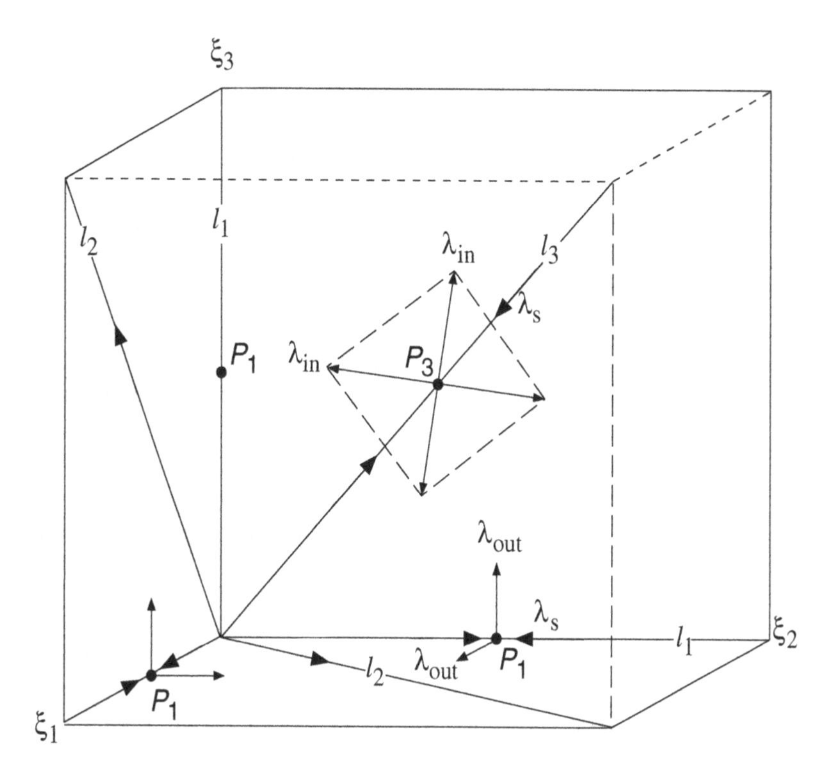

Figure 16.13 Phase space for a system of 3 interacting follicles ($N = 3$) with M_1 and M_2 chosen such that only two follicles ovulate. λ_{out} denotes eigenvalues with eigenvectors that point out of the coordinate hyperplane, while λ_{in} denotes eigenvalues with eigenvectors that are in the coordinate hyperplane. λ_s denotes the eigenvalue with eigenvector in the direction of the line of symmetry.

where A_1 and A_2 are diagonal matrices with diagonal elements a_1 and a_2 respectively, and B_1 and B_2 are matrices with all elements equal to b_1,

$$a_1 = \xi_M \frac{\partial \phi}{\partial \xi_i}\bigg|_{(\xi_M, M\xi_M)}, \qquad a_2 = \phi(0, M\xi_M), \qquad b_1 = \xi_M \frac{\partial \phi}{\partial \xi}\bigg|_{(\xi_M, M\xi_M)}. \tag{16.35}$$

The stability of P_M is determined by the eigenvalues of A, which, because of its block structure, are the eigenvalues of the two diagonal block matrices $A_1 + B_1$ and A_2. Hence, A has an eigenvalue $\lambda_{out} = a_2$ of multiplicity $N - M$ (from A_2), an eigenvalue $\lambda_s = a_1 + Mb_1 = -2$ of multiplicity 1, and an eigenvalue $\lambda_{in} = a_1$ of multiplicity $M - 1$, both coming from A_1. To verify the multiplicities of these eigenvalues, note that $A - a_1 I$ is a matrix of rank 1, so has a nullspace of dimension $M - 1$.

In the following discussion we let $Z = (\delta\xi_1, \ldots, \delta\xi_N)$ denote an eigenvector at P_M, and use subscripts to denote the different eigenvectors.

Perturbations Along l_M. Corresponding to the simple eigenvalue λ_s is the eigenvector Z_s whose components satisfy $\delta\xi_i = 1, i = 1, \ldots, M$, $\delta\xi_i = 0, i = M + 1, \ldots, N$. Hence,

Z_s is in the direction of l_M. Since $\lambda_s < 0$, it follows that l_M is on the stable manifold of P_M. Since symmetry is preserved along l_M, any solution that starts on l_M goes to P_M as $t \to \infty$.

Perturbations Orthogonal to l_M in the Coordinate Hyperplane. Corresponding to the eigenvalue λ_{in} are the eigenvectors $Z_1, \ldots, Z_{M-1}$ whose components are $\sum_{i=1}^{M} \delta\xi_i = 0, \delta\xi_i = 0, i = M+1, \ldots, N$. Z_1 to Z_{M-1} are independent vectors that lie in the coordinate hyperplane (since all have their last $M - N$ components equal to 0). Since they are also orthogonal to l_M, they span the orthogonal complement of l_M in the coordinate hyperplane.

Perturbations Orthogonal to l_M and the Coordinate Hyperplane. The eigenvalue λ_{out} has the corresponding eigenvectors $Z_{M+1}, \ldots, Z_{N-1}$ whose components satisfy $\sum_{i=M+1}^{N} \delta\xi_i = 0, \delta\xi_i = 0, i = 1, \ldots, M$. Finally, there is also the eigenvector Z_N with components $\delta\xi_i = (M-N)b_1, i = 1, \ldots, M$ and $\delta\xi_i = (a_1 - a_2) + Mb_1, i = M+1, \ldots, N$. All the eigenvectors corresponding to λ_{out} are orthogonal to both l_M and the coordinate hyperplane and span the orthogonal complement of the coordinate hyperplane.

These eigenvectors are illustrated in Fig. 16.13. At the critical point P_1, situated on the ξ_2 axis, there are two independent eigenvectors corresponding to λ_{out}, and these are both orthogonal to l_1, the ξ_2 axis. Note that as the coordinate hyperplane is a line in this case, there are no eigenvectors corresponding to λ_{in}. At P_3 the converse is true. Here there are no eigenvectors corresponding to λ_{out}, as the coordinate hyperplane is the entire space. In three dimensions, the only critical point that could have eigenvectors corresponding to all the eigenvalues λ_s, λ_{in}, and λ_{out} would be P_2. However, for these parameter values P_2 does not exist as $M_1 < 2 < M_2$.

It remains to determine the stability of each critical point P_M. This is easily done by direct computation of the eigenvalues, which gives

$$\lambda_s = a_1 + Mb_1 = -2, \tag{16.36}$$

$$\lambda_{in} = a_1 = \frac{(M_1 + M_2)M - 2M_1M_2}{(M - M_1)(M - M_2)}, \tag{16.37}$$

$$\lambda_{out} = a_2 = -\frac{(M_1 + M_2)M - M_1M_2}{(M - M_1)(M - M_2)}. \tag{16.38}$$

Plots of λ_{in} and λ_{out} as functions of M are shown in Fig. 16.14. For M between M_1 and M_2, P_M does not exist, but for $M > M_2$ and for $M < M^* = M_1M_2/(M_1 + M_2)$, λ_{in} and λ_{out} are of opposite signs. For $M^* < M < M_1$, the eigenvalues are both negative. It follows that if there are integers in the interval (M^*, M_1), then there are stable critical points P_M, with $M^* < M < M_1$. All other symmetric critical points are unstable.

Finally, we note (without proof) that there are critical points other than the symmetric ones, but they are all unstable.

In summary, when there are no integers in the interval (M^*, M_1) all the critical points are unstable and all the symmetric critical points are saddle points. In fact, from any starting point, all solutions approach infinity along one of the symmetric

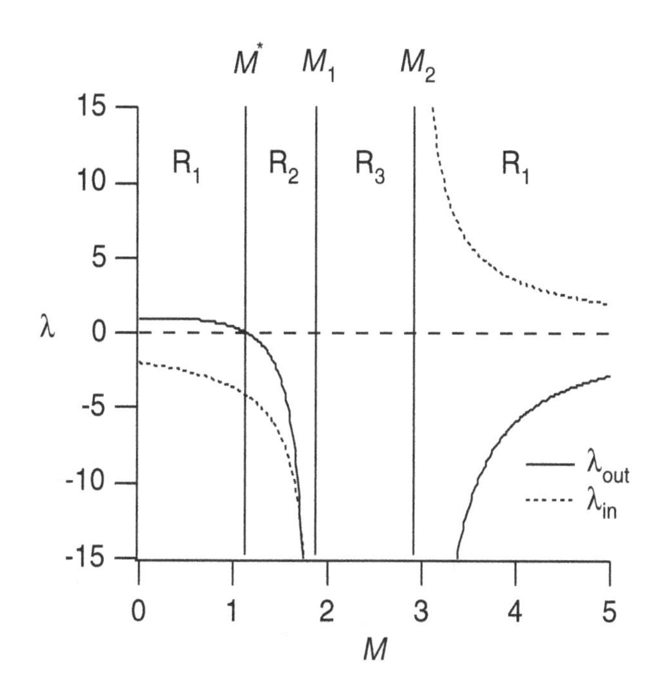

Figure 16.14 The eigenvalues of A as functions of M, calculated with the parameter values $M_1 = 1.9$, $M_2 = 2.9$. In the regions labeled R_1 the steady state P_M is a saddle point; in the region labeled R_2, P_M is stable; and in the region labeled R_3, there are no symmetric critical points, i.e., P_M does not exist for those values of M. Note that only integer values for M have any physical meaning. As described in the text, the eigenvalues λ_{out} and λ_{in} correspond, respectively, to eigenvectors pointing out and in of the symmetric hyperplane.

trajectories, l_M, where $M_1 < M < M_2$. These trajectories become infinite in finite time and are interpreted as ovulatory solutions. However, if there are integers in the interval (M^*, M_1), there are corresponding stable critical points. Any solution that starts in the domain of attraction of one of these stable critical points, P_{M_s} say, approaches P_{M_s} as time increases, and the system becomes stuck there. No follicles ovulate, but M_s follicles remain fixed at an intermediate maturity.

Stability of l_M

Although one might expect to observe ovulation numbers anywhere in the range M_1 to M_2, numerical simulations show that only some of these actually occur. This is illustrated in Fig. 16.15, where $M_1 = 3.85$ and $M_2 = 15.15$. In the previous numerical simulations (Fig. 16.11) we started with a fixed number of follicles with random initial maturities normally distributed; in Fig. 16.15, however, follicles mature at random times (generated by a Poisson process), so that the simulation more accurately reflects the physiological situation.

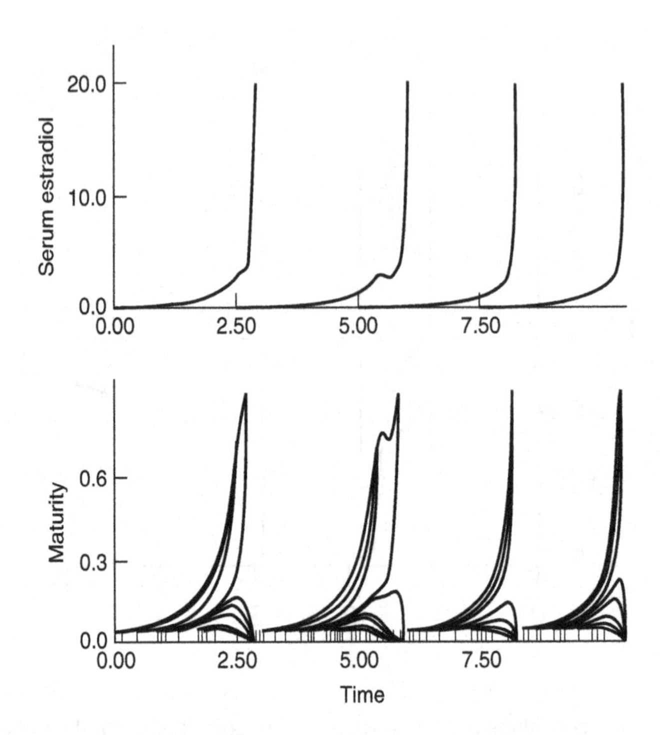

Figure 16.15 Typical solutions when follicles begin to develop at random times, generated by a Poisson process. Each small tick on the horizontal axis marks the initiation of development in a single follicle. Although the parameter values are $M_1 = 3.85$, $M_2 = 15.15$, and thus one might expect to observe ovulation numbers ranging from 4 to 15, only the ovulation numbers 4 and 5 are observed. (Lacker and Peskin, 1981, Fig. 11.)

Despite the random entry of follicles into the maturing pool, ovulation occurs at regular intervals, and the ovulation number varies little. Hence, the model generates periodic behavior from stochastic input. Furthermore, although we might expect to see ovulation numbers anywhere in the range 4 to 15, only the ovulation numbers 4 and 5 are observed. An explanation of this observation is found by examining the stability of the symmetric ovulatory solutions. This is done by transforming to a new coordinate system in which the ovulatory solutions, which become infinite in finite time, are transformed into finite critical points. The stability of these finite critical points can then be analyzed using standard linear stability methods.

We begin by noting that the initial ordering of a solution can never change. That is, if ξ_i starts above ξ_j, it remains above ξ_j for all time. This is true because

$$\frac{d}{dt}(\xi_i - \xi_j) = \xi_i\phi(\xi_i, \xi) - \xi_j\phi(\xi_j, \xi) = h(\xi_i, \xi_j, \xi)(\xi_i - \xi_j) \tag{16.39}$$

for some function h, and as long as ξ_i and ξ_j are bounded, so also is $h(\xi_i, \xi_j, \xi)$. Clearly,

$$\ln(\xi_i(t) - \xi_j(t)) = \ln(\xi_i(0) - \xi_j(0)) + \int_0^t h(\xi_i, \xi_j, \xi)\, dt. \tag{16.40}$$

If the right-hand side of this expression is bounded, so also is the left-hand side, so that $\xi_i(t) \neq \xi_j(t)$.

Since the original ordering of the maturities is preserved, we arrange the N follicles in order of maturity, with ξ_1 denoting the follicle with the greatest maturity, and define a new time scale by

$$\tau(t) = \int_0^t \xi_1^2(s)\, ds. \tag{16.41}$$

As $t \to T$, the finite time of ovulation, $\tau(t) \to \infty$. For as ξ_1 gets large, $d\xi_1/dt \approx \xi_1^3$, and hence ξ_1^2 behaves like $1/(T-t)$ as $t \to T$. Furthermore, ξ_1^2 is positive, and so τ is an increasing function of t that is therefore invertible. We use the inverse function to define new variables

$$\gamma_i(\tau) = \frac{\xi_i(t(\tau))}{\xi_1(t(\tau))}, \tag{16.42}$$

$$\Gamma(\tau) = \frac{\xi(t(\tau))}{\xi_1(t(\tau))}. \tag{16.43}$$

In terms of these new variables (16.26)–(16.27) become

$$\frac{d\gamma_i}{d\tau} = \gamma_i \Phi(\gamma_i, \Gamma), \qquad i = 1, \ldots, N, \tag{16.44}$$

$$\Gamma = \sum_{j=1}^{N} \gamma_j, \tag{16.45}$$

$$\Phi(\gamma_i, \Gamma) = (1-\gamma_i)[M_1 M_2(1+\gamma_i) - \Gamma(M_1+M_2)]. \tag{16.46}$$

Note that $\gamma_1(\tau) \equiv 1$, and $0 \le \gamma_i(\tau) \le 1$ for each i.

All ovulatory and anovulatory solutions correspond to critical points of (16.44)–(16.46) of the form

$$\gamma_i = \begin{cases} 1, & i = 1, \ldots, M, \\ 0, & i = M+1, \ldots, N. \end{cases} \tag{16.47}$$

Although ovulatory and anovulatory solutions look the same, they can be distinguished by determining whether the original variable ξ is finite. If so, the critical point corresponds to an anovulatory solution.

Equations (16.44)–(16.46) have only two distinct eigenvalues,

$$\lambda_1 = (M_1+M_2)M - 2M_1M_2, \tag{16.48}$$

$$\lambda_2 = -(M_1+M_2)M + M_1M_2, \tag{16.49}$$

which are plotted in Fig. 16.16. If M lies between M^* and $2M^*$, then both λ_1 and λ_2 are negative, so that the critical point is stable. Otherwise, the critical point is unstable. It follows that only ovulation numbers between M_1 and $2M^*$ are stable, and are therefore observable.

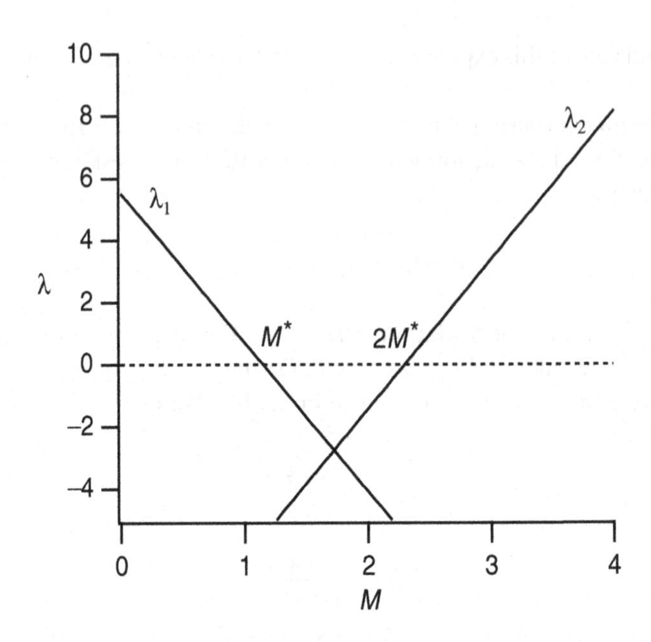

Figure 16.16 The eigenvalues λ_1 and λ_2 of the ovulation model in the transformed variables γ_i and τ.

In the numerical simulations shown in Fig. 16.15, $M_1 = 3.85$ and $M_2 = 15.15$, in which case $2M^* = 6.14$. This is consistent with the numerical simulations in which only ovulation numbers 4 and 5 were observed. One possible reason why ovulation number 6 is not observed is that it lies close to the stability boundary. Thus, its domain of attraction is relatively small, and therefore the probability that a random process finds this domain of attraction is also small.

The Effect of Population Size

With this model we can suggest an answer to the question of efficiency, namely, why do so many follicles begin the maturation process, only to atrophy and die? The answer appears to be that the mean time to ovulation is controlled more precisely by a large population than by a small one. This is illustrated in Fig. 16.17. For this figure, the model was simulated for 80 cycles, and the distribution of ovulation numbers and times was plotted for three different population sizes. Each population had the parameter values $M_1 = 6.1, M_2 = 5000$, and thus the expected ovulation numbers lie in the range 7 to 12. As the population size increases, the mean ovulation number decreases, but the shape of the distribution does not change a great deal. However, although the mean ovulation time (shown here centered at 0) does not change as the population size is increased, the distribution sharpens dramatically, and the range of observed ovulation times is dramatically reduced. Thus, while the majority of follicles atrophy and die, they have an important, although not immediately obvious, function: helping to regulate the timing

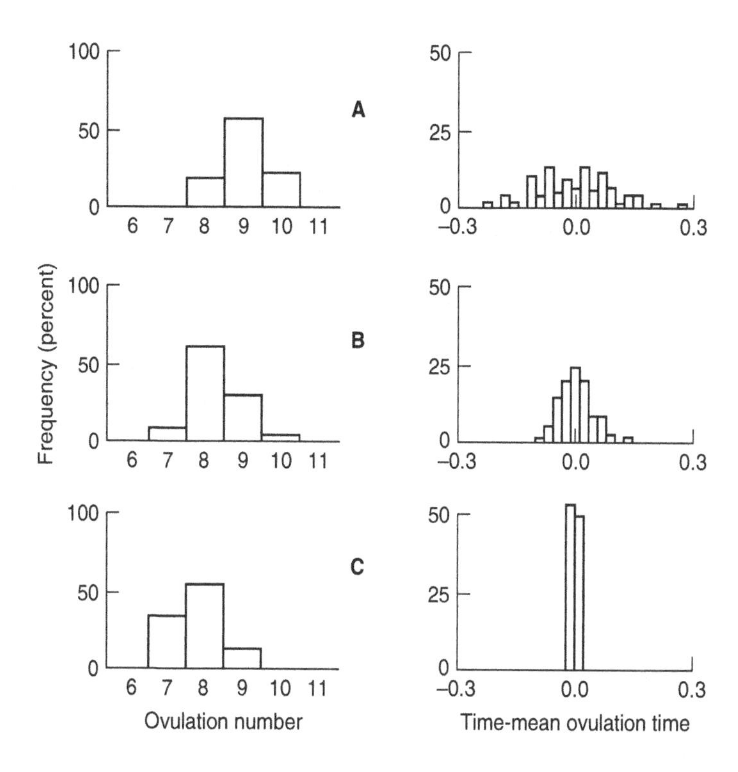

Figure 16.17 The effect of the population size on the distribution of ovulation numbers. In panels A, B, and C, there are, respectively, 10, 100, and 1000 follicles interacting. Parameter values are $M_1 = 6.1$, $M_2 = 5000$. As the follicle population gets larger, the mean ovulation time decreases only slightly, while the standard deviation of the distribution of ovulation times decreases. Thus, larger populations allow more precise control over the ovulation time. (Lacker, 1981, Fig. 13.)

of ovulation. This provides a possible explanation of why women near menopause (i.e., with fewer available oocytes) typically experience menstrual irregularities.

An Application of the Lacker Model

The Lacker model has been used as a tool to predict the response to ovarian superstimulation (Sarty and Pierson, 2005). A possible treatment for infertility is superstimulation of the ovaries by exogenous doses of hormones; an equivalent to FSH during the first portion of the ovulation cycle, followed by an equivalent to LH. In one treatment program, the size of the follicles and the level of estradiol were measured over the course of the treatment, with a number of individual follicles followed from day to day. For each follicle followed, this gave a vector $\xi_i(t), \xi(t), \dot{\xi}_i(t)$, where the derivative is calculated numerically from the data. Model parameters were then chosen so that $\dot{\xi}_i - \xi_i\phi(\xi_i, \xi)$ is as small as possible in the least squares sense.

Because the experimental data were collected during a period of exogenous superstimulation of the ovaries, one should not expect the Lacker model to give a close fit to the data, and indeed it does not. However, the model is a useful predictor of the ratio of ovulating follicles to dominant follicles (defined to be those with a diameter of 19 mm or greater). In particular, the model predicts that faster-growing follicles are less likely to ovulate, i.e., the speed with which the largest follicle reaches the target size of 19 mm is inversely related to probable ovulatory success. This result is similar to others which have shown that follicles from women who did not ovulate show more rapid growth patterns than follicles from women who do. Although the reasons for this are unclear, it gives a clear indication that to increase the number of ovulating follicles, it might be advantageous to slow the rate of follicle growth, possibly by reducing the size of the exogenous FSH dose.

16.2.3 Other Models of Ovulation

Although this model of ovulation is one of the simplest and most elegant, other, more complex, models have been constructed. For example, Schwarz (1969) and Bogumil et al. (1972) have proposed models that incorporate large numbers of parameters and are based more directly on experimental data.

The Lacker model was extended and generalized by Chávez-Ross et al. (1997) in a study of the possible mechanisms underlying polycystic ovary syndrome, a condition in which many large follicles develop but none becomes predominant, leading to failure of ovulation.

A model of a different type is due to Faddy and Gosden (1995). They constructed a compartmental model of follicle dynamics over the lifetime of an individual female and fit their model to experimental data to obtain follicle growth and death rates as functions of the individual's age. Although this compartmental model does not provide insight into the mechanisms underlying periodic ovulation and a constant ovulation number, it provides an understanding of follicle dynamics over a larger time span.

Most recently, Clément, Monniaux and their colleagues (Clément et al., 1997, 2001, 2002; Clément, 1998; Echenim et al., 2005) have modeled the follicle selection process, in a manner quite different from that of Lacker. In the Clément model, the progress of the granulosa cells through the cell cycle (see Chapter 10) is monitored, with some cells becoming atretic (i.e., dying) and others becoming differentiated. For each follicle, the maturity of the granulosa cells is described by a density function which satisfies a conservation PDE. The rates of entry into apoptosis and differentiation are governed by the blood concentration of FSH, where the rate of FSH production is governed by the total maturity of all the follicles, thus giving feedback between the follicles.

So far (i.e., by 2007), this model has been solved only for a population of five follicles, so it is too early to know whether or not it will prove to be an accurate and useful model of the follicle selection process.

16.3 Insulin and Glucose

Hormones secreted from cells in the pancreas are responsible for the control of glucose, amino acids, and other molecules that are necessary for metabolism. The pancreas contains a large number of secretory cells, grouped into about one million *islets of Langerhans* consisting of approximately 2,500 cells each. There are three principal secretory cell types: the α-cells secrete glucagon, the β-cells secrete insulin (see Chapter 9), and the δ-cells secrete somatostatin. Glucagon and insulin have complementary actions. A high concentration of glucose in the bloodstream (corresponding to an overabundance of nutrients) stimulates the production of insulin, which in turn induces storage of excess nutrient and decreases the rate at which nutrients are mobilized from storage areas such as adipose tissue or the liver. Insulin acts principally on three tissues: striated muscle (including the heart), liver, and adipose tissue. All the actions of insulin apparently stem from its interaction with a specific receptor in the plasma membrane of insulin-sensitive cells. How this interaction leads to the many actions of insulin on the cell is not fully understood. In striated muscle and adipose tissue, one important action of insulin is to stimulate the transport of glucose into the cell by a specific carrier (or carriers) in the plasma membrane. It appears to do this by recruiting glucose carriers to the plasma membrane from intracellular sites where they are inactive. Insulin thus increases the V_{max} of transport, often as much as 10- to 20-fold. When glucose enters the cell it is rapidly phosphorylated and metabolized.

In the case of the liver, insulin does not increase the rate of transport of glucose into the cell (although it increases the net uptake of glucose). In the liver, insulin acts on a number of intracellular enzymes to increase glucose storage and decrease mobilization of glucose stores. The details of how insulin does this are far from clear.

Glucagon raises the concentration of glucose in the bloodstream. It acts mainly but not entirely on the liver, where it stimulates glycogen breakdown and the formation of glucose from noncarbohydrate precursors such as lactate, glycerol, and amino acids. Glucagon released in the islets stimulates the β cells in the vicinity to secrete insulin.

Insulin Units

Historically, a unit (U) of insulin was defined to be that amount of insulin (in cubic centimeters) that lowers the percentage of blood sugar in a normal rabbit to 0.045 in 2 to 6 hours. The crudity of such a unit was the result of the fact that it was not possible to purify insulin until relatively recently, and thus a bioassay was the only way of determining the amount. An excellent discussion of historical insulin units is given by Lacy (1967).

Later, mouse units became more convenient, and a unit was defined to be the amount of insulin required to produce convulsions in half the mice under standard conditions. A mouse unit is about 1/600 of a rabbit unit. Fortunately, insulin extracted from most animals has equivalent activity in rabbits, mice, and men, although the guinea pig and capybara are exceptions to this rule. Various modifications were made to the conditions of these assays, but with the advent of reasonably pure preparations of insulin the unit has been redefined as 1/24 milligrams.

16.3.1 Insulin Sensitivity

Insulin resistance (i.e., a lowered ability of insulin to control blood glucose) plays a crucial role in diabetes, and has been implicated in a host of other diseases. Thus there have been a large number of studies of how insulin sensitivity can be measured experimentally (Mari, 2002). In general, these methods involve the addition of exogenous glucose and/or insulin, and then measurement of the time course of their blood concentrations. One of the most accurate methods is the euglycemic hyperinsulinemic glucose tolerance test (deFronzo et al., 1979). In this test a steady intravenous infusion of insulin is administered, to bring the blood concentration to a steady high level, while simultaneously the blood glucose concentration is clamped by continuous infusion of glucose through a second intravenous infusion. Once a steady state has been reached (which can take over an hour), the amount of glucose needed to maintain the glucose clamp is a measure of insulin sensitivity. At high insulin levels there is little glucose production by the liver, and thus the rate of glucose removal from the blood is due almost entirely to glucose metabolism, including uptake into the periphery. Thus, if insulin sensitivity is higher, it takes more glucose added continuously in the clamp to maintain a steady state; conversely, if insulin sensitivity is low, then the high insulin levels will not stimulate much glucose removal, and less added glucose is needed to maintain the steady state. However, since this procedure requires simultaneous IV infusions over a long time, and numerous blood samples, it is difficult to implement.

Subsequent methods have tried to determine insulin sensitivity in ways that are easier to apply. Some, such as the intravenous glucose tolerance test (IVGTT; for a mathematical study of this test see Gaetano and Arino, 2000), examine the response to a single intravenous bolus of glucose, while others such as the oral glucose tolerance test (OGTT; Mari et al., 2001) consider the response to a bolus of glucose taken orally. These methods depend crucially on mathematical models of the underlying processes, and most use, in one form or another, the so-called *minimal model* of Bergman et al. (1979; Bergman, 1989).

The meal tolerance test of Caumo et al. (2000), which is similar to the OGTT, provides a typical example of this kind of approach. In this test, glucose is administered orally and then the blood concentration of both glucose and insulin are measured as functions of time. (As a matter of interest, the meal consisted of 15 g Weetabix, 10 g skimmed milk, 250 ml pineapple juice, 50 g white-meat chicken, 60 g wholemeal bread and 10 g polyunsaturated margarine.)

To build a model of insulin sensitivity, we let $G(t)$, in units of mg/dl, denote the blood concentration of glucose, and let $X(t)$, in units of min^{-1}, denote insulin activity. We assume that G has a base value, G_b, to which it returns in the absence of perturbation, and that G is decreased at a rate proportional to both X and G. Hence,

$$\frac{dG}{dt} = p_1(G_b - G) - XG + \frac{r(t)}{V}, \tag{16.50}$$

where $r(t)$ is the rate at which glucose enters the bloodstream as a result of the meal, and is not known a priori. Here V is the glucose distribution volume per unit body weight, in units of ml/kg.

Similarly, insulin activity is assumed to increase whenever the concentration of insulin, $I(t)$ (in units of μU/ml), is above its base level, I_b, and is also assumed to decay spontaneously. Thus,

$$\frac{dX}{dt} = p_3(I(t) - I_b) - p_2 X. \tag{16.51}$$

The parameter p_2 has units min^{-1}, while p_3 has units ml μU^{-1}min^{-2}. Notice that at steady state, $X = \frac{p_3}{p_2}(I - I_b)$, so that the ratio $\frac{p_3}{p_2}$ is a measure of insulin sensitivity. That is, the larger the value of $\frac{p_3}{p_2}$, the greater the insulin activity for a given insulin concentration.

The test for insulin sensitivity is an inverse problem. That is, the glucose and insulin concentrations, $G(t)$ and $I(t)$, are measured quantities, and we wish to determine the ratio $\frac{p_3}{p_2}$.

First, notice that by integrating (16.51) from $t = 0$ to $t = \infty$, we find

$$\frac{p_3}{p_2} = \frac{\int_0^\infty X\, dt}{\int_0^\infty (I(t) - I_b)\, dt}, \tag{16.52}$$

where we have assumed that X is the same at the beginning ($t = 0$) as at the end ($t = \infty$) of the experiment, so that $\int_0^\infty \frac{dX}{dt}\, dt = 0$.

Since $I(t)$ is known, it remains to express $\int_0^\infty X\, dt$ in terms of known quantities. Dividing (16.50) by G and integrating from $t = 0$ to $t = \infty$, we learn that

$$\int_0^\infty \frac{1}{G}\frac{dG}{dt}\, dt = p_1 \int_0^\infty \frac{G_b - G}{G}\, dt - \int_0^\infty X\, dt + \frac{1}{V}\int_0^\infty \frac{r(t)}{G}\, dt. \tag{16.53}$$

However, since G is also the same at the beginning and end of the experiment, so that $\int_0^\infty \frac{1}{G}\frac{dG}{dt}\, dt = 0$, it follows that

$$\int_0^\infty X\, dt = p_1 \int_0^\infty \frac{G_b - G}{G}\, dt + \frac{1}{V}\int_0^\infty \frac{r(t)}{G}\, dt. \tag{16.54}$$

If we can estimate $r(t)$, the problem is solved. One option is to assume some functional form for $r(t)$, ensuring that it peaks at approximately the correct time and has the correct integral (since the total amount of added glucose is known). However, the approach used by Caumo et al. (2000) was to estimate the integral of r/G based on knowledge of G. An alternative model of G is

$$\frac{dG}{dt} = a(G_b - G) + br(t), \tag{16.55}$$

for some unknown constants a and b. The assumption here is that both (16.55) and (16.50) provide reasonable descriptions of the data for $G(t)$, provided a and b are chosen appropriately.

However, we can eliminate a and b from the problem as follows. First we integrate (16.55) directly, to get

$$\frac{a}{b} = \frac{\int_0^\infty r(t)\,dt}{\int_0^\infty (G - G_b)\,dt}. \tag{16.56}$$

Next we divide (16.55) by G and integrate, to get

$$\int_0^\infty \frac{r(t)}{G}\,dt = \frac{a}{b}\int_0^\infty \left(\frac{G - G_b}{G}\right)\,dt. \tag{16.57}$$

Substituting for $\frac{a}{b}$, we find that

$$\int_0^\infty \frac{r(t)}{G}\,dt = \left(\frac{\int_0^\infty r(t)\,dt}{\int_0^\infty (G - G_b)\,dt}\right)\int_0^\infty \left(\frac{G - G_b}{G}\right)\,dt. \tag{16.58}$$

Since $\int_0^\infty r(t)\,dt$ is the total amount of glucose ingested in the meal, which is known, the combination of (16.52), (16.54) and (16.58) gives an expression for p_3/p_2 entirely in terms of known quantities, as desired.

16.3.2 Pulsatile Insulin Secretion

Insulin secretion oscillates on a number of different time scales, ranging from tens of seconds to more than 100 minutes. The fast oscillations are caused (at least in part) by bursting electrical activity described in Chapter 9. During each burst of action potentials the cytoplasmic Ca^{2+} concentration rises as Ca^{2+} flows in through voltage-gated Ca^{2+} channels, and this rise in Ca^{2+} stimulates insulin secretion. Oscillations with a much larger period of around 100 minutes are also observed, and are called *ultradian oscillations* (Fig. 16.18). Finally, oscillations with intermediate frequencies of around 10 minutes or so also occur. One of the earliest observations of these oscillations was made in the rhesus monkey by Goodner et al. (1977), and some of their results are reproduced in Fig. 16.19. Glucagon and insulin oscillate out of phase, while insulin and glucose are in phase, with the increase of glucose leading the increase of insulin by an average of about one minute. Oscillations with intermediate frequency are also observed in isolated rat islets (Bergstrom et al., 1989; Berman et al., 1993), although, as can be seen from Fig. 16.20, spectral analysis is usually necessary to determine the principal underlying frequency. Once the underlying trend has been removed, a spectral decomposition of the data shows a frequency peak at about 0.07 min^{-1}, corresponding to a period of 14.5 minutes.

Ultradian insulin oscillations have a number of observable features. First, oscillations occur during constant intravenous glucose infusion and are not dependent on periodic nutrient absorption from the gut. However, damped oscillations occur after a

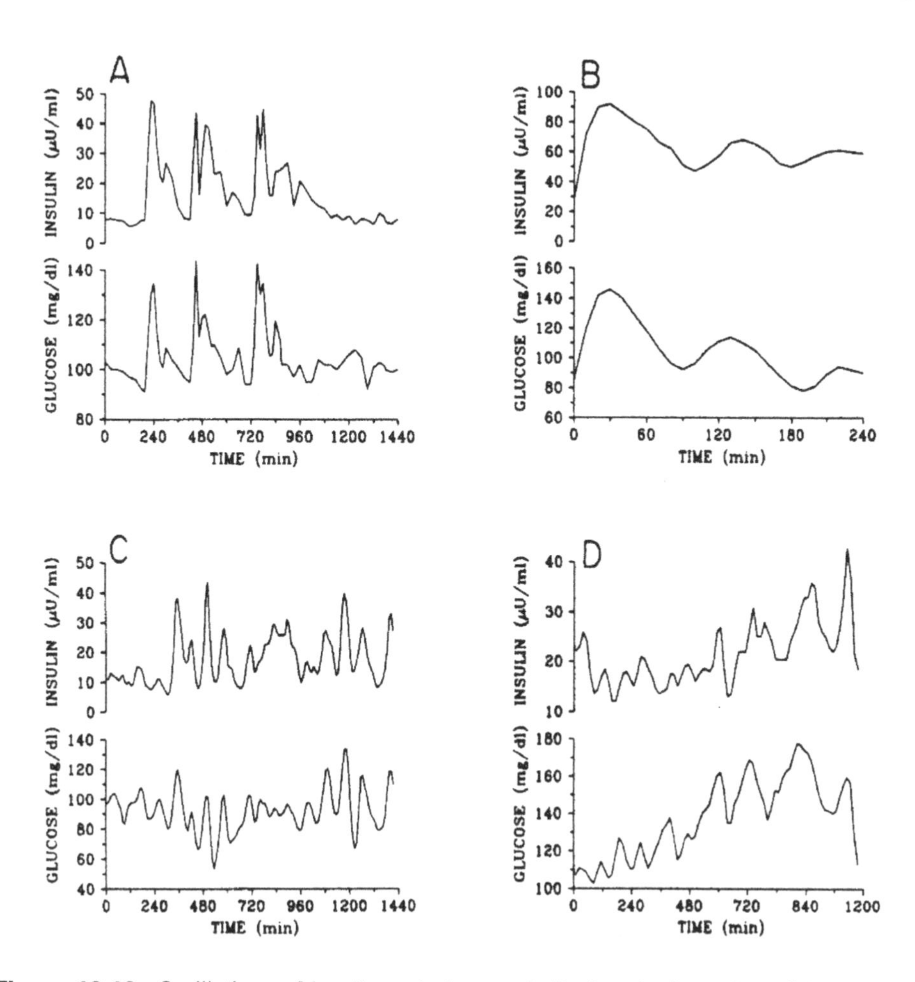

Figure 16.18 Oscillations of insulin and glucose. A: During the ingestion of three meals. B: During oral glucose. C: During continuous nutrition. D: During constant glucose infusion. Oscillations with a period of around 120 minutes occur even during constant stimulation (i.e., constant glucose infusion), and occur in a damped manner after a single stimulus such as ingestion of a meal. (Sturis et al., 1991, Fig. 1.)

single stimulus such as a meal. Second, glucose and insulin concentrations are highly correlated, with the glucose peak occurring about 10–20 minutes earlier than that of insulin. Third, the amplitude of the oscillations is an increasing function of glucose concentration, while the frequency is not; and fourth, the oscillations do not appear to depend on glucagon.

Although there are many possible mechanisms that are consistent with the above observations, they can all be explained by a relatively simple model (Sturis et al., 1991) in which the oscillations are produced by interactions between glucose and insulin.

A schematic diagram of the model is shown in Fig. 16.21. There are three pools in the model, representing remote insulin storage in the interstitial fluid, insulin in the

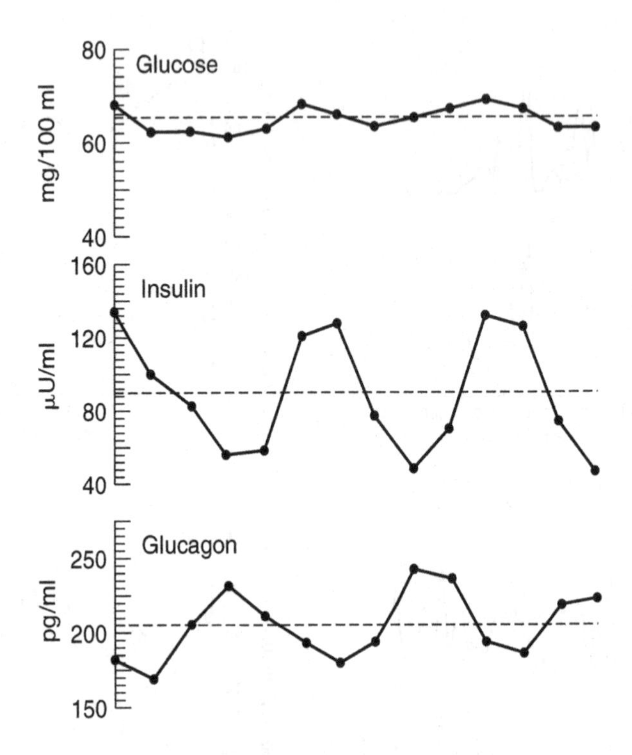

Figure 16.19 Intermediate frequency oscillations of glucose, insulin, and glucagon in monkeys. (Goodner et al., 1977, Fig. 1A.)

blood, and blood glucose. It turns out that two insulin pools are necessary, which is, by itself, an interesting model prediction. There are two delays, one explicit and the other implicit. Although plasma insulin regulates glucose production, it does so only after a delay of about 36 minutes. This delay is incorporated explicitly as a three-stage linear filter. An additional implicit delay arises because glucose utilization is regulated by the remote (interstitial) insulin, and not by the plasma insulin, while glucose has a direct effect (through insulin secretion from the pancreas) on plasma insulin levels.

We let I_p and I_i, in units of mU, denote the amounts of plasma insulin and remote insulin, respectively, and we let G, in units of mg, denote the total amount of glucose. Then the model equations follow from the following assumptions:

1. Plasma insulin is produced at a rate $f_1(G)$ that is dependent on plasma glucose. The insulin exchange with the remote pool is a linear function of the concentration difference between the pools $I_p/V_p - I_i/V_i$ with rate constant E, where V_p is the plasma volume and V_i is the interstitial volume. In addition, there is linear removal of insulin from the plasma by the kidneys and the liver, with rate constant $1/t_p$. Thus,

$$\frac{dI_p}{dt} = f_1(G) - \left(\frac{I_p}{V_p} - \frac{I_i}{V_i}\right)E - \frac{I_p}{t_p}. \tag{16.59}$$

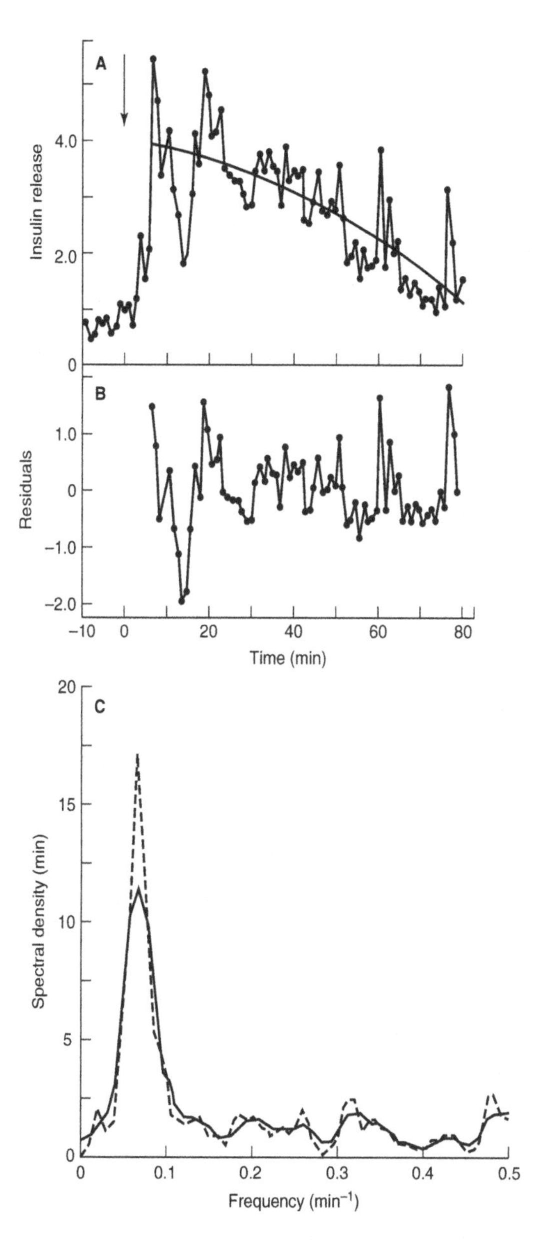

Figure 16.20 A: Oscillations of insulin release in perifused islets. The data indicate a slow time scale decreasing trend (the smooth line) on which are superimposed faster time scale oscillations. B: When the slow decrease is removed from the data, the residuals exhibit oscillations around 0. C: Spectral analysis of the residuals shows a frequency peak at about 0.07 min^{-1}, corresponding to oscillations with a period of 14.5 minutes. The dashed and continuous lines correspond to two different filters used in the spectral analysis. (Bergstrom et al., 1989. Figs. 1A, C, and 3.)

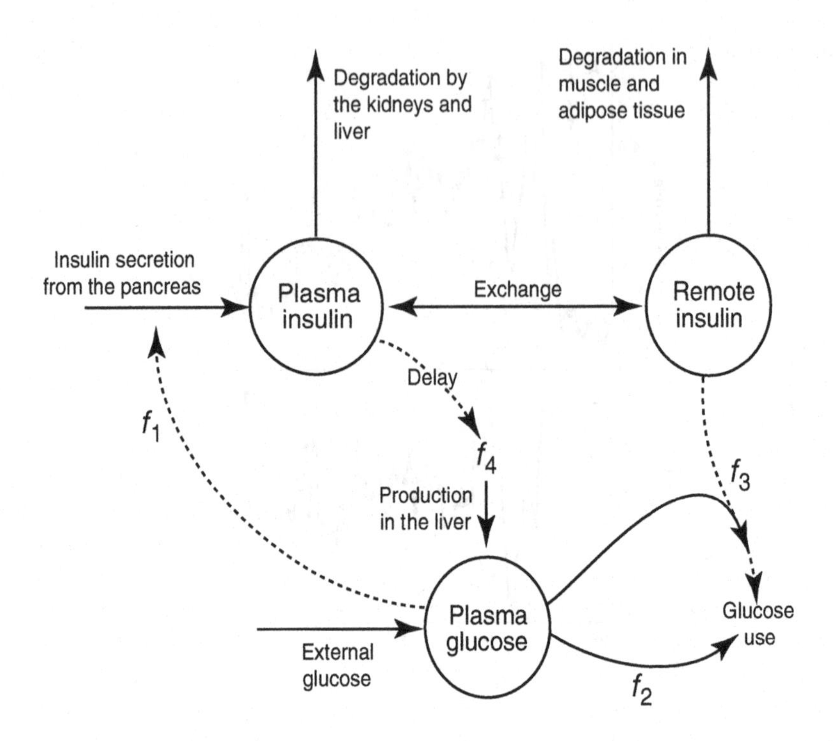

Figure 16.21 Schematic diagram of the model of ultradian insulin oscillations.

Note that this equation and the two that follow are written in terms of total amounts of insulin and glucose, rather than concentrations. Formulations using concentrations or total quantities are equivalent, provided that the blood and interstitial volumes remain constant, which we assume.

2. Remote insulin accumulates via exchange with the plasma pool and is degraded in muscle and adipose tissue at rate $1/t_i$:

$$\frac{dI_i}{dt} = \left(\frac{I_p}{V_p} - \frac{I_i}{V_i}\right)E - \frac{I_i}{t_i}. \tag{16.60}$$

3. Plasma glucose is produced at a rate f_4 that is dependent on plasma insulin, but only indirectly, as f_4 is a function of h_3, the output of a three-stage linear filter. The input to the filter is I_p, so glucose production is regulated by plasma insulin but delayed by the filter. There is input $I_G(t)$ from the addition of glucose from outside the system, by eating a meal, say. Finally, glucose is removed from the plasma by two processes. Thus,

$$\frac{dG}{dt} = f_4(h_3) + I_G(t) - f_2(G) - f_3(I_i)G. \tag{16.61}$$

Glucose utilization is described by two terms: $f_2(G)$ describes utilization of glucose that is independent of insulin, as occurs, for example, in the brain, and is

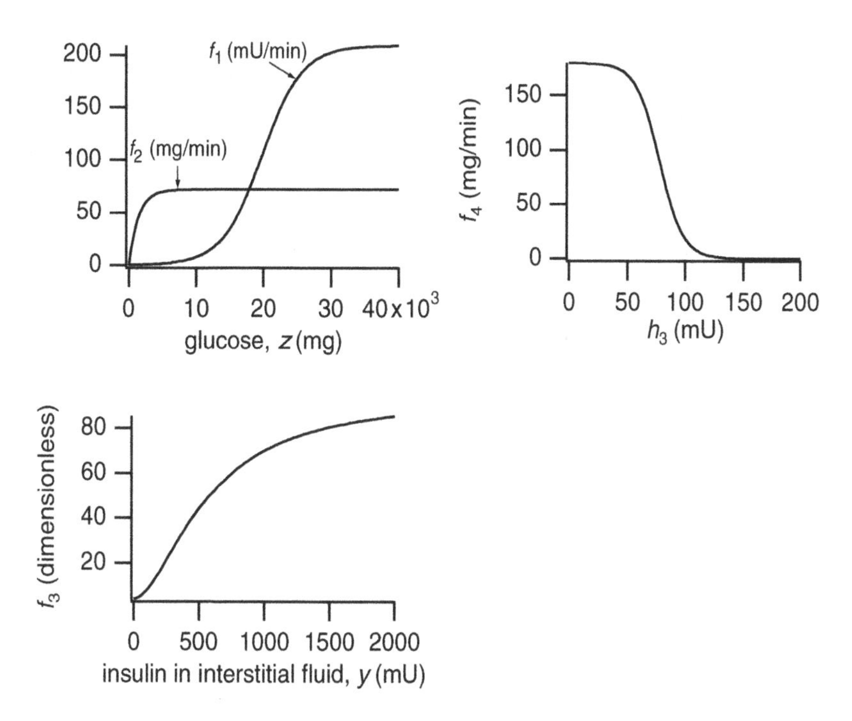

Figure 16.22 Graphs of $f_1, \ldots, f_4$ in the model of ultradian insulin oscillations. The exact forms of these functions are not physiologically significant, but are chosen to give the correct qualitative behavior.

an increasing function that saturates quickly. The second removal term, $f_3(I_i)G$, describes insulin-dependent utilization of glucose. f_3 is an increasing, sigmoidal function of I_i.

4. The three-stage linear filter satisfies the system of differential equations

$$t_d \frac{dh_1}{dt} = I_p - h_1, \tag{16.62}$$

$$t_d \frac{dh_2}{dt} = h_1 - h_2, \tag{16.63}$$

$$t_d \frac{dh_3}{dt} = h_2 - h_3. \tag{16.64}$$

The specific functional forms used for $f_1, \ldots, f_4$ are

$$f_1(G) = \frac{R_m}{1 + \exp(\frac{-G}{V_g C_1} + a_1)}, \tag{16.65}$$

$$f_2(G) = U_b \left[1 - \exp\left(\frac{-G}{C_2 V_g} \right) \right], \tag{16.66}$$

Table 16.4 Parameter values for the model of ultradian insulin oscillations.

V_p	3 l	C_2	144 mg l^{-1}
V_i	11 l	C_3	100 mg l^{-1}
V_g	10 l	C_4	80 mU l^{-1}
E	0.2 l min^{-1}	C_5	26 mU l^{-1}
t_p	6 min	U_b	72 mg min^{-1}
t_i	100 min	U_0	4 mg min^{-1}
t_d	12 min	U_m	94 mg min^{-1}
R_m	209 mU min^{-1}	R_g	180 mg min^{-1}
a_1	6.67	α	7.5
C_1	300 mg l^{-1}	β	1.77

$$f_3(I_i) = \frac{1}{C_3 V_g}\left(U_0 + \frac{U_m - U_0}{1 + (\kappa I_i)^{-\beta}}\right), \tag{16.67}$$

$$f_4(h_3) = \frac{R_g}{1 + \exp[(\alpha(\frac{h_3}{C_5 V_p} - 1)]}, \tag{16.68}$$

where $\kappa = \frac{1}{C_4}\left(\frac{1}{V_i} + \frac{1}{Et_i}\right)$, and these are graphed in Fig. 16.22. The parameter values are given in Table 16.4.

Numerical solution of the model equations shows that a constant infusion of glucose causes oscillations in insulin and glucose. As I increases, the oscillation period remains practically unchanged, but the amplitude increases (Fig. 16.23), in good qualitative agreement with experimental data. However, it is interesting that these oscillations disappear if the compartment of remote insulin is removed from the model. This indicates that the division of insulin into two functionally separate stores could play an important role in the dynamic control of insulin levels. Another prediction of the model is that the oscillations are dependent on the delay in the regulation of glucose production. If the delay caused by the three-stage filter is either too large or too small, the oscillations disappear (Exercise 8).

The Effect of Oscillatory Insulin Release

Sturis et al. (1995) found experimentally that an oscillatory insulin supply is more efficient at increasing glucose usage than is a constant insulin supply. The experiments were done by the addition of exogenous insulin, after somatostatin was used to suppress endogenous insulin production. Tolić et al. (2000) then used a simplified version of the above model of ultradian insulin oscillations to explain how such dependency on insulin oscillations could arise.

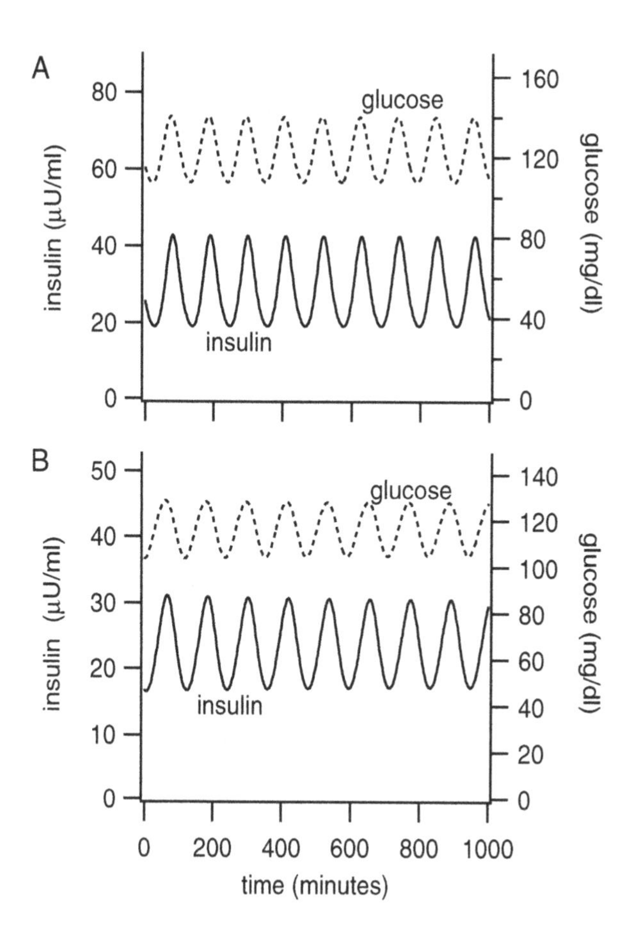

Figure 16.23 Ultradian insulin oscillations in the model. The glucose infusion rates are A: $I = 108$ mg/min, and B: $I = 216$ mg/min. Note that insulin and glucose are expressed in units of concentration. An amount is easily converted to a concentration by dividing by the volume of the appropriate compartment.

16.4 Adaptation of Hormone Receptors

It remains to answer the question of why hormone secretion is pulsatile in the first place. As with many oscillatory physiological systems, there is no completely satisfactory answer to this question. However, one plausible hypothesis has been proposed by Li and Goldbeter (1989). Based on a model of a hormone receptor first constructed by Segel, Goldbeter, and their coworkers (Segel et al., 1986; Knox et al., 1986), Li and Goldbeter constructed a model of a hormone receptor that responds best to stimuli of a certain frequency, thus providing a possible reason for the importance of pulsatility.

Closely linked to this hypothesis is the phenomenon of receptor adaptation. Often, the response to a constant hormone stimulus is much smaller than the response to a

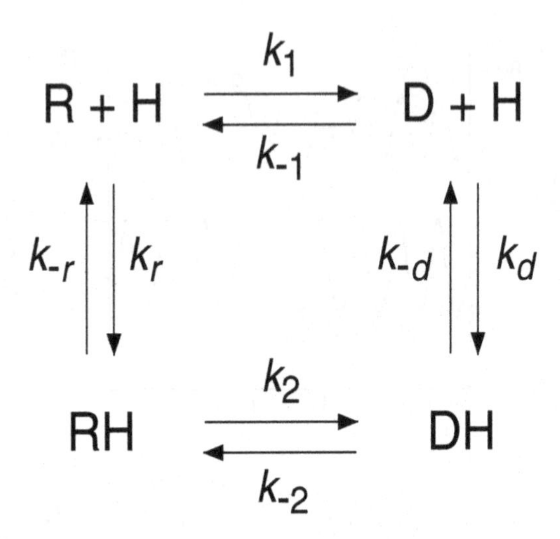

Figure 16.24 Schematic diagram of a model of a hormone receptor.

time-varying stimulus. In the extreme case, the receptor responds to a time-varying input, but has no response to a steady input, regardless of the input magnitude, a phenomenon called *exact adaptation*. A number of examples of adaptation are described in this book; for example, the models of the IP_3 receptor discussed in Chapter 7 show adaptation in their response to a step-function increase in Ca^{2+} concentration; i.e., their response is an initial peak in the Ca^{2+} release, followed by a decrease to a lower plateau as the receptor is slowly inactivated by Ca^{2+}. Similarly, as described in Chapter 19 biochemical feedback in photoreceptors can result in a system that displays remarkably precise adaptational properties, as embodied in Weber's law. Because of the importance of adaptation in physiological systems, it is interesting to study how adaptation arises in a simple receptor model.

The key assumption is that the hormone receptor can exist in two different conformational states, R and D, and each conformational state can have hormone bound or unbound (Fig. 16.24). For simplicity we assume that the active form of the receptor has hormone bound to the receptor in state R. The addition of hormone to the receptor system causes a change in the proportion of each receptor state, but the total receptor concentration is fixed.

Letting r, x, y, d denote $[R]/R_T, [RH]/R_T, [DH]/R_T$ and $[D]/R_T$ respectively, where R_T is the total receptor concentration, we find the following equations for the receptor system:

$$\frac{dr}{dt} = -[k_1 + k_r H(t)]r + k_{-r}x + k_{-1}d, \tag{16.69}$$

$$\frac{dx}{dt} = k_r H(t) r - (k_2 + k_{-r})x + k_{-2}y, \tag{16.70}$$

$$\frac{dy}{dt} = k_2x - (k_{-2} + k_{-d})y + k_dH(t)d. \tag{16.71}$$

Because of the conservation condition $r+x+y+d = 1$ there are only three independent variables, so only three equations are needed. The function $H(t)$ denotes the hormone concentration as a function of time, and is assumed to be known.

Each state of the receptor is assumed to have an intrinsic activity, and the total activity of the receptor is given by the sum over all the receptor states, weighted by the intrinsic activity of the state. Thus, the total activity A of the receptor is

$$A = a_1r + a_2x + a_3y + a_4d, \tag{16.72}$$

for some constants $a_1, \ldots, a_4$.

For simplicity, we assume that the binding of the ligand is essentially instantaneous, and thus

$$x = \frac{H(t)r}{K_r}, \tag{16.73}$$

$$y = \frac{H(t)d}{K_d}, \tag{16.74}$$

where $K_r = k_{-r}/k_r$ and $K_d = k_{-d}/k_d$. Using a standard quasi-equilibrium reduction (see Chapters 1 and 2), and assuming that $H(t)$ is slowly varying on the time scale of receptor binding, we find the single differential equation for the receptor,

$$\left(1 + \frac{H}{K_r}\right)\frac{dr}{dt} = \frac{k_{-1}K_d + k_{-2}H}{K_d + H}\left(1 - r\left(\frac{K_d + H}{K_dK_1} + \frac{K_r + H}{K_r}\right)\right), \tag{16.75}$$

where $K_1 = k_{-1}/k_1$, and $k_2 = \frac{k_{-2}K_r}{K_1K_d}$ in order to satisfy detailed balance. The steady states are given by

$$r_0 = \frac{1}{\frac{K_d+H}{K_1K_d} + \frac{K_r+H}{K_r}}, \tag{16.76}$$

$$x_0 = \frac{Hr_0}{K_r}, \tag{16.77}$$

$$y_0 = \frac{H\left(1 - \frac{K_r+H}{K_r}r_0\right)}{K_d + H}. \tag{16.78}$$

If $a_4 = 0$, so that the state d is completely inactive, the steady-state activity of the receptor is

$$\begin{aligned} A &= a_1r_0 + a_2x_0 + a_3y_0 \\ &= \frac{a_1K_1K_dK_r + H(a_2K_1K_d + a_3K_r)}{K_rK_d(K_1 + 1) + H(K_r + K_1K_d)}. \end{aligned} \tag{16.79}$$

In general, this is a saturating curve as a function of H. However, exact adaptation occurs if A is independent of H, in which case it must be that

$$A = A|_{H=0} = \lim_{H\to\infty} A, \tag{16.80}$$

so that

$$\frac{K_1 a_1}{1+K_1} = \frac{a_3 K_r + K_1 K_d a_2}{K_r + K_1 K_d}. \tag{16.81}$$

Note that since the right-hand side of (16.81) is the weighted average of a_2 and a_3, exact adaptation is possible only when a_1 is greater than the smaller of a_2 and a_3 (more precisely when $\frac{K_1 a_1}{1+K_1}$ lies between a_2 and a_3). In general, one expects a_2 to be larger than a_3 (as the RH form of the receptor has a greater intrinsic activity than its inactivated form DH), and thus $a_1 > a_3$ is required for exact adaptation. In other words, the intrinsic activity of the unbound receptor (R form) must be higher than the intrinsic activity of the inactivated receptor, even when the hormone is bound.

In response to a step increase in hormone concentration, the receptor state is first quickly converted to the RH form, which has a high activity, and thus the overall activity initially increases. However, over a longer time period, the RH form gradually converts to the DH form, which has a lower activity than the R (unbound) form. Thus, receptor inactivation decreases the activity back to the basal level. Thus, exact adaptation arises from a process of fast activation and slow inactivation, a mechanism that has appeared in many forms throughout this book.

16.5 EXERCISES

1. By taking partial derivatives of $f(\xi, \xi_i)$ with respect to ξ_i confirm that the model (16.25), when linearized about P_M, takes the form given in (16.33)–(16.35). Calculate the eigenvalues and eigenvectors of the matrix A.

2. This exercise works through the derivation of a Lyapunov function for the Lacker model (Akin and Lacker, 1984). Define $\delta(\xi) = \xi^2$, $\rho(\xi) = \xi^{-2} - 1$, and $\phi(p_i) = p_i(M_1 + M_2 - M_1 M_2 p_i)$, where $p_i = \xi_i/\xi$. Show that

$$\frac{d\xi_i}{dt} = \delta(\xi)\xi_i[\rho(\xi) + \phi(p_i)], \tag{16.82}$$

$$\frac{d\xi}{dt} = \delta(\xi)\xi[\rho(\xi) + \bar{\phi}], \tag{16.83}$$

$$\frac{dp_i}{dt} = \delta(\xi)p_i[\phi(p_i) - \bar{\phi}], \tag{16.84}$$

where

$$\bar{\phi} = \sum_{i=1}^{n} p_i \phi(p_i). \tag{16.85}$$

Define a new time scale τ by

$$\frac{d\tau}{dt} = \delta(\xi), \tag{16.86}$$

and show that

$$\frac{dp_i}{d\tau} = p_i[\phi(p_i) - \bar{\phi}]. \tag{16.87}$$

Finally, show that

$$V(p_1, \ldots, p_n) = \sum_{i=1}^{n} \int_0^{p_i} \phi(s)\, ds \tag{16.88}$$

is a Lyapunov function for the model by showing that

$$\frac{dV}{d\tau} = \sum_{i=1}^{n} p_i[\xi(p_i) - \bar{\phi}]^2 \geq 0. \tag{16.89}$$

Hint: Derive and use the fact that $\sum p_i(\xi(p_i) - \bar{\phi}) = 0$.

3. Since $\gamma_1 \equiv 1$, (16.44)–(16.46) provide no information about ξ_1. Use the original variables to show that $\bar{\xi}_1(\tau) = \xi(t(\tau))$ satisfies the differential equation

$$\frac{1}{2}\frac{d}{d\tau}\bar{\xi}_1^2 = 1 - \bar{\xi}_1^2(\Gamma - M_1)(\Gamma - M_2). \tag{16.90}$$

Find t as a function of τ. Describe how the original variables $\xi_i(t)$ may be obtained once the $\gamma_i(\tau)$ have been obtained by numerical solution of (16.44)–(16.46). Why is it preferable to numerically simulate the model in the transformed variables γ_i rather than the original variables ξ_i?

4. For the model of a hormone receptor assuming fast ligand binding (Section 16.4), calculate the response to a step function, and then to a stimulus of the form $H(t) = 1$ for $0 < t < t_0$, $H(t) = 0$ otherwise (call this stimulus a step pulse). Calculate the response to a series of step pulses, and calculate the width of the pulse and the time between pulses that gives the greatest average activity. Li and Goldbeter (1989) give the details.

5. (From Loeb and Strickland, 1987.) Many cells respond maximally to a hormone concentration that is much too low to saturate the hormone receptors. This can be explained by assuming that the response is dependent on a secondary mediator. Suppose that the hormone, H, combines reversibly with its receptor, R^o, to form the complex HR^o. Suppose that the secondary mediator M is formed at a rate proportional to $[HR^o]$ and is degraded with linear kinetics. Finally, suppose that M combines reversibly with its own receptor, R, to form MR, and that the cellular response is linearly proportional to [MR]. What is the fractional receptor occupancy as a function of [H]? Show that the fractional response (as a function of [H]) has the same shape as the fractional receptor occupancy curve, shifted to the left by a constant factor (when plotted against log[H]). Give a biological interpretation.

6. Consider the model of the activity of the LHRH neurons described by (16.4)–(16.6). Let a be a constant, instead of a function of time.

   (a) How does the steady state depend on a? What are the different possibilities?

   (b) Show that, depending on the values of the other parameters, an increase in a can lead either to an increase or a decrease in excitability.

   (c) Suppose that a is suddenly increased, held constant for a short period of time, and then returned to zero. Plot the solution in the phase plane. What happens as the period of time for which a is raised becomes shorter?

   (d) If $a(t)$ is the delta function, sketch the solution in the phase plane.

7. Find the response of the three-stage linear filter (16.62)–(16.64) to the sinusoidal input $I_p = I_0 + I_1 \sin(\omega t)$.

8. Consider the model of insulin oscillations described in Section 16.3.2.

   (a) Plot the steady states and periodic solutions of the model as, a function of I, and as a function of t_d. Show that, as t_d gets large or small, the oscillations disappear.

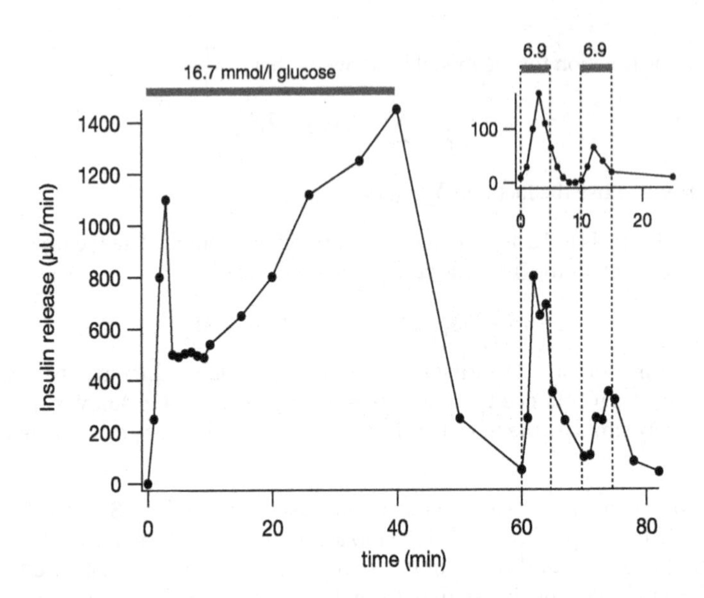

Figure 16.25 The response to glucose of the rate of insulin release in the perfused rat pancreas. The inset at the top right show the response to two glucose steps (each of size 6.9 mmol/l) while directly below the inset is shown the response to two identical glucose steps, but after the pancreas has been primed by the prior addition of 16.7 mmol/l glucose for 40 minutes (Nesher and Cerasi, 2002). We thank Professor Erol Cerasi for providing the original data for this figure.

(b) How can the model be modified to omit the remote insulin compartment? How does this change the oscillatory behavior in the model? (Hint: Combine both insulin compartments into a single compartment, with all the insulin inputs and outputs going to and from this single compartment, and no exchange term.) How is this approach different from assuming that the exchange between the plasma and remote insulin compartments is fast?

9. In response to a pulse of glucose, the rate of insulin release shows a biphasic response, as shown in Fig. 16.25. First, consider the responses shown in the inset at the top right. These are the responses to two identical glucose steps, each of magnitude 6.9 mmol/l, and each 5 minutes long. Clearly, the response during each glucose step first rises and then decays, while, in addition, the response to the second step is smaller than to the first.

Next, consider the curve in the main part of the figure. Initially, a step of 16.7 mmol/l glucose was added for 40 minutes; during this 40 minutes the rate of insulin release shows a clear triphasic behavior, first rising, then falling, then rising again. After this "priming" by glucose, the same two steps as in the inset were applied. Now, the response to these two steps is greatly enhanced, although, as in the inset, the response to the second step is smaller.

Clearly, there are a number of dynamic processes controlling the rate of insulin release, each operating on a different time scale. For instance, addition of glucose clearly first stimulates insulin release, then inhibits then, then stimulates it again. Furthermore, the data in the inset show that addition of glucose initiates an inhibiting mechanism that is still in operation 5 minutes after cessation of the initial stimulus.

Construct a mathematical model of this behavior. Do not worry about relating your model to specific known physiological mechanisms; construct a system of equations that exhibits approximately the correct behavior. Although there are many possible answers, your model will probably have a number of different variables, one for each of the dynamic processes discussed above and each with a different time constant, with the rate of insulin release modeled as a product of these variables.

CHAPTER 17

Renal Physiology

The kidneys perform two major functions. First, they excrete most of the end products of bodily metabolism, and second, they control the concentrations of most of the constituents of the body fluids. The main goal of this chapter is to gain some understanding of the processes by which the urine is formed and waste products removed from the bloodstream. The control of the constituents of the body fluids is discussed only secondarily.

The primary operating unit of the kidney is called a *nephron,* of which there are about a million in each kidney (Figs. 17.1 and 17.2). Each nephron is capable of forming urine by itself. The entrance of blood into the nephron is by the *afferent arteriole,* located in the renal cortex, and the tubules of the nephron and the associated peritubular capillaries extend deep into the renal medulla. The principal functional units of the nephron are the *glomerulus,* through which fluid is filtered from the blood; the *juxtaglomerular apparatus,* by which glomerular flow is controlled; and the *long tubule,* in which the filtered fluid is converted into urine.

17.1 The Glomerulus

The first stage of urine formation is the production of a filtrate of the blood plasma. The glomerulus, the primary filter, is a network of up to 50 parallel branching and anastomosing (rejoining) capillaries covered by epithelial cells and encased by *Bowman's capsule*. Blood enters the glomerulus by way of the afferent arteriole and leaves through the *efferent arteriole*. Pressure of the blood in the glomerulus makes the fluid filter into Bowman's capsule, carrying with the filtrate all the dissolved substances of small molecular weight. The glomerular membrane is almost completely impermeable

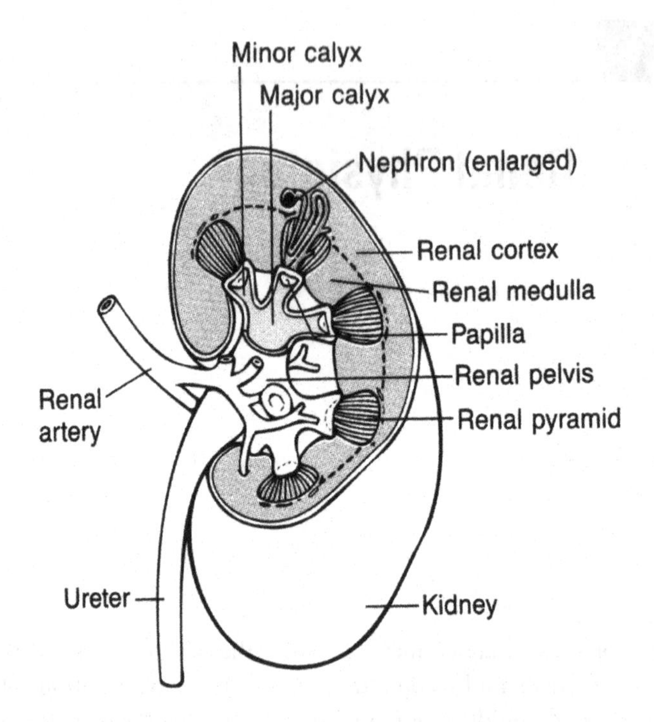

Figure 17.1 The kidney. (Guyton and Hall, 1996, Fig. 26-2, p. 317.)

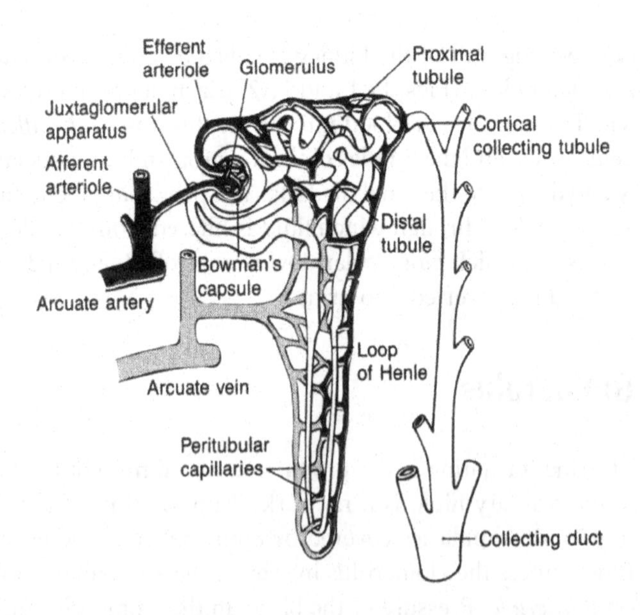

Figure 17.2 The nephron. (Guyton and Hall, 1996, Fig. 26-3, p. 318.)

to all plasma proteins, the smallest of which is albumin (molecular weight 69,000). As a result, the glomerular filtrate is identical to plasma except that it contains no significant amount of protein.

The quantity of filtrate formed each minute is called the *glomerular filtration rate*, and in a normal person averages about 125 ml/min. The filtration fraction is the fraction of renal plasma flow that becomes glomerular filtrate and is typically about 20 percent. Over 99 percent of the filtrate is reabsorbed in the tubules, with the remaining small portion passing into the urine.

There are three pressures that affect the rate of glomerular filtration. These are the pressure inside the glomerular capillaries that promote filtration, the pressure inside Bowman's capsule that opposes filtration, and the colloidal osmotic pressure (Section 2.7) of the plasma proteins inside the capillaries that opposes filtration.

A mathematical model of the glomerular filter can be described simply as follows. We assume that the glomerular capillaries comprise a one-dimensional tube with flow q_1 and that the surrounding Bowman's capsule is also effectively a one-dimensional tube with flow q_2 (Fig. 17.3). Since the flow across the glomerular capillaries is proportional to the pressure difference across the capillary wall, at steady state the rate of change of the flow in the capillary is

$$\frac{dq_1}{dx} = K_f(P_2 - P_1 + \pi_c), \tag{17.1}$$

where P_1 and P_2 are the hydrostatic fluid pressures in tubes 1 and 2, respectively, π_c is the osmotic pressure of suspended proteins and formed elements of blood, and K_f is the capillary filtration rate. The osmotic pressure of the suspended proteins is given by

$$\pi_c = RTc, \tag{17.2}$$

where c, the concentration expressed in moles per liter, is a function of x, since the suspension becomes more concentrated as it moves through the glomerulus. Since the large proteins bypass the filter, we have the conservation equation

$$c_i Q_i = c q_1, \tag{17.3}$$

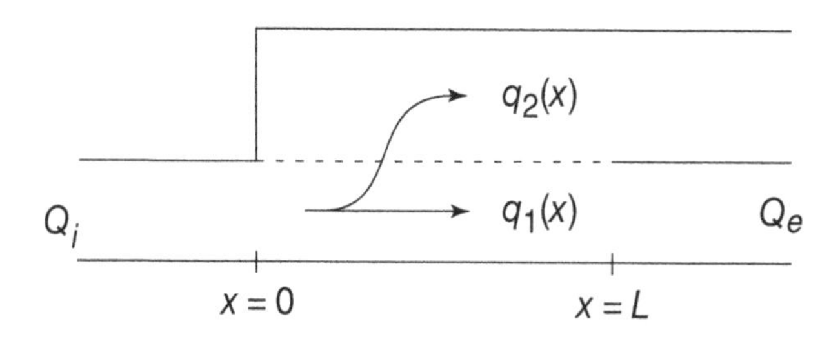

Figure 17.3 Schematic diagram of the glomerular filtration.

where c_i is the input concentration and Q_i is the input flux. It follows that

$$\pi_c = \pi_i \frac{Q_i}{q_1}, \tag{17.4}$$

where $\pi_i = RTc_i$ is the input osmotic pressure. Since the hydrostatic pressure drop in the glomerulus is small compared to the pressure drop in the efferent and afferent arterioles, we take P_1 and P_2 to be constants.

Equation (17.1) along with (17.4) gives a first-order differential equation for q_1, which is easily solved. Setting $q_1(L) = Q_e$ we find that

$$\frac{Q_e}{Q_i} + \alpha \ln \left(\frac{\frac{Q_e}{Q_i} - \alpha}{1 - \alpha} \right) = 1 - K_f L \frac{\pi_i}{\alpha Q_i}, \tag{17.5}$$

where Q_e is the efflux through the efferent arterioles, L is the length of the filter, and $\alpha = \pi_i/(P_1 - P_2)$.

Finally, we assume that the pressures and flow rates are controlled by the input and output arterioles, via

$$P_a - P_1 = R_a Q_i, \tag{17.6}$$

$$P_1 - P_e = R_e Q_e, \tag{17.7}$$

and that the flow out of the glomerulus into the proximal tubule is governed by

$$P_2 - P_d = R_d(Q_i - Q_e), \tag{17.8}$$

where P_a, P_e, and P_d are the afferent arteriole, efferent arteriole, and descending tubule pressures, respectively, and R_a, R_e, and R_d are the resistances of the afferent and efferent arterioles and proximal tubule, respectively. Typical values are $P_1 = 60, P_2 = 18, P_a = 100, P_e = 18, P_d = 14 - 18, \pi_i = 25$ mm Hg, with $Q_i = 650, Q_d = Q_i - Q_e = 125$ ml/min.

The flow rates and pressures vary as functions of the arterial pressure. To understand something of this variation, in Fig. 17.4 is shown the renal blood flow rate Q_i and the glomerular filtration flow rate as functions of the arterial pressure. It is no surprise that both of these are increasing functions of arterial pressure P_a.

The strategy for numerically computing this curve is as follows: with resistances R_a and R_e and pressures P_e, P_d, and π_i specified and fixed at typical levels, we pick a value for glomerular filtrate $Q_d = Q_i - Q_e$. For this value, we solve (17.5) (using a simple bisection algorithm) to find both Q_i and Q_e. From these, the corresponding pressures P_a, P_1, and P_2 are determined from (17.6) and (17.7), and plotted.

For this model, the filtration rate varies substantially as a function of arterial pressure. However, in reality (according to data shown in Fig. 17.5), the glomerular filtration rate remains relatively constant even when the arterial pressure varies between 75 to 160 mm Hg, suggesting that there is some autoregulation of the flow rate.

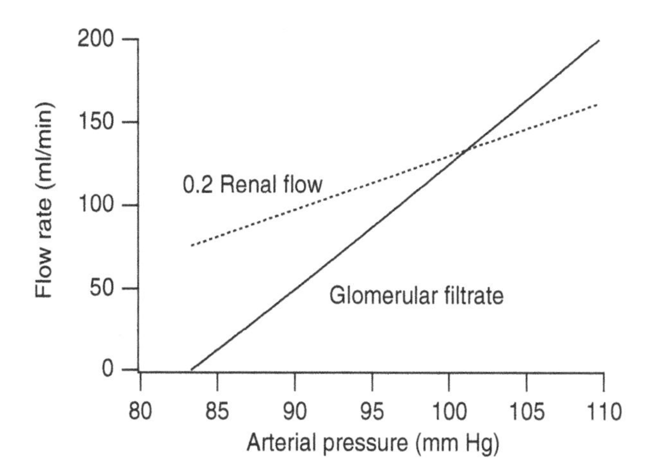

Figure 17.4 Unregulated glomerular filtration and renal blood flow plotted as functions of arterial pressure, with $P_d = 18$, $P_e = 0$ mm Hg.

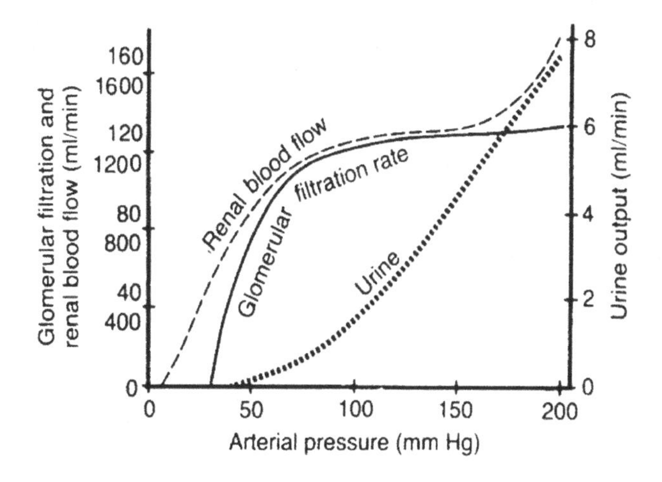

Figure 17.5 Autoregulation of renal blood flow and glomerular filtration rate but lack of autoregulation of urine flow during changes in renal arterial pressure. (Guyton and Hall, 1996, Fig. 26-13, p. 327.)

17.1.1 Autoregulation and Tubuloglomerular Oscillations

The need for autoregulation of the glomerular filtration rate is apparent. If the flow rate of filtrate is too slow, then we expect reabsorption to be too high, and the kidney fails to eliminate necessary waste products. On the other hand, at too high a flow rate, the tubules are unable to reabsorb those substances that need to be preserved and not eliminated, so that valuable substances are lost into the urine.

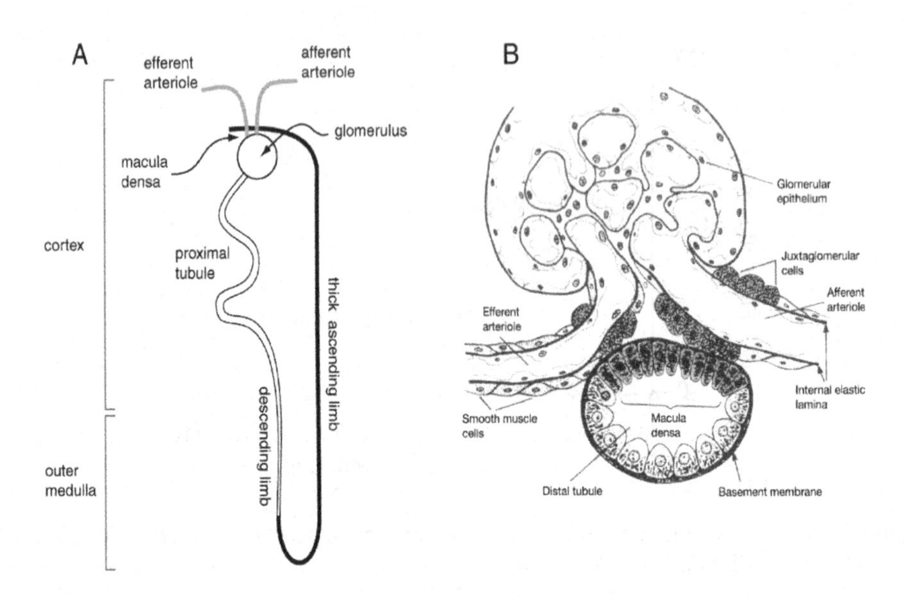

Figure 17.6 A: Schematic diagram of a short nephron, showing how the thick ascending limb ascends to the region of the glomerulus, enabling autoregulation of the glomerular filtration rate. Adapted from Pitman et al. (2004), Fig. 1. B: Structure of the juxtaglomerular apparatus. (Guyton and Hall, 1996, Fig. 26-14, p. 328.)

The glomerulus controls the rate of filtration by detecting the concentration of NaCl at the top of the thick ascending limb, most probably by detecting the chloride concentration. After its descent into the renal medulla, the loop of Henle returns to the proximity of the afferent and efferent arterioles at the glomerulus (Fig. 17.6A). The *juxtaglomerular complex* consists of *macula densa* cells in the distal tubule and *juxtaglomerular cells* in the walls of the afferent and efferent arterioles (as depicted in Fig. 17.6B).

As described in Section 17.2, the principal function of the thick ascending limb of the loop of Henle is the pumping of Na^+ out of the tubule into the interstitial space; it is almost entirely impermeable to water. Thus, the slower the rate of flow through the tubule, the greater the amount of Na^+ that can be transported out of the thick ascending limb, with Cl^- following passively. A low flow rate causes excessive reabsorption of Na^+ and Cl^- ions, resulting in too large a decrease of these ionic concentrations at the end of the loop. Conversely, a high flow rate results in a higher concentration of Na^+ and Cl^- at the macula densa.

The macula densa cells respond to decreases of NaCl concentration (by a mechanism not completely understood), by releasing a vasodilator that decreases the resistance of the afferent arterioles. Simultaneously, the juxtaglomerular cells release renin, an enzyme that enables the formation of angiotensin II, which constricts the efferent arterioles. The simultaneous effect of these is to increase the flow of filtrate through the glomerulus. Conversely, an increase in the concentration of NaCl

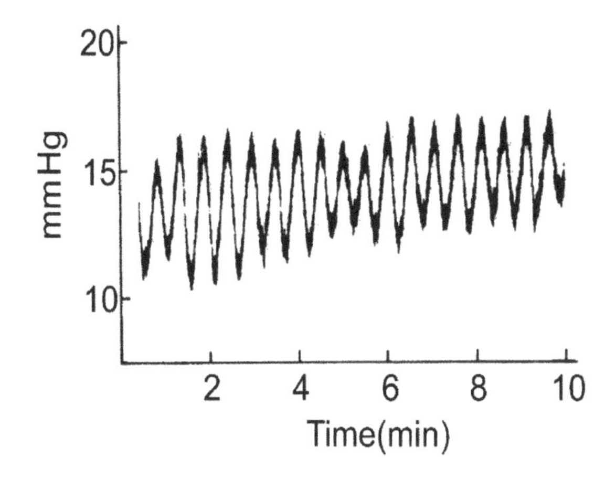

Figure 17.7 Spontaneous oscillations in proximal tubule pressure in a Sprague-Dawley rat. From Holstein-Rathlou and Leyssac, 1987, Fig. 5.

at the macula densa has the effect of decreasing the resistance of the afferent arteriole, and increasing the resistance of the efferent arteriole, thus decreasing the rate of filtration and the flow along the tubule.

It comes as no surprise that a control system such as this can result in oscillations in the rate of fluid flow along the loop of Henle (Fig. 17.7). These oscillations were first described in halothane-anesthetized Sprague-Dawley rats by Leyssac and Baumbach (1983), and have since been studied in detail by both modelers and experimentalists (Holstein-Rathlou and Leyssac, 1987; Holstein-Rathlou and Marsh, 1989, 1990, 1994; Holstein-Rathlou et al., 1991; Layton et al., 1991, 1995b, 1997, 2000; Pitman et al., 2004). A useful recent review is that of Just (2006).

A simple model of tubuloglomerular oscillations (Layton et al., 1991) focuses on the role of the chloride concentration in the thick ascending limb. We model the thick ascending limb as a one-dimensional tube, through which Cl^- is transported by fluid flow, while being removed actively (of course, it is the Na^+ that is removed actively, but since Cl^- follows passively, the effect is the same).

If we let $y = 0$ and $y = L$ denote, respectively, the beginning and end of the thick ascending limb of length L, and let $C(y, \tau)$ denote the Cl^- concentration at y at time τ, then conservation of Cl^- gives

$$\frac{\partial C}{\partial \tau} + \phi \frac{\partial C}{\partial y} = -R(C), \tag{17.9}$$

where $R(C)$ denotes the rate of removal of Cl^- from the tube via the pumping of Na^+, and ϕ denotes the velocity of fluid flow along the tube. For our simple model we assume that Cl^- is removed by a first-order process,

$$R(C) = -r_c C. \tag{17.10}$$

When the NaCl concentration at the macula densa is changed, the resistance of the afferent arteriole changes only after a time delay. Thus, ϕ is a function of the Cl^- concentration at the macula densa at some previous time, i.e., a function of $C(L, \tau - \bar{\tau})$.

The functional form of ϕ has been established by experimental studies to be of the form

$$\phi(C) = F_{\text{op}} + \Delta F \tanh(\alpha(\bar{C} - C)). \tag{17.11}$$

Notice that ϕ is a decreasing function of C. The constant $\bar{C}$ is the Cl^- concentration for which ϕ has an inflection point.

Finally, we assume that the input Cl^- concentration is known, $C(0,t) = C_0$.

We now rescale y, τ, and C, setting $y = Lx$, $\tau = \frac{L}{F_{\text{op}}}t$, $\bar{\tau} = \frac{L}{F_{\text{op}}}\bar{t}$, $C = C_0 c$, $K_1 = \Delta F/F_{\text{op}}$, and $K_2 = C_0$, so that the conservation law becomes the delayed partial differential equation

$$\frac{\partial c}{\partial t} + F(c(1, t - \bar{t}))\frac{\partial c}{\partial x} = -\mu c, \tag{17.12}$$

where

$$F(c) = 1 + K_1 \tanh(K_2(\bar{c} - c)), \tag{17.13}$$

and $\mu = r_C \frac{L}{F_{\text{op}}}$, subject to the boundary condition, $c(0,t) = 1$.

The first step in our analysis is to examine the steady-state solution. The steady-state solution, $s(x)$, is the solution of

$$F(s(1))\frac{\partial s}{\partial x} = -\mu s, \tag{17.14}$$

and so

$$s(x) = s(0)e^{-kx}, \tag{17.15}$$

where $k = \frac{\mu}{F(s(1))}$. Since $s(0) = 1$, and $s(1) = e^{-k}$, for consistency it must be that

$$kF(e^{-k}) = \mu. \tag{17.16}$$

Since F is a decreasing function of its argument, there is a unique value of k for which (17.16) holds, and hence a unique steady-state solution.

Layton et al. made the additional assumption that in normal, steady, operating conditions, $c(1,t) = \bar{c}$ and they chose the Cl^- uptake rate to be $\mu = -\ln \bar{c} = k$. In order to allow for abnormal operating conditions, we do not make this additional assumption here.

To understand something about this control mechanism, we calculate the sensitivity of $s(1)$ to changes in $s(0)$. The sensitivity σ is defined as

$$\sigma = \frac{s(0)}{s(1)}\frac{ds(1)}{ds(0)}. \tag{17.17}$$

Since

$$s(1)\exp\left(\frac{\mu}{F(s(1))}\right) = s(0), \tag{17.18}$$

it follows (after implicit differentiation) that

$$\sigma = \frac{1}{1 - \frac{k^2}{\mu}e^{-k}F'(e^{-k})} = \frac{1}{1 + \frac{k}{\mu}\gamma}, \tag{17.19}$$

where $\gamma = -ke^{-k}F'(e^{-k})$. Since, $F'(e^{-k}) < 0$, this implies that $s(1)$ becomes less sensitive to changes in input $s(0)$ as $F'(s(1))$ decreases (i.e., as F becomes steeper). Notice that with no control, $F'(c) = 0$, so that $\sigma = 1$.

The sensitivity is also changed by changes in the Cl^- removal rate. Sensitivity is smallest (and γ is largest) when $\mu = \ln \bar{c}$. (This is an approximate statement because the derivative of γ with respect to μ is nearly, but not exactly, zero at $\mu = \ln \bar{c}$.) However, as μ deviates from this optimal value, the steady solution $s(1)$ deviates from $\bar{c}$, leading to a flatter feedback response. With F' smaller in magnitude, so also is γ.

Oscillations in the macula densa Cl^- concentration, and thus in the other variables such as flow rate and glomerular pressure, can occur if the steady-state solution becomes unstable via a Hopf bifurcation. This is not the only possible instability mechanism, but it is a reasonable one to look for. The stability of $s(x)$ can be determined by writing $c(x,t) = s(x) + \epsilon u(x,t)$ and then expanding (17.12) to first order in ϵ. This gives

$$\frac{\partial u}{\partial t} + F(e^{-k})\frac{\partial u}{\partial x} = -\mu u + kF'(e^{-k})e^{-kx}u(1, t - \bar{t}), \tag{17.20}$$

with the boundary condition $u(0,t) = 0$.

We now look for solutions for u of the form $u(x,t) = f(x)e^{\lambda t}$, where $f(0) = 0$. Substituting this into (17.20) gives

$$\frac{\mu}{k}f'(x) = -(\mu + \lambda)f(x) + kF'(e^{-k})e^{-kx}f(1)e^{-\lambda\bar{t}}, \tag{17.21}$$

which can be solved to give

$$f(x) = \frac{1}{\lambda}kF'(e^{-k})f(1)e^{-\lambda\bar{t}}\left[e^{-kx} - e^{-\frac{k}{\mu}(\mu+\lambda)x}\right]. \tag{17.22}$$

For consistency, it must be that

$$f(1) = \frac{1}{\lambda}kF'(e^{-k})f(1)e^{-\lambda\bar{t}}\left[e^{-k} - e^{-\frac{k}{\mu}(\mu+\lambda)}\right], \tag{17.23}$$

from which it follows that λ must be a root of

$$\lambda = \gamma e^{-\lambda\bar{t}}(e^{-\frac{k}{\mu}\lambda} - 1) = -2\gamma e^{-\lambda(\bar{t}+\frac{k}{2\mu})}\sinh\left(\frac{k\lambda}{2\mu}\right), \tag{17.24}$$

where $\gamma = -kF'(e^{-k})e^{-k}$. Note that $\gamma > 0$ is related to the sensitivity of the feedback regulation in (17.19).

The only parameter values for which a Hopf bifurcation can occur are those for which λ has zero real part and nonzero imaginary part. If λ has zero real part we can take $\lambda = i\omega$. Substituting $\lambda = i\omega$ into (17.24) and equating real and imaginary parts gives the two equations

$$0 = 2\gamma \sin\left(\omega\left(\bar{t} + \frac{k}{2\mu}\right)\right)\sin\left(\frac{k\omega}{2\mu}\right), \tag{17.25}$$

$$\omega = -2\gamma \cos\left(\omega\left(\bar{t} + \frac{k}{2\mu}\right)\right)\sin\left(\frac{k\omega}{2\mu}\right). \tag{17.26}$$

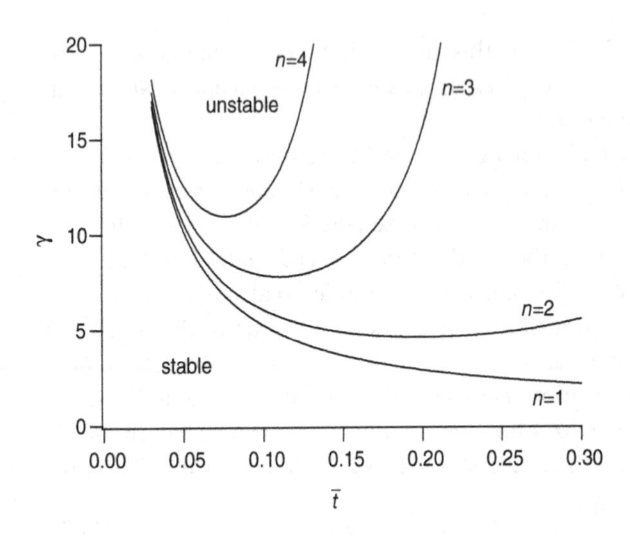

Figure 17.8 Plot of (17.28) for $n = 1, 2, 3, 4$. The steady-state solution is stable below the $n =$ 1 curve, and unstable above it.

The only nontrivial solution of the first of these has

$$\omega = \frac{n\pi}{\bar{t} + \frac{k}{2\mu}}, \quad n = 1, 2, 3, \ldots . \tag{17.27}$$

Thus, the curves in parameter space at which there are Hopf bifurcations are given by

$$\gamma = (-1)^{n+1} \frac{\omega}{2\sin(\frac{k\omega}{2\mu})}, \qquad \omega = \frac{n\pi}{\bar{t} + \frac{k}{2\mu}}, \quad n = 1, 2, 3, \ldots . \tag{17.28}$$

A plot of the curves (17.28) is given in Fig. 17.8 for $n = 1, 2, 3, 4$ with $\frac{k}{\mu} = 1$. The primary instability occurs at the curve corresponding to $n = 1$, while higher-frequency instabilities occur at the other curves. Typical corresponding solutions (calculated numerically) are shown in Fig. 17.9.

The main point of this plot is to demonstrate the importance of the two parameters, the time delay, $\bar{t}$, and the sensitivity feedback parameter, γ, on the onset of oscillations. Since the primary critical curve is monotone decreasing, oscillations occur if either of these parameters is sufficiently large (but both must be nonzero). The role of the other parameter of the problem, the rate of Cl^- removal from the thick ascending limb, is more subtle to unravel. As can be seen from (17.28), the primary critical curve is moved around a bit if $\frac{k}{\mu}$ is changed, but the basic shape of the curve is unchanged. What is more significant is that changes in μ lead to changes in the parameter γ. In fact, if μ deviates from its optimal value at $\mu = -\ln \bar{c}$, then γ decreases. Thus, the primary effect of changes of μ from its optimal value is to make oscillations less likely. Another way to express this is that if the delay $\bar{t}$ is sufficiently large, then there are two values of μ that are Hopf bifurcation points, one below and one above the optimal removal rate at $\mu = -\ln \bar{c}$. A plot of this critical curve is shown in Fig. 17.10.

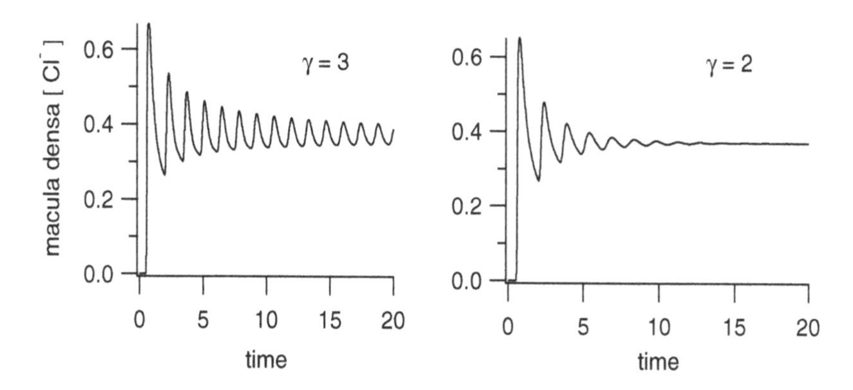

Figure 17.9 Typical oscillatory solutions of the model of tubuloglomerular oscillations, for two different feedback strengths. For both panels, $\bar{t} = 0.2$, $\frac{k}{\mu} = 1$. The Cl^- concentration is plotted in dimensionless units, where the concentration at the start of the thick ascending limb is 1. The time is also dimensionless; to get seconds, multiply by 15.5.

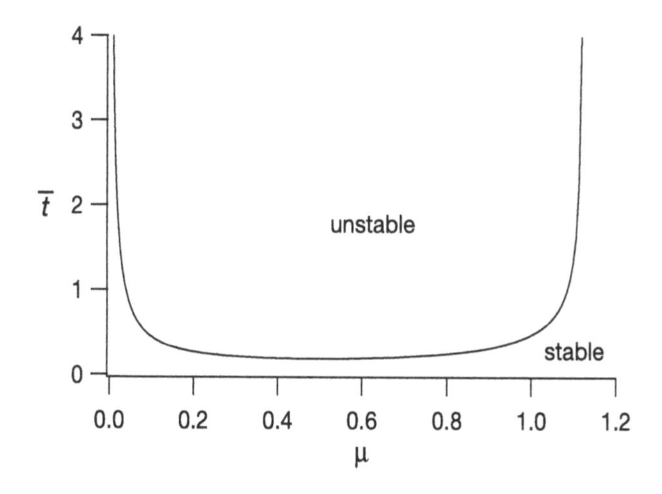

Figure 17.10 Critical curve in the $\mu, \bar{t}$ parameter space along which there are Hopf bifurcations in the glomerular feedback model. Parameter values are $K_1 = 1$, $K_2 = 10$, $\bar{c} = e^{-0.5}$.

A much more complex model of tubuloglomerular oscillations was constructed by Holstein-Rathlou and Marsh (1990), and the simple model above was compared in detail to this more complex model by Layton et al. (1991).

17.2 Urinary Concentration: The Loop of Henle

The challenge of any model of urine formation is to see how concentrating and diluting mechanisms work together to determine the composition of the urine and to regulate the interstitial contents, and then to account quantitatively for the concentrating ability of particular species. The challenge is substantial. For example, for humans, the maximal urine concentrating ability is 1200 mOsm/liter, while some desert animals, such as the Australian hopping mouse, can concentrate urine to as high as 10,000 mOsm/liter. It is not understood how such high urine concentrations can be obtained. It is also necessary that the kidney be able to produce a dilute urine under conditions of high fluid intake.

A normal 70 kg human must excrete about 600 mOsm of solute (waste products of metabolism and ingested ions) every day. The minimal amount of urine to transport these solutes, called the *obligatory urine volume* is

$$\text{obligatory volume} = \frac{\text{total solute/day}}{\text{maximal urine concentration}} \tag{17.29}$$

$$= \frac{600 \text{ mOsm/day}}{1200 \text{ mOsm/L}} = 0.5\text{L/day}. \tag{17.30}$$

This explains why severe dehydration occurs from drinking seawater. The concentration of salt in the oceans averages 3% sodium chloride, with osmolarity between 2000 and 2400 mOsm/liter. Drinking 1 liter of water with a concentration of 2400 mOsm/liter provides 2400 mOsm of solute that must be excreted. If the maximal urine concentration is 1200 mOsm/liter, then 2 liters of urine are required to rid the body of this ingested solute, a deficit of 1 liter, which must be drawn from the interstitial fluid. This explains why shipwreck victims who drink seawater are rapidly dehydrated, while (as Guyton and Hall have kindly pointed out) the victim's pet Australian hopping mouse can drink all the seawater it wants with impunity.

Urinary concentration or dilution is accomplished primarily in the loop of Henle. After leaving Bowman's capsule, the glomerular filtrate flows into a tubule having five sections: the *proximal tubule*, the *descending limb of the loop of Henle*, the *ascending limb of the loop of Henle*, the *distal tubule*, and, finally, the *collecting duct*. These tubules are surrounded by capillaries, called the *peritubular capillaries*, that reabsorb the fluid that has been extracted from the tubules. In Fig. 17.11 are shown the relative concentrations of various substances at different locations along the tubular system.

The purpose of the proximal tubule is to extract much of the water and dissolved chemicals (electrolytes, glucose, various amino acids, etc.) to be reabsorbed into the bloodstream while concentrating the waste products of metabolism. It is this concentrate that eventually flows as urine into the bladder. The proximal tubular cells have large numbers of mitochondria to support rapid active transport processes. Indeed, about 65 percent of the glomerular filtrate is reabsorbed before reaching the descending limb of the loop of Henle. Furthermore, glucose, proteins, amino acids, acetoacetate ions, and the vitamins are almost completely reabsorbed by active cotransport processes through the epithelial cells that line the proximal tubule.

Any substance that is reabsorbed into the bloodstream must first pass through the tubular membrane into the interstitium and then into peritubular capillaries. There are three primary mechanisms by which this transport takes place, all of which we have seen before (Fig. 17.12). First, there is active transport of Na^+ from the interior

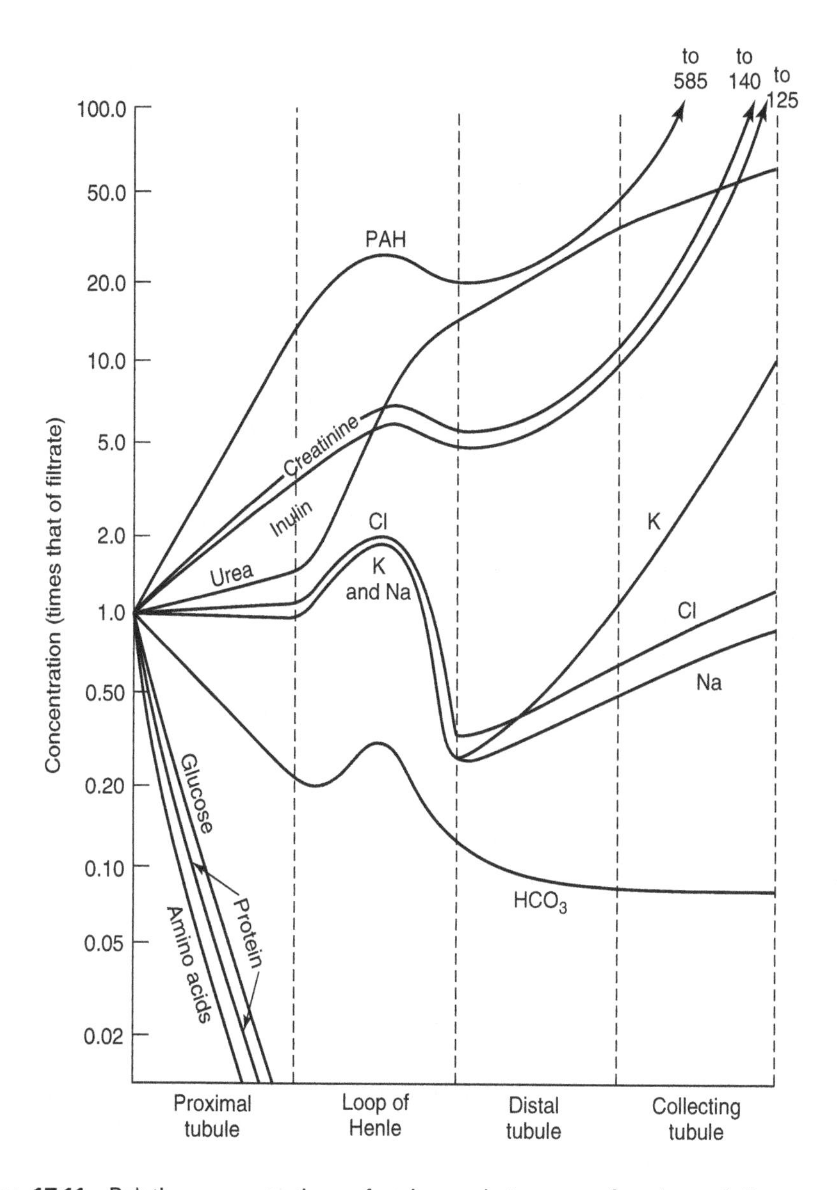

Figure 17.11 Relative concentrations of various substances as functions of distance along the renal tubule system. (Guyton and Hall, 1996, Fig. 27-11, p. 341.)

of the epithelial cells into the interstitium, mediated by a Na^+–K^+ ATPase. Although this pump actively pumps K^+ into the cell from the interstitium, both sides of the tubular epithelial cells are so permeable to K^+ that virtually all of the K^+ leaks back out of the cell almost immediately.

There are secondary transporters that use the gradient of Na^+ ions (established by the ATPase) to transport other substances from the tubular lumen into the interior

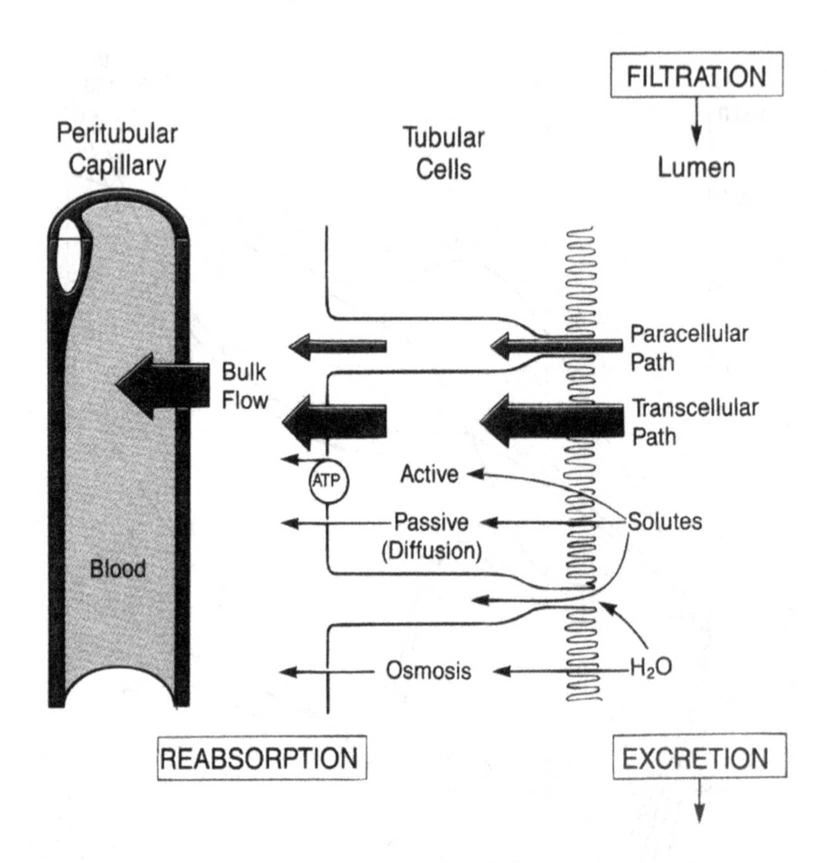

Figure 17.12 Schematic diagram of the reabsorption of water and solutes in the proximal tubule. (Guyton and Hall, 1996, Fig. 27-1, p. 332.)

of the epithelial cell. The most important of these are cotransporters of glucose and amino acid ions, but the epithelial cells of the proximal tubule also contain transporters of phosphate, Ca^{2+}, and Mg^{2+}. There is also a transporter that exchanges H^+ for Na^+ across the membrane of the epithelial cell membrane into the tubule. The third mechanism of transport is that of water across cell membranes, mediated by osmotic pressure (see Chapters 2 and 18).

The descending limb of the loop of Henle is lined with thin epithelial cells with few mitochondria, indicating minimal metabolic activity; it is highly permeable to water and moderately permeable to Na^+, urea, and most ions. The ascending limb of the loop of Henle begins with a thin wall but then about halfway up becomes grossly thickened. In contrast to the descending limb, the ascending limb is highly impermeable to water and urea. The cells of the thick ascending limb are similar to those of the proximal tubule, suited for strong active transport of Na^+ and Cl^- ions from the tubular lumen into the interstitial fluid.

The thick segment travels back to the region of the glomerulus, where it passes between the afferent and efferent arterioles, forming the juxtaglomerular apparatus, where much of the feedback control of the flow rate takes places. Passing beyond this point, the tubule becomes the distal tubule, the function of which is similar to that of the ascending limb of the loop of Henle.

Finally, the flow enters the descending collecting duct, which gathers the flow from several nephrons and descends back through the cortex and into the outer and inner zones of the medulla. The flow from the collecting duct then flows out of the kidney through the ureter on the way to the bladder. The cells lining the collecting duct are sensitive to a number of hormones that act to regulate their function as well as the final chemical composition of the urine. Primary among these hormones are *aldosterone* and *antidiuretic hormone* (ADH). Aldosterone determines the rate at which Na^+ ions are transported out of the tubular lumen, and ADH determines the permeability of the collecting duct to water, and thereby determines the final concentration of the urine. When there is no ADH present, the collecting duct is impermeable to water, but with ADH present, the permeability of the collecting duct allows water to be reabsorbed out of the collecting duct, leaving behind a more highly concentrated urine.

Putting this all together, we arrive at a qualitative summary of how a nephron operates. Along the ascending limb of the loop of Henle Na^+ is absorbed into the interstitium, either passively (in the thin ascending limb) or actively (in the thick ascending limb). This creates a high Na^+ concentration in the interstitium, which then serves to draw water out of the descending limb and allows Na^+ to reenter the descending limb. Hence, fluid entering the descending limb is progressively concentrated until, at the turning point of the loop, the fluid osmolarity is about 1200 mOsm/liter (the entering fluid is about 300 mOsm/liter). Clearly, because the fluid entering the ascending limb is so concentrated, Na^+ extraction from the ascending limb is enhanced, which further enhances water extraction from the descending limb, and so on. This positive feedback process is at the heart of the countercurrent mechanism, to be discussed in more detail below. As the fluid ascends the ascending limb, Na^+ is continually extracted until, at the level of the juxtaglomerular apparatus, the fluid in the tubule is considerably more dilute than the original filtrate. However (and this is the crucial part), the dilution process results in a steep gradient of Na^+ concentration in the interstitium, a gradient that can, when needed, concentrate the urine.

When there is no ADH present, the dilute urine formed by the loop of Henle proceeds through the collecting duct essentially unchanged, resulting in a large quantity of dilute urine. In the presence of large amounts of ADH, the collecting duct is highly permeable to water, so that by the time the filtrate reaches the level of the turning point of the loop of Henle, it is essentially at the same concentration as the interstitium, about 1200 mOsm/liter, thus giving a small quantity of concentrated urine.

It is important to emphasize that the principal functions of the loop of Henle are, first, the formation of dilute urine, which allows water to be excreted when necessary, and, second, the formation of the interstitial gradient in Na^+ concentration, which allows for the formation of a concentrated urine when necessary. The importance of the loop of Henle in creating the interstitial gradient of Na^+ concentration is underlined by the fact that although all vertebrates can produce dilute urine, only birds and mammals can produce hyperosmotic urine, and it is the kidneys of only these animals that contain loops of Henle.

17.2.1 The Countercurrent Mechanism

Solutes are exchanged between liquids by diffusion across their separating membranes. Since the rate of exchange is affected by the concentration difference across the membrane, the exchange rate is increased if large concentration differences can be maintained. One important way that large concentration differences can be maintained is by the *countercurrent mechanism*. As is described below, the countercurrent mechanism is important to renal function. Other examples of the countercurrent mechanism include the exchange of oxygen from water to blood through fish gills and the exchange of oxygen in the placenta between mother and fetus.

Suppose that two gases or liquids containing a solute flow along parallel tubes of length L, separated by a permeable membrane. We model this in the simplest possible way as a one-dimensional problem, and we assume that solute transport is a linear function of the concentration difference. Then the concentrations in the two one-dimensional tubes are given by

$$\frac{\partial C_1}{\partial t} + q_1 \frac{\partial C_1}{\partial x} = d(C_2 - C_1), \tag{17.31}$$

$$\frac{\partial C_2}{\partial t} + q_2 \frac{\partial C_2}{\partial x} = d(C_1 - C_2). \tag{17.32}$$

The mathematical problem is to find the outflow concentrations, given that the inflow concentrations, the length of the exchange chamber, and the flow velocities are known. It is a relatively easy matter to generalize this model to allow for an interstitium (see Exercise 5).

We assume that the flows are in steady state and that the input concentrations are C_1^0 and C_2^0. Then, if we add the two governing equations and integrate, we find that

$$q_1 C_1 + q_2 C_2 = k \text{ (a constant)}. \tag{17.33}$$

Pretending that k is known, we eliminate C_2 from (17.32) and find the differential equation for C_1,

$$\frac{dC_1}{dx} = \frac{d}{q_1 q_2}\left(k - (q_1 + q_2)C_1\right), \tag{17.34}$$

from which we learn that

$$C_1(x) = \kappa + (C_1(0) - \kappa)e^{-\lambda x}, \tag{17.35}$$

where $\kappa = \frac{k}{q_1+q_2}$ and $\lambda = d\left(\frac{q_1+q_2}{q_1 q_2}\right)$.

There are two cases to consider, namely when q_1 and q_2 are of the same sign and when they have different signs. If they have the same signs, say positive, then the input is at $x = 0$, and it must be that $C_1(0) = C_1^0, C_2(0) = C_2^0$, from which, using (17.33), it follows that

$$\frac{C_1(L)}{C_1^0} = \frac{1+\gamma\rho}{1+\rho} + \rho\frac{1-\gamma}{1+\rho}e^{-\lambda L}, \tag{17.36}$$

where $\gamma = C_2^0/C_1^0, \rho = q_2/q_1, \lambda = \frac{d}{q_1}(1+\frac{1}{\rho})$.

Suppose that the goal is to transfer material from vessel 1 to vessel 2, so that $\gamma < 1$. We learn from (17.36) that the output concentration from vessel 1 is an exponentially decreasing function of the residence length dL/q_1. Furthermore, the best that can be done (i.e., as $dL/q_1 \to \infty$) is $\frac{1+\gamma\rho}{1+\rho}$.

In the case that q_1 and q_2 are of opposite sign, say $q_1 > 0, q_2 < 0$, the inflow for vessel 1 is at $x = 0$, but the inflow for vessel 2 is at $x = L$. In this case we calculate that

$$\frac{C_1(L)}{C_1^0} = \frac{-\gamma\rho + (1-\rho+\gamma\rho)e^{-\lambda L}}{e^{-\lambda L} - \rho}, \tag{17.37}$$

where $\gamma = C_2(L)/C_1^0 = C_2^0/C_1^0, \rho = -q_2/q_1 > 0, \lambda = \frac{d}{q_1}(1-\frac{1}{\rho})$, provided that $\rho \neq 1$. In the special case $\rho = 1$, we have

$$\frac{C_1(L)}{C_1^0} = \frac{q_1 + \gamma dL}{q_1 + dL}. \tag{17.38}$$

Now we can see the substantial difference between a cocurrent (q_1 and q_2 of the same sign) and a countercurrent (q_1 and q_2 with the opposite sign). At fixed parameter values, if $\gamma < 1$, the expression for $C_1(L)/C_1^0$ in (17.36) is always larger than that in (17.37), implying that the total transfer of solute is always more efficient with a countercurrent than with a cocurrent.

In Fig. 17.13 is shown a comparison between a countercurrent and a cocurrent. The dashed curves show the transfer fraction $C_1(L)/C_1^0$ for a cocurrent, plotted as a function of the residence length dL/q_1, with input in tube 2, $C_2^0 = C_2(0) = 0$. The solid curves show the same quantity for a countercurrent, with input concentration $C_2^0 = C_2(L) = 0$.

In the limit of a long residence time (large dL/q_1), the transfer fraction becomes $1-\rho+\gamma\rho$ if $\rho < 1$, and γ if $\rho > 1$. Indeed, this is always smaller than the result for a cocurrent, $\frac{1+\gamma\rho}{1+\rho}$.

17.2.2 The Countercurrent Mechanism in Nephrons

The countercurrent mechanism works slightly differently in nephrons because the two parallel tubes, the descending branch and the ascending branch of the loop of Henle, are connected at their bottom end. Thus the flow and concentration of solute out of the descending tube must match the flow and concentration of solute into the ascending tube.

Mathematical models of the urine-concentrating mechanism have been around for some time, but all make use of the same basic physical principles, namely, the establishment of chemical gradients via active transport processes, the movement of ions via diffusion, and the transport of water by osmosis. The unique feature of the nephron is

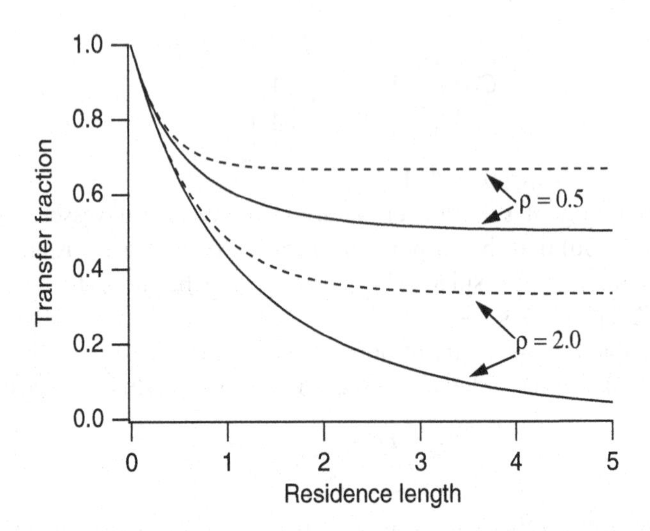

Figure 17.13 Transfer fraction for a cocurrent (dashed) and a countercurrent (solid) when $\gamma = 0$ plotted as a function of residence length.

its physical organization, which allows it to eliminate waste products while controlling other quantities. In what follows we present a model similar to that of Stephenson (1972, 1992) of urinary concentration that represents the gross organizational features of the loop of Henle. A number of other models are discussed in a special issue of the *Bulletin of Mathematical Biology* (volume 56, number 3, May 1994), while two useful reviews of mathematical work on the kidney are Knepper and Rector (1991) and Roy et al. (1992).

We view the loop of Henle as consisting of four compartments, including three tubules, the descending limb, the ascending limb, and the collecting duct, and a single compartment for the interstitium and peritubular capillaries (Fig. 17.14). The interstitium/capillary bed is treated as a one-dimensional tubule that accepts fluid from the other three tubules and loses it to the venules. It is an easy generalization to separate the peritubular capillaries and interstitium into separate compartments, but little is gained by doing so. In each of these compartments, one must keep track of the flow of water and the concentration of solutes. For the model presented here, we track only one solute, Na^+, because it is believed that the concentration of Na^+ in the interstitium determines over 90 percent of the osmotic pressure.

We assume that the flow in each of the tubes is a simple plug flow (positive in the positive x direction) with flow rates q_d, q_a, q_c, q_s for descending, ascending, collecting, and interstitial tubules, respectively. Similarly, the concentration of solute in each of these is denoted by c_d, c_a, c_c, c_s. The tubules are assumed to be one-dimensional, with glomerular filtrate entering the descending limb at $x = 0$, turning from the descending limb to the ascending limb at $x = L$, turning from the ascending limb to the collecting duct at $x = 0$, and finally exiting the collecting duct at $x = L$. We assume that the interstitium/capillary compartment drains at $x = 0$ with no flow at $x = L$.

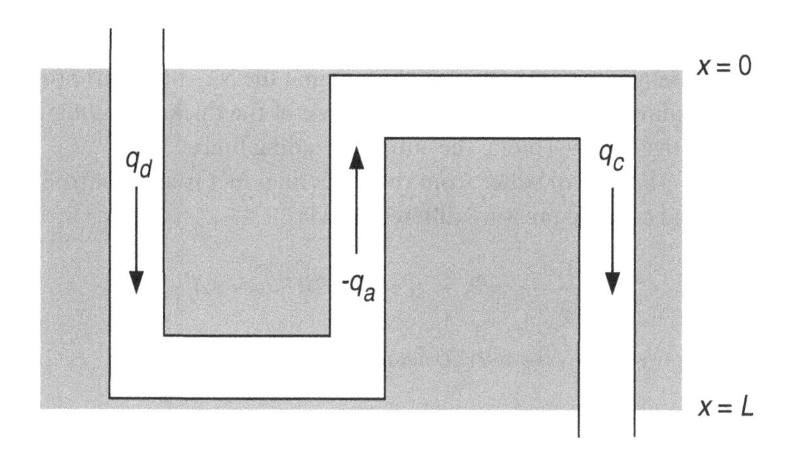

Figure 17.14 Diagram of the simple four-compartment model of the loop of Henle.

Descending Limb: The flux of water from the descending limb to the interstitium is controlled by the pressure difference and the osmotic pressure difference; hence

$$\frac{1}{k_d}\frac{dq_d}{dx} = P_s - \pi_s - P_d + 2RT(c_d - c_s), \tag{17.39}$$

where P_d and P_s are the hydrostatic pressures in the descending tubule and interstitium, π_s is the colloidal osmotic pressure of the interstitium, and k_d is the filtration rate for the descending tubule. The factor two multiplying the osmotic pressure due to the solute is to take into account the fact that the fluid is electrically neutral, and the flow of Na^+ ions is followed closely by the flow of chloride ions, both of which contribute to the osmotic pressure. The transport of Na^+ ions from the descending limb is governed by simple diffusion, so that at steady state we have

$$\frac{d(q_d c_d)}{dx} = h_d(c_s - c_d), \tag{17.40}$$

where h_d is the permeability of the descending limb to Na^+ ions.

Ascending Limb: The ascending limb is assumed to be impermeable to water, so that

$$\frac{dq_a}{dx} = 0, \tag{17.41}$$

and the flow of Na^+ out of the ascending limb is by an active process, so that

$$\frac{d(q_a c_a)}{dx} = -p. \tag{17.42}$$

The pump rate p certainly depends in nontrivial ways on the local concentrations of various ions. However, for this model we take p to be a constant. This simplifying assumption causes problems with the behavior of the model at low Na^+ concentrations, because there is nothing preventing the Na^+ concentration to become negative. Although the Na^+ ATPase is actually a Na^+–K^+ ATPase, the epithelial cells are highly permeable to K^+, and so we assume that K^+ can be safely ignored. For simplicity, we also ignore the fact that the Na^+ transport properties of the thin ascending limb are different from those of the thick ascending limb, and we assume active removal along the entire ascending limb.

Collecting Duct: The flow of water from the collecting duct is also controlled by the hydrostatic and osmotic pressure differences, via

$$\frac{1}{k_c}\frac{dq_c}{dx} = P_s - \pi_s - P_c + 2RT(c_c - c_s), \tag{17.43}$$

and the transport of Na^+ from the collecting duct is governed by

$$\frac{d(q_c c_c)}{dx} = h_c(c_s - c_c). \tag{17.44}$$

Here, k_c and h_c are the permeability of the collecting duct to water and Na^+, and are controlled by ADH and aldosterone, respectively.

Conservation Equations: Finally, because total fluid is conserved,

$$\frac{dq_s}{dx} = -\frac{d}{dx}(q_d + q_a + q_c), \tag{17.45}$$

and because total solute is conserved,

$$\frac{d(q_s c_s)}{dx} = -\frac{d}{dx}(q_d c_d + q_a c_a + q_c c_c). \tag{17.46}$$

To complete the description, we have the relationship between pressure and flow in a tube,

$$\frac{dP_j}{dx} = -R_j q_j, \tag{17.47}$$

for $j = d,a,c,s$. However, for renal modeling it is typical to take each pressure to be constant. Typical values for the pressures are P_d = 14–18 mm Hg, P_a = 10–14 mm Hg, P_c = 0–10 mm Hg, P_s = 6 mm Hg, and π_s = 17 mm Hg.

This description of the nephron consists of eight first-order differential equations in the eight unknowns q_j and c_j, for $j = d,a,c,s$. To complete the description, we need boundary conditions. We assume that the inputs $q_d(0)$ and $c_d(0)$ are known and given. Then, because the flow from the descending limb enters the ascending limb, $q_d(L) = -q_a(L)$ and $c_d(L) = c_a(L)$. Furthermore, $q_s(L) = 0$. At $x = 0$, flow from the ascending limb enters the collecting duct, so that $q_a(0) = -q_c(0)$ and $c_a(0) = c_c(0)$. Finally, since total fluid must be conserved, what goes in must go out, so that $q_d(0) + q_s(0) = q_c(L)$.

It is useful to nondimensionalize the equations by normalizing the flows and solute concentrations. Thus, we let

$$x = Ly, Q_j = \frac{q_j}{q_d(0)}, C_j = \frac{c_j}{c_d(0)} \text{ for } j = d,a,c,s,$$

and the dimensionless parameters are

$$\rho_j = \frac{q_d(0)}{2LRTc_d(0)k_j}, \Delta P_j = \frac{P_j + \pi_s - P_s}{RT2c_d(0)}, H_j = \frac{Lh_j}{q_d(0)}, \text{ for } j = d, c.$$

In this scaling $Q_d(0) = C_d(0) = 1$.

Three of these equations are trivially solved. In fact, it follows easily from (17.41), (17.45), and (17.46) that

$$Q_a = Q_a(0) = Q_a(L), \tag{17.48}$$

$$Q_d + Q_a + Q_c + Q_s = Q_c(L), \tag{17.49}$$

$$Q_dC_d + Q_aC_a + Q_cC_c + Q_sC_s = Q_c(L)C_c(L). \tag{17.50}$$

Two more identities can be found. If we use (17.40) to eliminate $c_d - c_s$ from (17.39), we obtain

$$\rho_d \frac{dQ_d}{dy} + \Delta P_d = C_d - C_s = -\frac{1}{H_d}\frac{d(Q_dC_d)}{dy}, \tag{17.51}$$

from which it follows that

$$\rho_d(Q_d - 1) + \frac{1}{H_d}(Q_dC_d - 1) = -\Delta P_d y. \tag{17.52}$$

Similarly, we use (17.44) to eliminate $c_c - c_s$ from (17.43) to obtain

$$\rho_c \frac{dQ_c}{dy} + \frac{1}{H_c}\frac{d(Q_cC_c)}{dy} = -\Delta P_c, \tag{17.53}$$

which integrates to

$$\rho_c(Q_c - Q_c(0)) + \frac{1}{H_c}(Q_cC_c - Q_c(0)C_c(0)) = -\Delta P_c y. \tag{17.54}$$

As discussed above, we assume that the Na^+ concentration in the ascending limb is always sufficiently high so that the Na^+–K^+ pump is saturated and the pump rate is independent of concentration, in which case the solution of (17.42) (in nondimensional variables) is

$$Q_aC_a = Q_aC_a(0) - Py, \tag{17.55}$$

where $P = \frac{pL}{c_d(0)q_d(0)}$ is the dimensionless Na^+ pump rate.

Having solved six of the original eight differential equations, we are left with a system of two first-order equations in two unknowns. The two equations are

$$\rho_d \frac{dQ_d}{dy} = -\Delta P_d + C_d - C_s, \tag{17.56}$$

$$\rho_c \frac{dQ_c}{dy} = -\Delta P_c + C_c - C_s, \tag{17.57}$$

subject to boundary conditions $Q_d = 1, Q_c = -Q_a$ at $y = 0$, and $Q_d = -Q_a$ at $y = 1$, where C_c, C_s, and C_d are functions of Q_d and Q_c. Although there are three boundary conditions for two first-order equations, the number Q_a is also unknown, so that this problem is well posed. Our goal in what follows is to understand the behavior of the solution of this system.

Formation of Urine Without ADH

The primary control of renal dialysis is accomplished in the collecting duct, where the amount of ADH determines the permeability of the collecting duct to water and the amount of aldosterone determines the permeability of the collecting duct to Na^+. Impairment of normal kidney function is often related to ADH. For example, the inability of the pituitary to produce adequate amounts of ADH is called *central diabetes insipidus*, and results in the formation of large amounts of dilute urine. On the other hand, with *nephrogenic diabetes insipidus*, the abnormality resides in the kidney, either as a failure of the countercurrent mechanism to produce an adequately hyperosmotic interstitium, or as the inability of the collecting ducts to respond to ADH. In either case, large volumes of dilute urine are formed.

Various drugs and hormones can have similar effects. For example, alcohol, clonidine (an antihypertensive drug), and haloperidol (a dopamine blocker) are known to inhibit the release of ADH. Other drugs such as nicotine and morphine stimulate the release of ADH. Drugs such as lithium (used to treat manic-depressives) and the antibiotic tetracyclines impair the ability of the collecting duct to respond to ADH.

The second important controller of urine formation is the hormone aldosterone. Aldosterone, secreted by zona glomerulosa cells in the adrenal cortex, works by diffusing into the epithelial cells, where it interacts with several receptor proteins and diffuses into the cell nucleus. In the cell nucleus it induces the production of the messenger RNA associated with several important proteins that are ingredients of Na^+ channels. The net effect is that (after about an hour) the number of Na^+ channels in the cell membrane increases, with a consequent increase of Na^+ conductance. Aldosterone is also known to increase the Na^+–K^+ ATPase activity in the collecting duct, as well as in other places in the nephron (a feature not included in this model), thereby increasing Na^+ removal and also K^+ excretion into the urine. For persons with *Addison's disease* (severely impaired or total lack of aldosterone), there is tremendous loss of Na^+ by the kidneys and accumulation of K^+. Conversely, excess aldosterone secretion, as occurs in patients with adrenal tumors (Conn's syndrome), is associated with Na^+ retention and K^+ depletion.

To see the effect of these controls we examine the behavior of our model in two limiting cases. In the first case, we assume that there is no ADH present, so that $\rho_c = \infty$, and that there is no aldosterone present, so that $H_c = 0$. In this case it follows from (17.53) that $Q_c = Q_c(0) = -Q_a$ and that $C_c = C_c(0) = C_a(0)$. In other words, there is no loss of either water or Na^+ from the collecting duct: the collecting duct has effectively been removed from the model.

It remains to determine what happens in the descending and ascending tubules. The flow is governed by the single differential equation

$$\rho_d \frac{dQ_d}{dy} = C_d - C_s - \Delta P_d = f(Q_d, Q_a, y), \tag{17.58}$$

where, from (17.50), (17.52), and (17.55),

$$C_d = \frac{1}{Q_d}(1 + \rho_d H_d(1 - Q_d) - \Delta P_d H_d y), \tag{17.59}$$

$$C_s = \frac{(P + \Delta P_d H_d)(1 - y)}{Q_d + Q_a} - \rho_d H_d, \tag{17.60}$$

subject to the boundary conditions $Q_d(0) = 1, Q_d(1) = -Q_a$. As before, Q_a is a constant, as the ascending limb is impermeable to water, and C_a is a linearly decreasing function of y.

We view this problem as a nonlinear eigenvalue problem, since it is a single first-order differential equation with two boundary conditions. The unknown parameter Q_a is the parameter that we adjust to make the solution satisfy the two boundary conditions. It is reasonable to take ρ_d to be small, since the descending tubule is quite permeable to water. In this case, however, the differential equation (17.58) is singular, since a small parameter multiplies the derivative. We overcome this difficulty by seeking a solution in the form $y = y(Q_d, \rho_d)$ satisfying the differential equation

$$f(Q_d, Q_a, y)\frac{dy}{dQ_d} = \rho_d \tag{17.61}$$

subject to boundary conditions $y = 0$ at $Q_d = 1$ and $y = 1$ at $Q_d = -Q_a$.

With ρ_d small we have a regular perturbation problem in which we seek y as a function of Q_d as a power series of ρ_d, which is solved as follows. We assume that y has a power series representation of the form

$$y = y_0 + \rho_d y_1 + \rho_d^2 y_2 + O(\rho_d^2), \tag{17.62}$$

substitute into (17.61), expand in powers of ρ_d, collect like powers of ρ_d, and then solve these sequentially. We find that

$$y = 1 - \frac{Q_a + Q_d}{PQ_d - \Delta P_d H_d Q_a}\left[1 - \Delta P_d(Q_d + H_d)\right] + O(\rho_d). \tag{17.63}$$

Notice that $y = 1$ at $Q_d = -Q_a$. Now we determine Q_a by setting $y = 0$, $Q_d = 1$ in (17.63), and solving for Q_a. To leading order in ρ_d we find that

$$-Q_a = 1 - \frac{P + H_d \Delta P_d}{1 - \Delta P_d} + O(\rho_d). \tag{17.64}$$

It is now a straightforward matter to plot y as a function of Q_d, and then rotate the axes so that we see Q_d as a function of y. This is depicted with a dashed curve in Fig. 17.15 (using formulas that include higher-order correction terms for ρ_d). For comparison we also include the curves calculated for the case where ADH is present;

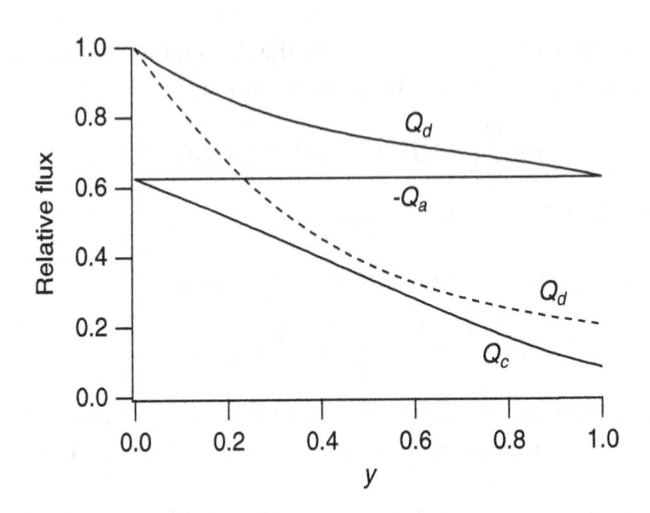

Figure 17.15 The flux of fluid in the loop of Henle, with ADH present (solid curve, $\rho_c = 2.0$) and without ADH present (dashed curve, $\rho_c = \infty$). Parameter values are $P = 0.9$, $\Delta P_d = 0.15$, $H_d = 0.1$, $\rho_d = 0.15$, $H_c = 0$.

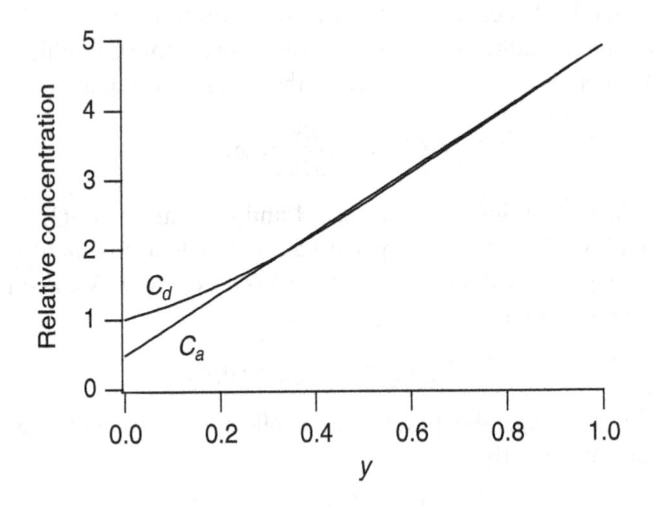

Figure 17.16 The solute concentration in the descending (C_d) and ascending (C_a) tubules with no ADH present ($\rho_c = \infty$), plotted as a function of distance y for the parameter set as in Fig. 17.15.

the details of that calculation are given below. Note that in either the presence or absence of ADH, Q_a is always independent of y, while in the absence of ADH, Q_c is also independent of y. Once Q_d is determined as a function of y, it is an easy matter to plot the concentrations C_d and C_a as functions of y, as shown in Fig. 17.16.

From these we can draw some conclusions about how the loop of Henle works in this mode. Sodium is extracted from the descending limb by simple diffusion and from the ascending loop by an active process. The Na^+ that is extracted from the ascending loop creates a large osmotic pressure in the interstitial region that serves to enhance the extraction of water from the descending loop. This emphasizes the importance of the countercurrent mechanism in the concentrating process. As the fluid proceeds down the descending loop, its Na^+ concentration is continually increasing, and during its passage along the ascending loop, its Na^+ concentration falls. At the lower end of the loop the relative concentration of the formed urine (i.e., of substances that are impermeable, such as creatinine) is $\frac{1}{Q_d(1)}$. This quantity represents the concentrating ability of the nephron in this mode. Since $C_a(0) < C_d(0)$, as can be seen from Fig. 17.16, by the time the fluid reaches the top of the ascending loop, it has been diluted. Furthermore, comparing the value of $Q_d(1)(= Q_c)$ in the absence of ADH (dashed curve in Fig. 17.15) to the value of $Q_c(1)$ in the presence of ADH (solid curve in Fig. 17.15) shows that the flux out of the collecting duct is higher in the absence of ADH. Hence, combining these two observations, we conclude that in the absence of ADH, the nephron produces a large quantity of dilute urine, while in the presence of ADH, it produces a smaller quantity of concentrated urine. This is consistent with the qualitative explanation of nephron function given earlier in the chapter.

In Fig. 17.17 are shown the solute concentration C_a and the flow rate Q at the upper end of the ascending tubule as functions of dimensionless pump rate P. The formed urine is dilute whenever this solute concentration is less than one. The fact that this concentration can become negative at larger pump rates is a failure of the model, since the pump rate in the model is not concentration dependent.

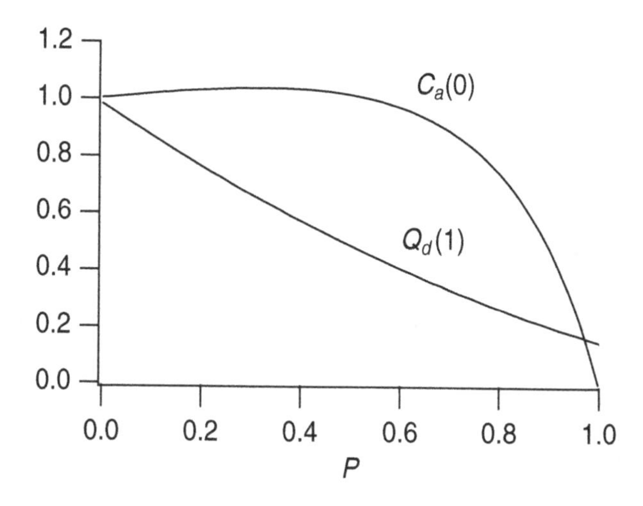

Figure 17.17 The solute concentration and the flow rate at the upper end of the ascending tubule plotted as functions of pump rate P when there is no ADH or aldosterone present ($\rho_c = \infty$, $H_c = 0$). Dilution occurs if the solute concentration is less than one. Parameter values are $\Delta P_d = 0.15$, $H_d = 0.1$, $\rho_d = 0.15$.

Formation of Urine With ADH

In the presence of ADH, the collecting tube is highly permeable to water, so that, since the concentration of Na^+ in the interstitium at the lower end of the tube is high, additional water can be extracted from the collecting duct, thereby concentrating the dilute urine formed by the loop of Henle.

To solve the governing equations in this case is much harder than in the case with no ADH. This is because the equations governing the flux (17.56) and (17.57) are both singular in the limit of zero ρ_d and ρ_c. Furthermore, one can show that the quasi-steady solution (found by setting $\rho_d = \rho_c = 0$ in (17.56) and (17.57)) cannot be made to satisfy the boundary conditions at $y = 1$, suggesting that the solution has a boundary layer. To avoid the difficulties associated with boundary layers, it is preferable to formulate the problem in terms of the solute flux $S_d = Q_d C_d$, because according to (17.51) this function is nearly linear and does not change rapidly when ρ_d is small.

In the case that ADH is present but there is no aldosterone ($H_c = 0$), the governing equations are

$$\frac{dS_d}{dy} = H_d\left(\frac{S_s}{Q_s} - \frac{S_d}{Q_d}\right) = H_d F_d(S_d, Q_c), \tag{17.65}$$

$$\rho_c \frac{dQ_c}{dy} = -\Delta P_c + \frac{S_c(0)}{Q_c} - \frac{S_s}{Q_s} = F_c(S_d, Q_c), \tag{17.66}$$

where

$$S_s = P(y-1) + S_d(1) - S_d, \tag{17.67}$$

$$Q_s = -1 - Q_a - Q_c + Q_c(1) - \frac{1 - S_d - \Delta P_d H_d y}{\rho_d H_d}, \tag{17.68}$$

$$Q_d = 1 + \frac{1 - S_d - \Delta P_d H_d y}{\rho_d H_d}, \tag{17.69}$$

subject to boundary conditions $S_d(0) = 1, Q_c(0) = 1 + \frac{1 - S_d - \Delta P_d H_d}{\rho_d H_d}$, and $Q_d(1) = -Q_a$.

These equations are difficult to solve because there are two unknown functions, S_d and Q_c, and an unknown constant Q_a, subject to three boundary conditions. One way to solve them is to introduce the constants Q_a and $Q_c(1)$ as unknown variables satisfying the obvious differential equations $\frac{dQ_a}{dy} = \frac{dQ_c(1)}{dy} = 0$, and to solve the expanded fourth-order system of equations in the four unknowns $S_d, Q_c, Q_a, Q_c(1)$ with four corresponding boundary conditions (adding the requirement that $Q_c = Q_c(1)$ at $y = 1$).

These equations were solved numerically using a centered difference scheme for the discretization and Newton's method to find a solution of the nonlinear equations (see Exercise 9). Typical results are shown in Fig. 17.18. Here we see what we expected (or hoped), namely that the collecting duct concentrates the dilute urine by extracting water. In fact, we see that the concentration increases on its path through the descending loop, decreases in the ascending loop, and then increases again in the collecting duct. This behavior is similar to the data for Na^+ concentration shown in Fig. 17.11.

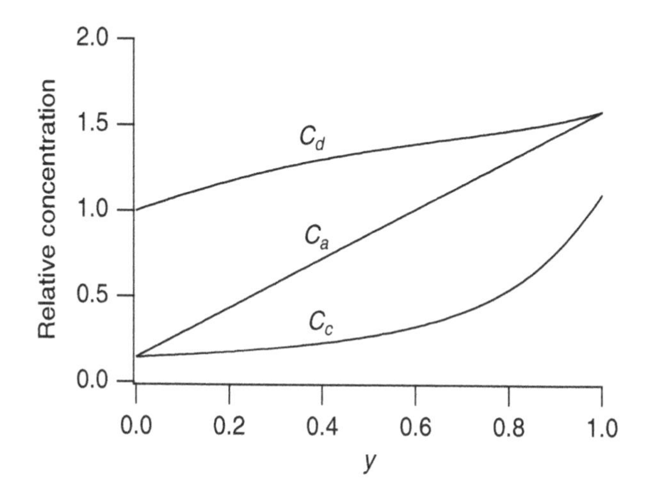

Figure 17.18 Solute concentrations in the loop of Henle and the collecting duct, plotted as functions of y for $P = 0.9, \Delta P_d = 0.15, \Delta P_c = 0.22, H_d = 0.1, \rho_d = 0.15$ and with $\rho_c = 2.0$, $H_c = 0$.

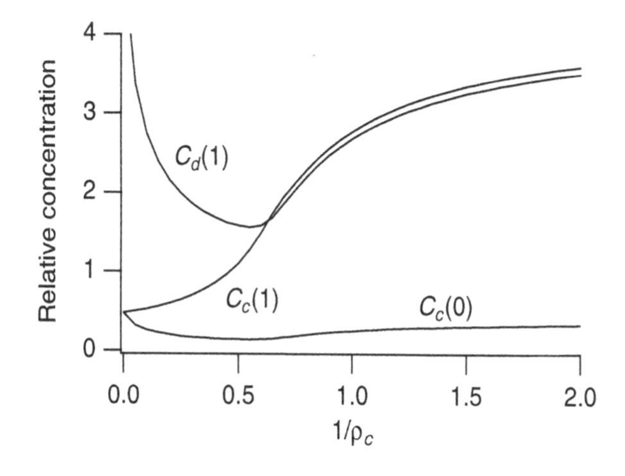

Figure 17.19 Solute concentrations at the bottom and top of the loop of Henle and at the end of the collecting duct plotted as functions of inverse permeability $\frac{1}{\rho_c}$, with $P = 0.9$, $\Delta P_d = 0.15$, $\Delta P_c = 0.22$, $H_d = 0.1$, $\rho_d = 0.15$, and $H_c = 0$.

The effect of the parameter ρ_c is shown in Figs. 17.19 and 17.20. In these figures are shown the solute concentrations and the flow rates at the bottom and top of the loop of Henle and at the end of the collecting duct. Here we see that the effect of ADH is, as expected, to reconcentrate the solute and to further reduce the loss of water.

The asymptotic value of $C_c(1)$ as $\rho_c \to 0$ is the maximal solute concentration possible and determines, for example, whether or not the individual can safely drink seawater without dehydration. The asymptotic value of $1/Q_c(1)$ represents the highest possible relative concentration of impermeable substances such as creatinine.

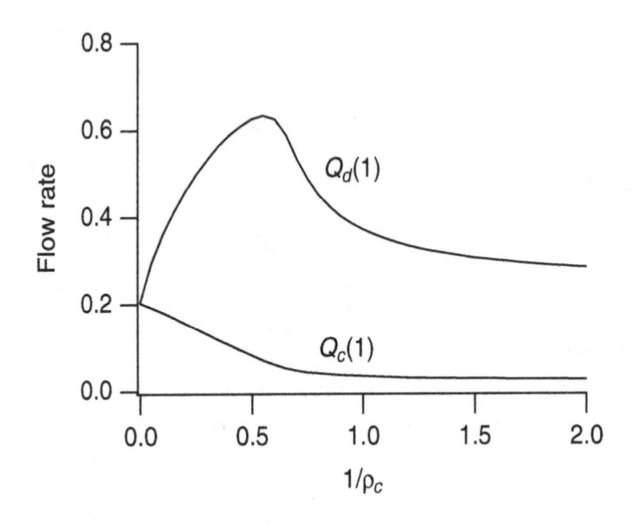

Figure 17.20 Fluid flow rates at the bottom of the loop of Henle and at the end of the collecting duct plotted as functions of inverse permeability $\frac{1}{\rho_c}$, with $P = 0.9, \Delta P_d = 0.15, \Delta P_c = 0.22, H_d = 0.1, \rho_d = 0.15$, and $H_c = 0$.

Further Generalizations

This model shows the basic principles behind nephron function, but the model is qualitative at best, and there are many questions that remain unanswered and many generalizations that might be pursued. For example, the model could be improved by incorporating a better representation of the interstitial/capillary bed flow, taking into account that the peritubular capillaries issue directly from the efferent arteriole of the glomerulus, thus determining the hydrostatic and osmotic pressures in the capillary bed. The model is also incorrect in that the active pumping of Na^+ out of the ascending limb is not concentration dependent, and as a result negative concentrations can occur for certain parameter values.

It is a fairly easy matter to add equations governing the flux of solutes other than Na^+, as the principles governing their flux are the same. One can also consider a time-dependent model in which the flow of water is not steady, by allowing the cross-sectional area of the tubules to vary. Nonsteady models are difficult to solve because they are stiff, and there is a substantial literature on the numerical analysis and simulation of time-dependent models (Layton et al., 1991).

Nephrons occur in a variety of lengths, and models describing kidney function have been devised that recognize that nephrons are distributed both in space and in length (Layton et al., 2004; Layton and Layton, 2003, 2005a,b). These models are partial differential equations, and again, because of inherent stiffness, their simulation requires careful choice of numerical algorithms (Layton et al., 1995a; Layton and Layton, 2002, 2003).

17.3 Models of Tubular Transport

So far in this chapter we have discussed glomerular filtration and the formation of urine. The third major group of models in renal physiology consists of those of the transport of water and solutes by the various epithelial cells that line the tubules. The most important of these models are undoubtedly those of Alan Weinstein; see, for example, Weinstein (1994, 1998a,b, 2000, 2003).

Two of the most significant questions addressed by these models is how the epithelial cell layer can transport water against an osmotic gradient, so-called *uphill* transport, or under conditions of little or no osmotic gradient, so-called *isotonic* transport. These questions were the subject of some of the earliest models (see, for instance, Weinstein and Stephenson, 1981), and remain important today. Discussion of simple models of uphill and isotonic water transport are deferred to Section 18.1.2, since these questions are also important in the gastrointestinal system.

Models of tubular transport usually also include descriptions of solute transport by both active and passive mechanisms. Typical solutes include Na^+, K^+, HCO_3^-, Cl^-, H, HPO4, NH4, glucose, and PO4, giving rise to complex models with a multitude of variables and parameters. Because the epithelial cells in different parts of the nephron have different permeabilities, pumps, and exchangers, there are different models of the collecting duct, the ascending and descending limbs, and the proximal tubule. As is described in Chapter 18, the lateral intercellular spaces (the spaces between the epithelial cells) play a crucial role in water and solute transport in models of this type, and are usually included explicitly in the models.

17.4 EXERCISES

1. The flow of glomerular filtrate and the total renal blood flow increase by 20 to 30 percent within 1 to 2 hours following a high-protein meal. How can this feature be incorporated into a model of renal function and regulation of glomerular function?

   Hint: Amino acids, which are released into the blood after a high-protein meal, are cotransported with Na^+ ions from the filtrate in the proximal tubule. Thus, high levels of amino acids leads to high reabsorption of Na^+ in the proximal tubule, and therefore lower than normal levels of Na^+ at the macula densa.

2. Show that (17.16) has a unique positive root.

3. How much water must one drink to prevent any dehydration after eating a 1.5 oz bag of potato chips? (See Exercise 18 in Chapter 2.) Remark: A mole of NaCl is 58.5 grams and it dissociates in water into 2 osmoles.

4. Why is alcohol a diuretic? What is the combined effect on urine formation of drinking beer (instead of water) while eating potato chips? What is the combined effect on urine formation of drinking beer while smoking cigarettes?

   Hint: Alcohol inhibits the release of ADH, while nicotine stimulates ADH release.

5. Construct a simple model of the countercurrent mechanism that includes an interstitial compartment (Fig. 17.21). Show that inclusion of the interstitium has no effect on the overall rates of transport. Allow the solute to diffuse in the interstitium, but not escape the boundaries.

   Hint: View the interstitium as a tube with zero flow rate.

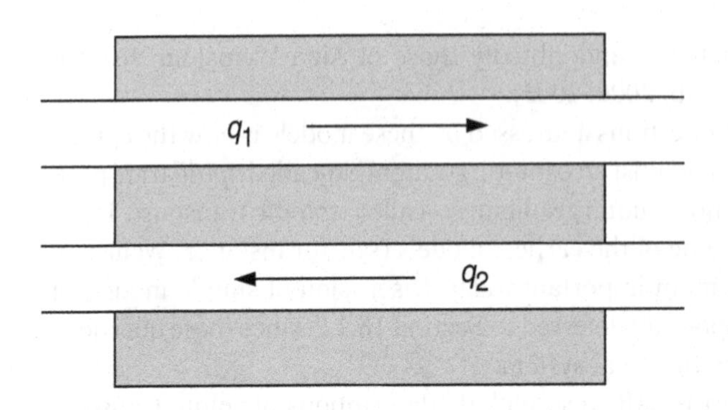

Figure 17.21 Diagram of a countercurrent flow mediated by an interstitium, for Exercise 5.

6. Generalize the four-compartment model of the loop of Henle by separating the interstitium and peritubular capillaries into separate compartments, allowing no flow across $x = 0$ or $x = L$ for the interstitium.

7. What changes in the exchange rates of the four-compartment model of the loop of Henle might better represent the geometry of the loop of Henle, as depicted in Fig. 17.2?

   Remark: Some features to consider include the location of the thickening of the ascending and descending limbs and the location of the junction of the peritubular capillaries with the arcuate vein.

8. Formulate a time-dependent four-compartment model of urine concentration that tracks the concentration of both Na^+ ions and urea.

9. Develop a numerical computer program to solve the equations of renal flow in the case that both ADH and aldosterone are present. It is preferable to formulate the problem in terms of the unknowns S_d and S_c and to expand the system of equations to a fourth-order system by allowing $S_d(1)$ and $S_c(1)$ to be unknowns that satisfy the simple differential equations $\frac{dS_d(1)}{dy} = 0$ and $\frac{dS_c(1)}{dy} = 0$. With the four unknowns, $S_d(y), S_c(y), S_d(1)$, and $S_c(1)$, the Jacobian matrix is a banded matrix, and numerical algorithms to solve banded problems are faster and more efficient than full matrix solvers.

10. Generalize the renal model to include a concentration-dependent Na^+ pump in the ascending tubule. Does this change in the model guarantee that the flux and concentrations are nowhere negative?

CHAPTER 18

The Gastrointestinal System

Although the detailed structure of the gastrointestinal tract varies from region to region, there is a common basic structure, outlined in the cross-section shown in Fig. 18.1. It is surrounded by a number of heavily innervated muscle layers, arranged both circularly and longitudinally. Contraction of these muscle layers can mix the contents of the tract and move food in a controlled manner in the appropriate direction. Beneath the muscle layer is the *submucosa*, consisting mostly of connective tissue, and beneath that is a thin layer of smooth muscle called the *muscularis mucosae*. Finally, there is the *lamina propria*, a layer of connective tissue containing capillaries and many kinds of secreting glands, and then a layer of epithelial cells, whose nature varies in different regions of the tract.

In addition to the muscle layers, there are two principal layers of neurons; the myenteric plexus, between the longitudinal and circular muscle layers, and the submucosal plexus, which lies in the submucosa. In general, stimulation of the myenteric plexus increases the rate and intensity of the rhythmic contractions of the gut, and increases the velocity of conduction of waves of excitation along the gut wall. The submucosal plexus is mainly sensory, receiving signals from stretch receptors in the gut wall, and from the gut epithelium. The gastrointestinal tract is also heavily innervated, which can control the activity of the entire gut, or part of it.

18.1 Fluid Absorption

The primary function of the gastrointestinal tract is to absorb nutrients from the mix of food and liquid that moves through it. To accomplish this, the absorptive surface of the intestines consists of many folds and bends called *valvulae conniventes*, which increase

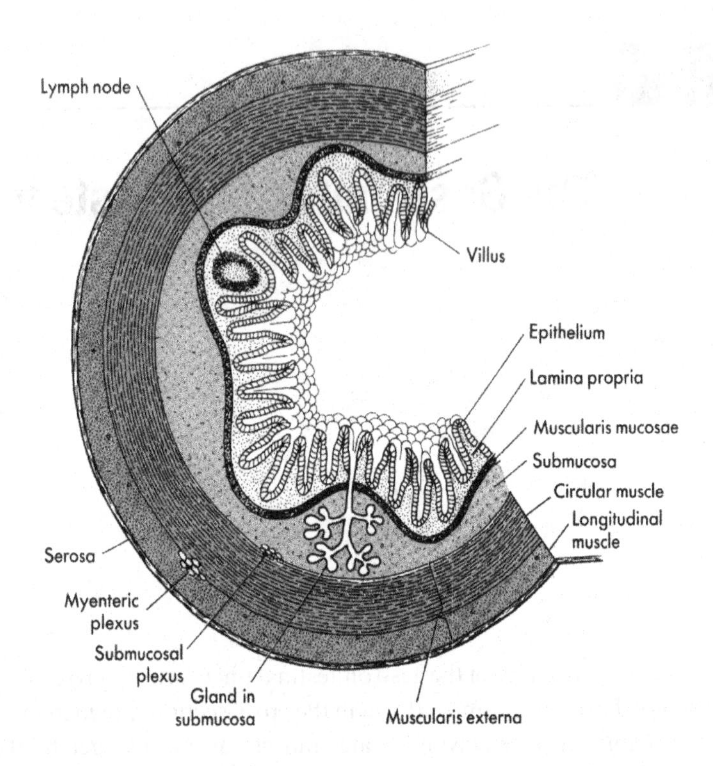

Figure 18.1 Cross-section of the gastrointestinal tract. The outermost layers of the tract consist of smooth muscle, while the innermost layer consists of epithelial cells. The epithelial cell layer contains many gastric pits, glands that secrete hydrochloric acid, and thus the stomach lumen is highly acidic. (Berne and Levy, 1993, Fig. 38-1, p. 616.)

the surface area of the absorptive mucosa about threefold. Located over the entire surface of the mucosa of the small intestine are millions of *villi*, which project about 1 mm from the surface of the mucosa and enhance the absorptive area another tenfold. The absorptive surface of the villi consists of epithelial cells that are characterized by a brush border, consisting of as many as 1000 *microvilli* 1 μm in length and 0.1 μm in diameter. The brush border increases the surface area exposed to the intestinal material by another twentyfold. The combination of all surface protrusions yields an absorptive surface area of about 250 square meters—about the surface area of a tennis court.

Epithelial cells are responsible for the absorption of nutrients and water from the intestine. The absorption of chemical nutrients, for example glucose and amino acids, is by the same process as in the kidney, via cotransporters with Na^+.

For many years, it was believed that water is absorbed, or transported through cells, simply because the membrane is porous to water. In this scenario, the movement of ions, either by passive diffusion or by active transport, sets up osmotic gradients which, in turn, drive the flow of water across the membrane. Recall that this is the basic assumption underlying the cell volume control models described in Chapter 2. However, the discovery in 1992 of transmembrane proteins that allow only (or at least mostly) water to permeate, called *aquaporins*, has changed this view dramatically (Reuss and Hirst, 2002). The first aquaporin to be discovered was AQP1, discovered in 1992 by Agre and his colleagues (Preston et al., 1992). Agre received the 2003 Nobel Prize in Chemistry for this discovery. Aquaporins have since been identified in practically every living organism, and have such high selectivity to water that they repel even H_3O^+ and hydrogen ions. The fact that water flow is through transmembrane protein structures means that, although it is driven by osmotic gradients, it is genetically regulated, and that malfunctioning channels may be associated with many diseases of the kidneys, skeletal muscle and other organs. A review by Agre et al. (2002) discusses a number of areas in which the discovery of aquaporins has had a major impact on our understanding of physiology and pathophysiology.

To complicate matters further, there is evidence that water can be transported actively, by cotransport with a solute (Loo et al., 2002; Zeuthen, 2000). This view has been challenged by other researchers, who claim that the water transport is osmotically driven but that the solute accumulates in spatially restricted areas not amenable to experimental access (Lapointe et al., 2002). However, it remains an intriguing possibility that, if true, would necessitate significant rethinking of models of water transport.

Although the control of absorption and transport of water is perhaps more complicated than first thought, osmosis appears to be the primary driving force. This being the case, one particularly important question arises; how can an epithelial cell layer transport water against its osmotic gradient (*uphill* transport)? Water transport under such conditions is observed in both the gastrointestinal system and the kidney. A related question is the mechanism of *isotonic* transport, in which the transported fluid has the same osmolality of the fluid into which it is transported.

Here we present two models of water transport via osmosis; the first is a simpler model that includes no spatial information, while the second, due to Diamond and Bossert (1967), shows how the maintenance of a standing gradient of solute concentration in a restricted spatial area can lead to isotonic and uphill water transport.

18.1.1 A Simple Model of Fluid Absorption

In general, gastrointestinal tract epithelial cells are not permeable to water on their lumenal side. However, there are 0.7–1.5 nm pores through the *tight junctions* between epithelial cells that permit water to diffuse readily between the lumen and the interstitium. The absorption of water through these pores is driven primarily by the Na^+ gradient between the lumen and the interstitium. Sodium is transported to the interior of the epithelial cell by passive transport and then is removed from the interior to the interstitium by a Na^+–K^+ ATPase. The Na^+ is transported from the interstitium by capillary blood flow.

To model the transport of water by the epithelial cell lining, we consider a small section of the epithelial gastrointestinal tract as two well-mixed compartments, the lumen and the interstitium, separated by a membrane (Fig. 18.2). We suppose that the Na^+ concentration in the lumen is n_l and in the cell interior is n_i, and that the

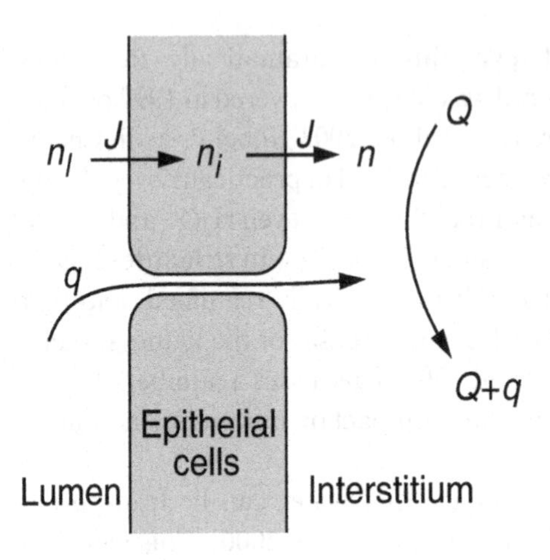

Figure 18.2 Diagram for osmotic transport of water across the epithelial cell wall. J denotes the flow of Na^+, q denotes the flow of water, and Q denotes capillary blood flow.

concentration of all osmolites in the interstitium is n. The flow, J, of Na^+ from the lumen to the interior of the cells is assumed to be passive (i.e., we ignore the effects of the membrane potential; see Exercise 1), and so

$$J = g(n_i - n_l), \tag{18.1}$$

for some constant g. Sodium flux from the cell interior to the interstitium is via an active Na^+–K^+ ATPase,

$$J = f(n_i), \tag{18.2}$$

for some saturating function f. The flow of water q through the tight junctions is driven by the osmotic pressure difference between the lumen and the interstitium, so that

$$Rq = n - n_l, \tag{18.3}$$

where R is the resistance (in appropriate units) of the tight junctions. Finally, we assume that there is a flow into and out of the interstitium provided by capillary flow. The influx of fluid is Q with an incoming concentration of osmolites n_0, while the outflow of osmolites is $Q+q$ at concentration n. At steady state, the conservation of Na^+ implies that

$$g(n_l - n_i) = f(n_i) \tag{18.4}$$

and

$$(Q + q)n - Qn_0 = f(n_i). \tag{18.5}$$

The behavior of this system of three algebraic equations is relatively easy to sort out. Since f is a positive, monotone increasing function of its argument, there is a one-to-one relationship between n_l and n_i,

$$n_l = n_i + \frac{1}{g}f(n_i). \tag{18.6}$$

We can use (18.3) to eliminate n from (18.5) and obtain

$$Rq^2 + (RQ + n_l)q + Q(n_l - n_0) - f(n_i) = 0. \tag{18.7}$$

Because the rate of Na^+ removal is dependent on the Na^+ concentration, we take $f(n) = \frac{Q_f n^3}{N^3+n^3}$, for some constants Q_f and N.

It is valuable to nondimensionalize this problem by scaling all concentrations by N, setting $u_j = n_j/N$ and $y = q/Q$. Then (18.7) becomes

$$\rho y^2 + (\rho + u_l)y + \kappa = 0, \tag{18.8}$$

where $\kappa = u_i - u_0 + (1-\gamma)\beta F(u_i) = 0, \rho = RQ/N, \gamma = g/Q$, and $\beta = \frac{Q_f}{gN}$, and (18.6) becomes

$$u_l = u_i + \beta F(u_i), \tag{18.9}$$

where $F(u) = \frac{u^3}{1+u^3}$. There are four nondimensional parameters, namely u_0, the (relative) concentration of incoming interstitial osmolites; ρ, the resistance of the tight junctions to water; γ, the relative permeability of the lumenal cell wall to Na^+; and β, the maximal velocity of active Na^+ transport (which depends primarily on the density of Na^+ pumps).

The easiest way to solve these equations for y as a function of u_l is to view the solution as the curve $y = y(u_i)$, $u_l = u_l(u_i)$, parameterized by u_i, since for each u_i, u_l is readily determined from (18.9), and then y is determined from (18.8). Observe that (18.8) is a quadratic polynomial in y that has at most one positive root. In fact, the larger root of this polynomial is positive if and only if $\kappa < 0$. Furthermore, the positive root is a monotone decreasing function of κ.

There are several behaviors of the solution depending on the parameter values. However, the behavior that is of most interest here occurs when $\beta(\gamma - 1)$ is a large positive number. In this case, κ is an N-shaped function of u_i, negative at $u_i = 0$, increasing for small values of u_i, then decreasing and finally increasing and eventually becoming positive for large u_i.

For much of parameter space this N-shaped behavior for κ translates into N-shaped behavior for the positive root of (18.8). That is, with $u_l = 0$, there is a positive root. This root initially decreases to a minimal value and then increases to a maximal value, whereupon it decreases and eventually becomes negative, as a function of u_l. This behavior is depicted in Fig. 18.3, with parameter values $\rho = u_0 = \beta = 1, \gamma = 10$.

The implications of this are interesting. It implies that one can maximize the absorption of water by adjusting the Na^+ level of the lumenal water. Thus, hydration occurs more quickly with fluids containing electrolytes than with pure water, as many high-performance athletes (such as road cyclists and long-distance runners) already know. However, too much Na^+ has the opposite effect of dehydrating the interstitium. This is a local effect only, as water is reabsorbed further along the tract.

When a person becomes dehydrated, large amounts of aldosterone are secreted by the adrenal glands. Aldosterone greatly enhances the transport of Na^+ by epithelial

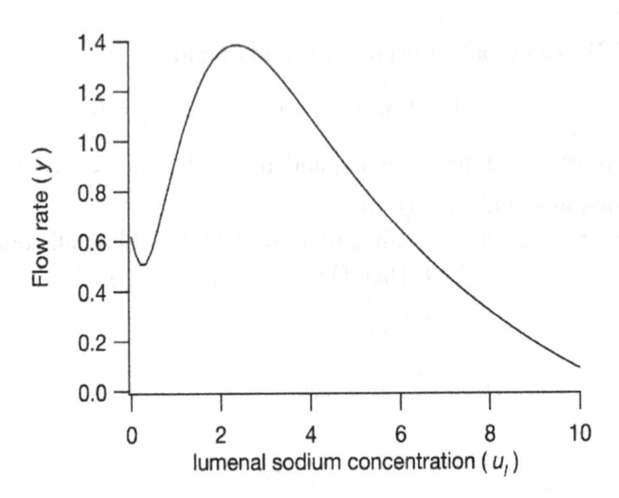

Figure 18.3 Flux of water through the epithelial membrane plotted as a function of lumenal Na^+ concentration, with parameter values $\rho = u_0 = \beta = 1$, $\gamma = 10$.

cells by activating the production of channel and pump proteins, which increases the passive and active transport of Na^+. Indeed, a person can acclimatize to heavy exercise in hot weather, as over a period of weeks increased aldosterone secretion from the adrenal cortex prevents excessive Na^+ loss in sweat, thus dispensing with the need for dietary Na^+ supplements. Loss of K^+ can still, however, be a problem.

In this model the presence of aldosterone can be modeled by increasing g, the conductivity of Na^+ transport from the lumen, and/or by increasing Q_f, the maximal rate of active Na^+ pumping. It is easy to see that the total flux of Na^+ $J = f(n_i)$ and the flux of water q both increase if either g or Q_f (or both) are increased. However, this increase is not without bound, since in the limit $g \to \infty$, we have $n_i \to n_l$, so that

$$\lim_{g\to\infty} J = f(n_l), \tag{18.10}$$

and

$$\lim_{g\to\infty} q = Q\left(\frac{n_0}{n_l} - 1\right) + \frac{f(n_l)}{n_l}, \tag{18.11}$$

when $R = 0$. Thus, if a person is dehydrated, aldosterone production works to increase Na^+ absorption and decrease water loss.

Now we can construct a simple model of water content and Na^+ concentration as a function of distance along the intestinal length. We suppose that the *chyme* (the mixture of food, water, and digestive secretions entering from the stomach) moves as a plug flow with constant velocity. Water is removed from the chyme by osmosis and Na^+ is removed by the epithelial cells at local rates determined by the local Na^+ concentration. In steady state,

$$\frac{dQ_w}{dx} = -q(n_l), \tag{18.12}$$

$$\frac{d(n_l Q_w)}{dx} = -J(n_l), \tag{18.13}$$

where Q_w is the flow of water in the intestine, x is the distance along the intestine, and $q(n_l)$ and $J(n_l)$ are the removal rates of water and Na^+, such as those suggested above. The analysis of this system of equations is straightforward and is left as an exercise (Exercise 2).

There are two common abnormalities that can occur in this process. *Constipation* occurs if the movement of feces through the large intestine is abnormally slow, allowing more time for the removal of water and therefore hardening and drying of the feces. Any pathology of the intestines that obstructs normal movement, including tumors, ulcers, or forced inhibition of normal defecation reflexes, can cause constipation.

The opposite condition, in which there is rapid movement of the feces through the large intestine, is known as *diarrhea*. There are several causes of diarrhea, the most common of which is infectious diarrhea, in which a viral or bacterial infection causes an inflammation of the mucosa. Wherever it is infected, the rate of secretion of the mucosa is greatly increased, with the net effect that large quantities of fluid are made available to aid in the elimination of the infectious agent.

For example, the toxins of cholera and other diarrheal bacteria stimulate immature epithelial cells (which are constantly being produced) to release large amounts of Na^+ and water, presumably to combat the disease by washing away the bacteria. However, if this excess secretion of Na^+ and water cannot be overcome by the absorption by mature, healthy cells, the result can be lethal because of serious dehydration. In most instances, the life of a cholera victim can be saved by intravenous administration of large amounts of NaCl solution to make up for the loss.

18.1.2 Standing-Gradient Osmotic Flow

In order for water transport to occur by osmosis, it appears at first glance that the extracellular fluid must be hypertonic, i.e., have a higher solute concentration than the cell cytoplasm. For, if it were not, how could water be transported osmotically out of the cell? However, it has been known for many years that many types of epithelia (such as in the intestine and the kidney) are capable of transporting water in the absence of an osmotic gradient, or even against an osmotic gradient. Furthermore, the transported fluid itself is very close to isotonic. Both of these observations are difficult to reconcile with simple theories of solute-linked transport.

To explain such observations, Diamond and Bossert (1967; their model was an extension of the original idea of Curran and MacIntosh, 1962) proposed the standing-gradient osmotic flow model, a model that was, until recently, the basis for most quantitative explanations of solute-linked water transport. The standing-gradient

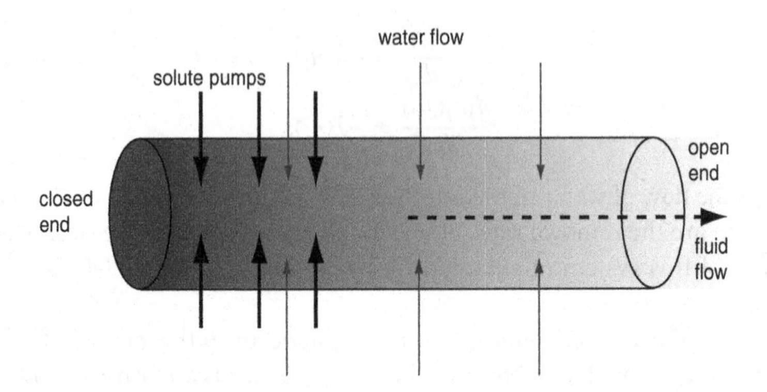

Figure 18.4 Schematic diagram of a long, narrow, tube in which solute-linked water transport occurs. The open end is the side to which fluid is secreted.

model was analyzed mathematically by Segel (1970; see also Lin and Segel, 1988), and Weinstein and Stephenson (1981) analyze a slight variant of this model.

One important characteristic of solute-linked water transport by epithelia is that, at the ultrastructural level, such epithelia have long, narrow channels, open at the end out of which water is flowing, and closed at the other end. This channel is sometimes the lateral intercellular space (long, narrow channels between neighboring cells), but it can also be infoldings of the basilar membrane. Because this is such a common feature of transporting epithelial cells, Diamond and Bossert proposed that it was this anatomical feature that was responsible for the ability of epithelia to secrete (almost) isotonic fluid, and they constructed a mathematical model to show how this could be.

Before embarking on the mathematical presentation, it is helpful first to gain an intuitive feel for how such a process works. Suppose we have a long, narrow tube, closed at one end and open at the other, as shown in Fig. 18.4. When solute is actively pumped into the closed end of the tube, the higher solute concentration in this restricted space is not immediately dissipated by diffusion, and thus a localized concentration gradient is set up across the tube wall, drawing water into the tube from the cytoplasm. However, water transport into the tube sets up a flow of water, which washes the solute out of the tube into the extracellular medium. Thus, a localized high solute concentration can be used to drive water flow even in the absence of concentration gradients on a larger spatial scale.

Suppose that solute-linked water transport occurs in a long, narrow, cylindrical tube of length L and radius r, closed at one end (Fig. 18.4). Solute is actively transported across the border of the tube, from the cell cytoplasm into the tube, and water follows passively, driven by the osmotic pressure difference.

Let $N(x)$ be the rate of active solute transport, with units of moles per unit area per unit time, across the boundary of the cylindrical tube. Usually, N is nonzero only at that portion of the channel close to the closed end. Also, let $v(x)$ be the velocity of the fluid flow at point x, where a positive velocity denotes flow in the direction of increasing x, and let $c(x,t)$ be the concentration of the solute in the tube at position x.

The differential equations for c and v are derived by standard conservation arguments (see, for example, Section 14.1.2, or Section 15.3). For a circular tube of radius r,

$$\pi r^2 \frac{\partial c}{\partial t} = -\pi r^2 \frac{\partial J}{\partial x} + 2\pi r N(x)$$
$$= -\pi r^2 \frac{\partial}{\partial x}\left(vc - D\frac{\partial c}{\partial x}\right) + 2\pi r N(x), \tag{18.14}$$

since $J = vc - D\frac{\partial c}{\partial x}$ is the flux per unit area of solute in the tube.

In steady state, this gives

$$D\frac{d^2c}{dx^2} - \frac{d}{dx}(vc) + \frac{2N(x)}{r} = 0. \tag{18.15}$$

Similarly, the velocity of water along the tube is v, and the flow per unit area of water entering the tube driven by osmotic pressure is $P(c(x) - c_0)$, where c_0 is the concentration of solute everywhere outside the cylindrical tube. Thus, in steady state, conservation of water implies that

$$\frac{dv}{dx} = \frac{2P}{r}(c(x) - c_0). \tag{18.16}$$

At the closed end there is no flow of water or solute, and thus

$$v(0) = \frac{dc(0)}{dx} = 0. \tag{18.17}$$

However, because this is a third-order system, we need a third boundary condition. For this we assume that the open end of the channel opens into a large space for which c is everywhere c_0, so take

$$c(L) = c_0. \tag{18.18}$$

Notice that this boundary condition is an approximation only; a more accurate model would determine the movement of solute and water in the tube as well as in the surrounding space. However, $c = c_0$ is a reasonable approximation if solute is quickly transported away from the end of the tube, and is thus forced to remain at a constant concentration there. This is the same approximation that is made in all introductory textbooks that discuss the heat equation for a one-dimensional conducting rod, with ends held at a fixed temperature by a large heat bath. Notice also that a similar problem occurs at the external sides of the tube walls, since anywhere solute is removed, the concentration cannot be identically c_0 there, even though that is the assumption we make.

Several features of the solution can be deduced without completely solving the equations. First, integrating (18.15) and (18.16) over the length of the tube, and applying the boundary conditions, one finds that

$$v(L)c_0 - D\frac{dc(L)}{dx} = \int_0^L \frac{2N(x)}{r}\,dx, \tag{18.19}$$

and

$$v(L) = \frac{2LP}{r}(\bar{c} - c_0), \qquad \bar{c} = \frac{1}{L}\int_0^L c\,dx. \tag{18.20}$$

The second of these tells us that the higher the average concentration of solute in the tube, the higher the efflux velocity.

The concentration of solute in the efflux, c_e, is defined to be the rate of total solute pumped into the tube, divided by the total flow out of the tube:

$$c_e = \frac{2\pi r \int_0^L N(x)\,dx}{\pi r^2 v(L)} = \frac{2\int_0^L N(x)\,dx}{rv(L)}. \tag{18.21}$$

Clearly, c_e is not the same as c_0, In fact, from (18.19),

$$c_e = c_0 - \frac{D}{v(L)}\frac{dc(L)}{dx}. \tag{18.22}$$

Thus, the greater the fluid velocity at the mouth of the channel, the lower the efflux concentration, and the closer the efflux fluid is to isotonic.

Some typical solutions are shown in Fig. 18.5. We take $N(x)$ to be the step function $N(x) = N_0 H(aL-x)$, so that solute is pumped into the tube only in the region $0 < x < aL$, with $a = 0.1$. For $x > aL$, the concentration of solute decreases along the length of the tube, and, depending on the parameters, the efflux can be almost isotonic with the background solution. For the parameters given in the caption to Fig. 18.5, the efflux concentration is 0.434 μM when $N_0 = 0.3$, and is 0.344 when $N_0 = 0.1$. As the length of the channel decreases, but with solute actively pumped into only the 10% of the channel closest to the closed end ($a = 0.1$), the efflux concentration increases; for a short channel (with these same parameter values) the efflux concentration can be almost ten times the background concentration. This is because in a short channel the solute diffuses out of the tube relatively rapidly, so that the osmotic driving force is decreased, and there is less space over which osmotic balance can be achieved.

The solutions shown in Fig. 18.5 were computed numerically using a shooting method. Since we know $v(0) = 0$ and $\frac{dc(0)}{dx} = 0$, we can guess a value of $c(0)$, and then solve the initial value problem numerically to determine $c(L)$. If $c(L)$ is smaller than c_0, we increase the initial value for c, while if $c(L)$ is too large, we reduce the initial value for c. Thus, by trial and error (or more systematically, using bisection), we can determine the value $c(0)$ that gives the correct boundary value at L.

Although this numerical method is direct, there are difficulties because the problem is stiff. Near $x = L$, the solution of the differential equations is close to a saddle point, and shooting into a saddle point is notoriously unstable. Said another way, if D is small (in a way made precise below), then this problem is singular, since a small number multiplies the largest derivative.

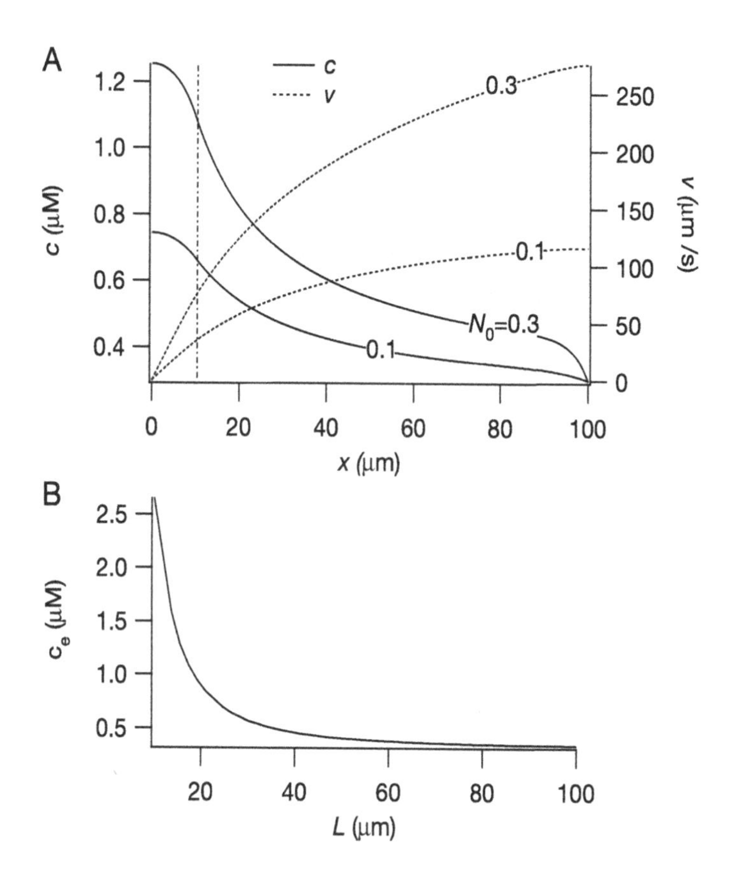

Figure 18.5 A: The standing concentration gradient of solute in the tube and the associated fluid velocity, computed numerically. Parameters are $D = 1000\ \mu\text{m}^2\,\text{s}^{-1}$, $r = 0.05\ \mu\text{m}$, $c_0 = 0.3\ \mu\text{M}$, $P = 0.2\ \mu\text{m}\,\text{s}^{-1}\,\mu\text{M}^{-1}$, and $L = 100\ \mu\text{m}$. Furthermore, $N(x) = N_0$ on the region $0 \le x \le aL$, $a = 0.1$ (i.e., to the left of the vertical dashed line) and is zero elsewhere. Two different values were used for N_0 (in units of nmoles per square meter per second); $N_0 = 0.3$ and $N_0 = 0.1$. B: The efflux solute concentration as a function of L with other parameters the same as in A ($N_0 = 0.1$).

Thus, the second way to find solutions of this problem is with perturbation arguments, taking advantage of small parameters. To study the problem in the small diffusion limit (which is the limit in which numerical solutions are most difficult to obtain), we nondimensionalize the problem, setting $y = \frac{x}{L}$, $u = \frac{c}{c_0}$ and $w = \frac{vr}{c_0 PL}$, and substituting into the equations to get

$$\epsilon \frac{d^2u}{dy^2} - \frac{d}{dy}(wu) = -2n(y), \tag{18.23}$$

$$\frac{dw}{dy} = 2(u-1), \tag{18.24}$$

where $\epsilon = \frac{Dr}{L^2 c_0 P}$ and $n(x) = \frac{N(x)}{c_0^2 P}$, with the corresponding boundary conditions

$$w(0) = \frac{du}{dy}(0) = 0, \tag{18.25}$$

$$u(1) = 1. \tag{18.26}$$

For the parameter values shown in Fig. 18.5, $\epsilon = 0.08$. Notice that while we call this the small diffusion limit, the diffusion coefficient is not really a parameter that can be freely varied, but is fixed for the particular solute. For the analysis that follows ϵ is required to be small, and this can be accomplished by many combinations of parameters, including cytoplasmic concentration, permeability, tube length or tube radius, which are not a priori fixed. Thus, we could equally well identify this as the long tube limit or the high permeability limit.

To find the outer solution we set $\epsilon = 0$, which gives the reduced system of equations

$$w_o \frac{du_o}{dy} + 2(u_o - 1) = 2n(y), \tag{18.27}$$

$$\frac{dw_o}{dy} = 2(u_o - 1). \tag{18.28}$$

Since this is a second-order system, rather than a third-order system, we can satisfy only two of the three boundary conditions. Which one of the three to drop is not apparent until one examines the boundary layer equations, as is done next. However, in anticipation of that result, we solve the reduced system subject to the conditions $w_o(0) = \frac{du_o(0)}{dy} = 0$.

Now, since $\frac{du_o(0)}{dy} = 0$, for consistency it must be that

$$u_o(0)(u_o(0) - 1) = n(0). \tag{18.29}$$

The unique positive root of this quadratic equation, say u_d (which is also greater than 1), is the initial value for $u_o(y)$. The solution of the outer problem is now readily determined. If $n(y)$ is a step function, $n(y) = n_0 H(a - y)$, with $0 < a < 1$, then for $0 < y < a$, u_o is constant, and $w_o(y)$ is linear in y,

$$u_o(y) = u_d, \qquad w_o(y) = 2(u_d - 1)y. \tag{18.30}$$

For $a < y < 1$, $w_o(y)u_o(y) = w_a u_d$, where $w_a = w(a) = 2a(u_d - 1)$. Then,

$$\frac{dw_o}{dy} = 2\left(\frac{w_a u_d}{w_o} - 1\right), \tag{18.31}$$

which can be integrated to find

$$-w_o + w_a - w_a u_d \ln\left(\frac{w_o - w_a u_d}{w_a - w_a u_d}\right) = 2(y - a), \tag{18.32}$$

so that $w_o(y)$ and $u_o(y)$ are determined implicitly. A plot of this outer solution is shown in Fig. 18.6.

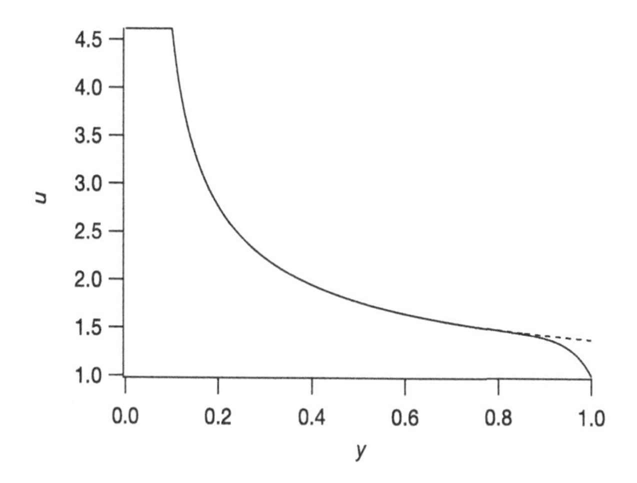

Figure 18.6 The outer solution $u_0(y)$ (shown dashed) and uniformly valid composite solution, computed with the same parameter values as in Fig. 18.5, with $N_0 = 0.3$. Note that the two solutions agree well everywhere except in the boundary layer close to $y = 1$, where the boundary layer correction changes rapidly in order to satisfy the boundary condition at $y = 1$.

The outer solution cannot satisfy the boundary condition at $y = 1$. In fact, $u_o(1)$ is certain to be greater than one. To satisfy the condition $u(1) = 1$, we seek a boundary layer correction to the outer solution. We introduce the scaled variable

$$\xi = \frac{y-1}{\epsilon}, \tag{18.33}$$

and in terms of this variable the original system of equations becomes

$$\frac{d^2u}{d\xi^2} - \frac{d}{d\xi}(wu) = -2\epsilon n(y), \tag{18.34}$$

$$\frac{dw}{d\xi} = 2\epsilon(u-1). \tag{18.35}$$

To leading order in ϵ, we have the boundary layer equations

$$\frac{d^2u_i}{d\xi^2} - \frac{d}{d\xi}(w_iu_i) = 0, \tag{18.36}$$

$$\frac{dw_i}{d\xi} = 0, \tag{18.37}$$

where the subscript i is intended to denote that this is the inner solution. This system is also easily solved: w_i is constant and clearly must be equal to $w_o(1)$, so that the boundary layer equation for u_i reduces to

$$\frac{d^2u_i}{d\xi^2} - w_o(1)\frac{du_i}{d\xi}u_i = 0, \tag{18.38}$$

with general solution

$$u_i(\xi) = A + B\exp(w_o(1)\xi). \tag{18.39}$$

Now we see why this boundary layer is located at $y = 1$. Since the solution $u_i(\xi)$ is bounded in the limit $\xi \to -\infty$, but not in the limit $\xi \to \infty$, it can be a boundary layer solution only on the right, not the left. We pick the constants A and B so that $u_i(0) = 1$ and so that it matches the outer solution, i.e., $\lim_{\xi\to-\infty} = u_o(1)$. Thus,

$$u_i(\xi) = u_o(1) + (1 - u_o(1))\exp(w_o(1)\xi) + O(\epsilon). \tag{18.40}$$

Finally, the composite solution, or uniformly valid solution, is given by

$$u(y) = u_o(y) + (1 - u_o(1))\exp\left(\frac{1}{\epsilon}w_o(1)(y-1)\right) + O(\epsilon). \tag{18.41}$$

A plot of this composite solution is also shown in Fig. 18.6.

A different scaling is necessary if the tube is short or permeability is small. As before, we let $y = \frac{x}{L}$, and $u = \frac{c}{c_0}$ but take $W = \frac{vL}{D}$, and find

$$\frac{d^2u}{dy^2} - \frac{d}{dy}(Wu) + 2m(x) = 0, \tag{18.42}$$

where $m(x) = \frac{L^2N(x)}{Drc_0}$, and

$$\frac{dW}{dx} = 2\eta(u-1), \tag{18.43}$$

where $\eta = \frac{c_0PL^2}{Dr} = \frac{1}{\epsilon}$. The boundary conditions are unchanged from before.

With $\eta \ll 1$, this is a regular perturbation problem whose solution is readily found as a power series in η. We leave the details of this calculation as an exercise. However, the result is that to leading order in η, but expressed in terms of original dimensional variables,

$$\bar{c} = c_0 + \frac{N_0}{Dr}\left(1 - \frac{1}{3}a^2\right)aL^2. \tag{18.44}$$

We recover the efflux velocity from (18.20).

18.1.3 Uphill Water Transport

In the model of the previous section, the background concentration everywhere outside the cylindrical tube was assumed to be the same, c_0. Thus, the model demonstrated that a standing gradient of solute concentration can transport water between regions of identical osmolality, and do so in such a way that the transported fluid is as close to this background osmolality as desired.

However, in many cases an epithelial cell layer can do something even more surprising; it can transport water against an osmotic pressure difference, yet using only osmosis to do so. How this works can be seen by a simple modification of the standing-gradient model (Weinstein and Stephenson, 1981).

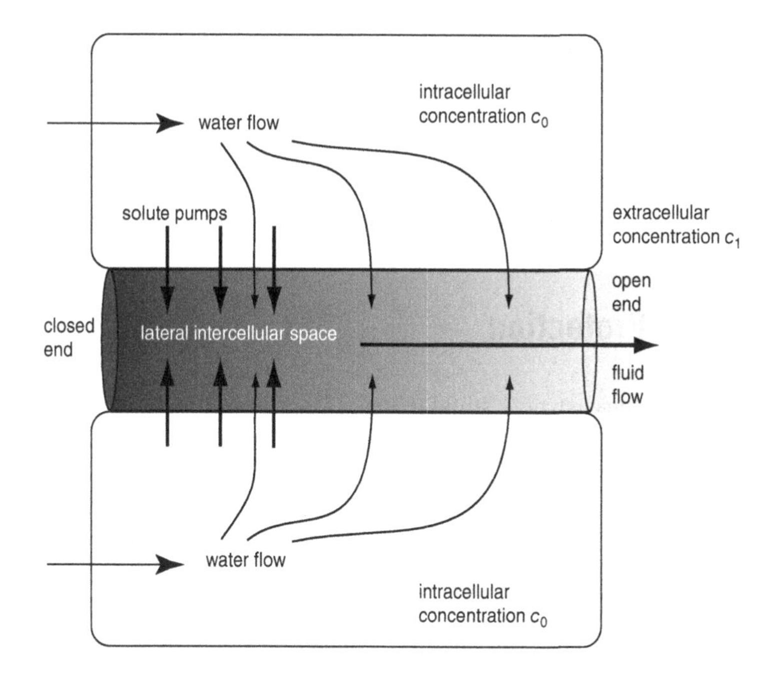

Figure 18.7 Schematic diagram of the standing-gradient model with separated intracellular and extracellular regions.

Instead of assuming that the concentration everywhere outside the cylindrical tube is the same, we divide the external region into intracellular and extracellular regions, with concentrations c_0 and c_1, respectively (Fig. 18.7), with $c_0 > c_1$. The model equations remain the same as previously, with the one exception that the boundary condition for c at $x = l$ is $c(L) = c_1$.

The solution of this system of equations has the same general features as before, namely solute concentration is a decreasing function of x. Anywhere that $c > c_0$, water is drawn from the cell into the tube, while anywhere that $c < c_0$, water is drawn from the tube into the cell. The net flux is determined by (18.20). Accordingly, if the average concentration in the tube is greater than that in the cell, there is a net positive fluid velocity at the open end. Conversely, if the average concentration in the tube is less than that in the cell, there is a net negative fluid velocity at the open end, drawing water into the cell.

So an interesting question is to determine under what conditions there is net fluid outflow when $c_1 < c_0$.

In the small diffusion (long tube length) limit, the net flow is always positive, regardless of how small c_1 might be. This is because, in the small diffusion limit, the concentration drops below c_0 only in a thin boundary layer, and the net velocity is affected by this only slightly.

In the small tube length (large diffusion, small permeability) limit, the flow can go either way. It is easy to determine that

$$\bar{c} = c_1 + a\frac{N_0}{r}\frac{L^2}{D}\left(1 - \frac{1}{3}a^2\right), \tag{18.45}$$

from which the direction of net flow is also easily determined.

18.2 Gastric Protection

The inner surface of the gastrointestinal tract is a layer of columnar epithelial cells that actively secrete mucus and a fluid rich in bicarbonate. The mucus is highly viscous and coats the cells with a 0.5–1.0 mm thick layer that is insoluble by other gastric secretions and creates a lubricating boundary for the intestinal wall. In addition, this layer of cells is studded with a large number of gastric pits (Fig. 18.8). Each gastric pit contains *parietal cells* that secrete hydrochloric acid through an active transport process, leading to a pH of about 1 in the stomach lumen. Since the pH of the blood supplying the surface epithelium is about 7.4, there is a large H^+ concentration gradient (approximately a millionfold increase in concentration of hydrogen) across each epithelial cell. Clearly, the epithelial cells must be protected from the high lumenal acidity. It is believed that

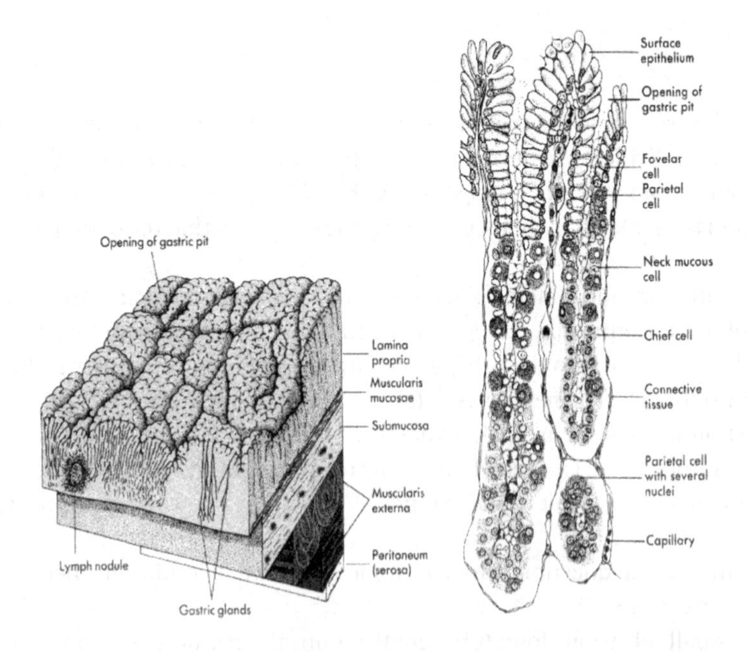

Figure 18.8 Closeup view of the gastric mucosa and two gastric pits. The epithelium of the gastric wall contains large numbers of gastric pits, each of which is lined by parietal cells that secrete HCl. (Berne and Levy, 1993, Fig. 39-9, p. 659.)

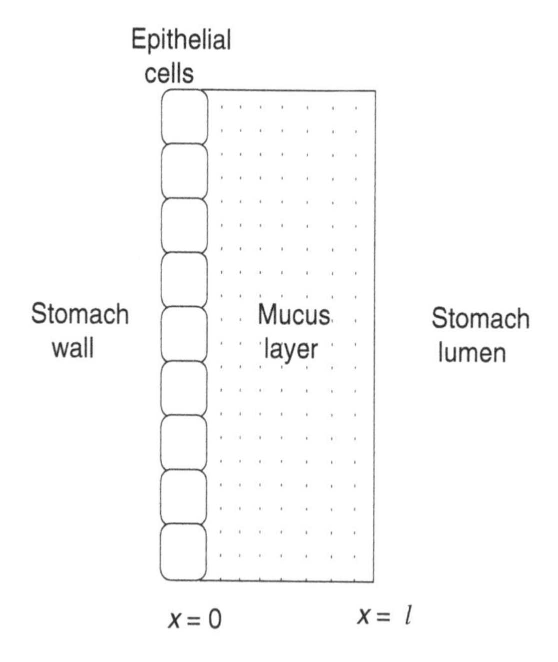

Figure 18.9 Schematic diagram of the mucus layer in the model of gastric protection.

the secretion of mucus and bicarbonate by epithelial cells plays an important role in gastric protection.

18.2.1 A Steady-State Model

To model gastric protection (following Engel et al., 1984) we assume that the lumenal surface of the gastric mucosa is a plane located at $x = 0$, where x is a coordinate measured perpendicular to the mucosal wall, while the mucus layer is of uniform thickness l. Thus, the mucus–lumen interface lies at $x = l$, as illustrated in Fig. 18.9. Inside the mucus layer H^+ and HCO_3^- react according to

$$H^+ + HCO_3^- \underset{k_-}{\overset{k_+}{\rightleftarrows}} H_2O + CO_2. \tag{18.46}$$

This bicarbonate buffering system is one of the most important buffering systems in the body, and its role in the transport of carbon dioxide is discussed in Chapter 14.

In the time-dependent problem, each species obeys a reaction–diffusion equation, such as

$$\frac{\partial [H^+]}{\partial t} = D_{H^+} \frac{\partial^2 [H^+]}{\partial x^2} - k_+ [H^+][HCO_3^-] + k_- [CO_2], \tag{18.47}$$

where D_{H^+} is the diffusion coefficient of H^+ in the mucus layer. However, at steady state the time derivatives are zero, and the partial derivatives with respect to x become ordinary derivatives. Thus, at steady state,

$$D_{\mathrm{H}^+}\frac{d^2[\mathrm{H}^+]}{dx^2} = D_{\mathrm{HCO}_3^-}\frac{d^2[\mathrm{HCO}_3^-]}{dx^2} = k_+[\mathrm{H}^+][\mathrm{HCO}_3^-] - k_-[\mathrm{CO}_2], \tag{18.48}$$

$$D_{\mathrm{CO}_2}\frac{d^2[\mathrm{CO}_2]}{dx^2} = -k_+[\mathrm{H}^+][\mathrm{HCO}_3^-] + k_-[\mathrm{CO}_2]. \tag{18.49}$$

To complete the formulation of the problem we add boundary conditions at the epithelial and lumenal boundaries of the mucus layer. On the lumenal side we assume that $[\mathrm{H}^+] = [\mathrm{H}^+]_l$ and $[\mathrm{CO}_2] = [\mathrm{CO}_2]_l$ are constant and known, determined by the concentration of the contents of the gastrointestinal tract, while $[\mathrm{HCO}_3^-]$ is given by the equilibrium relation

$$[\mathrm{HCO}_3^-]_l = \frac{k_-}{k_+}\frac{[\mathrm{CO}_2]_l}{[\mathrm{H}^+]_l}. \tag{18.50}$$

On the epithelial side we assume that the fluxes of HCO_3^- and CO_2 are known, as these chemicals are actively secreted by the epithelial cells, and thus, from Fick's law,

$$D_{\mathrm{HCO}_3^-}\frac{d[\mathrm{HCO}_3^-]}{dx} = -\bar{J}, \tag{18.51}$$

$$D_{\mathrm{CO}_2}\frac{d[\mathrm{CO}_2]}{dx} = -\bar{I}, \tag{18.52}$$

at $x = 0$, for some known constants $\bar{J}$ and $\bar{I}$. Finally, we assume that the flux of H^+ across the boundary at $x = 0$ is proportional to the concentration difference across the boundary; i.e.,

$$D_{\mathrm{H}^+}\frac{d[\mathrm{H}^+]}{dx} = P_{\mathrm{H}^+}\left([\mathrm{H}^+] - [\mathrm{H}^+]_{\mathrm{epi}}\right), \tag{18.53}$$

where P_{H^+} is the permeability and $[\mathrm{H}^+]_{\mathrm{epi}}$ is the concentration of H^+ in the epithelial cells. Since the concentration of H^+ in the epithelial cells is low compared to the concentration external to the cell, we set $[\mathrm{H}^+]_{\mathrm{epi}}$ to zero, and thus require

$$D_{\mathrm{H}^+}\frac{d[\mathrm{H}^+]}{dx} = P_{\mathrm{H}^+}[\mathrm{H}^+] \tag{18.54}$$

at $x = 0$.

To study this system of equations, we introduce nondimensional variables $y = x/l, u = [\mathrm{H}^+]/[\mathrm{H}^+]_l, v = [\mathrm{HCO}_3^-]/[\mathrm{H}^+]_l, w = [\mathrm{CO}_2]/[\mathrm{H}^+]_l$, in terms of which the model becomes

$$\epsilon\frac{d^2u}{dy^2} = uv - \zeta w, \tag{18.55}$$

$$\frac{d^2u}{dy^2} = \gamma\frac{d^2v}{dy^2} = -\beta\frac{d^2w}{dy^2}, \tag{18.56}$$

where $\beta = D_{CO_2}/D_{H^+}, \gamma = D_{HCO_3^-}/D_{H^+}, \epsilon = \frac{D_{H^+}}{k_+ l^2 [H^+]_l}, \zeta = \frac{k_-}{k_+ [H^+]_l}$. The boundary conditions at $y = 1$ $(x = l)$ are

$$u(1) = 1, \qquad v(1) = \zeta\alpha, \qquad w(1) = \alpha, \tag{18.57}$$

and the boundary conditions at $y = 0$ are

$$\frac{du}{dy}(0) = \lambda u(0), \qquad \gamma\frac{dv}{dy}(0) = -J, \qquad \beta\frac{dw}{dy}(0) = -I, \tag{18.58}$$

where $\alpha = \frac{[CO_2]_l}{[H^+]_l}, J = \frac{\bar{J}l}{D_{H^+}[H^+]_l}, I = \frac{\bar{I}l}{D_{H^+}[H^+]_l}, \lambda = \frac{P_{H^+}l}{D_{H^+}}$. Integrating (18.56) from 0 to y and using the boundary conditions (18.58) we obtain

$$\frac{du}{dy} - \lambda u(0) = \gamma\frac{dv}{dy} + J = -\beta\frac{dw}{dy} - I. \tag{18.59}$$

Integrating (18.58) from y to 1 and applying the boundary conditions (18.57) gives

$$u - 1 - \lambda u(0)(y-1) = \gamma(v - \zeta\alpha) + J(y-1) = -\beta(w - \alpha) - I(y-1). \tag{18.60}$$

From this we obtain v and w as functions of u and y:

$$v(y) = \zeta\alpha + \frac{1}{\gamma}[u(y) - 1 - (\lambda u(0) + J)(y-1)], \tag{18.61}$$

and

$$w(y) = \alpha - \frac{1}{\beta}[u(y) - 1 + (I - \lambda u(0))(y-1)]. \tag{18.62}$$

Thus, we can write the model as

$$\epsilon\frac{d^2u}{dy^2} = uv - \zeta w = f(u(y), y), \tag{18.63}$$

$$\frac{du}{dy}(0) = \lambda u(0), \qquad u(1) = 1. \tag{18.64}$$

From the molecular weights of the chemicals, we estimate $\beta \approx 0.14$ and $\gamma \approx 0.13$. The forward and reverse rates of the bicarbonate reaction are, respectively, $k_- = 11\ s^{-1}$ and $k_+ = 2.6 \times 10^{10}\ cm^3 \cdot mol^{-1} \cdot s^{-1}$. Other experimentally determined quantities include $\bar{J} = 1.4 \times 10^{-10}\ mol \cdot cm^{-2} \cdot s^{-1}$, $[H^+]_l = 140$ mM, $l = 0.05$ cm, $D_{H^+} = 1.75 \times 10^{-5} cm^2 \cdot s^{-1}$, and $P_{H^+} = 1.3 \times 10^{-5} cm \cdot s^{-1}$. From these parameter values we see that $\epsilon = O(10^{-7})$ and $\zeta = O(10^{-6})$ are small parameters, while $\lambda = 0.037$ and $J = 0.0003$.

We now use singular perturbation theory to solve this two-point boundary value problem. This approach is possible because ϵ, which is the ratio of the rate of diffusion through the mucus to the rate of reaction, is small. Outside of a thin layer the bicarbonate reaction is in a pseudo-steady state at each point in space; in this region, diffusion of hydrogen ions or bicarbonate plays little role. It is only within the thin layer that the bicarbonate concentration is determined by the balance of reaction and diffusion. This allows the representation of the solution in two different spatial variables, one describing the solution outside this thin layer, and one describing the solution inside it. The solutions are then matched to obtain a uniformly valid solution. As described below, although the bicarbonate reaction is in local chemical equilibrium outside the thin layer, the bulk of the reaction actually occurs within the thin layer. For the parameter values used here, the thin layer occurs at $y = 0$, but this need not necessarily be so. If the acidity of the lumen is low enough, the thin reaction layer occurs within the mucus layer (see Exercise 4).

The Outer Solution

We look for a solution of the form

$$u = u_0 + \epsilon u_1 + \cdots, \tag{18.65}$$

substitute into the differential equation, and equate coefficients of powers of ϵ. This gives a hierarchy of equations for the outer solution. To lowest order in ϵ we have

$$0 = f(u_0, y), \tag{18.66}$$

$$\frac{du_0}{dy}(0) = \lambda u_0(0), \qquad u_0(1) = 1. \tag{18.67}$$

Obviously, both boundary conditions cannot be satisfied, so we drop the boundary condition at $y = 0$ and keep the boundary condition at $y = 1$. There are good physical reasons for this choice. As discussed above, the balance of reaction and diffusion is important only in a thin layer around $y = 0$. Thus, if we ignore diffusion (by setting $\epsilon = 0$) we do not expect to be able to satisfy the boundary condition at $y = 0$. (This is also the correct mathematical choice, because, as is discussed below, there is a "corner layer" at $y = 0$; the other choice, ignoring the boundary condition at $y = 1$, fails to produce a valid solution.)

The equation $f(u_0, y) = 0$ is the quadratic polynomial in u_0,

$$\begin{aligned} &\beta u_0^2 + [\zeta\gamma(\alpha\beta + 1) - \beta + \beta(J + \lambda u_0(0))(1 - y)]u_0 \\ &\quad + \zeta\gamma[(\lambda u_0(0) - I)(1 - y) - \alpha\beta - 1] = 0, \end{aligned} \tag{18.68}$$

so it can be solved exactly. The easiest way to represent this solution is to find y as a function of u_0, since (18.68) is linear in y. However, because ζ is small, we find that

$$u_0(y) = 1 + (\lambda u_0(0) + J)(y - 1) + O(\zeta), \tag{18.69}$$

and

$$v = O(\zeta). \tag{18.70}$$

Next we set $y = 0$ in (18.68) and solve for $u_0(0)$ to get

$$u_0(0) = \frac{1 - J}{1 + \lambda} + O(\zeta), \tag{18.71}$$

from which it follows that

$$u_0(y) = 1 - \frac{\lambda + J}{\lambda + 1}(1 - y) + O(\zeta), \tag{18.72}$$

$$w(y) = \alpha + \frac{J + I}{\beta}(1 - y) + O(\zeta). \tag{18.73}$$

Hence, to leading order, there is no HCO_3^- in the mucus layer, and H^+ and CO_2 vary linearly with distance through the mucus layer.

The Inner Solution

The outer solution (18.72) does not satisfy the boundary condition at $y = 0$. A uniformly valid solution of this problem must include a "corner layer," that is, a solution with large second derivative, which therefore changes slope, but not value (at least to lowest order), in a small region close to $y = 0$. The corner layer here results from the fact that the boundary condition at $y = 0$ is expressed in terms of the derivative of u at 0. Hence, to satisfy the boundary condition, the derivative of u must change quickly. It is beyond the scope of this book to give a detailed description of the construction of this corner layer (see Engel et al., 1984, or, for a more general description, Keener, 1998, or Holmes, 1995). Suffice it to say that the corner layer is found by introducing a scaled variable $\tilde{y} = y/\sqrt{\epsilon}$ (which eliminates ϵ from the second derivative term in (18.63)) and then seeking a power series solution in powers of $\sqrt{\epsilon}$. The result is a modification of the outer solution by the addition of a term of the form

$$\sqrt{\epsilon} e^{-\mu y/\sqrt{\epsilon}}, \tag{18.74}$$

which is of small amplitude and satisfies the boundary condition at the origin. As a result, to leading order in ϵ, the outer solution provides a uniformly valid representation of the solution on the entire interval $0 < y < 1$. A sketch of the solution is given in Fig. 18.10.

Physical Interpretation of the Corner Layer

There is an interesting interpretation of the corner layer in terms of the physiology of the problem. Recall that the boundary conditions at $y = 0$ for the original problem are

$$\frac{du}{dy}(0) = \lambda u(0), \qquad \gamma \frac{dv}{dy}(0) = -J, \qquad \beta \frac{dw}{dy}(0) = -I. \tag{18.75}$$

The outer solution does not satisfy these boundary conditions. Instead, if we evaluate the derivatives of the outer solution to leading order, we find that

$$\frac{du_0}{dy}(0) = \lambda u_0(0) + J, \qquad \frac{dv_0}{dy}(0) = 0, \qquad \beta \frac{dw_0}{dy}(0) = -I + J. \tag{18.76}$$

In other words, to lowest order, the bicarbonate flux J can be replaced by a flux of hydrogen ions in the opposite direction with the same magnitude, and the CO_2 flux must be altered to compensate. Thus the original problem, which has a bicarbonate source at $y = 0$, is replaced by a simpler problem that has a H^+ sink and a CO_2 source

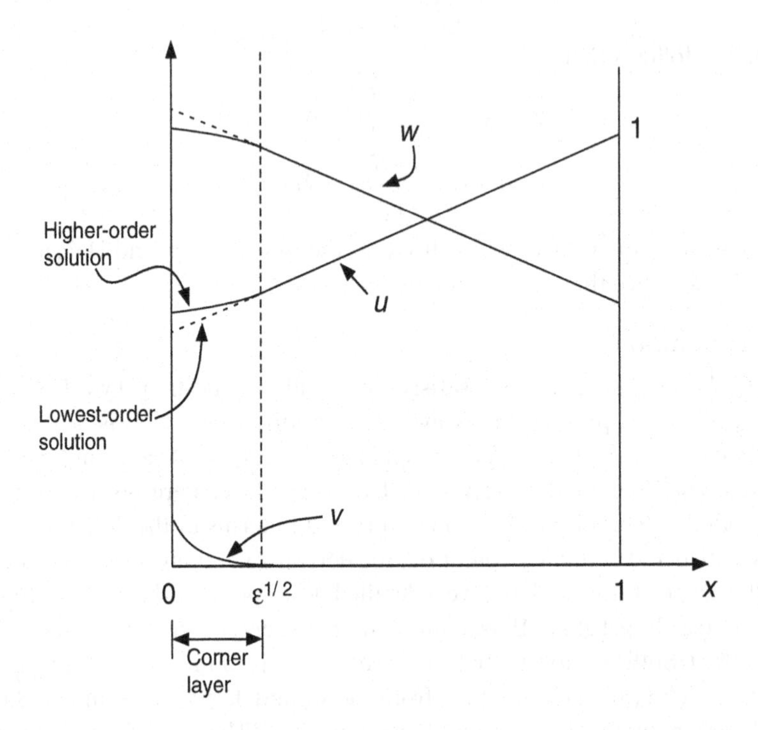

Figure 18.10 Sketch (not to scale) of the solution to the model of gastric protection. (Adapted from Engel et al., 1984, Fig. 10.)

at $y = 0$ and no H^+–HCO_3^- reaction. This implies that each bicarbonate molecule that exits the epithelium reacts immediately with a hydrogen ion, with the consequent disappearance of the hydrogen ion. Hence, to lowest order, all the chemical reaction occurs within the corner layer.

In dimensional variables, the outer solution is

$$[H^+] = [H^+]_l - \frac{\bar{J} + P_{H^+}[H^+]_l}{D_{H^+} + P_{H^+}l}(l - x), \tag{18.77}$$

$$[CO_2] = [CO_2]_l + \frac{\bar{J} + \bar{I}}{D_{CO_2}}(l - x), \tag{18.78}$$

and thus, at the epithelial surface,

$$[H^+]_0 = \frac{D_{H^+}[H^+]_l - \bar{J}l}{D_{H^+} + P_{H^+}l}. \tag{18.79}$$

Using experimentally determined values for the parameters, we find that $[H^+]_0 = 135$ mM, a decrease of only 3.5%, which is too small to protect the epithelial cells from high lumenal acidity. Thus, this simple model of the mucus layer is insufficient to explain how the epithelial layer is protected.

18.2.2 Gastric Acid Secretion and Neutralization

The primary difficulty with the above model of gastric protection is that the flux of bicarbonate, J, is too small to cause a sufficient reduction of hydrogen ions at the surface of the epithelial cells. A model that addresses this shortcoming by examining the relationship between hydrochloric acid secretion and the release of bicarbonate was constructed by Lacker and his coworkers (de Beus et al., 1993).

Hydrochloric acid is secreted from the parietal cells of the oxyntic glands using a number of reactions. First, water in the cells is dissociated into hydrogen and hydroxyl ions in the cell cytoplasm. The hydrogen ions are actively secreted via a H^+–K^+ ATPase. In addition, chloride ions are actively secreted and Na^+ ions are actively absorbed, via separate ATPases. The result is a high concentration of hydrochloric acid in the lumen. At the same time, carbon dioxide combines with hydroxyl ions (catalyzed by carbonic anhydrase) to form carbonic acid and thence bicarbonate. This bicarbonate diffuses out of the cell into the extracellular medium and is transported by the capillary blood flow. The direction of capillary blood flow is from the oxyntic cell in the gastric pit to the epithelial lining of the lumen. Since the epithelial cells are downstream of the oxyntic cells, they absorb bicarbonate from the blood and then secrete it into the mucus. Thus, as acid production increases, so does the rate at which bicarbonate is secreted into the lumen by the epithelial cells. According to de Beus et al., the lack of this feature in the Engel model caused an underestimation of the rate of bicarbonate secretion from the epithelial layer.

De Beus et al. estimated the model parameters from the available experimental literature and showed that analytic solutions in certain simplified cases agreed well with the full solution. Of particular interest is their reproduction of the *alkaline tide*. As the rate of H^+ secretion into the lumen increases, the downstream $[H^+]$ (i.e., the gastric venous $[H^+]$) *decreases*. This reinforces the major idea behind this model, that secretion of HCO_3^- by the epithelial cells is driven by H^+ secretion by the oxyntic cells, so that gastric protection is automatically increased as the lumenal $[H^+]$ increases.

Over longer time scales the secretion of gastric acid is controlled by a complex network of positive and negative feedback processes involving the enteric and central nervous systems and a number of different cell types. For example, food stimulates the production of gastrin by G cells, which in turn stimulates the production of H^+ from parietal cells. However, gastrin also stimulates the production of somatostatin from D cells (in the corpus of the stomach) which inhibits the production of H^+ from the parietal cells. Gastrin also acts on enterochromaffin-like cells (ECL cells) in the corpus to stimulate the secretion of histamine, which enhances acid secretion and potentiates the effect of gastrin on parietal cells. To further complicate matters, D cells in the antrum produce somatostatin in response to high H^+ concentrations to form another negative feedback loop by inhibiting the production of gastrin by the antral G cells.

In addition to these control mechanisms, the numbers of G, ECL, D and parietal cells are controlled by additional feedback mechanisms that operate over a time scale of days to weeks, so that gastric acid secretion can be modulated by long-term changes in food ingestion.

A complicated model of this system has been constructed and analyzed by Joseph et al., (2002). Their principal conclusion was that the action of somatostatin was the most crucial feedback mechanism for maintaining a stable acid balance in the stomach.

18.3 Coupled Oscillators in the Small Intestine

One principal function of the gastrointestinal tract is to mix ingested food and move it through the tract in the appropriate direction. It does this by contraction of the layers of smooth muscle illustrated in Fig. 18.1, contractions that are controlled on a number of different levels. At the lowest level, each smooth muscle cell has intrinsic electrical activity, which can be oscillatory in nature. At higher levels, the properties of the local oscillations are modified by extrinsic and intrinsic neuronal stimulation, or chemical stimuli. Different parts of the tract have different kinds of contractile behavior. Here, we focus on the electrical activity of the smooth muscle of the small intestine. The small intestine is itself divided into three different sections: the first 25 cm or so after the pylorus (the passage from the stomach to the small intestine, controlled by the pyloric sphincter) is called the *duodenum*; the next section, comprising about 40% of the length of the small intestine, is called the *jejunum*; while the remainder is called the *ileum*. However, although this nomenclature is useful for understanding some of the experimental results we present here, we do not distinguish between the electrical activity of different sections of the small intestine.

18.3.1 Temporal Control of Contractions

Smooth muscle cells throughout the gastrointestinal tract exhibit oscillations in their membrane potential, with periods ranging from 2 to 40 cycles/min. A typical example of this *electrical control activity*, or ECA, is shown in Fig. 18.11. Although depolarization of the cell membrane potential can cause muscular contractions, this happens only if the membrane potential is depolarized past a threshold, in which case the potential begins to oscillate, or burst, at a much higher frequency. Whether or not bursting occurs depends on the level of neuronal or chemical stimulation. In this way, contractile activity depends on the local oscillatory properties of the smooth muscle cells, as well as on the higher-level control processes. Electrical bursts are termed *electrical response activity*, or ERA. Muscular contractions cannot occur with a frequency greater than that of the ECA, and thus the properties of the local ECA constrain the possible types of muscular contraction.

Often, when faced with cellular oscillators, modelers seek to understand the cellular mechanisms that cause such behavior; this book contains many examples of this approach. Here, by contrast, we examine what happens when a large number of oscillators are coupled to one another, without concern for the exact mechanisms underlying each

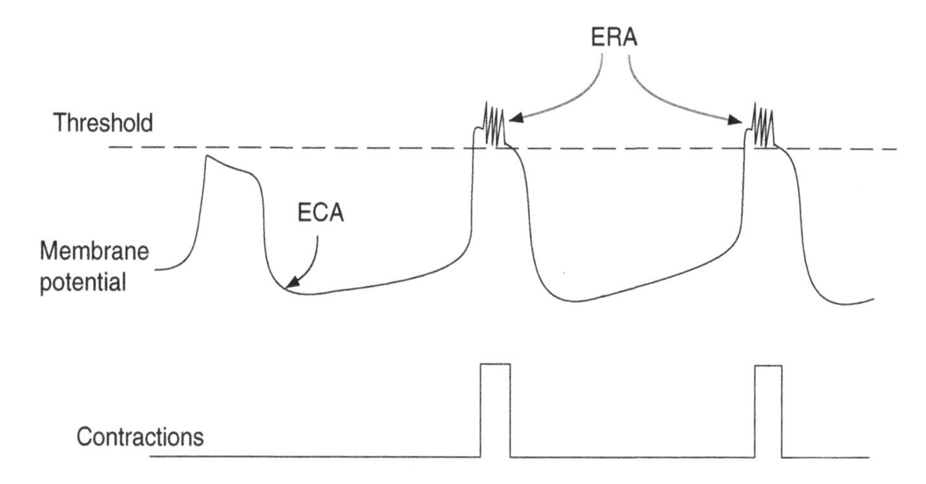

Figure 18.11 Schematic diagram of electrical control activity (ECA), electrical response activity (ERA), and muscular contraction. (Adapted from Sarna, 1989, Fig. 2.)

oscillation. Although this approach cannot determine a direct relationship between cellular properties and global behavior, it provides greater insight into how coupled oscillators can give rise to organized wave activity of the type that is frequently seen in the stomach and small intestine.

18.3.2 Waves of Electrical Activity

If each local oscillator were uncoupled from its neighbors, we expect there would be no organized waves of contraction moving along the intestine. However, the main point of this section is that weak coupling between the oscillators causes the propagation of waves of ECA and ERA along the intestine.

The importance of coupling between the local oscillators is demonstrated in the top panel of Fig. 18.12, where is shown the experimentally measured frequency of segments of the small intestine in the intact intestine, and in segments that have been dissociated from one another by circumferential cuts across the intestine. In the intact intestine, the frequency of the ECA is constant over the entire region close to the pylorus, even though the intrinsic frequency is steadily decreasing over this region. At approximately 60 cm from the pylorus the ECA frequency begins to decrease. In the frequency plateau (the region of constant frequency) region, each oscillator is phase-locked to its neighbor, resulting in organized waves that move along the intestine away from the pylorus. This is illustrated in the top panel of Fig. 18.13, where the oscillation peaks in neighboring parts of the intestine are connected by solid lines. The slope of the solid line gives the speed of the wave along the intestine, and the fact that subsequent lines are regularly spaced, parallel and straight, shows that the waves are repetitive and highly organized. Following the frequency plateau is a region where the ECA frequency decreases along

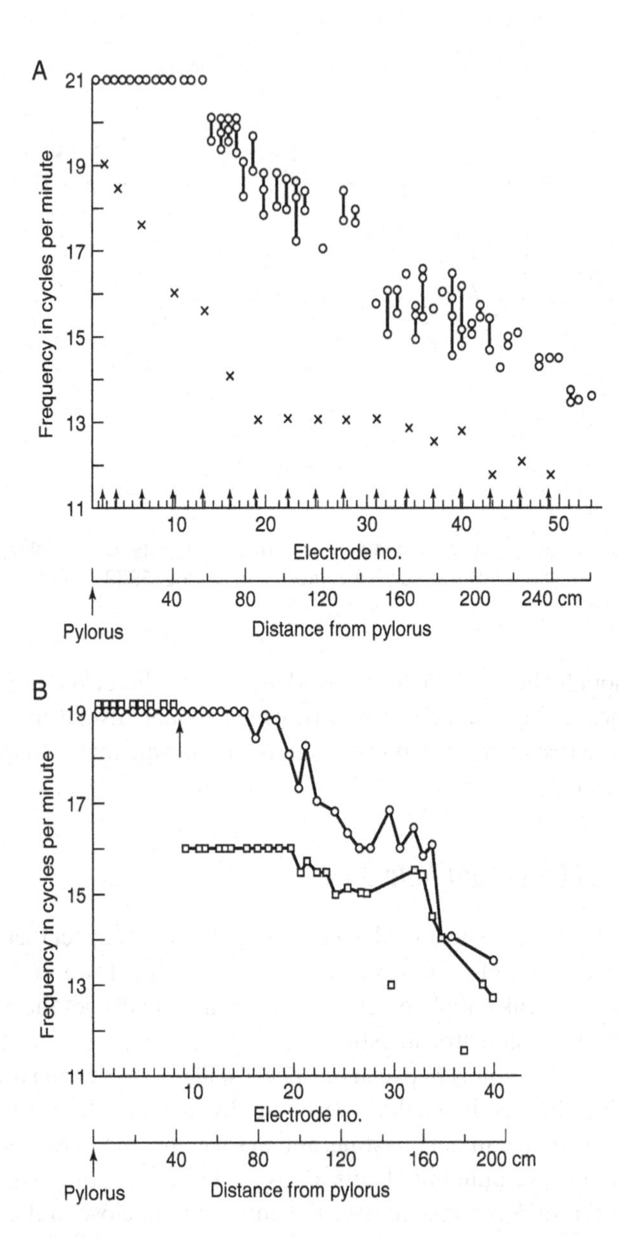

Figure 18.12 A: Intact (circles) and intrinsic frequency (crosses) of ECA in dog small intestine. The intrinsic frequencies were obtained by cutting across the small intestine at the places indicated by the arrows so as to disrupt oscillator coupling. B: The effect of a single cut (at the arrow) across the small intestine. To the right of the cut a frequency plateau still occurs, but now at a lower frequency than in the intact intestine. To the left of the cut the frequencies are unchanged. (Sarna, 1989, Fig. 10.)

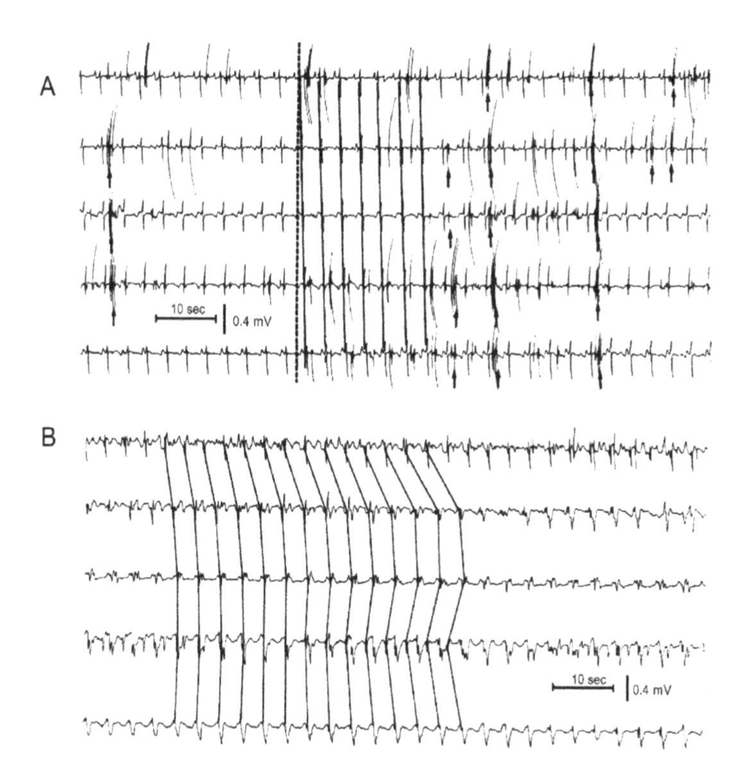

Figure 18.13 Experimental recordings from dog small intestine. A: Recordings taken from the frequency plateau region close to the pylorus. Solid lines connect the peaks of the ERA; these lines are straight and parallel, indicating the propagation of regular wave trains in this region. B: Recordings taken from the variable frequency region. The peaks of the ERA are not well organized, indicating that regular wave propagation has broken down. (Sarna, 1989, Fig. 11 A and B.)

the intestine, and the corresponding waves are not phase-locked and therefore much less regular (lower panel of Fig. 18.13).

Note that phase locking (i.e., oscillation with the same frequency) does not necessarily imply that there is wave-like behavior. A *phase wave* occurs when there is a constant advance (or delay) of phase from one point to the next along the length of the intestine.

When the segments are uncoupled, each shows oscillatory ECA, but with an intrinsic frequency that decreases with distance from the pylorus; the frequency plateau disappears in the isolated segments. It appears that in the intact intestine, the highest-frequency segment closest to the pylorus entrains the nearby oscillators, which have similar but lower frequencies. However, when the difference in intrinsic frequency is too large, entrainment is not possible: the frequency plateau breaks down, and the waves lose regularity. This is illustrated further in the lower panel of Fig. 18.12, which

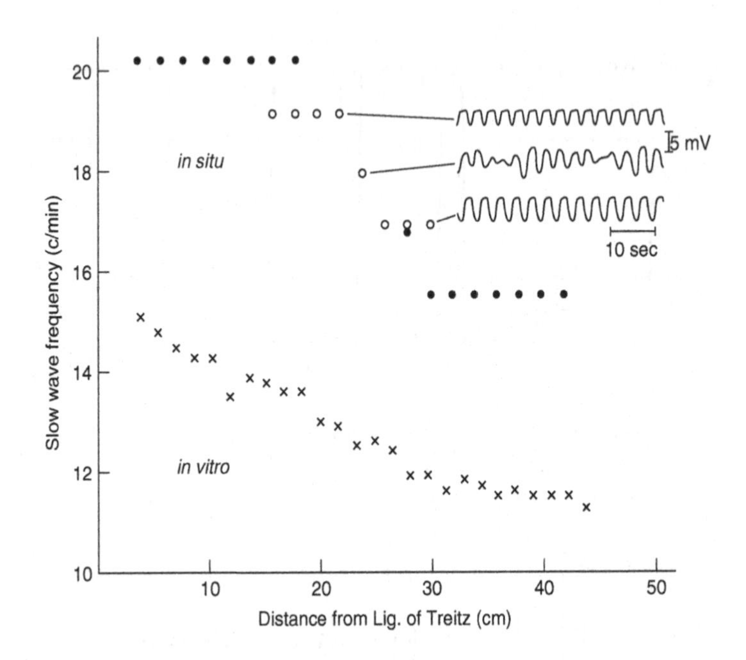

Figure 18.14 Frequency of the ECA in the cat. As the distance from the ligament of Treitz increases, the *in vitro* measurements (crosses) show a steady decline in frequency, while the *in vivo* measurements (open and filled circles) show clear frequency plateaus. (Diamant and Bortoff, 1969, Fig. 2.) (The ligament of Treitz marks the beginning of the jejunum.)

shows the effect of a single cut in the intestine part of the way along the frequency plateau. To the left of the cut, the ECA frequency is entrained to the same high frequency as that of the frequency plateau in the intact intestine. To the right of the cut, a new frequency plateau emerges as the highest-frequency oscillator again entrains its neighbors. In this case the frequency of the second plateau is lower than that of the first, as it is entrained to an oscillator with a lower frequency, but it extends further to the right, into the region where the intact intestine has a variable ECA frequency.

There is some evidence to suggest that ECA frequency decreases along the intestine in a stepwise fashion, and this is illustrated in Fig. 18.14. The frequency plateaus are separated by regions where the amplitude of the oscillation is variable. Often, however, the wave activity in subsequent plateaus is less organized than in the first, as the oscillations are not so closely phase-locked.

18.3.3 Models of Coupled Oscillators

The two primary means by which the waves of electrical activity in the intestine have been studied are with numerical simulations of large coupled systems of oscillators and with rigorous mathematical analysis of approximating "phase equations."

Numerical Investigations

A number of investigators have used numerical simulations to study the behavior of chains of coupled oscillators in the small intestine (Nelsen and Becker, 1968; Diamant et al., 1970; Sarna et al., 1971; Robertson-Dunn and Linkens, 1974; Brown et al., 1975; Patton and Linkens, 1978). As a typical example, Diamant et al. (1970) coupled from 5 to 25 van der Pol oscillators with frequencies that decreased along the chain. Each oscillator was coupled to its nearest neighbor with the lower frequency in a procedure called forward coupling, and the coupling was assumed to be resistive. Numerical simulations showed that the oscillators are organized into frequency plateaus, whose lengths increased as the coupling strength increased. Because the coupling was in the forward direction only, the frequency plateaus lay above the intrinsic frequencies of the individual oscillators. The frequency plateaus were separated by regions in which the local frequency waxed and waned.

The Phase Equations

The mathematical study of waves of electrical activity on the small intestine begins with the assumption that there are $n+1$ coupled oscillators, described by the system of equations

$$\frac{du_i}{dt} = F_i(u_i) + \epsilon \sum_{j=1}^{n+1} a_{ij} H(u_j), \tag{18.80}$$

where u_i is the vector of independent variables describing the ith oscillator, and where a_{ij} are the coupling coefficients. The oscillators are assumed to be nearly identical, so that the behavior of each oscillator is described approximately by some periodic function, denoted by $u_i = U(\omega(\epsilon)t + \delta\theta_i(t))$, where $\delta\theta_i$ is the *phase shift* of the oscillator and is presumed to be slowly varying. Then, the equations (18.80) can be reduced (using multiscale or averaging techniques) to equations describing the phase shifts of the individual oscillators, of the form

$$\frac{d}{d\tau}\delta\theta_i = \xi_i - \Omega_1 + \sum_{j=1}^{n+1} a_{ij} h(\delta\theta_j - \delta\theta_i) + O(\epsilon^2), \qquad i = 1, \dots, n+1, \tag{18.81}$$

for some periodic function h, where $\tau = \epsilon t$ is a slow time. The phase equations are asymptotically valid in the limit that the coupling is weak and the oscillators are similar. A derivation of the phase equations is given in Section 12.7, where the function h and the constants ξ_i and Ω_1 are determined. As a reminder, recall that $2\pi(1+\epsilon\xi_i)$ is the natural (uncoupled) frequency of the ith oscillator, and that $\omega(\epsilon) = 1 + \epsilon\Omega_1 + O(\epsilon^2)$.

When each oscillator is coupled only to its nearest neighbors in a linear chain, the equations are

$$\frac{du_i}{dt} = F_i(u_i) + \epsilon(u_{i+1} - u_i) + \epsilon(u_{i-1} - u_i). \tag{18.82}$$

Here, the term $\epsilon(u_{i+1} - u_i)$ is deleted if $i = n + 1$, and the term $\epsilon(u_{i-1} - u_i)$ is deleted if $i = 1$. Then, the phase equations are of the form (18.81), where $a_{ij} = 1$ if $j = i + 1$ or if $j = i - 1$, $a_{ii} = -a_{i,i+1} - a_{i,i-1}$, and all other elements of a_{ij} are zero. We find equations for the consecutive phase differences $\phi_i = \delta\theta_{i+1} - \delta\theta_i$ to be

$$\frac{d\phi_i}{d\tau} = [\Delta_i + h(\phi_{i+1}) + h(-\phi_i) - h(\phi_i) - h(-\phi_{i-1})] + O(\epsilon^2), \tag{18.83}$$

where $\Delta_i = \xi_{i+1} - \xi_i$ is a measure of the amount of detuning of the oscillators, i.e., how much the natural frequencies vary along the chain. The term $h(-\phi_{i-1})$ is omitted if $i = 1$, and the term $h(\phi_{i+1})$ is omitted if $i = n$. Finally, we take h to be odd, in which case the phase difference equation becomes

$$\dot{\phi} = \beta\Delta + K\mathbf{H}(\phi), \tag{18.84}$$

where $\phi = (\phi_1, \ldots, \phi_n)$, $\beta\Delta = (\Delta_1, \ldots, \Delta_n)$, and $\mathbf{H} = (h(\phi_1), \ldots, h(\phi_n))$, and K is a tridiagonal matrix with -2 on the diagonal and 1 above and below the diagonal. Here, the dot denotes differentiation with respect to the slow time $\tau = \epsilon t$. The parameter β has been introduced as a control parameter for the amplitude of the gradient of the uncoupled oscillator frequency (i.e., the strength of the detuning).

Some Simple Solutions

Before discussing how frequency plateaus arise in the phase equation, it is useful to consider the solution in some simpler cases.

Two Coupled Oscillators

For two coupled oscillators there is only a single phase equation,

$$\frac{d\phi}{d\tau} = \beta\Delta - 2h(\phi). \tag{18.85}$$

A phase-locked solution is one for which the phase difference between neighboring oscillators does not change, i.e., ϕ is constant. Thus, phase-locked solutions are found by setting $d\phi/d\tau = 0$ and solving for Δ. This gives

$$\beta\Delta = 2h(\phi). \tag{18.86}$$

Since h is 2π-periodic and odd, we can solve (18.86) only if $\beta\Delta$ is not too large, as otherwise it would be greater than the maximum value of $2h$. In a common example, $h(\phi)$ is taken to be $\sin(\phi)$, in which case $|\beta\Delta|$ must be less than two for a phase-locked solution to exist. Since $\beta\Delta$ measures the amount of detuning, a phase-locked solution exists if and only if the two oscillators have natural frequencies that are not too different. If $\beta\Delta$ is small enough, the phase difference established between the oscillators (at least to lowest order in ϵ) is given by the solution of (18.86).

Three Coupled Oscillators

When three oscillators are coupled, the two phase difference equations are

$$\frac{d\phi_1}{d\tau} = \beta\Delta_1 - 2h(\phi_1) + h(\phi_2), \tag{18.87}$$

$$\frac{d\phi_2}{d\tau} = \beta\Delta_2 - 2h(\phi_2) + h(\phi_1), \tag{18.88}$$

and so a phase-locked solution occurs if

$$\frac{2\beta\Delta_1 + \beta\Delta_2}{3} = h(\phi_1), \tag{18.89}$$

$$\frac{2\beta\Delta_2 + \beta\Delta_1}{3} = h(\phi_2). \tag{18.90}$$

Clearly, this can be solved only if $|\beta(2\Delta_1 + \Delta_2)|$ and $|\beta(2\Delta_2 + \Delta_1)|$ are small enough.

It is important to note that if solutions for ϕ_1 and ϕ_2 exist, ϕ_1 does not necessarily equal ϕ_2. Thus, although the oscillators are phase-locked, the phase difference between the first and second oscillators is not necessarily the same as the phase difference between the second and third. If the phase differences are unequal, there is not a regular (constant speed) phase wave moving along the chain. Hence, phase locking does not necessarily imply a wavelike behavior.

To have a regular phase wave, the phase differences must be both constant and *equal* along the chain of oscillators. In this case, the peak of the wave moves at a constant speed down the oscillator chain. For the case of three coupled oscillators, phase wave solutions exist if $\Delta_1 = \Delta_2$. In other words, if the frequency difference between the first and second oscillators is the same as the difference between the second and third, and if this difference is not too large, then a phase wave solution exists.

This highlights the fact that in general, we can specify the frequency gradient along the oscillator chain and then solve for the phase differences, or we can specify the phase differences and then solve for the required frequency gradient, but we cannot specify the phase difference and the frequency gradient, expecting a phase wave.

A Chain of Coupled Oscillators

The equations for a phase-locked steady-state solution for a chain of $n + 1$ coupled oscillators are given by

$$\beta\Delta + K\mathbf{H}(\phi) = 0. \tag{18.91}$$

This is a system of n equations in n unknowns. In general, we can view the frequencies as given and the phase differences as unknown, or we can specify the phase differences and view the frequency differences as unknown. For example, if we seek a solution that is both phase-locked and has a phase wave, we need $\phi_1 = \phi_2 = \cdots = \phi_n$. Letting $h(\phi_i) = \eta$, we obtain $\Delta_1 = \Delta_n = \eta/\beta$ and $\Delta_i = 0$ for $i = 2, \ldots, n-1$. Thus, a phase wave solution exists only if all the middle oscillators have the same frequency, ω say, while the first oscillator is tuned to $\omega - \eta/\beta$ and the last oscillator to $\omega + \eta/\beta$. Note that

η can be either positive or negative, as different signs correspond to waves moving in opposite directions.

This observation poses a dilemma for the application of the phase equation to the electrical waves in the small intestine. Recall that ECA in the small intestine has a frequency plateau in the region close to the pylorus, and in this plateau, waves appear to be traveling away from the pylorus. These are phase-locked, with constant phase difference along the intestine. However, each segment of the small intestine has a natural oscillation frequency that decreases with distance from the pylorus. These two observations are inconsistent with the phase equations, for which a constant phase difference implies a constant natural frequency on the interior of the chain.

Frequency Plateaus

One partial solution to this dilemma was given by Ermentrout and Kopell (1984). They showed that on each plateau, the phase differences are not exactly constant, but make small oscillations, being locked only in an average sense, a phenomenon sometimes called *phase trapping*.

For simplicity, we assume that $h(\phi)$ is odd and 2π-periodic, with a maximum M at ϕ_M and a minimum m at ϕ_m, qualitatively like $\sin\phi$ (Fig. 18.15). Critical points of the differential equation (18.84) are solutions of

$$\mathbf{H}(\phi) = K^{-1}(-\beta\Delta), \tag{18.92}$$

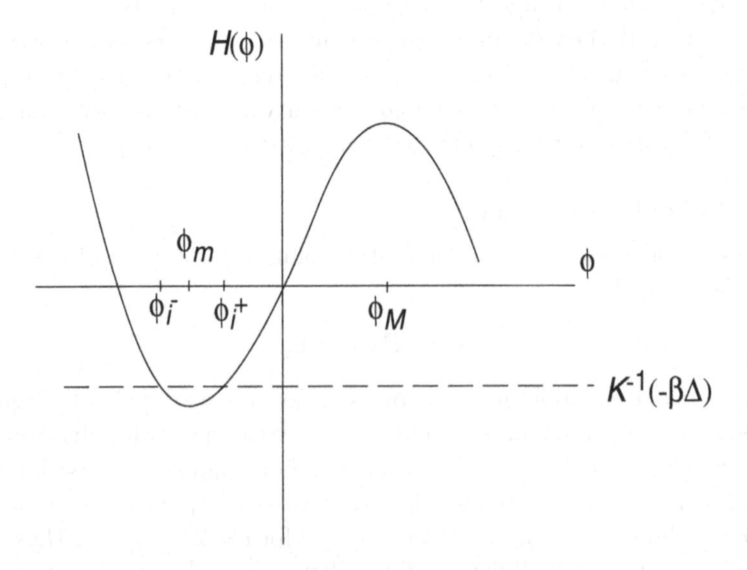

Figure 18.15 The roots of the ith component of the phase difference equation.

which has a solution if and only if every component of $K^{-1}(-\beta\Delta)$ lies between m and M. Let

$$\beta_0 = \max\{\beta : m \le (K^{-1}(-\beta\Delta))_i \le M, \text{ for every } i\}. \tag{18.93}$$

When $\beta < \beta_0$, for every i there are two solutions to the scalar equation $h(\phi_i) = (K^{-1}(-\beta\Delta))_i$. These solutions, which we denote by ϕ_i^+ and ϕ_i^-, are shown in Fig. 18.15.

Since each component of ϕ can have one of two values, there are 2^n possible steady states when $\beta < \beta_0$. Because of the definition of β_0, there is some value j such that the roots ϕ_j^+ and ϕ_j^- coalesce and disappear as β crosses β_0. Thus, as β crosses β_0, all the critical points coalesce in pairs and disappear. This follows because every critical point can be matched with another that agrees with it in every component $i \neq j$, one having jth component ϕ_j^+, the other having jth component ϕ_j^-. When $\beta < \beta_0$ the members of the pair differ only in the jth component; when $\beta = \beta_0$ the members of the pair are identical, and when $\beta > \beta_0$ there is no solution for the jth component, and so both solutions fail to exist.

There is one particular pair of critical points that is of interest for reasons that are explained below. Let $\xi_j(\beta)$ denote the critical point whose kth component is ϕ_k^- for all $k \neq j$, and whose jth component is ϕ_j^+. Also, let $\xi_0(\beta)$ denote the critical point with the kth component equal to ϕ_k^- for all k. Clearly, as $\beta \to \beta_0$, ξ_j coalesces with ξ_0 and the two critical points disappear.

Finally, to complete the preliminaries, we restrict Δ to be of a particular form. Since experimental data suggest that the natural frequencies of the oscillators decrease approximately linearly along the small intestine, it is reasonable to take $\Delta = (-1, \ldots, -1)$, corresponding to a linear decrease in frequency along the oscillator chain. We also assume that there is an even number of oscillators in the chain (an odd number of phase differences), i.e., that n is odd, with $n = 2j-1$, and thus the central phase difference is at position j. For this choice of Δ and n, the solution of (18.92) fails first at the jth component; that is, if j is the position of the middle phase difference, then $\phi_j^+ \to \phi_j^-$ as $\beta \to \beta_0$. This is easily seen by noting that $K^{-1}(\beta\Delta)$ has kth component $-\beta k(n+1-k)/2 < 0$ (Exercise 5). Hence $\phi_k^{\pm}(\beta) < 0$ for all k as long as $\beta < \beta_0$. Further, when $n = 2j-1$, $k(n+1-k) = k(2j-k)$, which is greatest when $k = j$. Since the jth component of the solution to (18.92) is the one with the greatest modulus, it follows that the jth component is the first to "hit" the minimum and disappear.

We now return to the particular pair of steady states, ξ_0 and ξ_j, defined above. Linear stability analysis of the system (18.84) shows that ξ_0 is a stable node, while ξ_j is a saddle point with one positive and $n-1$ negative eigenvalues. Further, both branches of the unstable manifold at ξ_j tend to ξ_0 as $\tau \to \infty$, and thus the closure of the unstable manifold forms a closed loop.

We can get some insight into the meaning of this last statement and why it is true by considering the special case $n = 2$. It is convenient to introduce the change of variables $\psi = K^{-1}\phi$, in which case the system of differential equations (18.84) becomes

$$\dot{\psi} = K^{-1}\beta\Delta + \mathbf{H}(K\psi), \tag{18.94}$$

or, in the specific case that $n = 2$,

$$\dot{\psi}_1 = \beta + h(\psi_2 - 2\psi_1), \tag{18.95}$$

$$\dot{\psi}_2 = \beta + h(\psi_1 - 2\psi_2). \tag{18.96}$$

This is a two-dimensional system whose phase portrait is easily studied. First, note that since (18.84) is a flow on a torus, so also is this system. The torus for (18.84) is the domain $0 \le \phi_i \le 2\pi, i = 1, \ldots, n$, with the boundary at $\phi_i = 0$ "identified" with, or equivalent to, the boundary at $\phi_i = 2\pi$. Here, however, the boundaries of the torus are modified, being the four straight lines

$$\psi_1 - 2\psi_2 = 0, -2\pi, \tag{18.97}$$

$$\psi_2 - 2\psi_1 = 0, -2\pi. \tag{18.98}$$

These bounding lines are shown dashed in Fig. 18.16. Now, the flow on this torus can be understood by first examining the nullclines $\dot{\psi}_1 = 0$ and $\dot{\psi}_2 = 0$. There are four such curves,

$$\psi_2 - 2\psi_1 = -\phi_\pm, \tag{18.99}$$

$$\psi_1 - 2\psi_2 = -\phi_\pm, \tag{18.100}$$

where the numbers $-\phi_\pm$ satisfy $h(-\phi_\pm) = -\beta$, as depicted in Fig. 18.15. The nullclines are shown in Fig. 18.16 by solid lines.

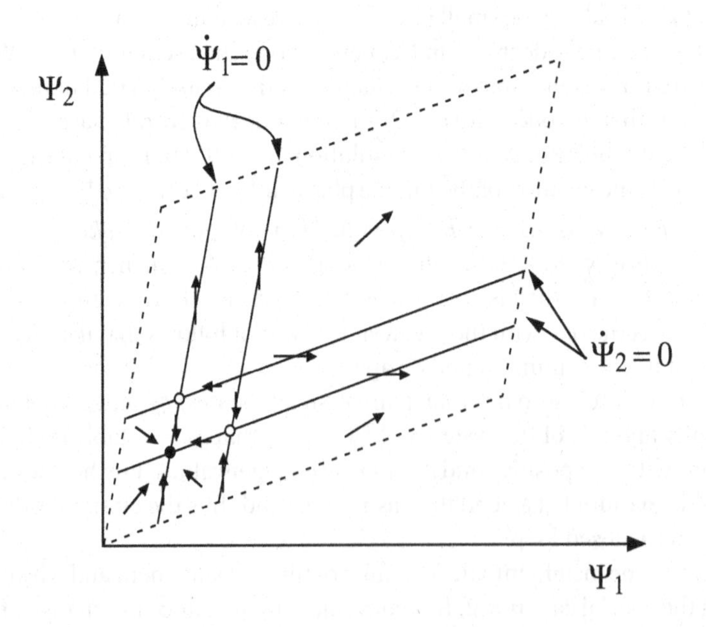

Figure 18.16 Phase portrait for the two-dimensional system of equations (18.97)–(18.98). The stable critical point is denoted by a filled circle, and the two saddle points by open circles.

Clearly, there are four critical points. By sketching in a few elements of the vector field, we can see that the critical point at the leftmost and lowest position is the only stable critical point (denoted by a filled circle), two of the critical points are saddle points (denoted by open circles), and the fourth is an unstable node. We also see that the stable critical point is a global attractor. That is, every trajectory, excluding the other critical points, tends to the unique stable critical point as time goes to infinity. It follows that the unstable manifold of each saddle point forms a closed loop; both closed loops are therefore invariant manifolds.

In general, for arbitrary n, when $\beta < \beta_0$ there is a closed invariant manifold containing the two steady states, ξ_j and ξ_0. Furthermore, this loop is a smooth invariant attracting cycle on which ϕ_j completes a full rotation from 0 to 2π but on which the other ϕ's do not. In other words, as ϕ_j moves through the 2π-cycle, all the other $\phi_k, k \neq j$, vary without making a full cycle. This is illustrated in Fig. 18.17 by orbit A. Orbit B, however, experiences a full 2π cycle in angle θ_2 for every 2π cycle of θ_1. It follows that the invariant attracting manifold formed by the two branches of the unstable manifold of ξ_j is homotopic to (i.e., is continuously deformable into) the circle $\phi_k = 0, k \neq j$, $0 \leq \phi_j \leq 2\pi$.

The crucial result proved by Ermentrout and Kopell is that this invariant attracting manifold exists even when $\beta > \beta_0$. Thus, although steady states of the phase difference equations disappear when $\beta > \beta_0$, a smooth, invariant, attracting manifold that is homotopic to the circle $\phi_k = 0, k \neq j, 0 \leq \phi_j \leq 2\pi$, persists. Since it contains no critical points, this manifold is an attracting limit cycle. This stable limit cycle corresponds to a pair of frequency plateaus in the chain of coupled oscillators. To see this, define the average frequency of the kth oscillator to be

$$\omega(\epsilon) + \lim_{T\to\infty} \frac{1}{T} \int_0^T \delta\theta_k'(t)\, dt, \tag{18.101}$$

provided that the limit exists. Here, a prime denotes differentiation with respect to t. Note that if $\delta\theta_k'$ is constant, the frequency is exactly $\omega(\epsilon) + \delta\theta_k'$, as expected. Subtracting $\delta\theta_k'$ from $\delta\theta_{k+1}'$ gives the average phase difference as

$$\lim_{T\to\infty} \frac{\epsilon}{T} \int_0^T \phi_k'(\tau)\, d\tau. \tag{18.102}$$

Around the attracting limit cycle, this simplifies to

$$\frac{\epsilon}{T_0} \int_0^{T_0} \phi_k'(\tau)\, d\tau, \tag{18.103}$$

where $T_0(\beta)$ is the period of the limit cycle on the torus. However, we readily calculate that

$$\int_0^{T_0} \phi_k'(\tau)\, d\tau = \begin{cases} 0, & k \neq j, \\ 2\pi, & k = j. \end{cases} \tag{18.104}$$

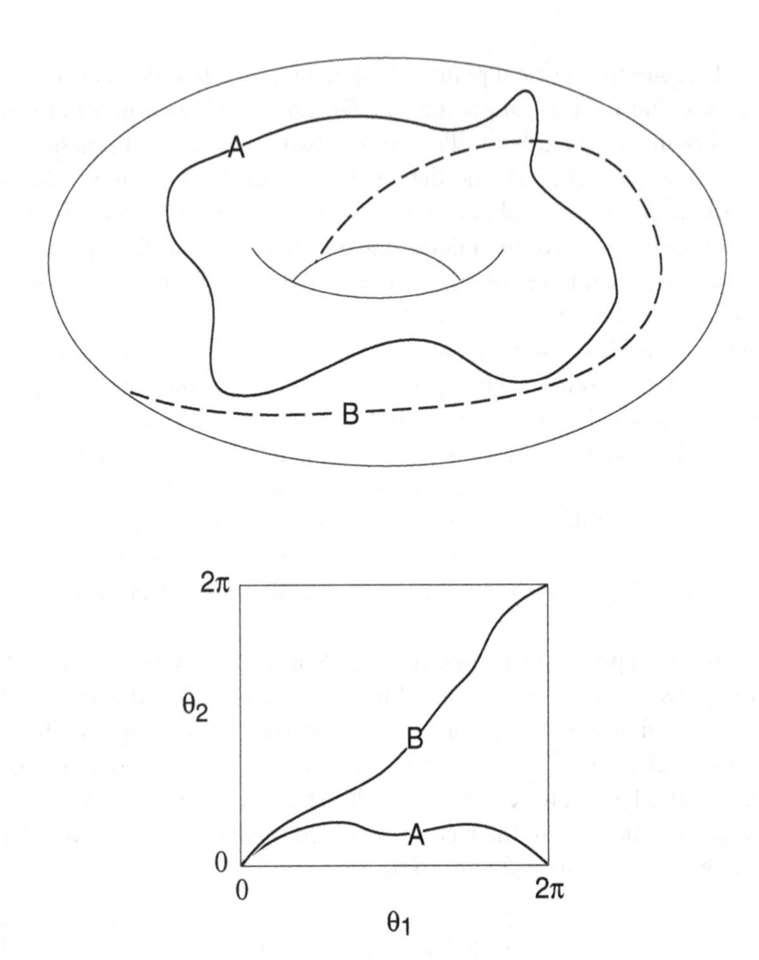

Figure 18.17 Schematic diagram of two orbits on a torus, drawn two different ways. The upper panel shows the orbits on a torus, while the lower panel has unfolded the torus into a square, periodic in both directions. Orbit A is homotopic to the circle $\theta_2 = 0$, while orbit B is not.

It follows that the first j oscillators all have the same average frequency, as do the oscillators from $j+1$ to n. Between the jth and the $(j+1)$st oscillators there is a frequency jump of $2\pi\epsilon/T_0(\beta)$.

It is important to note that the phase differences $\phi_k, k \neq j$, make small oscillations about the constant ϕ_k^-, but are not identically constant. Thus, on each frequency plateau, the phases are not locked, but only "trapped" on average over each cycle. In contrast to some experimental data, one therefore does not expect to see exactly regular propagating waves appearing on each plateau, since such waves require phase locking at each instant and a constant phase difference along the plateau. The precise reasons for this discrepancy are, as yet, unknown.

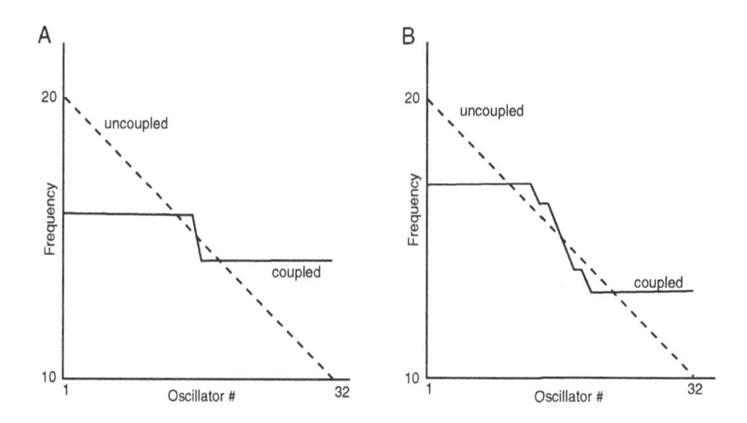

Figure 18.18 Frequency plateaus in a chain of 32 coupled oscillators. For these numerical simulations the phase equation was chosen to be $\dot{\phi}_k = -10/31 + \delta(\sin\phi_{k+1} - 2\sin\phi_k + \sin\phi_{k-1})$. (A): $\delta = 32$, (B): $\delta = 18$. Note that decreasing δ while leaving the intrinsic oscillator frequencies unchanged is equivalent to increasing β. (Ermentrout and Kopell, 1984, Fig. 4.1.)

Numerical Solutions

Numerical solutions of the phase equation for two different values of β are shown in Fig. 18.18. Once β is greater than the critical value β_0 a pair of plateaus emerges, and as β is increased further, multiple plateaus appear. Comparison of Fig. 18.12 with Fig. 18.18 shows that the above model differs from experimental results in an important way. In the model, the frequency of the plateau lies between the maximum and minimum natural frequencies of the oscillators in the plateau; a plateau cannot have a higher frequency than all of its constituent oscillators, as is seen in the experimental data. However, the above simple model can be extended to obtain better qualitative agreement with experiment. For example, the coupling between cells can be made stronger in one direction than the other (nonisotropic coupling) or it can be made nonuniform along the oscillator chain, in which case the phase difference equation can reproduce the asymmetrical behavior exhibited by the experimental system.

However, despite this qualitative agreement, it must be admitted that the simple model presented here does not give a quantitative explanation of the properties of frequency plateaus in the small intestine. It is an excellent example of how, to obtain an analytical understanding of a particular phenomenon, it is often necessary to study a model that has been reduced to caricature by successive approximations. Although hope of quantitative agreement is thereby lost, such simple models often permit a substantial understanding of the underlying structure.

18.3.4 Interstitial Cells of Cajal

There is, of course, much more than coupled oscillators involved in the control of the gastrointestinal electrical activity. In particular, in recent years it has become clear that the interstitial cells of Cajal (ICC) play an extremely important role in the regulation of ECA, although we still have only a limited understanding of exactly what that role is.

ICC were first discovered by Santiago Ramon y Cajal in 1911 (Cajal, 1893, 1911) who called them "primitive neurons" and theorized that they were modulators of smooth muscle contraction, modulated by the nervous system in their turn. ICC can be divided into two groups. In most regions of the gastrointestinal tract, ICC form a thin layer between the longitudinal and circular layers of smooth muscle. These are called *myenteric* ICC. The second group of ICC, the *intramuscular* ICC, are distributed through the smooth muscle cells themselves. We now know that ICC are the pacemaker cells that drive the electrical slow wave in gastrointestinal smooth muscle, and that isolated smooth muscle cells cannot oscillate independently (Sanders, 1996; Čamborová et al., 2003; Takaki, 2003; Hirst and Ward, 2003; Hirst and Edwards, 2004). Although far more is known about the biophysical properties of myenteric ICC than of intramuscular ICC, the mechanisms controlling pacemaker activity are still far from clear. IP_3-dependent release of Ca^{2+} from internal stores such as the endoplasmic reticulum (see Chapter 7) is clearly one important mechanism, and this released Ca^{2+} acts on Ca^{2+}-sensitive ionic channels, particularly a Ca^{2+}-sensitive Cl^- current. It is likely that Ca^{2+} release from the mitochondria also plays an important role. However, no more definite mechanism has yet been proposed and confirmed in detail. ICC are also modulated by the enteric nervous system; it seems that innervation of the gastrointestinal tract is directed more to the ICC than to the smooth muscle itself (Hirst and Edwards, 2004). The question of how ICC operate, how they are modulated by the nervous system, and how they in turn control smooth muscle contractility, is a fascinating area in which modeling is likely to play an increasingly important role over the next few years.

The importance of ICC for generating pacemaker potentials underlines the important similarities between the gastrointestinal tract and the heart. In both organs, neurons innervate pacemaker cells, which in turn coordinate their activity in some way to control muscle motility. In the pacemaker cells and the muscle cells of both the gastrointestinal tract and the heart, the control of internal Ca^{2+}, via IP_3 receptors, ryanodine receptors, and Ca^{2+}-sensitive ion channels, plays a crucial role. These similarities are shown schematically in Fig. 18.19.

18.3.5 Biophysical and Anatomical Models

The most detailed biophysical model of gastrointestinal electrical activity to date is that of Miftakhov et al. (1999a,b), who constructed a model of a "functional unit" of the small intestine. In addition to the smooth muscle cells themselves, this functional unit included a number of different neurons in the enteric nervous system (each with dendritic, somatic and axonal components), as well as the dynamics of neurotransmitters and Ca^{2+} in the synapses. In this model local distention of the gut wall causes a response in mechanoreceptors (dendritic, somatic and axonal in sequence), stimulation of a response in a secondary neuron, and thence stimulation of the smooth muscle cell by opening of L-type Ca^{2+} channels and a consequent increase of Ca^{2+} concentration

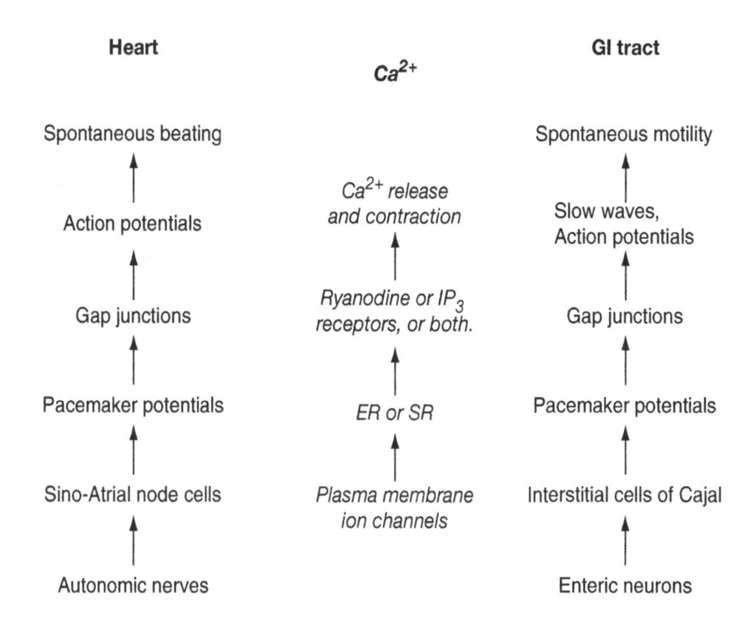

Figure 18.19 Schematic diagram showing the similarities between the gastrointestinal tract and the heart. Adapted from Takaki, 2003, Fig. 15.

inside the smooth muscle cell, causing contraction. The generated force itself feeds back to stimulate the mechanoreceptors again, resulting in multiple contractions in response to the initial isolated stimulus.

The enormous number of parameters and equations required to construct this model, the almost complete lack of experimental data about most of these parameters, and the relative lack of experimental data against which such a model can be tested, raises a number of thorny questions about model identifiability. How much model complexity can be justified by experimental data, and how much is absolutely necessary to answer the scientific question under investigation, are difficult questions that have no generic answers; each model must be treated on its own merits. Although there is no doubt that complex biophysical models will be required before we can have a complete understanding of how a syncytium of smooth muscle and its associated innervation works to propel food along the gut, it is perhaps premature to place too much reliance on such complex models at this stage.

A different approach is taken by Pullan et al. (2004) in their detailed anatomical model of the stomach. Instead of studying a highly detailed biophysical model on a simple geometry, they constructed a complex geometry but used only highly simplified biophysical models. Using data from the visible human project (Spitzer et al., 1996) they constructed a geometrically accurate stomach on which a finite element grid could be superimposed and the model equations solved. However, they model the electrical activity using the FitzHugh–Nagumo equations (Chapter 5). Thus, although the model shows how a wave of electrical excitation can spread around the stomach, as yet it offers little insight into the detailed mechanisms or significance of ECA, or into the behavior of a syncytium of coupled smooth muscle oscillators. However, as more detailed and accurate biophysical models become available they can (relatively) easily be incorporated into the detailed geometrical model.

The electrical activity of the stomach can be measured by an electrogastrogram (EGG) in the same way that the electrical activity of the heart can be measured by an ECG (Chapter 12). The earliest such recordings were those of Alvarez (1922) and Davis et al. (1957). However, even today EGG recordings, particularly those aimed at measuring electrical activity in the intestines, are difficult to make and to interpret (Bradshaw et al., 1997, 1999; Allescher et al., 1998). The difficulties are clear; each heartbeat is a single event, localized, relatively precise, and repeated many times in succession. Gastrointestinal electrical activity has none of these attributes, being diffuse, changing in nature and frequency along the length of the stomach and intestines, and continuously occurring.

18.4 EXERCISES

1. Modify (18.1) to account for the effects of the membrane potential on Na^+ flux. How does membrane potential affect the transport of water?

2. Use the model of local Na^+ removal and osmotic transport of water to analyze the removal of water and Na^+ along the length of the intestinal tract.

   (a) Give a phase-plane analysis for the system. What is the trajectory of Na^+ and water if Na^+ is initially quite high? What is the trajectory of Na^+ and water if Na^+ is initially quite low? How does this compare with the trajectory when there is no Na^+?

   (b) If the flow of water from the intestine is assumed to depend solely on the Na^+ concentration, then the flow can become negative, which is clearly unphysiological. How might this assumption be modified and justified on physical grounds?

   Hint: As the chyme dries, one expects the continued extraction of water to become more difficult.

3. For the the model of Section 18.1.2, find the average concentration of solute in a tube for which $\eta = \frac{c_0 PL^2}{Dr}$ is small. Use this to find the efflux velocity and efflux concentration. Under what conditions on parameters can water be pumped into pure water?

4. (a) Find two terms of the power series representation of the solution of (18.68) in powers of ζ. Find $u(0)$ to the same order in ζ.

   (b) It appears from the leading-order solution that $u(0)$ is negative if $J > 1$. Show that this is not correct, but that $u(0) > 0$ for all positive values of J. Show that when $J > 1$, the bulk of the reaction occurs in a thin layer contained within the mucus layer, and that the epithelial surface is completely protected. What is the physical interpretation of the condition $J > 1$?

5. Show that the kth component of $K^{-1}(\beta\Delta)$ in (18.92) is $\alpha_k = -\beta k(n+1-k)/2$.

   Hint: Verify that $(K\alpha)_i = \beta$.

6. What steady phase-locked solutions are possible for two coupled identical oscillators? What stable, steady, phase-locked solutions are possible?

7. Describe the behavior expected from two coupled oscillators when $\beta\Delta = -2-\epsilon$ and $h(\phi) = \sin\phi$ for $\epsilon \ll 1$. Check your prediction numerically. This behavior is called *rhythm splitting*, and is discussed in more detail in Murray (2002).

8. In Exercise 22 of Chapter 12 the coupling function $h(\theta)$ is calculated for a collection of coupled FitzHugh–Nagumo equations. Solve the phase equation with this coupling function numerically with arbitrary initial data to determine $\phi(t)$. How quickly do identical FitzHugh–Nagumo oscillators synchronize?

9. Extend the previous question (and refer to Chapters 7 and 8) to study the synchronization of intracellular Ca^{2+} oscillations. Suppose two cells, each with a well-mixed interior, are coupled by a membrane through which Ca^{2+} can diffuse through gap junctions. Suppose further that each cell exhibits intracellular Ca^{2+} oscillations of slightly different periods (i.e., each cell has a slightly different background IP_3 concentration). Using your favorite model of Ca^{2+} oscillations, determine the coupling function numerically, and thus determine how fast such synchronization occurs, as a function of the intercellular permeability of Ca^{2+}.

CHAPTER 19

The Retina and Vision

The visual system is arguably the most important system through which our brain gathers information about our surroundings, and forms one of our most complex physiological systems. In vertebrates, light entering the eye through the lens is detected by photosensitive pigment in the photoreceptors, converted to an electrical signal, and passed back through the layers of the retina to the optic nerve, and from there, through the visual nuclei, to the visual cortex of the brain. At each stage, the signal passes through an elaborate system of biochemical and neural feedbacks, the vast majority of which are poorly, if at all, understood.

Although there is great variety in detail between the eyes of different species, a number of important features are qualitatively conserved. Perhaps the most striking of these features is the ability of the visual system to adapt to background light. As the background light level increases, the sensitivity of the visual system is decreased, which allows for operation over a huge range of light levels. From a dim starlit night to a bright sunny day, the background light level varies over 10 orders of magnitude (Hood and Finkelstein, 1986), and yet our eyes continue to operate across all these levels without becoming saturated with light. The visual system accomplishes this by ensuring that its sensitivity varies approximately inversely with the background light, a relationship known as *Weber's law* (Weber, 1834) and one that we discuss in detail in the next section.

Because of this adaptation, the eye is more sensitive to changes in light level than to a steady input. When a space-independent pulse of light is shone on the entire eye, the retina responds with a large-amplitude signal at the beginning followed by a decrease to a lower plateau. Similarly, at the end of the pulse, a large negative transient is followed by a return to a plateau. Response to transients with adaptation in steady conditions is characteristic of inhibitory feedback, or *self-inhibition*, which occurs at a number of levels in the retina. Furthermore, the retina is sensitive to particular frequencies of flashing light, with the most sensitive frequency being a function of the background light level.

Adaptation is also manifested spatially. When a time-independent strip of light is applied to the retina, the response is greatest at the edges of the pattern. These response variations are known as *Mach bands* and are due to *lateral inhibition*, which plays a similar role in space as self-inhibition plays in time. For example, in the interior of a uniformly bright part of the visual field, neurons are inhibited from all sides, while regions near the edge receive little inhibition from their dimly illuminated neighbors and therefore appear brighter. The result is contour enhancement. The effect of lateral inhibition can be seen in the white intersections of Fig. 19.1. In particular, if one looks intently at one of the white intersections, the remaining intersections appear to have a gray or darkened interior, and the center of the white strips appear slightly darkened compared to their edges, because of lateral inhibition.

The visual system operates on many levels, ranging from the biochemistry of the photopigments, to the cellular electrophysiology of the individual retinal cells, to the neural pathways responsible for image processing, to the large-scale structure of the visual cortex. Because of this complexity, there is still much debate over the mechanisms of adaptation, and the place where it occurs. Certainly, adaptation occurs in psy-

Figure 19.1 Test pattern with which to observe the effects of self-inhibition and lateral inhibition.

chophysical experiments (i.e., with human subjects who report what they detect); one example of this has already been seen in Fig. 19.1. As another example, psychophysical measurement of the threshold-versus-intensity function shows Weber's law behavior. Plots of the ratio $\delta I/I$, where δI is the brightness of the smallest detectable flash superimposed on the background I, shows that $\delta I/I$ is approximately constant over a range of background light levels. Although relating these psychophysical results to specific mechanisms is not an easy task, it is clear that adaptation occurs at the very lowest level of the visual system, inside the photoreceptors themselves.

Obviously, there is insufficient space here for a detailed study of all these aspects of the visual system. Here we concentrate on retinal mechanisms and omit discussion of mechanisms at the level of the visual cortex. The latter questions are better addressed in specialist books on neuroscience. For a more comprehensive view of the visual system, the reader is referred to the excellent book by Nicholls, Martin, and Wallace (1992); other discussions of visual processing can be found in Graham (1989), Blakemore (1990), Landy and Movshon (1991), and Spilmann and Werner (1990). A review of the connection between psychophysical experiments and the underlying physiology is given by Hood (1998), while an excellent earlier review of psychophysical experiments is Barlow (1972).

19.1 Retinal Light Adaptation

The first stage of visual processing occurs in the retina, a structure consisting of at least five major neuronal cell types (Fig. 19.2). After entering the eye, light passes through all the cell layers of the retina before being absorbed by photosensitive pigments in the photoreceptors in the final layer of cells. (A functional reason for this arrangement is not known.) Photoreceptors come in two varieties: rods, which operate in conditions of low light, and cones, which operate in bright light conditions and detect color. In the dark, photoreceptors have a resting membrane potential of around −40 mV, and they hyperpolarize in response to light. The light response is graded, with larger light stimuli resulting in larger hyperpolarizations. Note that this is different behavior from typical neurons, in which the action potential is a depolarization and is all-or-nothing, as described in Chapter 5. Photoreceptors make connections to both horizontal cells and bipolar cells. Each horizontal cell makes connections to many photoreceptors (and, often, to bipolar cells), and is coupled to other horizontal cells by gap junctions. The bipolar cells form a more direct pathway, coupling photoreceptor responses to ganglion cells, but this is also a simplification. Amacrine cells connect only to bipolar cells and ganglion cells, and their precise function is unknown. Ganglion cells (which fire action potentials, unlike photoreceptors and horizontal cells) are the output stage of the retina and form the optic nerve. The interconnections among the retinal cells are complex and not well understood; there has been a great deal of work done on how the retina detects features such as moving edges and orientation, while ignoring much of the information presented to it. Here, we describe only the simplest models.

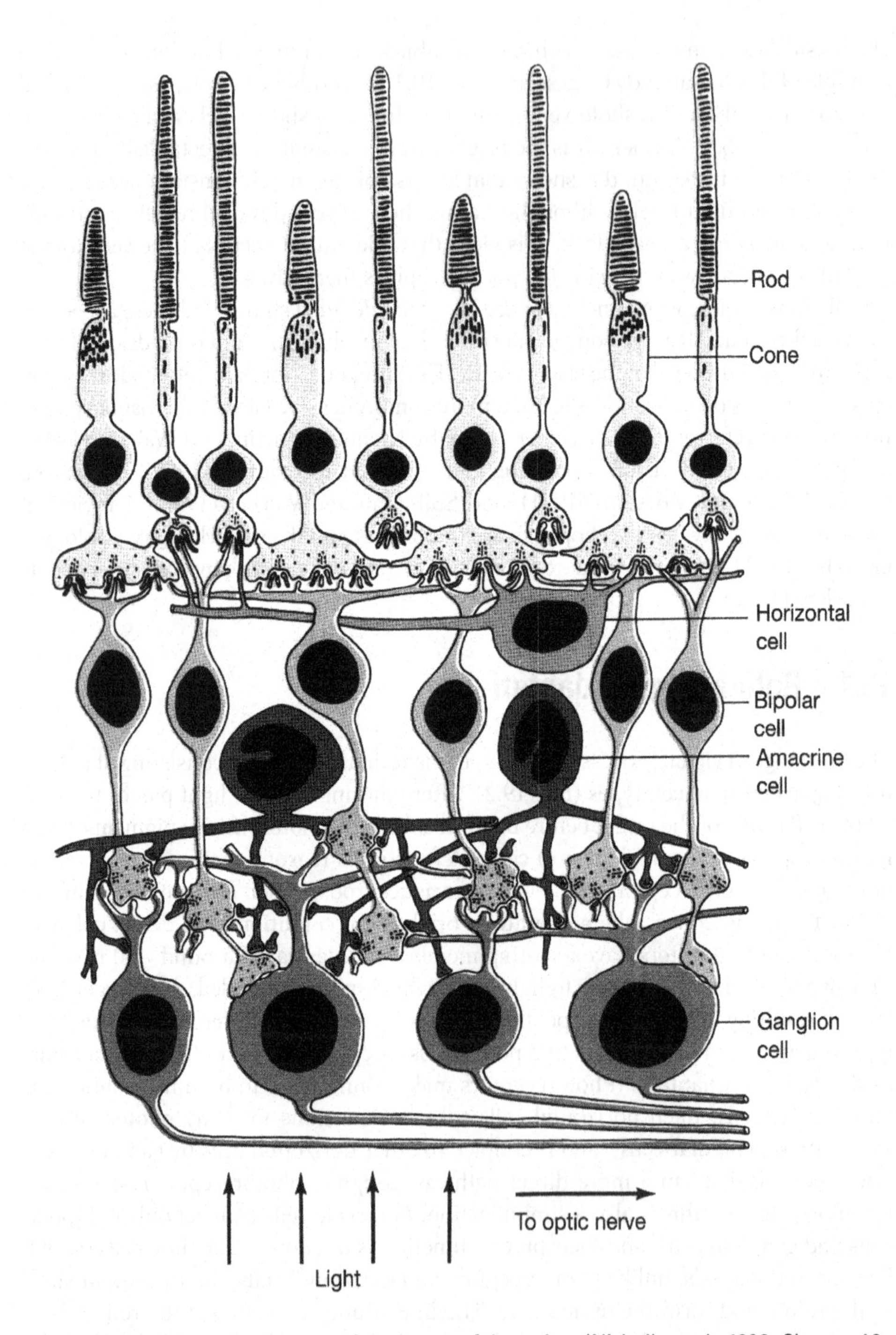

Figure 19.2 Schematic diagram of the layers of the retina. (Nicholls et al., 1992, Chapter 16, Fig. 14, p. 583.)

19.1.1 Weber's Law and Contrast Detection

One of the basic features of the retina is *light adaptation*, the ability to adapt to varying levels of background light. Over a wide range of light levels, the sensitivity of the retina is observed to be approximately inversely proportional to the background light level. This is an example of Weber's law, as discussed above, and is sometimes called the *Weber–Fechner law*. There are three common definitions of sensitivity. It can mean *psychophysical sensitivity*, defined as 1/threshold, where the threshold is the minimal stimulus necessary to elicit an observable response when superimposed on a given background. Weber's law describes the fact that in psychophysical experiments the threshold increases as the background light level increases.

A second definition of sensitivity is the one used most in this chapter. In response to a small light flash, the membrane potential of a photoreceptor (or horizontal cell) first decreases and then returns to rest (recall that the voltage hyperpolarizes in response to a light flash). If $V(I, I_0)$ is the maximum deviation of the membrane potential in response to a light flash of magnitude I on a background of I_0, then the *peak sensitivity* is

$$S(I_0) = \left.\frac{\partial V}{\partial I}\right|_{I=I_0}. \tag{19.1}$$

In the physiology literature (for example, Fain et al., 2001) the sensitivity is defined as the amplitude of the voltage response to a small flash divided by the intensity of the flash (in units of photons per area). Since $V(I, I_0) \approx \frac{\partial V(I,I_0)}{\partial I}|_{I=I_0} I$, these two definitions are equivalent as long as one uses only small flashes.

The third definition of sensitivity is the *steady-state sensitivity*. If $V_0(I_0)$ is the steady response as a function of the background light level, then the steady-state sensitivity is defined to be dV_0/dI_0.

Light adaptation serves two fundamentally important purposes. First, it helps the retina handle the wide range of light levels in which the eye must operate. The eye functions in a range of light levels that spans about 10 log units, from a starlit night to bright sunlight. (Light intensities are typically plotted on a dimensionless logarithmic scale. For example, if I_0 is a standard unit of light intensity, and the intensity of the light stimulus is I, then $\log(\frac{I}{I_0}) = \log I - \log I_0$, so that on a logarithmic scale, the unit scale I_0 only shifts $\log \frac{I}{I_0}$). Further, the retina is so sensitive that it can reliably detect as few as 20 photons, and can even, although less reliably, detect single photons.

The two requirements of operation over a wide range of light levels and high sensitivity in the dark are in conflict. Without control mechanisms, a retina that can detect single photons would be saturated, and hence blinded, by bright light. In bright light, there is a *saturation catastrophe*, in which every photoreceptor is saturated, each sending the same signal to the brain, so that no contrast in the scene can be detected. However, for the human retina this saturation catastrophe is about 10 log units above the level of no response. This range of light sensitivity is achieved partly by the use of two different types of photoreceptors, rods and cones, having different sensitivities, rods operating in dim light and cones in bright light. However, by itself, two types of photoreceptors are inadequate to account for the observed range of light sensitivity.

The second effect of light adaptation is to send a signal to the brain that is dependent only on the contrast in the scene, not on the background light level. When a scene is observed with different background light levels, the amount of light reflected from an object in that scene varies considerably. For example, if you read a book inside where the light is relatively dim, it looks the same as if you were to read it outside in bright light. This happens despite the fact that the amount of light reflecting off the black letters outside is considerably more than the amount of light reflecting off the white page inside. Nevertheless, in both situations we see a white page and black text. Clearly, the eye is measuring something other than the total amount of light coming from an object. In fact, the eye measures the contrast in the scene, which, since it is dependent only on the reflectances of the objects, is independent of the background light level. As another striking example of the importance of contrast detection, consider a black and white television screen. When the television is switched off the screen is uniformly gray. When the television is switched on there is no mechanism for making parts of the screen darker. Nevertheless, we still see deep blacks on the screen. Of course, those parts of the screen that look black do so only in contrast to those other parts of the screen which are brighter.

Contrast detection is a consequence of Weber's law, as can be seen from the following argument of Shapley and Enroth-Cugell (1984). Suppose we observe an object superimposed on a background, where the background reflectance is R_b, the object reflectance is R_o, and the background light level is I. As the receptive field of a retinal neuron moves across the boundary of the object, the stimulus it receives changes from IR_b to IR_o, a difference of $IR_o - IR_b$. Since, according to Weber's law, the sensitivity of the cell is inversely proportional to IR_b (the amount of light reaching the cell from the background), the cell's response will be approximately proportional to $(IR_o - IR_b)/(IR_b) = (R_o - R_b)/R_b$, which is dependent only on the contrast in the scene.

19.1.2 Intensity–Response Curves and the Naka–Rushton Equation

Light adaptation in photoreceptors can be clearly seen in the experimental results reproduced in Fig. 19.3 (also discussed in detail by Fain et al., 2001). In that experiment, a salamander rod was presented with a series of light flashes superimposed on backgrounds of different intensities and the resultant photocurrent was measured. (Although we have not yet discussed the photocurrent in receptors, it is reasonable to assume that the voltage shows similar qualitative responses. The photocurrent is discussed in detail in Section 19.2.) Three important features can be easily seen. First, as the intensity of the flash increases (for any fixed background), the magnitude of the response increases with the intensity of the flash, but eventually saturates. Second, as the background light level increases, a larger flash is needed to evoke a response of a given size. Thus, the rod is adapting. Third, in response to the step of light at the beginning of the lowest trace (background light level 1.57), the response shows a clear

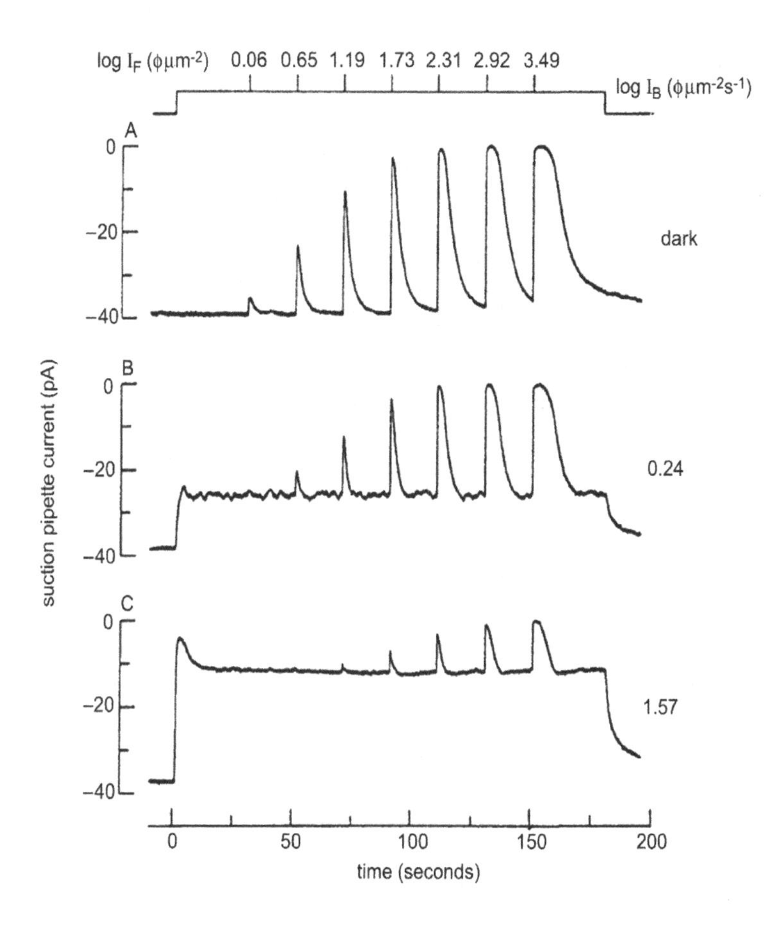

Figure 19.3 Light adaptation in a salamander rod. I_F is the intensity of the light flash and I_B is the intensity of the background, both in units of photons per μm^2. The suction pipette current is the total photocurrent through the rod; models of the photocurrent are discussed in Section 19.2. However, note that a decrease in photocurrent corresponds to a hyperpolarization of the photoreceptor. The timing of the flashes is shown at the top of the figure. Fain et al., (2001), Fig. 1.

initial peak followed by relaxation to a lower steady level. This relaxation to a lower steady level is a consequence of the adaptation of the receptor.

Around each background light level the peak of the flash response is approximately described by the Naka–Rushton equation (Naka and Rushton, 1966; yet another incarnation of the Michaelis–Menten equation described in detail in Chapter 1)

$$\frac{R_{\text{peak}}}{R_{\max}} = \frac{I}{I + \sigma(I_0)}, \tag{19.2}$$

where R_{peak} is the peak of the response, $R_{\max}$ is the maximal response to a saturating flash, I is the intensity of the flash, and I_0 is the background light level. The response is

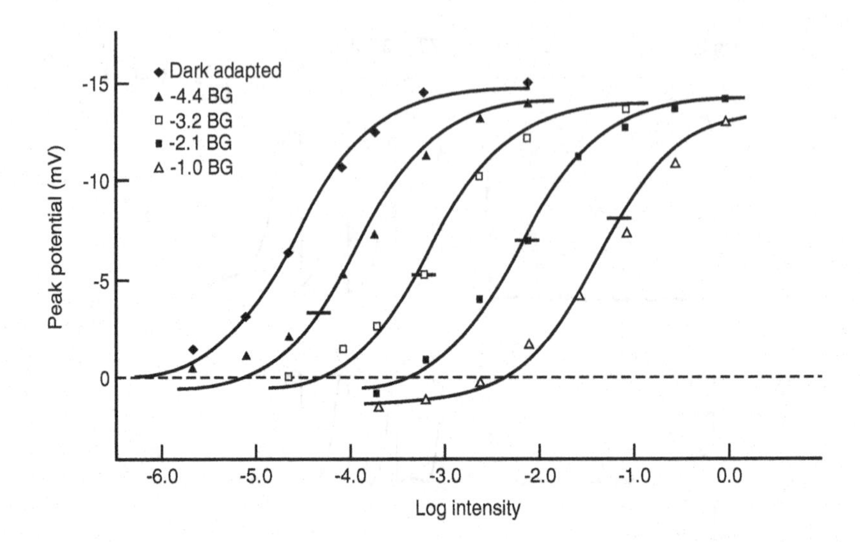

Figure 19.4 Intensity response curves measured in a red-sensitive cone of the turtle. The peak of the response to a light flash (either increasing or decreasing) is plotted against the log of the intensity of flash. Each data set corresponds to a different background light level, and the smooth curves are drawn by using the Naka–Rushton equation, (19.3), as a template, and shifting it across and down for the higher background light levels. The short horizontal lines denote the resting hyperpolarization for each background light level. The membrane potential has been scaled to be zero in the dark (Normann and Perlman, 1979, Fig. 7).

half maximal when $I = \sigma$. It is experimentally observed that as I_0 increases, so does σ. Thus at higher background light levels it takes a larger flash to obtain a half-maximal response, as would be expected from an adapting receptor.

Much of this information can be elegantly summarized by intensity–response curves (Fig. 19.4), a set of curves that repays careful consideration. For a fixed background light level I_0, we consider the response to a family of superimposed light flashes, positive or negative, and plot the peak of the flash response against the intensity of the flash. Note that Fig. 19.4 shows photoreceptor voltage plotted against flash intensity (instead of photocurrent as in Fig. 19.3) and the vertical axis is reversed.

Let $V(I, I_0)$ denote the peak response to a flash of intensity I on a background of I_0. It turns out that the entire set of flash responses can be described by using the Naka–Rushton equation

$$\frac{V(I, I_0)}{V_{\max}} = \frac{I}{I + \sigma(I_0)} \tag{19.3}$$

as a template and moving it across and down slightly to describe the higher background light levels (see Exercise 1). From Fig. 19.4 it can be seen that σ is an increasing function of I_0. Thus, as the background light level increases, the response curve maintains its shape, but shifts to higher light levels and moves down slightly (although this downward shift is not accounted for in (19.3)). Note that since I and σ are positive, the Naka–Rushton equation is always well defined.

In contrast to the peak response, the steady response, depicted by the small horizontal lines in Fig. 19.4, is a much shallower function of the background light level. In this way, retinal neurons can detect contrast over a wide range of light levels without saturating. The steep Naka–Rushton intensity–response curves around each background light level give a high sensitivity to changes superimposed on that background, but the shallower dependence of the steady response on the background light level postpones saturation.

In the experimental literature, adaptation is often discussed in terms of changes in sensitivity, so it is useful to present a brief discussion of this approach here also. Sensitivity (at least in the way we use it here) is defined by

$$S(I_0) = \left.\frac{dV(I, I_0)}{dI}\right|_{I=I_0}. \tag{19.4}$$

Note that the sensitivity, S, is a function of the background light level, which is the whole point of adaptation. If V satisfies the Naka–Rushton equation then

$$S = V_{\max}\frac{\sigma(I_0)}{(I_0 + \sigma(I_0))^2}. \tag{19.5}$$

Defining S_D to be the sensitivity in the dark, it follows that $S_D = V_{\max}/\sigma(0)$ and thus

$$\frac{S(I_0)}{S_D} = \frac{\sigma(I_0)\sigma_D}{[I_0 + \sigma(I_0)]^2}. \tag{19.6}$$

We know that σ is an increasing function of I_0, but until now have not specified what the function is. If we specify σ to be a linear increasing function of I_0, i.e.,

$$\sigma = \sigma_D + kI_0, \tag{19.7}$$

for some constant, k, then

$$\frac{S(I_0)}{S_D} = \frac{(\sigma_D + kI_0)\sigma_D}{[I_0(1+k) + \sigma_D]^2}. \tag{19.8}$$

Notice that as I_0 gets large, so that $kI_0 \gg \sigma_D$, then S/S_D is proportional to $1/I_0$, which is Weber's law. Fortunately, $\sigma(I_0)$ measured experimentally is close to linear for many photoreceptors (Perlman and Normann, 1998).

Many experimental measurements have shown that the relative sensitivity can be well fit by the expression

$$\frac{S}{S_D} = \frac{A}{A + I_0}, \tag{19.9}$$

where A is some constant to be determined by fitting to the data. Because this expression for the sensitivity satisfies all the necessary criteria (it is equal to 1 when $I_0 = 0$, it goes to zero as $I_0 \to \infty$, and it is approximately proportional to $1/I_0$ for a range of I_0) and is very simple, it has been used often in the literature (for example Baylor and Hodgkin, 1974; Baylor et al., 1980; Fain, 1976). Although (19.9) differs from the expression for the sensitivity (19.8) given by the Naka–Rushton equation, it is left as an exercise (Exercise 2) to show that the two expressions for the sensitivity are practically indistinguishable.

19.2 Photoreceptor Physiology

Since we know from the results in Fig. 19.3 that adaptation occurs at the level of the individual photoreceptor, we construct a model of the biochemistry and electrophysiology of a photoreceptor to see how such adaptation can come about. As a preliminary to the construction of such a model, we present a brief discussion of the physiology of the vertebrate photoreceptor. (Because of the significant differences between vertebrate and invertebrate photoreceptors we consider only the former here.) More detailed discussions are given in Fain and Matthews (1990), McNaughton (1990), and Pugh and Lamb (1990). A selection of detailed articles is given in Hargrave et al. (1992) and an excellent review of adaptation in vertebrate photoreceptors is given by Fain et al. (2001).

Vertebrate photoreceptors are composed of two principal segments: an outer segment that contains the photosensitive pigment, and an inner segment that contains the necessary cellular machinery. A connecting process, called an axon, connects the inner segment to a *synaptic pedicle*, which communicates with neurons (such as horizontal and bipolar cells) in the inner layers of the retina. In rods, the photosensitive pigment is located on a stack of membrane-enclosed disks that take up the majority of the space in the outer segment, while in cones, the pigment is located on invaginations of the outer segment membrane. The connecting process does not transmit action potentials, and hence the name "axon" is somewhat misleading. Photoreceptors respond in a graded manner to light, and give an analog, rather than a digital, output.

In the dark, the resting membrane potential is about −40 mV. Current, carried by Na^+ and Ca^{2+} ions, flows into the cell through light-sensitive channels in the outer segment, and is balanced by current, carried mostly by K^+ ions, flowing out through K^+ channels in the inner segment. Thus, in the dark there is a circulating current of about 35–60 pA. In the dark, the light-sensitive channels are held open by the binding of three molecules of cGMP. Ionic balance is maintained by a Na^+–K^+ pump in the inner segment that removes three Na^+ ions for the entry of two K^+ ions, and a Na^+–Ca^{2+}, K^+ exchanger in the outer segment that removes one Ca^{2+} and one K^+ for the entry of four Na^+ ions. The Na^+–Ca^{2+}, K^+ exchanger is the principal method for Ca^{2+} extrusion from the cytoplasm.

The light response begins when a photon of light strikes the photosensitive pigment, initiating a series of reactions (described in more detail below) that results in the activation of rhodopsin, and its subsequent binding to a G-protein, transducin. The bound transducin exchanges a molecule of GDP for GTP and then binds to cGMP-phosphodiesterase (PDE), thereby activating PDE to PDE*. Since the rate of hydrolysis

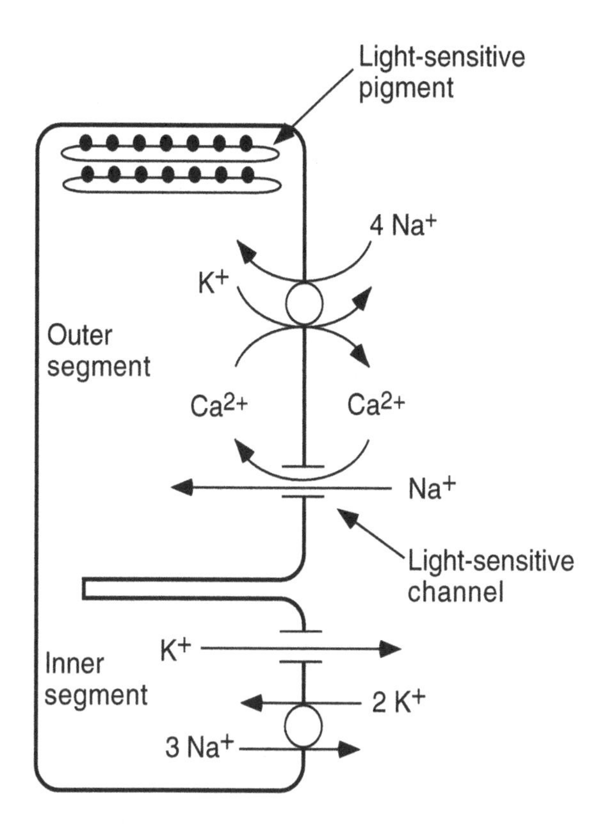

Figure 19.5 Schematic diagram of a photoreceptor, showing the major ionic currents and pumps that regulate phototransduction. (Adapted from McNaughton, 1990.)

of cGMP by PDE* is greater than by PDE, this leads to a decline in [cGMP] and subsequent closure of some of the light-sensitive channels. As the light-sensitive conductance decreases, the membrane potential moves closer to the reversal potential of the inner segment K^+ conductance (about −65 mV), hyperpolarizing the membrane.

Light adaptation is a highly complex process, resulting from a large number of interacting mechanisms, but there is little doubt that the most important player is the cytoplasmic free Ca^{2+} concentration. When the light-sensitive channels close, the entry of Ca^{2+} is restricted, as about 20–30% of the light-sensitive current is carried by Ca^{2+}. However, since the Na^+–Ca^{2+}, K^+ exchanger continues to operate, the intracellular $[Ca^{2+}]$ falls. This decrease in $[Ca^{2+}]$ increases the activity of an enzyme called guanylate cyclase that makes cGMP from GTP. Thus, a decrease in $[Ca^{2+}]$ results in an increase in the rate of production of cGMP, reopening the light-sensitive channels, completing the feedback loop. A schematic diagram of the reactions involved in adaptation is given in Fig. 19.6. Although it is likely that there are other important reactions involved in phototransduction (for example, Ca^{2+} may affect the activity of PDE), the above scheme incorporates many essential features of the light response.

The mechanisms of phototransduction are similar in rods and cones, with one important difference being the light-sensitive pigment contained in the cell. In rods, rhodopsin consists of retinal and a protein called *scotopsin*, while in cones, the

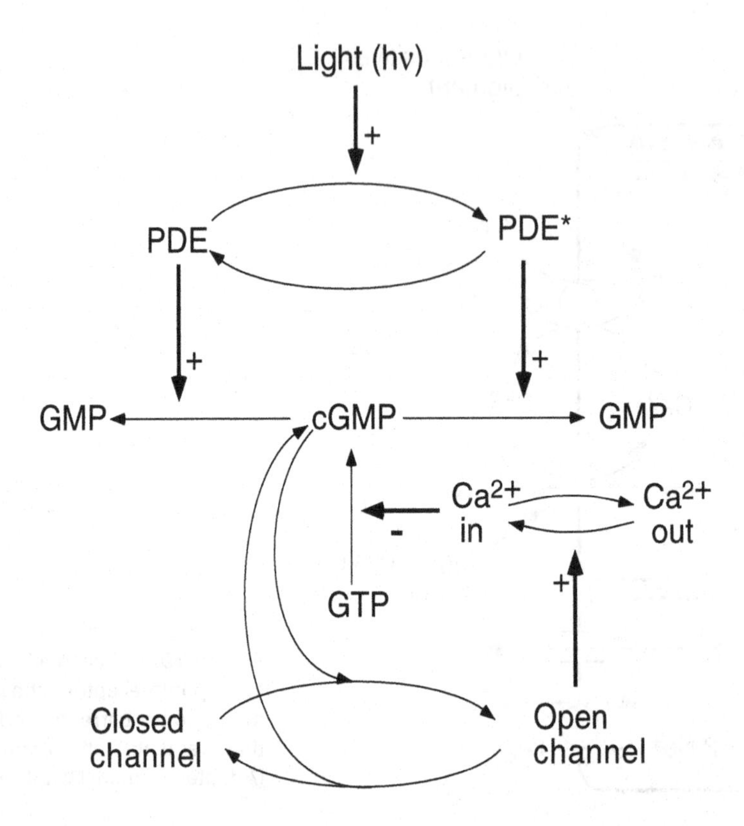

Figure 19.6 Schematic diagram of the major reactions involved in light adaptation. The initial cascade, leading to the activation of PDE, is condensed into a single step here, but is considered in more detail in Fig. 19.7. (Adapted from Sneyd and Tranchina, 1989.)

rhodopsin consists of retinal and different proteins, called *photopsins*. The primary effect of this compositional difference is that in rods, rhodopsin absorbs light in a range of wavelengths centered at 505 nm, while the rhodopsin in cones absorbs light in a range of wavelengths centered at 445 (blue cones), 535 (green cones), and 570 (red cones) nm. Night blindness is caused by insensitivity of rods because of inadequate amounts of rhodopsin, often associated with vitamin A deficiency. Colorblindness, on the other hand, occurs when green or red cones are missing or when blue cones are underrepresented. Colorblindness is a genetically inherited disorder. Another important difference between rods and cones is that the light response of a cone is much faster than that of a rod, due principally to its smaller size.

There are a number of models of phototransduction, some of which (Baylor et al., 1974a,b; Carpenter and Grossberg, 1981) were constructed before the molecular events underlying adaptation were well known. More detailed models include those of Tranchina and his colleagues for turtle cones (Sneyd and Tranchina, 1989; Tranchina et al., 1991), Forti et al. (1989) for newt rods, and Tamura et al. (1991) for primate rods. These models have confirmed that feedback of Ca^{2+} on the activity of guanylate cyclase is indeed sufficient to explain many features of the light response in both rods and cones. Detailed models of the initial cascade and the activation process have been constructed by Cobbs and Pugh (1987) and Lamb and Pugh (1992). The recent models of Hamer (Hamer 2000a,b; Hamer et al., 2005), building upon the earlier model of Nikonov et al. (1998), provide by far the most detailed quantitative fit of a model to experimental data, while a detailed stochastic view of the initial reactions in the cascade is given by Hamer et al. (2003).

19.2.1 The Initial Cascade

Although the main consequence of the absorption of light by a photoreceptor is the conversion of PDE to a more active form, and a resultant decline in the concentration of cGMP, there are many biochemical steps between these events (Fig. 19.7). Absorption of a photon causes the isomerization of 11-cis retinal to the all-trans form, and this in turn causes a series of isomerizations of rhodopsin, ending with metarhodopsin II.

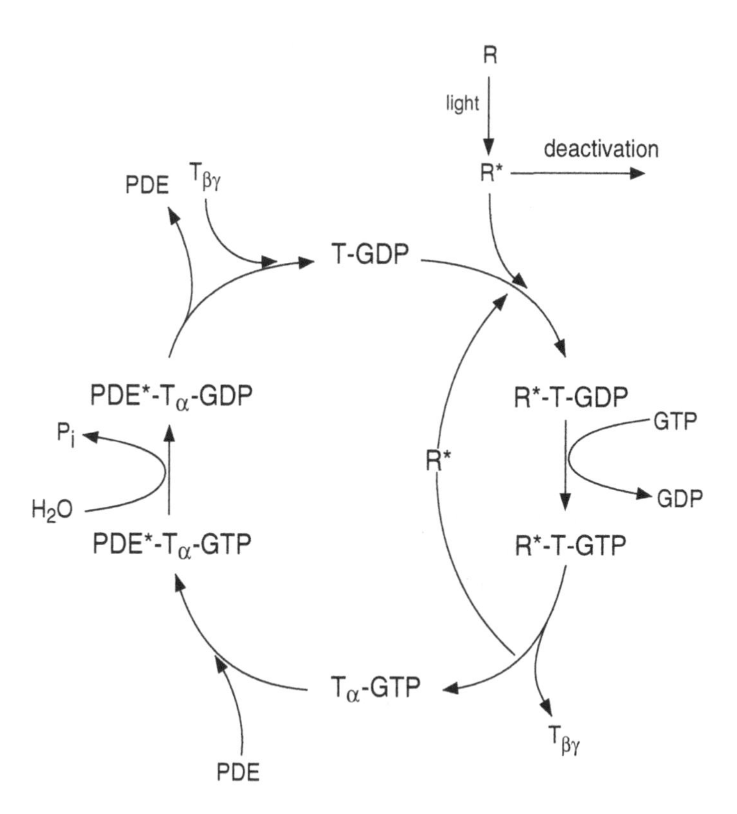

Figure 19.7 Schematic diagram of the biochemical reactions involved in the initial steps of phototransduction. (Adapted from Stryer, 1986, Fig. 6.)

Metarhodopsin II is converted to metarhodopsin III, which is, in turn, hydrolyzed to opsin and all-trans retinal. The activated form of rhodopsin is metarhodopsin II, called R* here. In the dark, the G-protein transducin is in its deactivated form, T-GDP. After absorption of a photon, R* binds to T-GDP and catalyzes the exchange of GDP for GTP. This exchange reduces the affinity of R* for transducin, and also causes transducin to split into an α subunit, T_α-GTP, and a $\beta\gamma$ subunit, $T_{\beta\gamma}$. It is T_α-GTP that then binds to PDE, forming the complex PDE*-T_α-GTP, which is the activated form of the PDE. The cycle is completed when the GTP is dephosphorylated to GDP, the PDE leaves the complex, and the T_α subunit recombines with the $T_{\beta\gamma}$ subunit forming the deactivated form of transducin again. All of these reactions occur in the membrane that contains rhodopsin, and are thus influenced by the speed at which the various proteins can diffuse within the membrane, an aspect that we do not consider here (but see Lamb and Pugh, 1992).

Although it is possible to model this sequence of reactions in detail (Exercise 4), we do not do so here, as it appears that nonlinearities in the initial stages of the light response have little effect on adaptation at low light levels. This allows the considerable simplification of modeling the initial stages of the light response as a simple multistage linear system.

We model the initial cascade as a sequence of linear reactions,

$$\frac{dr}{dt} = l_1 I(t) - l_2 r, \tag{19.10}$$

$$\frac{dg_1}{dt} = l_3 r - l_4 g_1, \tag{19.11}$$

$$\frac{dg_2}{dt} = l_5 g_1 - l_6 g_2, \tag{19.12}$$

$$\frac{dg}{dt} = l_7 g_2 - l_8 g, \tag{19.13}$$

where $l_1, \ldots, l_8$ are rate constants, r is the concentration of R*, g is the concentration of transducin, and g_1 and g_2 are hypothetical intermediate states between the formation of activated rhodopsin and the activation of transducin. These intermediate states need not occur in this specific location, but could be included anywhere preceding the activation of PDE. Two intermediate states are used because there is evidence that to get acceptable agreement with data, at least four stages are needed before the activation of PDE (Cobbs and Pugh, 1987; Hamer and Tyler, 1995). The transfer function (see Appendix 19.8) of this linear system is

$$H(\omega) = \frac{\eta}{(1 + i\omega\tau_1)^4}, \tag{19.14}$$

where $\eta = l_1 l_3 l_5 l_7 \tau_1^4$, and where we have assumed that $l_2 = l_4 = l_6 = l_8 = 1/\tau_1$. The impulse response $K(t)$ of the system is given by

$$K(t) = \frac{\eta}{\tau_1 3!}\left(\frac{t}{\tau_1}\right)^3 e^{-t/\tau_1}. \tag{19.15}$$

Finally, we let p denote the concentration of PDE*, and let P_0 denote the total concentration of PDE, to get

$$\frac{dp}{dt} = s(t)(P_0 - p) - k_1 p, \tag{19.16}$$

where

$$s(t) = \int_{-\infty}^{t} I(\tau)K(t-\tau)\,d\tau. \tag{19.17}$$

If $P_0/p \ll 1$ then $P_0 - p \approx P_0$, in which case the deactivation of PDE* is linear (Hodgkin and Nunn, 1988).

19.2.2 Light Adaptation in Cones

We can now incorporate the model of the initial cascade into a complete model of excitation and adaptation by including equations for the concentrations of cGMP, Ca^{2+}, and Na^+, as well as the membrane potential. First, we scale p by P_0, and then let $x = [\text{cGMP}]/[\text{cGMP}]_{\text{dark}}$, $y = [Ca^{2+}]/[Ca^{2+}]_{\text{dark}}$, and $z = [Na^+]/[Na^+]_{\text{dark}}$, so that $x = y = z = 1$ in the dark. We also shift the membrane potential V so that $V = 0$ in the dark.

cGMP is produced at some rate dependent upon Ca^{2+} concentration, given by $g(y)$, an unknown function to be determined. cGMP is hydrolyzed by both the active (at rate proportional to xp) and the inactive (at rate proportional to $x(1-p)$) forms of PDE, although the rate of hydrolysis by PDE* is faster than by PDE. Thus,

$$\frac{dx}{dt} = g(y) - \gamma xp - \delta x(1-p) = g(y) - (\gamma - \delta)xp - \delta x. \tag{19.18}$$

Note that the units of $g(y)$, γ, and δ are s^{-1}. (The assumptions behind (19.18) are not as simple as they may seem; see Exercise 5.)

The light-sensitive channel is held open by three cGMP molecules, and, in the physiological regime, has a current–voltage relation proportional to e^{-V/V^*}, for some constant V^*. In general, one expects the number of open light-sensitive channels to be a sigmoidal function of x, as in (1.70), with a Hill coefficient of 3. However, since x is very small in the dark (and becomes even smaller in the presence of light), few light-sensitive channels ever open, and thus the light-sensitive current J_{ls} is well represented by

$$J_{\text{ls}} = Jx^3 e^{-V/V^*}, \tag{19.19}$$

for some constant J, with units of current.

Calcium enters the cell via the light-sensitive current, of which approximately 15% is carried by Ca^{2+}, and is pumped out by the Na^+–Ca^{2+},K^+ exchanger. Assuming that the Na^+–Ca^{2+},K^+ exchanger removes Ca^{2+} with first-order kinetics, the balance equation for Ca^{2+} is

$$\beta\frac{dy}{dt} = \frac{\kappa}{2Fv y_d} Jx^3 e^{-V/V^*} - k_2 y, \tag{19.20}$$

where ν is the cell volume, κ is the fraction of the light-sensitive current carried by Ca^{2+}, F is Faraday's constant, k_2 is the rate of the exchanger, and y_d denotes $[Ca^{2+}]_{dark}$. To incorporate Ca^{2+} buffering, we assume that the ratio of bound to free Ca^{2+} is β, and that the buffering is fast and linear. This means that the rate of change of $[Ca^{2+}]$ must be scaled by β (a detailed discussion of Ca^{2+} buffering is given in Section 7.4). Typically, β is approximately 99.

Similarly, the balance equation for Na^+ is derived by assuming that the exchanger brings 4 Na^+ ions in for each Ca^{2+} ion it pumps out, and that the rate of the Na^+–K^+ pump is a linear function of Na^+. Further, most of the light-sensitive current not carried by Ca^{2+} is carried by Na^+. Thus,

$$\frac{dz}{dt} = \frac{1-\kappa}{F\nu z_d} J x^3 e^{-V/V^*} + \frac{4k_2 y_d}{z_d} y - k_3 z, \tag{19.21}$$

where z_d denotes $[Na^+]_{dark}$.

Some parameter relationships can be determined and the equations for y and z simplified by using the fact that $x = y = 1$, $V = 0$ must be a steady state. From this it follows that

$$k_2 = \frac{J\kappa}{2F\nu y_d} \tag{19.22}$$

and

$$k_3 = \frac{(1-\kappa)J}{F\nu z_d} + \frac{4k_2 y_d}{z_d} = \frac{J(1+\kappa)}{F\nu z_d}, \tag{19.23}$$

so that

$$\tau_y \frac{dy}{dt} = x^3 e^{-V/V^*} - y, \tag{19.24}$$

$$\tau_z \frac{dz}{dt} = \left(\frac{1-\kappa}{1+\kappa}\right) x^3 e^{-V/V^*} + \left(\frac{2\kappa}{1+\kappa}\right) y - z, \tag{19.25}$$

where $\tau_z = \frac{F\nu z_d}{J(1+\kappa)}$ and $1/\tau_y = \beta k_2$.

Finally, we derive an equation for the membrane potential. Since the exchangers and pumps transfer net charge across the cell membrane, there are four sources of transmembrane current: the light-sensitive current, the Na^+–K^+ pump current, the Na^+–Ca^{2+}, K^+ exchange current, and the light-insensitive K^+ current, which is modeled as an ohmic conductance. Also note that for every 1 Ca^{2+} ion pumped out of the cell, one positive charge enters, and for every 3 Na^+ ions pumped out, one positive charge leaves. Thus

$$C_m \frac{dV}{dt} = J x^3 e^{-V/V^*} - \frac{F k_3 z_d \nu}{3} z - G(V - E) + (F k_2 y_d \nu) y, \tag{19.26}$$

where G and E are, respectively the conductance and reversal potential of the light-insensitive K^+ channel, and C_m is the capacitance of the cell membrane. Recall that the potential V is measured relative to the potential in the dark. Using (19.22) and

(19.23), the voltage equation becomes

$$C_m \frac{dV}{dt} = Jx^3 e^{-V/V^*} - \frac{J(1+\kappa)}{3} z - G(V-E) + \frac{J\kappa}{2} y, \tag{19.27}$$

and then using that $V = 0$, $y = z = 1$ must be a steady state, we get

$$J = \frac{-6GE}{4+\kappa}. \tag{19.28}$$

Substituting this expression back into the voltage equation gives

$$\tau_m \frac{dV}{dt} = -\left(\frac{6E}{4+\kappa}\right) x^3 e^{-V/V^*} + 2\left(\frac{1+\kappa}{4+\kappa}\right) Ez - (V-E) - \left(\frac{3E\kappa}{4+\kappa}\right) y, \tag{19.29}$$

where $\tau_m = C_m/G$ is the membrane time constant.

In summary, the model equations are

$$\frac{dx}{dt} = g(y) - (\gamma - \delta)xp - \delta x, \tag{19.30}$$

$$\tau_y \frac{dy}{dt} = x^3 e^{-V/V^*} - y, \tag{19.31}$$

$$\tau_z \frac{dz}{dt} = \left(\frac{1-\kappa}{1+\kappa}\right) x^3 e^{-V/V^*} + \left(\frac{2\kappa}{1+\kappa}\right) y - z, \tag{19.32}$$

$$\tau_m \frac{dV}{dt} = -\left(\frac{6E}{4+\kappa}\right) x^3 e^{-V/V^*} + 2\left(\frac{1+\kappa}{4+\kappa}\right) Ez - (V-E) - \left(\frac{3E\kappa}{4+\kappa}\right) y. \tag{19.33}$$

Determination of the Unknowns

The unknown function $g(y)$ (the Ca^{2+}-dependent rate of cGMP production) is determined by requiring the steady-state membrane potential to be the logarithmic function

$$V_0 = -s_1 \log(1 + s_2 I_0), \tag{19.34}$$

for some constants s_1, s_2. The form of this steady-state relation gives very good agreement with experimental data from turtle cones (although it does not give exact Weber's law behavior). This results in a long and complicated expression for $g(y)$ that we do not give here, as its analytic form has no physiological significance (see Exercise 3). Its shape, however, is of interest, and that can be determined only after the parameters are determined by fitting to experimental data.

Some of the parameters are known from experiment. For example, κ, the proportion of the light-sensitive current carried by Ca^{2+}, is known to be 0.1–0.15, while τ_z, the time constant for Na^+ extrusion, is known to be around 0.04 s. Similarly, from measurements of the current/voltage relation of the light-sensitive channel, V^* is known to be 35.7 mV. The remaining unknown parameters (s_1, s_2, E, τ_y, k_4, γ, δ, η, τ_1, and τ_m) are determined by fitting the first-order transfer function of the model to experimental data (typical experimental data are shown in Fig. 19.22). The results of this parameter estimation are given in Table 19.1.

Table 19.1 Parameter values for the model of light adaptation in turtle cones. (Tranchina et al., 1991.)

s_1 = 1.59 mV	s_2 = 1130
E = −13 mV	V^* = 35.7 mV
τ_y = 0.07 s	k_1 = 35.4 s^{-1}
γ = 303 s^{-1}	δ = 5 s^{-1}
κ = 0.1	η = 52.5 s^{-1}
τ_1 = 0.012 s	τ_m = 0.016 s
τ_z = 0.04 s	

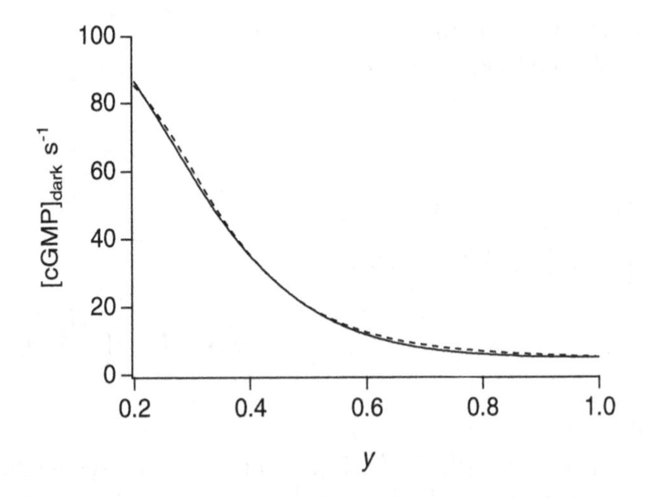

Figure 19.8 Predicted and measured activities of guanylate cyclase as functions of the Ca^{2+} concentration. The solid line denotes $g(y)$ (theoretical prediction), and the dotted line denotes $A(y)$ (experimental measurement). For convenience, the activity of guanylate cyclase is expressed in units of $[cGMP]_{dark}$ per second.

Model Predictions and Behavior

The most interesting prediction of the model is the shape of the feedback function $g(y)$ that mediates light adaptation. A plot of g is given in Fig. 19.8. In the physiological regime, $g(y)$ is well approximated by the function $A(y)$, where

$$A(y) = 4 + \frac{91}{1 + (y/0.34)^4}. \tag{19.35}$$

In other words, as $[Ca^{2+}]$ falls, the rate of cGMP production by guanylate cyclase rises along a sigmoidal curve, with a Hill coefficient of 4. This prediction of the model has been confirmed experimentally (Koch and Stryer, 1988), thus lending quantitative support to the hypothesis that the modulation of guanylate cyclase activity by $[Ca^{2+}]$ is sufficient to account for light adaptation in turtle cones. Although experimental data

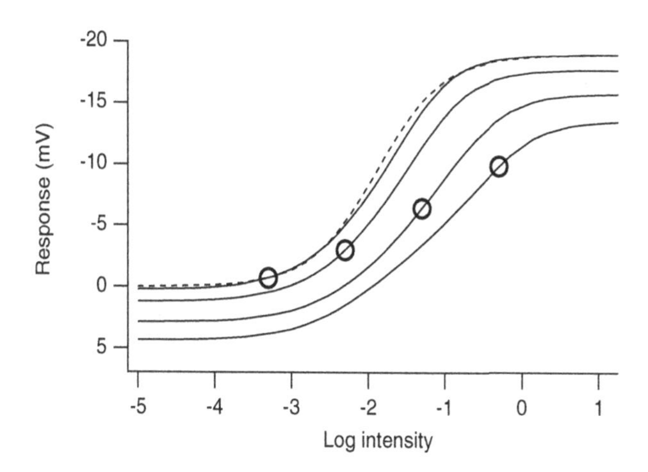

Figure 19.9 Intensity–response curves from the model of adaptation in turtle cones. The dotted line is the Naka–Rushton equation, and the solid lines are the model results. The open symbols denote the steady states for four different light levels.

are not entirely consistent, and there is thus still considerable debate about the correct Hill coefficient, more modern studies (Hamer, 2000b) have concluded that the Hill coefficient is possibly closer to 3 or even 2.

The model exhibits constant contrast sensitivity over a range of lower light levels. As I_0 increases, the contrast sensitivity first increases and then decreases slightly, in agreement with the results of Daly and Normann (1985); the impulse response becomes biphasic; and the time-to-peak decreases as the response speeds up. Further, the intensity–response curves agree well with the Naka–Rushton equation (19.3) and shift to the right and slightly down as I_0 increases (Fig. 19.9), again in good agreement with experimental data. Some of these properties of the model are explored further in Exercise 3.

An unexpected prediction of the model is that [cGMP] does not fall much as the background light level is increased. For example, if the background light level is changed so that the sensitivity decreases by a factor of 1000, [cGMP] decreases by less than a factor of two (see Sneyd and Tranchina, 1989, Fig. 4). This gives a possible explanation for the puzzling observation that even though a decrease in [cGMP] is believed to underlie light adaptation, such decreases are sometimes not experimentally observed (DeVries et al., 1979; Dawis et al., 1988). In other words, the model predicts that even though a decrease in [cGMP] may indeed mediate light adaptation, the actual decrease may be too small to measure reliably.

The model agrees quantitatively with experiment in a number of other ways (discussed in detail by Tranchina et al., 1991), lending further support to the hypothesis that it provides an excellent description of many features of light adaptation. Similar conclusions have been reached by Forti et al. (1989), who modeled phototransduction in newt rods, and Tamura et al. (1991), who modeled adaptation in primate rods. It thus appears that although Ca^{2+} feedback on the activity of guanylate cyclase cannot explain all features of light adaptation in rods and cones (Fain et al., 2001), it is one of the principal mechanisms.

19.3 A Model of Adaptation in Amphibian Rods

The most elegant models of phototransduction in amphibian and vertebrate rods are those of Hamer, Tranchina, and their colleagues. Hamer (2000a,b) fitted a modified version of the earlier model of Nikonov et al. (1998) to a set of flash responses from the larval tiger salamander rod. Hamer also attempted to reproduce in a qualitative way a wider array of experimental data, including step responses from newt rods and responses to highly saturating flashes. This model was then extended by Hamer et al. (2003, 2005) to consider the single-photon response.

Hamer's model differs from the above model of adaptation in cones in two principal ways. First, it includes a more detailed description of Ca^{2+} buffering, and, second, it models only the photocurrent, not the voltage. This latter feature makes the model simpler in some respects.

The initial cascade is modeled by two reactions only,

$$\frac{dR^*}{dt} = I(t) - \frac{1}{\tau_R}R^*, \tag{19.36}$$

$$\frac{dE^*}{dt} = \nu R^* - \frac{1}{\tau_E}E^*, \tag{19.37}$$

where R^* and E^* denote, respectively, the number of activated rhodopsin and PDE molecules, and as before, $I(t)$ is the light input, and ν, τ_R, and τ_E are constants.

Calcium (c) and cGMP (g) concentrations are the only other model variables. The rate of production of cGMP is dependent on c, and is given by an expression similar to (19.35). Also, cGMP is degraded by E^* as well as at some background rate. Thus,

$$\frac{dg}{dt} = \frac{A_{\max}}{1+(\frac{c}{K_{\mathrm{Ca}}})^{n_c}} - (\beta_{\mathrm{dark}} + \beta_E E^*)g. \tag{19.38}$$

The Ca^{2+} equations are similar to those discussed in Chapter 7. Calcium enters the cell via a light-sensitive current, is pumped out by membrane ATPases, and is buffered. Thus,

$$\frac{dc}{dt} = \alpha f J_{\mathrm{dark}} - \gamma(c - c_{\min}) - k_{\mathrm{on}}(b_t - b)c + k_{\mathrm{off}}b, \tag{19.39}$$

$$\frac{db}{dt} = k_{\mathrm{on}}(b_t - b)c - k_{\mathrm{off}}b. \tag{19.40}$$

Here, J_{dark} is the light-sensitive current in the dark, f is the fraction of open light-sensitive channels, and α is a constant that converts Ca^{2+} current to Ca^{2+} concentration flux. The model of Ca^{2+} pumping is highly simplified, since it is assumed that the pumps remove Ca^{2+} at a linear rate that is zero at $c_{\min}$, while the buffering model is the same

Table 19.2 Parameters of the model of light adaptation in amphibian rods (Hamer, 2000a).

τ_R	= 0.42 s	τ_E	= 1.2 s
ν	= 1735 s^{-1}	b_t	= 395 μM
γ	= 100 s^{-1}	k_{on}	= 0.17 $\mu M^{-1} s^{-1}$
k_{off}	= 2.35 s^{-1}	c_{min}	= 0.005 μM
K_c	= 0.219 μM	n_c	= 2.85
A_{max}	= 4.461 $\mu M\, s^{-1}$	β_E	= 1.68×10^{-5} s^{-1}
β_{dark}	= 0.136 s^{-1}	n_g	= 2.21
α	= 0.78 $\mu M\, s^{-1}\, pA^{-1}$	J_{dark}	= 72.3 pA

as that discussed in Section 7.4. Hence, b_t is the total concentration of buffer, and b is the concentration of buffer bound to Ca^{2+}.

It remains to specify the form of f, the fraction of open light-sensitive channels. As in the model of adaptation in cones, we assume that f is a power of g, and thus

$$f = \left(\frac{g}{g_{dark}}\right)^{n_g}, \tag{19.41}$$

where g_{dark} is the concentration of cGMP in the dark.

Panel A in Fig. 19.10 shows the data and the model fit. The model was fit to a set of flash responses from the larval tiger salamander, and excellent quantitative agreement was obtained. However, Hamer then proceeded to make a series of more ambitious comparisons wherein the model response was compared to a series of step responses (panel B) and a series of highly saturating flash responses (panel C). Step responses of the model (panel B) capture the salient features of amphibian rod step responses over a range of step intensities spanning 4.7 log units (Forti et al., 1989), including the initial nose on the responses, and the recovery to a steady-state plateau. This feature is due to active Ca^{2+}-mediated gain control, and is observed in virtually all rod and cone step responses.

Similarly, the model captures features of highly saturated amphibian rod flash responses over a 6.5 log unit range (panel C), including the empirically observed intensity-dependence of saturation period (Pepperberg et al., 1992).

Both the model step responses and saturated flash responses have a multiphase recovery that comprises an initial faster phase, and then a slower, late phase that only appears for intense steps and for flashes sufficiently intense to elicit very long saturation periods (Pepperberg et al., 1992). The slow components in the model recovery are, however, still too fast (and less oscillatory) compared to the physiology. The transition to the slow phase is a common feature of high-intensity flash and step rod responses in both cold-blooded and mammalian species (Forti et al., 1989; Torre et al., 1990; Pepperberg et al., 1992; Nakatani et al., 1991; Tamura et al., 1991). Forti et al. (1989) and Hamer et al. (2005) showed that the slow phase of the recovery could be accounted for by including in a model a slow back-reaction from inactivated to activated rhodopsin, as suggested earlier by Lamb (1981).

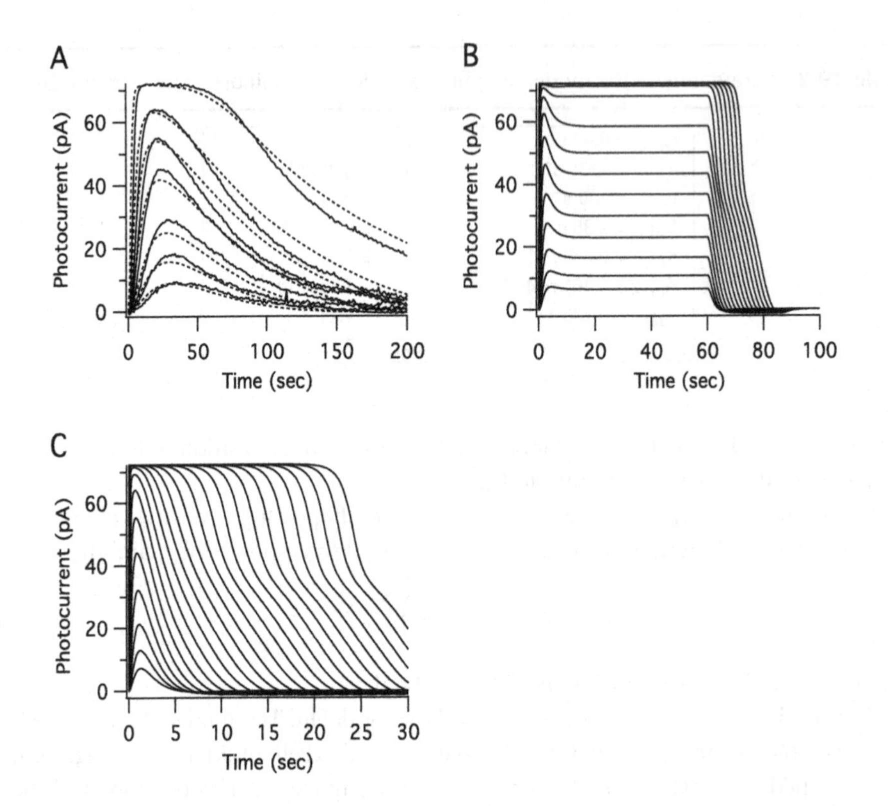

Figure 19.10 A: Experimental data and model fits from Hamer (2000a). Flash responses from larval tiger salamander rods (solid lines), and model fits (dotted lines). Responses were obtained under voltage-clamp, in the perforated patch mode, and provided to Hamer by J.I. Korenbrot. The intensities applied to the rod were 13.33, 27.35, 53.79, 147.6, 309.8, 619.7, and 3541 R*/20 ms flash. B: Model step responses, simulating a family of step responses from Newt rods as presented in Fig. 6A of Forti et al. (1989). The range of step intensities used was 5 to 220,000 R*/sec. C: Highly saturated flash responses generated by the model, simulating the salient features of salamander rod responses shown in Pepperberg et al. (1992). The range of flash strengths used was 8 to $> 2.9 \times 10^7$ R*/flash. We thank R.D. Hamer for providing the original data for this figure.

These disagreements between model and experiment in the bright flash and bright step regime are almost certainly the result of mechanisms that have been omitted from such simple models. For example, nonlinearities in the initial stages of the light response are likely to be important for bright flashes, a feature that is lacking in both models discussed above (but see Exercise 4). In addition, Ca^{2+}-dependent modulation of cGMP production is not the only mechanism for adaptation (Fain et al., 2001). For example, Ca^{2+} is also known to modulate the light-dependent channels directly, with a decrease in $[Ca^{2+}]$ causing an increase in the affinity of the channels for cGMP. It has also been proposed that Ca^{2+} modulates early steps in the phototransduction cascade, possibly by altering the gain of activation of PDE by rhodopsin in possible combination with modulation of the PDE activity. In a theoretical study, Koutalos et al. (1995) estimated that, in dim light, about 80% of adaptation in salamander rods is mediated by the action of Ca^{2+} on guanylate cyclase, while at bright light (6 log units brighter) about 60% is mediated by the action of Ca^{2+} on the activity of PDE. However, whatever the exact details, according to Fain et al. (2001) there is no compelling evidence for adaptation being mediated by any messenger other than Ca^{2+}.

19.3.1 Single-Photon Responses

One of the most striking features of phototransduction is that not only can a rod respond to a single photon, it does so in a reproducible way that varies little from photon to photon. At first glance this might seem uninteresting, until one realizes that a single photon response is mediated by a single rhodopsin molecule in a random location on any one of 1000–2000 disk surfaces. Since the lifetime of a single molecule must, necessarily, be stochastic in nature, how can this produce a reproducible single-photon response? Baylor et al. (1979) were the first to study this question, which has long been considered one of the most puzzling and important problems in photoreceptor physiology.

To consider the problem in slightly more detail, suppose that the decay of an activated rhodopsin molecule is a Poisson process, with rate k_0. We are interested in the probability distribution for how long the rhodopsin molecule stays in the activated state, since this determines how many PDE molecules (on average) are activated.

Following the methods described in Section 2.9.1, we know that if the rhodopsin molecule is activated at time $t = 0$, the probability that it remains activated at time t is given by P_{R^*}, which is the solution of the differential equation

$$\frac{dP_{R^*}}{dt} = -k_0 P_{R^*}, \tag{19.42}$$

where $P_{R^*}(0) = 1$. Thus, $P_{R^*}(t) = e^{-k_0 t}$. It follows that the probability distribution function for the time that the rhodopsin molecule stays activated is $k_0 P_{R^*}(t) = k_0 e^{-k_0 t}$. This *activated time distribution* has mean μ, and standard deviation σ, both equal to $1/k_0$, and thus

$$\frac{\sigma}{\mu} = 1. \tag{19.43}$$

This ratio, called the coefficient of variation, is a nondimensional measure of the spread of the distribution.

Thus, a single inactivation step results in an activation lifetime that is highly variable, much more variable than is seen in experimental data. However, in reality, rhodopsin is inactivated in a series of phosphorylation steps, which results in a greatly decreased variation (Baylor et al., 1979; Rieke and Baylor, 1998a,b; Whitlock and Lamb, 1999; Field and Rieke, 2002; Hamer et al., 2003).

To see this, suppose that upon light activation, rhodopsin starts in state R_0, where R_i denotes rhodopsin that has been phosphorylated i times. Suppose further that each state R_i activates PDE at the same rate. This second assumption is not necessary, but simplifies the analysis considerably (for a detailed analysis of a model that does not adopt this assumption, see Hamer et al., 2003). Finally, assume that there is a maximum of n phosphorylation sites and the rhodopsin is inactivated when all the sites are phosphorylated.

The corresponding Markov model is

$$R_0 \xrightarrow{k_0} R_1 \xrightarrow{k_1} R_2 \xrightarrow{k_2} \cdots \xrightarrow{k_{n-2}} R_{n-1} \xrightarrow{k_{n-1}} \text{inactive rhodopsin.}$$

The total time that rhodopsin is active is the time it takes to get from state R_0 to the inactive state R_n. From Section 2.9.1, we know that the probability distribution function for the time of inactivation is $\frac{dP_n}{dt} = k_{n-1}P_{n-1}$, where $P_k(t)$ is the probability of being in state k at time t, and is found as the solution of

$$\frac{dP_0}{dt} = -k_0 P_0, \tag{19.44}$$

$$\frac{dP_1}{dt} = k_0 P_0 - k_1 P_1, \tag{19.45}$$

$$\vdots \tag{19.46}$$

$$\frac{dP_{n-1}}{dt} = k_{n-2}P_{n-2} - k_{n-1}P_{n-1}, \tag{19.47}$$

with $P_0(0) = 1$, and $P_k(0) = 0$ for $k > 0$. In other words, P_{n-1} is the impulse response of an $(n-1)$-stage linear filter (Appendix 19.8). For any n this system is easily solved using Fourier transforms (or any other method suitable for linear systems), and the mean and standard deviation calculated. For example, when $n = 3$, the distribution of the activated time has coefficient of variation

$$\frac{\sigma}{\mu} = \frac{\sqrt{k_2^2k_1^2 + k_0^2k_1^2 + k_0^2k_2^2}}{k_2k_1 + k_1k_0 + k_2k_0}, \tag{19.48}$$

which is less than one, since all the rate constants are positive. In particular, if the reaction sites are independent so that $k_0 = 3k_2$, $k_1 = 2k_2$, then $\frac{\sigma}{\mu} = \frac{7}{11}$.

In the special case that all the k_i's are equal, inactivation is by an n-step Poisson process, and the coefficient of variation is given by the simple formula

$$\frac{\sigma}{\mu} = \frac{1}{\sqrt{n}}. \tag{19.49}$$

Clearly, as the number of phosphorylation steps increases, the coefficient of variation decreases. This has also been derived in Hamer et al. (2003), and verified empirically in transgenically modified mouse rods with varying numbers of functional phosphorylation sites on rhodopsin (Doan et al., 2006).

A Unified Model

Although multiple phosphorylation steps provide a plausible answer to how a single photon can generate a reproducible response, to understand how phototransduction works at all light levels, it is necessary to construct a single model that can reproduce not only the single-photon response, but also the responses to flashes and steps of light, both bright and dim. This constitutes a severe set of tests against which to validate any model of phototransduction, and it was not until 2005 that such a model appeared (Hamer et al., 2005). Hamer et al. elaborated their 2003 model and were able to capture many qualitative and quantitative features of vertebrate rod responses from several species. Using a single set of parameters (except for sensitivity adjustments to account for differences between species and experimental preparations), Hamer et al. were able to account for (among other things): (i) single-photon responses and reproducibility statistics, (ii) dark-adapted flash responses over a wide range of flash intensities; (iii) step responses over a 4.7 log unit range, and (iv) steady-state light-adapted responses under normal conditions and when Ca^{2+} feedback was prevented.

19.4 Lateral Inhibition

Thus far we have considered only the responses of individual photoreceptors. However, spatial interactions in the retina also play an important role in regulation of the light response. Some of the earliest studies of this were done in the retina of the horseshoe crab *Limulus polyphemus*. As a side issue, it is interesting to note that the photoreceptors of the horseshoe crab (and other invertebrates) operate along quite different lines than those of vertebrates (Dorlöchter and Stieve, 1997). In fact, the mechanisms underlying the photoresponse in invertebrates have a great deal in common with the mechanisms underlying Ca^{2+} dynamics in other cell types, as discussed in detail in Chapter 7. However, despite these intriguing similarities we do not discuss invertebrate phototransduction in any further detail.

In 1967 Haldan Hartline won (one third of) the Nobel Prize in Physiology or Medicine for his work on the retina of the horseshoe crab. Together with his colleague, Floyd Ratliff, he published in the 1950s a series of papers that showed how lateral inhibition played an important role in the horseshoe crab retina (Hartline et al., 1956; Hartline and Ratliff, 1957, 1958; Ratliff and Hartline, 1959).

The *Limulus* eye is a compound eye containing about 1000 ommatidia. Nerve fibers from each ommatidium come together to form the optic nerve, while lateral interconnections arise from a layer of nerve fibers directly behind the ommatidia. Two classic experiments are shown in Fig. 19.11. In panel A are shown the results of applying a light step (or ramp in the lower panel) to a single receptor. When lateral inhibition is removed (by the simple expedient of shining light only ever on a single receptor) then the receptor responds in a steplike fashion (open triangles). However, if the light stimulus is applied to the entire eye, thus allowing lateral inhibition, the receptor responds by accentuating the edges of the step or ramp (open circles). The temporal development of inhibition is shown in panel B. In a manner reminiscent of the responses of vertebrate photoreceptors (as is discussed above), receptors respond to a light stimulus with a transient peak followed by adaptation to a lower steady level. When the

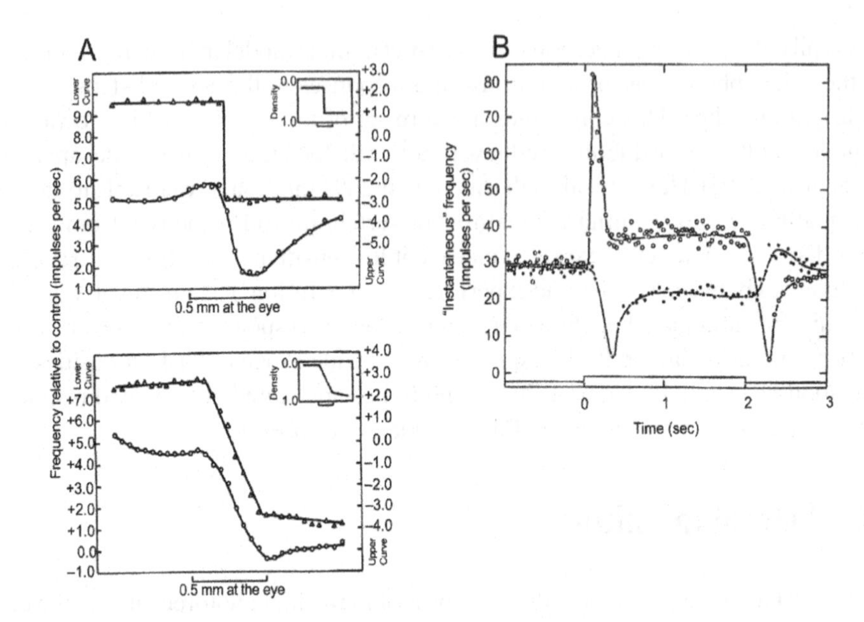

Figure 19.11 Lateral inhibition in the *Limulus* eye. A: the response to a step or a ramp. The open triangles show the response of a single receptor as a light step (or ramp) is moved across it, in conditions that eliminate lateral inhibition (i.e., only a single receptor was illuminated). The open circles correspond to the same experiment with lateral inhibition (i.e., when the step or ramp was applied to the entire eye and moved across the receptor). Reproduced from Ratliff and Hartline (1959), Figs. 4 and 5. B: simultaneous excitatory and inhibitory responses in neighboring receptors. One receptor (filled circles) was illuminated uniformly, while the other (open circles) was stimulated with light as indicated in the trace along the bottom of the graph. The stimulated receptor responds with a transient peak followed by adaptation and a return to a lower plateau (as in the vertebrate photoreceptors discussed above). The neighboring receptor, which is not directly stimulated by the light, is inhibited by the response of the first receptor. Reproduced from Ratliff (1961), Fig. 10.

light is removed the behavior is reversed, with a transient dip followed by recovery. A neighboring nonilluminated receptor (filled circles) responds in the opposite way as it is inhibited by the stimulated receptor.

To explain their results, Hartline and Ratliff formulated the Hartline–Ratliff equations which describe the steady response of a receptor subject to lateral inhibition. If r_p is the response of the pth receptor and e_p is the excitation applied to the pth receptor, then

$$r_p = e_p - \sum_{j=1}^{n} K_{p,j}(r_j - r_{p,j}^0), \tag{19.50}$$

where $K_{p,j}$ is the inhibitory coupling coefficient from the jth receptor to the pth receptor, and $r_{p,j}^0$ is the threshold below which the response of receptor j cannot inhibit receptor

p. $K_{p,j}$ decreases with distance from cell p, while r^0 increases with distance. Thus the decline of inhibition with distance is built into the model.

Although there have been a number of detailed studies of the Hartline–Ratliff equations and the *Limulus* eye (Knight et al., 1970; Hartline and Knight, 1974; Ratliff et al., 1974; Brodie et al., 1978a,b; Grzywacz et al., 1992), we do not discuss them here. Instead we discuss a simpler version of the model due to Peskin (1976).

19.4.1 A Simple Model of Lateral Inhibition

We suppose that E is the excitation of a receptor by light and that I is the inhibition of the receptor from its neighbors. The photoreceptor response is $R = E - I$. A light stimulus L causes an excitation E in the receptor, and E decays with time constant τ. The response of the receptor R provides an input into a layer of inhibitory cells, which are laterally connected, and so the inhibition spreads laterally by diffusion and decays with time constant 1. The model equations are

$$\tau \frac{\partial E}{\partial t} = L - E, \tag{19.51}$$

$$\frac{\partial I}{\partial t} = \nabla^2 I - I + \lambda R, \tag{19.52}$$

$$R = E - I. \tag{19.53}$$

Space-Independent Behavior

If the light stimulus is spatially uniform, then spatial dependence can be ignored, and the model equations reduce to the ordinary differential equations

$$\tau \frac{dE}{dt} = L - E, \tag{19.54}$$

$$\frac{dI}{dt} + (\lambda + 1)I = \lambda E. \tag{19.55}$$

If L is a unit step applied at time $t = 0$, then the response at subsequent times is

$$R = E - I \tag{19.56}$$

$$= \frac{1}{\lambda + 1} - \frac{k - 1}{k - \lambda - 1} e^{-kt} + \frac{\lambda k}{(k - \lambda - 1)(\lambda + 1)} e^{-(\lambda+1)t}, \tag{19.57}$$

where $k = 1/\tau$. The response R is graphed in Fig. 19.12A, whence it can be seen that the response is an initial peak followed by a decay to a plateau.

Time-Independent Behavior

If the input is steady, so that time derivatives vanish, then $E = L$ and

$$\nabla^2 I = (\lambda + 1)I - \lambda L. \tag{19.58}$$

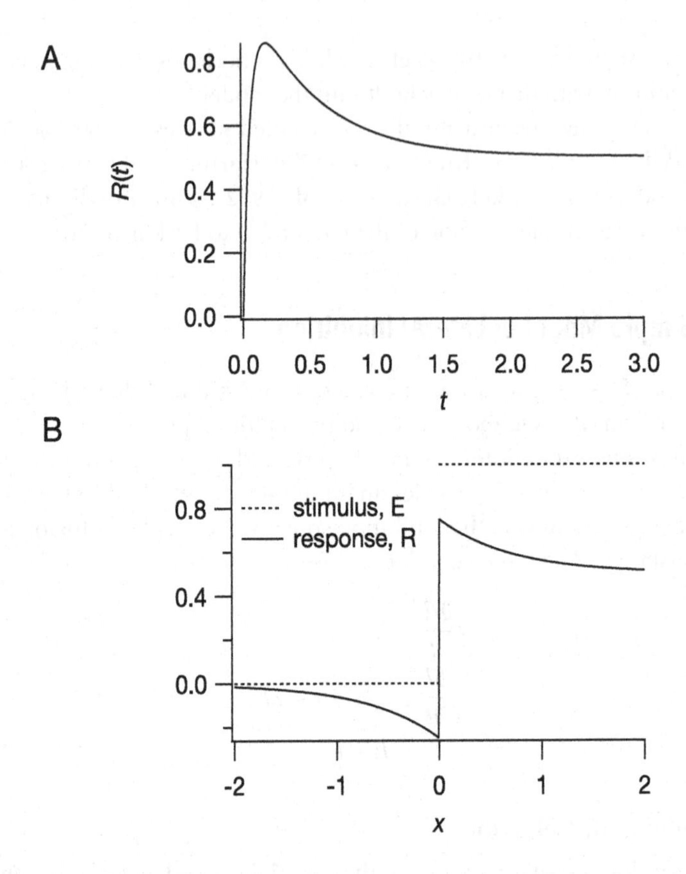

Figure 19.12 Solutions of the qualitative model of lateral inhibition in the retina, calculated with the parameters $\lambda = 1$, $k = 20$. A: The space-independent response. B: The steady response to a band of light extending from $x = 0$ to $x = \infty$.

Suppose there is an edge in the pattern of light, represented by

$$L(x,y) = \begin{cases} 1, & x > 0, \\ 0, & x < 0. \end{cases} \tag{19.59}$$

Then the solution for I is

$$I = \begin{cases} \frac{\lambda}{\lambda+1}\left(1 - \frac{1}{2}e^{-x\sqrt{\lambda+1}}\right), & x > 0, \\ \frac{\lambda}{\lambda+1}\frac{1}{2}e^{x\sqrt{\lambda+1}}, & x < 0, \end{cases} \tag{19.60}$$

where I and dI/dx are required to be continuous at $x = 0$. However, since $E = L$, $R = E - I$ is discontinuous at $x = 0$. Graphs of E and $R = E - I$ are shown in Fig. 19.12B: as can be seen, R exhibits Mach bands at the edge of the light stimulus.

19.4.2 Photoreceptor and Horizontal Cell Interactions

Lateral inhibition in the vertebrate retina results from different processes from those in the invertebrate. In the vertebrate retina photoreceptors and horizontal cells form layers of cells through which their potential can spread laterally. The output from the photoreceptors is directed toward the horizontal cells, but the response of the horizontal cells also influences the photoreceptors, forming a feedback loop with spatial interactions. A detailed model of receptor/horizontal cell interactions was constructed by Krausz and Naka (1980), and the model parameters were determined by fitting to experimental data from the catfish retina. The model is depicted in Fig. 19.13.

In this model, receptor and horizontal cells are assumed to form continuous sheets, within which voltage spreads continuously. The coupling coefficient for voltage spread in the sheet of receptors differs from that in the horizontal cell sheet. The receptors feed forward to the horizontal cells, with transfer function $\hat{A}$, and the horizontal cells feed back to the receptors with transfer function $\hat{k}$. The receptor response is the excitation due to light minus that due to horizontal cell feedback.

We first consider the model in which the voltage spreads laterally in the horizontal cell layer, but not in the photoreceptor layer. To specify this model we must first determine how voltage spreads within a cell layer. The primary assumption is that the horizontal cell layer is effectively a continuous two-dimensional sheet of cytoplasm, and spread of current within this layer can be modeled by the passive cable equation (Chapter 4), with a source term describing the current input from the photoreceptor layer. If the variations of light around the mean are small, it is reasonable to assume that the ionic currents are passive and that the governing equation is linear. Thus, from

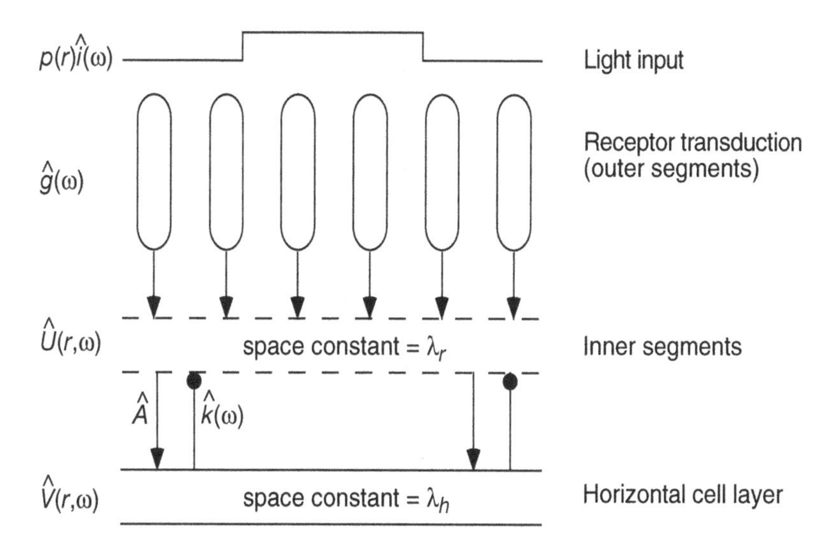

Figure 19.13 Schematic diagram of the lateral inhibition model of Krausz and Naka. (Adapted from Krausz and Naka, 1980.)

(4.14) we have

$$\tau_h \frac{\partial V}{\partial t} + V = \lambda_h^2 \nabla^2 V + R_h I_{\text{ph}}, \tag{19.61}$$

where τ_h is the membrane time constant, λ_h is the membrane space constant, R_h is the membrane resistivity, and I_{ph} is the current input from the photoreceptor layer.

To simplify the model, we assume that the light input, and all subsequent responses, are radially symmetric, functions only of the distance from the center of the stimulus. We suppose that the light input is $Ii(t)p(r)$, i.e., modulated temporally by $i(t)$ and spatially by $p(r)$. We let $\hat{U}(r,\omega)$ denote the Fourier transform of receptor potential at position r, and let $\hat{g}(\omega)$ denote the transfer function of the linear stages of receptor phototransduction. Then, in the frequency domain we have

$$\hat{U} = Ip(r)\hat{g}(\omega)\hat{i}(\omega) - \hat{k}(\omega)\hat{V}. \tag{19.62}$$

U is influenced by two terms, the first due to excitation by light, and the second due to inhibitory feedback from horizontal cells, with transfer function $\hat{k}(\omega)$. Finally, taking the feedforward transfer function to be $\hat{A}(\omega)$, and taking $V = \hat{V}e^{i\omega t}$, we obtain

$$\lambda_h^2 \nabla^2 \hat{V} - (1 + i\omega\tau_h)\hat{V} = -\hat{A}(\omega)\hat{U}. \tag{19.63}$$

Although we expect the qualitative behavior of the Krausz–Naka model to be similar to that of Peskin's model, the goal of this model is to obtain quantitative agreement with experiment by fitting it directly to data.

We simplify (19.62) and (19.63) by a change of variables. We set

$$\Phi = \frac{\hat{U}}{I\hat{g}(\omega)\hat{i}(\omega)}, \tag{19.64}$$

$$\Psi = \frac{\hat{V}}{I\hat{g}(\omega)\hat{i}(\omega)\hat{A}(\omega)}, \tag{19.65}$$

and from (19.63) find that

$$\nabla^2 \Psi - \frac{1}{\alpha^2(\omega)}\Psi = -\frac{p(r)}{\lambda_h^2}, \tag{19.66}$$

where

$$\alpha^2(\omega) = \frac{\lambda_h^2}{1 + i\omega\tau_h + \hat{A}(\omega)\hat{k}(\omega)}. \tag{19.67}$$

It is left as an exercise (Exercise 10) to show that the solution of (19.66), assuming an infinite domain, is

$$\Psi = \frac{1}{\lambda_h^2}\int_0^\infty p(s)G(r,s)\,s\,ds, \tag{19.68}$$

where G, the fundamental solution, is given by

$$G(r,s) = \begin{cases} I_0(r/\alpha)K_0(s/\alpha), & r < s, \\ K_0(r/\alpha)I_0(s/\alpha), & r > s. \end{cases} \tag{19.69}$$

Here I_0 and K_0 are modified Bessel functions of the first and second kind of order zero. (Unfortunately, I_0 is standard notation for the modified Bessel function of the first kind, but it should not be confused with the background light level.) Of particular interest is the case of a circular spot of light of radius R,

$$p(s) = \begin{cases} 1, & s < R, \\ 0 & s > R. \end{cases} \tag{19.70}$$

In this case, we can use the identities $\frac{d}{dz}(zK_1(z)) = zK_0(z)$ and $\frac{d}{dz}(zI_1(z)) = zI_0(z)$ to evaluate the integral (19.68) with the result that

$$\Psi(r,\omega) = \frac{1}{\lambda_h^2} F(r,R,\omega), \tag{19.71}$$

where

$$F(r,R,\omega) = \begin{cases} \alpha^2[1 - (R/\alpha)I_0(r/\alpha)K_1(R/\alpha)], & r < R, \\ \alpha R I_1(R/\alpha)K_0(r/\alpha), & r > R. \end{cases} \tag{19.72}$$

Fitting to Data

Krausz and Naka determined the model parameters by fitting the ratio of the uniform field response to the spot response to experimental data (Fig. 19.14). The field response (i.e., taking $R \to \infty$) can be calculated from (19.66) by setting $p(r) \equiv 1$, in which case the constant solution for Ψ is easily seen to be $\Psi(r,\omega)_{\text{field}} = \alpha^2/\lambda_h^2$. Thus,

$$\frac{\hat{V}_{\text{spot}}}{\hat{V}_{\text{field}}} = \frac{\Psi(r,\omega)_{\text{spot}}}{\Psi(r,\omega)_{\text{field}}} = \frac{1}{\alpha^2} F(r,R,\omega). \tag{19.73}$$

For fixed values of r and R, $\alpha(\omega)$ can be determined to give good agreement between model and experiment. Note that by taking response ratios, dependence on the feedforward steps within each photoreceptor is eliminated. Thus, attention is focused on the interactions between the horizontal cells and the photoreceptors. However, the model cannot distinguish between $\hat{A}$ and $\hat{k}$, since only the product of these terms appear. So, for simplicity, it is assumed that $\hat{A}$ is a constant gain, with no frequency dependence, and that $\hat{k}$ has unity gain. Values for $\hat{k}(\omega)$ are obtained at each frequency, and then $k(t)$ determined from an inverse Fourier transform. The result is well approximated by

$$k(t) = \frac{3}{\tau} e^{-(t-t_0)/\tau} [1 - e^{-(t-t_0)/\tau}]^2, \tag{19.74}$$

a sigmoidal-shaped rising curve followed by exponential decay. The parameters t_0 and τ are, respectively, the feedback delay and the feedback time constant. The function $k(t)$ has an important physiological interpretation, as it describes the feedback from horizontal cells to photoreceptors; in response to a delta function input from the horizontal cells, the photoreceptor response is given by $-k(t)$. Parameters resulting from the fit are given in Table 19.3. The membrane time constant of the horizontal cells was found to be small in all cases, and so was set to zero. Hence, the potential of the horizontal cell layer responds essentially instantaneously to a stimulus. However, the

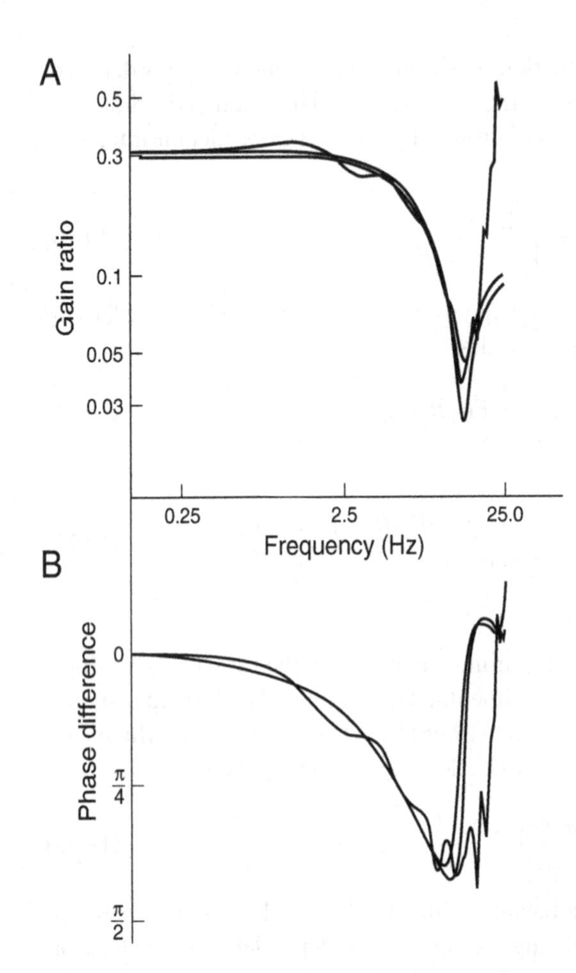

Figure 19.14 Spot-to-field transfer function measured in the catfish retina. The wavy traces are the experimental results, and the two smooth curves are results from the model, using two slightly different parameter sets, one of which is given in Table 19.3. (Krausz and Naka, 1980, Fig. 4.)

Table 19.3 Parameters of the Krausz–Naka model for catfish retinal neurons. These parameters correspond to the model in which there is no coupling between photoreceptors.

Parameter	Value
λ_h	= 0.267 mm
τ_h	= 0 ms
$\hat{A}$	= 3.77
τ	= 24.8 ms
t_0	= 0.022 ms

time constant for the response of the photoreceptor layer to horizontal cell feedback is significant.

Predicting the Response to a Moving Grating

As a test of the model, Krausz and Naka calculated the response to a 1-dimensional moving grating, and compared the result to experimental data. A moving grating provides a light stimulus of the form $I = I(t - x/c)$, where c is the speed of the grating. We look for solutions of the form $V = V(\xi)$, where $\xi = t - x/c$, in which case the differential equation (19.61) becomes

$$\frac{\lambda_h^2}{c^2}V'' - \tau_h V' - V = -R_h I_{\text{ph}}, \tag{19.75}$$

where a prime denotes differentiation with respect to ξ. Taking Fourier transforms with respect to ξ, and recalling that in the frequency domain the input to the horizontal cell layer is given by $\hat{A}\hat{U}$, we find that

$$(1 + i\omega\tau_h)\hat{V} = -\frac{\omega^2\lambda_h^2}{c^2}\hat{V} + \hat{A}\hat{U}. \tag{19.76}$$

Assuming that the light input is given by $Ie^{i\omega\xi}$, then $\hat{U}$ satisfies

$$\hat{U} = I\hat{g}(\omega) - \hat{k}(\omega)\hat{V}. \tag{19.77}$$

Hence,

$$\begin{aligned}\Psi(\omega) &= \frac{\hat{V}}{\bar{I}\hat{g}(\omega)\hat{A}}\\ &= \frac{1}{1 + i\omega\tau_h + \hat{A}\hat{k} + \omega^2\lambda_h^2/c^2}\\ &= \left(\frac{\alpha^2}{\lambda_h^2}\right)\frac{c^2}{c^2 + \omega^2\alpha^2}.\end{aligned} \tag{19.78}$$

Krausz and Naka showed that the model predictions for the rectilinear stimulus predict experimental results accurately, confirming that the model provides a general quantitative description of the horizontal cell response that is not limited to the data on which it was based.

Receptor Coupling

Receptors are electrically coupled by gap junctions, and the potential spreads through the receptor layer in a continuous fashion, as it does in the horizontal cell layer, but with a different space constant, λ_r (Lamb and Simon, 1977; Detwiler and Hodgkin, 1979). To incorporate receptor coupling, we need only add spatial coupling for the receptor layer. In the frequency domain, we have

$$\lambda_r^2\nabla^2\Phi - (1 + i\omega\tau_r)\Phi - \hat{k}(\omega)\hat{A}(\omega)\Psi = -p(r), \tag{19.79}$$

$$\lambda_h^2\nabla^2\Psi - (1 + i\omega\tau_h)\Psi = -\Phi, \tag{19.80}$$

where τ_r is the membrane time constant for the receptor cell layer. Writing $q_r = (1 + i\omega\tau_r)/\lambda_r^2$ and $q_h = (1 + i\omega\tau_h)/\lambda_h^2$ and substituting (19.80) into (19.79) gives

$$\left(\nabla^2 - \frac{1}{\gamma^2}\right)\left(\nabla^2 - \frac{1}{\delta^2}\right)\Psi = \frac{p(r)}{\lambda_h^2\lambda_r^2}, \tag{19.81}$$

where $1/\gamma^2$ and $1/\delta^2$ are defined by

$$\frac{1}{\gamma^2} + \frac{1}{\delta^2} = q_r + q_h, \tag{19.82}$$

$$\frac{1}{\gamma^2\delta^2} = q_r q_h + \frac{\hat{A}(\omega)\hat{k}(\omega)}{\lambda_h^2\lambda_r^2}. \tag{19.83}$$

Note that γ and δ are analogous to α in (19.66).

Now we define $\chi(r, \gamma)$ to satisfy

$$\left(\nabla^2 - \frac{1}{\gamma^2}\right)\chi(r, \gamma) = -p(r). \tag{19.84}$$

Since the operator $\left(\nabla^2 - \frac{1}{\gamma^2}\right)$ has a unique inverse, we use $\chi(r, \gamma)$ to eliminate p in (19.81) and find that

$$\left(\nabla^2 - \frac{1}{\delta^2}\right)\Psi = \frac{-\chi(r, \gamma)}{\lambda_h^2\lambda_r^2}. \tag{19.85}$$

Similarly, by symmetry

$$\left(\nabla^2 - \frac{1}{\gamma^2}\right)\Psi = \frac{-\chi(r, \delta)}{\lambda_h^2\lambda_r^2}. \tag{19.86}$$

Subtracting these two equations, we obtain

$$\Psi(r, \omega) = \frac{1}{\lambda_h^2\lambda_r^2}\left(\frac{\gamma^2\delta^2}{\gamma^2 - \delta^2}\right)[\chi(r, \gamma) - \chi(r, \delta)]. \tag{19.87}$$

Solving (19.85) for $\nabla^2\Psi$, substituting into (19.80), and using the expression for Ψ then gives

$$\Phi(r, \omega) = \frac{1}{\lambda_r^2}\left(\frac{\gamma^2\delta^2}{\gamma^2 - \delta^2}\right)\left[\left(q_h - \frac{1}{\gamma^2}\right)\chi(r, \gamma) - \left(q_h - \frac{1}{\delta^2}\right)\chi(r, \delta)\right]. \tag{19.88}$$

Since (19.84) for χ is of the same form as (19.66) for Ψ, its solution takes the same form. Thus, (19.87) and (19.88) give an explicit solution to the general problem of electrical flow in two coupled cell layers connected by reciprocal pathways.

19.5 Detection of Motion and Directional Selectivity

It seems intuitively obvious that the detection of motion is one of the most important jobs of the visual system of an animal, either prey or predator, and it is believed to be one of the oldest features of the visual system (Nakayama, 1985; a detailed review of motion detection is given by Clifford and Ibbotson, 2003). It is known that, in some vertebrates at least, directional selectivity arises in the retina itself, with directionally sensitive ganglion cells responding preferentially to motion in a certain direction (Barlow and Levick, 1965; Taylor et al., 2000; Vaney and Taylor, 2002). A qualitative model of how

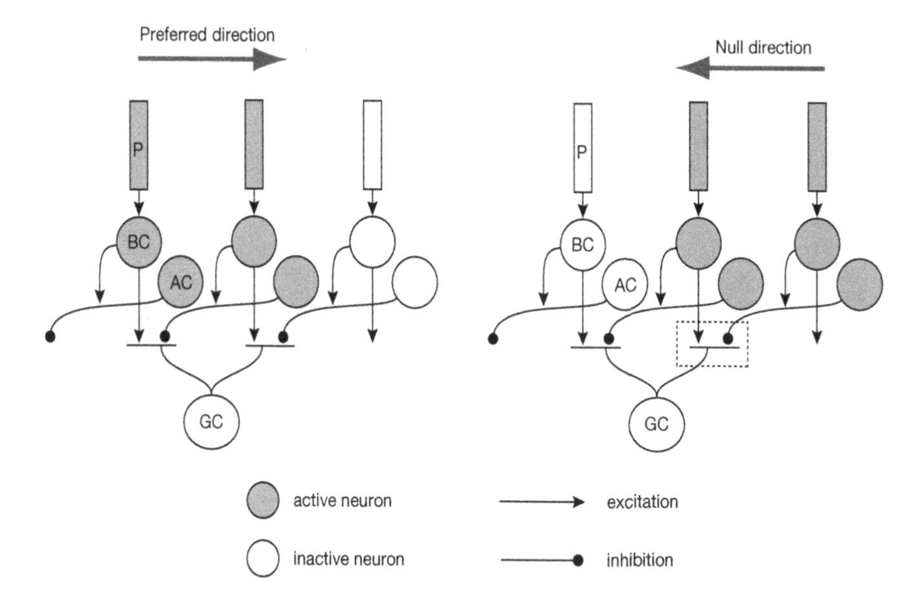

Figure 19.15 Proposed mechanism for directional selective ganglion cells. P – photoreceptor; BC – bipolar cell; AC – amacrine cell; GC – ganglion cell. When motion is in the null direction, the inhibitory and excitatory inputs at the ganglion cell cancel out (as shown by the dashed box), giving a smaller response. Thus, directional selectivity arises from the asymmetry of the connections from amacrine cells to ganglion cells. Adapted from Taylor et al., 2000, Fig. 1.

such direction selectivity arises is sketched in Fig. 19.15. According to this scheme, directional selectivity arises because of the asymmetry of the connections between amacrine and ganglion cells. When motion is in the preferred direction, the inhibitory effects of the amacrine cell layer are all in the reverse direction, and thus act too late to inhibit the ganglion cell responses. However, when motion is in the null direction, the amacrine cells feed forward making the excitation and inhibition happen at the same time at the same place, thus decreasing the ganglion cell response.

The earliest quantitative model of motion detection was that of Hassenstein and Reichardt (1956; Reichardt, 1961; Borst and Egelhaaf, 1989; an introduction for the nonexpert is given by Borst, 2000), a model that was based on experimental work on the beetle *Chlorophanus* and that has become a classic in the field. A schematic diagram of a simple Reichardt detector is shown in Fig. 19.16. Two photoreceptors (inputs I_1 and I_2), separated by a distance Δx, feed forward to the output stage (a ganglion cell, for example). Input I_1 is delayed by Δt and multiplied by I_2, while the mirror image procedure is applied to I_2. Finally, the results are subtracted to get the response, R.

For ease of notation, let $\Delta x = h$ and $\Delta t = k$. Then, if the light stimulus is $s(x,t)$, it follows that

$$\begin{aligned} R(t) &= I_1(t)I_2(t-k) - I_1(t-k)I_2(t) \\ &= s(0,t)s(h,t-k) - s(0,t-k)s(h,t), \end{aligned} \tag{19.89}$$

where for convenience we have assumed that the first photoreceptor, I_1, is at $x = 0$.

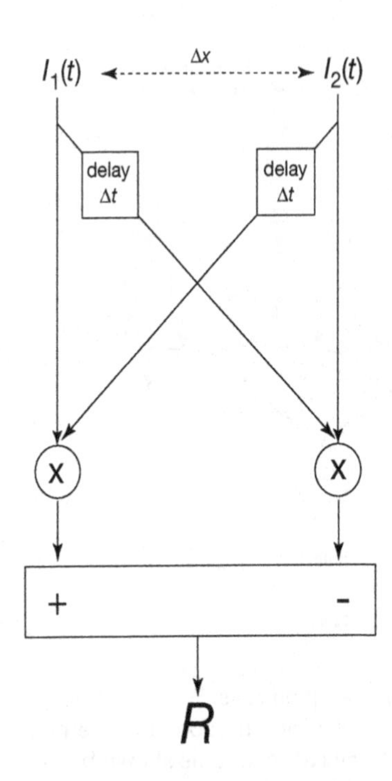

Figure 19.16 Schematic diagram of a simple Reichardt detector. Two spatially separated inputs are delayed, cross-multiplied and subtracted to get a response, R, whose sign depends on the direction of motion.

This motion detector has two important properties. First, when the response is integrated over a sufficiently long time, the steady components of I_1 and I_2 have no effect; only those portions of the input that correspond to a moving signal are detected. Although this is not true for an arbitrary input, it is true for realistic light stimuli. Second, when the velocity of the stimulus is reversed, the integral of the response changes sign. Thus, the sign of the integral of R determines the direction of motion.

Both these properties can be simply illustrated by using a stimulus of the form $s(x,t) = s_0 + s_1(x - ct)$, where c is the speed at which the stimulus is moved across the two receptors. Two basic forms for s_1 cover almost the entire range of physiological stimuli; periodic with mean zero, or with compact support (i.e., nonzero only on some finite region). An example of the first case is a flicker stimulus presented to the eye, while an example of the second case is a moving bar. Assuming either one of these two forms for s_1, we get

$$R(t) = s_0 \left[s_1(h - ct + ck) - s_1(h - ct) + s_1(-ct) - s_1(-ct + ck)\right] + s_1(-ct)s_1(h - ct + ck) - s_1(-ct + ck)s_1(h - ct). \tag{19.90}$$

When this is integrated from $t = -\infty$ to ∞, all the terms in the square brackets cancel out (since s_1 has either mean zero or compact support, by assumption). Thus

$$\int R(t)\,dt = \int s_1(-ct)s_1(h - ct + ck) - s_1(-ct + ck)s_1(h - ct)\,dt. \tag{19.91}$$

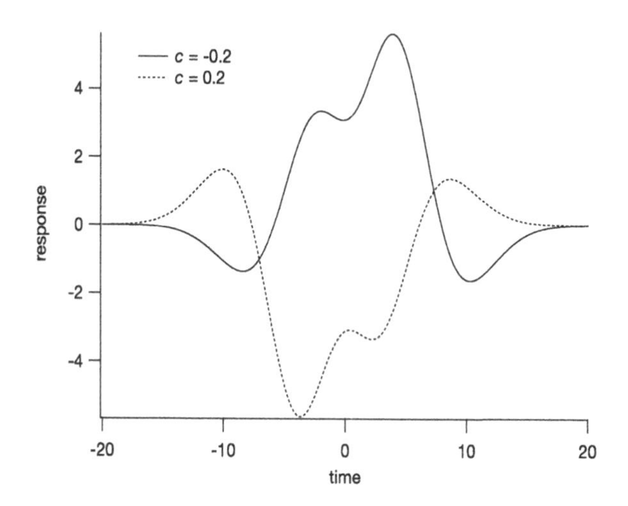

Figure 19.17 Response of a simple Reichardt detector to the stimulus $s(x,t) = 10[1 + e^{-(x-ct-1)^2} + 0.8e^{-(x-ct+0.2)^2}]$ (calculated with the values $\Delta x = 0.5$, $\Delta t = 0.2$). When c changes sign, the response is close to a reflection through the origin of the first response. In fact, it is not an exact reflection, but the integrals of the two responses are of the same amplitude with opposite sign.

Now denote the response for stimulus speed c, by $R_c(t)$. Since

$$\int R_{-c}(-t)\,dt = \int s_1(-ct)s_1(h-ct-ck) - s_1(-ct-ck)s_1(h-ct)\,dt$$
$$= -\int R_c(t)\,dt \tag{19.92}$$

(use the change of variables $\tau = t - k$), it follows that

$$\int [R_c(t) + R_{-c}(-t)]\,dt = 0. \tag{19.93}$$

Thus, if the velocity of the stimulus is reversed, the response is close to a reflection through the origin of the original response, having exactly the same integral, of opposite sign. We illustrate this in Fig. 19.17.

19.6 Receptive Fields

The output stage of the retina is the layer of ganglion cells, which extend through the optic nerve to the lateral geniculate nucleus, from there transmitting signals to the visual cortex. Each ganglion cell responds by a series of action potentials, with the information encoded in the frequency and duration of the wave train. Thus, ganglion cells transmit a digital signal typical of neurons. The input stage of ganglion cells is highly organized. Each ganglion cell responds only to light in a well-defined part of the retina, called the *receptive field* of the cell, and these receptive fields are organized into two concentric,

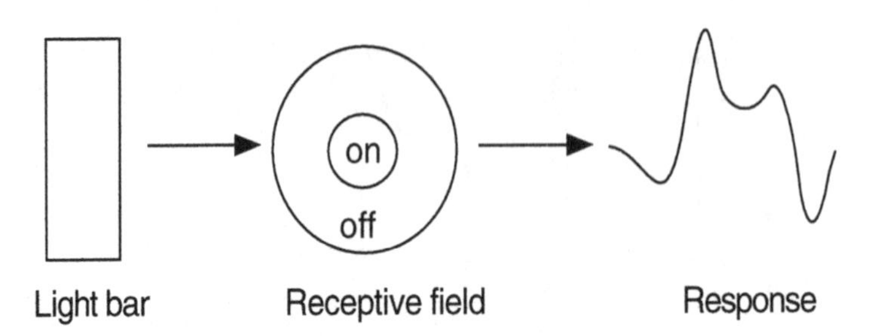

Figure 19.18 Diagram of the center-surround arrangement of the receptive field of an on-center retinal ganglion cell, and its response to a wide bar moving across the receptive field. (Adapted from Rodieck, 1965, Fig. 1.)

mutually antagonistic regions, the center and the surround (Fig. 19.18). Although in reality the story is not quite so simple (see Troy and Shou, 2002, for an excellent modern review) we discuss only simple receptive fields of this type. This center-surround organization was discovered by Kuffler (1953, 1973) and is considered one of the landmark discoveries in visual science. Interestingly, one of Kuffler's early colleagues was FitzHugh, of later FitzHugh–Nagumo fame (Barlow, FitzHugh and Kuffler, 1957).

The center can be either excitatory (*on-center*) or inhibitory (*off-center*). A white figure moved across the receptive field of an on-center cell gives the same response as a black figure moved across the receptive field of an off-center cell. A typical response curve for a bar moving across the receptive field is shown in Fig. 19.18. Note that the ganglion cell has a large response to the edges of the bar, but responds much less to the maintained stimulus in the middle of the bar. This is reminiscent of the Mach bands seen in the Krausz–Naka model. There are different types of on/off responses for ganglion cells. Some respond to both "on" and "off," while others respond only to one or the other. Some ganglion cells are directionally dependent, responding to a stimulus only if it enters the receptive field from a particular direction. Other ganglion cells are color dependent.

The different responses of the on-center and off-center ganglion cells are the result of their connections to different types of bipolar cells. On-center and off-center ganglion cells synapse to bipolar cells that respond, respectively, to light increments and decrements. These different bipolar cell responses are due, in turn, to their different responses to glutamate released at the synapse with the photoreceptor.

Recall that, in the dark the photoreceptor is depolarized and thus there is a continual release of glutamate at the synapse with the bipolar cell. A light stimulus causes hyperpolarization of the photoreceptor and a consequent decrease in glutamate release. To see what effect this decrease in glutamate has on the bipolar cell, consider first an on-center bipolar cell. These bipolar cells express glutamate receptors (metabotropic glutamate receptors) that, when bound to glutamate, activate intracellular pathways that result in the closure of cAMP-gated Na^+ channels and hyperpolarization of the cell (contrast this with the ACh-sensitive channels discussed in Section 8.1.5). Thus, in the dark, on-center bipolar cells are hyperpolarized due to the high levels of glutamate released by the photoreceptor. When the photoreceptor is stimulated by light glutamate levels fall, the Na^+ channels in the bipolar cells open and the bipolar cell depolarizes, sending a positive signal to the ganglion cell.

The off-center bipolar cell expresses a different type of glutamate receptor (an ionotropic, or AMPA, receptor) that responds to glutamate by depolarizing the cell. Thus, in the dark these bipolar cells are depolarized. When stimulated by light the photoreceptors release less glutamate and the off-center bipolar cell hyperpolarizes, thus sending a negative signal to the ganglion cell.

Although it is not yet entirely understood how the center-surround field occurs, it is likely that it is mediated by feedback from horizontal cells to photoreceptors. Horizontal cells synapse with photoreceptors over a considerable area and are connected to one another by gap junctions. It is thus plausible that negative feedback from the surround to the ganglion cell is mediated by feedback from the horizontal cell layer to the photoreceptors in the center, and thus to the bipolar cell and the ganglion cell.

One of the earliest models of ganglion cell behavior was constructed by Rodieck (1965). In this model it is assumed that the response of a ganglion cell is a weighted sum of the responses from each part of the receptive field, with negative weights for the inhibitory part of the field and positive weights for the excitatory part. Here we consider only on-center cells, as the model is the same for off-center cells, with reversed signs. We also consider the model in one spatial dimension only, as the extension to two dimensions introduces greater algebraic complexity, but no new concepts.

Suppose that the steady response to a step change in illumination of a small area dx centered at the point x is given by $f(x)dx$. From consideration of experimental data, Rodieck showed that $f(x)$ can be described as the sum of two Gaussians, one contributing a positive component from the center and the other contributing a negative component from the surround. Thus,

$$f(x) = \frac{g_1\sigma_1}{\sqrt{\pi}}e^{-\sigma_1^2x^2} - \frac{g_2\sigma_2}{\sqrt{\pi}}e^{-\sigma_2^2x^2}, \tag{19.94}$$

which is plotted in Fig. 19.19. The constants g_1 and g_2 are the gains of the excitatory center and inhibitory surround, respectively, and the parameters σ_1 and σ_2 control their radial size. Now suppose that the response of the ganglion cell is infinitely fast, and that the cell is stimulated with a semi-infinite bar, extending from $x = -\infty$ to $x = ct$, so that the edge of the bar is moving from left to right with speed c. Then, the response of the ganglion cell, $R(t)$, is

$$R(t) = \int_{-\infty}^{ct} f(x)\,dx \tag{19.95}$$

$$= g_1\left[\frac{1}{2} + \frac{1}{\sqrt{\pi}}\text{erf}(ct/\sigma_1)\right] + g_2\left[\frac{1}{2} + \frac{1}{\sqrt{\pi}}\text{erf}(ct/\sigma_2)\right], \tag{19.96}$$

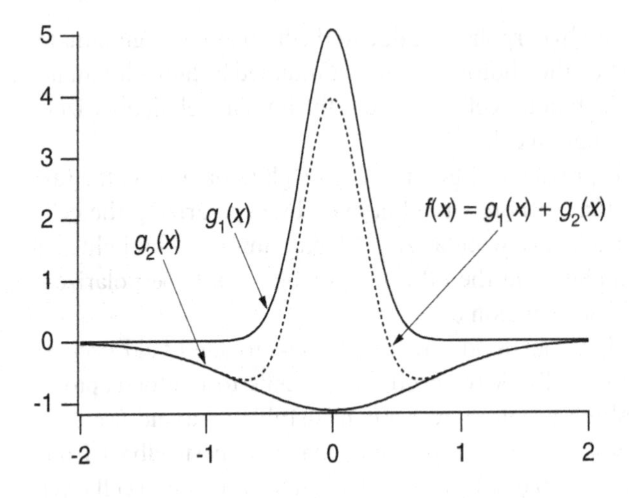

Figure 19.19 The addition of two Gaussian distributions, one with positive sign denoting the excitatory center, and one with negative sign denoting the inhibitory surround, gives the response function $f(x)$, which weights the light stimulus according to its position in space. Computed using $\sigma_1 = 3$, $\sigma_2 = 1$, $g_1 = 3$, $g_2 = 1$.

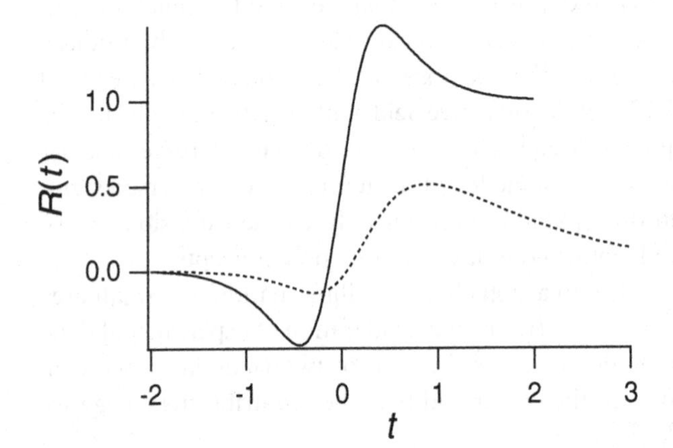

Figure 19.20 Response of a ganglion cell to a moving bar of semi-infinite width. Solid line assuming that the response at each point x is infinitely fast; dotted line assuming that each point x responds to the light stimulus according to (19.98). Both curves were calculated with $\sigma_1 = 3$, $\sigma_2 = 1$, $g_1 = 3$, $g_2 = 1$, $c = 1$.

where erf(x), the error function, is defined by

$$\operatorname{erf}(x) = \frac{2}{\sqrt{\pi}} \int_0^\infty e^{-x^2}\, dx. \tag{19.97}$$

From the plot of R given in Fig. 19.20 (solid line) it can be seen that the ganglion cell responds preferentially to the edge of the bar, as is seen in experimental data.

In reality, the response of the ganglion cell is not infinitely fast, but the response to a step of light has an initial peak followed by a decrease to a lower plateau. Thus, in Rodieck's model, the time-dependent response to a step input is taken to be

$$h(t) = \left[1 + te^{-t}\right] H(t), \tag{19.98}$$

where $H(t)$ is the Heaviside function. This is a simple way to incorporate the dynamic behavior of the earlier retinal stages. More general forms for the response produce little difference in the overall response.

To illustrate the temporal behavior of the overall response, we calculate the response to the moving bar, extending from $x = -\infty$ to $x = ct$. Since the edge of the moving bar reaches an element at position x at time $t = x/c$, the response of an element at position x is $f(x)h(t - x/c)dx$. Integrating over the entire domain gives

$$R(t) = \int_{-\infty}^{\infty} f(x)h(t - x/c)\,dx, \tag{19.99}$$

$$= \underbrace{\int_{-\infty}^{ct} f(x)\,dx}_{\text{steady term}} + \underbrace{\int_{-\infty}^{ct} f(x)(t - x/c)e^{x/c-t}\,dx}_{\text{transient term}}, \tag{19.100}$$

which is graphed in Fig. 19.20 (dotted line). If $f(x)$ decays sufficiently rapidly at $\pm\infty$, the transient term goes to zero as $t \to \infty$, leaving only the steady response. Of course, since retinas are not infinite in extent, f is zero outside a bounded domain, and so such decay is guaranteed. Further, in the limit as $c \to 0$, keeping ct fixed, the transient term again approaches zero. That is, if the bar moves slowly, the effect of $h(t)$ is small, again as expected. On the other hand, as $c \to \infty$, $R(t)$ approaches $h(t)\int_{-\infty}^{\infty} f(x)\,dx$, which is exactly the response to a space-independent flash.

Not only should a model incorporate the temporal dynamics of the receptive field responses, it should also incorporate the fact that the responses of the center and surround have different temporal kinetics. Thus, the most accurate models include an additional temporal filter in the response of the surround, the so-called Gaussian center-surround model (Troy and Shou, 2002). However, since these models quickly become quite complicated we do not discuss them in any further detail here.

Cells higher up in the visual pathway, in the lateral geniculate nucleus and the visual cortex, have progressively more complex receptive fields, designed to make particular cells respond maximally to stimuli of particular orientation or direction of movement. The above model serves as a brief introduction to the type of modeling involved in the analysis of receptive fields. A more detailed discussion of receptive fields is given by Kuffler et al. (1984) and Troy and Shou (2002).

19.7 The Pupil Light Reflex

The control of pupil size is yet another way in which the eye can adjust to varying levels of light intensity. While the adjustment of pupil size accounts for much less of visual adaptation than those mechanisms described earlier, it is nonetheless an important control mechanism.

The size of the pupil of the eye is determined by a balance between constricting and dilating mechanisms. Pupil constriction is caused by contraction of the circularly arranged pupillary constrictor muscle, which is innervated by parasympathetic fibers. The motor nucleus for this muscle is the Edinger–Westphal nucleus located in the oculomotor complex of the midbrain. Dilation is controlled by contraction of the radially arranged pupillary dilator muscle innervated by sympathetic fibers and by inhibition of the Edinger–Westphal nucleus.

The effect of the pupil light reflex is to control the retinal light flux

$$\phi = IA, \tag{19.101}$$

where I is the illuminance (lumen mm^{-1}) and A is the pupil area (mm^2). It performs this function by acting like the aperture of a camera. When light is shined on the retina, the pupil constricts, thereby decreasing ϕ. However, there is a latency of $\approx$180–400 ms following a change in light input before changes in pupil size are detected.

This combination of negative feedback with delay may lead to oscillations of pupil size. These oscillations were first observed by a British army officer, Major Stern, who noticed that pupil cycling could be induced by carefully focusing a narrow beam of light at the pupillary margin. Initially, the retina is exposed to light, causing the pupil to constrict, but this causes the iris to block the light from reaching the retina, so that the pupil subsequently dilates, re-exposing the retina to light, and so on indefinitely.

Longtin and Milton (1989; Milton, 2003) developed a model of the dynamics of pupil contraction and dilation. In their model it is assumed that the light flux ϕ is transformed after a time delay τ_r into neural action potentials that travel along the optic nerve. The frequency of these action potentials is related to ϕ by

$$N(t) = \eta F\left(\ln\left[\frac{\phi(t-\tau_r)}{\bar{\phi}}\right]\right), \tag{19.102}$$

where $F(x) = x$ for $x \geq 0$ and $F(x) = 0$ for $x < 0$, $\bar{\phi}$ is a threshold retinal light level (the light level below which there is no response), and η is a rate constant. The notation $\phi(t - \tau_r)$ is used to indicate dependence on the flux at time τ_r in the past.

This afferent neural action potential rate is used by the midbrain nuclei, after an additional time delay τ_t, to produce an efferent neural signal. This signal exits the midbrain along preganglionic parasympathetic nerve fibers, which terminate in the ciliary ganglion where the pupillary sphincter is innervated. Neural action potentials at the neuromuscular junction result in the release of neurotransmitter (ACh), which diffuses across the synaptic cleft, thus generating muscle action potentials and initiating muscle contraction. These events are assumed to require an additional time τ_m.

The relationship between iris muscle activity x and the rate of arriving action potentials $E(t)$ is not known. We take a simple differential relationship

$$\tau_x \frac{dx}{dt} + x = E(t), \tag{19.103}$$

where

$$E(t) = \gamma F\left(\ln\left[\frac{\phi(t-\tau)}{\bar{\phi}}\right]\right), \tag{19.104}$$

and $\tau = \tau_r + \tau_t + \tau_m$ is the total time delay in the system.

Finally, we close the model by assuming some relationship between iris muscle activity x and pupil area A as $A = f(x)$. For example, one reasonable possibility is the

Hill equation

$$A = f(x) = \Lambda_0 + \frac{\Lambda\theta^n}{x^n + \theta^n}, \tag{19.105}$$

for which area is a decreasing function of activity, with maximal area $\Lambda + \Lambda_0$ and minimal area Λ_0. It follows that the differential equation governing iris muscle activity is

$$\tau_x \frac{dx}{dt} + x = \gamma F\left(\ln\left[\frac{I(t-\tau)f(x(t-\tau))}{\bar{\phi}}\right]\right) \tag{19.106}$$

$$= g(x(t-\tau), I(t-\tau)). \tag{19.107}$$

19.7.1 Linear Stability Analysis

Because $f(x)$ is a decreasing function of x, a steady solution of (19.107) is assured when the input $I(t)$ is constant. We identify this value of x as x^*, satisfying $x^* = g(x^*, I)$. Linearized about x^*, the delay differential equation becomes

$$\tau_x \frac{dX}{dt} + X = -GX(t-\tau), \tag{19.108}$$

where $G = -g_x(x^*, I) = -\gamma \frac{f'(x^*)}{f(x^*)}$ is called the *gain* of this negative feedback system. If we set $X = X_0 e^{\mu t}$, we find the characteristic equation for μ to be

$$\tau_x \mu + 1 = -Ge^{-\mu\tau}. \tag{19.109}$$

If $|G| < 1$, there are no roots of this equation with positive real part; the solution is linearly stable. Since $G > 0$, there are no positive real roots of this characteristic equation. The only possible way for the solution to become unstable is through a Hopf bifurcation, whereby a root of (19.109) with nonzero imaginary part changes the sign of its real part. If we set $\mu = i\omega$, we can separate (19.109) into real and imaginary parts, obtaining

$$G\cos\omega\tau = -1, \tag{19.110}$$

$$G\sin\omega\tau = \tau_x\omega. \tag{19.111}$$

From these two expressions, we readily find a parametric representation of the critical stability curve to be

$$G = \frac{-1}{\cos\eta}, \qquad \frac{\tau}{\tau_x} = \frac{-\eta}{\tan\eta}. \tag{19.112}$$

The first instability curve is plotted in Fig. 19.21, with the gain G plotted as a function of the dimensionless delay τ/τ_x. It is easily seen that on the critical curve G is a decreasing function of τ/τ_x. If the delay is larger than the critical delay, the steady solution is unstable and there is a stable periodic solution of the full differential delay equation (19.107), corresponding to periodic cycling of pupil size with constant light stimulus.

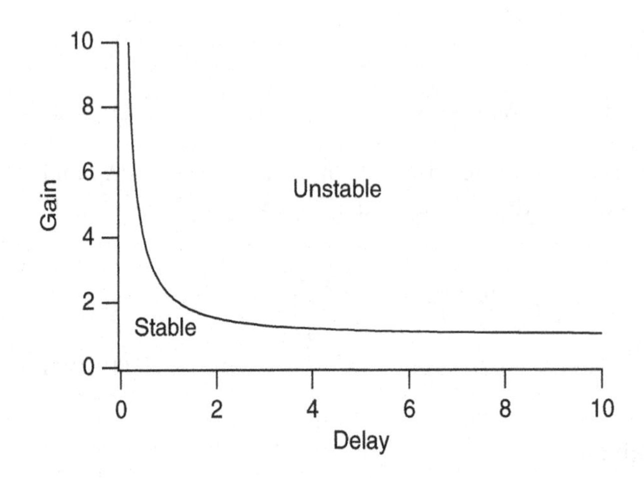

Figure 19.21 Critical stability curve for the pupil light reflex.

19.8 Appendix: Linear Systems Theory

One of the most widely used tools in the study of the visual system is linear systems analysis. Here we have assumed that the basic tools of linear function theory, such as Fourier transforms, delta functions, and the convolution theorem, are familiar to readers. There are numerous books that provide the necessary background, for example, Papoulis (1962), Stakgold (1998) and Haberman (2004).

The essential idea of linear systems theory is that for any linear differential equation $L[u] = f(t)$, where $L[\cdot]$ is a time-autonomous differential operator, the Fourier transform of the solution can be written as

$$\hat{u}(\omega) = T(\omega)\hat{f}(\omega), \tag{19.113}$$

where $\hat{f}(\omega)$ is the Fourier transform of the input function $f(t)$, and $T(\omega)$ is called the *transfer function* for this linear system. Note that if $f(t)$ is the delta function, then $\hat{u} = T(\omega)$, and thus the transfer function is the Fourier transform of the impulse response. The transfer function can also be found by assuming an input of the form $f = e^{i\omega t}$ and looking for an output of the form $u = T(\omega)e^{i\omega t}$. Thus, the amplitude and phase of the sinusoidal input are modulated by the amplitude and phase of the transfer function. Such sinusoidally varying inputs are commonly used in experimental studies of the visual system; by varying the frequency of the stimulus and measuring (at each fixed frequency) the amplitude and phase of the output, the transfer function can be experimentally determined.

Once the transfer function (and thus the impulse response) of a linear system is known, its response to any input can be calculated. Intuitively, any function $f(t)$ can be considered a superposition of impulses, and thus the response to the input $f(t)$ is the sum of the responses to these impulses. Thus, if $K(t)$ is the impulse response of the linear operator L, the solution to $L[u] = f(t)$ is $u(t) = \int_{-\infty}^{t} f(\tau)K(t-\tau)\,d\tau$. Although this exposition is far too brief for a complete presentation of the theory, the main point is that, for a linear system, experimental measurement of the transfer function suffices for a complete determination of the system. For a detailed discussion of impulse responses, Green's functions, and fundamental solutions of differential equations, the reader is referred to Stakgold (1997), Haberman (2003) or Kevorkian (2000).

One of the simplest mathematical realizations of a linear system is the linear differential equation

$$\frac{dx}{dt} + ax = f(t), \qquad a > 0, \tag{19.114}$$

where $f(t)$ is the input to the system. If we assume that $f(t) = e^{i\omega t}$, and look for solutions of the form $x = A(\omega)e^{i\omega t}$, we get

$$A(\omega) = \frac{1}{a + i\omega}. \tag{19.115}$$

This is the transfer function of (19.114). It can also be calculated by taking the Fourier transform of (19.114). The inverse Fourier transform of the transfer function is $H(t)e^{-at}$, where H is the Heaviside function, which is the impulse response of (19.114) (Exercise 12). If we now solve (19.114) directly by using the integrating factor e^{at} we get

$$\left(x(t)e^{at}\right)' = f(t)e^{at}, \tag{19.116}$$

and thus, integrating from $-\infty$ to t and assuming the solution is bounded at $-\infty$, we get

$$x(t) = \int_{-\infty}^{t} f(\tau)e^{-a(t-\tau)}\,d\tau. \tag{19.117}$$

Thus the response to a general input is the convolution of the input with the impulse response, as claimed above.

This approach is easily extended to multiple linear differential equations, as in (19.10)–(19.13). The transfer function and impulse response of those equations is calculated in exactly the same way; we assume an input of the form $I(t) = e^{i\omega t}$ and look for solutions of the form $g(t) = H(\omega)e^{i\omega t}$.

Typical experimental data are shown in Fig. 19.22. The light input to the system is modulated around a mean level, and the output is the membrane potential of a turtle horizontal cell. Here there are five different background light levels, corresponding to the five curves in the figure. Gain is measured in units of mV photon^{-1} and plotted relative to gain in the dark.

As can be seen from Fig. 19.22, when the background light level increases by one log unit in intensity (indicated by moving from filled squares to open circles, or from open circles to filled triangles, etc.), the relative gain at low frequency decreases by approximately one log unit. Hence, over a range of light levels the low-frequency gain is inversely proportional to I_0, and thus the steady-state sensitivity obeys Weber's law. At high frequencies and low background light levels, Weber's law breaks down, and the gain becomes nearly independent of the background light level. It is important to note that the steady-state sensitivity is not the same as the peak sensitivity. In photoreceptors,

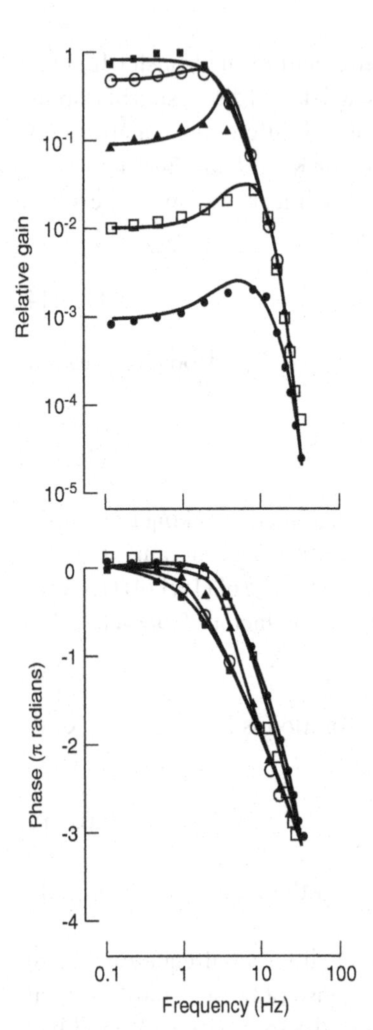

Figure 19.22 A family of temporal frequency responses from turtle horizontal cells, measured at five different mean light levels. (From top to bottom, the filled square, open circle, filled triangle, open square, and filled circle denote background lights of intensity, respectively, −4, −3, −2, −1, and 0 log units.) Symbols are experimental data, and the smooth curves are from the model of Tranchina et al. (1984), which is not described here. The data are presented in a typical Bode plot format, with the amplitude plotted in the upper panel, the phase difference in the lower. (Tranchina et al., 1984, Fig. 1.)

however, for reasons that are not clear, both the steady-state and peak sensitivity follow Weber's law approximately. From these data it can also be seen that turtle horizontal cells respond best to stimuli of a given frequency. At the two highest light levels the data are clearly band-pass in nature, with a maximal response at around 10 Hz.

One significant problem is that most realistic systems are nonlinear, so a transfer function cannot be defined (since solutions are not the linear superposition of fundamental solutions). However, if the amplitude of the sinusoidal input is small, so that $I(t) = \epsilon e^{i\omega t}$, then the response should also be small, of the form $\epsilon T(\omega) e^{i\omega t} + O(\epsilon^2)$. If ϵ is small enough, and higher-order terms can be neglected, the response of the system is well described by the *first-order transfer function* $T(\omega)$.

Of course, the response of retinal cells to stimuli is not linear, and therefore one can determine only their first-order frequency response. However, measurement of the amplitudes of higher harmonics ($e^{2i\omega t}, e^{3i\omega t}, e^{4i\omega t}$, etc.) in response to an input of the form $e^{i\omega t}$ indicates that nonlinearities have little effect for the light stimuli used in the experiments. Thus, the behavior of retinal cells can be described well by their first-order frequency responses.

Suppose $V_0(x)$ denotes the response to a steady input x. If the input is of the form $I(t) = I_0 + \epsilon e^{i\omega t}$, then the output is of the form $V(t) = V_0(I_0) + \epsilon V_1(w; I_0)e^{i\omega t} + O(\epsilon^2)$. The function $V_1(w; I_0)$ is the first-order transfer function, or first-order frequency response. When $\omega = 0$, this becomes $V_0(I_0 + \epsilon) = V_0(I_0) + \epsilon V_1(0; I_0) + O(\epsilon^2)$. However, expanding $V_0(I_0 + \epsilon)$ in a Taylor series around I_0 gives $V_0(I_0 + \epsilon) = V_0(I_0) + \epsilon V_0'(I_0) + O(\epsilon^2)$, from which it follows that the steady-state sensitivity is

$$\frac{dV_0}{dI_0} = V_1(0; I_0), \tag{19.118}$$

an identity that is of considerable use.

19.9 EXERCISES

1. Show that changing σ in the Naka–Rushton equation (19.3) translates the curve along the $\log I$ axis.

2. Compare the two expressions (19.8) and (19.9) for the sensitivity. (Hint: Choose the arbitrary values $\sigma_D = 0.1$, $k = 1$ and plot (19.8), then adjust the value of A until the curves look the same.)

3. In this exercise, some of the solutions of the model of light adaptation in turtle cones are obtained (Section 19.2.2).

   (a) Before one can calculate solutions of the model, the expression for $g(y)$ must be derived. By requiring the steady-state solution of the model to be

   $$V_0 = -s_1 \log(1 + s_2 I_0), \tag{19.119}$$

   show that

   $$g(y) = \left(ye^{E(1-y)/V^*}\right)^{1/3}\left(\delta + \frac{(\gamma - \delta)\eta(e^{-E(1-y)/s_1} - 1)}{s_2 k_1 + \eta(e^{-E(1-y)/s_1} - 1)}\right). \tag{19.120}$$

   Plot $g(y)$ (using the parameters given in Table 19.1) and compare to the functional form found experimentally (19.35), i.e.,

   $$A(y) = 4 + \frac{84}{1 + (y/0.34)^4}. \tag{19.121}$$

   (b) Calculate the impulse responses of the model for a range of background light levels. For each background light level I_0, let the magnitude of the impulse be I_0 also. Do this for $I_0 = 0.5, 0.05, 0.005, 0.0005$ and 0.00005. How close to linear are the responses? What can be said about the contrast sensitivity?

   (c) Scale each of the impulse responses calculated above so that each has a maximum value of 1, and plot them all on the same graph. How do the impulse responses change as the light level increases?

(d) Replace $g(y)$ by $A(y)$ and recalculate the steady state V_0 and the impulse responses (at the same background light levels as in part (b) above). Are there any significant differences?

4. In 1989, Forti et al. published a model of phototransduction in newt rods. In many respects their model is similar to the model discussed here. However, one major difference is in how they chose to model the initial stages of the light response. Instead of using a linear filter (which is, to be sure, a bit simplistic), Forti et al. assumed that the initial stages of the light response could be modeled by the reactions

$$\mathrm{R} \overset{h\nu}{\longrightarrow} \mathrm{R}^* \overset{\alpha_1}{\longrightarrow} \tag{19.122}$$

$$\mathrm{R}^* + \mathrm{T} \overset{\epsilon}{\longrightarrow} \mathrm{R}^* + \mathrm{T}^* \tag{19.123}$$

$$\mathrm{T}^* \overset{\beta}{\longrightarrow} \tag{19.124}$$

$$\mathrm{T}^* + \mathrm{P} \underset{\tau_2}{\overset{\tau_1}{\rightleftarrows}} \mathrm{T}^* + \mathrm{P}^* \tag{19.125}$$

Here, R denotes rhodopsin, T denotes transducin and P denotes PDE. The * superscript denotes the activated form, and $h\nu$ denotes the action of light. The parameter values of the model are given in Table 19.4. (These are not exactly the reactions assumed by Forti et al. as we have omitted the inactivated rhodopsin state. However, this makes almost no difference to the results presented here.)

(a) Write down the differential equations for this system of reactions, assuming the amount of rhodopsin is not limiting, and solve them numerically. Plot $P^*(t)$ for a variety of stimuli, each 10 ms in duration, with magnitudes 0.2, 0.5, 1, 2, 5, 10, 20, 50, 100 and 500. What can you conclude about the linearity of the response when the magnitude of the stimulus is small? What can you conclude about the linearity of the response as the magnitude of the stimulus increases?

(b) Compare the model in part (a) with the detailed reaction scheme in Fig. 19.7. Where do they disagree? Modify the model so that it agrees better with the more detailed reaction scheme. (Hint: The conservation equation for transducin must be considered more carefully, as must the differential equation for T*.) Solve the new model numerically for the same stimuli as part (a), and compare with the solutions of the Forti et al. model. Explain the differences.

(c) Compare the model of part (a) to the linear cascade model in the book. Where are they similar? Where are they not similar? Why?

Table 19.4 Parameter values for the model of the initial stages of the light response in newt rods (Forti et al., 1989). P_0 is the total amount of PDE, and T_0 is the total amount of transducin.

$T_0 = 1000\ \mu$M	$P_0 = 100\ \mu$M
$\alpha_1 = 20$ s^{-1}	$\epsilon = 0.5$ s^{-1} μM^{-1}
$\tau_1 = 0.1$ s^{-1} μM^{-1}	$\tau_2 = 10$ s^{-1}
$\beta_1 = 10.6$ s^{-1}	

5. cGMP is hydrolyzed to GMP by PDE or PDE* in an enzymatic reaction, and as was described in Chapter 1, such reactions do not necessarily follow the law of mass action. Assuming that cGMP reacts with PDE according to

$$\text{cGMP} + \text{PDE} \underset{k_{-1}}{\overset{k_1}{\rightleftarrows}} \text{complex} \overset{k_2}{\rightarrow} \text{GMP} + \text{PDE}, \tag{19.126}$$

derive the conditions under which (19.18) may be expected to apply, keeping in mind that [PDE] is much larger than [cGMP], and thus the usual approximation of enzyme kinetics does not apply. (See also Exercise 14 of Chapter 1.)

6. Show that the current generated by the electrogenic exchange pumps should not be ignored in the modeling of a photoreceptor. (Hint: By writing down the balance equations for Na^+ and Ca^{2+} show that the outward current generated by the electrogenic Na^+–K^+ pump in the inner segment must be approximately a third of the total inward light-sensitive current.) How significant is the current generated by the Na^+–Ca^{2+}, K^+ exchanger in the outer segment? Should the electrogenic pumps be included in the model if the model is compared to photocurrent measurements and not to voltage measurements?

7. Find the solution of (19.44)–(19.47) in the case that $k_i = k$ for all i. Find the mean and variance for the time of inactivation.

8. Calculate the response of the lateral inhibition model (Section 19.4) when the light stimulus is a strip of width $2a$, modulated sinusoidally with frequency ω. Show that as a increases (i.e., as the stimulus goes from a spot to a uniform field), the gain at $x = 0$ becomes more band-pass in nature.

   Hint: Show that the solution for $\hat{I}$ is of the form

$$\hat{I} = \begin{cases} \frac{\lambda\hat{L}}{(i\omega\tau+1)(1+\lambda+i\omega)}[1 + A\cosh(x\sqrt{1+\lambda+i\omega})], & |x| < a, \\ B\exp(-|x|\sqrt{1+\lambda+i\omega}), & |x| > a, \end{cases} \tag{19.127}$$

   and require that $\hat{I}$ and its derivative be continuous at $|x| = a$. Define the gain as the response divided by the stimulus, and calculate the amplitude of the gain at $x = 0$ as a function of ω and a. (The details are given in Peskin (1976)).

9. Calculate the response of the lateral inhibition model (Section 19.4) to a moving step of light. Show that as the speed tends to zero, the response approaches the response to a steady step, while as the speed goes to infinity, the response behaves like the response to a step of light presented simultaneously to the entire retina. (This question is taken from Peskin (1976)).

   Hint: To exhibit the space-like limit write R as a function of $x + ct$, where c is the speed of the moving step. To exhibit the timelike limit, write R as a function of $t + x/c$.

10. Derive (19.69) and verify (19.71). Hints: Show first that everywhere except $r = s$, G satisfies the modified Bessel equation of order zero, $z^2\frac{d^2u}{dz^2} + z\frac{du}{dz} - z^2u = 0$. Two independent solutions of the modified Bessel equation are the modified Bessel functions of the first and second kind. Then show that the jump condition at $r = s$ is

$$\left.\frac{dG}{dr}\right|_{r=s^+} - \left.\frac{dG}{dr}\right|_{r=s^-} = -\frac{1}{s}, \tag{19.128}$$

   and use the fact that the Wronskian W is

$$W(K_\nu, I_\nu) = K_\nu(z)\frac{d}{dz}I_\nu(z) - I_\nu(z)\frac{d}{dz}K_\nu(z) = \frac{1}{z}. \tag{19.129}$$

11. Show that the Krausz–Naka model (Section 19.4.2) is similar to the Peskin model (Section 19.4.1), with a few more details. Thus, investigate the response of the Krausz–Naka model to a space-independent light step and a time-independent bar of light. Demonstrate Mach bands and adaptation, as in the Peskin model.

12. (a) By taking the Fourier transform of (19.114) show that the transfer function is given by (19.115).

    (b) Calculate the impulse response of (19.114) by solving the differential equation directly. (*Hint*: First integrate from $t = -\epsilon$ to $t = \epsilon$ and let $\epsilon \to 0$ to obtain a jump condition at $t = 0$. Then use that jump condition as an initial condition for the differential equation.)

CHAPTER 20

The Inner Ear

The mammalian ear has three major components: the outer, middle, and inner ears (Fig. 20.1A). The outer ear consists of a cartilaginous flange, the *pinna*, incorporating a resonant cavity that connects to the *ear canal* and finally to the *tympanic membrane*. It performs an initial filtering of the sound waves, increasing the sound pressure gain at the tympanic membrane in the 2 to 7 kHz region. It also aids sound localization. Bats, for example, have highly developed pinnae, with a high degree of directional selectivity. Although less efficient in humans, the outer ear accounts for our ability to distinguish whether sounds come from above or below, in front or behind.

The function of the middle ear is to transmit the sound vibrations from the tympanic membrane to the cochlea. Because of the much higher impedance of the cochlear fluid, the middle ear also functions as an impedance-matching device, focusing the energy of the tympanic membrane on the oval window of the cochlea. If not for impedance matching, much of the energy of the sound waves in air would be reflected by the cochlear fluid. This impedance matching is carried out by the *ossicles*, three small bones, the *malleus, incus*, and *stapes*, that connect the tympanic membrane to the oval window. The tympanic membrane has a much higher surface area than the oval window, and the ossicles act as levers that increase the force at the expense of velocity, resulting in the required concentration of energy at the oval window.

Most of the events central to hearing occur in the inner ear, in particular the *cochlea*. The vestibular apparatus (the semicircular canals and the otolith organs) are also in the inner ear, but their principal function is the detection of movement and acceleration, not sound. The cochlea is a tube, about 35 mm long, divided longitudinally into three compartments and twisted into a spiral (Fig. 20.1B). The three compartments are the *scala vestibuli*, the *scala tympani*, and the *scala media*, and they wind around the spiral together, preserving their spatial orientation. Reissner's membrane separates the scala

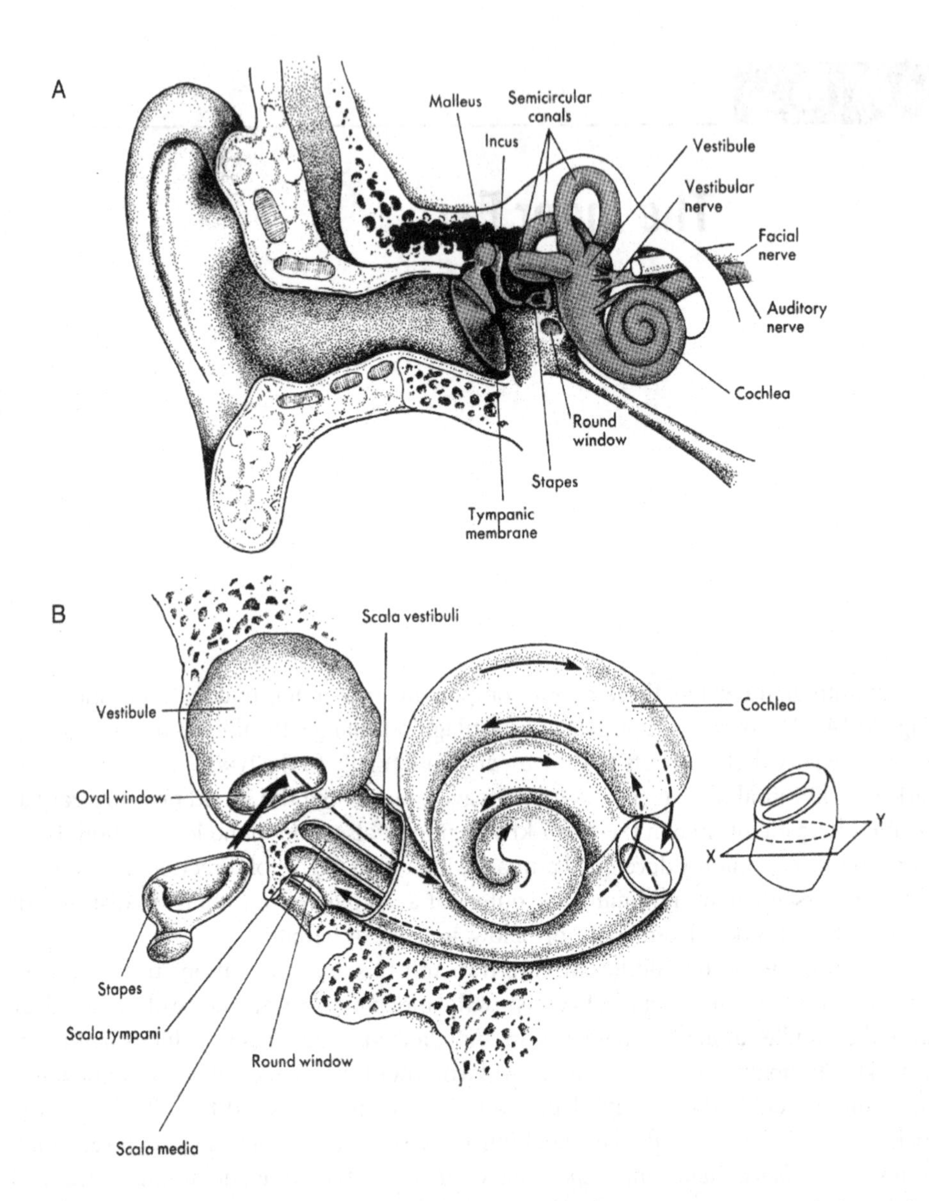

Figure 20.1 Location and structure of the cochlea. A: Location of the cochlea in relation to the middle ear, the tympanic membrane, and the outer ear. B: Diagram of the cochlea at increased magnification, showing its spiral structure and the relative positions of the two larger internal compartments, the scala vestibuli and the scala tympani. (Berne and Levy, 1993, Fig. 10-6.)

vestibuli from the scala media, which in turn is separated from the scala tympani by the *spiral lamina* and the *basilar membrane* (Fig. 20.2A). The scala vestibuli and the scala tympani are filled with *perilymph*, a fluid similar to extracellular fluid, while the scala media is filled with *endolymph*, a fluid with a high K^+ concentration and a low Na^+ concentration. Sound waves transmitted through the middle ear are focused by the

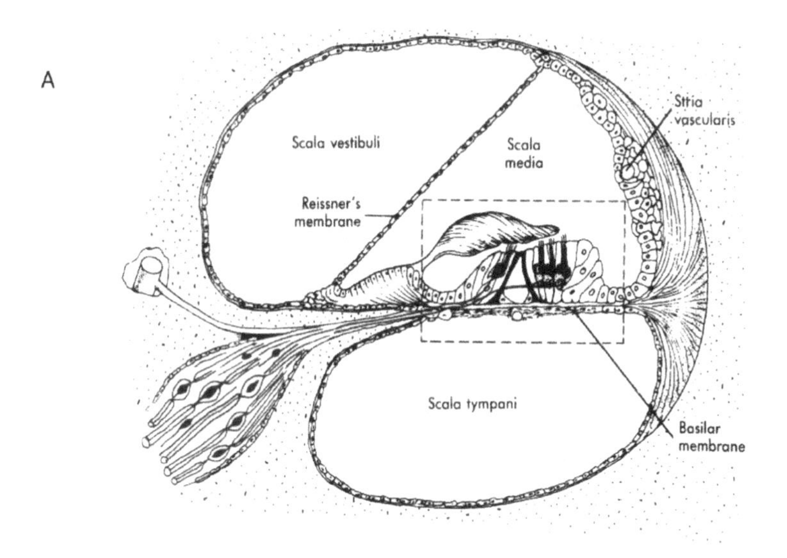

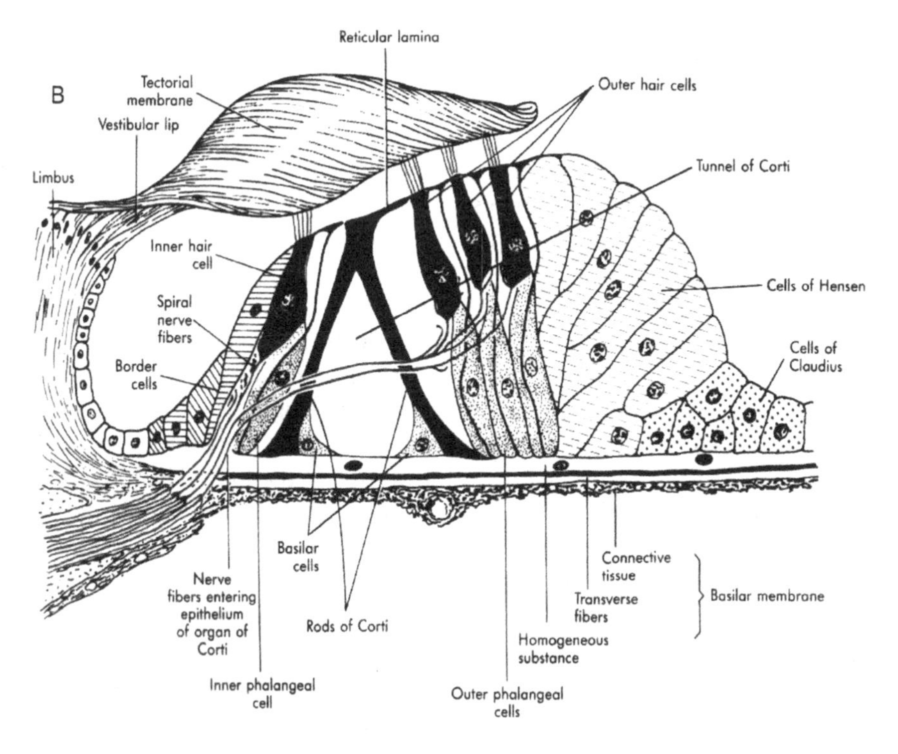

Figure 20.2 Location and structure of the cochlea, continued. A: Cross-section of the cochlea in the plane indicated by the inset in Fig. 23.1B. B: Enlarged view of the organ of Corti, including the basilar membrane, the tectorial membrane, and the hair cells. (Berne and Levy, 1993, Fig. 10-6.)

stapes onto the oval window, an opening into the scala vestibuli. The resultant waves in the perilymph travel along the length of the scala vestibuli, creating complementary waves in the basilar membrane and the scala tympani. At the end of the cochlea, an opening between the scala vestibuli and the scala tympani, the *helicotrema*, equalizes the local pressure in the two compartments. Because the perilymph is essentially incompressible, it is necessary for the scala tympani also to have an opening analogous to the oval window; otherwise, conservation of mass would preclude movement of the stapes. The opening in the scala tympani is called the *round window*. Inward motion of the stapes at the oval window is compensated for by the corresponding outward motion of fluid at the round window.

Transduction of sound into electrical impulses is carried out by the *organ of Corti* (Fig. 20.2B), which sits on top of the basilar membrane. *Hair cells* in the organ of Corti have hairs projecting out the top, and these hairs touch a flap called the *tectorial membrane* that sits over the organ of Corti. Waves in the basilar membrane create a shear force on these hairs, which in turn causes a change in the membrane potential of the hair cell. This is transmitted to nerve cells, and from there to the brain.

20.1 Frequency Tuning

The task of the cochlea is to identify the constituent frequencies of a sound wave, and thus identify the sound. The different ways in which this is accomplished in different animals fall into three principal groupings: mechanical tuning of the hair cells, mechanical tuning of the basilar membrane, and electrical tuning of the hair cells (Hudspeth, 1985; Eatock, 2000).

One of the earliest theories of frequency tuning was that of Helmholtz (1875) who proposed that the ear consists of an array of sharply tuned elements, each resonating with a particular frequency. Although this is far from the full story, resonance plays an important role. For example, in many lizards, the length of the hair bundles on the hair cells increases systematically from the base to the apex. In much the same way that a longer string produces notes of lower pitch, the longer hair cell bundles respond preferentially to inputs of lower frequency, while the short bundles are tuned to higher frequencies. Thus, the input frequency can be determined by the position of maximal stimulation. In mammals, the basilar membrane itself acts as a frequency analyzer, and this is discussed in the next section. The third tuning mechanism results from the properties of ionic channels in the hair cell membrane. Each hair cell is an electrical resonator, with a band-pass frequency response. The input frequency that gives the greatest response is a function of the biophysical properties of the hair cell, and the systematic variation of these properties along the length of the cochlea allows the cochlea to distinguish between frequencies based on the position of maximal response.

We do not discuss the first tuning mechanism further, but concentrate on the remaining two. We begin by looking at models of the basilar membrane that demonstrate

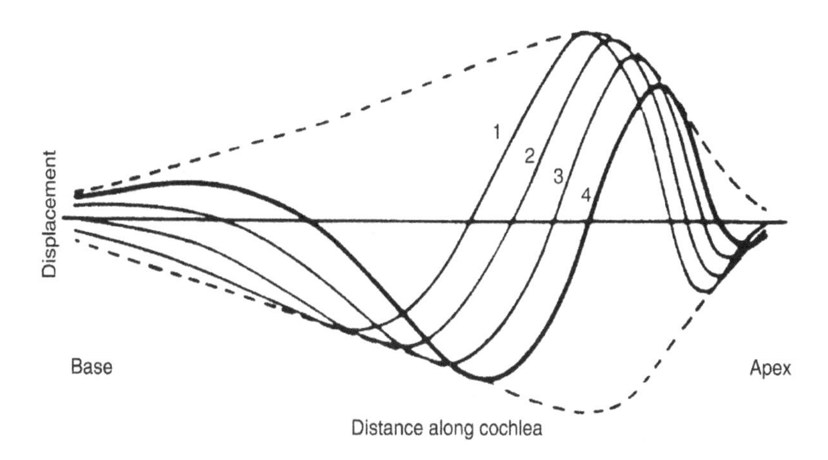

Figure 20.3 Membrane waves and their envelope in the cochlea. The solid lines show the deflection of the basilar membrane at successive times, denoted (in order of increasing time) by 1, 2, 3, 4. The dashed line is the envelope of the membrane wave, and remains constant over time. (von Békésy, 1960, Fig. 12-17.)

mechanical tuning, and then discuss models of resonance of the hair cell membrane potential.

20.1.1 Cochlear Macromechanics

In mammals, vibrations of the stapes set up a wave with a particular shape on the basilar membrane. The amplitude envelope of the wave is first increasing, then decreasing, and the position of the peak of the envelope is dependent on the frequency of the stimulus (von Békésy, 1960), as illustrated in Fig. 20.3. The wave speed decreases as it moves along the membrane, resulting in a continual decrease in phase, and an apparent increase in frequency. Low-frequency stimuli have a wave envelope that peaks closer to the apex of the cochlea (i.e., near the helicotrema), and as the frequency of the stimulus increases, the envelope peak moves toward the base of the cochlea, as illustrated in Fig. 20.4.

The amplitude of the envelope is a two-dimensional function of distance from the stapes and frequency of stimulation; the curves shown in Fig. 20.4 are cross-sections of the function for fixed frequency. Another way to present the data is to give cross-sections for a fixed distance. This gives the envelope amplitude as a function of frequency, for a fixed distance from the stapes, i.e., the frequency response of the basilar membrane for that fixed distance. Frequency responses measured by von Békésy are shown in Fig. 20.5, from which it can be seen that each part of the basilar membrane responds maximally to a certain frequency, and as the frequency increases, the site of maximum response moves toward the stapes. In this way the cochlea determines the frequency of the incoming signal from the place on the basilar membrane of maximal amplitude, the so-called *place theory* of hearing.

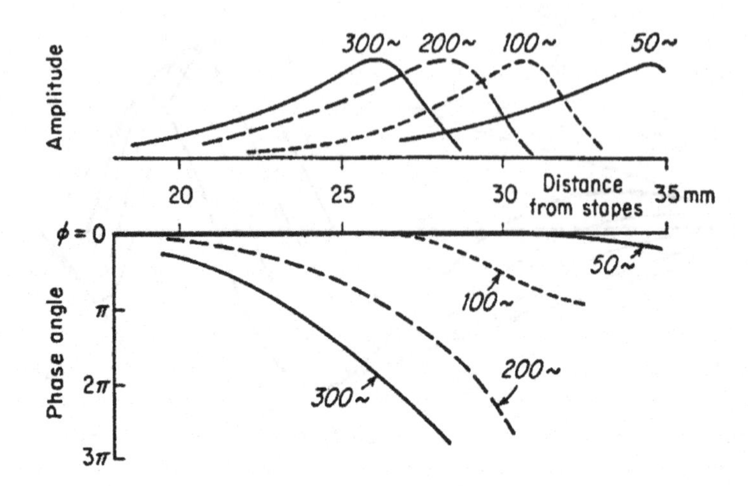

Figure 20.4 Amplitude and phase of the cochlear membrane wave for four different frequencies. As the frequency of the wave increases, the peak of the wave envelope moves toward the base of the cochlea (i.e., toward the oval and round windows). (von Békésy, 1960, Fig. 11-58.)

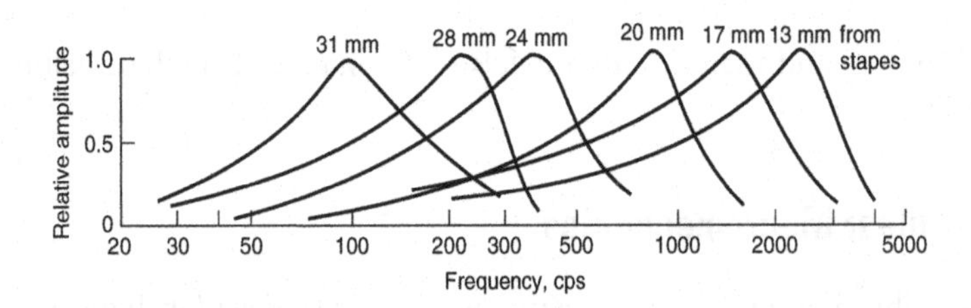

Figure 20.5 Frequency responses of the basilar membrane measured at different distances from the stapes. Only the amplitude is shown. Close to the stapes, the basilar membrane responds preferentially to tones of high frequency, while farther away from the stapes, the membrane responds preferentially to tones of lower frequencies. (von Békésy, 1960, Fig. 11-49.)

Although many of von Békésy's results have since been superseded (he performed his experiments, somewhat gruesomely, on cadavers, but it is now known that the properties of the basilar membrane and hair cells in a living person are different), the experimental results of von Békésy, and the associated place theory, were some of the most important studies of the cochlea in the 20th century, and have had a huge effect on theoretical studies.

The name "basilar membrane" is misleading, as it is not a true membrane. This is shown by the fact that when it is cut, the edges do not retract. Thus it is not under tension; resistance to movement comes from the bending elasticity. The stiffness of the basilar membrane decreases exponentially from the base to the apex, with a length constant of about 7 mm. Although the width of the cochlea decreases from the base to the apex, the width of the basilar membrane increases in this direction.

Models of waves on the basilar membrane can be distinguished by the types of equations used for the membrane and the fluid. Early models by Ranke (1950) and Zwislocki (1965) assumed the perilymph to be incompressible and inviscid, and modeled the basilar membrane as a damped, forced harmonic oscillator, with no elastic coupling along the length of the membrane. Ranke used deep-water wave theory, while Zwislocki used shallow-water wave theory, leading to considerable controversy over which was the best approach. These models were developed by many authors, the best known being due to Peterson and Bogert (1950), Fletcher (1951), Lesser and Berkley (1972), and Siebert (1974). Subsequent models by Steele (1974), Inselberg and Chadwick (1976), Chadwick et al. (1976), Chadwick (1980), and Holmes (1980a,b, 1982) used more sophisticated representations of the basilar membrane as an elastic plate and incorporated fluid viscosity and the geometry of the plate. Surveys of experimental and theoretical results can be found in Dallos et al. (1990, 1996). Here we give an overview of some of the earlier and simpler models, as they provide elegant demonstrations of how the basilar membrane and the perilymph can interact to give the types of waves observed by von Békésy.

20.2 Models of the Cochlea

20.2.1 Equations of Motion for an Incompressible Fluid

The fluid in the cochlea surrounding the basilar membrane is incompressible, and assumed to be inviscid. The equations of motion of this fluid are well known, and are derived in many places (e.g., Batchelor, 1967).

We let $\mathbf{u} = (u_1, u_2, u_3)$ be the fluid velocity, p the pressure, and ρ the density of the fluid. The mass of fluid in a fixed volume V can change only in response to fluid flux across the boundary of the volume. Thus,

$$\frac{d}{dt}\int_V \rho \, dV = -\int_S \rho(\mathbf{u}\cdot\mathbf{n})\, dS, \tag{20.1}$$

where S is the surface of V, and $\mathbf{n} = (n_1, n_2, n_3)$ is the outward unit normal to V. Similarly, the momentum of the fluid in a fixed domain V can change only in response to applied forces or to the flux of momentum across the boundary of the domain. Thus (for an inviscid fluid) conservation of momentum implies that

$$\frac{d}{dt}\int_V \rho u_i \, dV = -\int_S [(\mathbf{u}\cdot\mathbf{n})\rho u_i + p n_i]\, dS. \tag{20.2}$$

Using the divergence theorem to convert surface integrals to volume integrals, we obtain

$$\int_V \left(\frac{\partial \rho u_i}{\partial t} + \rho\nabla\cdot(u_i\mathbf{u}) + \frac{\partial p}{\partial x_i}\right) dV = 0, \tag{20.3}$$

$$\int_V \frac{\partial \rho}{\partial t} + \nabla\cdot\mathbf{u}\, dV = 0. \tag{20.4}$$

Finally, since V is arbitrary, and assuming that the density ρ is constant, it follows that

$$\rho\frac{\partial \mathbf{u}}{\partial t} + \rho(\nabla\cdot\mathbf{u})\mathbf{u} + \nabla p = 0, \tag{20.5}$$

$$\nabla\cdot\mathbf{u} = 0. \tag{20.6}$$

When the fluid motions are of small amplitude, as is expected to be true in the cochlea, the nonlinear terms may be ignored, yielding

$$\rho\frac{\partial \mathbf{u}}{\partial t} + \nabla p = 0, \tag{20.7}$$

$$\nabla\cdot\mathbf{u} = 0. \tag{20.8}$$

An important special case is when $\mathbf{u} = \nabla\phi$ for some potential ϕ (an irrotational flow), in which case (20.7) and (20.8) become

$$\rho\frac{\partial\phi}{\partial t} + p = 0, \tag{20.9}$$

$$\nabla^2\phi = 0, \tag{20.10}$$

where p is normalized so that the constant of integration is zero.

20.2.2 The Basilar Membrane as a Harmonic Oscillator

One of the simplest models of the cochlea combines (20.9) and (20.10) with the equation of a damped, forced harmonic oscillator. One of the clearest presentations of this model is due to Lesser and Berkley (1972). In their model, the cochlea is assumed to have a configuration as shown in Fig. 20.6. Thus, letting subscripts 1 and 2 denote quantities in the upper and lower compartments, respectively, we have two copies of (20.9) and (20.10),

$$\rho\frac{\partial\phi_1}{\partial t} + p_1 = \rho\frac{\partial\phi_2}{\partial t} + p_2 = 0, \tag{20.11}$$

$$\nabla^2\phi_1 = \nabla^2\phi_2 = 0, \tag{20.12}$$

where the pressure is determined only up to an arbitrary constant.

Each point of the basilar membrane is modeled as a simple damped harmonic oscillator with mass, damping, and stiffness that vary along the length of the membrane. Thus, the movement of any part of the membrane is assumed to be independent of the movement of neighboring parts of the membrane, as there is no direct lateral coupling. The position of the basilar membrane, $y = \eta(x,t)$, is specified by

$$m(x)\frac{\partial^2\eta}{\partial t^2} + r(x)\frac{\partial\eta}{\partial t} + k(x)\eta = p_2(x,\eta(x,t),t) - p_1(x,\eta(x,t),t), \tag{20.13}$$

where $m(x)$ is the mass per unit area of the basilar membrane, $r(x)$ is its damping coefficient, and $k(x)$ is its stiffness (Hooke's constant) per unit area.

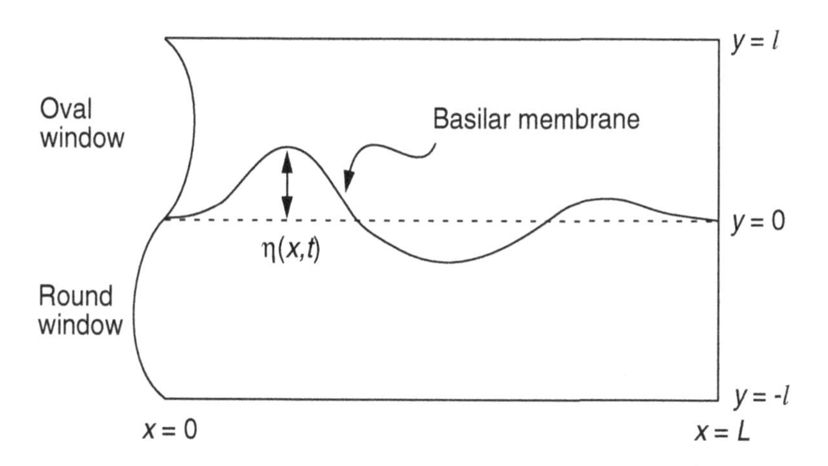

Figure 20.6 Schematic diagram of the cochlea, adapted from the model of Lesser and Berkley (1972). The cochlea is modeled as having two rectangular compartments filled with fluid, separated by the basilar membrane. The upper compartment corresponds to the scala vestibuli, and the lower compartment to the scala tympani. For simplicity, the scala media, shown in Fig. 20.1, is omitted from the model.

Boundary conditions are determined by requiring that the fluid velocity at the boundary matches the velocity of the boundary. Thus,

$$\frac{\partial \eta}{\partial t} = \frac{\partial \phi_1}{\partial y} = \frac{\partial \phi_2}{\partial y}, \qquad y = \eta(x,t),\ 0 < x < L. \tag{20.14}$$

We further assume that there is no vertical motion at the top, so that

$$\frac{\partial \phi_1}{\partial y} = 0, \qquad y = l,\ 0 < x < L. \tag{20.15}$$

As formulated, this is a very difficult free boundary problem. A significant simplification occurs if we assume that the displacement is small. Then, the leading-order linear problem has a driving force which is the pressure difference evaluated at $y = 0$, rather than at $y = \eta$, so that

$$m(x)\frac{\partial^2 \eta}{\partial t^2} + r(x)\frac{\partial \eta}{\partial t} + k(x)\eta = p_2(x,0,t) - p_1(x,0,t). \tag{20.16}$$

The boundary conditions also simplify to

$$\frac{\partial \eta}{\partial t} = \frac{\partial \phi_1}{\partial y} = \frac{\partial \phi_2}{\partial y}, \qquad y = 0,\ 0 < x < L. \tag{20.17}$$

There are a number of ways to specify how the system is externally forced. One way, due to Lesser and Berkley, is to assume that the motion of the stapes in contact with the oval window determines the position of the oval window. Since $\partial\phi/\partial x$ is the x

component of the fluid velocity, the boundary condition at $x = 0$ is

$$\frac{\partial \phi_1}{\partial x} = \frac{\partial F(y,t)}{\partial t}, \qquad 0 < y < l, \tag{20.18}$$

where $F(y,t)$ is the specified horizontal displacement of the oval window. Further, we assume that there is no horizontal motion at the far end, so that at $x = L$

$$\frac{\partial \phi_1}{\partial x} = 0, \qquad 0 < y < l. \tag{20.19}$$

20.2.3 An Analytical Solution

Because of the inherent symmetry of the problem, we seek solutions that are odd in y (Exercise 1). Thus, we consider only the problem in the upper region and drop the subscript 1.

When the input has a single frequency, $F(y,t) = \hat{F}(y)e^{i\omega t}$, then $\phi(x,y,t)$ is of the form $\hat{\phi}(x,y;\omega)e^{i\omega t}$ and similarly for the other variables. Looking for solutions of this form for all the variables, we obtain the equations

$$\nabla^2\hat{\phi} = 0, \qquad \hat{p} + i\omega\rho\hat{\phi} = 0, \tag{20.20}$$

$$\frac{\partial \hat{\phi}}{\partial y} = i\omega\hat{\eta}, \qquad i\omega\hat{\eta}Z = -\hat{p}, \qquad \text{on } y = 0, \tag{20.21}$$

$$\frac{\partial \hat{\phi}}{\partial x} = U_0, \qquad \text{on } x = 0, \tag{20.22}$$

$$\frac{\partial \hat{\phi}}{\partial x} = 0, \qquad \text{on } x = L, \tag{20.23}$$

$$\frac{\partial \hat{\phi}}{\partial y} = 0, \qquad \text{on } y = l, \tag{20.24}$$

where $Z = i\omega m + r + k/(i\omega)$ and $U_0 = i\omega\hat{F}$. By looking for solutions in the frequency domain, we have transformed the differential equations on the basilar membrane into algebraic equations. The term $i\omega Z$ is the frequency response of the damped harmonic oscillator, and Z, the *impedance*, is a function of x. Also note that in (20.21) the pressure is assumed to be an odd function of y.

Finally, we nondimensionalize the model equations by scaling x and y by L, Z by $i\omega\rho L$, and $\hat{\phi}$ by U_0L; rearranging; and dropping the hats we get

$$\nabla^2\phi = 0, \tag{20.25}$$

$$\frac{\partial \phi}{\partial y} = \frac{2\phi}{Z}, \qquad \text{on } y = 0, \tag{20.26}$$

$$\frac{\partial \phi}{\partial x} = 1 \qquad \text{on } x = 0, \tag{20.27}$$

$$\frac{\partial \phi}{\partial x} = 0 \qquad \text{on } x = 1, \tag{20.28}$$

$$\frac{\partial \phi}{\partial y} = 0 \qquad \text{on } y = \sigma, \tag{20.29}$$

where $\sigma = l/L$.

The analytical solution of this problem can be found using standard Fourier series. We look for solutions of the form

$$\phi = x\left(1 - \frac{x}{2}\right) - \sigma y\left(1 - \frac{y}{2\sigma}\right) + \sum_{n=0}^{\infty} A_n \cosh[n\pi(\sigma - y)] \cos(n\pi x), \tag{20.30}$$

for some, as yet undetermined, constants A_n. Since ϕ satisfies all the boundary conditions except (20.26), we use (20.26) to determine the unknown coefficients A_n. This gives

$$\begin{aligned} &\sigma + \sum_{n=0}^{\infty} n\pi A_n \sinh(n\pi\sigma)\cos(n\pi x) \\ &\quad - \frac{2}{Z}\left[x(1 - \frac{x}{2}) + \sum_{n=0}^{\infty} A_n \cosh(n\pi\sigma)\cos(n\pi x)\right] = 0. \end{aligned} \tag{20.31}$$

Multiplying by $\cos(m\pi x)$, and integrating from 0 to 1, we find

$$A_m \alpha_m = f_m, \tag{20.32}$$

where

$$\alpha_m = \frac{1}{Z}\cosh(m\pi\sigma) - \frac{1}{2} n\pi \sinh(m\pi\sigma) \tag{20.33}$$

and

$$f_m = \sigma\delta_{m0} - \int_0^1 \frac{x(2-x)\cos(m\pi x)}{Z}\, dx = -\frac{2}{m^2\pi^2}. \tag{20.34}$$

The coefficients A_n can now be evaluated explicitly, and substituted into (20.30). Typical results are shown in Fig. 20.7. The wave envelope has the same qualitative shape as von Békésy's results (Fig. 20.3), and the peak of the wave envelope moves toward the base of the cochlea as the frequency is increased.

20.2.4 Long-Wave and Short-Wave Models

Although the Fourier solution shows that the behavior of the Lesser and Berkley model is qualitatively correct, it would be nice to get a better analytical understanding of the behavior of the basilar membrane. There are two classic approximations of the model equations that allow further analytical investigation. The long-wave approximation, studied by Zwislocki and others, assumes that the wavelength is long compared to the depth of the cochlea, and the short-wave approximation of Ranke assumes the opposite,

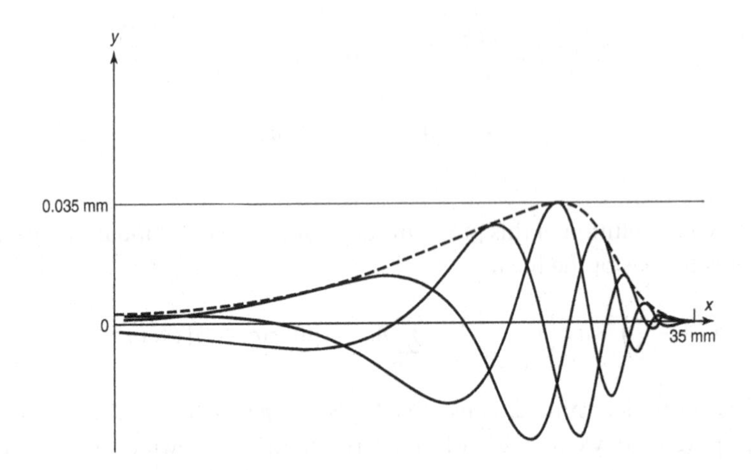

Figure 20.7 Results from the Lesser and Berkley model, showing a typical wave on the basilar membrane and the wave envelope. Parameters are $m = 0.05$ g/cm^2, $k = 10^7 e^{-1.5x}$ dynes/cm^3, $r = 3000e^{-1.5x}$ dynes sec/cm^3, $\omega = 1000$/sec. The perilymph was assumed to have the same density as water, 1 g/cm^3 (Lesser and Berkley, 1972, Fig. 6.)

that the cochlea is effectively infinitely deep. Experiments suggest that the depth of the cochlea has little effect on the cochlear wave, supporting the short-wave theory. Indeed, even if one side of the cochlea is completely removed, there is little effect on the wave. However, neither model gives a complete description of cochlear behavior (Zwislocki, 1953).

Both short-wave and long-wave models can be derived as approximate cases of the model described in Section 20.2.2. To show this, we use a generalized form of the previous model (Siebert, 1974), as illustrated in Fig. 20.8. The only change is to assume that there is a direct mechanical forcing at the two ends of the basilar membrane. Modifying the equation of membrane motion (20.16) to include this direct forcing gives

$$m(x)\frac{\partial^2 \eta}{\partial t^2} + r(x)\frac{\partial \eta}{\partial t} + k(x)\eta = p_2(x,0,t) - p_1(x,0,t) + F_0(t)\delta(x) - F_L(t)\delta(x-L). \quad (20.35)$$

As before, we assume that the forcing is at a single frequency with $F_0(t) = F_0 e^{i\omega t}$ and $F_L(t) = F_L e^{i\omega t}$. It follows from (20.17), (20.20), and (20.35) that

$$\nabla^2 p(x,y) = 0, \quad (20.36)$$

$$-i\omega\eta = \frac{1}{i\omega\rho}\frac{\partial p(x,0)}{\partial y}, \quad (20.37)$$

$$Y(x)p(x,0) = -i\omega\eta(x) + \eta_0\delta(x) - \eta_L\delta(x-L), \quad (20.38)$$

where $Y = 2/Z$, $\eta_0 = \frac{F_0}{Z(0)}$, $\eta_L = \frac{F_L}{Z(L)}$, and where we have dropped the hats associated with the Fourier transform.

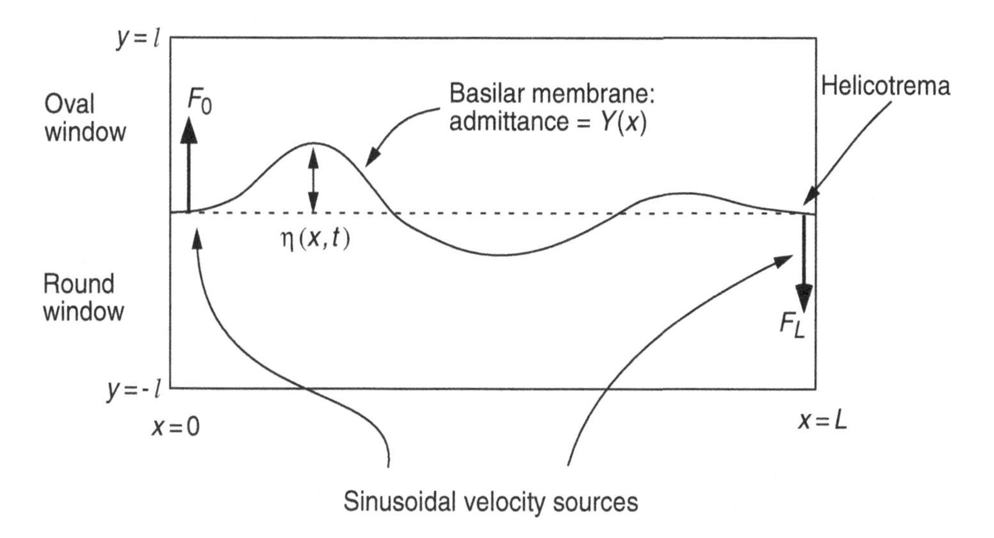

Figure 20.8 Schematic diagram of the cochlea model of Siebert (1974). It differs from the model of Lesser and Berkley in the boundary conditions at $x = 0$ and $x = L$, where it is assumed that there is direct mechanical forcing at both ends of the membrane.

Because Laplace's equation is separable on a rectangular domain, we use Fourier series to write the solution of (20.36) as

$$p(x, y) = \sum_{n=-\infty}^{\infty} \alpha_n \frac{\cosh[2\pi n(y-l)/L]}{\cosh(2\pi nl/L)} e^{2\pi inx/L}, \tag{20.39}$$

where we have used the boundary condition $\partial p/\partial y = 0$ on $y = l$. It follows that

$$-i\omega\eta = \frac{1}{i\omega\rho}\frac{\partial p(x,0)}{\partial y} = -\frac{1}{i\omega\rho}\sum_{n=-\infty}^{\infty} \alpha_n \frac{2\pi n}{L}\tanh(2\pi nl/L)e^{2\pi inx/L}. \tag{20.40}$$

Since our only interest is in the behavior of the basilar membrane, from now on we restrict our attention to $p(x, 0)$, which we denote by $p(x)$.

The Shallow-Water Approximation

In the shallow-water approximation, the wavelengths of the waves on the basilar membrane are assumed to be greater than the depth of the cochlea. As a consequence, we assume that $\alpha_n = 0$ for all $n > N$, for some integer N such that $Nl/L \ll 1$. Since the sum over n includes only those terms with $nl/L \ll 1$, it follows that for each term in the sum, $\tanh(2\pi nl/L)$ can be approximated by the lowest-order term in its Taylor expansion. Thus, $\tanh(2\pi nl/L) \approx 2\pi nl/L$, and so (20.40) becomes

$$-i\omega\eta \approx -\frac{l}{i\omega\rho}\sum_{n=-\infty}^{\infty} \alpha_n \left(\frac{2\pi n}{L}\right)^2 e^{2\pi inx/L}. \tag{20.41}$$

However, a similar argument applied to (20.39) implies that

$$p(x) = \sum_{n=-\infty}^{\infty} \alpha_n e^{2\pi inx/L}, \tag{20.42}$$

and thus, combining this with (20.41), we have

$$-i\omega\eta \approx \frac{l}{i\omega\rho}\frac{d^2p}{dx^2}. \tag{20.43}$$

Combining this with (20.37) and (20.38), we get a single equation for $p(x)$,

$$Y(x)p(x) = \frac{l}{i\omega\rho}\frac{d^2p(x)}{dx^2} + \eta_0\delta_0(x) - \eta_L\delta_L(x). \tag{20.44}$$

To convert the delta functions in this equation into boundary conditions, we integrate the differential equation (20.44) from $x = -\epsilon$ to $x = +\epsilon$ and let $\epsilon \to 0$ and find that

$$\frac{dp}{dx} = -i\omega\rho\eta_0 \qquad \text{at } x = 0, \tag{20.45}$$

where we have assumed that $dp/dx = 0$ at $x = 0^-$, which is outside the boundaries of the cochlea. Similarly, integrating from $x = L - \epsilon$ to $x = L + \epsilon$ and letting $\epsilon \to 0$ gives

$$\frac{dp}{dx} = i\omega\rho\eta_L \qquad \text{at } x = L. \tag{20.46}$$

Note that when $\eta_0 = 1$ and $\eta_L = 0$ (and since $p + i\omega\rho\phi = 0$), these boundary conditions are the same as (20.27) and (20.28) used in the Lesser–Berkley model.

The analysis of this equation exploits the fact that $Y(x)$ is a slowly varying function. To see what this means mathematically, note that $i\omega\rho Y(x)/l$ has dimensional units of length^{-2}, which determines the length scale (wavelength) of the spatial oscillations of $p(x)$. On the other hand, $Y(x)$ varies exponentially with a length constant of $\lambda^{-1} \approx 0.7$ cm. If the ratio of these two length constants is small, then we assert that $Y(x)$ is a slowly varying function. Furthermore, there is a rescaling of space, $x = z/q$, of (20.44), putting it into the dimensionless form

$$\frac{d^2p}{dz^2} + g^2(\epsilon z)p(z) = 0, \tag{20.47}$$

where ϵ is a small positive number and $g^2(\epsilon z) = \frac{-i\omega\rho Y(z/q)}{lq^2}$ is of order one in amplitude and slowly varying in z. Note that q is an arbitrary length scale, chosen so that g^2 is of order one in amplitude; by assumption, $\lambda/q \ll 1$.

As a specific example, suppose that $m = 0, k(x) = k_0e^{-\lambda x}, r(x) = r_0e^{-\lambda x}$, in which case

$$\frac{i\omega\rho Y(x)}{l} = \frac{-2\omega^2\rho}{lk_0}\frac{e^{\lambda x}}{1 + i\omega r_0/k_0}. \tag{20.48}$$

We set $q^2 = \frac{2\omega^2\rho}{lk_0}$, and then define ϵ by

$$\epsilon = \frac{\lambda}{2q}. \tag{20.49}$$

If the parameters are such that $\epsilon \ll 1$, we have a slowly varying oscillation.

Problems of this type are well known in the theory of oscillations and can be solved approximately using multiscale analysis (Kevorkian and Cole, 1996; Keener, 1998). We wish to find approximate solutions of (20.47). If g were a constant ($\epsilon = 0$), the solution of (20.47) would be simply

$$p(z) = Ae^{igz} + Be^{-igz}. \tag{20.50}$$

However, since g is assumed to be slowly varying, we expect this basic solution to be a reasonable local (but not global) approximation. To find a solution that has a longer range of validity, we introduce two scales, a slow scale variable $\sigma = \epsilon z$ and a fast variable τ for which $\frac{d\tau}{dz} = f(\epsilon z)$, where f is a function to be determined. It follows that the derivative $\frac{d}{dz}$ must be replaced by partial derivatives

$$\frac{d}{dz} = f(\sigma)\frac{\partial}{\partial \tau} + \epsilon\frac{\partial}{\partial \sigma}. \tag{20.51}$$

In terms of these two variables the original ordinary differential equation (20.47) becomes the partial differential equation

$$f^2(\sigma)\frac{\partial^2 p}{\partial \tau^2} + \epsilon f(\sigma)\frac{\partial^2 p}{\partial \sigma \partial \tau} + \epsilon\frac{\partial}{\partial \sigma}\left(f(\sigma)\frac{\partial p}{\partial \tau}\right) + \epsilon^2\frac{\partial}{\partial \sigma}\left(f(\sigma)\frac{\partial p}{\partial \sigma}\right) + g^2(\sigma)p = 0. \tag{20.52}$$

The obvious choice for f is $f = g$, because then the equation to leading order in ϵ is

$$\frac{\partial^2 p}{\partial \tau^2} + p = 0 \tag{20.53}$$

with general solution

$$P_0 = A(\sigma)e^{i\tau} + B(\sigma)e^{-i\tau}. \tag{20.54}$$

Notice that A and B are functions of the slow variable σ, since (20.53) is a partial differential equation.

To determine the functions A and B, we set $p = P_0 + \epsilon P_1 + O(\epsilon^2)$, collect like powers of ϵ, and determine that the equation for P_1 is

$$g^2(\sigma)\left(\frac{\partial^2 P_1}{\partial \tau^2} + P_1\right) = -g(\sigma)\frac{\partial^2 P_0}{\partial \sigma \partial \tau} - \frac{\partial}{\partial \sigma}\left(g(\sigma)\frac{\partial P_0}{\partial \tau}\right). \tag{20.55}$$

Since we want P_1 to be a periodic function of τ, we require that P_1 be "nonsecular," that is, that the right-hand side of (20.55) contain no terms proportional to $e^{i\tau}$ or $e^{-i\tau}$. It follows that

$$\frac{\partial}{\partial \sigma}(gA^2) = 0, \qquad \frac{\partial}{\partial \sigma}(gB^2) = 0, \tag{20.56}$$

or that

$$A(\sigma) = \frac{A_0}{\sqrt{g(\sigma)}}, \qquad B(\sigma) = \frac{B_0}{\sqrt{g(\sigma)}}, \tag{20.57}$$

from which we obtain, to lowest order in ϵ,

$$p = \frac{1}{\sqrt{g(\sigma)}}(A_0 e^{iG(z)} + B_0 e^{-iG(z)}), \tag{20.58}$$

where $G(z) = \int_0^z g(\epsilon z)\,dz$.

In terms of the original dimensioned variables, this is

$$p(x) = \phi^{-1/2}\left(A_1 \exp\left[i\int_0^x \phi(s)\,ds\right] + B_1 \exp\left[-i\int_0^x \phi(s)\,ds\right]\right), \tag{20.59}$$

where

$$\phi(x) = \sqrt{\frac{-i\omega\rho Y(x)}{l}}. \tag{20.60}$$

The constants A_1 and B_1 are determined from boundary conditions (20.45) and (20.46), and then the membrane displacement is found from the identity $i\omega\eta(x) = -Y(x)p(x)$.

The key feature of this solution is that it is oscillatory with an envelope, whose maximal amplitude and position are determined by the frequency ω. We get some idea of this behavior in the special case $m = 0$, $k(x) = k_0 e^{-\lambda x}$, and $r(x) = r_0 e^{-\lambda x}$ in which case $\phi(x) = \alpha e^{\lambda x/2}$, where $\alpha^2 = \frac{2\omega^2\rho}{l(k_0+i\omega r_0)}$. If we let $\alpha = \alpha_r + i\alpha_i$ and suppose that $\frac{\alpha_i}{\lambda} \gg 1$ (not valid at low frequencies), then with $\eta_L = 0$, we find that

$$\eta(x) = -\frac{1}{i\omega}Y(x)p(x) \approx \hat{A}\exp\left(\frac{3\lambda x}{4} - \frac{2\alpha_i}{\lambda}e^{\lambda x/2} + \frac{2i\alpha_r}{\lambda}e^{\lambda x/2}\right). \tag{20.61}$$

This represents an oscillation with exponentially increasing phase and amplitude

$$|\eta| \approx |\hat{A}|\exp\left(\frac{3\lambda x}{4} - \frac{2\alpha_i}{\lambda}e^{\lambda x/2}\right). \tag{20.62}$$

The maximal value of this envelope occurs at

$$x_p = -\frac{2}{\lambda}\ln\left(\frac{4\alpha_i}{3\lambda}\right). \tag{20.63}$$

The location of this maximum is dependent on frequency, as

$$\alpha_i \approx \sqrt{\frac{\rho\omega}{lr_0}}, \tag{20.64}$$

provided that ω is sufficiently large. Thus, for large ω we have

$$x_p = -\frac{1}{\lambda}\ln\left(\frac{16\rho\omega}{9l\lambda^2 r_0}\right). \tag{20.65}$$

The Deep-Water Approximation

The second approach, the *short-wave*, or *deep-water*, approximation, assumes that the wavelength of the membrane waves is short compared to the cochlear depth. In this case the Fourier expansion of $p(x)$ includes only high frequencies, and so $\alpha_n = 0$ whenever $|n| < N$ for some large integer N. However, when $|n| > N$ and $l \gg L$, then $\tanh(2\pi nl/L) \approx \text{sign}(n)$. Thus, (20.40) becomes

$$-i\omega\eta \approx -\frac{1}{i\omega\rho}\sum_{n=-\infty}^{\infty}\alpha_n\frac{2\pi}{L}|n|e^{2\pi inx/L}. \tag{20.66}$$

Now we separate the sum into two pieces by defining two functions,

$$p_+(x) = \sum_{n=0}^{\infty}\alpha_n e^{2\pi inx/L} \tag{20.67}$$

and

$$p_-(x) = \sum_{n=-\infty}^{-1}\alpha_n e^{2\pi inx/L}, \tag{20.68}$$

and then observe that (20.38) becomes

$$Yp = Y(p_+ + p_-) \approx \frac{1}{\omega\rho}\left[\frac{dp_+}{dx} - \frac{dp_-}{dx}\right] + \eta_0\delta_0(x) - \eta_L\delta_L(x), \tag{20.69}$$

which we take to be the governing equation for p.

We can remove the delta function influence from this equation by integrating across the boundaries at $x = 0$ and $x = L$, and assuming that outside the cochlea, $p = 0$. This gives

$$\frac{1}{\omega\rho}[p_+(0) - p_-(0)] = \eta_0, \tag{20.70}$$

$$\frac{1}{\omega\rho}[p_+(L) - p_-(L)] = \eta_L. \tag{20.71}$$

Although p_+ is a linear combination of only positive (spatial) frequencies, the same is not true of Yp_+. However, if we assume that Y is a slowly varying function of x, then the Fourier series of Y with Fourier coefficients b_k has $b_k \approx 0$ whenever $|k| > k_0$, for some number k_0 that is small compared to the dominant frequency of p_+. It follows that

$$Yp_+ = \sum_{k=-\infty}^{\infty} c_k e^{2\pi ikx/L}, \tag{20.72}$$

where $c_k = \sum_{j=0}^{\infty}\alpha_j b_{k-j}$. If the dominant frequencies of p_+ and Y are separated, as stated above, then c_k is small for $k \leq 0$. Thus we can approximate Yp_+ by its Fourier series with positive frequencies. A similar argument applies for Yp_-. With these approximations,

(20.69) separates into a pair of differential equations for the positive and negative frequencies separately,

$$Y(x)p_\pm(x) \approx \frac{\pm 1}{\omega\rho}\frac{dp_\pm}{dx}, \qquad 0 < x < L. \tag{20.73}$$

These first-order linear equations can be integrated directly to get

$$p_\pm(x) = A_\pm \exp\left[\pm\omega\rho\int_0^x Y(\zeta)\,d\zeta\right] \tag{20.74}$$

for some constants $A_\pm$, so that

$$p = A_+ \exp\left[\omega\rho\int_0^x Y(\zeta)\,d\zeta\right] + A_- \exp\left[-\omega\rho\int_0^x Y(\zeta)\,d\zeta\right]. \tag{20.75}$$

We use the boundary conditions at $x = 0$ and $x = L$ to determine the constants $A_\pm$. From (20.70) and (20.71) it follows that these constants must satisfy the equations

$$A_+ - A_- = \omega\rho\eta_0, \qquad \gamma A_+ - \frac{1}{\gamma}A_- = \omega\rho\eta_L, \tag{20.76}$$

where $\gamma = \exp[\omega\rho\int_0^L Y(\zeta)\,d\zeta]$, from which it follows that

$$A_+ = \frac{\omega\rho}{\gamma^2 - 1}[\gamma\eta_L - \eta_0], \tag{20.77}$$

$$A_- = \frac{\omega\rho}{\gamma^2 - 1}[\gamma\eta_L - \gamma^2\eta_0]. \tag{20.78}$$

For physiological values of Y, $|\gamma| \gg 1$ for all except the lowest frequencies; for example, for the parameter values in Fig. 20.7, $|\gamma| = 40$ when $\omega = 800$, and $|\gamma| = 10^9$ when $\omega = 1500$. Since on physical grounds η_0 and η_L are not large, it follows that $|A_-| \approx -\omega\rho\eta_0 \gg |A_+|$. Finally, from (20.38) and (20.75), we find that the membrane displacement is given by

$$\eta = -i\rho\eta_0 Y(x) \exp\left[-\omega\rho\int_0^x Y(\zeta)\,d\zeta\right]. \tag{20.79}$$

The amplitude of η can be plotted as a function of x to give the envelope of the wave on the basilar membrane. The frequency response is similarly obtained, by fixing x and plotting $|\eta|$ as a function of ω. Typical results are shown in Fig. 20.9. The qualitative agreement with data is good, with the peak of the wave envelope moving toward the stapes as the frequency increases.

In the special case $m = 0, k(x) = k_0 e^{-\lambda x}, r(x) = r_0 e^{-\lambda x}$, we can calculate the waveform (20.79) to be

$$\eta = 2\eta_0\xi \exp[\lambda x + \beta(1 - e^{\lambda x})], \tag{20.80}$$

where

$$\beta = \frac{2\omega^3\rho r_0}{\lambda(k_0^2 + \omega^2 r_0^2)} + i\frac{2\omega^2\rho k_0}{\lambda(k_0^2 + \omega^2 r_0^2)} = \beta_r + i\beta_i \tag{20.81}$$

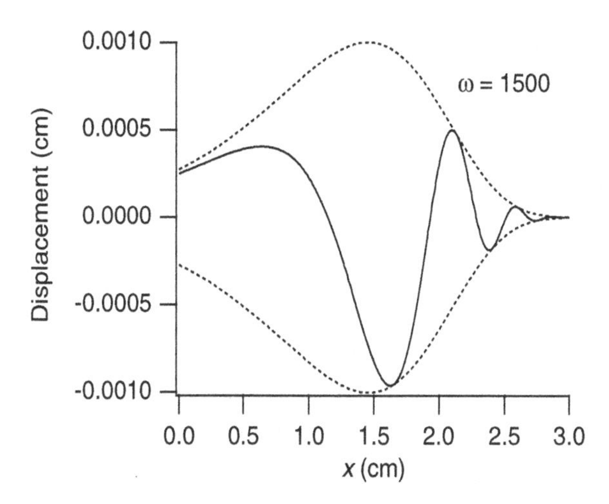

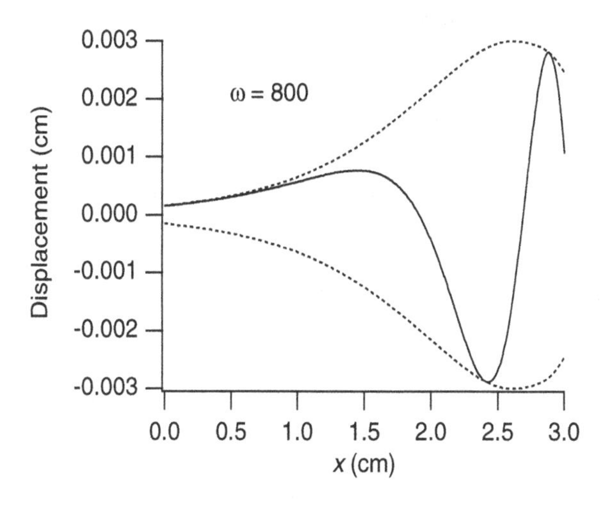

Figure 20.9 Plots of the amplitude of the wave on the basilar membrane for two different frequencies. The envelope of the wave is shown as a dotted line. Calculated from the deep-water approximation (20.75) using the same parameter values as in the Lesser and Berkley model (given in the caption to Fig. 20.7).

and

$$\xi = \frac{\omega\rho}{i\omega\rho + k_0}. \tag{20.82}$$

Here we again see an oscillatory waveform with an envelope of amplitude

$$|\eta| = 2\eta_0|\xi|\exp[\lambda x + \beta_r(1 - e^{\lambda x})], \tag{20.83}$$

the maximum of which occurs at

$$x_p = \frac{-1}{\lambda}\ln\left(\frac{2\omega^3\rho r_0}{\lambda(k_0^2 + \omega^2 r_0^2)}\right). \tag{20.84}$$

According to this expression, the peak of the envelope moves to the left (toward the base of the cochlea) as ω increases. The principal fault of the short-wave model is that the phase of the model waves increases much more than is observed experimentally.

A similar solution was found by Peskin (1976, 1981), who calculated an exact solution to a special case of the cochlear model. In his model the cochlear membrane was taken to be infinitely long, with $r(x)$ and $k(x)$ chosen to be decaying exponential functions with decay rate λ and a fluid container of height $\lambda l = \pi/2$. With these assumptions and simplifications, Peskin found the exact solution using conformal mapping and contour integration techniques.

20.2.5 More Complex Models

In this chapter we have concentrated on the simpler models of the basilar membrane that assume that the cochlea is two-dimensional and that the basilar membrane can be described by a point impedance function, i.e., that each point of the basilar membrane acts as a damped harmonic oscillator, with no coupling along the length of the membrane except for that imposed indirectly via the fluid motion.

Although the wave motion on the basilar membrane is an important component of the hearing process, many other factors are involved (Pickles, 1982; Rhode, 1984; Hudspeth, 1985). Nonlinearities in the cochlear response and acoustic emissions suggest the presence of active feedback processes that modulate the waveform (Section 20.4). This feedback may occur in the outer hair cells and the organ of Corti. Simple hydrodynamic models do not reproduce the degree of tuning observed in the mammalian cochlea, and the precise tuning mechanism is still controversial. Many other, more complex, models have been constructed (see, for example, Steele, 1974; Steele and Tabor, 1979a,b; Inselberg and Chadwick, 1976; Chadwick et al., 1976; Chadwick, 1980; Holmes, 1980a,b, 1982). In general, these models use similar equations for the fluid flow, but model the basilar membrane in greater detail, including spatial coupling in the membrane. The resultant membrane equations are of fourth order in space, and heavy use is made of asymptotic expansions in the solution of the model equations. A particularly detailed study was performed by Steele (1974), who constructed a series of models ranging from a plate in an infinite body of fluid up to a tapered elastic basilar membrane, a cochlea with rigid walls, and flexible arches of Corti. Recent models are based on finite element descriptions of the basilar membrane, organ of Corti, and the outer hair cells, and incorporate acoustic, electrical and mechanical elements (Ramamoorthy et al., 2007), while the three-dimensional spiral nature of the cochlea has been modeled by Givelberg and Bunn (2003; Givelberg, 2004).

20.3 Electrical Resonance in Hair Cells

In many lower vertebrates frequency decomposition is performed, not by a wave on the basilar membrane, but by the hair cells themselves. Hair cells in the turtle cochlea and the bullfrog sacculus (to name but two examples) respond preferentially to stimuli of a certain frequency, and this band-pass response is mediated by the ionic channels

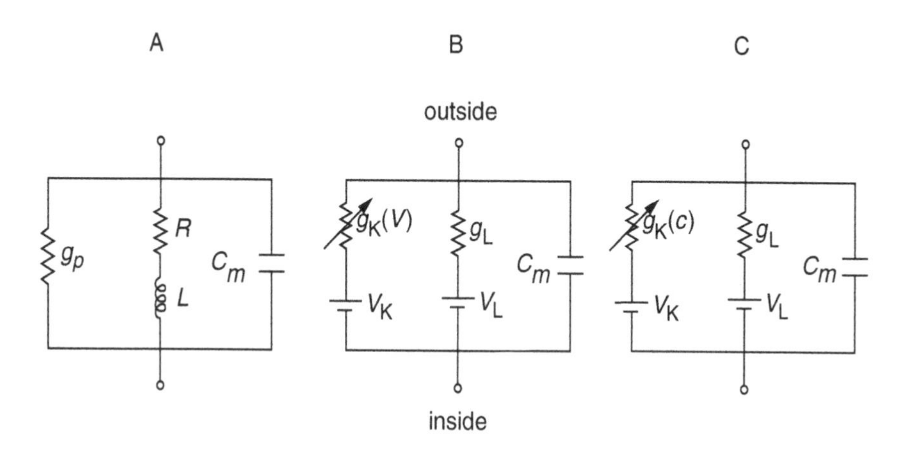

Figure 20.10 Electrical circuits for electrical tuning, adapted from Ashmore and Attwell (1985). A: The basic model, with an inductance in place of ionic currents. B: Voltage-gated K^+ current, with a conductance that is increased by membrane depolarization. C: Ca^{2+}-gated current, with the K^+ conductance controlled by the intracellular concentration of Ca^{2+}, denoted by c.

in the hair cell membrane. At the top of each hair cell is the hair bundle, a group of stereocilia connected to each other at the tips by a thin fiber, called a tip link. Each stereocilium is rigid and, in response to a force applied at the tip, pivots around its base rather than bending. It is postulated that the tip links act like elastic springs connected directly to ionic channels such that when the hair bundle is deflected in one direction, the tip links pull channels open, while when the hair bundle is deflected in the opposite direction, the tip links relax and allow channels to close. The mechanically sensitive ion channels are nonselective, and the modulation of current flow through these channels results in hyperpolarization or depolarization of the hair cell membrane. The membrane potential is then modulated by other ionic channels, including K^+ channels, Ca^{2+}-sensitive K^+ channels, and voltage-sensitive Ca^{2+} channels. The structure, tuning, sensitivity, and function of hair cells are reviewed by Hudspeth (1985, 1989; Hudspeth and Gillespie, 1994), and these papers give a readable summary of the field.

In response to a step current input, the membrane potential of hair cells exhibits damped oscillations, with a period and amplitude dependent on the size of the step. Thus, each cell has a natural frequency of oscillation and responds best to a stimulus at a similar frequency. Crawford and Fettiplace (1981) and Ashmore and Attwell (1985) have developed simple models of electrical resonance that while not based on the details of known mechanisms, provide a good description of the experimental results. Later work by Hudspeth and Lewis (1988a,b), using a more detailed model, showed that the measured properties of the ionic conductances are sufficient to explain resonance in the hair cells of the bullfrog sacculus.

20.3.1 An Electrical Circuit Analogue

The models of Crawford and Fettiplace (1981) and Ashmore and Attwell (1985) are based on the electrical circuit shown in Fig. 20.10A. In response to a current input I, the voltage V is given by

$$\frac{d^2V}{dt^2} + \gamma\frac{dV}{dt} + \omega_0^2 V = f(t), \tag{20.85}$$

where

$$\gamma = \frac{g_p}{C_m} + \frac{R}{L}, \tag{20.86}$$

$$\omega_0^2 = \frac{g_p R + 1}{LC_m}, \tag{20.87}$$

$$f(t) = \frac{1}{C_m}\frac{dI}{dt} + \frac{IR}{LC_m}. \tag{20.88}$$

It is simplest to demonstrate resonance when $f(t) = e^{i\omega t}$, in which case $V = V_1(\omega)e^{i\omega t}$, where

$$V_1(\omega) = \frac{1}{\omega_0^2 - \omega^2 + i\gamma\omega}. \tag{20.89}$$

Thus, $|V_1|$ has a band-pass frequency response, with a maximum at $\hat{\omega}$, where $\hat{\omega}^2 = \omega_0^2 - \gamma^2/2$. Solutions of (20.85) are of the form $\exp(-\gamma t/2)\exp\left(\pm i\sqrt{\omega_0^2 - \gamma^2/4}\right)$, and thus $\hat{\omega}$ is slightly smaller than the natural frequency of oscillation of the system. However, if damping is small (i.e., if γ is small), the maximum amplitude of the frequency response occurs at approximately the natural frequency of oscillation. The sharpness of the peak of $|V_1|$ is a measure of the degree of tuning of the electrical circuit, with sharper peaks giving greater frequency selectivity. Since

$$\frac{d^2}{d\omega^2}\left[(\omega_0^2 - \omega^2)^2 + \gamma^2\omega^2\right]\Big|_{\omega=\hat{\omega}} = 4\gamma^2(2Q^2 - 1), \tag{20.90}$$

where

$$Q = \frac{\omega_0}{\gamma}, \tag{20.91}$$

it follows that Q, often called the *quality factor*, is a useful measure of the degree of tuning. As Q increases, so does the frequency selectivity of the circuit.

We now consider the response of the circuit when the input is a sinusoidally varying current. When $I = e^{i\omega t}$,

$$V_1(\omega) = \frac{\frac{R}{LC_m} + \frac{i\omega}{C_m}}{\omega_0^2 - \omega^2 + i\gamma\omega}, \tag{20.92}$$

which again corresponds to a band-pass filter, with the maximum response occurring at $\hat{\omega}$, where

$$(\hat{\omega}^2)^2 + 2\left(\frac{R}{L}\right)^2 \hat{\omega}^2 + \left(\frac{R}{L}\right)^2 (\gamma^2 - 2\omega_0^2) = 0. \qquad (20.93)$$

Crawford and Fettiplace (1981) used a model of this type (without the leak conductance g_p) to determine the electrical tuning characteristics of hair cells from the turtle cochlea. By comparison of these tuning curves with tuning curves obtained by acoustic stimulation of the hair cells they were able to determine that electrical resonance can account for most of the frequency selectivity of the hair cell.

Although the above circuit exhibits the required resonance, it would be much more satisfactory to explain electrical resonance in terms of components that have a more direct connection to the hair cell. This can be done in at least two ways. In Fig. 20.10B and C two circuits are shown, one involving a voltage-sensitive K^+ conductance, the other a Ca^{2+}-sensitive K^+ conductance, that, formally at least, are equivalent to the circuit in Fig. 20.10A.

Consider first the circuit in Fig. 20.10B. If the leak has a constant conductance, and the K^+ conductance is a function of time and voltage, then

$$I = C_m \frac{dV}{dt} + g_{\mathrm{L}}(V - V_{\mathrm{L}}) + f g_{\mathrm{K}}(V - V_{\mathrm{K}}), \qquad (20.94)$$

$$\tau \frac{df}{dt} = f_\infty - f, \qquad (20.95)$$

where we take a linear approximation for f_∞,

$$f_\infty = f_r + \mu(V - V_r). \qquad (20.96)$$

Here, V_r is assumed to be the resting membrane potential, f_r is the value of f when $V = V_r$, and μ is the slope of the activation curve at the steady state. Note that at steady state,

$$0 = g_{\mathrm{L}}(V_r - V_{\mathrm{L}}) - g_{\mathrm{K}} f_r (V_{\mathrm{K}} - V_r), \qquad (20.97)$$

and thus V_{L} can be eliminated.

It follows that

$$\frac{dI}{dt} + \frac{I}{\tau} = C_m \frac{d^2\tilde{V}}{dt^2} + \left(g_{\mathrm{L}} + g_{\mathrm{K}} f_r + \frac{C_m}{\tau}\right)\frac{d\tilde{V}}{dt} + \left(\frac{g_{\mathrm{L}} + g_{\mathrm{K}} f_r + g_{\mathrm{K}}(V_r - V_{\mathrm{K}})\mu}{\tau}\right)\tilde{V}, \qquad (20.98)$$

where $\tilde{V} = V - V_r$ and where we have linearized the equation around V_r by assuming that $V \approx V_r$. Equation (20.98) is equivalent to (20.85)–(20.88) if

$$L = \frac{\tau}{g_{\mathrm{K}}(V_r - V_{\mathrm{K}})\mu}, \qquad (20.99)$$

$$R = \frac{1}{g_{\mathrm{K}}(V_r - V_{\mathrm{K}})\mu}, \qquad (20.100)$$

$$g_p = g_{\mathrm{L}} + g_{\mathrm{K}} f_r. \qquad (20.101)$$

A similar procedure can be followed for the circuit in Fig. 20.10C, in which the K^+ conductance is Ca^{2+}-dependent rather than voltage-dependent, but extra assumptions about the Ca^{2+} kinetics must be made. As a first approximation, it is assumed that Ca^{2+} enters the cell through channels at a rate that is a linear function of voltage, with slope θ, and is removed with first-order kinetics, i.e., at a rate proportional to its concentration. Finally, it is assumed that the proportion of open K^+ channels is linearly related to the Ca^{2+} concentration. Thus,

$$I = C_m \frac{dV}{dt} + g_L(V - V_L) + g_K kc(V - V_K), \tag{20.102}$$

$$W\frac{dc}{dt} = \frac{I_r + \theta(V - V_r)}{F} - pc, \tag{20.103}$$

where c denotes Ca^{2+} concentration, F is Faraday's constant, W is the cell volume, p is the rate of Ca^{2+} pumping, and I_r is the steady Ca^{2+} current when $V = V_r$. The constant k is the rate at which Ca^{2+} activates the K^+ current. Again, linearizing this system about the steady state gives a system that is equivalent to (20.85)–(20.88), provided that

$$L = \frac{WF}{g_K(V_r - V_K)k\theta}, \tag{20.104}$$

$$R = \frac{pF}{g_K(V_r - V_K)k\theta}, \tag{20.105}$$

$$g_p = g_L + g_K kc_r, \tag{20.106}$$

where c_r is the steady Ca^{2+} concentration at the resting potential.

Ashmore and Attwell showed that although the model with the voltage-sensitive K^+ conductance can generate a wide range of optimal frequencies, physiological values for the parameters result in values for the quality factor Q that are an order of magnitude too low. Thus, for reasonable parameters, the model can distinguish between frequencies, but not sharply enough. Experimental values for Q are often 5 or more, while Q values in the model are not above 0.7. This, they argue, is the result of the low value of μ: a physiological value for μ is about 0.33 mV^{-1}, but Q is large enough in the model with μ about 3 mV^{-1}. Thus, it appears that the activation of the K^+ current by voltage is not steep enough to account for the observed resonance in hair cells.

In the third model, however, the activation of the K^+ current by Ca^{2+} can be made much steeper. Here, the effective activation slope of the K^+ channel is $k\theta/(pF)$, which can be made large by decreasing the pump rate p or by increasing the sensitivity of the K^+ channel to Ca^{2+}. Ashmore and Attwell conclude that frequency tuning in hair cells is more likely the result of a Ca^{2+}-sensitive K^+ conductance than of a voltage-sensitive conductance.

20.3.2 A Mechanistic Model of Frequency Tuning

This conclusion has been upheld by the work of Hudspeth and Lewis (1988a,b). Based on a series of experiments in which they measured the kinetic properties of the ionic

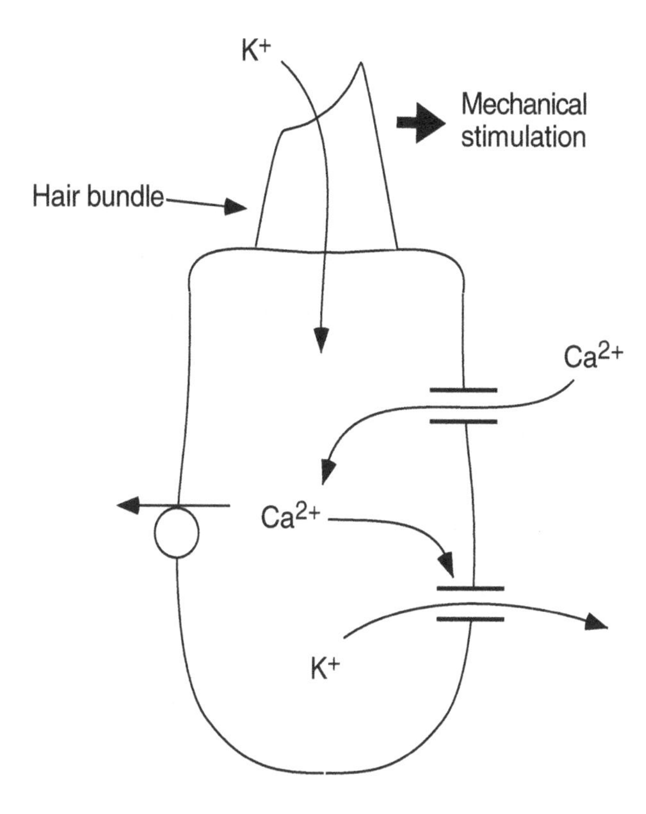

Figure 20.11 Schematic diagram of a model of electrical tuning in hair cells, adapted from Hudspeth (1985).

conductances in saccular hair cells of the bullfrog, Hudspeth and Lewis constructed a detailed model of electrical resonance in these cells. They concluded that the observed properties of the Ca^{2+}-sensitive K^+ conductance, in concert with a voltage-sensitive Ca^{2+} conductance and a leak, are a sufficient quantitative explanation of frequency tuning in these cells.

A schematic diagram of their model is given in Fig. 20.11. Mechanical deflection of the hair bundle opens transduction channels in the hair bundle allowing the entry of positive ions, mostly K^+. The consequent depolarization of the cell activates voltage-gated Ca^{2+} channels, and the intracellular Ca^{2+} concentration rises. This, in turn, opens Ca^{2+}-sensitive K^+ channels. K^+ ions flow out of the cell, and the cell repolarizes. Ca^{2+} balance is maintained by pumps that remove Ca^{2+} from the hair cell. One crucial, and rather unusual, feature of the model is that K^+ can both enter and leave the cell passively. Since the hair bundle projects into the scala media, the fluid surrounding the hair bundle (the endolymph in the case of hair cells in the cochlea) is of different composition from that surrounding the base of the hair cell, having a high K^+ and a low Na^+ concentration.

We do not present all the many details of the model here. Suffice it to say that it is assumed that there are three significant ionic currents contributing to resonance in the hair cell: a voltage-gated Ca^{2+} current, a Ca^{2+}-activated K^+ current, and a leak current.

Thus, for an applied current I,

$$I = C_m \frac{dV}{dt} + I_c + I_{kc} + I_L. \tag{20.107}$$

The voltage-gated Ca^{2+} current I_c and the leak currents are described by similar equations as in the Hodgkin–Huxley model (Chapter 5). The model of the Ca^{2+}-activated K^+ channel, I_{kc}, is considerably more complicated. It is assumed that the channel has three closed states and two open states: binding of two Ca^{2+} ions converts the channel into a state in which it can spontaneously open, while the binding of an additional Ca^{2+} ion can prolong opening. The transition rate constants are dependent on Ca^{2+} and voltage. Finally, Ca^{2+} handling is treated simply by assuming that Ca^{2+} enters through the Ca^{2+} channel and is removed by a first-order process.

The parameters (of which there are about 30) were determined by constraining the model to agree with voltage-clamp data from a single cell, and then the response of the model to current pulses was investigated. It was found that depolarizing current steps induced damped membrane potential oscillations in the model, with a frequency and amplitude dependent on the magnitude of the current step, in close agreement with experimental data (Fig. 20.12).

To simulate a transduction current, a term $I_T = g_T(V - V_T)$ is added to the right-hand side of (20.107). The transduction conductance is assumed to be a function of hair cell displacement, which, in turn, is assumed to vary sinusoidally. The resultant model frequency response is band-pass in nature, with the maximal response at frequency 112 Hz and a quality factor of 3. The frequency at which the response is maximal (the

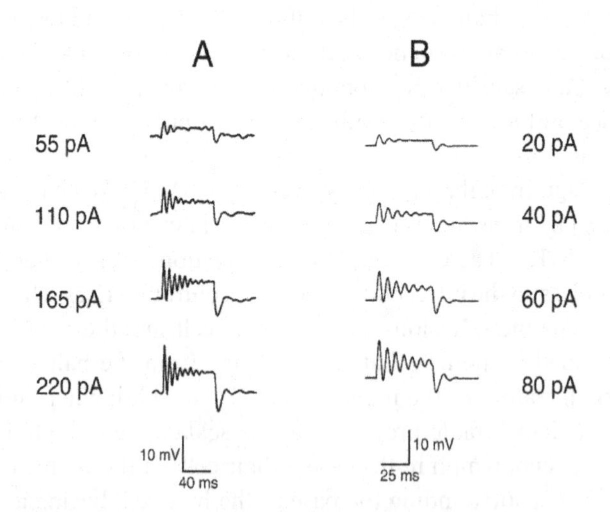

Figure 20.12 A: Responses of bullfrog saccular hair cells to depolarizing current steps. As the current step increases in size, the hair cells show more pronounced oscillatory behavior. (Adapted from Hudspeth and Lewis, 1988b, Fig. 3.) B: Responses of the model to current steps. (Adapted from Hudspeth and Lewis, 1988b, Fig. 6.)

resonant frequency) is a function of the model parameters, and realistic changes in the model parameters can account for the range of experimentally observed resonant frequencies in the bullfrog sacculus. In particular, because the resonant frequency is sensitive to g_{kc}, the model predicts that controlling the number of Ca^{2+}-sensitive K^+ channels is one simple way in which cells could tune their frequency response.

20.4 The Nonlinear Cochlear Amplifier

In Chapter 19 we saw that photoreceptors are able to respond in a reliable and reproducible way to single photons; the visual system thus attains the maximum possible sensitivity. Similarly, one of the most extraordinary things about the cochlea is its extreme sensitivity, which appears to be limited only by thermal noise (Manley, 2001).

The ability of the cochlea to detect low-amplitude signals is surprising, given that the cochlear fluid would be expected to damp out signals with low power. As Hudspeth (2005) has so graphically expressed it, asking the basilar membrane to vibrate under such conditions is akin to asking a tuning fork to vibrate in honey. As early as 1948, Gold pointed out this difficulty, and proposed the existence of some active mechanism amplifying the response to small sounds.

Such high sensitivity (as well as sharp frequency tuning) is believed to result from active mechanical amplification by hair cells. The exact molecular basis for such amplification is unclear, and almost certainly differs in different species. In mammals, outer hair cells contract and elongate in response to depolarization and hyperpolarization respectively. Vibrations of the basilar membrane cause lateral movement of the hair bundle on the top of the outer hair cell, which in turn opens and closes ion channels in the stereocilia, leading to changes in membrane potential (Hudspeth, 1997; Nobili et al., 1998; Manley, 2001; Hudspeth et al., 2000). It is fascinating to note that the opening and closing of ion channels in response to movement of the hair bundle is a direct mechanical interaction, and thus extremely fast. Elastic springs (tip links) connect the tip of one stereocilium directly to the ion channel; when the stereocilium moves, the tip link is pulled, thus opening the ion channel. It seems that these changes in the length of the outer hair cell result from the contraction or expansion of billions of proteins within the outer membrane, and that these proteins react to the membrane potential directly, without any need for ATP. However, the exact identity and properties of these proteins remains elusive (Ashmore et al., 2000).

20.4.1 Negative Stiffness, Adaptation, and Oscillations

Hair bundles of the bullfrog sacculus display spontaneous twitches, or oscillations that, when entrained by a stimulus of the resonant frequency, respond with an amplitude about twice that of the initial stimulus. One possible mechanism for such spontaneous oscillations has been proposed by Martin et al. (2000), and is based on the fact that the hair bundle has a negative stiffness for small displacements. This is an extraordinary

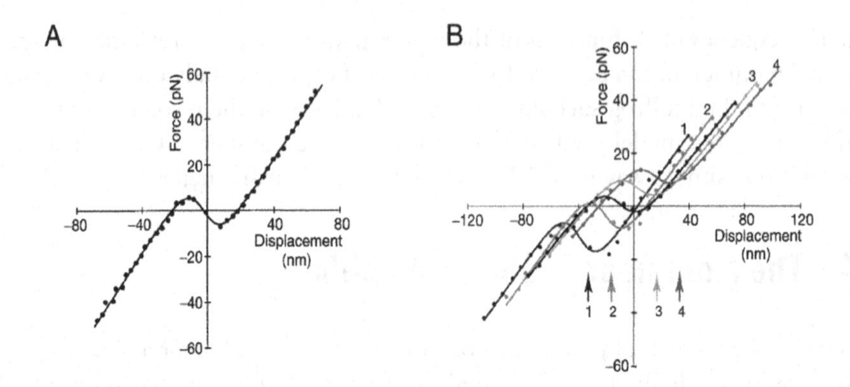

Figure 20.13 A: Negative stiffness in hair bundles from the bullfrog sacculus. Small perturbations from the zero-displacement resting state cause greater displacements, and thus zero displacement is an unstable steady state (Martin et al., 2000, Fig. 1C). B: Adaptation of the force–displacement curve. When a fixed displacement is externally imposed, the force–displacement curve shifts in the direction of the applied displacement, and also moves vertically. The black curve is the force–displacement curve in the absence of any externally imposed displacement, while the other curves, in varying shades of gray (numbered for easier identification), were measured after the imposed displacement had been maintained for 100 ms. The arrows along the bottom show the positions of the externally imposed offset displacements (Martin et al., 2000, Fig. 2).

result. A typical spring (as taught in first year math and physics classes) has a positive stiffness; when perturbed from its rest position, the spring exerts a force that tries to restore the rest position, i.e., a positive displacement gives a positive force, while a negative displacement gives a negative force.

But hair bundles from the bullfrog sacculus have a very much more interesting response (Fig. 20.13A). A small positive displacement results in a negative force, tending to *increase* the displacement, leading to positive feedback and an unstable rest state. Similarly, a negative displacement causes a positive force. Hence there are two stable rest states; one with a positive displacement, one with a negative displacement.

Furthermore, the hair bundle displays adaptation, in a manner reminiscent of photoreceptor adaptation (Chapter 19). This is illustrated in Fig. 20.13B. If a fixed displacement is externally imposed, the force–displacement curve shifts both horizontally and vertically so as to bring the position of the imposed displacement within the region of negative stiffness. Hence, the response to a step increase in load is an initial large increase in force, followed by a slower relaxation. If the maintained displacement is large enough, this shift of the force–displacement curve results in the disappearance of two steady states in a saddle-node bifurcation.

Thus, spontaneous oscillations can arise. When the hair bundle tries to sit at the stable steady state with positive displacement, the force–displacement curve gradually adapts, shifting up and to the right, until this stable steady state no longer exists.

The hair bundle is then forced to flip to the other stable steady state (at a negative displacement), whereupon the process repeats itself in reverse. This results in a spontaneous oscillation where the hair bundle is alternately flipping between the two stable steady states, and the force–displacement curve is continually adapting. It is left as an exercise (Exercise 8) to formulate a mathematical model of this process.

20.4.2 Nonlinear Compression and Hopf Bifurcations

Another important feature of the cochlea is that low-amplitude signals are more sharply tuned than signals of high amplitude, and thus show greater frequency selectivity. Furthermore, there is no audible sound for which the cochlear response is linear; cochlear responses are essentially nonlinear.

Typical experimental data (from a living chinchilla) are shown in Fig. 20.14. At low signal amplitudes the frequency–response curve is sharply peaked, but becomes less so at high amplitudes. Also, at the resonant frequency (shown by the vertical dashed line), the response quickly saturates as the stimulus amplitude increases. This is called *nonlinear response compression*. Stated another way, the velocity of the response to a small-amplitude stimulus is much greater than would be expected were the cochlea to behave in a linear manner.

Eguíluz et al. (2000) have shown that both these properties of the response could result from the proximity of a Hopf bifurcation. Since we have already seen from

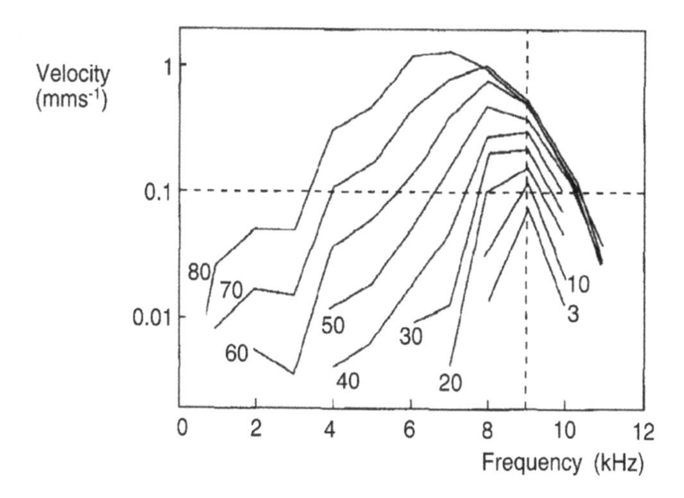

Figure 20.14 The velocity of one point on the basilar membrane of a live chinchilla, plotted as a function of frequency. Each curve corresponds to a different stimulus amplitude, labeled in decibels. At low stimulus strength the frequency–response curve is sharply peaked, but becomes less so as the stimulus amplitude increases. In addition, the response at the resonant frequency is not a linear function of stimulus amplitude, but saturates (Reprinted from Ruggero, 1992, Fig. 1b, with permission from Elsevier).

biophysical models that bullfrog saccular hair cells appear to be dynamically close to a Hopf bifurcation, this provides a plausible and appealing explanation of the responses seen in Fig. 20.13.

In normal form (Wiggins, 2003), the equations for a Hopf bifurcation are

$$\frac{dx}{dt} = \mu x - \omega_0 y - (x^2 + y^2)x, \tag{20.108}$$

$$\frac{dy}{dt} = \omega_0 x + \mu y - (x^2 + y^2)y, \tag{20.109}$$

or, in complex form,

$$\frac{dz}{dt} = (\mu + i\omega_0)z - |z|^2 z, \tag{20.110}$$

where $z = x + iy$. This system has a Hopf bifurcation at $\mu = 0$, and the oscillations that appear at the Hopf bifurcation have period $2\pi/\omega_0$, and have the form $z = \sqrt{\mu}e^{i\omega_0 t}$.

If we drive this oscillator with a forcing term $\alpha e^{i\omega t}$, and look for solutions of the form $z = Ae^{i(\omega t+\phi)}$, we get

$$(iA(\omega - \omega_0) - \mu A + A^3)e^{i\phi} = \alpha. \tag{20.111}$$

Since we are interested only in the amplitude, A, of the response, it suffices to take the modulus of both sides of (20.111), from which it follows that

$$\alpha^2 = A^6 - 2\mu A^4 + [\mu^2 + (\omega - \omega_0)^2]A^2. \tag{20.112}$$

For a given stimulus strength (i.e., a given α), we can plot A as a function of stimulus frequency, ω. The results are shown in Fig. 20.15. When α is small, the response has a sharp frequency tuning, and, at the resonant frequency $\omega = \omega_0$, the amplitude shows nonlinear compression. This behavior is qualitatively the same as that seen

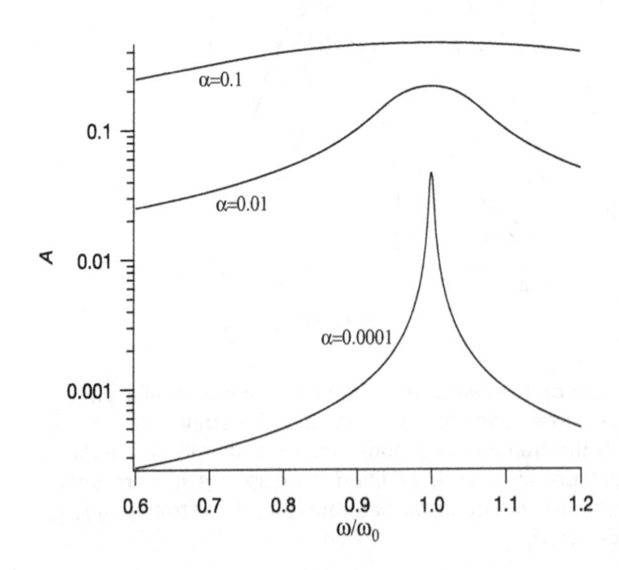

Figure 20.15 The amplitude, A, of the response of (20.108)–(20.109) to an input of the form $\alpha e^{i\omega t}$.

experimentally (Fig. 20.14). At the resonant frequency, $\omega = \omega_0$, (20.112) becomes

$$\alpha = A^3 - \mu A, \tag{20.113}$$

from which we see that, at the Hopf bifurcation point (i.e., when $\mu = 0$), A depends on the one-third power of the input amplitude, and thus the response is nonlinearly compressed, as seen in the experimental data.

It is still uncertain whether the hair cells of any particular animal actually tune their responses in this way by sitting close to a Hopf bifurcation, and the exact mechanical forces that are responsible for nonlinear compression, sharp tuning, and amplification remain unclear. Nevertheless, this simple theory provides an intriguing explanation of this phenomenon.

20.5 EXERCISES

1. Show that if $F(y,t)$ in the Lesser and Berkley model is an odd function of y, it is sufficient to consider only the solution in the region $0 < y < l$. Hint: Show that $\phi_2(y,t) = -\phi_1(-y,t)$ and $p_2(y,t) = -p_1(-y,t)$ satisfy the differential equations for ϕ_2 and p_2. Hence the potential and pressure are odd functions of y.

2. Formulate a model of the cochlea in which the basilar membrane is spatially coupled in the x direction as if it were a damped string. How does the Fourier transform of this problem differ from the Lesser–Berkley model (20.25)–(20.29)?

3. Show that (20.75) describes two waves moving in opposite directions. Hint: Y has an imaginary component, and thus the terms in (20.75) have phases with opposite signs. What are the envelopes of the waves?

4. If the functions $m(x), r(x)$, and $k(x)$ are proportional to the same exponential, then the solution of (20.44) can be found exactly.

   (a) Suppose that $m = m_0 e^{-\lambda x}$, $k(x) = k_0 e^{-\lambda x}$, $r(x) = r_0 e^{-\lambda x}$. Show that (20.44) becomes
   $$\frac{d^2 p}{dx^2} + \alpha^2 e^{\lambda x} p = 0, \tag{20.114}$$
   where $\alpha^2 = \frac{2\omega^2 \rho}{l(i\omega r_0 + k_0 - \omega^2 m_0)}$.

   (b) Show that the transformation $s = \frac{2\alpha}{\lambda} e^{\lambda x/2}$ transforms (20.114) into
   $$s\frac{d}{ds}\left(s\frac{dp}{ds}\right) + s^2 p = 0, \tag{20.115}$$
   which is Bessel's equation of order zero. Thus, the general solution of (20.114) is
   $$p(x) = AJ_0\left(\frac{2\alpha}{\lambda} e^{\lambda x/2}\right) + BY_0\left(\frac{2\alpha}{\lambda} e^{\lambda x/2}\right), \tag{20.116}$$
   where J_0 and Y_0 are the zeroth-order Bessel functions of the first and second kind, or equivalently,
   $$p(x) = \tilde{A} H_0^{(1)}(s) + \tilde{B} H_0^{(2)}(s), \tag{20.117}$$
   where $H_0^{(1)}(s)$ and $H_0^{(2)}(s)$ are the zeroth-order Hankel functions of the first and second kind. Use the boundary conditions (20.45) and (20.46) to determine the coefficients $\tilde{A}$ and $\tilde{B}$.

(c) Use the asymptotic behavior of the Hankel functions (Keener, 1998)

$$H_0^{(1)}(s) \sim \left(\frac{2}{\pi s}\right)^{1/2} e^{i(s-\pi/4)}, \qquad H_0^{(2)}(s) \sim \left(\frac{2}{\pi s}\right)^{1/2} e^{-i(s-\pi/4)} \tag{20.118}$$

to approximate $p(x)$ in the case that $\tilde{B} = 0$. Show that $\eta(x)$ is approximated by (20.61). Under what conditions is this approximation valid?

5. For the long-wave model it was claimed that the phase of η, where η is given by (20.80), does not agree with experimental data. Confirm this by showing that for a fixed ω, the phase increases exponentially with x, and that the phase does not approach zero as ω approaches zero.

6. Compare the location of the envelope maximum x_p for the shallow-water approximation (20.65) in the case $l\lambda = \pi/2$ with that for the deep-water approximation (20.84), for large ω.

7. Let $\eta_0 = 1$, $\eta_L = 0$, and $m = 0.05e^{-1.5x}$, and let the other parameters be the same as in the Lesser and Berkley model. Use the boundary conditions to solve for A_1 and B_1 in (20.59), and thus calculate the displacement of the basilar membrane for the shallow-water model. Compare the shallow-water and deep-water models by plotting the displacements for high and low frequencies. How does the behavior of the long-wave model change as l is decreased?

8. Formulate a mathematical model of spontaneous oscillations that incorporates the force–displacement adaptation described in Section 20.4.1, and show that the model can exhibit relaxation oscillations. (Hint: Use a second-order differential equation to model the hair bundle as a damped spring, and at least one additional first-order equation to model adaptation of the spring. The model equations need not be biophysically realistic, but merely a qualitative caricature of the process.)

Appendix: Units and Physical Constants

Quantity	Name	Symbol	Units
Amount	mole	mol	
Electric charge	coulomb	C	
Mass	gram	g	
Temperature	kelvin	K	
Time	second	s	
Length	meter	m	
Force	newton	N	$kg \cdot m \cdot s^{-2}$
Energy	joule	J	$N \cdot m$
Pressure	pascal	Pa	$N \cdot m^{-2}$
Capacitance	farad	F	$A \cdot s \cdot V^{-1}$
Resistance	ohm	Ω	$V \cdot A^{-1}$
Electric current	ampere	A	$C \cdot s^{-1}$
Conductance	siemen	S	$A \cdot V^{-1} = \Omega^{-1}$
Potential difference	volt	V	$N \cdot m \cdot C^{-1}$
Concentration	molar	M	$mol \cdot L^{-1}$
Atomic Mass	dalton	D	$g\, N_A^{-1}$

Unit Scale Factors

Name	Prefix	Scale factor
femto	f	$\times 10^{-15}$
pico	p	$\times 10^{-12}$
nano	n	$\times 10^{-9}$
micro	μ	$\times 10^{-6}$
milli	m	$\times 10^{-3}$
centi	c	$\times 10^{-2}$
deci	d	$\times 10^{-1}$
kilo	k	$\times 10^{3}$
mega	M	$\times 10^{6}$
giga	G	$\times 10^{9}$

Physical Constant	Symbol	Value
Boltzmann's constant	k	1.381×10^{-23} J $\cdot$ K^{-1}
Planck's constant	h	6.626×10^{-34}J $\cdot$ s
Avogadro's number	N_A	6.02257×10^{23} mol^{-1}
unit charge	q	1.6×10^{-19} C
gravitational constant	g	9.78049 m/s^2
Faraday's constant	F	9.649×10^{4}C $\cdot$ mol^{-1}
permittivity of free space	ϵ_0	8.854×10^{-12} F/m
universal gas constant	R	8.315 J mol^{-1} $\cdot$ K^{-1}
atmosphere	atm	1.01325×10^{5} N $\cdot$ m^{-2}
insulin unit	U	$\frac{1}{24000}$g

Other Useful Identities and Quantities

1 atm = 760 mmHg
$R = kN_A$
$F = qN_A$
pH $= -\log_{10}[H^+]$ with $[H^+]$ in moles per liter
273.15 K = 0°C (ice point)
$T_{\text{Kelvin}} = T_{\text{centigrade}} - 273.15$
$T_{\text{Farenheit}} = \frac{9}{5}T_{\text{centigrade}} + 32$
$\frac{RT}{F} = 25.8$ mV at 27°C
density of pure water at 4°C = 1 gm/cm^3
C_m; capacitance of the cell membrane $\approx 1\ \mu$F/cm^2
Liter; 1 liter = 10^{-3} m^3
ϵ; dielectric constant for water = 80.4 ϵ_0

Lumen: 1 lm = quantity of light emitted by $\frac{1}{60}$ cm^2 surface area of pure platinum at its melting temperature (1770°C), within a solid angle of 1 steradian.

References

Adair, G. S. (1925) A critical study of the direct method of measuring osmotic pressure of hemoglobin, *Proceedings of the Royal Society of London A*. 108: 627–637.

Agre, P., L. S. King, M. Yasui, W. B. Guggino, O. P. Ottersen, Y. Fujiyoshi, A. Engel and S. Nielsen (2002) Aquaporin water channels–from atomic structure to clinical medicine, *Journal of Physiology*. 542: 3–16.

Aharon, S., H. Parnas and I. Parnas (1994) The magnitude and significance of Ca^{2+} domains for release of neurotransmitter, *Bulletin of Mathematical Biology*. 56: 1095–1119.

Akin, E. and H. M. Lacker (1984) Ovulation control: the right number or nothing, *Journal of Mathematical Biology*. 20: 113–132.

Albers, R. W., S. Fahn and G. J. Koval (1963) The role of sodium ions in the activation of electrophorus electric organ adenosine triphosphatase, *Proceedings of the National Academy of Sciences USA*. 50: 474–481.

Alberts, B., D. Bray, J. Lewis, M. Raff, K. Roberts and J. D. Watson (1994) *Molecular Biology of the Cell (Third Edition)*: Garland Publishing, Inc., New York, London.

Aldrich, R. W., D. P. Corey and C. F. Stevens (1983) A reinterpretation of mammalian sodium channel gating based on single channel recording, *Nature*. 306: 436–441.

Allbritton, N. L., T. Meyer and L. Stryer (1992) Range of messenger action of calcium ion and inositol 1,4,5-trisphosphate, *Science*. 258: 1812–1815.

Allen, N. A., K. C. Chen, C. A. Shaffer, J. J. Tyson and L. T. Watson (2006) Computer evaluation of network dynamics models with application to cell cycle control in budding yeast, *IEE Proceedings Systems Biology*. 153: 13–21.

Allescher, H. D., K. Abraham-Fuchs, R. E. Dunkel and M. Classen (1998) Biomagnetic 3-dimensional spatial and temporal characterization of electrical activity of human stomach, *Digestive Diseases and Sciences*. 43: 683–693.

Alt, W. and D. A. Lauffenberger (1987) Transient behavior of a chemotaxis system modelling certain types of tissue inflammation, *Journal of Mathematical Biology*. 24: 691–722.

Alvarez, W. C. (1922) The electrogastrogram and what it shows, *Journal of the American Medical Association*. 78: 1116–1119.

Anand, M., K. Rajagopal and K. R. Rajagopal (2003) A model incorporating some of the mechanical and biochemical factors underlying clot formation and dissolution in flowing blood, *Journal of Theoretical Medicine*. 5: 183–218.

Anand, M., K. Rajagopal and K. R. Rajagopal (2005) A model for the formation and lysis of blood clots, *Pathophysiology of Haemostasis and Thrombosis*. 34: 109–120.

Anderson, C., N. Trayanova and K. Skouibine (2000) Termination of spiral waves with biphasic shocks: role of virtual electrode polarization, *Journal of Cardiovascular Electrophysiology*. 11: 1386–1396.

Anliker, M., R. L. Rockwell and E. Ogden (1971a) Nonlinear analysis of flow pulses and shock waves in arteries. Part I: derivation and properties of mathematical model, *Zeitschrift für Angewandte Mathematik und Physik*. 22: 217–246.

Anliker, M., R. L. Rockwell and E. Ogden (1971b) Nonlinear analysis of flow pulses and shock waves in arteries. Part II: parametric study related to clinical problems., *Zeitschrift für Angewandte Mathematik und Physik*. 22: 563–581.

Apell, H. J. (1989) Electrogenic properties of the Na,K pump, *Journal of Membrane Biology*. 110: 103–114.

Apell, H. J. (2004) How do P-type ATPases transport ions?, *Bioelectrochemistry*. 63: 149–156.

Armstrong, C. M. (1981) Sodium channels and gating currents, *Physiological Reviews*. 61: 644–683.

Armstrong, C. M. and F. Bezanilla (1973) Currents related to movement of the gating particles of the sodium channels, *Nature*. 242: 459–461.

Armstrong, C. M. and F. Bezanilla (1974) Charge movement associated with the opening and closing of the activation gates of the Na channels, *Journal of General Physiology*. 63: 533–552.

Armstrong, C. M. and F. Bezanilla (1977) Inactivation of the sodium channel II Gating current experiments, *Journal of General Physiology*. 70: 567–590.

Arnold, V. I. (1983) *Geometric Methods in the Theory of Ordinary Differential Equations*: Springer-Verlag, New York.

Aronson, D. G. and H. F. Weinberger (1975) *Nonlinear diffusion in population genetics, combustion and nerve pulse propagation*: Springer-Verlag, New York.

Arrhenius, S. (1889) On the reaction velocity of the inversion of cane sugar by acids, *Zeitschrift fr physikalische Chemie*. 4: 226 ff.

Asdell, S. A. (1946) *Patterns of Mammalian Reproduction*: Comstock Publishing Company, New York.

Ashmore, J. F. and D. Attwell (1985) Models for electrical tuning in hair cells, *Proceedings of the Royal Society of London B*. 226: 325–344.

Ashmore, J. F., G. S. Geleoc and L. Harbott (2000) Molecular mechanisms of sound amplification in the mammalian cochlea, *Proceedings of the National Academy of Sciences USA*. 97: 11759–11764.

Ataullakhanov, F. I., Y. V. Krasotkina, V. I. Sarbash, R. I. Volkova, E. I. Sinauridse and A. Y. Kondratovich (2002a) Spatio-temporal dynamics of blood coagulation and pattern formation. An experimental study., *International Journal of Bifurcation and Chaos*. 12: 1969–1983.

Ataullakhanov, F. I., V. I. Zarnitsina, A. V. Pokhilko, A. I. Lobanov and O. L. Morozova (2002b) Spatio-temporal dynamics of blood coagulation and pattern formation. A theoretical approach., *International Journal of Bifurcation and Chaos*. 12: 1985–2002.

Atri, A., J. Amundson, D. Clapham and J. Sneyd (1993) A single-pool model for intracellular calcium oscillations and waves in the *Xenopus laevis* oocyte, *Biophysical Journal*. 65: 1727–1739.

Atwater, I., C. M. Dawson, A. Scott, G. Eddlestone and E. Rojas (1980) The nature of the oscillatory behavior in electrical activity for pancreatic β-cell, *Hormone and Metabolic Research*. 10 (suppl.): 100–107.

Atwater, I. and J. Rinzel (1986) *The β-cell bursting pattern and intracellular calcium*. In: Ionic Channels in Cells and Model Systems, Ed: R. Latorre, Plenum Press, New York, London.

Bélair, J., M. C. Mackey and J. M. Mahaffy (1995) Age-structured and two-delay models for erythropoiesis, *Mathematical Biosciences*. 128: 317–346.

Backx, P. H., P. P. d. Tombe, J. H. K. V. Deen, B. J. M. Mulder and H. E. D. J. t. Keurs (1989) A model of propagating calcium-induced calcium release mediated by calcium diffusion, *Journal of General Physiology*. 93: 963–977.

Ball, F. G., Y. Cai, J. B. Kadane and A. O'Hagan (1999) Bayesian inference for ion-channel gating mechanisms directly from single-channel recordings, using Markov chain Monte Carlo, *Proceedings of the Royal Society of London A*. 455: 2879–2932.

Bar, M. and M. Eiswirth (1993) Turbulence due to spiral breakup in a continuous excitable medium, *Physical Review E*. 48: 1635–1637.

Barcilon, V. (1992) Ion flow through narrow membrane channels: Part I, *SIAM Journal on Applied Mathematics*. 52: 1391–1404.

Barcilon, V., D. P. Chen and R. S. Eisenberg (1992) Ion flow through narrow membrane channels: Part II, *SIAM Journal on Applied Mathematics*. 52: 1405–1425.

Barkley, D. (1994) Euclidean symmetry and the dynamics of rotating spiral waves, *Physical Review Letters*. 72: 164–167.

Barlow, H. B. (1972) Single units and sensation: a neuron doctrine for perceptual psychology?, *Perception*. 1: 371–394.

Barlow, H. B., R. Fitzhugh and S. W. Kuffler (1957) Change of organization in the receptive fields of the cat's retina during dark adaptation, *Journal of Physiology*. 137: 338–354.

Barlow, H. B. and W. R. Levick (1965) The mechanism of directionally selective units in rabbit's retina, *Journal of Physiology*. 178: 477–504.

Batchelor, G. K. (1967) *An Introduction to Fluid Dynamics*: Cambridge University Press, Cambridge.

Batzel, J. J., F. Kappel, D. Schneditz and H. T. Tran (2007) *Cardiovascular and Respiratory System: Modeling, Analysis and Control*: SIAM, Philadelphia.

Batzel, J. J. and H. T. Tran (2000a) Stability of the human respiratory control system. I. Analysis of a two-dimensional delay state-space model, *Journal of Mathematical Biology*. 41: 45–79.

Batzel, J. J. and H. T. Tran (2000b) Stability of the human respiratory control system. II. Analysis of a three-dimensional delay state-space model, *Journal of Mathematical Biology*. 41: 80–102.

Batzel, J. J. and H. T. Tran (2000c) Modeling instability in the control system for human respiration: applications to infant non-REM sleep, *Applied Mathematics and Computation*. 110: 1–51.

Baylor, D. A. and A. L. Hodgkin (1974) Changes in time scale and sensitivity in turtle photoreceptors, *Journal of Physiology*. 242: 729–758.

Baylor, D. A., A. L. Hodgkin and T. D. Lamb (1974a) The electrical response of turtle cones to flashes and steps of light, *Journal of Physiology*. 242: 685–727.

Baylor, D. A., A. L. Hodgkin and T. D. Lamb (1974b) Reconstruction of the electrical responses of turtle cones to flashes and steps of light, *Journal of Physiology*. 242: 759–791.

Baylor, D. A., T. D. Lamb and K. W. Yau (1979) Responses of retinal rods to single photons, *Journal of Physiology*. 288: 613–634.

Baylor, D. A., G. Matthews and K. W. Yau (1980) Two components of electrical dark noise in toad retinal rod outer segments, *Journal of Physiology*. 309: 591–621.

Beck, J. S., R. Laprade and J.-Y. Lapointe (1994) Coupling between transepithelial Na transport and basolateral K conductance in renal proximal tubule, *American Journal of Physiology — Renal, Fluid and Electrolyte Physiology*. 266: F517–527.

Beeler, G. W. and H. J. Reuter (1977) Reconstruction of the action potential of ventricular myocardial fibers, *Journal of Physiology*. 268: 177–210.

Begenisich, T. B. and M. D. Cahalan (1980) Sodium channel permeation in squid axons I: reversal potential experiments, *Journal of Physiology*. 307: 217–242.

Beltrami, E. and J. Jesty (1995) Mathematical analysis of activation thresholds in enzyme-catalyzed positive feedbacks: application to the feedbacks of blood coagulation, *Proceedings of the National Academy of Sciences USA*. 92: 8744–8748.

Ben-Tal, A. (2006) Simplified models for gas exchange in the human lungs, *Journal of Theoretical Biology*. 238: 474–495.

Ben-Tal, A. and J. C. Smith (2008) A model for control of breathing in mammals: coupling neural dynamics to peripheral gas exchange and transport, *Journal of Theoretical Biology*. 251: 480–497.

Bennett, M. R., L. Farnell and W. G. Gibson (2004) The facilitated probability of quantal secretion within an array of calcium channels of an active zone at the amphibian neuromuscular junction, *Biophysical Journal*. 86: 2674–2690.

Bennett, M. R., L. Farnell and W. G. Gibson (2005) A quantitative model of purinergic junctional transmission of calcium waves in astrocyte networks, *Biophysical Journal*. 89: 2235–2250.

Bennett, M. R., L. Farnell, W. G. Gibson and P. Dickens (2007) Mechanisms of calcium sequestration during facilitation at active zones of an amphibian neuromuscular junction, *Journal of Theoretical Biology*. 247: 230–241.

Bensoussan, A., J.-L. Lions and G. Papanicolaou (1978) *Asymptotic Analysis for Periodic Structures*: North-Holland, Amsterdam, New York.

Bentele, K. and M. Falcke (2007) Quasi-steady approximation for ion channel currents, *Biophysical Journal*. 93: 2597–2608.

Bergman, R. N. (1989) Toward physiological understanding of glucose tolerance: minimal-model approach, *Diabetes*. 38: 1512–1527.

Bergman, R. N., Y. Z. Ider, C. R. Bowden and C. Cobelli (1979) Quantitative estimation of insulin sensitivity, *American Journal of Physiology — Endocrinology and Metabolism*. 236: E667–677.

Bergstrom, R. W., W. Y. Fujimoto, D. C. Teller and C. D. Haën (1989) Oscillatory insulin secretion in perifused isolated rat islets, *American Journal of Physiology — Endocrinology and Metabolism*. 257: E479–485.

Berman, N., H.-F. Chou, A. Berman and E. Ipp (1993) A mathematical model of oscillatory insulin secretion, *American Journal of Physiology — Regulatory, Integrative and Comparative Physiology*. 264: R839–851.

Berne, R. M. and M. N. Levy, Eds. (1993) *Physiology (Third Edition)*. Mosby Year Book, St. Louis.

Berne, R. M. and M. N. Levy, Eds. (1998) *Physiology (Fourth Edition)*. Mosby Year Book, St. Louis.

Bernstein, J. (1902) Untersuchungen zur Thermodynamik der bioelektrischen Ströme. Erster Theil., *Pflügers Archive*. 82: 521–562.

Berridge, M. J. (1997) Elementary and global aspects of calcium signalling, *Journal of Physiology*. 499: 291–306.

Berridge, M. J., M. D. Bootman and H. L. Roderick (2003) Calcium signalling: dynamics, homeostasis and remodelling, *Nature Reviews Molecular Cell Biology*. 4: 517–529.

Berridge, M. J. and A. Galione (1988) Cytosolic calcium oscillators, *FASEB Journal*. 2: 3074–3082.

Bers, D. M. (2001) *Excitation–Contraction Coupling and Cardiac Contractile Force (Second edition)*: Kluwer, New York.

Bers, D. M. (2002) Cardiac excitation–contraction coupling, *Nature*. 415: 198–205.

Bers, D. M. and E. Perez-Reyes (1999) Ca channels in cardiac myocytes: structure and function in Ca influx and intracellular Ca release, *Cardiovascular Research*. 42: 339–360.

Bertram, R., M. J. Butte, T. Kiemel and A. Sherman (1995) Topological and phenomenological classification of bursting oscillations, *Bulletin of Mathematical Biology*. 57: 413–439.

Bertram, R., M. Egli, N. Toporikova and M. E. Freeman (2006b) A mathematical model for the mating-induced prolactin rhythm of female rats, *American Journal of Physiology — Endocrinology and Metabolism*. 290: E573–582.

Bertram, R., M. Gram Pedersen, D. S. Luciani and A. Sherman (2006a) A simplified model for mitochondrial ATP production, *Journal of Theoretical Biology*. 243: 575–586.

Bertram, R., L. Satin, M. Zhang, P. Smolen and A. Sherman (2004) Calcium and glycolysis mediate multiple bursting modes in pancreatic islets, *Biophysical Journal*. 87: 3074–3087.

Bertram, R., L. S. Satin, M. G. Pedersen, D. S. Luciani and A. Sherman (2007b) Interaction of glycolysis and mitochondrial respiration in metabolic oscillations of pancreatic islets, *Biophysical Journal*. 92: 1544–1555.

Bertram, R. and A. Sherman (2004a) A calcium-based phantom bursting model for pancreatic islets, *Bulletin of Mathematical Biology*. 66: 1313–1344.

Bertram, R. and A. Sherman (2004b) Filtering of calcium transients by the endoplasmic reticulum in pancreatic beta-cells, *Biophysical Journal*. 87: 3775–3785.

Bertram, R., A. Sherman and L. S. Satin (2007a) Metabolic and electrical oscillations: partners in controlling pulsatile insulin secretion, *American Journal of Physiology — Endocrinology and Metabolism*. 293: E890–900.

Bertram, R., A. Sherman and E. F. Stanley (1996) Single-domain/bound calcium hypothesis of transmitter release and facilitation, *Journal of Neurophysiology*. 75: 1919–1931.

Beuter, A., L. Glass, M. C. Mackey and M. S. Titcombe (2003) *Nonlinear Dynamics in Physiology and Medicine*: Springer-Verlag, New York.

Bezprozvanny, I., J. Watras and B. E. Ehrlich (1991) Bell-shaped calcium-response curves of Ins(1,4,5)P_3- and calcium-gated channels from endoplasmic reticulum of cerebellum, *Nature*. 351: 751–754.

Bier, M., B. Teusink, B. N. Kholodenko and H. V. Westerhoff (1996) Control analysis of glycolytic oscillations, *Biophysical Chemistry*. 62: 15–24.

Blakemore, C., Ed. (1990) *Vision: Coding and Efficiency*. Cambridge University Press, Cambridge, UK.

Blatow, M., A. Caputi, N. Burnashev, H. Monyer and A. Rozov (2003) Ca^{2+} buffer saturation underlies paired pulse facilitation in calbindin-D28k-containing terminals, *Neuron*. 38: 79–88.

Bliss, R. D., P. R. Painter and A. G. Marr (1982) Role of feedback inhibition in stabilizing the classical operon, *Journal of Theoretical Biology*. 97: 177–193.

Blum, J. J., D. D. Carr and M. C. Reed (1992) Theoretical analysis of lipid transport in sciatic nerve, *Biochimica et Biophysica Acta*. 1125: 313–320.

Blum, J. J. and M. C. Reed (1985) A model for fast axonal transport, *Cell Motility*. 5: 507–527.

Blum, J. J. and M. C. Reed (1989) A model for slow axonal transport and its application to neurofilamentous neuropathies, *Cell Motility and the Cytoskeleton*. 12: 53–65.

Bluman, G. W. and H. C. Tuckwell (1987) Methods for obtaining analytical solutions for Rall's model neuron, *Journal of Neuroscience Methods*. 20: 151–166.

Bogumil, R. J., M. Ferin, J. Rootenberg, L. Speroff and R. L. Van De Wiele (1972) Mathematical studies of the human menstrual cycle. I. formulation of a mathematical model., *Journal of Clinical Endocrinology and Metabolism*. 35: 126–156.

Bohr, C., K. A. Hasselbach and A. Krogh (1904) Über einen in biologischen Beziehung wichtigen Einfluss, den die Kohlensauerspannung des Blutes auf dessen Sauerstoffbindung übt., *Skand Arch Physiol*. 15: 401–412.

Boitano, S., E. R. Dirksen and M. J. Sanderson (1992) Intercellular propagation of calcium waves mediated by inositol trisphosphate, *Science*. 258: 292–294.

Bootman, M., E. Niggli, M. Berridge and P. Lipp (1997a) Imaging the hierarchical Ca^{2+} signalling system in HeLa cells, *Journal of Physiology*. 499: 307–314.

Bootman, M. D., M. J. Berridge and P. Lipp (1997b) Cooking with calcium: the recipes for composing global signals from elementary events, *Cell*. 91: 367–373.

Borghans, J. A. M., R. J. De Boer and L. A. Segel (1996) Extending the quasi-steady state approximation by changing variables, *Bulletin of Mathematical Biology*. 58: 43–63.

Borghans, J. M., G. Dupont and A. Goldbeter (1997) Complex intracellular calcium oscillations. A theoretical exploration of possible mechanisms, *Biophysical Chemistry*. 66: 25–41.

Borisuk, M. T. and J. J. Tyson (1998) Bifurcation analysis of a model of mitotic control in frog eggs, *Journal of Theoretical Biology*. 195: 69–85.

Borst, A. (2000) Models of motion detection, *Nature Neuroscience*. 3 Suppl: 1168.

Borst, A. and M. Egelhaaf (1989) Principles of visual motion detection, *Trends in Neurosciences*. 12: 297–306.

Bowen, J. R., A. Acrivos and A. K. Oppenheim (1963) Singular perturbation refinement to quasi-steady state approximation in chemical kinetics, *Chemical Engineering Science*. 18: 177–188.

Boyce, W. E. and R. C. DiPrima (1997) *Elementary Differential Equations and Boundary Value Problems (Sixth Edition)*: John Wiley and Sons, New York.

Boyd, I. A. and A. R. Martin (1956) The end-plate potential in mammalian muscle, *Journal of Physiology*. 132: 74–91.

Brabant, G., K. Prank and C. Schöfl (1992) Pulsatile patterns in hormone secretion, *Trends in Endocrinology and Metabolism*. 3: 183–190.

Bradshaw, L. A., S. H. Allos, J. P. Wikswo, Jr. and W. O. Richards (1997) Correlation and comparison of magnetic and electric detection of small intestinal electrical activity, *American Journal of Physiology — Gastrointestinal and Liver Physiology*. 272: G1159–1167.

Bradshaw, L. A., J. K. Ladipo, D. J. Staton, J. P. Wikswo, Jr. and W. O. Richards (1999) The human vector magnetogastrogram and magnetoenterogram, *IEEE Transactions in Biomedical Engineering*. 46: 959–970.

Braun, M. (1993) *Differential Equations and their Applications (Fourth Edition)*: Springer-Verlag, New York.

Briggs, G. E. and J. B. S. Haldane (1925) A note on the kinematics of enzyme action, *Biochemical Journal*. 19: 338–339.

Brink, P. R. and S. V. Ramanan (1985) A model for the diffusion of fluorescent probes in the septate giant axon of earthworm: axoplasmic diffusion and junctional membrane permeability, *Biophysical Journal*. 48: 299–309.

Britton, N. F. (1986) *Reaction–Diffusion Equations and their Applications to Biology*: Academic Press, London.

Brodie, S. E., B. W. Knight and F. Ratliff (1978a) The response of the Limulus retina to moving stimuli: a prediction by Fourier synthesis, *Journal of General Physiology*. 72: 129–166.

Brodie, S. E., B. W. Knight and F. Ratliff (1978b) The spatiotemporal transfer function of the Limulus lateral eye, *Journal of General Physiology*. 72: 167–202.

Brokaw, C. J. (1976) Computer simulation of movement-generating cross-bridges, *Biophysical Journal*. 16: 1013–1027.

Brown, B. H., H. L. Duthie, A. R. Horn and R. H. Smallwood (1975) A linked oscillator model of electrical activity of human small intestine, *American Journal of Physiology*. 229: 384–388.

Brown, D., A. E. Herbison, J. E. Robinson, R. W. Marrs and G. Leng (1994) Modelling the luteinizing hormone-releasing hormone pulse generator, *Neuroscience*. 63: 869–879.

Brown, D., E. A. Stephens, R. G. Smith, G. Li and G. Leng (2004) Estimation of parameters for a mathematical model of growth hormone secretion, *Journal of Neuroendocrinology*. 16: 936–946.

Bugrim, A., R. Fontanilla, B. B. Eutenier, J. Keizer and R. Nuccitelli (2003) Sperm initiate a Ca^{2+} wave in frog eggs that is more similar to Ca^{2+} waves initiated by IP_3 than by Ca^{2+}, *Biophysical Journal*. 84: 1580–1590.

Bungay, S. D., P. A. Gentry and R. D. Gentry (2003) A mathematical model of lipid-mediated thrombin generation, *Mathematical Medicine and Biology*. 20: 105–129.

Burger, H. C. and J. B. Van Milaan (1946) Heart-vector and leads, *British Heart Journal*. 8: 157–161.

Burger, H. C. and J. B. Van Milaan (1947) Heart-vector and leads. Part II, *British Heart Journal*. 9: 154–160.

Burger, H. C. and J. B. Van Milaan (1948) Heart-vector and leads. Part III: geometrical representation, *British Heart Journal*. 10: 229–233.

Bursztyn, L., O. Eytan, A. J. Jaffa and D. Elad (2007) Modeling myometrial smooth muscle contraction, *Annals of the New York Academy of Sciences*. 1101: 110–138.

Burton, W. K., N. Cabrera and F. C. Frank (1951) The growth of crystals and the equilibrium structure of their surfaces, *Philosophical Transactions of the Royal Society of London A*. 243: 299–358.

Butera, R. J., Jr., J. Rinzel and J. C. Smith (1999a) Models of respiratory rhythm generation in the pre-Bötzinger complex. I. Bursting pacemaker neurons, *Journal of Neurophysiology*. 82: 382–397.

Butera, R. J., Jr., J. Rinzel and J. C. Smith (1999b) Models of respiratory rhythm generation in the pre-Bötzinger complex. II. Populations Of coupled pacemaker neurons, *Journal of Neurophysiology*. 82: 398–415.

Cajal, S. R. (1893) Sur les ganglions et plexus nerveux de lintestin, *Comptes Rendus des Seances de la Société de Biologie et de ses Filiales*. 45: 217–223.

Cajal, S. R. (1911) *Histologie du Systéme Nerveux de l'Homme et des Vertébrés*: Maloine, Paris.

Callamaras, N., J. S. Marchant, X. P. Sun and I. Parker (1998) Activation and co-ordination of $InsP_3$-mediated elementary Ca^{2+} events during global Ca^{2+} signals in Xenopus oocytes, *Journal of Physiology*. 509: 81–91.

Čamborová, P., P. Hubka, I. Šulková and I. Hulín (2003) The pacemaker activity of interstitial cells of Cajal and gastric electrical activity, *Physiological Research*. 52: 275–284.

Campbell, D. T. and B. Hille (1976) Kinetic and pharmacological properties of the sodium channel of frog skeletal muscle, *Journal of General Physiology*. 67: 309–323.

Carafoli, E. (2002) Calcium signaling: a tale for all seasons, *Proceedings of the National Academy of Sciences USA*. 99: 1115–1122.

Carafoli, E., L. Santella, D. Branca and M. Brini (2001) Generation, control, and processing of cellular calcium signals, *Critical Reviews in Biochemistry and Molecular Biology*. 36: 107–260.

Carpenter, G. (1977) A geometric approach to singular perturbation problems with applications to nerve impulse equations, *Journal of Differential Equations*. 23: 335–367.

Carpenter, G. A. and S. Grossberg (1981) Adaptation and transmitter gating in vertebrate photoreceptors, *Journal of Theoretical Neurobiology*. 1: 1–42.

Cartwright, M. and M. Husain (1986) A model for the control of testosterone secretion, *Journal of Theoretical Biology*. 123: 239–250.

Castellan, G. W. (1971) *Physical Chemistry (Second Edition)*: Addison-Wesley, Reading,

Casten, R. G., H. Cohen and P. A. Lagerstrom (1975) Perturbation analysis of an approximation to the Hodgkin–Huxley theory, *Quarterly of Applied Mathematics*. 32: 365–402.

Caulfield, J. B. and T. K. Borg (1979) The collagen network of the heart, *Laboratory Investigation*. 40: 364–372.

Caumo, A., R. N. Bergman and C. Cobelli (2000) Insulin sensitivity from meal tolerance tests in normal subjects: a minimal model index, *Journal of Clinical Endocrinology and Metabolism*. 85: 4396–4402.

Chávez-Ross, A., S. Franks, H. D. Mason, K. Hardy and J. Stark (1997) Modelling the control of ovulation and polycystic ovary syndrome, *Journal of Mathematical Biology*. 36: 95–118.

Chadwick, R. (1980) *Studies in cochlear mechanics*. In: Mathematical Modeling of the Hearing Process Lecture Notes in Biomathematics, 43, Ed: M. H. Holmes and L. A. Rubenfeld, Springer-Verlag, Berlin, Heidelberg, New York.

Chadwick, R. S., A. Inselberg and K. Johnson (1976) Mathematical model of the cochlea. II: results and conclusions, *SIAM Journal on Applied Mathematics*. 30: 164–179.

Champneys, A. R., V. Kirk, E. Knobloch, B. E. Oldeman and J. Sneyd (2007) When Shil'nikov meets Hopf in excitable systems, *SIAM Journal on Applied Dynamical Systems*. 6: 663–693.

Changeux, J. P. (1965) The control of biochemical reactions, *Scientific American*. 212: 36–45.

Chapman, R. A. and C. H. Fry (1978) An analysis of the cable properties of frog ventricular myocardium, *Journal of Physiology*. 283: 263–282.

Charles, A. (1998) Intercellular calcium waves in glia, *Glia*. 24: 39–49.

Charles, A. C., J. E. Merrill, E. R. Dirksen and M. J. Sanderson (1991) Intercellular signaling in glial cells: calcium waves and oscillations in response to mechanical stimulation and glutamate, *Neuron*. 6: 983–992.

Charles, A. C., C. C. G. Naus, D. Zhu, G. M. Kidder, E. R. Dirksen and M. J. Sanderson (1992) Intercellular calcium signaling via gap junctions in glioma cells, *Journal of Cell Biology*. 118: 195–201.

Charnock, J. S. and R. L. Post (1963) Studies of the mechanism of cation transport. I. The preparation and properties of a cation-stimulated adenosine-triphosphatase from guinea pig kidney cortex, *Australian Journal of Experimental Biology and Medical Science*. 41: 547–560.

Chay, T. R. (1986) On the effect of the intracellular calcium-sensitive K^+ channel in the bursting pancreatic β-cell, *Biophysical Journal*. 50: 765–777.

Chay, T. R. (1987) The effect of inactivation of calcium channels by intracellular Ca^{2+} ions in the bursting pancreatic β-cell, *Cell Biophysics*. 11: 77–90.

Chay, T. R. (1996a) Modeling slowly bursting neurons via calcium store and voltage-independent calcium current, *Neural Computation*. 8: 951–978.

Chay, T. R. (1996b) Electrical bursting and luminal calcium oscillation in excitable cell models, *Biological Cybernetics*. 75: 419–431.

Chay, T. R. (1997) Effects of extracellular calcium on electrical bursting and intracellular and luminal calcium oscillations in insulin secreting pancreatic beta-cells, *Biophysical Journal*. 73: 1673–1688.

Chay, T. R. and D. L. Cook (1988) Endogenous bursting patterns in excitable cells, *Mathematical Biosciences*. 90: 139–153.

Chay, T. R. and H. S. Kang (1987) *Multiple oscillatory states and chaos in the endogeneous activity of excitable cells: pancreatic β-cell as an example*. In: Chaos in Biological Systems, Ed: H. Degn, A. V. Holden and L. F. Olsen, Plenum Press, New York.

Chay, T. R. and J. Keizer (1983) Minimal model for membrane oscillations in the pancreatic β-cell, *Biophysical Journal*. 42: 181–190.

Cheer, A., R. Nuccitelli, G. F. Oster and J.-P. Vincent (1987) Cortical waves in vertebrate eggs I: the activation waves, *Journal of Theoretical Biology*. 124: 377–404.

Chen, D. and R. Eisenberg (1993) Charges, currents, and potentials in ionic channels of one conformation, *Biophysical Journal*. 64: 1405–1421.

Chen, D. P., V. Barcilon and R. S. Eisenberg (1992) Constant fields and constant gradients in open ionic channels, *Biophysical Journal*. 61: 1372–1393.

Chen, K. C., L. Calzone, A. Csikasz-Nagy, F. R. Cross, B. Novak and J. J. Tyson (2004) Integrative analysis of cell cycle control in budding yeast, *Molecular Biology of the Cell*. 15: 3841–3862.

Chen, K. C., A. Csikasz-Nagy, B. Gyorffy, J. Val, B. Novak and J. J. Tyson (2000) Kinetic analysis of a molecular model of the budding yeast cell cycle, *Molecular Biology of the Cell*. 11: 369–391.

Chen, L. and M. Q. Meng (1995) Compact and scattered gap junctions in diffusion mediated cell-cell communication, *Journal of Theoretical Biology*. 176: 39–45.

Chen, P. S., P. D. Wolf, E. G. Dixon, N. D. Danieley, D. W. Frazier, W. M. Smith and R. E. Ideker (1988) Mechanism of ventricular vulnerability to single premature stimuli in open-chest dogs, *Circulation Research*. 62: 1191–1209.

Cheng, H., M. Fill, H. Valdivia and W. J. Lederer (1995) Models of Ca^{2+} release channel adaptation, *Science*. 267: 2009–2010.

Cheng, H., W. J. Lederer and M. B. Cannell (1993) Calcium sparks: elementary events underlying excitation–contraction coupling in heart muscle, *Science*. 262: 740–744.

Cherniack, N. S. and G. S. Longobardo (2006) Mathematical models of periodic breathing and their usefulness in understanding cardiovascular and respiratory disorders, *Experimental Physiology*. 91: 295–305.

Cheyne, J. (1818) A case of apoplexy, in which the fleshy part of the heart was converted into fat, *Dublin Hospital Reports*. 2: 216.

Christ, G. J., P. R. Brink and S. V. Ramanan (1994) Dynamic gap junctional communication: a delimiting model for tissue responses, *Biophysical Journal*. 67: 1335–1344.

Ciliberto, A., B. Novak and J. J. Tyson (2003) Mathematical model of the morphogenesis checkpoint in budding yeast, *Journal of Cell Biology*. 163: 1243–1254.

Civan, M. M. and R. J. Bookman (1982) Transepithelial Na^+ transport and the intracellular fluids: a computer study, *Journal of Membrane Biology*. 65: 63–80.

Civan, M. M. and R. J. Podolsky (1966) Contraction kinetics of striated muscle fibres following quick changes in load, *Journal of Physiology*. 184: 511–534.

Clément, F., M. A. Gruet, P. Monget, M. Terqui, E. Jolivet and D. Monniaux (1997) Growth kinetics of the granulosa cell population in ovarian follicles: an approach by mathematical modelling, *Cell Proliferation*. 30: 255–270.

Clément, F., D. Monniaux, J. Stark, K. Hardy, J. C. Thalabard, S. Franks and D. Claude (2001) Mathematical model of FSH-induced cAMP production in ovarian follicles, *American Journal of Physiology — Endocrinology and Metabolism*. 281: E35–53.

Clément, F., D. Monniaux, J. C. Thalabard and D. Claude (2002) Contribution of a mathematical modelling approach to the understanding of the ovarian function, *Comptes Rendus des Seances de la Société de Biologie et de ses Filiales*. 325: 473–485.

Clancy, C. E. and Y. Rudy (1999) Linking a genetic defect to its cellular phenotype in a cardiac arrhythmia, *Nature*. 400: 566–569.

Clancy, C. E. and Y. Rudy (2001) Cellular consequences of HERG mutations in the long QT syndrome: precursors to sudden cardiac death, *Cardiovascular Research*. 50: 301–313.

Clapham, D. (1995) Intracellular calcium–replenishing the stores, *Nature*. 375: 634–635.

Clark, A. J. (1933) *The Mode of Action of Drugs on Cells*: Edward Arnold and Co., London.

Clark, L. H., P. M. Schlosser and J. F. Selgrade (2003) Multiple stable periodic solutions in a model for hormonal control of the menstrual cycle, *Bulletin of Mathematical Biology*. 65: 157–173.

Clifford, C. W. and M. R. Ibbotson (2003) Fundamental mechanisms of visual motion detection: models, cells and functions, *Progress in Neurobiology*. 68: 409–437.

Cobbs, W. H. and E. N. Pugh (1987) Kinetics and components of the flash photocurrent of isolated retinal rods of the larval salamander *Ambystoma Tigrinum*, *Journal of Physiology*. 394: 529–572.

Coddington, E. A. and N. Levinson (1984) *Ordinary Differential Equations*: Robert E. Krieger Publishing Company, Malabar, Florida.

Cohen, M. A. and J. A. Taylor (2002) Short-term cardiovascular oscillations in man: measuring and modelling the physiologies, *Journal of Physiology*. 542: 669–683.

Cole, K. S., H. A. Antosiewicz and P. Rabinowitz (1955) Automatic computation of nerve excitation, *SIAM Journal on Applied Mathematics*. 3: 153–172.

Cole, K. S. and H. J. Curtis (1940) Electric impedance of the squid giant axon during activity, *Journal of General Physiology*. 22: 649–670.

Colegrove, S. L., M. A. Albrecht and D. D. Friel (2000) Quantitative analysis of mitochondrial Ca^{2+} uptake and release pathways in sympathetic neurons. Reconstruction of the recovery after depolarization-evoked $[Ca^{2+}]_i$ elevations, *Journal of General Physiology*. 115: 371–388.

Colijn, C. and M. C. Mackey (2005a) A mathematical model of hematopoiesis–I. Periodic chronic myelogenous leukemia, *Journal of Theoretical Biology*. 237: 117–132.

Colijn, C. and M. C. Mackey (2005b) A mathematical model of hematopoiesis: II. Cyclical neutropenia, *Journal of Theoretical Biology*. 237: 133–146.

Colli-Franzone, P., L. Guerri and S. Rovida (1990) Wavefront propagation in an activation model of the anisotropic cardiac tissue: asymptotic analysis and numerical simulations, *Journal of Mathematical Biology*. 28: 121–176.

Colli-Franzone, P., L. Guerri and B. Taccardi (1993) Spread of excitation in a myocardial volume: simulation studies in a model of anisotropic ventricular muscle activated by a point stimulation, *Journal of Cardiovascular Electrophysiology*. 4: 144–160.

Colquhoun, D. (1994) *Practical analysis of single-channel records*. In: Microelectrode Techniques: The Plymouth Workshop Handbook (Second Edition), Ed: D. Ogden, Mill Hill, London.

Colquhoun, D. (2006) The quantitative analysis of drug-receptor interactions: a short history, *Trends in Pharmacological Sciences*. 27: 149–157.

Colquhoun, D. and A. G. Hawkes (1977) Relaxation and fluctuations of membrane currents that flow through drug-operated channels, *Proceedings of the Royal Society of London B*. 199: 231–262.

Colquhoun, D. and A. G. Hawkes (1981) On the stochastic properties of single ion channels, *Proceedings of the Royal Society of London B*. 211: 205–235.

Colquhoun, D. and A. G. Hawkes (1982) On the stochastic properties of bursts of single ion channel openings and of clusters of bursts, *Philosophical Transactions of the Royal Society of London B*. 300: 1–59.

Colquhoun, D. and A. G. Hawkes (1994) *The interpretation of single channel recordings*. In: Microelectrode Techniques: The Plymouth Workshop Handbook (Second Edition), Ed: D. Ogden, Mill Hill, London.

Coombes, S. (2001) The effect of ion pumps on the speed of travelling waves in the fire-diffuse-fire model of Ca^{2+} release, *Bulletin of Mathematical Biology*. 63: 1–20.

Coombes, S. and P. C. Bressloff (2003) Saltatory waves in the spike-diffuse-spike model of active dendritic spines, *Physical Review Letters*. 91: 028102.

Coombes, S., R. Hinch and Y. Timofeeva (2004) Receptors, sparks and waves in a fire-diffuse-fire framework for calcium release, *Progress in Biophysics and Molecular Biology*. 85: 197–216.

Coombes, S. and Y. Timofeeva (2003) Sparks and waves in a stochastic fire-diffuse-fire model of Ca^{2+} release, *Physical Review E — Statistical, Nonlinear and Soft Matter Physics*. 68: 021915.

Cornell-Bell, A. H., S. M. Finkbeiner, M. S. Cooper and S. J. Smith (1990) Glutamate induces calcium waves in cultured astrocytes: long-range glial signaling, *Science*. 247: 470–473.

Cornish-Bowden, A. and R. Eisenthal (1974) Statistical considerations in the estimation of enzyme kinetic parameters by the direct linear plot and other methods, *Biochemical Journal*. 139: 721–730.

Costa, K. D., P. J. Hunter, J. M. Rogers, J. M. Guccione, L. K. Waldman and A. D. McCulloch (1996a) A three-dimensional finite element method for large elastic deformations of ventricular myocardium: I–Cylindrical and spherical polar coordinates, *Journal of Biomechanical Engineering*. 118: 452–463.

Costa, K. D., P. J. Hunter, J. S. Wayne, L. K. Waldman, J. M. Guccione and A. D. McCulloch (1996b) A three-dimensional finite element method for large elastic deformations of ventricular myocardium: II–Prolate spheroidal coordinates, *Journal of Biomechanical Engineering*. 118: 464–472.

Courant, R. and D. Hilbert (1953) *Methods of Mathematical Physics*: Wiley-Interscience, New York.

Courtemanche, M., J. P. Keener and L. Glass (1993) Instabilities of a propagating pulse in a ring of excitable media, *Physical Review Letters*. 70: 2182–2185.

Courtemanche, M., J. P. Keener and L. Glass (1996) A delay equation representation of pulse circulation on a ring in excitable media, *SIAM Journal on Applied Mathematics*. 56: 119–142.

Courtemanche, M., R. J. Ramirez and S. Nattel (1998) Ionic mechanisms underlying human atrial action potential properties: insights from a mathematical model, *American Journal of Physiology — Heart and Circulatory Physiology*. 275: H301–321.

Courtemanche, M. and A. T. Winfree (1991) Re-entrant rotating waves in a Beeler–Reuter-based model of 2-dimensional cardiac electrical activity, *International Journal of Bifurcation and Chaos*. 1: 431–444.

Cox, S. J. (2004) Estimating the location and time course of synaptic input from multi-site potential recordings, *Journal of Computational Neuroscience*. 17: 225–243.

Crawford, A. C. and R. Fettiplace (1981) An electrical tuning mechanism in turtle cochlear hair cells, *Journal of Physiology*. 312: 377–412.

Curran, P. F. and J. R. MacIntosh (1962) A model system for biological water transport, *Nature*. 193: 347–348.

Cuthbertson, K. S. R. and T. R. Chay (1991) Modelling receptor-controlled intracellular calcium oscillators, *Cell Calcium*. 12: 97–109.

Cytrynbaum, E. and J. P. Keener (2002) Stability conditions for the traveling pulse: Modifying the restitution hypothesis, *Chaos*. 12: 788–799.

Dörlochter, M. and H. Stieve (1997) The Limulus ventral photoreceptor: light response and the role of calcium in a classic preparation, *Progress in Neurobiology*. 53: 451–515.

Dallos, P., C. D. Geisler, J. W. Matthews, M. A. Ruggero and C. R. Steele, Eds. (1990)

The Mechanics and Biophysics of Hearing. Lecture Notes in Biomathematics, 87: Springer-Verlag, Berlin, Heidelberg, New York.

Dallos, P., A. N. Popper and R. R. Fay, Eds. (1996) *The Cochlea*. Springer-Verlag, New York.

Daly, S. J. and R. I. Normann (1985) Temporal information processing in cones: effects of light adaptation on temporal summation and modulation, *Vision Research*. 25: 1197–1206.

Danø, S., P. G. Sorensen and F. Hynne (1999) Sustained oscillations in living cells, *Nature*. 402: 320–322.

Dani, J. A. and D. G. Levitt (1990) Diffusion and kinetic approaches to describe permeation in ionic channels, *Journal of Theoretical Biology*. 146: 289–301.

Danielsen, M. and J. T. Ottesen (2001) Describing the pumping heart as a pressure source, *Journal of Theoretical Biology*. 212: 71–81.

Dargan, S. L. and I. Parker (2003) Buffer kinetics shape the spatiotemporal patterns of IP_3-evoked Ca^{2+} signals, *Journal of Physiology*. 553: 775–788.

Davis, B. O., N. Holtz and J. C. Davis (1985) *Conceptual Human Physiology*: C.E. Merrill Pub. Co., Columbus.

Davis, R. C., L. Garafolo and F. P. Gault (1957) An exploration of abdominal potential, *Journal of Comparative and Physiological Psychology*. 52: 519–523.

Dawis, S. M., R. M. Graeff, R. A. Heyman, T. F. Walseth and N. D. Goldberg (1988) Regulation of cyclic GMP metabolism in toad photoreceptors, *Journal of Biological Chemistry*. 263: 8771–8785.

Dawson, D. C. (1992) *Water transport: principles and perspectives*. In: The Kidney: Physiology and Pathophysiology, Ed: D. W. Seldin and G. Giebisch, Raven Press, New York.

Dawson, D. C. and N. W. Richards (1990) Basolateral K conductance: role in regulation of NaCl absorption and secretion, *American Journal of Physiology — Cell Physiology*. 259: C181–195.

Dayan, P. and L. F. Abbott (2001) *Theoretical Neuroscience: Computational and Mathematical Modeling of Neural Systems*: MIT Press, Cambridge, MA.

de Beus, A. M., T. L. Fabry and H. M. Lacker (1993) A gastric acid secretion model, *Biophysical Journal*. 65: 362–378.

De Gaetano, A. and O. Arino (2000) Mathematical modelling of the intravenous glucose tolerance test, *Journal of Mathematical Biology*. 40: 136–168.

de Schutter, E., Ed. (2000) *Computational Neuroscience: Realistic Modeling for Experimentalists*. CRC Press, Boca Raton, Florida.

de Vries, G. (1995) *Analysis of models of bursting electrical activity in pancreatic beta cells. PhD Thesis, Department of Mathematics*: University of British Columbia, Vancouver.

De Young, G. W. and J. Keizer (1992) A single pool IP_3-receptor based model for agonist stimulated Ca^{2+} oscillations, *Proceedings of the National Academy of Sciences USA*. 89: 9895–9899.

deBoer, R. W., J. M. Karemaker and J. Strackee (1987) Hemodynamic fluctuations and baroreflex sensitivity in humans: a beat-to-beat model, *American Journal of Physiology — Heart and Circulatory Physiology*. 253: H680–689.

DeFronzo, R. A., J. D. Tobin and R. Andres (1979) Glucose clamp technique: a method for quantifying insulin secretion and resistance, *American Journal of Physiology — Endocrinology and Metabolism*. 237: E214–223.

del Castillo, J. and B. Katz (1954) Quantal components of the end-plate potential, *Journal of Physiology*. 124: 560–573.

del Castillo, J. and B. Katz (1957) Interaction at end-plate receptors between different choline derivatives, *Proceedings of the Royal Society of London B*. 146: 369–381.

Del Negro, C. A., C. G. Wilson, R. J. Butera, H. Rigatto and J. C. Smith (2002) Periodicity, mixed-mode oscillations, and quasiperiodicity in a rhythm-generating neural network, *Biophysical Journal*. 82: 206–214.

Demer, L. L., C. M. Wortham, E. R. Dirksen and M. J. Sanderson (1993) Mechanical stimulation induces intercellular calcium signalling in bovine aortic endothelial cells, *American Journal of Physiology — Heart and Circulatory Physiology*. 33: H2094–2102.

Deonier, R. C., S. Tavaré and M. S. Waterman (2004) *Computational Genome Analysis: An Introduction*: Springer-Verlag, New York.

Destexhe, A., Z. F. Mainen and T. J. Sejnowski (1998) *Kinetic models of synaptic transmission*. In: Methods in Neuronal Modeling, Ed: C. Koch and I. Segev, MIT Press, Cambridge, MA.

Detwiler, P. B. and A. L. Hodgkin (1979) Electrical coupling between cones in the turtle retina, *Journal of Physiology*. 291: 75–100.

DeVries, G. W., A. I. Cohen, O. H. Lowry and J. A. Ferendelli (1979) Cyclic nucleotides in the cone-dominant ground squirrel retina, *Experimental Eye Research*. 29: 315–321.

Diamant, N. E. and A. Bortoff (1969) Nature of the intestinal slow-wave frequency gradient, *American Journal of Physiology*. 216: 301–307.

Diamant, N. E., P. K. Rose and E. J. Davison (1970) Computer simulation of intestinal slow-wave frequency gradient, *American Journal of Physiology*. 219: 1684–1690.

Diamond, J. M. and W. H. Bossert (1967) Standing-gradient osmotic flow. A mechanism for coupling of water and solute transport in epithelia, *Journal of General Physiology*. 50: 2061–2083.

DiFrancesco, D. and D. Noble (1985) A model of cardiac electrical activity incorporating ionic pumps and concentration changes, *Philosophical Transactions of the Royal Society of London B*. 307: 353–398.

Dixon, M. and E. C. Webb (1979) *Enzymes (Third Edition)*: Academic Press, New York.

Doan, T., A. Mendez, P. B. Detwiler, J. Chen and F. Rieke (2006) Multiple phosphorylation sites confer reproducibility of the rod's single-photon responses, *Science*. 313: 530–533.

Dockery, J. D. and J. P. Keener (1989) Diffusive effects on dispersion in excitable media, *SIAM Journal on Applied Mathematics*. 49: 539–566.

Dodd, A. N., J. Love and A. A. Webb (2005a) The plant clock shows its metal: circadian regulation of cytosolic free Ca^{2+}, *Trends in Plant Science*. 10: 15–21.

Dodd, A. N., N. Salathia, A. Hall, E. Kevei, R. Toth, F. Nagy, J. M. Hibberd, A. J. Millar and A. A. Webb (2005b) Plant circadian clocks increase photosynthesis, growth, survival, and competitive advantage, *Science*. 309: 630–633.

Doedel, E. (1986) *Software for continuation and bifurcation problems in ordinary differential equations*: California Institute of Technology, Pasadena.

Dolmetsch, R. E., K. Xu and R. S. Lewis (1998) Calcium oscillations increase the efficiency and specificity of gene expression, *Nature*. 392: 933–936.

Domijan, M., R. Murray and J. Sneyd (2006) Dynamic probing of the mechanisms underlying calcium oscillations, *Journal of Nonlinear Science*. 16: 483–506.

Duffy, M. R., N. F. Britton and J. D. Murray (1980) Spiral wave solutions of practical reaction–diffusion systems, *SIAM Journal on Applied Mathematics*. 39: 8–13.

Dunlap, J. C. (1998) Circadian rhythms; an end in the beginning, *Science*. 280: 1548–1549.

Dunlap, J. C. (1999) Molecular bases for circadian clocks, *Cell*. 96: 271–290.

Dupont, G. and C. Erneux (1997) Simulations of the effects of inositol 1,4,5-trisphosphate 3-kinase and 5-phosphatase activities on Ca^{2+} oscillations, *Cell Calcium*. 22: 321–331.

Dupont, G., O. Koukoui, C. Clair, C. Erneux, S. Swillens and L. Combettes (2003) Ca^{2+} oscillations in hepatocytes do not require the modulation of $InsP_3$ 3-kinase activity by Ca^{2+}, *FEBS Letters*. 534: 101–105.

Dupont, G., T. Tordjmann, C. Clair, S. Swillens, M. Claret and L. Combettes (2000) Mechanism of receptor-oriented intercellular calcium wave propagation in hepatocytes, *FASEB Journal*. 14: 279–289.

Durand, D. (1984) The somatic shunt cable model for neurons, *Biophysical Journal*. 46: 645–653.

Durrett, R. (2002) *Probability Models for DNA Sequence Evolution*: Springer Verlag, New York.

Eason, J. and N. Trayanova (2002) Phase singularities and termination of spiral wave reentry, *Journal of Cardiovascular Electrophysiology*. 13: 672–679.

Eatock, R. A. (2000) Adaptation in hair cells, *Annual Review of Neuroscience*. 23: 285–314.

Eaton, W. A., E. R. Henry, J. Hofrichter and A. Mozzarelli (1999) Is cooperative oxygen binding by hemoglobin really understood?, *Nature Structural Biology*. 6: 351–358.

Ebihara, L. and E. A. Johnson (1980) Fast sodium current in cardiac muscle, a quantitative description, *Biophysical Journal*. 32: 779–790.

Echenim, N., D. Monniaux, M. Sorine and F. Clément (2005) Multi-scale modeling of the follicle selection process in the ovary, *Mathematical Biosciences*. 198: 57–79.

Edelstein-Keshet, L. (1988) *Mathematical Models in Biology*: McGraw-Hill, New York.

Efimov, I. R., F. Aguel, Y. Cheng, B. Wollenzier and N. Trayanova (2000a) Virtual electrode polarization in the far field: implications for external defibrillation, *American Journal of Physiology — Heart and Circulatory Physiology*. 279: H1055–1070.

Efimov, I. R., Y. Cheng, D. R. Van Wagoner, T. Mazgalev and P. J. Tchou (1998) Virtual electrode-induced phase singularity: a basic mechanism of defibrillation failure, *Circulation Research*. 82: 918–925.

Efimov, I. R., R. A. Gray and B. J. Roth (2000b) Virtual electrodes and deexcitation: new insights into fibrillation induction and defibrillation, *Journal of Cardiovascular Electrophysiology*. 11: 339–353.

Eguiluz, V. M., M. Ospeck, Y. Choe, A. J. Hudspeth and M. O. Magnasco (2000) Essential nonlinearities in hearing, *Physical Review Letters*. 84: 5232–5235.

Einstein, A. (1906) Eine neue Bestimmung der Moleküldimensionen, *Annals of Physics*. 19: 289.

Eisenberg, E. and L. E. Greene (1980) The relation of muscle biochemistry to muscle physiology, *Annual Review of Physiology*. 42: 293–309.

Eisenberg, E. and T. L. Hill (1978) A cross-bridge model of muscle contraction, *Progress in Biophysics and Molecular Biology*. 33: 55–82.

Eisenthal, R. and A. Cornish-Bowden (1974) A new graphical method for estimating enzyme kinetic parameters, *Biochemical Journal*. 139: 715–720.

Elston, T., H. Wang and G. Oster (1998) Energy transduction in ATP synthase, *Nature*. 391: 510–513.

Elston, T. C. (2000) Models of post-translational protein translocation, *Biophysical Journal*. 79: 2235–2251.

Elston, T. C. and C. S. Peskin (2000) The role of protein flexibility in molecular motor function: coupled diffusion in a tilted periodic potential, *SIAM Journal on Applied Mathematics*. 60: 842–867.

Endo, M., M. Tanaka and Y. Ogawa (1970) Calcium-induced release of calcium from the sarcoplasmic reticulum of skinned skeletal muscle fibres, *Nature*. 228: 34–36.

Engel, E., A. Peskoff, G. L. Kauffman and M. I. Grossman (1984) Analysis of hydrogen ion concentration in the gastric gel mucus layer, *American Journal of Physiology — Gastrointestinal and Liver Physiology*. 247: G321–338.

Ermentrout, B. (2002) *Simulating, Analyzing, and Animating Dynamical Systems: A Guide to Xppaut for Researchers and Students*: SIAM, Philadelphia.

Ermentrout, G. B. and N. Kopell (1984) Frequency plateaus in a chain of weakly coupled oscillators, *SIAM Journal on Mathematical Analysis*. 15: 215–237.

Ewens, W. J. and G. Grant (2005) *Statistical Methods in Bioinformatics: An Introduction*: Springer Verlag, New York.

Eyring, H., R. Lumry and J. W. Woodbury (1949) Some applications of modern rate theory to physiological systems, *Record of Chemical Progress*. 10: 100–114.

Fabiato, A. (1983) Calcium-induced release of calcium from the cardiac sarcoplasmic reticulum, *American Journal of Physiology — Cell Physiology*. 245: C1–14.

Faddy, M. J. and R. G. Gosden (1995) A mathematical model of follicle dynamics in the human ovary, *Human Reproduction*. 10: 770–775.

Fain, G. and H. R. Matthews (1990) Calcium and the mechanism of light adaptation in vertebrate photoreceptors, *Trends in Neuroscience*. 13: 378–384.

Fain, G. L. (1976) Sensitivity of toad rods: Dependence on wave-length and background illumination, *Journal of Physiology*. 261: 71–101.

Fain, G. L., H. R. Matthews, M. C. Cornwall and Y. Koutalos (2001) Adaptation in vertebrate photoreceptors, *Physiological Reviews*. 81: 117–151.

Fajmut, A., M. Brumen and S. Schuster (2005) Theoretical model of the interactions between Ca^{2+}, calmodulin and myosin light chain kinase, *FEBS Letters*. 579: 4361–4366.

Falcke, M. (2003a) Buffers and oscillations in intracellular Ca^{2+} dynamics, *Biophysical Journal*. 84: 28–41.

Falcke, M. (2003b) On the role of stochastic channel behavior in intracellular Ca^{2+} dynamics, *Biophysical Journal*. 84: 42–56.

Falcke, M. (2004) Reading the patterns in living cells–the physics of Ca^{2+} signaling, *Advances in Physics*. 53: 255–440.

Falcke, M., J. L. Hudson, P. Camacho and J. D. Lechleiter (1999) Impact of mitochondrial Ca^{2+} cycling on pattern formation and stability, *Biophysical Journal*. 77: 37–44.

Falcke, M., M. Or-Guil and M. Bar (2000) Dispersion gap and localized spiral waves in a model for intracellular Ca^{2+} dynamics, *Physical Review Letters*. 84: 4753–4756.

Fall, C. P., E. S. Marland, J. M. Wagner and J. J. Tyson, Eds. (2002) *Computational Cell Biology*: Springer-Verlag, New York.

Farkas, I., D. Helbing and T. Vicsek (2002) Mexican waves in an excitable medium, *Nature*. 419: 131–132.

Fast, V. G. and A. G. Kleber (1993) Microscopic conduction in cultured strands of neonatal rat heart cells measured with voltage-sensitive dyes, *Circulation Research*. 73: 914–925.

Fatt, P. and B. Katz (1952) Spontaneous subthreshold activity at motor nerve endings, *Journal of Physiology*. 117: 109–128.

Feldman, J. L. and C. A. Del Negro (2006) Looking for inspiration: new perspectives on respiratory rhythm, *Nature Reviews Neuroscience*. 7: 232–242.

Felmy, F., E. Neher and R. Schneggenburger (2003) The timing of phasic transmitter release is Ca^{2+}-dependent and lacks a direct influence of presynaptic membrane potential, *Proceedings of the National Academy of Sciences USA*. 100: 15200–15205.

Fenn, W. O., H. Rahn and A. B. Otis (1946) A theoretical study of the composition of the alveolar air at altitude, *American Journal of Physiology*. 146: 637–653.

Field, G. D. and F. Rieke (2002) Mechanisms regulating variability of the single photon responses of mammalian rod photoreceptors, *Neuron*. 35: 733–747.

Fields, R. D. and B. Stevens-Graham (2002) New insights into neuron–glia communication, *Science*. 298: 556–562.

Fife, P. (1979) *Mathematical Aspects of Reacting and Diffusing Systems*: Springer-Verlag, Berlin.

Fife, P. C. and J. B. McLeod (1977) The approach of solutions of nonlinear diffusion equations to travelling front solutions, *Archive for Rational Mechanics and Analysis*. 65: 335–361.

Fill, M. and J. A. Copello (2002) Ryanodine receptor calcium release channels, *Physiological Reviews*. 82: 893–922.

Fill, M., A. Zahradnikova, C. A. Villalba-Galea, I. Zahradnik, A. L. Escobar and S. Gyorke (2000) Ryanodine receptor adaptation, *Journal of General Physiology*. 116: 873–882.

Finkelstein, A. and C. S. Peskin (1984) Some unexpected consequences of a simple physical mechanism for voltage-dependent gating in biological membranes, *Biophysical Journal*. 46: 549–558.

Fishman, M. A. and A. S. Perelson (1993) Modeling T cell-antigen presenting cell interactions, *Journal of Theoretical Biology*. 160: 311–342.

Fishman, M. A. and A. S. Perelson (1994) Th1/Th2 cross regulation, *Journal of Theoretical Biology*. 170: 25–56.

FitzHugh, R. (1960) Thresholds and plateaus in the Hodgkin–Huxley nerve equations, *Journal of General Physiology*. 43: 867–896.

FitzHugh, R. (1961) Impulses and physiological states in theoretical models of nerve membrane, *Biophysical Journal*. 1: 445–466.

FitzHugh, R. (1969) *Mathematical models of excitation and propagation in nerve*. In: Biological Engineering, Ed: H. P. Schwan, McGraw-Hill, New York.

Fletcher, H. (1951) On the dynamics of the cochlea, *Journal of the Acoustical Society of America*. 23: 637–645.

Foerster, P., S. Muller and B. Hess (1989) Critical size and curvature of wave formation in an excitable chemical medium, *Proceedings of the National Academy of Sciences USA*. 86: 6831–6834.

Fogelson, A. L. (1992) Continuum models of platelet aggregation: Formulation and mechanical properties, *SIAM Journal on Applied Mathematics*. 52: 1089–1110.

Fogelson, A. L. and R. D. Guy (2004) Platelet–wall interactions in continuum models of platelet thrombosis: formulation and numerical solution, *Mathematical Medicine and Biology*. 21: 293–334.

Fogelson, A. L. and A. L. Kuharsky (1998) Membrane binding-site density can modulate activation thresholds in enzyme systems, *Journal of Theoretical Biology*. 193: 1–18.

Fogelson, A. L. and N. Tania (2005) Coagulation under flow: the influence of flow-mediated transport on the initiation and inhibition of coagulation, *Pathophysiology of Haemostasis and Thrombosis*. 34: 91–108.

Fogelson, A. L. and R. S. Zucker (1985) Presynaptic calcium diffusion from various arrays of single channels, *Biophysical Journal*. 48: 1003–1017.

Forger, D. B. and C. S. Peskin (2003) A detailed predictive model of the mammalian circadian clock, *Proceedings of the National Academy of Sciences USA*. 100: 14806–14811.

Forger, D. B. and C. S. Peskin (2004) Model based conjectures on mammalian clock controversies, *Journal of Theoretical Biology*. 230: 533–539.

Forger, D. B. and C. S. Peskin (2005) Stochastic simulation of the mammalian circadian clock, *Proceedings of the National Academy of Sciences USA*. 102: 321–324.

Forti, S., A. Menini, G. Rispoli and V. Torre (1989) Kinetics of phototransduction in retinal rods of the newt *Triturus Cristatus*, *Journal of Physiology*. 419: 265–295.

Fowler, A. C. and G. P. Kalamangalam (2000) The role of the central chemoreceptor in causing periodic breathing, *IMA Journal of Mathematics Applied in Medicine and Biology*. 17: 147–167.

Fowler, A. C. and G. P. Kalamangalam (2002) Periodic breathing at high altitude, *IMA*

Journal of Mathematics Applied in Medicine and Biology. 19: 293–313.

Fowler, A. C., G. P. Kalamangalam and G. Kember (1993) A mathematical analysis of the Grodins model of respiratory control, *IMA Journal of Mathematics Applied in Medicine and Biology*. 10: 249–280.

Fowler, A. C. and M. J. McGuinness (2005) A delay recruitment model of the cardiovascular control system, *Journal of Mathematical Biology*. 51: 508–526.

Fox, R. F. and Y. Lu (1994) Emergent collective behavior in large numbers of globally coupled independently stochastic ion channels, *Physical Review E*. 49: 3421–3431.

Frank, O. (1899) Die Grundform des Arteriellen Pulses (see translation by Sagawa et al., 1990), *Zeitschrift für Biologie*. 37: 483–526.

Frankel, M. L. and G. I. Sivashinsky (1987) On the nonlinear diffusive theory of curved flames, *Journal de Physique (Paris)*. 48: 25–28.

Frankel, M. L. and G. I. Sivashinsky (1988) On the equation of a curved flame front, *Physica D*. 30: 28–42.

Frankenhaeuser, B. (1960a) Quantitative description of sodium currents in myelinated nerve fibres of *Xenopus laevis*, *Journal of Physiology*. 151: 491–501.

Frankenhaeuser, B. (1960b) Sodium permeability in toad nerve and in squid nerve, *Journal of Physiology*. 152: 159–166.

Frankenhaeuser, B. (1963) A quantitative description of potassium currents in myelinated nerve fibres of *Xenopus laevis*, *Journal of Physiology*. 169: 424–430.

Frazier, D. W., P. D. Wolf and R. E. Ideker (1988) Electrically induced reentry in normal myocardium–evidence of a phase singularity, *PACE — Pacing and Clinical Electrophysiology*. 11: 482.

Frazier, D. W., P. D. Wolf, J. M. Wharton, A. S. Tang, W. M. Smith and R. E. Ideker (1989) Stimulus-induced critical point. Mechanism for electrical initiation of reentry in normal canine myocardium, *Journal of Clinical Investigation*. 83: 1039–1052.

Fredkin, D. R. and J. A. Rice (1992) Bayesian restoration of single-channel patch clamp recordings, *Biometrics*. 48: 427–448.

Frenzen, C. L. and P. K. Maini (1988) Enzyme kinetics for a two-step enzymic reaction with comparable initial enzyme-substrate ratios, *Journal of Mathematical Biology*. 26: 689–703.

Friel, D. (1995) $[Ca^{2+}]_i$ oscillations in sympathetic neurons: an experimental test of a theoretical model, *Biophysical Journal*. 68: 1752–1766.

Friel, D. (2000) Mitochondria as regulators of stimulus-evoked calcium signals in neurons., *Cell Calcium*. 28: 307–316.

Fye, W. B. (1986) Carl Ludwig and the Leipzig Physiological Institute: "a factory of new knowledge", *Circulation*. 74: 920–928.

Gardiner, C. W. (2004) *Handbook of Stochastic Methods for Physics, Chemistry and the Natural Sciences (Third Edition)*: Springer-Verlag, New York.

Garfinkel, A., Y. H. Kim, O. Voroshilovsky, Z. Qu, J. R. Kil, M. H. Lee, H. S. Karagueuzian, J. N. Weiss and P. S. Chen (2000) Preventing ventricular fibrillation by flattening cardiac restitution, *Proceedings of the National Academy of Sciences USA*. 97: 6061–6066.

Gatti, R. A., W. A. Robinson, A. S. Denaire, M. Nesbit, J. J. McCullogh, M. Ballow and R. A. Good (1973) Cyclic leukocytosis in chronic myelogenous leukemia, *Blood*. 41: 771–782.

Gerhardt, M., H. Schuster and J. J. Tyson (1990) A cellular automaton model of excitable media including curvature and dispersion, *Science*. 247: 1563–1566.

Gilkey, J. C., L. F. Jaffe, E. B. Ridgway and G. T. Reynolds (1978) A free calcium wave traverses the activating egg of the medaka, *Oryzias latipes*, *Journal of Cell Biology*. 76: 448–466.

Givelberg, E. (2004) Modeling elastic shells immersed in fluid, *Communications on Pure and Applied Mathematics*. 57: 283–309.

Givelberg, E. and J. Bunn (2003) A comprehensive three-dimensional model of the cochlea, *Journal of Computational Physics*. 191: 377–391.

Glass, L., A. L. Goldberger, M. Courtemanche and A. Shrier (1987) Nonlinear dynamics, chaos and complex cardiac arrhythmias, *Proceedings of the Royal Society of London A*. 413: 9–26.

Glass, L. and D. Kaplan (1995) *Understanding Nonlinear Dynamics*: Springer-Verlag, New York.

Glass, L. and M. C. Mackey (1988) *From Clocks to Chaos*: Princeton University Press, Princeton.

Glynn, I. M. (2002) A hundred years of sodium pumping, *Annual Review of Physiology*. 64: 1–18.

Goel, P., J. Sneyd and A. Friedman (2006) Homogenization of the cell cytoplasm: the calcium bidomain equations, *SIAM Journal on*

Multiscale Modeling and Simulation. 5: 1045–1062.

Gold, T. (1948) Hearing. II. The physical basis of the action of the cochlea, *Proceedings of the Royal Society of London B*. 135: 492–498.

Goldberger, A. L. and E. Goldberger (1994) *Clinical Electrocardiography: a Simplified Approach (Fifth Edition)*: Mosby, St. Louis.

Goldbeter, A. (1995) A model for circadian oscillations in the Drosophila period protein (PER), *Proceedings of the Royal Society of London B*. 261: 319–324.

Goldbeter, A. (1996) *Biochemical Oscillations and Cellular Rhythms: the Molecular Bases of Periodic and Chaotic Behaviour*: Cambridge University Press, Cambridge.

Goldbeter, A., G. Dupont and M. J. Berridge (1990) Minimal model for signal-induced Ca^{2+} oscillations and for their frequency encoding through protein phosphorylation, *Proceedings of the National Academy of Sciences USA*. 87: 1461–1465.

Goldbeter, A. and D. E. Koshland, Jr. (1981) An amplified sensitivity arising from covalent modification in biological systems, *Proceedings of the National Academy of Sciences USA*. 78: 6840–6844.

Goldbeter, A. and R. Lefever (1972) Dissipative structures for an allosteric model; application to glycolytic oscillations, *Biophysical Journal*. 12: 1302–1315.

Gomatam, J. and P. Grindrod (1987) Three dimensional waves in excitable reaction–diffusion systems, *Journal of Mathematical Biology*. 25: 611–622.

Goodner, C. J., B. C. Walike, D. J. Koerker, J. W. Ensinck, A. C. Brown, E. W. Chideckel, J. Palmer and L. Kalnasy (1977) Insulin, glucagon, and glucose exhibit synchronous, sustained oscillations in fasting monkeys, *Science*. 195: 177–179.

Goodwin, B. C. (1965) Oscillatory behavior in enzymatic control processes, *Advances in Enzyme Regulation*. 3: 425–438.

Gordon, A. M., A. F. Huxley and F. J. Julian (1966) The variation in isometric tension with sarcomere length in vertebrate muscle fibres, *Journal of Physiology*. 184: 170–192.

Graham, N. (1989) *Visual Pattern Analyzers*: Oxford University Press, New York.

Greenstein, J. L., R. Hinch and R. L. Winslow (2006) Mechanisms of excitation–contraction coupling in an integrative model of the cardiac ventricular myocyte, *Biophysical Journal*. 90: 77–91.

Greenstein, J. L. and R. L. Winslow (2002) An integrative model of the cardiac ventricular myocyte incorporating local control of Ca^{2+} release, *Biophysical Journal*. 83: 2918–2945.

Griffith, J. S. (1968a) Mathematics of cellular control processes. I. Negative feedback to one gene, *Journal of Theoretical Biology*. 20: 202–208.

Griffith, J. S. (1968b) Mathematics of cellular control processes. II. Positive feedback to one gene, *Journal of Theoretical Biology*. 20: 209–216.

Griffith, J. S. (1971) *Mathematical Neurobiology*: Academic Press, London.

Grindrod, P. (1991) *Patterns and Waves: the Theory and Application of Reaction-Diffusion Equations*: Clarendon Press, Oxford.

Grodins, F. S., J. Buell and A. J. Bart (1967) Mathematical analysis and digital simulation of the respiratory control system, *Journal of Applied Physiology*. 22: 260–276.

Grubelnik, V., A. Z. Larsen, U. Kummer, L. F. Olsen and M. Marhl (2001) Mitochondria regulate the amplitude of simple and complex calcium oscillations, *Biophysical Chemistry*. 94: 59–74.

Grzywacz, N. M., P. Hillman and B. W. Knight (1992) Response transfer functions of Limulus ventral photoreceptors: interpretation in terms of transduction mechanisms, *Biological Cybernetics*. 66: 429–435.

Guccione, J. M. and A. D. McCulloch (1993) Mechanics of active contraction in cardiac muscle: Part I–Constitutive relations for fiber stress that describe deactivation, *Journal of Biomechanical Engineering*. 115: 72–81.

Guccione, J. M., L. K. Waldman and A. D. McCulloch (1993) Mechanics of active contraction in cardiac muscle: Part II–Cylindrical models of the systolic left ventricle, *Journal of Biomechanical Engineering*. 115: 82–90.

Guckenheimer, J. and P. Holmes (1983) *Nonlinear Oscillations, Dynamical Systems, and Bifurcations of Vector Fields*: Springer-Verlag, New York, Heidelberg, Berlin.

Guevara, M. R. and L. Glass (1982) Phase locking, period doubling bifurcations and chaos in a mathematical model of a periodically driven oscillator: a theory for the entrainment of biological oscillators and the generation of cardiac dysrhythmias, *Journal of Mathematical Biology*. 14: 1–23.

Guneroth, W. G. (1965) *Pediatric Electrocardiography*: W.B. Saunders Co., Philadelphia.

Guo, J. S. and J. C. Tsai (2006) The asymptotic behavior of solutions of the buffered bistable system, *Journal of Mathematical Biology*. 53: 179–213.

Guyton, A. C. (1963) *Circulatory Physiology: Cardiac Output and its Regulation*: W.B. Saunders, Philadelphia.

Guyton, A. C. and J. E. Hall (1996) *Textbook of Medical Physiology (Ninth Edition)*: W.B. Saunders, Philadelphia.

Höfer, T. (1999) Model of intercellular calcium oscillations in hepatocytes: synchronization of heterogeneous cells, *Biophysical Journal*. 77: 1244–1256.

Höfer, T., H. Nathansen, M. Lohning, A. Radbruch and R. Heinrich (2002) GATA-3 transcriptional imprinting in Th2 lymphocytes: a mathematical model, *Proceedings of the National Academy of Sciences USA*. 99: 9364–9368.

Höfer, T., A. Politi and R. Heinrich (2001) Intercellular Ca^{2+} wave propagation through gap-junctional Ca^{2+} diffusion: a theoretical study, *Biophysical Journal*. 80: 75–87.

Höfer, T., L. Venance and C. Giaume (2002) Control and plasticity of intercellular calcium waves in astrocytes: a modeling approach, *Journal of Neuroscience*. 22: 4850–4859.

Haberman, R. (2004) *Applied Partial Differential Equations : with Fourier Series and Boundary Value Problems*: Pearson Prentice Hall, Upper Saddle River, N.J.

Hai, C. M. and R. A. Murphy (1989a) Ca^{2+}, crossbridge phosphorylation, and contraction, *Annual Review of Physiology*. 51: 285–298.

Hai, C. M. and R. A. Murphy (1989b) Cross-bridge dephosphorylation and relaxation of vascular smooth muscle, *American Journal of Physiology — Cell Physiology*. 256: C282–287.

Hajnóczky, G. and A. P. Thomas (1997) Minimal requirements for calcium oscillations driven by the IP_3 receptor, *Embo Journal*. 16: 3533–3543.

Hale, J. K. and H. Koçak (1991) *Dynamics and Bifurcations*: Springer-Verlag, New York, Berlin, Heidelberg.

Hall, G. M., S. Bahar and D. J. Gauthier (1999) Prevalence of rate-dependent behaviors in cardiac muscle, *Physical Review Letters*. 82: 2995–2998.

Hamer, R. D. (2000a) Computational analysis of vertebrate phototransduction: combined quantitative and qualitative modeling of dark- and light-adapted responses in amphibian rods, *Visual Neuroscience*. 17: 679–699.

Hamer, R. D. (2000b) Analysis of Ca^{++}-dependent gain changes in PDE activation in vertebrate rod phototransduction, *Molecular Vision*. 6: 265–286.

Hamer, R. D., S. C. Nicholas, D. Tranchina, T. D. Lamb and J. L. Jarvinen (2005) Toward a unified model of vertebrate rod phototransduction, *Visual Neuroscience*. 22: 417–436.

Hamer, R. D., S. C. Nicholas, D. Tranchina, P. A. Liebman and T. D. Lamb (2003) Multiple steps of phosphorylation of activated rhodopsin can account for the reproducibility of vertebrate rod single-photon responses, *Journal of General Physiology*. 122: 419–444.

Hamer, R. D. and C. W. Tyler (1995) Phototransduction: modeling the primate cone flash response, *Visual Neuroscience*. 12: 1063–1082.

Hamill, O. P., A. Marty, E. Neher, B. Sakmann and F. J. Sigworth (1981) Improved patch-clamp techniques for high-resolution current recording from cells and cell-free membrane patches, *Pflügers Arch*. 391: 85–100.

Hargrave, P. A., K. P. Hoffman and U. B. Kaupp, Eds. (1992) *Signal Transduction in Photoreceptor Cells*: Springer-Verlag, Berlin.

Harootunian, A. T., J. P. Kao and R. Y. Tsien (1988) Agonist-induced calcium oscillations in depolarized fibroblasts and their manipulation by photoreleased Ins(1,4,5)P_3, Ca^{++}, and Ca^{++} buffer, *Cold Spring Harbor Symposia on Quantitative Biology*. 53: 935–943.

Hartline, H. K. and B. W. Knight, Jr. (1974) The processing of visual information in a simple retina, *Annals of the New York Academy of Sciences*. 231: 12–18.

Hartline, H. K. and F. Ratliff (1957) Inhibitory interaction of receptor units in the eye of Limulus, *Journal of General Physiology*. 40: 357–376.

Hartline, H. K. and F. Ratliff (1958) Spatial summation of inhibitory influences in the eye of Limulus, and the mutual interaction of receptor units, *Journal of General Physiology*. 41: 1049–1066.

Hartline, H. K., H. G. Wagner and F. Ratliff (1956) Inhibition in the eye of Limulus, *Journal of General Physiology*. 39: 651–673.

Hassenstein, B. and W. Reichardt (1956) Systemtheorische analyse der zeit-, reihenfolgen- und vorzeichenauswertung bei der bewegungsperzeption des rüsselkäfers

Chlorophanus, *Zeitschrift für Naturforschung B*. 11: 513–524.

Hastings, S., J. Tyson and D. Webster (1977) Existence of periodic solutions for negative feedback cellular control systems, *Journal of Differential Equations*. 25: 39–64.

Hastings, S. P. (1975) The existence of progressive wave solutions to the Hodgkin–Huxley equations, *Archive for Rational Mechanics and Analysis*. 60: 229–257.

Haurie, C., D. C. Dale and M. C. Mackey (1998) Cyclical neutropenia and other periodic hematological disorders: a review of mechanisms and mathematical models, *Blood*. 92: 2629–2640.

Haurie, C., D. C. Dale, R. Rudnicki and M. C. Mackey (2000) Modeling complex neutrophil dynamics in the grey collie, *Journal of Theoretical Biology*. 204: 505–519.

Haurie, C., R. Person, D. C. Dale and M. C. Mackey (1999) Hematopoietic dynamics in grey collies, *Experimental Hematology*. 27: 1139–1148.

Hearn, T., C. Haurie and M. C. Mackey (1998) Cyclical neutropenia and the peripheral control of white blood cell production, *Journal of Theoretical Biology*. 192: 167–181.

Hearon, J. Z. (1948) The kinetics of blood coagulation, *Bulletin of Mathematical Biology*. 10: 175–186.

Heineken, F. G., H. M. Tsuchiya and R. Aris (1967) On the mathematical status of the pseudo-steady state hypothesis of biochemical kinetics, *Mathematical Biosciences*. 1: 95–113.

Helmholtz, H. L. F. (1875) *On the Sensations of Tone as a Physiological Basis for the Theory of Music*: Longmans, Green and Co., London.

Henriquez, C. S. (1993) Simulating the electrical behavior of cardiac tissue using the bidomain model, *CRC Critical Reviews in Biomedical Engineering*. 21: 1–77.

Henriquez, C. S., A. L. Muzikant and C. K. Smoak (1996) Anisotropy, fiber curvature, and bath loading effects on activation in thin and thick cardiac tissue preparations: simulations in a three-dimensional bidomain model, *Journal of Cardiovascular Electrophysiology*. 7: 424–444.

Hering, E. (1869) Uber den einfluss der atmung auf den kreislauf i. Uber athenbewegungen des gefasssystems., *Sitzungsberichte Kaiserlich Akad Wissenschaft Mathemat-Naturwissenschaft Classe*. 60: 829–856.

Hess, B. and A. Boiteux (1973) *Substrate Control of Glycolytic Oscillations*. In: Biological and Biochemical Oscillators, Ed: B. Chance, E. K. Pye, A. K. Ghosh and B. Hess, Academic Press, New York.

Hilgemann, D. W. (2004) New insights into the molecular and cellular workings of the cardiac Na^+/Ca^{2+} exchanger, *American Journal of Physiology — Cell Physiology*. 287: C1167–1172.

Hill, A. V. (1938) The heat of shortening and the dynamic constants of muscle, *Proceedings of the Royal Society of London B*. 126: 136–195.

Hill, T. L. (1974) Theoretical formalism for the sliding filament model of contraction of striated muscle. Part I., *Progress in Biophysics and Molecular Biology*. 28: 267–340.

Hill, T. L. (1975) Theoretical formalism for the sliding filament model of contraction of striated muscle. Part II., *Progress in Biophysics and Molecular Biology*. 29: 105–159.

Hille, B. (2001) *Ionic Channels of Excitable Membranes (Third Edition)*: Sinauer, Sunderland, MA.

Hille, B. and W. Schwartz (1978) Potassium channels as multi-ion single-file pores, *Journal of General Physiology*. 72: 409–442.

Himmel, D. M. and T. R. Chay (1987) Theoretical studies on the electrical activity of pancreatic β-cells as a function of glucose, *Biophysical Journal*. 51: 89–107.

Hindmarsh, J. L. and R. M. Rose (1982) A model of the nerve impulse using two first order differential equations, *Nature*. 296: 162–164.

Hindmarsh, J. L. and R. M. Rose (1984) A model of neuronal bursting using three coupled first order differential equations, *Proceedings of the Royal Society of London B*. 221: 87–102.

Hirose, K., S. Kadowaki, M. Tanabe, H. Takeshima and M. Iino (1999) Spatiotemporal dynamics of inositol 1,4,5-trisphosphate that underlies complex Ca^{2+} mobilization patterns, *Science*. 284: 1527–1530.

Hirsch, M. W., C. C. Pugh and M. Shub (1977) *Invariant Manifolds*: Springer-Verlag, New York.

Hirsch, M. W. and S. Smale (1974) *Differential Equations, Dynamical Systems and Linear Algebra*: Academic Press, New York.

Hirst, G. D. and F. R. Edwards (2004) Role of interstitial cells of Cajal in the control of gastric motility, *Journal of Pharmacological Sciences*. 96: 1–10.

Hirst, G. D. and S. M. Ward (2003) Interstitial cells: involvement in rhythmicity and neural control of gut smooth muscle, *Journal of Physiology*. 550: 337–346.

Hodgkin, A. L. (1976) Chance and design in electrophysiology: an informal account of certain experiments on nerve carried out between 1934 and 1952, *Journal of Physiology*. 263: 1–21.

Hodgkin, A. L. and A. F. Huxley (1952a) Currents carried by sodium and potassium ions through the membrane of the giant axon of *Loligo*, *Journal of Physiology*. 116: 449–472.

Hodgkin, A. L. and A. F. Huxley (1952b) The components of membrane conductance in the giant axon of *Loligo*, *Journal of Physiology*. 116: 473–496.

Hodgkin, A. L. and A. F. Huxley (1952c) The dual effect of membrane potential on sodium conductance in the giant axon of *Loligo*, *Journal of Physiology*. 116: 497–506.

Hodgkin, A. L. and A. F. Huxley (1952d) A quantitative description of membrane current and its application to conduction and excitation in nerve, *Journal of Physiology*. 117: 500–544.

Hodgkin, A. L., A. F. Huxley and B. Katz (1952) Measurement of current-voltage relations in the membrane of the giant axon of *Loligo*, *Journal of Physiology*. 116: 424–448.

Hodgkin, A. L. and B. Katz (1949) The effect of sodium ions on the electrical activity of the giant axon of the squid, *Journal of Physiology*. 108: 37–77.

Hodgkin, A. L. and R. D. Keynes (1955) The potassium permeability of a giant nerve fibre, *Journal of Physiology*. 128: 61–88.

Hodgkin, A. L. and B. J. Nunn (1988) Control of light-sensitive current in salamander rods, *Journal of Physiology*. 403: 439–471.

Hodgkin, A. L. and W. A. H. Rushton (1946) The electrical constants of a crustacean nerve fibre, *Proceedings of the Royal Society of London B*. 133: 444–479.

Hodgson, M. E. A. and P. J. Green (1999) Bayesian choice among Markov models of ion channels using Markov chain Monte Carlo, *Proceedings of the Royal Society of London A*. 455: 3425–3448.

Hofer, A. M. and E. M. Brown (2003) Extracellular calcium sensing and signalling, *Nature Reviews Molecular Cell Biology*. 4: 530–538.

Holmes, M. H. (1980a) An analysis of a low-frequency model of the cochlea, *Journal of the Acoustical Society of America*. 68: 482–488.

Holmes, M. H. (1980b) Low frequency asymptotics for a hydroelastic model of the cochlea, *SIAM Journal on Applied Mathematics*. 38: 445–456.

Holmes, M. H. (1982) A mathematical model of the dynamics of the inner ear, *Journal of Fluid Mechanics*. 116: 59–75.

Holmes, M. H. (1995) *Introduction to Perturbation Methods*: Springer-Verlag, New York.

Holstein-Rathlou, N. H. (1993) Oscillations and chaos in renal blood flow control, *Journal of the American Society of Nephrology*. 4: 1275–1287.

Holstein-Rathlou, N. H. and P. P. Leyssac (1987) Oscillations in the proximal intratubular pressure: a mathematical model, *American Journal of Physiology — Renal, Fluid and Electrolyte Physiology*. 252: F560–572.

Holstein-Rathlou, N. H. and D. J. Marsh (1989) Oscillations of tubular pressure, flow, and distal chloride concentration in rats, *American Journal of Physiology — Renal, Fluid and Electrolyte Physiology*. 256: F1007–1014.

Holstein-Rathlou, N. H. and D. J. Marsh (1990) A dynamic model of the tubuloglomerular feedback mechanism, *American Journal of Physiology — Renal, Fluid and Electrolyte Physiology*. 258: F1448–1459.

Holstein-Rathlou, N. H. and D. J. Marsh (1994) A dynamic model of renal blood flow autoregulation, *Bulletin of Mathematical Biology*. 56: 411–429.

Holstein-Rathlou, N. H., A. J. Wagner and D. J. Marsh (1991) Tubuloglomerular feedback dynamics and renal blood flow autoregulation in rats, *American Journal of Physiology — Renal, Fluid and Electrolyte Physiology*. 260: F53–68.

Hood, D. C. (1998) Lower-level visual processing and models of light adaptation, *Annual Review of Psychology*. 49: 503–535.

Hood, D. C. and M. A. Finkelstein (1986) *Sensitivity to light*. In: Handbook of Perception and Human Performance, Volume 1: Sensory Processes and Perception (Chapter 5), Ed: K. R. Boff, L. Kaufman and J. P. Thomas, Wiley, New York.

Hooks, D. A., K. A. Tomlinson, S. G. Marsden, I. J. LeGrice, B. H. Smaill, A. J. Pullan and P. J. Hunter (2002) Cardiac microstructure: implications for electrical propagation and defibrillation in the heart, *Circulation Research*. 91: 331–338.

Hoppensteadt, F. C. and J. P. Keener (1982) Phase locking of biological clocks, *Journal of Mathematical Biology*. 15: 339–349.

Hoppensteadt, F. C. and C. S. Peskin (2001) *Modeling and Simulation in Medicine and the Life Sciences*: Springer-Verlag, New York.

Houart, G., G. Dupont and A. Goldbeter (1999) Bursting, chaos and birhythmicity originating from self-modulation of the inositol 1,4,5-trisphosphate signal in a model for intracellular Ca^{2+} oscillations, *Bulletin of Mathematical Biology*. 61: 507–530.

Howard, J. (2001) *Mechanics of Motor Proteins and the Cytoskeleton*: Sinauer Associates, Sunderland, MA.

Howard, J., A. J. Hudspeth and R. D. Vale (1989) Movement of microtubules by single kinesin molecules, *Nature*. 342: 154–158.

Hudspeth, A. (1997) Mechanical amplification of stimuli by hair cells, *Current Opinion in Neurobiology*. 7: 480–486.

Hudspeth, A. J. (1985) The cellular basis of hearing: the biophysics of hair cells, *Science*. 230: 745–752.

Hudspeth, A. J. (1989) How the ear's works work, *Nature*. 341: 397–404.

Hudspeth, A. J. (2005) How the ear's works work: mechanoelectrical transduction and amplification by hair cells, *Comptes Rendus des Seances de la Société de Biologie et de ses Filiales*. 328: 155–162.

Hudspeth, A. J., Y. Choe, A. D. Mehta and P. Martin (2000) Putting ion channels to work: mechanoelectrical transduction, adaptation, and amplification by hair cells, *Proceedings of the National Academy of Sciences USA*. 97: 11765–11772.

Hudspeth, A. J. and P. G. Gillespie (1994) Pulling springs to tune transduction: adaptation by hair cells, *Neuron*. 12: 1–9.

Hudspeth, A. J. and R. S. Lewis (1988a) Kinetic analysis of voltage- and ion-dependent conductances in saccular hair cells of the bull-frog, *Rana Catesbeiana*, *Journal of Physiology*. 400: 237–274.

Hudspeth, A. J. and R. S. Lewis (1988b) A model for electrical resonance and frequency tuning in saccular hair cells of the bull-frog, *Rana Catesbeiana*, *Journal of Physiology*. 400: 275–297.

Huertas, M. A. and G. D. Smith (2007) The dynamics of luminal depletion and the stochastic gating of Ca^{2+}-activated Ca^{2+} channels and release sites, *Journal of Theoretical Biology*. 246: 332–354.

Hunter, P. J. (1995) Myocardial constitutive laws for continuum mechanics models of the heart, *Advances in Experimental Medicine and Biology*. 382: 303–318.

Hunter, P. J., A. D. McCulloch and H. E. ter Keurs (1998) Modelling the mechanical properties of cardiac muscle, *Progress in Biophysics and Molecular Biology*. 69: 289–331.

Hunter, P. J., A. J. Pullan and B. H. Smaill (2003) Modeling total heart function, *Annual Review of Biomedical Engineering*. 5: 147–177.

Huntsman, L. L., E. O. Attinger and A. Noordergraaf (1978) *Metabolic autoregulation of blood flow in skeletal muscle*. In: Cardiovascular System Dynamics, Ed: J. Baan, A. Noordergraaf and J. Raines, MIT Press, Cambridge MA.

Huxley, A. F. (1957) Muscle structure and theories of contraction, *Progress in Biophysics*. 7: 255–318.

Huxley, A. F. and R. M. Simmons (1971) Proposed mechanism of force generation in striated muscle, *Nature*. 233: 533–538.

Iacobas, D. A., S. O. Suadicani, D. C. Spray and E. Scemes (2006) A stochastic two-dimensional model of intercellular Ca^{2+} wave spread in glia, *Biophysical Journal*. 90: 24–41.

Imredy, J. P. and D. T. Yue (1994) Mechanism of Ca^{2+}-sensitive inactivation of L-type Ca^{2+} channels, *Neuron*. 12: 1301–1318.

Inselberg, A. and R. S. Chadwick (1976) Mathematical model of the cochlea. I: formulation and solution, *SIAM Journal on Applied Mathematics*. 30: 149–163.

Irving, M., J. Maylie, N. L. Sizto and W. K. Chandler (1990) Intracellular diffusion in the presence of mobile buffers: application to proton movement in muscle, *Biophysical Journal*. 57: 717–721.

Iyer, A. N. and R. A. Gray (2001) An experimentalist's approach to accurate localization of phase singularities during reentry, *Annals of Biomedical Engineering*. 29: 47–59.

Izhikevich, E. M. (2000) Neural excitability, spiking and bursting, *International Journal of Bifurcation and Chaos*. 10: 1171–1266.

Izu, L. T., W. G. Wier and C. W. Balke (2001) Evolution of cardiac calcium waves from stochastic calcium sparks, *Biophysical Journal*. 80: 103–120.

Jack, J. J. B., D. Noble and R. W. Tsien (1975) *Electric Current Flow in Excitable Cells*: Oxford University Press, Oxford.

Jacob, F. and J. Monod (1961) Genetic regulatory mechanisms in the synthesis of proteins, *Journal of Molecular Biology*. 3: 318–356.

Jacob, F., D. Perrin, C. Sanchez and J. Monod (1960) L'opéron : groupe de gène à expression par un opérateur, *Comptes Rendus des*

Sceances de la Société de Biologie et de ses Filiales. 250: 1727–1729.

Jaffrin, M.-Y. and C. G. Caro (1995) *Biological flows*: Plenum Press, New York.

Jafri, M. S. and J. Keizer (1995) On the roles of Ca^{2+} diffusion, Ca^{2+} buffers and the endoplasmic reticulum in IP_3-induced Ca^{2+} waves, *Biophysical Journal*. 69: 2139–2153.

Jafri, M. S., J. J. Rice and R. L. Winslow (1998) Cardiac Ca^{2+} dynamics: the roles of ryanodine receptor adaptation and sarcoplasmic reticulum load, *Biophysical Journal*. 74: 1149–1168.

Jahnke, W., C. Henze and A. T. Winfree (1988) Chemical vortex dynamics in three-dimensional excitable media, *Nature*. 336: 662–665.

Jahnke, W. and A. T. Winfree (1991) A survey of spiral-wave behaviors in the Oregonator model, *International Journal of Bifurcation and Chaos*. 1: 445–466.

Jakobsson, E. (1980) Interactions of cell volume, membrane potential, and membrane transport parameters, *American Journal of Physiology — Cell Physiology*. 238: C196–206.

Janeway, C. A., P. Travers, M. Walport and M. Shlomchik (2001) *Immunobiology: The Immune System in Health and Disease (Fifth Edition)*: Garland Publishing, New York.

Jelić, S., Z. Čupić and L. Kolar-Anić (2005) Mathematical modeling of the hypothalamic–pituitary–adrenal system activity, *Mathematical Biosciences*. 197: 173–187.

Jesty, J., E. Beltrami and G. Willems (1993) Mathematical analysis of a proteolytic positive-feedback loop: dependence of lag time and enzyme yields on the initial conditions and kinetic parameters, *Biochemistry*. 32: 6266–6274.

Jewell, B. R. and D. R. Wilkie (1958) An analysis of the mechanical components in frog's striated muscle, *Journal of Physiology*. 143: 515–540.

Jones, C. K. R. T. (1984) Stability of the traveling wave solutions of the FitzHugh–Nagumo system, *Transactions of the American Mathematical Society*. 286: 431–469.

Jones, K. C. and K. G. Mann (1994) A model for the tissue factor pathway to thrombin. II. A mathematical simulation, *Journal of Biological Chemistry*. 269: 23367–23373.

Joseph, I. M., Y. Zavros, J. L. Merchant and D. Kirschner (2003) A model for integrative study of human gastric acid secretion, *Journal of Applied Physiology*. 94: 1602–1618.

Julian, F. J. (1969) Activation in a skeletal muscle contraction model with a modification for insect fibrillar muscle, *Biophysical Journal*. 9: 547–570.

Julien, C. (2006) The enigma of Mayer waves: Facts and models, *Cardiovascular Research*. 70: 12–21.

Jung, P., A. Cornell-Bell, K. S. Madden and F. Moss (1998) Noise-induced spiral waves in astrocyte syncytia show evidence of self-organized criticality, *Journal of Neurophysiology*. 79: 1098–1101.

Jung, P., A. Cornell-Bell, F. Moss, S. Kadar, J. Wang and K. Showalter (1998) Noise sustained waves in subexcitable media: From chemical waves to brain waves, *Chaos*. 8: 567–575.

Just, A. (2006) Mechanisms of renal blood flow autoregulation: dynamics and contributions, *American Journal of Physiology — Regulatory Integrative and Comparative Physiology*. 292: 1–17.

Kang, T. M. and D. W. Hilgemann (2004) Multiple transport modes of the cardiac Na^+/Ca^{2+} exchanger, *Nature*. 427: 544–548.

Kaplan, W. (1981) *Advanced Engineering Mathematics*: Addison-Wesley, Reading, MA.

Karma, A. (1993) Spiral breakup in model equations of action potential propagation in cardiac tissue, *Physical Review Letters*. 71: 1103–1106.

Karma, A. (1994) Electrical alternans and spiral wave breakup in cardiac tissue, *Chaos*. 4: 461–472.

Katz, B. and R. Miledi (1968) The role of calcium in neuromuscular facilitation, *Journal of Physiology*. 195: 481–492.

Keener, J. P. (1980a) Waves in excitable media, *SIAM Journal on Applied Mathematics*. 39: 528–548.

Keener, J. P. (1980b) Chaotic behavior in piecewise continuous difference equations, *Transactions of the American Mathematical Society*. 261: 589–604.

Keener, J. P. (1981) On cardiac arrhythmias: AV conduction block, *Journal of Mathematical Biology*. 12: 215–225.

Keener, J. P. (1983) Analog circuitry for the van der Pol and FitzHugh–Nagumo equation, *IEEE Transactions on Systems, Man and Cybernetics*. SMC-13: 1010–1014.

Keener, J. P. (1986) A geometrical theory for spiral waves in excitable media, *SIAM Journal on Applied Mathematics*. 46: 1039–1056.

Keener, J. P. (1987) Propagation and its failure in coupled systems of discrete excitable cells, *SIAM Journal on Applied Mathematics*. 47: 556–572.

Keener, J. P. (1988) The dynamics of three dimensional scroll waves in excitable media, *Physica D*. 31: 269–276.

Keener, J. P. (1991a) An eikonal–curvature equation for action potential propagation in myocardium, *Journal of Mathematical Biology*. 29: 629–651.

Keener, J. P. (1991b) The effects of discrete gap junctional coupling on propagation in myocardium, *Journal of Theoretical Biology*. 148: 49–82.

Keener, J. P. (1992) The core of the spiral, *SIAM Journal on Applied Mathematics*. 52: 1372–1390.

Keener, J. P. (1994) Symmetric spirals in media with relaxation kinetics and two diffusing species, *Physica D*. 70: 61–73.

Keener, J. P. (1998) *Principles of Applied Mathematics, Transformation and Approximation (Second Edition)*: Perseus Books, Cambridge, Massachusetts.

Keener, J. P. (2000a) Homogenization and propagation in the bistable equation, *Physica D*. 136: 1–17.

Keener, J. P. (2000b) Propagation of waves in an excitable medium with discrete release sites, *SIAM Journal on Applied Mathematics*. 61: 317–334.

Keener, J. P. (2004) The topology of defibrillation, *Journal of Theoretical Biology*. 230: 459–473.

Keener, J. P. (2006) Stochastic calcium oscillations, *Mathematical Medicine and Biology*. 23: 1–25.

Keener, J. P. and E. Cytrynbaum (2003) The effect of spatial scale of resistive inhomogeneity on defibrillation of cardiac tissue, *Journal of Theoretical Biology*. 223: 233–248.

Keener, J. P. and L. Glass (1984) Global bifurcations of a periodically forced oscillator, *Journal of Mathematical Biology*. 21: 175–190.

Keener, J. P., F. C. Hoppensteadt and J. Rinzel (1981) Integrate and fire models of nerve membrane response to oscillatory input, *SIAM Journal on Applied Mathematics*. 41: 503–517.

Keener, J. P. and A. V. Panfilov (1995) *Three-dimensional propagation in the heart: the effects of geometry and fiber orientation on propagation in myocardium*. In: Cardiac Electrophysiology From Cell to Bedside, Ed: D. P. Zipes and J. Jalife, Saunders, Philadelphia PA.

Keener, J. P. and A. V. Panfilov (1996) A biophysical model for defibrillation of cardiac tissue, *Biophysical Journal*. 71: 1335–1345.

Keener, J. P. and A. V. Panfilov (1997) *The effects of geometry and fibre orientation on propagation and extracellular potentials in myocardium*. In: Computational Biology of the Heart, Ed: A. V. Panfilov and A. V. Holden, John Wiley and Sons, New York.

Keener, J. P. and J. J. Tyson (1986) Spiral waves in the Belousov–Zhabotinsky reaction, *Physica D*. 21: 307–324.

Keener, J. P. and J. J. Tyson (1992) The dynamics of scroll waves in excitable media, *SIAM Review*. 34: 1–39.

Keizer, J. and G. DeYoung (1994) Simplification of a realistic model of IP_3-induced Ca^{2+} oscillations, *Journal of Theoretical Biology*. 166: 431–442.

Keizer, J. and L. Levine (1996) Ryanodine receptor adaptation and Ca^{2+}-induced Ca^{2+} release-dependent Ca^{2+} oscillations, *Biophysical Journal*. 71: 3477–3487.

Keizer, J. and G. Magnus (1989) ATP-sensitive potassium channel and bursting in the pancreatic beta cell, *Biophysical Journal*. 56: 229–242.

Keizer, J. and G. D. Smith (1998) Spark-to-wave transition: saltatory transmission of calcium waves in cardiac myocytes, *Biophysical Chemistry*. 72: 87–100.

Keizer, J., G. D. Smith, S. Ponce-Dawson and J. E. Pearson (1998) Saltatory propagation of Ca^{2+} waves by Ca^{2+} sparks, *Biophysical Journal*. 75: 595–600.

Keizer, J. and P. Smolen (1991) Bursting electrical activity in pancreatic β-cells caused by Ca^{2+} and voltage-inactivated Ca^{2+} channels, *Proceedings of the National Academy of Sciences USA*. 88: 3897–3901.

Keller, E. F. and L. A. Segel (1971) Models for chemotaxis, *Journal of Theoretical Biology*. 30: 225–234.

Kernevez, J.-P. (1980) *Enzyme Mathematics*: North-Holland Publishing Company, Amsterdam, New York.

Kessler, D. A. and R. Kupferman (1996) Spirals in excitable media: the free-boundary limit with diffusion, *Physica D*. 97: 509–516.

Kevorkian, J. (2000) *Partial Differential Equations: Analytical Solution Techniques*: Springer, New York.

Kevorkian, J. and J. D. Cole (1996) *Multiple Scale and Singular Perturbation Methods*: Springer-Verlag, New York.

Khoo, M. C., R. E. Kronauer, K. P. Strohl and A. S. Slutsky (1982) Factors inducing periodic breathing in humans: a general model, *Journal of Applied Physiology*. 53: 644–659.

Kidd, J. F., K. E. Fogarty, R. A. Tuft and P. Thorn (1999) The role of Ca^{2+} feedback in shaping $InsP_3$-evoked Ca^{2+} signals in mouse pancreatic acinar cells, *Journal of Physiology*. 520: 187–201.

Kim, W. T., M. G. Rioult and A. H. Cornell-Bell (1994) Glutamate-induced calcium signaling in astrocytes, *Glia*. 11: 173–184.

Klingauf, J. and E. Neher (1997) Modeling buffered Ca^{2+} diffusion near the membrane: implications for secretion in neuroendocrine cells, *Biophysical Journal*. 72: 674–690.

Kluger, Y., Z. Lian, X. Zhang, P. E. Newburger and S. M. Weissman (2004) A panorama of lineage-specific transcription in hematopoiesis, *Bioessays*. 26: 1276–1287.

Knepper, M. A. and F. C. Rector, Jr. (1991) *Urinary concentration and dilution*. In: The Kidney (4th edition) Volume 1, Ed: B. M. Brenner and F. C. Rector, Jr., Saunders, Philadelphia.

Knight, B. W. (1972) Dynamics of encoding a population of neurons, *Journal of General Physiology*. 59: 734–766.

Knight, B. W., J. I. Toyoda and F. A. Dodge, Jr. (1970) A quantitative description of the dynamics of excitation and inhibition in the eye of Limulus, *Journal of General Physiology*. 56: 421–437.

Knisley, S. B., T. F. Blitchington, B. C. Hill, A. O. Grant, W. M. Smith, T. C. Pilkington and R. E. Ideker (1993) Optical measurements of transmembrane potential changes during electric field stimulation of ventricular cells, *Circulation Research*. 72: 255–270.

Knobil, E. (1981) Patterns of hormonal signals and hormone action, *New England Journal of Medicine*. 305: 1582–1583.

Knox, B. E., P. N. Devreotes, A. Goldbeter and L. A. Segel (1986) A molecular mechanism for sensory adaptation based on ligand-induced receptor modification, *Proceedings of the National Academy of Sciences USA*. 83: 2345–2349.

Koch, C. (1999) *Biophysics of Computation: Information Processing in Single Neurons*: Oxford University Press, Oxford.

Koch, C. and I. Segev, Eds. (1998) *Methods in Neuronal Modeling: From Ions to Networks (Third Edition)*. MIT Press, Cambridge, MA.

Koch, K.-W. and L. Stryer (1988) Highly cooperative feedback control of retinal rod guanylate cyclase by calcium ions, *Nature*. 334: 64–66.

Koefoed-Johnsen, V. and H. H. Ussing (1958) The nature of the frog skin potential, *Acta Physiologica Scandinavica*. 42: 298–308.

Koenigsberger, M., R. Sauser, J. L. Beny and J. J. Meister (2006) Effects of arterial wall stress on vasomotion, *Biophysical Journal*. 91: 1663–1674.

Koenigsberger, M., R. Sauser, M. Lamboley, J. L. Beny and J. J. Meister (2004) Ca^{2+} dynamics in a population of smooth muscle cells: modeling the recruitment and synchronization, *Biophysical Journal*. 87: 92–104.

Koenigsberger, M., R. Sauser and J. J. Meister (2005) Emergent properties of electrically coupled smooth muscle cells, *Bulletin of Mathematical Biology*. 67: 1253–1272.

Kohler, H.-H. and K. Heckman (1979) Unidirectional fluxes in saturated single-file pores of biological and artificial membranes I: pores containing no more than one vacancy, *Journal of Theoretical Biology*. 79: 381–401.

Kopell, N. and G. B. Ermentrout (1986) Subcellular oscillations and bursting, *Mathematical Biosciences*. 78: 265–291.

Kopell, N. and L. N. Howard (1973) Plane wave solutions to reaction–diffusion equations, *Studies in Applied Mathematics*. 52: 291–328.

Koshland, D. E., Jr. and K. Hamadani (2002) Proteomics and models for enzyme cooperativity, *Journal of Biological Chemistry*. 277: 46841–46844.

Koshland, D. E., Jr., G. Nemethy and D. Filmer (1966) Comparison of experimental binding data and theoretical models in proteins containing subunits, *Biochemistry*. 5: 365–385.

Koutalos, Y., K. Nakatani and K. W. Yau (1995) The cGMP–phosphodiesterase and its contribution to sensitivity regulation in retinal rods, *Journal of General Physiology*. 106: 891–921.

Kramers, H. A. (1940) Brownian motion in a field of force and the diffusion model of chemical reactions, *Physica*. 7: 284–304.

Krane, D. E. and M. L. Raymer (2003) *Fundamental Concepts of Bioinformatics*: Benjamin Cummings, San Francisco.

Krassowska, W., T. C. Pilkington and R. E. Ideker (1987) Periodic conductivity as a mechanism for cardiac stimulation and defibrillation, *IEEE Transactions in Biomedical Engineering*. 34: 555–560.

Krausz, H. I. and K.-I. Naka (1980) Spatiotemporal testing and modeling of catfish retinal neurons, *Biophysical Journal*. 29: 13–36.

Kreyszig, E. (1994) *Advanced Engineering Mathematics (Seventh Edition)*: John Wiley and Sons, New York.

Kucera, J. P., S. Rohr and Y. Rudy (2002) Localization of sodium channels in intercalated disks modulates cardiac conduction, *Circulation Research*. 91: 1176–1182.

Kuffler, S. W. (1953) Discharge patterns and functional organization of the mammalian retina, *Journal of Neurophysiology*. 16: 37–68.

Kuffler, S. W. (1973) The single-cell approach in the visual system and the study of receptive fields, *Investigative Ophthalmology*. 12: 794–813.

Kuffler, S. W., J. G. Nicholls and R. Martin (1984) *From Neuron to Brain (Second Edition)*: Sinaeur Associates, Sunderland, MA.

Kuramoto, Y. and T. Tsuzuki (1976) Persistent propagation of concentration waves in dissipative media far from thermal equilibrium, *Progress of Theoretical Physics*. 55: 356–369.

Kuramoto, Y. and T. Yamada (1976) Pattern formation in oscillatory chemical reactions, *Progress of Theoretical Physics*. 56: 724–740.

Läuger, P. (1973) Ion transport through pores: a rate-theory analysis, *Biochimica et Biophysica Acta*. 311: 423–441.

Lacker, H. M. (1981) Regulation of ovulation number in mammals: a follicle interaction law that controls maturation, *Biophysical Journal*. 35: 433–454.

Lacker, H. M. and C. S. Peskin (1981) *Control of ovulation number in a model of ovarian follicular maturation*. In: Lectures on Mathematics in the Life Sciences, Ed: S. Childress, American Mathematical Society, Providence.

Lacker, H. M. and C. S. Peskin (1986) A mathematical method for unique determination of cross-bridge properties from steady-state mechanical and energetic experiments on macroscopic muscle, *Lectures on Mathematics in the Life Sciences*. 16: 121–153.

Lacy, A. H. (1967) The unit of insulin, *Diabetes*. 16: 198–200.

Laidler, K. J. (1969) *Theories of Chemical Reaction Rates*: McGraw-Hill, New York.

Lamb, T. D. (1981) The involvement of rod photoreceptors in dark adaptation, *Vision Research*. 21: 1773–1782.

Lamb, T. D. and E. N. Pugh (1992) A quantitative account of the activation steps involved in phototransduction in amphibian photoreceptors, *Journal of Physiology*. 449: 719–758.

Lamb, T. D. and E. J. Simon (1977) Analysis of electrical noise in turtle cones, *Journal of Physiology*. 272: 435–468.

Landy, M. S. and J. A. Movshon, Eds. (1991) *Computational Models of Visual Processing*: MIT Press, Cambridge, MA.

Lane, D. C., J. D. Murray and V. S. Manoranjan (1987) Analysis of wave phenomena in a morphogenetic mechanochemical model and an application to post-fertilisation waves on eggs, *IMA Journal of Mathematics Applied in Medicine and Biology*. 4: 309–331.

Lange, R. L. and H. H. Hecht (1962) The mechanism of Cheyne–Stokes respiration, *Journal of Clinical Investigation*. 41: 42–52.

Langer, G. A. and A. Peskoff (1996) Calcium concentration and movement in the diadic cleft space of the cardiac ventricular cell, *Biophysical Journal*. 70: 1169–1182.

Lapointe, J. Y., M. Gagnon, S. Poirier and P. Bissonnette (2002) The presence of local osmotic gradients can account for the water flux driven by the Na^+-glucose cotransporter, *Journal of Physiology*. 542: 61–62.

Layton, A. T. and H. E. Layton (2002) A numerical method for renal models that represent tubules with abrupt changes in membrane properties, *Journal of Mathematical Biology*. 45: 549–567.

Layton, A. T. and H. E. Layton (2003) A region-based model framework for the rat urine concentrating mechanism, *Bulletin of Mathematical Biology*. 65: 859–901.

Layton, A. T. and H. E. Layton (2005a) A region-based mathematical model of the urine concentrating mechanism in the rat outer medulla. I. Formulation and base-case results, *American Journal of Physiology — Renal, Fluid and Electrolyte Physiology*. 289: F1346–1366.

Layton, A. T. and H. E. Layton (2005b) A region-based mathematical model of the urine concentrating mechanism in the rat outer medulla. II. Parameter sensitivity and tubular inhomogeneity, *American Journal of Physiology — Renal, Fluid and Electrolyte Physiology*. 289: F1367–1381.

Layton, A. T., T. L. Pannabecker, W. H. Dantzler and H. E. Layton (2004) Two modes for concentrating urine in rat inner medulla, *American Journal of Physiology — Renal, Fluid and Electrolyte Physiology*. 287: F816–839.

Layton, H. E., E. B. Pitman and M. A. Knepper (1995a) A dynamic numerical method for models of the urine concentrating mechanism, *SIAM Journal on Applied Mathematics*. 55: 1390–1418.

Layton, H. E., E. B. Pitman and L. C. Moore (1991) Bifurcation analysis of TGF-mediated oscillations in SNGFR, *American Journal of Physiology — Renal, Fluid and Electrolyte Physiology*. 261: F904–919.

Layton, H. E., E. B. Pitman and L. C. Moore (1995b) Instantaneous and steady-state gains in the tubuloglomerular feedback system, *American Journal of Physiology — Renal, Fluid and Electrolyte Physiology*. 268: F163–174.

Layton, H. E., E. B. Pitman and L. C. Moore (1997) Spectral properties of the tubuloglomerular feedback system, *American Journal of Physiology — Renal, Fluid and Electrolyte Physiology*. 273: F635–649.

Layton, H. E., E. B. Pitman and L. C. Moore (2000) Limit-cycle oscillations and tubuloglomerular feedback regulation of distal sodium delivery, *American Journal of Physiology — Renal, Fluid and Electrolyte Physiology*. 278: F287–301.

LeBeau, A. P., A. B. Robson, A. E. McKinnon, R. A. Donald and J. Sneyd (1997) Generation of action potentials in a mathematical model of corticotrophs, *Biophysical Journal*. 73: 1263–1275.

LeBeau, A. P., A. B. Robson, A. E. McKinnon and J. Sneyd (1998) Analysis of a reduced model of corticotroph action potentials, *Journal of Theoretical Biology*. 192: 319–339.

Lechleiter, J. and D. Clapham (1992) Molecular mechanisms of intracellular calcium excitability in *X. laevis* oocytes, *Cell*. 69: 283–294.

Lechleiter, J., S. Girard, D. Clapham and E. Peralta (1991a) Subcellular patterns of calcium release determined by G protein-specific residues of muscarinic receptors, *Nature*. 350: 505–508.

Lechleiter, J., S. Girard, E. Peralta and D. Clapham (1991b) Spiral calcium wave propagation and annihilation in *Xenopus laevis* oocytes, *Science*. 252: 123–126.

Leloup, J. C. and A. Goldbeter (1998) A model for circadian rhythms in Drosophila incorporating the formation of a complex between the PER and TIM proteins, *Journal of Biological Rhythms*. 13: 70–87.

Leloup, J. C. and A. Goldbeter (2003) Toward a detailed computational model for the mammalian circadian clock, *Proceedings of the National Academy of Sciences USA*. 100: 7051–7056.

Leloup, J. C. and A. Goldbeter (2004) Modeling the mammalian circadian clock: sensitivity analysis and multiplicity of oscillatory mechanisms, *Journal of Theoretical Biology*. 230: 541–562.

Lenbury, Y. and P. Pornsawad (2005) A delay-differential equation model of the feedback-controlled hypothalamus-pituitary-adrenal axis in humans, *Mathematical Medicine and Biology*. 22: 15–33.

Lesser, M. B. and D. A. Berkley (1972) Fluid mechanics of the cochlea. Part I., *Journal of Fluid Mechanics*. 51: 497–512.

Levine, I. N. (2002) *Physical Chemistry (Fifth Edition)*: McGraw-Hill, Tokyo.

Levine, S. N. (1966) Enzyme amplifier kinetics, *Science*. 152: 651–653.

Lew, V. L., H. G. Ferreira and T. Moura (1979) The behaviour of transporting epithelial cells. I. Computer analysis of a basic model, *Proceedings of the Royal Society of London B*. 206: 53–83.

Lewis, M. (2005) The *lac* repressor, *CR Biologies*. 328: 521–548.

Lewis, T. J. and J. P. Keener (2000) Wave-blocking in excitable media due to regions of depressed excitability, *SIAM Journal on Applied Mathematics*. 61: 293–396.

Leyssac, P. P. and L. Baumbach (1983) An oscillating intratubular pressure response to alterations in Henle loop flow in the rat kidney, *Acta Physiologica Scandinavica*. 117: 415–419.

Li, W., J. Llopis, M. Whitney, G. Zlokarnik and R. Y. Tsien (1998) Cell-permeant caged $InsP_3$ ester shows that Ca^{2+} spike frequency can optimize gene expression, *Nature*. 392: 936–941.

Li, Y.-X. and A. Goldbeter (1989) Frequency specificity in intercellular communication: influence of patterns of periodic signaling on target cell responsiveness, *Biophysical Journal*. 55: 125–145.

Li, Y.-X., J. Keizer, S. S. Stojilkovic and J. Rinzel (1995) Ca^{2+} excitability of the ER membrane: an explanation for IP_3-induced Ca^{2+} oscillations, *American Journal of Physiology — Cell Physiology*. 269: C1079–1092.

Li, Y.-X. and J. Rinzel (1994) Equations for $InsP_3$ receptor-mediated $[Ca^{2+}]$ oscillations derived from a detailed kinetic model: a Hodgkin–Huxley like formalism, *Journal of Theoretical Biology*. 166: 461–473.

Li, Y.-X., J. Rinzel, J. Keizer and S. S. Stojilkovic (1994) Calcium oscillations in pituitary gonadotrophs: comparison of experiment and theory, *Proceedings of the National Academy of Sciences USA*. 91: 58–62.

Li, Y.-X., S. S. Stojilkovic, J. Keizer and J. Rinzel (1997) Sensing and refilling calcium stores in an excitable cell, *Biophysical Journal*. 72: 1080–1091.

Lighthill, J. (1975) *Mathematical Biofluiddynamics*: SIAM, Philadelphia, PA.

Lin, C. C. and L. A. Segel (1988) *Mathematics Applied to Deterministic Problems in the Natural Sciences*: SIAM, Philadelphia.

Lin, S. C. and D. E. Bergles (2004) Synaptic signaling between neurons and glia, *Glia*. 47: 290–298.

Liu, B.-Z. and G.-M. Deng (1991) An improved mathematical model of hormone secretion in the hypothalamo–pituitary–gonadal axis in man, *Journal of Theoretical Biology*. 150: 51–58.

Llinás, R., I. Z. Steinberg and K. Walton (1976) Presynaptic calcium currents and their relation to synaptic transmission: voltage clamp study in squid giant synapse and theoretical model for the calcium gate, *Proceedings of the National Academy of Sciences USA*. 73: 2918–2922.

Loeb, J. N. and S. Strickland (1987) Hormone binding and coupled response relationships in systems dependent on the generation of secondary mediators, *Molecular Endocrinology*. 1: 75–82.

Lombard, W. P. (1916) The Life and Work of Carl Ludwig, *Science*. 44: 363–375.

Longobardo, G., C. J. Evangelisti and N. S. Cherniack (2005) Introduction of respiratory pattern generators into models of respiratory control, *Respiratory Physiology and Neurobiology*. 148: 285–301.

Longtin, A. and J. G. Milton (1989) Modelling autonomous oscillations in the human pupil light reflex using non-linear delay-differential equations, *Bulletin of Mathematical Biology*. 51: 605–624.

Loo, D. D., E. M. Wright and T. Zeuthen (2002) Water pumps, *Journal of Physiology*. 542: 53–60.

Lugosi, E. and A. T. Winfree (1988) Simulation of wave propagation in three dimensions using Fortran on the Cyber 205, *Journal of Computational Chemistry*. 9: 689–701.

Luo, C. H. and Y. Rudy (1991) A model of the ventricular cardiac action potential; depolarization, repolarization and their interaction, *Circulation Research*. 68: 1501–1526.

Luo, C. H. and Y. Rudy (1994a) A dynamic model of the cardiac ventricular action potential; I: Simulations of ionic currents and concentration changes, *Circulation Research*. 74: 1071–1096.

Luo, C. H. and Y. Rudy (1994b) A dynamic model of the cardiac ventricular action potential; II: Afterdepolarizations, triggered activity and potentiation, *Circulation Research*. 74: 1097–1113.

Lytton, J., M. Westlin, S. E. Burk, G. E. Shull and D. H. MacLennan (1992) Functional comparisons between isoforms of the sarcoplasmic or endoplasmic reticulum family of calcium pumps, *Journal of Biological Chemistry*. 267: 14483–14489.

MacGregor, D. J. and G. Leng (2005) Modelling the hypothalamic control of growth hormone secretion, *Journal of Neuroendocrinology*. 17: 788–803.

Mackey, M. C. (1978) Unified hypothesis for the origin of aplastic anemia and periodic hematopoiesis, *Blood*. 51: 941–956.

Mackey, M. C. (1979) Periodic auto-immune hemolytic anemia: an induced dynamical disease, *Bulletin of Mathematical Biology*. 41: 829–834.

Mackey, M. C. and L. Glass (1977) Oscillation and chaos in physiological control systems, *Science*. 197: 287–289.

Mackey, M. C., C. Haurie and J. Bélair (2003) *Cell replication and control*. In: Nonlinear Dynamics in Physiology and Medicine, Ed: A. Beuter, L. Glass, M. C. Mackey and M. S. Titcombe, Springer-Verlag, New York.

Mackey, M. C. and J. G. Milton (1987) Dynamical diseases, *Annals of the New York Academy of Sciences*. 504: 16–32.

Mackey, M. C., M. Santillán and N. Yildirim (2004) Modeling operon dynamics: the tryptophan and lactose operons as paradigms, *Comptes Rendus des Seances de la Société de Biologie et de ses Filiales*. 327: 211–224.

Macknight, A. D. C. (1988) Principles of cell volume regulation, *Renal Physiology and Biochemistry*. 3–5: 114–141.

MacLennan, D. H. and E. G. Kranias (2003) Phospholamban: a crucial regulator of cardiac contractility, *Nature Reviews Molecular Cell Biology*. 4: 566–577.

MacLennan, D. H., W. J. Rice and N. M. Green (1997) The mechanism of Ca^{2+} transport by sarco(endo)plasmic reticulum Ca^{2+}-ATPases, *Journal of Biological Chemistry*. 272: 28815–28818.

Madsen, M. F., S. Dano and P. G. Sorensen (2005) On the mechanisms of glycolytic oscillations in yeast, *FEBS Journal*. 272: 2648–2660.

Maginu, K. (1985) Geometrical characteristics associated with stability and bifurcations of periodic travelling waves in reaction–diffusion equations, *SIAM Journal on Applied Mathematics*. 45: 750–774.

Magleby, K. L. and C. F. Stevens (1972) A quantitative description of end-plate currents, *Journal of Physiology*. 223: 173–197.

Magnus, G. and J. Keizer (1997) Minimal model of beta-cell mitochondrial Ca^{2+} handling, *American Journal of Physiology — Cell Physiology*. 273: C717–733.

Magnus, G. and J. Keizer (1998a) Model of beta-cell mitochondrial calcium handling and electrical activity. I. Cytoplasmic variables, *American Journal of Physiology — Cell Physiology*. 274: C1158–1173.

Magnus, G. and J. Keizer (1998b) Model of beta-cell mitochondrial calcium handling and electrical activity. II. Mitochondrial variables, *American Journal of Physiology — Cell Physiology*. 274: C1174–1184.

Mallik, R. and S. P. Gross (2004) Molecular motors: strategies to get along, *Current Biology*. 14: R971–982.

Manley, G. A. (2001) Evidence for an active process and a cochlear amplifier in nonmammals, *Journal of Neurophysiology*. 86: 541–549.

Marchant, J., N. Callamaras and I. Parker (1999) Initiation of IP_3-mediated Ca^{2+} waves in Xenopus oocytes, *Embo Journal*. 18: 5285–5299.

Marchant, J. S. and I. Parker (2001) Role of elementary Ca^{2+} puffs in generating repetitive Ca^{2+} oscillations, *Embo Journal*. 20: 65–76.

Marchant, J. S. and C. W. Taylor (1998) Rapid activation and partial inactivation of inositol trisphosphate receptors by inositol trisphosphate, *Biochemistry*. 37: 11524–11533.

Marhl, M., T. Haberichter, M. Brumen and R. Heinrich (2000) Complex calcium oscillations and the role of mitochondria and cytosolic proteins., *Biosystems*. 57: 75–86.

Mari, A. (2002) Mathematical modeling in glucose metabolism and insulin secretion, *Current Opinion in Clinical Nutrition and Metabolic Care*. 5: 495–501.

Mari, A., G. Pacini, E. Murphy, B. Ludvik and J. J. Nolan (2001) A model-based method for assessing insulin sensitivity from the oral glucose tolerance test, *Diabetes Care*. 24: 539–548.

Mariani, L., M. Lohning, A. Radbruch and T. Hofer (2004) Transcriptional control networks of cell differentiation: insights from helper T lymphocytes, *Progress in Biophysics and Molecular Biology*. 86: 45–76.

Marland, E. (1998) *The Dynamics of the Sarcomere*: PhD Thesis, Department of Mathematics, University of Utah, Salt Lake City.

Martin, P., A. D. Mehta and A. J. Hudspeth (2000) Negative hair-bundle stiffness betrays a mechanism for mechanical amplification by the hair cell, *Proceedings of the National Academy of Sciences USA*. 97: 12026–12031.

Matveev, V., R. Bertram and A. Sherman (2006) Residual bound Ca^{2+} can account for the effects of Ca^{2+} buffers on synaptic facilitation, *Journal of Neurophysiology*. 96: 3389–3397.

Matveev, V., A. Sherman and R. S. Zucker (2002) New and corrected simulations of synaptic facilitation, *Biophysical Journal*. 83: 1368–1373.

Matveev, V., R. S. Zucker and A. Sherman (2004) Facilitation through buffer saturation: constraints on endogenous buffering properties, *Biophysical Journal*. 86: 2691–2709.

Mayer, S. (1877) Studien zur physiologie des herzens und der blutgefasse. V. Uber spontane blutdruckschwankungen, *Sitzungsberichte Kaiserlich Akad Wissenschaft Mathemat-Naturwissenschaft Classe*. 74: 281–307.

McAllister, R. E., D. Noble and R. W. Tsien (1975) Reconstruction of the electrical activity of cardiac Purkinje fibres, *Journal of Physiology*. 251: 1–59.

McCulloch, A., L. Waldman, J. Rogers and J. Guccione (1992) Large-scale finite element analysis of the beating heart, *Critical Reviews in Biomedical Engineering*. 20: 427–449.

McCulloch, A. D. (1995) *Cardiac biomechanics*. In: Biomedical Engineering Handbook: The Electrical Engineering Handbook Series, Ed: J. D. Branzino, CRC Press, Boca Raton.

McCulloch, A. D. and G. Paternostro (2005) Cardiac systems biology, *Annals of the New York Academy of Sciences*. 1047: 283–295.

McDonald, D. A. (1974) *Blood Flow in Arteries (Second Edition)*: Arnold, London.

McKean, H. P. (1970) Nagumo's equation, *Advances in Mathematics*. 4: 209–223.

McKenzie, A. and J. Sneyd (1998) On the formation and breakup of spiral waves of calcium, *International Journal of Bifurcation and Chaos*. 8: 2003–2012.

McLachlan, R. I., N. L. Cohen, K. D. Dahl, W. J. Bremner and M. R. Soules (1990) Serum inhibin levels during the periovulatory interval in normal women: relationships with sex steroid and gonadotrophin levels, *Clinical Endocrinology*. 32: 39–48.

McNaughton, P. A. (1990) Light response of vertebrate photoreceptors, *Physiological Reviews*. 70: 847–883.

McQuarrie, D. A. (1967) *Stochastic Approach to Chemical Kinetics*: Methuen and Co., London.

Meinrenken, C. J., J. G. Borst and B. Sakmann (2003) Local routes revisited: the space and time dependence of the Ca^{2+} signal for phasic transmitter release at the rat calyx of Held, *Journal of Physiology*. 547: 665–689.

Meyer, T. and L. Stryer (1988) Molecular model for receptor-stimulated calcium spiking, *Proceedings of the National Academy of Sciences USA*. 85: 5051–5055.

Meyer, T. and L. Stryer (1991) Calcium spiking, *Annual Review of Biophysics and Biophysical Chemistry*. 20: 153–174.

Michaelis, L. and M. I. Menten (1913) Die Kinetik der Invertinwirkung, *Biochemische Zeitschrift*. 49: 333–369.

Miftakhov, R. N., G. R. Abdusheva and J. Christensen (1999a) Numerical simulation of motility patterns of the small bowel. 1. formulation of a mathematical model, *Journal of Theoretical Biology*. 197: 89–112.

Miftakhov, R. N., G. R. Abdusheva and J. Christensen (1999b) Numerical simulation of motility patterns of the small bowel. II. Comparative pharmacological validation of a mathematical model, *Journal of Theoretical Biology*. 200: 261–290.

Mijailovich, S. M., J. P. Butler and J. J. Fredberg (2000) Perturbed equilibria of myosin binding in airway smooth muscle: bond-length distributions, mechanics, and ATP metabolism, *Biophysical Journal*. 79: 2667–2681.

Milhorn, H. T., Jr. and P. E. Pulley, Jr. (1968) A theoretical study of pulmonary capillary gas exchange and venous admixture, *Biophysical Journal*. 8: 337–357.

Miller, R. N. and J. Rinzel (1981) The dependence of impulse propagation speed on firing frequency, dispersion, for the Hodgkin–Huxley model, *Biophysical Journal*. 34: 227–259.

Milton, J. (2003) *Pupil light reflex: delays and oscillations*. In: Nonlinear Dynamics in Physiology and Medicine, Ed: A. Beuter, L. Glass, M. C. Mackey and M. S. Titcombe, Springer-Verlag, New York.

Milton, J. G. and M. C. Mackey (1989) Periodic haematological diseases: mystical entities or dynamical disorders?, *Journal of the Royal College of Physicians of London*. 23: 236–241.

Mines, G. R. (1914) On circulating excitations in heart muscle and their possible relation to tachycardia and fibrillation, *Transactions of the Royal Society of Canada*. 4: 43–53.

Minorsky, N. (1962) *Nonlinear Oscillations*: Van Nostrand, New York.

Miura, R. M. (1981) *Nonlinear waves in neuronal cortical structures*. In: Nonlinear Phenomena in Physics and Biology, Ed: R. H. Enns, B. L. Jones, R. M. Miura and S. S. Rangnekar, Plenum Press, New York.

Moe, G. K., W. C. Rheinboldt and J. A. Abildskov (1964) A computer model of atrial fibrillation, *American Heart Journal*. 67: 200–220.

Mogilner, A., T. C. Elston, H. Wang and G. Oster (2002) *Molecular motors: examples*. In: Computational Cell Biology, Ed: C. P. Fall, E. S. Marland, J. M. Wagner and J. J. Tyson, Springer-Verlag, New York.

Mogilner, A. and G. Oster (1996) Cell motility driven by actin polymerization, *Biophysical Journal*. 71: 3030–3045.

Mogilner, A. and G. Oster (1999) The polymerization ratchet model explains the force–velocity relation for growing microtubules, *European Journal of Biophysics*. 28: 235–242.

Mogilner, A. and G. Oster (2003) Force generation by actin polymerization II: the elastic ratchet and tethered filaments, *Biophysical Journal*. 84: 1591–1605.

Monod, J., J. Wyman and J. P. Changeux (1965) On the nature of allosteric transition: A plausible model, *Journal of Molecular Biology*. 12: 88–118.

Morley, A. (1979) Cyclic hemopoiesis and feedback control, *Blood Cells*. 5: 283–296.

Morris, C. and H. Lecar (1981) Voltage oscillations in the barnacle giant muscle fiber, *Biophysical Journal*. 35: 193–213.

Mount, D. W. (2001) *Bioinformatics: Sequence and Genome Analysis*: Cold Spring Harbor Laboratory Press, Cold Spring Harbor, N.Y.

Mountcastle, V. B., Ed. (1974) *Medical Physiology (Thirteenth Edition)*: C.V. Mosby Co., Saint Louis.

Murphy, R. A. (1994) What is special about smooth muscle? The significance of covalent crossbridge regulation, *FASEB Journal*. 8: 311–318.

Murray, J. D. (1971) On the molecular mechanism of facilitated oxygen diffusion by haemoglobin and myoglobin, *Proceedings of the Royal Society of London B*. 178: 95–110.

Murray, J. D. (1984) *Asymptotic Analysis*: Springer-Verlag, New York.

Murray, J. D. (2002) *Mathematical Biology (Third Edition)*: Springer-Verlag, New York.

Murray, J. D. and J. Wyman (1971) Facilitated diffusion: the case of carbon monoxide, *Journal of Biological Chemistry*. 246: 5903–5906.

Nagumo, J., S. Arimoto and S. Yoshizawa (1964) An active pulse transmission line simulating nerve axon, *Proceedings of the Institute of Radio Engineers*. 50: 2061–2070.

Naka, K. I. and W. A. Rushton (1966) S-potentials from luminosity units in the retina of fish (Cyprinidae), *Journal of Physiology*. 185: 587–599.

Nakatani, K., T. Tamura and K. W. Yau (1991) Light adaptation in retinal rods of the rabbit and two other nonprimate mammals, *Journal of General Physiology*. 97: 413–435.

Nakayama, K. (1985) Biological image motion processing: a review, *Vision Research*. 25: 625–660.

Naraghi, M., T. H. Muller and E. Neher (1998) Two-dimensional determination of the cellular Ca^{2+} binding in bovine chromaffin cells, *Biophysical Journal*. 75: 1635–1647.

Naraghi, M. and E. Neher (1997) Linearized buffered Ca^{2+} diffusion in microdomains and its implications for calculation of $[Ca^{2+}]$ at the mouth of a calcium channel, *Journal of Neuroscience*. 17: 6961–6973.

Nash, M. S., K. W. Young, R. A. Challiss and S. R. Nahorski (2001) Intracellular signalling. Receptor-specific messenger oscillations, *Nature*. 413: 381–382.

Nasmyth, K. (1995) Evolution of the cell cycle, *Philosophical Transactions of the Royal Society of London B*. 349: 271–281.

Nasmyth, K. (1996) Viewpoint: putting the cell cycle in order, *Science*. 274: 1643–1645.

Nathanson, M. H., A. D. Burgstahler, A. Mennone, M. B. Fallon, C. B. Gonzalez and J. C. Saez (1995) Ca^{2+} waves are organized among hepatocytes in the intact organ, *American Journal of Physiology — Gastrointestinal and Liver Physiology*. 269: G167–171.

Nedergaard, M. (1994) Direct signaling from astrocytes to neurons in cultures of mammalian brain cells, *Science*. 263: 1768–1771.

Neher, E. (1998a) Usefulness and limitations of linear approximations to the understanding of Ca^{++} signals, *Cell Calcium*. 24: 345–357.

Neher, E. (1998b) Vesicle pools and Ca^{2+} microdomains: new tools for understanding their roles in neurotransmitter release, *Neuron*. 20: 389–399.

Nelsen, T. S. and J. C. Becker (1968) Simulation of the electrical and mechanical gradient of the small intestine, *American Journal of Physiology*. 214: 749–757.

Nesheim, M. E., R. P. Tracy and K. G. Mann (1984) "Clotspeed," a mathematical simulation of the functional properties of prothrombinase, *Journal of Biological Chemistry*. 259: 1447–1453.

Nesheim, M. E., R. P. Tracy, P. B. Tracy, D. S. Boskovic and K. G. Mann (1992) Mathematical simulation of prothrombinase, *Methods in Enzymology*. 215: 316–328.

Nesher, R. and E. Cerasi (2002) Modeling phasic insulin release: immediate and time-dependent effects of glucose, *Diabetes*. 51 Suppl 1: S53–59.

Neu, J. C. (1979) Chemical waves and the diffusive coupling of limit cycle oscillators, *SIAM Journal on Applied Mathematics*. 36: 509–515.

Neu, J. C. and W. Krassowska (1993) Homogenization of syncytial tissues, *Critical Reviews in Biomedical Engineering*. 21: 137–199.

Nicholls, J. G., A. R. Martin and B. G. Wallace (1992) *From Neuron to Brain (Third Edition)*: Sinauer Associates, Inc., Sunderland, MA.

Niederer, S. A., P. J. Hunter and N. P. Smith (2006) A quantitative analysis of cardiac myocyte relaxation: a simulation study, *Biophysical Journal*. 90: 1697–1722.

Nielsen, K., P. G. Sørensen and F. Hynne (1997) Chaos in glycolysis, *Journal of Theoretical Biology*. 186: 303–306.

Nielsen, P. M. F., I. J. LeGrice and B. H. Smaill (1991) A mathematical model of geometry and fibrous structure of the heart, *American Journal of Physiology — Heart and Circulatory Physiology*. 260: H1365–1378.

Nikonov, S., N. Engheta and E. N. Pugh, Jr. (1998) Kinetics of recovery of the dark-adapted salamander rod photoresponse, *Journal of General Physiology*. 111: 7–37.

Nobili, R., F. Mammano and J. Ashmore (1998) How well do we understand the cochlea?, *Trends in Neuroscience*. 21: 159–167.

Noble, D. (1962) A modification of the Hodgkin–Huxley equations applicable to Purkinje fiber action and pacemaker potential, *Journal of Physiology*. 160: 317–352.

Noble, D. (2002a) Modelling the heart: insights, failures and progress, *Bioessays*. 24: 1155–1163.

Noble, D. (2002b) Modeling the heart–from genes to cells to the whole organ, *Science*. 295: 1678–1682.

Noble, D. and S. J. Noble (1984) A model of sino-atrial node electrical activity using a modification of the DiFrancesco–Noble (1984) equations, *Proceedings of the Royal Society of London B*. 222: 295–304.

Nolasco, J. B. and R. W. Dahlen (1968) A graphic method for the study of alternation in cardiac action potentials, *Journal of Applied Physiology*. 25: 191–196.

Norman, R. A. and I. Perlman (1979) The effects of background illumination on the photoresponses of red and green cones, *Journal of Physiology*. 286: 491–507.

Novak, B., Z. Pataki, A. Ciliberto and J. J. Tyson (2001) Mathematical model of the cell division cycle of fission yeast, *Chaos*. 11: 277–286.

Novak, B. and J. J. Tyson (1993a) Numerical analysis of a comprehensive model of M-phase control in *Xenopus* oocyte extracts and intact embryos, *Journal of Cell Science*. 106: 1153–1168.

Novak, B. and J. J. Tyson (1993b) Modeling the cell division cycle: M phase trigger oscillations and size control, *Journal of Theoretical Biology*. 165: 101–134.

Novak, B. and J. J. Tyson (2004) A model for restriction point control of the mammalian cell cycle, *Journal of Theoretical Biology*. 230: 563–579.

Nowak, M. A. and R. M. May (2000) *Virus Dynamics: Mathematical Principles of Immunology and Virology*: Oxford University Press, Oxford.

Nowycky, M. C. and M. J. Pinter (1993) Time courses of calcium and calcium-bound buffers following calcium influx in a model cell, *Biophysical Journal*. 64: 77–91.

Nuccitelli, R., D. L. Yim and T. Smart (1993) The sperm-induced Ca^{2+} wave following fertilization of the Xenopus egg requires the production of Ins(1,4,5)P_3, *Developmental Biology*. 158: 200–212.

Nunemaker, C. S., R. Bertram, A. Sherman, K. Tsaneva-Atanasova, C. R. Daniel and L. S. Satin (2006) Glucose modulates $[Ca^{2+}]_i$ oscillations in pancreatic islets via ionic and glycolytic mechanisms, *Biophysical Journal*. 91: 2082–2096.

O'Neill, P. V. (1983) *Advanced Engineering Mathematics*: Wadsworth, Belmont CA.

Ohta, T., M. Mimura and R. Kobayashi (1989) Higher dimensional localized patterns in excitable media, *Physica D*. 34: 115–144.

Olufsen, M. S., A. Nadim and L. A. Lipsitz (2002) Dynamics of cerebral blood flow regulation explained using a lumped parameter model, *American Journal of Physiology — Regulatory, Integrative, and Comparative Physiology*. 282: R611–622.

Olufsen, M. S., C. S. Peskin, W. Y. Kim, E. M. Pedersen, A. Nadim and J. Larsen (2000) Numerical simulation and experimental validation of blood flow in arteries with structured-tree outflow conditions, *Annals of Biomedical Engineering*. 28: 1281–1299.

Orrenius, S., B. Zhivotovsky and P. Nicotera (2003) Regulation of cell death: the calcium–apoptosis link, *Nature Reviews Molecular Cell Biology*. 4: 552–565.

Ortoleva, P. and J. Ross (1973) Phase waves in oscillatory chemical reactions, *Journal of Chemical Physics*. 58: 5673–5680.

Ortoleva, P. and J. Ross (1974) On a variety of wave phenomena in chemical reactions, *Journal of Chemical Physics*. 60: 5090–5107.

Osher, S. and J. A. Sethian (1988) Fronts propagating with curvature-dependent speed: algorithms based on Hamilton–Jacobi formulations, *Journal of Computational Physics*. 79: 12–49.

Otani, N. F. and R. F. Gilmour, Jr. (1997) Memory models for the electrical properties of local cardiac systems, *Journal of Theoretical Biology*. 187: 409–436.

Othmer, H. G. (1976) The qualitative dynamics of a class of biochemical control circuits, *Journal of Mathematical Biology*. 3: 53–78.

Ottesen, J. T. (1997) Modelling of the baroreflex-feedback mechanism with time-delay, *Journal of Mathematical Biology*. 36: 41–63.

Ottesen, J. T., M. S. Olufsen and J. K. Larsen (2004) *Applied Mathematical Models in Human Physiology*: SIAM, Philadelphia.

Ozbudak, E. M., M. Thattai, H. N. Lim, B. I. Shraiman and A. Van Oudenaarden (2004) Multistability in the lactose utilization network of Escherichia coli, *Nature*. 427: 737–740.

Pace, N., E. Strajman and E. L. Walker (1950) Acceleration of carbon monoxide elimination in man by high pressure oxygen, *Science*. 111: 652–654.

Panfilov, A. V. and P. Hogeweg (1995) Spiral break-up in a modified FitzHugh–Nagumo model, *Physics Letters A*. 176: 295–299.

Panfilov, A. V. and A. V. Holden (1990) Self-generation of turbulent vortices in a two-dimensional model of cardiac tissue, *Physics Letters A*. 151: 23–26.

Panfilov, A. V. and J. P. Keener (1995) Re-entry in an anatomical model of the heart, *Chaos, Solitons and Fractals*. 5: 681–689.

Papoulis, A. (1962) *The Fourier Integral and its Applications*: McGraw-Hill, New York.

Parker, I., J. Choi and Y. Yao (1996b) Elementary events of $InsP_3$-induced Ca^{2+} liberation in Xenopus oocytes: hot spots, puffs and blips, *Cell Calcium*. 20: 105–121.

Parker, I. and Y. Yao (1996) Ca^{2+} transients associated with openings of inositol trisphosphate-gated channels in Xenopus oocytes, *Journal of Physiology*. 491: 663–668.

Parker, I., Y. Yao and V. Ilyin (1996a) Fast kinetics of calcium liberation induced in *Xenopus* oocytes by photoreleased inositol trisphosphate, *Biophysical Journal*. 70: 222–237.

Parker, I., W. J. Zang and W. G. Wier (1996c) Ca^{2+} sparks involving multiple Ca^{2+} release sites along Z-lines in rat heart cells, *Journal of Physiology*. 497: 31–38.

Parnas, H., G. Hovav and I. Parnas (1989) Effect of Ca^{2+} diffusion on the time course of neurotransmitter release, *Biophysical Journal*. 55: 859–874.

Parnas, H. and L. A. Segel (1980) A theoretical explanation for some effects of calcium on the facilitation of neurotransmitter release, *Journal of Theoretical Biology*. 84: 3–29.

Parnas, H., J. C. Valle-Lisboa and L. A. Segel (2002) Can the Ca^{2+} hypothesis and the Ca^{2+}-voltage hypothesis for neurotransmitter release be reconciled?, *Proceedings of the National Academy of Sciences USA*. 99: 17149–17154.

Pate, E. (1997) *Mathematical modeling of muscle crossbridge mechanics*. In: Case Studies in Mathematical Biology, Ed: H. Othmer, F. Adler, M. Lewis and J. Dallon, Prentice Hall, Upper Saddle River, New Jersey.

Pate, E. and R. Cooke (1989) A model of crossbridge action: the effects of ATP, ADP and Pi, *Journal of Muscle Research and Cell Motility*. 10: 181–196.

Pate, E. and R. Cooke (1991) Simulation of stochastic processes in motile crossbridge systems, *Journal of Muscle Research and Cell Motility*. 12: 376–393.

Patlak, J. (1991) Molecular kinetics of voltage-dependent Na^+ channels., *Physiological Reviews*. 71: 1047–1080.

Patneau, D. K. and M. L. Mayer (1991) Kinetic analysis of interactions between kainate and AMPA: evidence for activation of a single receptor in mouse hippocampal neurons, *Neuron*. 6: 785–798.

Patton, R. J. and D. A. Linkens (1978) Hodgkin–Huxley type electronic modelling of gastrointestinal electrical activity, *Medical and Biological Engineering and Computing*. 16: 195–202.

Pauling, L. (1935) The oxygen equilibrium of hemoglobin and its structural interpretation, *Proceedings of the National Academy of Sciences USA*. 21: 186–191.

Pauwelussen, J. P. (1981) Nerve impulse propagation in a branching nerve system: a simple model, *Physica D*. 4: 67–88.

Payne, S. and C. Stephens (2005) The response of the cross-bridge cycle model to oscillations in intracellular calcium: A mathematical analysis, *Conference proceedings: Annual International Conference of the IEEE Engineering in Medicine and Biology Society*. 7: 7305–7308.

Pearson, J. E. and S. Ponce-Dawson (1998) Crisis on skid row, *Physica A*. 257: 141–148.

Pedley, T. J. (1980) *The Fluid Mechanics of Large Blood Vessels*: Cambridge University Press, Cambridge.

Pelce, P. and J. Sun (1991) Wave front interaction in steadily rotating spirals, *Physica D*. 48: 353–366.

Pepperberg, D. R., M. C. Cornwall, M. Kahlert, K. P. Hofmann, J. Jin, G. J. Jones and H. Ripps (1992) Light-dependent delay in the falling phase of the retinal rod photoresponse, *Visual Neuroscience*. 8: 9–18.

Perelson, A. S. (2002) Modelling viral and immune system dynamics, *Nature Reviews Immunology*. 2: 28–36.

Perlman, I. and R. A. Normann (1998) Light adaptation and sensitivity controlling mechanisms in vertebrate photoreceptors, *Progress in Retinal and Eye Research*. 17: 523–563.

Pernarowski, M. (1994) Fast subsystem bifurcations in a slowly varying Liénard system exhibiting bursting, *SIAM Journal on Applied Mathematics*. 54: 814–832.

Pernarowski, M., R. M. Miura and J. Kevorkian (1991) *The Sherman–Rinzel–Keizer model for bursting electrical activity in the pancreatic β-cell*. In: Differential Equations Models in Biology, Epidemiology and Ecology, Ed: S. Busenberg and M. Martelli, Springer-Verlag, New York.

Pernarowski, M., R. M. Miura and J. Kevorkian (1992) Perturbation techniques for models of bursting electrical activity in pancreatic β-cells, *SIAM Journal on Applied Mathematics*. 52: 1627–1650.

Perutz, M. F. (1970) Stereochemistry of cooperative effects in haemoglobin, *Nature*. 228: 726–739.

Perutz, M. F., W. Bolton, R. Diamond, H. Muirhead and H. Watson (1964) Structure of haemoglobin. An X-ray examination of reduced horse haemoglobin, *Nature*. 203: 687–690.

Peskin, C. S. (1975) *Mathematical Aspects of Heart Physiology*: Courant Institute of Mathematical Sciences Lecture Notes, New York.

Peskin, C. S. (1976) *Partial Differential Equations in Biology*: Courant Institute of Mathematical Sciences Lecture Notes, New York.

Peskin, C. S. (1981) Lectures on mathematical aspects of physiology, *AMS Lectures in Applied Mathematics*. 19: 38–69.

Peskin, C. S. (1991) *Mathematical Aspects of Neurophysiology*: Courant Institute of Mathematical Sciences Lecture Notes, New York.

Peskin, C. S. (2002) The immersed boundary method, *Acta Numerica*. 11: 479–517.

Peskin, C. S. and D. M. McQueen (1989) A three-dimensional computational method for blood flow in the heart. I. Immersed elastic fibers in a viscous incompressible fluid., *Journal of Computational Physics*. 81: 372–405.

Peskin, C. S. and D. M. McQueen (1992) Cardiac fluid dynamics, *Critical Reviews in Biomedical Engineering*. 20: 451–459.

Peskin, C. S., G. M. Odell and G. F. Oster (1993) Cellular motions and thermal fluctuations: the Brownian ratchet, *Biophysical Journal*. 65: 316–324.

Peskin, C. S. and G. Oster (1995) Coordinated hydrolysis explains the mechanical behavior of kinesin, *Biophysical Journal*. 68: 202S–210.

Peskin, C. S. and G. F. Oster (1995) Force production by depolymerizing microtubules: load–velocity curves and run-pause statistics, *Biophysical Journal*. 69: 2268–2276.

Peskoff, A. and G. A. Langer (1998) Calcium concentration and movement in the ventricular cardiac cell during an excitation–contraction cycle, *Biophysical Journal*. 74: 153–174.

Peskoff, A., J. A. Post and G. A. Langer (1992) Sarcolemmal calcium binding sites in heart: II. Mathematical model for diffusion of calcium released from the sarcoplasmic reticulum into the diadic region, *Journal of Membrane Biology*. 129: 59–69.

Peterson, L. C. and B. P. Bogert (1950) A dynamical theory of the cochlea, *Journal of the Acoustical Society of America*. 22: 369–381.

Pickles, J. O. (1982) *An Introduction to the Physiology of Hearing*: Academic Press, London.

Pitman, E. B., R. M. Zaritski, K. J. Kesseler, L. C. Moore and H. E. Layton (2004) Feedback-mediated dynamics in two coupled nephrons, *Bulletin of Mathematical Biology*. 66: 1463–1492.

Plant, R. E. (1981) Bifurcation and resonance in a model for bursting nerve cells, *Journal of Mathematical Biology*. 11: 15–32.

Podolsky, R. J. and A. C. Nolan (1972) *Cross-bridge properties derived from physiological studies of frog muscle fibres*. In: Contractility of Muscle Cells and Related Processes, Ed: R. J. Podolsky, Prentice Hall, Englewood Cliffs, NJ.

Podolsky, R. J. and A. C. Nolan (1973) *Muscle Contraction Transients, Cross-Bridge Kinetics and the Fenn Effect.*: 37th Cold Spring Harbor Symposium of Quantitative Biology, Cold Spring Harbor, New York.

Podolsky, R. J., A. C. Nolan and S. A. Zaveler (1969) Cross-bridge properties derived from muscle isotonic velocity transients, *Proceedings of the National Academy of Sciences USA*. 64: 504–511.

Politi, A., L. D. Gaspers, A. P. Thomas and T. Hofer (2006) Models of IP_3 and Ca^{2+} oscillations: frequency encoding and identification of underlying feedbacks, *Biophysical Journal*. 90: 3120–3133.

Pollack, G. H. (1976) Intercellular coupling in the atrioventricular node and other tissues of the rabbit heart, *Journal of Physiology*. 255: 275–298.

Ponce-Dawson, S., J. Keizer and J. E. Pearson (1999) Fire-diffuse-fire model of dynamics of intracellular calcium waves, *Proceedings of the National Academy of Sciences USA*. 96: 6060–6063.

Preston, G. M., T. P. Carroll, W. B. Guggino and P. Agre (1992) Appearance of water channels in *Xenopus* oocytes expressing red cell CHIP28 protein, *Science*. 256: 385–387.

Pries, A. R. and T. W. Secomb (2000) Microcirculatory network structures and models, *Annals of Biomedical Engineering*. 28: 916–921.

Pries, A. R. and T. W. Secomb (2005) Control of blood vessel structure: insights from theoretical models, *American Journal of Physiology — Heart and Circulatory Physiology*. 288: H1010–1015.

Pries, A. R., T. W. Secomb and P. Gaehtgens (1996) Biophysical aspects of blood flow in the microvasculature, *Cardiovascular Research*. 32: 654–667.

Pugh, E. N. and T. D. Lamb (1990) Cyclic GMP and calcium: messengers of excitation and adaptation in vertebrate photoreceptors, *Vision Research*. 30: 1923–1948.

Pullan, A., L. Cheng, R. Yassi and M. Buist (2004) Modelling gastrointestinal bioelectric activity, *Progress in Biophysics and Molecular Biology*. 85: 523–550.

Qian, H. (2000) The mathematical theory of molecular motor movement and chemomechanical energy transduction, *Journal of Mathematical Chemistry*. 27: 219–234.

Qu, Z., A. Garfinkel, P. S. Chen and J. N. Weiss (2000) Mechanisms of discordant alternans and induction of reentry in simulated cardiac tissue, *Circulation*. 102: 1664–1670.

Röttingen, J. and J. G. Iversen (2000) Ruled by waves? Intracellular and intercellular calcium signalling, *Acta Physiologica Scandinavica*. 169: 203–219.

Rahn, H. (1949) A concept of mean alveolar air and the ventilation–bloodflow relationships during pulmonary gas exchange, *American Journal of Physiology*. 158: 21–30.

Rall, W. (1957) Membrane time constant of motoneurons, *Science*. 126: 454.

Rall, W. (1959) Branching dendritic trees and motoneuron membrane resistivity, *Experimental Neurology*. 2: 491–527.

Rall, W. (1960) Membrane potential transients and membrane time constant of motoneurons, *Experimental Neurology*. 2: 503–532.

Rall, W. (1969) Time constants and electrotonic length of membrane cylinders and neurons, *Biophysical Journal*. 9: 1483–1508.

Rall, W. (1977) *Core conductor theory and cable properties of neurons*. In: Handbook of Physiology The Nervous System I, Ed: J. M. Brookhart and V. B. Mountcastle, American Physiological Society, Bethesda, MD.

Ramamoorthy, S., N. V. Deo and K. Grosh (2007) A mechano-electro-acoustical model for the cochlea: response to acoustic stimuli, *Journal of the Acoustical Society of America*. 121: 2758–2773.

Ramanan, S. V. and P. R. Brink (1990) Exact solution of a model of diffusion in an infinite chain or monolayer of cells coupled by gap junctions, *Biophysical Journal*. 58: 631–639.

Ramirez, J. M. and D. W. Richter (1996) The neuronal mechanisms of respiratory rhythm generation, *Current Opinion in Neurobiology*. 6: 817–825.

Rand, R. H. and P. J. Holmes (1980) Bifurcation of periodic motions in two weakly coupled van der Pol oscillators, *Journal of Non-linear Mechanics*. 15: 387–399.

Ranke, O. F. (1950) Theory of operation of the cochlea: A contribution to the hydrodynamics of the cochlea, *Journal of the Acoustical Society of America*. 22: 772–777.

Rapp, P. E. (1975) A theoretical investigation of a large class of biochemical oscillations, *Mathematical Biosciences*. 25: 165–188.

Rapp, P. E. (1976) Mathematical techniques for the study of oscillations in biochemical control loops, *Bulletin of the Institute of Mathematics and its Applications*. 12: 11–21.

Rapp, P. E. and M. J. Berridge (1977) Oscillations in calcium–cyclic AMP control loops form the basis of pacemaker activity and other high frequency biological rhythms, *Journal of Theoretical Biology*. 66: 497–525.

Ratliff, F. (1961) *Inhibitory interaction and the detection and enhancement of contours*. In: Sensory Communication, Ed: W. A. Rosenblith, MIT Press, Cambridge, MA.

Ratliff, F. and H. K. Hartline (1959) The responses of Limulus optic nerve fibers to patterns of illumination on the receptor mosaic, *Journal of General Physiology*. 42: 1241–1255.

Ratliff, F., B. W. Knight, Jr., F. A. Dodge, Jr. and H. K. Hartline (1974) Fourier analysis of dynamics of excitation and inhibition in the eye of Limulus: amplitude, phase and distance, *Vision Research*. 14: 1155–1168.

Rauch, J. and J. Smoller (1978) Qualitative theory of the FitzHugh–Nagumo equations, *Advances in Mathematics*. 27: 12–44.

Reed, M. C. and J. J. Blum (1986) Theoretical analysis of radioactivity profiles during fast axonal transport: effects of deposition and turnover, *Cell Motility and the Cytoskeleton*. 6: 620–627.

Reeve, E. B. and A. C. Guyton, Eds. (1967) *Physical Bases of Circulatory Transport: Regulation and Exchange*: W.B. Saunders, Philadelphia.

Reichardt, W. (1961) *Autocorrelation, a principle for the evaluation of sensory information by the central nervous system*. In: Sensory Communication, Ed: W. A. Rosenblith, Cambridge, MA.

Reijenga, K. A., H. V. Westerhoff, B. N. Kholodenko and J. L. Snoep (2002) Control analysis for autonomously oscillating biochemical networks, *Biophysical Journal*. 82: 99–108.

Reimann, P. (2002) Brownian motors: Noisy transport far from equilibrium, *Physics Reports*. 361: 57–265.

Reuss, L. and B. H. Hirst (2002) Water transport controversies–an overview, *Journal of Physiology*. 542: 1–2.

Rhode, W. S. (1984) Cochlear mechanics, *Annual Review of Physiology*. 46: 231–246.

Richter, D. W. (1996) *Neural regulation of respiration: rhythmogenesis and afferent control*. In: Comprehensive Human Physiology, Ed: R. Gregor and U. Windhorst, Springer-Verlag, Berlin.

Ridgway, E. B., J. C. Gilkey and L. F. Jaffe (1977) Free calcium increases explosively in activating medaka eggs, *Proceedings of the National Academy of Sciences USA*. 74: 623–627.

Rieke, F. and D. A. Baylor (1998a) Origin of reproducibility in the responses of retinal rods to single photons, *Biophysical Journal*. 75: 1836–1857.

Rieke, F. and D. A. Baylor (1998b) Single-photon detection by rod cells of the retina, *Reviews of Modern Physics*. 70: 1027–1036.

Riley, R. L. and A. Cournand (1949) " Ideal" alveolar air and the analysis of ventilation–perfusion relationships in the lungs, *Journal of Applied Physiology*. 1: 825–847.

Riley, R. L. and A. Cournand (1951) Analysis of factors affecting partial pressures of oxygen and carbon dioxide in gas and blood of lungs; theory, *Journal of Applied Physiology*. 4: 77–101.

Rinzel, J. (1978) On repetitive activity in nerve, *Federation Proceedings*. 37: 2793–2802.

Rinzel, J. (1985) *Bursting oscillations in an excitable membrane model*. In: Ordinary and Partial Differential Equations, Ed: B. D. Sleeman and R. J. Jarvis, Springer-Verlag, New York.

Rinzel, J. (1987) *A formal classification of bursting mechanisms in excitable systems*. In: Mathematical Topics in Population Biology, Morphogenesis, and Neurosciences, Lecture Notes in Biomathematics, Vol 71, Ed: E. Teramoto and M. Yamaguti, Springer-Verlag, Berlin.

Rinzel, J. (1990) Electrical excitability of cells, theory and experiment: review of the Hodgkin–Huxley foundation and an update, *Bulletin of Mathematical Biology*. 52: 5–23.

Rinzel, J. and J. P. Keener (1983) Hopf bifurcation to repetitive activity in nerve, *SIAM Journal on Applied Mathematics*. 43: 907–922.

Rinzel, J. and J. B. Keller (1973) Traveling wave solutions of a nerve conduction equation, *Biophysical Journal*. 13: 1313–1337.

Rinzel, J. and Y. S. Lee (1986) *On different mechanisms for membrane potential bursting*. In: Nonlinear Oscillations in Biology and Chemistry, Lecture Notes in Biomathematics, Vol 66, Ed: H. G. Othmer, Springer-Verlag, New York.

Rinzel, J. and Y. S. Lee (1987) Dissection of a model for neuronal parabolic bursting, *Journal of Mathematical Biology*. 25: 653–675.

Rinzel, J. and K. Maginu (1984) *Kinematic analysis of wave pattern formation in excitable media*. In: Non-equilibrium Dynamics in Chemical Systems, Ed: A. Pacault and C. Vidal, Springer-Verlag, Berlin.

Robb-Gaspers, L. D. and A. P. Thomas (1995) Coordination of Ca^{2+} signaling by intercellular propagation of Ca^{2+} waves in the intact liver, *Journal of Biological Chemistry*. 270: 8102–8107.

Roberts, D. and A. M. Scher (1982) Effect of tissue anisotropy on extracellular potential fields in canine myocardium *in situ*, *Circulation Research*. 50: 342–351.

Robertson-Dunn, B. and D. A. Linkens (1974) A mathematical model of the slow-wave electrical activity of the human small intestine, *Medical and Biological Engineering*. 12: 750–758.

Robinson, T. F., L. Cohen-Gould and S. M. Factor (1983) Skeletal framework of mammalian heart muscle. Arrangement of inter- and pericellular connective tissue structures, *Laboratory Investigation*. 49: 482–498.

Rodieck, R. W. (1965) Quantitative analysis of cat retinal ganglion cell response to visual stimuli, *Vision Research*. 5: 583–601.

Rooney, T. A. and A. P. Thomas (1993) Intracellular calcium waves generated by Ins(1,4,5)P_3-dependent mechanisms, *Cell Calcium*. 14: 674–690.

Roper, P., J. Callaway and W. Armstrong (2004) Burst initiation and termination in phasic vasopressin cells of the rat supraoptic nucleus: a combined mathematical, electrical, and calcium fluorescence study, *Journal of Neuroscience*. 24: 4818–4831.

Roper, P., J. Callaway, T. Shevchenko, R. Teruyama and W. Armstrong (2003) AHP's, HAP's and DAP's: how potassium currents regulate the excitability of rat supraoptic neurones, *Journal of Computational Neuroscience*. 15: 367–389.

Rorsman, P. and G. Trube (1986) Calcium and delayed potassium currents in mouse pancreatic β-cells under voltage clamp conditions, *Journal of Physiology*. 375: 531–550.

Roth, B. J. (1992) How the anisotropy of the intracellular and extracellular conductivities influences stimulation of cardiac muscle, *Journal of Mathematical Biology*. 30: 633–646.

Roughton, F. J. W., E. C. DeLand, J. C. Kernohan and J. W. Severinghaus (1972) *Some recent studies of the oxyhaemoglobin dissociation curve of human blood under physiological conditions and the fitting of the Adair equation to the standard curve*. In: Oxygen Affinity of Hemoglobin and Red Cell Acid Base States, Ed: M. Rorth and P. Astrup, Academic Press, New York.

Roy, D. R., H. E. Layton and R. L. Jamison (1992) *Countercurrent mechanism and its regulation*. In: The Kidney: Physiology and Pathophysiology, Ed: D. W. Seldin and G. Giebisch, Raven Press, New York.

Rubinow, S. I. (1973) *Mathematical Problems in the Biological Sciences*: SIAM, Philadelphia.

Rubinow, S. I. (1975) *Introduction to Mathematical Biology*: John Wiley and Sons, New York.

Rubinow, S. I. and M. Dembo (1977) The facilitated diffusion of oxygen by hemoglobin and myoglobin, *Biophysical Journal*. 18: 29–42.

Ruggero, M. A. (1992) Responses to sound of the basilar membrane of the mammalian cochlea, *Current Opinion in Neurobiology*. 2: 449–456.

Rushmer, R. F. (1976) *Structure and Function of the Cardiovascular System (Second Edition)*: W.B. Saunders Co., Philadelphia.

Rybak, I. A., J. F. Paton and J. S. Schwaber (1997a) Modeling neural mechanisms for genesis of respiratory rhythm and pattern. I. Models of respiratory neurons, *Journal of Neurophysiology*. 77: 1994–2006.

Rybak, I. A., J. F. Paton and J. S. Schwaber (1997b) Modeling neural mechanisms for genesis of respiratory rhythm and pattern. II. Network models of the central respiratory pattern generator, *Journal of Neurophysiology*. 77: 2007–2026.

Rybak, I. A., N. A. Shevtsova, J. F. Paton, T. E. Dick, W. M. St-John, M. Morschel and M. Dutschmann (2004) Modeling the ponto-medullary respiratory network, *Respiratory Physiology and Neurobiology*. 143: 307–319.

Sabah, N. H. and R. A. Spangler (1970) Repetitive response of the Hodgkin–Huxley model for the squid giant axon, *Journal of Theoretical Biology*. 29: 155–171.

Sachs, F., F. Qin and P. Palade (1995) Models of Ca^{2+} release channel adaptation, *Science*. 267: 2010–2011.

Sagawa, K., R. K. Lie and J. Schaefer (1990) Translation of Otto Frank's Paper "Die Grundform des Arteriellen Pulses", Zeitschrift für Biologie, 37:483-526 (1899), *Journal of Molecular and Cellular Cardiology*. 22: 253–277.

Sagawa, K., H. Suga and K. Nakayama (1978) *Instantaneous pressure–volume ratio of the left ventricle versus instantaneous force–length relation of papillary muscle*. In: Cardiovascular System Dynamics, Ed: J. Baan, A. Noordergraaf and J. Raines, MIT Press, Cambridge, MA.

Sakmann, B. and E. Neher (1995) *Single-Channel Recording (Second Edition)*: Plenum Press, New York.

Sala, F. and A. Hernández-Cruz (1990) Calcium diffusion modeling in a spherical neuron: relevance of buffering properties, *Biophysical Journal*. 57: 313–324.

Sanders, K. M. (1996) A case for interstitial cells of Cajal as pacemakers and mediators of neurotransmission in the gastrointestinal tract, *Gastroenterology*. 111: 492–515.

Sanderson, M. J., A. C. Charles, S. Boitano and E. R. Dirksen (1994) Mechanisms and function of intercellular calcium signaling, *Molecular and Cellular Endocrinology*. 98: 173–187.

Sanderson, M. J., A. C. Charles and E. R. Dirksen (1990) Mechanical stimulation and intercellular communication increases intracellular Ca^{2+} in epithelial cells, *Cell Regulation*. 1: 585–596.

Santillán, M. and M. C. Mackey (2001a) Dynamic behavior in mathematical models of the tryptophan operon, *Chaos*. 11: 261–268.

Santillán, M. and M. C. Mackey (2001b) Dynamic regulation of the tryptophan operon: a modeling study and comparison with experimental data, *Proceedings of the National Academy of Sciences USA*. 98: 1364–1369.

Santillán, M. and M. C. Mackey (2004) Influence of catabolite repression and inducer exclusion on the bistable behavior of the lac operon, *Biophysical Journal*. 86: 1282–1292.

Sarna, S. K. (1989) *In vivo myoelectric activity: methods, analysis and interpretation*. In: Handbook of Physiology. Section 6: The Gastrointestinal System, Ed: S. G. Schultz, J. D. Wood and B. B. Rauner, American Physiological Society, Bethesda, Maryland.

Sarna, S. K., E. E. Daniel and Y. J. Kingma (1971) Simulation of the slow wave electrical activity of small intestine, *American Journal of Physiology*. 221: 166–175.

Sarty, G. E. and R. A. Pierson (2005) An application of Lacker's mathematical model for the prediction of ovarian response to superstimulation, *Mathematical Biosciences*. 198: 80–96.

Schiefer, A., G. Meissner and G. Isenberg (1995) Ca^{2+} activation and Ca^{2+} inactivation of canine reconstituted cardiac sarcoplasmic reticulum Ca^{2+}-release channels, *Journal of Physiology*. 489: 337–348.

Schlosser, P. M. and J. F. Selgrade (2000) A model of gonadotropin regulation during the menstrual cycle in women: qualitative features, *Environmental Health Perspectives*. 108 Suppl 5: 873–881.

Schmitz, S., H. Franke, J. Brusis and H. E. Wichmann (1993) Quantification of the cell kinetic effects of G-CSF using a model of human granulopoiesis, *Experimental Hematology*. 21: 755–760.

Schmitz, S., M. Loeffler, J. B. Jones, R. D. Lange and H. E. Wichmann (1990) Synchrony of bone marrow proliferation and maturation as the origin of cyclic haemopoiesis, *Cell and Tissue Kinetics*. 23: 425–442.

Schultz, S. G. (1981) Homocellular regulatory mechanisms in sodium-transporting epithelia: avoidance of extinction by "flush-through", *American Journal of Physiology — Renal, Fluid and Electrolyte Physiology*. 241: F579–590.

Schumaker, M. F. and R. MacKinnon (1990) A simple model for multi-ion permeation, *Biophysical Journal*. 58: 975–984.

Schuster, S., M. Marhl and T. Höfer (2002) Modelling of simple and complex calcium oscillations. From single-cell responses to intercellular signalling, *European Journal of Biochemistry*. 269: 1333–1355.

Schwartz, N. B. (1969) A model for the regulation of ovulation in the rat, *Recent Progress in Hormone Research*. 25: 1–53.

Segel, I. H. (1975) *Enzyme Kinetics: Behavior and Analysis of Rapid Equilibrium and Steady-State Enzyme Systems*: John Wiley & Sons. Republished in the Wiley Classics Library Edition, 1993. Wiley, Hoboken, New Jersey.

Segel, L. and A. Goldbeter (1994) Scaling in biochemical kinetics: dissection of a relaxation oscillator, *Journal of Mathematical Biology*. 32: 147–160.

Segel, L. A. (1970) Standing-gradient flows driven by active solute transport, *Journal of Theoretical Biology*. 29: 233–250.

Segel, L. A. (1977) *Mathematics Applied to Continuum Mechanics*: MacMillan, New York.

Segel, L. A. (1988) On the validity of the steady state assumption of enzyme kinetics, *Bulletin of Mathematical Biology*. 50: 579–593.

Segel, L. A., I. Chet and Y. Henis (1977) A simple quantitative assay for bacterial motility, *Journal of General Microbiology*. 98: 329–337.

Segel, L. A., A. Goldbeter, P. N. Devreotes and B. E. Knox (1986) A mechanism for exact sensory adaptation based on receptor modification, *Journal of Theoretical Biology*. 120: 151–179.

Segel, L. A. and A. S. Perelson (1992) Plasmid copy number control: a case study of the quasi-steady state assumption, *Journal of Theoretical Biology*. 158: 481–494.

Segel, L. A. and M. Slemrod (1989) The quasi-steady state assumption: a case study in perturbation, *SIAM Review*. 31: 446–447.

Segev, I., J. Rinzel and G. M. Shepherd (1995) *The Theoretical Foundation of Dendritic Function*: MIT Press, Cambridge, MA.

Sel'kov, E. E. (1968) Self-oscillations in glycolysis, *European Journal of Biochemistry*. 4: 79–86.

Selgrade, J. F. and P. M. Schlosser (1999) A model for the production of ovarian hormones during the menstrual cycle, *Fields Institute Communications*. 21: 429–446.

Selivanov, V. A., F. Ichas, E. L. Holmuhamedov, L. S. Jouaville, Y. V. Evtodienko and J. P. Mazat (1998) A model of mitochondrial Ca^{2+}-induced Ca^{2+} release simulating the Ca^{2+} oscillations and spikes generated by mitochondria, *Biophysical Chemistry*. 72: 111–121.

Sha, W., J. Moore, K. Chen, A. D. Lassaletta, C. S. Yi, J. J. Tyson and J. C. Sible (2003) Hysteresis drives cell-cycle transitions in Xenopus laevis egg extracts, *Proceedings of the National Academy of Sciences USA*. 100: 975–980.

Shannon, T. R., F. Wang, J. Puglisi, C. Weber and D. M. Bers (2004) A mathematical treatment of integrated Ca dynamics within the ventricular myocyte, *Biophysical Journal*. 87: 3351–3371.

Shapley, R. M. and C. Enroth-Cugell (1984) *Visual adaptation and retinal gain controls*. In: Progress in Retinal Research, Ed: N. Osborne and G. Chader, Pergamon Press, London.

Sheetz, M. P. and J. A. Spudich (1983) Movement of myosin-coated fluorescent beads on actin cables in vitro, *Nature*. 303: 31–35.

Shen, P. and R. Larter (1995) Chaos in intracellular Ca^{2+} oscillations in a new model for non-excitable cells, *Cell Calcium*. 17: 225–232.

Sherman, A. (1994) Anti-phase, asymmetric and aperiodic oscillations in excitable cells–I. coupled bursters, *Bulletin of Mathematical Biology*. 56: 811–835.

Sherman, A. and J. Rinzel (1991) Model for synchronization of pancreatic β-cells by gap junction coupling, *Biophysical Journal*. 59: 547–559.

Sherman, A., J. Rinzel and J. Keizer (1988) Emergence of organized bursting in clusters of pancreatic β-cells by channel sharing, *Biophysical Journal*. 54: 411–425.

Shorten, P. R., A. B. Robson, A. E. McKinnon and D. J. Wall (2000) CRH-induced electrical activity and calcium signalling in pituitary corticotrophs, *Journal of Theoretical Biology*. 206: 395–405.

Shorten, P. R. and D. J. Wall (2000) A Hodgkin–Huxley model exhibiting bursting oscillations, *Bulletin of Mathematical Biology*. 62: 695–715.

Shotkin, L. M. (1974a) A model for LH levels in the recently-castrated adult rat and its comparison with experiment, *Journal of Theoretical Biology*. 43: 1–14.

Shotkin, L. M. (1974b) A model for the effect of daily injections of gonadal hormones on LH levels in recently-castrated adult rats and its comparison with experiment, *Journal of Theoretical Biology*. 43: 15–28.

Shuai, J. W. and P. Jung (2002a) Optimal intracellular calcium signaling, *Physical Review Letters*. 88: 068102.

Shuai, J. W. and P. Jung (2002b) Stochastic properties of Ca^{2+} release of inositol 1,4,5-trisphosphate receptor clusters, *Biophysical Journal*. 83: 87–97.

Shuai, J. W. and P. Jung (2003) Optimal ion channel clustering for intracellular calcium signaling, *Proceedings of the National Academy of Sciences USA*. 100: 506–510.

Shuttleworth, T. J. (1999) What drives calcium entry during $[Ca^{2+}]_i$ oscillations?–challenging the capacitative model, *Cell Calcium*. 25: 237–246.

Siebert, W. M. (1974) Ranke revisited–a simple short-wave cochlear model, *Journal of the Acoustical Society of America*. 56: 594–600.

Simon, S. M., C. S. Peskin and G. F. Oster (1992) What drives the translocation of proteins?, *Proceedings of the National Academy of Sciences USA*. 89: 3770–3774.

Sinha, S. (1988) Theoretical study of tryptophan operon: application in microbial technology, *Biotechnology and Bioengineering*. 31: 117–124.

Smart, J. L. and J. A. McCammon (1998) Analysis of synaptic transmission in the neuromuscular junction using a continuum finite element model, *Biophysical Journal*. 75: 1679–1688.

Smith, G. D. (1996) Analytical steady-state solution to the rapid buffering approximation near an open Ca^{2+} channel, *Biophysical Journal*. 71: 3064–3072.

Smith, G. D., L. Dai, R. M. Miura and A. Sherman (2001) Asymptotic analysis of buffered calcium diffusion near a point source, *SIAM Journal on Applied Mathematics*. 61: 1816–1838.

Smith, G. D., J. E. Keizer, M. D. Stern, W. J. Lederer and H. Cheng (1998) A simple numerical model of calcium spark formation and detection in cardiac myocytes, *Biophysical Journal*. 75: 15–32.

Smith, G. D., J. E. Keizer, M. D. Stern, W. J. Lederer and H. Cheng (1998) A simple numerical model of calcium spark formation and detection in cardiac myocytes, *Biophysical Journal*. 75: 15–32.

Smith, G. D., J. Wagner and J. Keizer (1996) Validity of the rapid buffering approximation near a point source of calcium ions, *Biophysical Journal*. 70: 2527–2539.

Smith, J. C., A. P. Abdala, H. Koizumi, I. A. Rybak and J. F. Paton (2007) Spatial and functional architecture of the Mammalian brain stem respiratory network: a hierarchy of three oscillatory mechanisms, *Journal of Neurophysiology*. 98: 3370–3387.

Smith, J. C., H. H. Ellenberger, K. Ballanyi, D. W. Richter and J. L. Feldman (1991) Pre-Bötzinger complex: a brainstem region that may generate respiratory rhythm in mammals, *Science*. 254: 726–729.

Smith, J. M. and R. J. Cohen (1984) Simple finite element model accounts for wide range of cardiac dysrhythmias, *Proceedings of the National Academy of Sciences USA*. 81: 233–237.

Smith, N. P. and E. J. Crampin (2004) Development of models of active ion transport for whole-cell modelling: cardiac sodium–potassium pump as a case study, *Progress in Biophysics and Molecular Biology*. 85: 387–405.

Smith, W. R. (1980) Hypothalamic regulation of pituitary secretion of luteinizing hormone–II. Feedback control of gonadotropin secretion, *Bulletin of Mathematical Biology*. 42: 57–78.

Smith, W. R. (1983) Qualitative mathematical models of endocrine systems, *American Journal of Physiology — Regulatory, Integrative and Comparative Physiology*. 245: R473–477.

Smolen, P. (1995) A model for glycolytic oscillations based on skeletal muscle phosphofructokinase kinetics, *Journal of Theoretical Biology*. 174: 137–148.

Smolen, P. and J. Keizer (1992) Slow voltage inactivation of Ca^{2+} currents and bursting mechanisms for the mouse pancreatic β-cell, *Journal of Membrane Biology*. 127: 9–19.

Smoller, J. (1994) *Shock Waves and Reaction-Diffusion Equations (Second Edition)*: Springer-Verlag, New York.

Sneyd, J., Ed. (2005) *Tutorials in Mathematical Biosciences II: Mathematical Modeling of Calcium Dynamics and Signal Transduction*. Springer-Verlag, New York.

Sneyd, J. and A. Atri (1993) Curvature dependence of a model for calcium wave propagation, *Physica D*. 65: 365–372.

Sneyd, J., A. C. Charles and M. J. Sanderson (1994) A model for the propagation of intercellular calcium waves, *American Journal of Physiology — Cell Physiology*. 266: C293–302.

Sneyd, J., P. D. Dale and A. Duffy (1998) Traveling waves in buffered systems: applications to calcium waves, *SIAM Journal on Applied Mathematics*. 58: 1178–1192.

Sneyd, J. and J. F. Dufour (2002) A dynamic model of the type-2 inositol trisphosphate receptor, *Proceedings of the National Academy of Sciences USA*. 99: 2398–2403.

Sneyd, J. and M. Falcke (2005) Models of the inositol trisphosphate receptor, *Progress in Biophysics and Molecular Biology*. 89: 207–245.

Sneyd, J., M. Falcke, J. F. Dufour and C. Fox (2004a) A comparison of three models of the inositol trisphosphate receptor, *Progress in Biophysics and Molecular Biology*. 85: 121–140.

Sneyd, J., J. Keizer and M. J. Sanderson (1995b) Mechanisms of calcium oscillations and waves: a quantitative analysis, *FASEB Journal*. 9: 1463–1472.

Sneyd, J. and J. Sherratt (1997) On the propagation of calcium waves in an inhomogeneous medium, *SIAM Journal on Applied Mathematics*. 57: 73–94.

Sneyd, J. and D. Tranchina (1989) Phototransduction in cones: an inverse problem in enzyme kinetics, *Bulletin of Mathematical Biology*. 51: 749–784.

Sneyd, J., K. Tsaneva-Atanasova, V. Reznikov, Y. Bai, M. J. Sanderson and D. I. Yule (2006) A method for determining the dependence of calcium oscillations on inositol trisphosphate oscillations, *Proceedings of the National Academy of Sciences USA*. 103: 1675–1680.

Sneyd, J., K. Tsaneva-Atanasova, D. I. Yule, J. L. Thompson and T. J. Shuttleworth (2004b) Control of calcium oscillations by membrane fluxes, *Proceedings of the National Academy of Sciences USA*. 101: 1392–1396.

Sneyd, J., B. Wetton, A. C. Charles and M. J. Sanderson (1995a) Intercellular calcium waves mediated by diffusion of inositol trisphosphate: a two-dimensional model, *American Journal of Physiology — Cell Physiology*. 268: C1537–1545.

Sneyd, J., M. Wilkins, A. Strahonja and M. J. Sanderson (1998) Calcium waves and oscillations driven by an intercellular gradient of inositol (1,4,5)-trisphosphate, *Biophysical Chemistry*. 72: 101–109.

Sobie, E. A., K. W. Dilly, J. dos Santos Cruz, W. J. Lederer and M. S. Jafri (2002)

Termination of cardiac Ca^{2+} sparks: an investigative mathematical model of calcium-induced calcium release, *Biophysical Journal*. 83: 59–78.

Soeller, C. and M. B. Cannell (1997) Numerical simulation of local calcium movements during L-type calcium channel gating in the cardiac diad, *Biophysical Journal*. 73: 97–111.

Soeller, C. and M. B. Cannell (2002a) Estimation of the sarcoplasmic reticulum Ca^{2+} release flux underlying Ca^{2+} sparks, *Biophysical Journal*. 82: 2396–2414.

Soeller, C. and M. B. Cannell (2002b) A Monte Carlo model of ryanodine receptor gating in the diadic cleft of cardiac muscle, *Biophysical Journal*. 82: 76a.

Soeller, C. and M. B. Cannell (2004) Analysing cardiac excitation–contraction coupling with mathematical models of local control, *Progress in Biophysics and Molecular Biology*. 85: 141–162.

Spach, M. S., W. T. Miller, 3rd, D. B. Geselowitz, R. C. Barr, J. M. Kootsey and E. A. Johnson (1981) The discontinuous nature of propagation in normal canine cardiac muscle. Evidence for recurrent discontinuities of intracellular resistance that affect the membrane currents, *Circulation Research*. 48: 39–54.

Spilmann, L. and J. S. Werner, Eds. (1990) *Visual Perception: The Neurophysiological Foundations*: Academic Press, London.

Spitzer, V., M. J. Ackerman, A. L. Scherzinger and R. M. Whitlock (1996) The visible human male: a technical report, *Journal of the American Medical Informatics Association*. 3: 118–130.

Spudich, J. A., S. J. Kron and M. P. Sheetz (1985) Movement of myosin-coated beads on oriented filaments reconstituted from purified actin, *Nature*. 315: 584–586.

Stakgold, I. (1998) *Green's Functions and Boundary Value Problems*: Wiley, New York.

Starmer, C. F., A. R. Lancaster, A. A. Lastra and A. O. Grant (1992) Cardiac instability amplified by use-dependent Na channel blockade, *American Journal of Physiology — Heart and Circulatory Physiology*. 262: H1305–1310.

Starmer, C. F., A. A. Lastra, V. V. Nesterenko and A. O. Grant (1991) Proarrhythmic response to sodium channel blockade. Theoretical model and numerical experiments, *Circulation*. 84: 1364–1377.

Steele, C. R. (1974) Behavior of the basilar membrane with pure-tone excitation, *Journal of the Acoustical Society of America*. 55: 148–162.

Steele, C. R. and L. Tabor (1979a) Comparison of WKB and finite difference calculations for a two-dimensional cochlear model, *Journal of the Acoustical Society of America*. 65: 1001–1006.

Steele, C. R. and L. Tabor (1979b) Comparison of WKB calculations and experimental results for three-dimensional cochlear models, *Journal of the Acoustical Society of America*. 65: 1007–1018.

Stephenson, J. L. (1972) Concentration of the urine in a central core model of the counterflow system, *Kidney International*. 2: 85–94.

Stephenson, J. L. (1992) *Urinary concentration and dilution: models*. In: Handbook of Physiology. Section 8: Renal Physiology, Ed: E. E. Windhager, American Physiological Society, Bethesda, Maryland.

Stern, M. D. (1992) Buffering of calcium in the vicinity of a channel pore, *Cell Calcium*. 13: 183–192.

Stern, M. D. (1992) Theory of excitation–contraction coupling in cardiac muscle, *Biophysical Journal*. 63: 497–517.

Stern, M. D., G. Pizarro and E. Ríos (1997) Local control model of excitation–contraction coupling in skeletal muscle, *Journal of General Physiology*. 110: 415–440.

Stern, M. D., L. S. Song, H. Cheng, J. S. Sham, H. T. Yang, K. R. Boheler and E. Rios (1999) Local control models of cardiac excitation–contraction coupling. A possible role for allosteric interactions between ryanodine receptors, *Journal of General Physiology*. 113: 469–489.

Stiles, J. R. and T. M. Bartol (2000) *Monte Carlo methods for simulating realistic synaptic microphysiology using MCell*. In: Computational Neuroscience: Realistic Modeling for Experimentalists, Ed: E. D. Schutter, CRC Press, New York.

Stojilkovic, S. S., J. Reinhart and K. J. Catt (1994) Gonadotropin-releasing hormone receptors: structure and signal transduction pathways, *Endocrine Reviews*. 15: 462–499.

Stoker, J. J. (1950) *Nonlinear Vibrations*: Interscience, New York.

Stokes, W. (1854) *The Diseases of the Heart and Aorta*: Hodges and Smith, Dublin.

Strang, G. (1986) *Introduction to Applied Mathematics*: Wellesley-Cambridge Press, Wellesley, MA.

Streeter, D. D. J. (1979) *Gross morphology and fiber geometry of the heart*. In: Handbook of Physiology. Section 2: The Cardiovascular System, Volume I: The Heart, Ed: American Physiological Society, Bethesda, MD.

Strieter, J., J. L. Stephenson, L. G. Palmer and A. M. Weinstein (1990) Volume-activated chloride permeability can mediate cell volume regulation in a mathematical model of a tight epithelium, *Journal of General Physiology*. 96: 319–344.

Strogatz, S. H. (1994) *Nonlinear Dynamics and Chaos*: Addison-Wesley, Reading, MA.

Stryer, L. (1986) Cyclic GMP cascade of vision, *Annual Review of Neuroscience*. 9: 87–119.

Stryer, L. (1988) *Biochemistry (Third Edition)*: W.H. Freeman, New York.

Sturis, J., K. S. Polonsky, E. Mosekilde and E. V. Cauter (1991) Computer model for mechanisms underlying ultradian oscillations of insulin and glucose, *American Journal of Physiology — Endocrinology and Metabolism*. 260: E801–809.

Sturis, J., A. J. Scheen, R. Leproult, K. S. Polonsky and E. van Cauter (1995) 24-hour glucose profiles during continuous or oscillatory insulin infusion. Demonstration of the functional significance of ultradian insulin oscillations, *Journal of Clinical Investigation*. 95: 1464–1471.

Sun, X. P., N. Callamaras, J. S. Marchant and I. Parker (1998) A continuum of $InsP_3$-mediated elementary Ca^{2+} signalling events in Xenopus oocytes, *Journal of Physiology*. 509: 67–80.

Sveiczer, A., A. Csikasz-Nagy, B. Gyorffy, J. J. Tyson and B. Novak (2000) Modeling the fission yeast cell cycle: quantized cycle times in *wee1⁻ cdc25Δ* mutant cells, *Proceedings of the National Academy of Sciences USA*. 97: 7865–7870.

Swillens, S., P. Champeil, L. Combettes and G. Dupont (1998) Stochastic simulation of a single inositol 1,4,5-trisphosphate-sensitive Ca^{2+} channel reveals repetitive openings during "blip-like" Ca^{2+} transients, *Cell Calcium*. 23: 291–302.

Swillens, S., G. Dupont, L. Combettes and P. Champeil (1999) From calcium blips to calcium puffs: theoretical analysis of the requirements for interchannel communication, *Proceedings of the National Academy of Sciences USA*. 96: 13750–13755.

Swillens, S. and D. Mercan (1990) Computer simulation of a cytosolic calcium oscillator, *Biochemical Journal*. 271: 835–838.

Taccardi, B., R. L. Lux and P. R. Erschler (1992) Effect of myocardial fiber direction on 3-dimensional shape of excitation wavefront and associated potential distributions in ventricular walls, *Circulation*. 86 (Suppl. I): 752.

Tai, K., S. D. Bond, H. R. MacMillan, N. A. Baker, M. J. Holst and J. A. McCammon (2003) Finite element simulations of acetylcholine diffusion in neuromuscular junctions, *Biophysical Journal*. 84: 2234–2241.

Takaki, M. (2003) Gut pacemaker cells: the interstitial cells of Cajal (ICC), *Journal of Smooth Muscle Research*. 39: 137–161.

Tameyasu, T. (2002) Simulation of Ca^{2+} release from the sarcoplasmic reticulum with three-dimensional sarcomere model in cardiac muscle, *Japanese Journal of Physiology*. 52: 361–369.

Tamura, T., K. Nakatani and K.-W. Yau (1991) Calcium feedback and sensitivity regulation in primate rods, *Journal of General Physiology*. 98: 95–130.

Tang, Y., T. Schlumpberger, T. Kim, M. Lueker and R. S. Zucker (2000) Effects of mobile buffers on facilitation: experimental and computational studies, *Biophysical Journal*. 78: 2735–2751.

Tang, Y. and J. L. Stephenson (1996) Calcium dynamics and homeostasis in a mathematical model of the principal cell of the cortical collecting tubule, *Journal of General Physiology*. 107: 207–230.

Tang, Y., J. L. Stephenson and H. G. Othmer (1996) Simplification and analysis of models of calcium dynamics based on IP_3-sensitive calcium channel kinetics, *Biophysical Journal*. 70: 246–263.

Taniguchi, K., S. Kaya, K. Abe and S. Mardh (2001) The oligomeric nature of Na/K-transport ATPase, *Journal of Biochemistry*. 129: 335–342.

Tawhai, M. H. and K. S. Burrowes (2003) Developing integrative computational models of pulmonary structure, *Anatomical record Part B, New anatomist*. 275: 207–218.

Tawhai, M. H., P. Hunter, J. Tschirren, J. Reinhardt, G. McLennan and E. A. Hoffman (2004) CT-based geometry analysis and finite element models of the human and ovine bronchial tree, *Journal of Applied Physiology*. 97: 2310–2321.

Tawhai, M. H., M. P. Nash and E. A. Hoffman (2006) An imaging-based computational approach to model ventilation distribution and soft-tissue deformation in the ovine lung, *Academic Radiology*. 13: 113–120.

Taylor, C. W. (1998) Inositol trisphosphate receptors: Ca^{2+}-modulated intracellular Ca^{2+} channels, *Biochimica et Biophysica Acta*. 1436: 19–33.

Taylor, W. R., S. He, W. R. Levick and D. I. Vaney (2000) Dendritic computation of direction selectivity by retinal ganglion cells, *Science*. 289: 2347–2350.

ten Tusscher, K. H., D. Noble, P. J. Noble and A. V. Panfilov (2004) A model for human ventricular tissue, *American Journal of Physiology — Heart and Circulatory Physiology*. 286: H1573–1589.

Thomas, A. P., G. S. Bird, G. Hajnoczky, L. D. Robb-Gaspers and J. W. Putney, Jr. (1996) Spatial and temporal aspects of cellular calcium signaling, *FASEB Journal*. 10: 1505–1517.

Thomas, D., P. Lipp, S. C. Tovey, M. J. Berridge, W. Li, R. Y. Tsien and M. D. Bootman (2000) Microscopic properties of elementary Ca^{2+} release sites in non-excitable cells, *Current Biology*. 10: 8–15.

Tolić, I. M., E. Mosekilde and J. Sturis (2000) Modeling the insulin–glucose feedback system: the significance of pulsatile insulin secretion, *Journal of Theoretical Biology*. 207: 361–375.

Topor, Z. L., M. Pawlicki and J. E. Remmers (2004) A computational model of the human respiratory control system: responses to hypoxia and hypercapnia, *Annals of Biomedical Engineering*. 32: 1530–1545.

Tordjmann, T., B. Berthon, M. Claret and L. Combettes (1997) Coordinated intercellular calcium waves induced by noradrenaline in rat hepatocytes: dual control by gap junction permeability and agonist, *Embo Journal*. 16: 5398–5407.

Tordjmann, T., B. Berthon, E. Jacquemin, C. Clair, N. Stelly, G. Guillon, M. Claret and L. Combettes (1998) Receptor-oriented intercellular calcium waves evoked by vasopressin in rat hepatocytes, *Embo Journal*. 17: 4695–4703.

Torre, V., S. Forti, A. Menini and M. Campani (1990) Model of phototransduction in retinal rods, *Cold Spring Harbor Symposia in Quantitative Biology*. 55: 563–573.

Tosteson, D. C. and J. F. Hoffman (1960) Regulation of cell volume by active cation transport in high and low potassium sheep red cells, *Journal of General Physiology*. 44: 169–194.

Tranchina, D., J. Gordon and R. Shapley (1984) Retinal light adaptation–evidence for a feedback mechanism, *Nature*. 310: 314–316.

Tranchina, D., J. Sneyd and I. D. Cadenas (1991) Light adaptation in turtle cones: testing and analysis of a model for phototransduction, *Biophysical Journal*. 60: 217–237.

Tranquillo, R. and D. Lauffenberger (1987) Stochastic models of leukocyte chemosensory movement, *Journal of Mathematical Biology*. 25: 229–262.

Traube, L. (1865) Ueber periodische thatigkeits-aeusserungen des vasomotorischen und hemmungs-nervencentrums, *Medizin Wissenschaft*. 56: 881–885.

Troy, J. B. and T. Shou (2002) The receptive fields of cat retinal ganglion cells in physiological and pathological states: where we are after half a century of research, *Progress in Retinal and Eye Research*. 21: 263–302.

Troy, W. C. (1976) Bifurcation phenomena in FitzHugh's nerve conduction equations, *Journal of Mathematical Analysis and Applications*. 54: 678–690.

Troy, W. C. (1978) The bifurcation of periodic solutions in the Hodgkin–Huxley equations, *Quarterly of Applied Mathematics*. 36: 73–83.

Tsai, J.-C. and J. Sneyd (2005) Existence and stability of traveling waves in buffered systems, *SIAM Journal on Applied Mathematics*. 66: 1675–1680.

Tsai, J.-C. and J. Sneyd (2007a) Are buffers boring?: uniqueness and asymptotical stability of traveling wave fronts in the buffered bistable system, *Journal of Mathematical Biology*. 54: 513–553.

Tsai, J. C. and J. Sneyd (2007b) Traveling waves in the discrete fast buffered bistable system, *Journal of Mathematical Biology*. 55: 605–652.

Tsaneva-Atanasova, K., D. I. Yule and J. Sneyd (2005) Calcium oscillations in a triplet of pancreatic acinar cells, *Biophysical Journal*. 88: 1535–1551.

Tsaneva-Atanasova, K., C. L. Zimliki, R. Bertram and A. Sherman (2006) Diffusion of calcium and metabolites in pancreatic islets: killing oscillations with a pitchfork, *Biophysical Journal*. 90: 3434–3446.

Tse, A., F. W. Tse, W. Almers and B. Hille (1993) Rhythmic exocytosis stimulated by GnRH-induced calcium oscillations in rat gonadotropes, *Science*. 260: 82–84.

Tuckwell, H. C. (1988) *Introduction to Theoretical Neurobiology*: Cambridge University Press, Cambridge.

Tuckwell, H. C. and R. M. Miura (1978) A mathematical model for spreading cortical depression, *Biophysical Journal*. 23: 257–276.

Tung, L. (1978) *A bi-domain model for describing ischemic myocardial D-C potentials. Ph.D. Thesis*: MIT, Cambridge, MA.

Tyson, J. J., A. Csikasz-Nagy and B. Novak (2002) The dynamics of cell cycle regulation, *Bioessays*. 24: 1095–1109.

Tyson, J. J. and P. C. Fife (1980) Target patterns in a realistic model of the Belousov–Zhabotinskii reaction, *Journal of Chemical Physics*. 73: 2224–2237.

Tyson, J. J., C. I. Hong, C. D. Thron and B. Novak (1999) A simple model of circadian rhythms based on dimerization and proteolysis of PER and TIM, *Biophysical Journal*. 77: 2411–2417.

Tyson, J. J. and J. P. Keener (1988) Singular perturbation theory of traveling waves in excitable media, *Physica D*. 32: 327–361.

Tyson, J. J. and B. Novak (2001) Regulation of the eukaryotic cell cycle: molecular antagonism, hysteresis, and irreversible transitions, *Journal of Theoretical Biology*. 210: 249–263.

Tyson, J. J. and H. G. Othmer (1978) The dynamics of feedback control circuits in biochemical pathways, *Progress in Theoretical Biology*. 5: 1–62.

Ullah, G., P. Jung and A. H. Cornell-Bell (2006) Anti-phase calcium oscillations in astrocytes via inositol (1,4,5)-trisphosphate regeneration, *Cell Calcium*. 39: 197–208.

Urban, B. W. and S. B. Hladky (1979) Ion transport in the simplest single file pore, *Biochimica et Biophysica Acta*. 554: 410–429.

Ursino, M. (1998) Interaction between carotid baroregulation and the pulsating heart: a mathematical model, *American Journal of Physiology — Heart and Circulatory Physiology*. 275: H1733–1747.

Ursino, M. (1999) A mathematical model of the carotid baroregulation in pulsating conditions, *IEEE Transactions in Biomedical Engineering*. 46: 382–392.

Ussing, H. H. (1949) The distinction by means of tracers between active transport and diffusion, *Acta Physiologica Scandinavica*. 19: 43–56.

Ussing, H. H. (1982) Volume regulation of frog skin epithelium, *Acta Physiologica Scandinavica*. 114: 363–369.

Uyeda, T. Q., S. J. Kron and J. A. Spudich (1990) Myosin step size. Estimation from slow sliding movement of actin over low densities of heavy meromyosin, *Journal of Molecular Biology*. 214: 699–710.

van der Pol, B. and J. van der Mark (1928) The heartbeat considered as a relaxation oscillation, and an electrical model of the heart, *Philosophical Magazine*. 6: 763–775.

van Kampen, N. G. (2007) *Stochastic Processes in Physics and Chemistry (Third Edition)*: Elsevier, Amsterdam.

van Milligen, B. P., P. D. Bons, B. A. Carreras and R. Sanchez (2006) On the applicability of Fick's law to diffusion in inhomogeneous systems, *European Journal of Physics*. 26: 913–925.

Vandenberg, C. A. and F. Bezanilla (1991) A sodium channel gating model based on single channel, macroscopic ionic, and gating currents in the squid giant axon, *Biophysical Journal*. 60: 1511–1533.

Vaney, D. I. and W. R. Taylor (2002) Direction selectivity in the retina, *Current Opinion in Neurobiology*. 12: 405–410.

Vesce, S., P. Bezzi and A. Volterra (1999) The active role of astrocytes in synaptic transmission, *Cellular and Molecular Life Sciences*. 56: 991–1000.

Vielle, B. (2005) Mathematical analysis of Mayer waves, *Journal of Mathematical Biology*. 50: 595–606.

Vodopick, H., E. M. Rupp, C. L. Edwards, F. A. Goswitz and J. J. Beauchamp (1972) Spontaneous cyclic leukocytosis and thrombocytosis in chronic granulocytic leukemia, *New England Journal of Medicine*. 286: 284–290.

von Békésy, V. (1960) *Experiments in Hearing*: McGraw-Hill, New York. Reprinted in 1989 by the Acoustical Society of America.

von Euler, C. (1980) Central pattern generation during breathing, *Trends in Neuroscience*. 3: 275–277.

Wagner, J. and J. Keizer (1994) Effects of rapid buffers on Ca^{2+} diffusion and Ca^{2+} oscillations, *Biophysical Journal*. 67: 447–456.

Wagner, J., Y.-X. Li, J. Pearson and J. Keizer (1998) Simulation of the fertilization Ca^{2+} wave in *Xenopus laevis* eggs, *Biophysical Journal*. 75: 2088–2097.

Waldo, A. L., A. J. Camm, et al. (1996) Effect of d-sotalol on mortality in patients with left ventricular dysfunction after recent and remote myocardial infarction. The SWORD

Investigators. Survival With Oral d-Sotalol, *Lancet*. 348: 7–12.

Wang, H. and G. Oster (1998) Energy transduction in the F1 motor of ATP synthase, *Nature*. 396: 279–282.

Wang, X.-J. and J. Rinzel (1995) *Oscillatory and bursting properties of neurons*. In: The Handbook of Brain Theory and Neural Networks, Ed: M. Arbib, MIT Press, Cambridge, MA.

Wang, I., A.Z. Politi, N. Tania, Y. Bai, M.J. Sanderson and J. Sneyd (2008) A mathematical model of airway and pulmonary arteriole smooth muscle, *Biophysical Journal*. 94: 2053–2064.

Watanabe, M., N. F. Otani and R. F. Gilmour, Jr. (1995) Biphasic restitution of action potential duration and complex dynamics in ventricular myocardium, *Circulation Research*. 76: 915–921.

Watanabe, M. A., F. H. Fenton, S. J. Evans, H. M. Hastings and A. Karma (2001) Mechanisms for discordant alternans, *Journal of Cardiovascular Electrophysiology*. 12: 196–206.

Waterman, M. S. (1995) *Introduction to Computational Biology: Maps, Sequences, and Genomes*: Chapman and Hall, Boca Raton, FL.

Webb, S. E. and A. L. Miller (2003) Calcium signalling during embryonic development, *Nature Reviews Molecular Cell Biology*. 4: 539–551.

Weber, E. H. (1834) *De pulsu, resorptione, auditu et tactu annotationes anatomicæ, et physiologicæ. Author's summary: Ueber den Tastsinn, Arch. Anat. u. Physiol., 1835, 152*: Leipzig.

Weinstein, A. (1992) Analysis of volume regulation in an epithelial cell model, *Bulletin of Mathematical Biology*. 54: 537–561.

Weinstein, A. (1996) Coupling of entry to exit by peritubular K^+ permeability in a mathematical model of rat proximal tubule, *American Journal of Physiology — Renal, Fluid and Electrolyte Physiology*. 271: F158–168.

Weinstein, A. M. (1994) Mathematical models of tubular transport, *Annual Review of Physiology*. 56: 691–709.

Weinstein, A. M. (1998a) A mathematical model of the inner medullary collecting duct of the rat: pathways for Na and K transport, *American Journal of Physiology — Renal, Fluid and Electrolyte Physiology*. 274: F841–855.

Weinstein, A. M. (1998b) A mathematical model of the inner medullary collecting duct of the rat: acid/base transport, *American Journal of Physiology — Renal, Fluid and Electrolyte Physiology*. 274: F856–867.

Weinstein, A. M. (2000) A mathematical model of the outer medullary collecting duct of the rat, *American Journal of Physiology — Renal, Fluid and Electrolyte Physiology*. 279: F24–45.

Weinstein, A. M. (2003) Mathematical models of renal fluid and electrolyte transport: acknowledging our uncertainty, *American Journal of Physiology — Renal, Fluid and Electrolyte Physiology*. 284: F871–884.

Weinstein, A. M. and J. L. Stephenson (1981) Coupled water transport in standing gradient models of the lateral intercellular space, *Biophysical Journal*. 35: 167–191.

Weiss, J. N., A. Karma, Y. Shiferaw, P. S. Chen, A. Garfinkel and Z. Qu (2006) From pulsus to pulseless: the saga of cardiac alternans, *Circulation Research*. 98: 1244–1253.

Wenckebach, K. F. (1904) *Arrhythmia of the Heart: a Physiological and Clinical Study*: Green, Edinburgh.

West, J. B. (1985) *Ventilation/Blood Flow and Gas Exchange*: Blackwell Scientific Publications, Oxford.

West, J. B. (2004) Understanding pulmonary gas exchange: ventilation–perfusion relationships, *American Journal of Physiology — Lung, Cellular and Molecular Physiology*. 287: L1071–1072.

Wetsel, W. C., M. M. Valenca, I. Merchenthaler, Z. Liposits, F. J. Lopez, R. I. Weiner, P. L. Mellon and A. Negro-Vilar (1992) Intrinsic pulsatile secretory activity of immortalized luteinizing hormone-releasing hormone-secreting neurons, *Proceedings of the National Academy of Sciences USA*. 89: 4149–4153.

White, D. C. S. and J. Thorson (1975) *The Kinetics of Muscle Contraction*: Pergamon Press. Originally published in Progress in Biophysics and Molecular Biology, volume 27, 1973, Oxford, New York.

White, J. B., G. P. Walcott, A. E. Pollard and R. E. Ideker (1998) Myocardial discontinuities: a substrate for producing virtual electrodes that directly excite the myocardium by shocks, *Circulation*. 97: 1738–1745.

Whiteley, J. P., D. J. Gavaghan and C. E. Hahn (2001) Modelling inert gas exchange in tissue and mixed-venous blood return to the lungs, *Journal of Theoretical Biology*. 209: 431–443.

Whiteley, J. P., D. J. Gavaghan and C. E. Hahn (2002) Mathematical modelling of oxygen transport to tissue, *Journal of Mathematical Biology*. 44: 503–522.

Whiteley, J. P., D. J. Gavaghan and C. E. Hahn (2003a) Periodic breathing induced by arterial oxygen partial pressure oscillations, *Mathematical Medicine and Biology*. 20: 205–224.

Whiteley, J. P., D. J. Gavaghan and C. E. Hahn (2003b) Mathematical modelling of pulmonary gas transport, *Journal of Mathematical Biology*. 47: 79–99.

Whitham, G. B. (1974) *Linear and Nonlinear Waves*: Wiley-Interscience, New York.

Whitlock, G. G. and T. D. Lamb (1999) Variability in the time course of single photon responses from toad rods: termination of rhodopsin's activity, *Neuron*. 23: 337–351.

Wichmann, H. E., M. Loeffler and S. Schmitz (1988) A concept of hemopoietic regulation and its biomathematical realization, *Blood Cells*. 14: 411–429.

Wiener, N. and A. Rosenblueth (1946) The mathematical formulation of the problem of conduction of impulses in a network of connected excitable elements, specifically in cardiac muscle, *Archivos del Instituto de Cardiologia de Mexico*. 16: 205–265.

Wier, W. G., T. M. Egan, J. R. Lopez-Lopez and C. W. Balke (1994) Local control of excitation–contraction coupling in rat heart cells, *Journal of Physiology*. 474: 463–471.

Wierschem, K. and R. Bertram (2004) Complex bursting in pancreatic islets: a potential glycolytic mechanism, *Journal of Theoretical Biology*. 228: 513–521.

Wiggins, S. (2003) *Introduction to Applied Nonlinear Dynamical Systems and Chaos*: Springer-Verlag, New York.

Wikswo, J. P., Jr., S. F. Lin and R. A. Abbas (1995) Virtual electrodes in cardiac tissue: a common mechanism for anodal and cathodal stimulation, *Biophysical Journal*. 69: 2195–2210.

Wildt, L., A. Häusler, G. Marshall, J. S. Hutchison, T. M. Plant, P. E. Belchetz and E. Knobil (1981) Frequency and amplitude of gonadotropin-releasing hormone stimulation and gonadotropin secretion in the Rhesus monkey, *Endocrinology*. 109: 376–385.

Willems, G. M., T. Lindhout, W. T. Hermens and H. C. Hemker (1991) Simulation model for thrombin generation in plasma, *Haemostasis*. 21: 197–207.

Williams, M. M. (1990) *Hematology*: McGraw-Hill, New York.

Winfree, A. T. (1967) Biological rhythms and the behavior of populations of coupled oscillators, *Journal of Theoretical Biology*. 16: 15–42.

Winfree, A. T. (1972) Spiral waves of chemical activity, *Science*. 175: 634–636.

Winfree, A. T. (1973) Scroll-shaped waves of chemical activity in three dimensions, *Science*. 181: 937–939.

Winfree, A. T. (1974) Rotating chemical reactions, *Scientific American*. 230: 82–95.

Winfree, A. T. (1980) *The Geometry of Biological Time*: Springer-Verlag, New York.

Winfree, A. T. (1987) *When Time Breaks Down*: Princeton University Press, Princeton, NJ.

Winfree, A. T. (1991) Varieties of spiral wave behavior: an experimentalist's approach to the theory of excitable media, *Chaos*. 1: 303–334 .

Winfree, A. T. and S. H. Strogatz (1983a) Singular filaments organize chemical waves in three dimensions: 1. Geometrically simple waves, *Physica D*. 8: 35–49.

Winfree, A. T. and S. H. Strogatz (1983b) Singular filaments organize chemical waves in three dimensions: 2. twisted waves, *Physica D*. 9: 65–80.

Winfree, A. T. and S. H. Strogatz (1983c) Singular filaments organize chemical waves in three dimensions: 3. knotted waves, *Physica D*. 9: 333–345.

Winfree, A. T. and S. H. Strogatz (1984) Singular filaments organize chemical waves in three dimensions: 4. wave taxonomy, *Physica D*. 13: 221–233.

Winslow, R. L., R. Hinch and J. L. Greenstein (2005) *Mechanisms and models of cardiac excitation–contraction coupling*. In: Tutorials in Mathematical Biosciences II: Mathematical Modeling of Calcium Dynamics and Signal Transduction, Ed: J. Sneyd, Springer-Verlag, New York.

Wittenberg, J. B. (1966) The molecular mechanism of haemoglobin-facilitated oxygen diffusion, *Journal of Biological Chemistry*. 241: 104–114.

Wong, P., S. Gladney and J. D. Keasling (1997) Mathematical model of the lac operon: inducer exclusion, catabolite repression, and diauxic growth on glucose and lactose, *Biotechnology Progress*. 13: 132–143.

Woodbury, J. W. (1971) *Eyring rate theory model of the current–voltage relationship of ion channels in excitable membranes*. In: Chemical Dynamics: Papers in Honor of Henry Eyring, Ed: J. Hirschfelder, John Wiley and Sons, Inc., New York.

Wyman, J. (1966) Facilitated diffusion and the possible role of myoglobin as a transport mechanism, *Journal of Biological Chemistry*. 241: 115–121.

Wyman, R. J. (1977) Neural generation of breathing rhythm, *Annual Review of Physiology*. 39: 417–448.

Yamada, W. M. and R. S. Zucker (1992) Time course of transmitter release calculated from simulations of a calcium diffusion model, *Biophysical Journal*. 61: 671–682.

Yanagida, E. (1985) Stability of fast travelling pulse solutions of the FitzHugh–Nagumo equation, *Journal of Mathematical Biology*. 22: 81–104.

Yanagihara, K., A. Noma and H. Irisawa (1980) Reconstruction of sino-atrial node pacemaker potential based on voltage clamp experiments, *Japanese Journal of Physiology*. 30: 841–857.

Yao, Y., J. Choi and I. Parker (1995) Quantal puffs of intracellular Ca^{2+} evoked by inositol trisphosphate in Xenopus oocytes, *Journal of Physiology*. 482: 533–553.

Yates, A., R. Callard and J. Stark (2004) Combining cytokine signalling with T-bet and GATA-3 regulation in Th1 and Th2 differentiation: a model for cellular decision-making, *Journal of Theoretical Biology*. 231: 181–196.

Yehia, A. R., D. Jeandupeux, F. Alonso and M. R. Guevara (1999) Hysteresis and bistability in the direct transition from 1:1 to 2:1 rhythm in periodically driven single ventricular cells, *Chaos*. 9: 916–931.

Yildirim, N. and M. C. Mackey (2003) Feedback regulation in the lactose operon: a mathematical modeling study and comparison with experimental data, *Biophysical Journal*. 84: 2841–2851.

Yildirim, N., M. Santillán, D. Horike and M. C. Mackey (2004) Dynamics and bistability in a reduced model of the lac operon, *Chaos*. 14: 279–292.

Young, K. W., M. S. Nash, R. A. Challiss and S. R. Nahorski (2003) Role of Ca^{2+} feedback on single cell inositol 1,4,5-trisphosphate oscillations mediated by G-protein-coupled receptors, *Journal of Biological Chemistry*. 278: 20753–20760.

Young, R. C. (1997) A computer model of uterine contractions based on action potential propagation and intercellular calcium waves, *Obstetrics and Gynecology*. 89: 604–608.

Young, R. C. and R. O. Hession (1997) Paracrine and intracellular signaling mechanisms of calcium waves in cultured human uterine myocytes, *Obstetrics and Gynecology*. 90: 928–932.

Yule, D. I., E. Stuenkel and J. A. Williams (1996) Intercellular calcium waves in rat pancreatic acini: mechanism of transmission, *American Journal of Physiology — Cell Physiology*. 271: C1285–1294.

Zahradnikova, A. and I. Zahradnik (1996) A minimal gating model for the cardiac calcium release channel, *Biophysical Journal*. 71: 2996–3012.

Zarnitsina, V. I., A. V. Pokhilko and F. I. Ataullakhanov (1996a) A mathematical model for the spatio-temporal dynamics of intrinsic pathway of blood coagulation. I. The model description, *Thrombosis Research*. 84: 225–236.

Zarnitsina, V. I., A. V. Pokhilko and F. I. Ataullakhanov (1996b) A mathematical model for the spatio-temporal dynamics of intrinsic pathway of blood coagulation. II. Results, *Thrombosis Research*. 84: 333–344.

Zeuthen, T. (2000) Molecular water pumps, *Reviews of Physiology, Biochemistry and Pharmacology*. 141: 97–151.

Zhang, M., P. Goforth, R. Bertram, A. Sherman and L. Satin (2003) The Ca^{2+} dynamics of isolated mouse beta-cells and islets: implications for mathematical models, *Biophysical Journal*. 84: 2852–2870.

Zhou, X., S. B. Knisley, W. M. Smith, D. Rollins, A. E. Pollard and R. E. Ideker (1998) Spatial changes in the transmembrane potential during extracellular electric stimulation, *Circulation Research*. 83: 1003–1014.

Zhou, Z. and E. Neher (1993) Mobile and immobile calcium buffers in bovine adrenal chromaffin cells, *Journal of Physiology*. 469: 245–273.

Zigmond, S. H. (1977) Ability of polymorphonuclear leukocytes to orient in gradients of chemotactic factors, *Journal of Cell Biology*. 75: 606–616.

Zigmond, S. H., H. I. Levitsky and B. J. Kreel (1981) Cell polarity: an examination of its behavioral expression and its consequences for polymorphonuclear leukocyte chemotaxis, *Journal of Cell Biology*. 89: 585–592.

Zinner, B. (1992) Existence of traveling wavefront solutions for the discrete Nagumo equation, *Journal of Differential Equations*. 96: 1–27.

Zipes, D. P. and J. Jalife (1995) *Cardiac Electrophysiology; From Cell to Bedside*

(Second Edition): W. B. Saunders Co., Philadelphia.

Zucker, R. S. and A. L. Fogelson (1986) Relationship between transmitter release and presynaptic calcium influx when calcium enters through discrete channels, *Proceedings of the National Academy of Sciences USA*. 83: 3032–3036.

Zucker, R. S. and L. Landò (1986) Mechanism of transmitter release: voltage hypothesis and calcium hypothesis, *Science*. 231: 574–579.

Zucker, R. S. and W. G. Regehr (2002) Short-term synaptic plasticity, *Annual Review of Physiology*. 64: 355–405.

Zwislocki, J. (1953) Review of recent mathematical theories of cochlear dynamics, *Journal of the Acoustical Society of America*. 25: 743–751.

Index

Printed in the United States
By Bookmasters